Advanced Steels

Yuqing Weng • Han Dong • Yong Gan
Editors

Advanced Steels

The Recent Scenario in Steel Science and Technology

Metallurgical Industry Press

Springer

Editors
Prof. Yuqing Weng
The Chinese Society for Metals
Beijing 100711
People's Republic of China
e-mail: weng@csm.org.cn

Prof. Yong Gan
Central Iron and Steel Research Institute
Chinese Academy of Engineering
Beijing 100081
People's Republic of China
e-mail: gany@cisri.com.cn

Prof. Han Dong
Central Iron and Steel Research Institute
National Engineering Research Center of Advanced Steel Technology
No. 76 Xue Yuan Nan Lu
Beijing 100081
People's Republic of China
e-mail: donghan@nercast.com

ISBN 978-3-642-17664-7 e-ISBN 978-3-642-17665-4
DOI 10.1007/978-3-642-17665-4
Springer Heidelberg Dordrecht London New York

Jointly published with Metallurgical Industry Press, Beijing and Springer-Verlag GmbH Berlin Heidelberg
ISBN 978-7-5024-5436-4 Metallurgical Industry Press – Not for sale outside the mainland of China

Cover design: eStudio Calamar S.L.

Printed on acid-free paper

Springer is part of Springer Science+Business Media (www.springer.com)

Preface

At present, Steel is one of the most common material widely used in the world, both for structural and functional applications. Steel has been the basic material for weaponry, agriculture, construction, etc. in the human society since the beginning of iron age, and now it is still playing very important roles in the world. It is generally believed that steel is really a kind of advanced materials due to its advantages during processing, fabrication, applications, and also recycling. People cannot image what the world would be if there be no steel around us.

Steels have been widely using for construction, automobile, machinery, energy, transportation, daily life, etc. in this special occasion that people take much more care with the climate change and global warming. Will steels still play an important role to our society in the future? Yes, it will be. More advanced steel products with the characteristics of high performance, low cost, easy fabrication, low tolerance, and environment benign have been developed to meet the demands from both market and environment protection. It seems there is no stop of this advancing trend.

The development of steel products is dependent on the steel knowledge we have. Although there have been a good accumulation of steel knowledge since the massive production of liquid steel, the new phenomena and roles in steels have still been investigated in recent years. Now people involved in steel research, steel processing and steel applications are concerned more and more with the progresses of steel science and technology than ever before, and have made great contributions to steel knowledge. This is one of the reasons why steel products change year by year. In order to illustrate the current status of steels, the editors of this book decided to ask outstanding professors and researchers all of the world to write a review on their research fields on the occasion of ICAS 2010.

The First International Conference on Advanced Steels was held at Guilin, China, November 8–11, 2010. The International Conference on Advanced Steels (ICAS) is the merging of two international series conferences: "International Symposium on Ultrafine Grained Structures (ISUGS)" and "International Conference on Advanced Structural Steels (ICASS)". Over 270 papers have been presented in the Conference. It was really a platform for people all over the world to share their contributed works in steels with their colleagues effectively. ICAS 2010 will cover almost every aspect of steels: physical metallurgy, steel grades, processing and fabrication, simulation, properties and applications, etc. It is a comprehensive conference on steel products and technologies. Plenary and keynote speakers are very active in the relative steel fields, and are invited to illustrate their works in this specific proceedings in detail.

The aim of this book is to introduce steel researchers and technologists to the understanding of present status of different kinds of steels and relative technologies. It covers general review on steel industry, physical metallurgy, HSLA steel, automobile steel, specialty steel, processing and fabrications. It is the summary of steels over past decades and also the forecast of advanced steels into the future. I believe physically that this specific book would help people to have the progresses of steels in hand.

Beijing, China Rang Cai

Contents

Part I General Review

Advanced Steel and Our Society: Better Steel, Better World 3
Yong Gan

Innovative Steels for Low Carbon Economy 9
Lejiang Xu

Development and Outlook of Advanced High Strength Steel in Ansteel 15
Xiaogang Zhang

Technical Progress and Product Development of TISCO Stainless Steel 19
Xiao Bo Li

The State of Steel Industry in India and its Future Prospects 27
Sanak Mishra

On the Performance Improvement of Steels through M^3 Structure Control.... 35
Han Dong, Xingjun Sun, Wenquan Cao, Zhengdong Liu, Maoqiu Wang, and Yuqing Weng

High-Strength Steels: Control of Structure and Properties 59
A. S. Oryshchenko and E. I. Khlusova

Ultra-high Strength Steel Treated by Using Quenching–Partitioning–Tempering Process 67
T. Y. Hsu (Zuyao Xu) and Xuejun Jin

Part II Physical Metallurgy Frontier

Long-term Stabilization of Steel Availability under Limited Resources 77
Kotobu Nagai

Grain Boundary Carbon Segregation Estimated by McLean and Seah-Hondros Models 81
Setsuo Takaki, Nobuo Nakada and Toshihiro Tsuchiyama

Nano-Preciptates Design with Hydrogen Trapping Character in High Strength Steel 87
Fu-Gao Wei, Toru Hara and Kaneaki Tsuzaki

Micro-Mechanical Behavior of Inclusions in Advanced Steels 93
Xishan Xie, Yanpin Zeng, Miaomiao Wang, and Hongmei Fan

Dislocation Assisted Phase Transformation Observed in Iron Alloys 103
Yoon-UK Heo, Masaki Takeuchi, Kazuo Furuya, and Hu-Chul Lee

Solution and Precipitation of Secondary Phase in Steels: Phenomenon, Theory, and Practice 109
Qilong Yong, Xinjun Sun, Gengwei Yang, and Zhengyan Zhang

Ways to Manage Both Strength and Ductility in Nanostructured Steels 119
Nobuhiro Tsuji

Steels: Data Exploration for Discovery and Data-Sharing 131
Guoquan Liu

Long Life High Strength Steels to Resist Fatigue Failure and Delayed Fracture 137
Weijun Hui, Han Dong, Yuqing Weng, Jie Shi, and Maoqiu Wang

Part III Auto Sheet Steels

Metallurgical Perspectives on Advanced Sheet Steels for Automotive Application 163
Debanshu Bhattacharya

Recent Development of Nb-Containing DP590, DP780 and DP980 Steels for Production on Continuous Galvanizing Lines 177
K. Cho, K. V. Redkin, M. Hua, C. I. Garcia, and A. J. DeArdo

Lightweight Car Body and Application of High Strength Steels 187
Mingtu Ma and Hongliang Yi

Design of Lean Maraging TRIP Steels 199
Dirk Ponge, Julio Millán, and Dierk Raabe

The 3rd Generation Automobile Sheet Steels Presenting with Ultrahigh Strength and High Ductility 209
Wenquan Cao, Jie Shi, Chang Wang, Cunyu Wang, Le Xu, Maoqiu Wang, Yuqing Weng, and Han Dong

Challenges Toward the Further Strengthening of Sheet Steel 229
K. Ushioda, J. Takahashi, S. Takebayashi, D. Maeda, K. Hayashi and Y. R. Abe

Developments in High Strength Steels with Duplex Microstructures of Bainite or Martensite with Retained Austenite: Progress with Quenching and Partitioning Heat Treatment 241
David Edmonds, David Matlock and John Speer

Development and Application of Q&P Sheet Steels 255
Li Wang and Weijun Feng

Microstructure and Mechanical Properties of Al-Added High Mn Austenitic Steel 259
Jae-Eun jin and Young-Kook Lee

Microstructure and Property Control of Advanced High Strength Automotive Steels 265
Lin Li

Microstructure and Mechanical Properties of a TRIP Steel Containing 7 Mass% Mn 275
Seong-Jun Park, Chang-Seok Oh, and Sung-Joon Kim

Part IV Advanced High Strength Low Alloy Steels

Development of High Strength and High Performance Linepipe and Shipbuilding Steels 281
Ki Kang Bong, Ju Seok Kang, Jang Yong Yoo, Dong Han Seo, In Shik Suh, and Gyu Baek An

MoNb-based Alloying Concepts for Low-Carbon Bainitic Steels 289
Hardy Mohrbacher, Xinjun Sun, Qilong Yong, and Han Dong

Vanadium in Bainitic Steels: A Review of Recent Developments 303
Yu Li and David Milbourn

Nanostructural Engineering of TMCP Steels 309
Peter D. Hodgson, Ilana B. Timokhina, Hossein Beladi, and Subrata Mukherjee

Research of Low Carbon Nb-Ti-B Microalloyed High Strength Hot Strip Steels with Yield Strength ≥700 MPa 317
Hongtao Zhang, Chengbin Liu, and Ganyun Pang

Mechanical Properties and Microstructure of X80 Hot-Rolled Steel Strip for the Second West-East Gas Pipeline 333
Junhua Kong, Lin Zheng, Lixin Wu, Xiaoguo Liu, and Liwei Li

Refinement of Prior Austenite Grain in Advanced Pipeline Steel 341
Chengjia Shang and Chengliang Miao

Part V Specialty Steels

Grain Boundary Hardening and Single Crystal Plasticity in High Nitrogen Austenitic Stainless Steels 359
Markus O. Speidel

Unexplored Possibilities of Nitrogen Alloying of Steel 363
Jacques Foct

High-Nitrogen Steels: The Current State and Development Trends 367
Anatoly G. Svyazhin, Jerzy Siwka, and Ludmila M. Kaputkina

Development of Stainless Steels with Superior Mechanical Properties: A Correlation Between Structure and Properties in Nanoscale/Sub-micron Grained Austenitic Stainless Steel 371
S. Rajasekhara, L. P. Karjalainen, A. Kyröläinen, and P. J. Ferreira

Advanced Heat Resistant Austenitic Stainless Steels 385
Guocai Chai, Jan-Olof Nilsson, Magnus Boström, Jan Högberg, and Urban Forsberg

Research and Development of Advanced Boiler Steel Tubes and Pipes Used for 600°C USC Power Plants in China 399
Z. D. Liu, S. C. Cheng, H. S. Bao, G. Yang, Y. Gan, S. Q. Xu, Q. J. Wang, Y. R. Guo, and S. P. Tan

Strengthening Mechanisms in Creep of Advanced Ferritic Power Plant Steels Based on Creep Deformation Analysis 409
Fujio Abe

New Products and Techniques of Mould Steels 423
Xiaochun Wu and Luoping Xu

Research on Large-size Pre-hardened Mould Blocks of Plastic Mould Steels 443
Dangshen Ma, Lin Wang, Aijun Kang, Qiang Guo, Yongwei Wang, Zaizhi Chen, Lihong Cao, Weiji Zhou and Nailu Chen

Developments and Challenges of China High-Speed Steel Industry over Last Decade 453
Lizhi Wu

Part VI Advanced Steel Processing and Fabrication

Study of Weldability of High Nitrogen Stainless Steel 465
Zhiling Tian, Yun Peng, Lin Zhao, Hongjun Xiao, and Chengyong Ma

Thermomechanical Processing and Role of Microalloying in Eutectoid Steels 475
J. M. Rodriguez-Ibabe and B. López

Study of Non-metallic Inclusions in High Strength Alloy Steel Refined by Using High Basicity and High Al_2O_3 Content Slag 485
Xinhua Wang, Min Jiang, Bing Chen, and Wanjun Wang

Formation of Ultrafine Grained Ferrite + Cementite Duplex Structure by Warm Deformation 495
Tadashi Furuhara and Behrang Poorganji

Pangang Rail Production System Innovation and New Products Development 501
Dongsheng Mei

The Influence of Strong Magnetic Field on Alloy Carbide Precipitation in Fe-C-Mo Alloy 509
Tingping Hou and Kaiming Wu

Part I

General Review

Advanced Steel and Our Society: Better Steel, Better World (Opening Address and the Introduction of the Specific Proceedings)

Yong Gan

Abstract
It has been generally believed that steel is a kind of advanced materials, presenting characteristics to meet a variety of requirements. They could be applied to the circumstances subject to the elevated temperature up to 650°C and cryogenic temperature down to −196°C, to the applied stresses from 100 up to 5,000 MPa, to the corrosion of atmosphere, acid, alkali, salt, etc. Steels has been widely used for construction, automobile, rails, shipbuilding, petrochemistry, machinery, weaponry, daily life, etc.

Keywords
Steels • Low alloy steels • Iron bridge

1 The Importance of Steels

It has been generally believed that steel is a kind of advanced materials, presenting characteristics to meet a variety of requirements. They could be applied to the circumstances subject to the elevated temperature up to 650°C and cryogenic temperature down to −196°C, to the applied stresses from 100 up to 5,000 MPa, to the corrosion of atmosphere, acid, alkali, salt, etc. Steels has been widely used for construction, automobile, rails, shipbuilding, petrochemistry, machinery, weaponry, daily life, etc. (Fig. 1). Thanks to the heaven that there have existed a vast resources of iron ores and human beings have accumulated the experiences to produce and to use steels, which have changed our world remarkably.

Steel industry is the basic link in the economic chain. It provides raw materials to the downstream sectors, such as machinery, automotive, shipbuilding, appliance, and construction (Fig. 2). And it also draws upstream sectors, such as coal mine, electricity, transportation, mineral ores, ferro-alloys, machinery, etc., through the consumption of their products. Steel industry is actually an index to evaluate the industrialization of a country and the comprehensive national power. Generally speaking, the major developed countries are almost stronger at steel industry. Thanks to the advantages of steel, they play very important roles in economy, sustainable society, public finance and tax, defense, and employment.

2 Historic Review of Steels

In the year of 1333 BC, Tutabkhamun's sarcophagus had both a gold and a steel dagger upon it (Fig. 3), signifying the importance of both metals. It was believed to be made from meteorites in Hittite, now Syria. In 1867, the essayist Thomas Carlyle declared: "the nation that gains control of iron soon gains control of gold." At least, it was really true from the beginning of iron age to the end of World War II.

It is obliging to illustrate the two sites of UNESCO World's Heritage to you for the evidence of steels for Industrial Revolution.

Volklingen Ironworks in Germany was an integrated ironworks that was built and equipped in the nineteenth and twentieth centuries and has remained intact (Fig. 4).

Y. Gan (✉)
Chinese Academy of Engineering, Central Iron and Steel Research Institute, Beijing, China
e-mail: gany@cisri.com.cn

Y. Weng et al. (eds.), *Advanced Steels*, DOI: 10.1007/978-3-642-17665-4_1,

Fig. 1 Steels accompany with us in every aspects

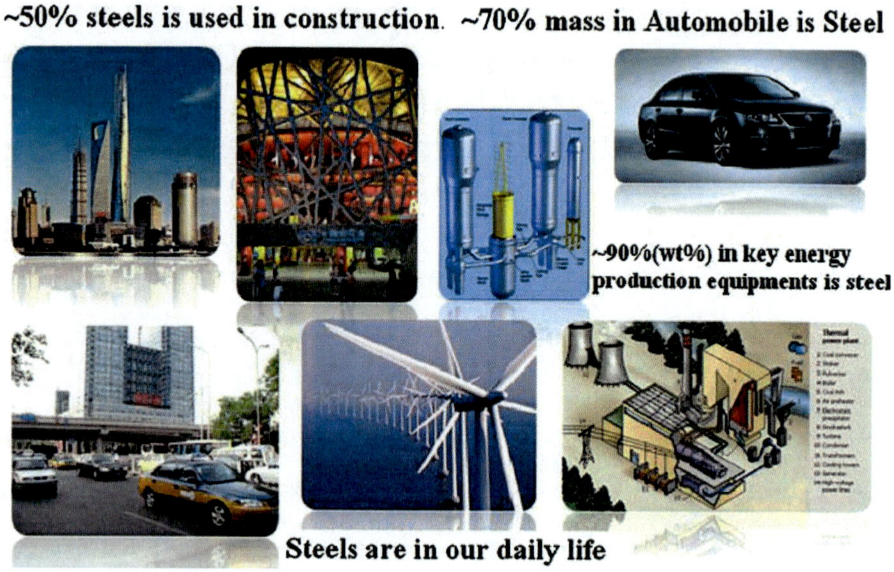

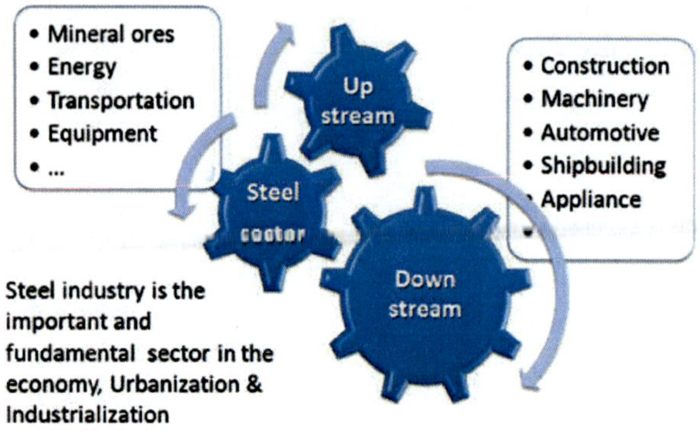

Fig. 2 The roles of steel industry in the economy

The ironworks, which cover some 6 ha, dominate the city of Völklingen. Although it has recently gone out of production, it is the only intact example, in the whole of western Europe and North America, of an integrated ironworks that was built and equipped in the nineteenth and twentieth centuries and has remained intact. And due to the production of steel in the history, it was listed as the site of UNESCO World's Heritage (http://whc.unesco.org/en/list/687/gallery/).

The world's first cast iron bridge was built over the River Severn at Coalbrookdale in 1779 (Fig. 5). Ironbridge Gorge in UK, the site of the world's first cast iron bridge, is known throughout the world as the symbol of the Industrial Revolution. Not only iron founders and industrial spies flocked to see this wondrous bridge, but also artists and travelers. The bridge had a far reaching impact: on local society and the economy, on bridge design and on the use of cast iron in building. The story of the bridge's conservation begins in 1784 with reports of cracks in the southern abutments, and is brought up to date with the English Heritage sponsored work of 1999 (http://www.ironbridge.org.uk/about_us/the_iron_bridge/index.asp).

Ironbridge is known throughout the world as the symbol of the Industrial Revolution. It contains all the elements of progress that contributed to the rapid development of this industrial region in the eighteenth century, from the mines themselves to the railway lines. Nearby, the blast furnace of Coalbrookdale, built in 1708, is a reminder of the discovery of coke. The bridge at Ironbridge, the world's first bridge constructed of iron, had a considerable influence on developments in the fields of technology and architecture (http://whc.unesco.org/en/list/371).

And a modern integrated steel plant, Caofeidian, has been constructed as one of the example of steel technology innovation in China (Fig. 6). Steels play a very important role in the urbanization and industrialization. There are strong demands for steel products not only in quantity but also quality, even for environment benign. For Caofeidian, the steel plant has been established to possess three fundamental roles, steel production, energy conversion, and waste treatment. It may as the model for newly constructed steel plants.

Low alloy steels are as approximately 30% of total steel products. The efforts on the increase of both strength and toughness (ductility) have not stopped over past 50 years. Although Q345 steel is widely produced and applied, higher strength steels are now preferred to construct high rise and large span building, long span bridges, high pressure large diameter pipelines, light weight vehicles, large ships, e.g. Q420 and Q460 steel plates used to construct of "Bird Nest", "Water Cube" and CCTV Station Building for Beijing Olympic Games (Fig. 7); Q420 steel plates for the construction of Dashengguan Bridge over Yangtze

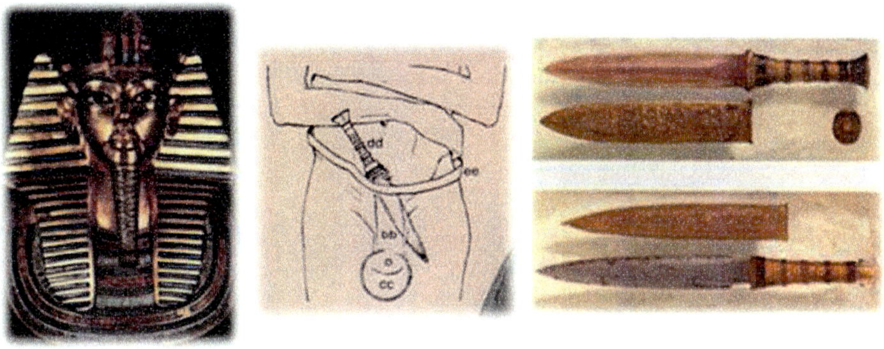

Fig. 3 Tutabkhamun's steel dagger in ancient Egypt over 3000 years

Fig. 4 The intact Volklingen Ironworks in Germany

Fig. 5 Ironbridge Gorge in UK, the site of the world's first cast iron bridge, is known throughout the world as the symbol of the Industrial Revolution

River of high speed railway from Beijing to Shanghai, X80 steel plates for west to east oil pipeline construction, 590 MPa steel plate to reduce the weight of vehicles, FH40, DH40 and EH40 steel plates for large ship building, etc.

Fig. 6 Caifeidian, a newly constructed steel plant in China, possesses three fundamental roles: steel production, energy conversion, and waste treatment

3 Future Perspective of Steels

Steel is the basic material for almost every sector such as construction, machinery, transportation, energy, utensil, etc. From ancient time to now, steel has been playing a very important role in the civilization of human beings. Our world has been changed significantly since the application of steel. Steel will lead us to be higher, faster, and stronger. The main topic of the first International Conference on Advanced Steel is: Better Steel, Better World.

There is no doubt that steel will still be the dominant material in the foreseeable future. Steel is really a type of advanced materials that changes day by day. This change is mainly due to the contribution of physical and chemical metallurgy, steel processing and facilities, market requirements, etc. The new constraints of environment protection and resource saving should be borne in mind in the development of steels in the future. It is noticeable that the requirements for steel products to be of high performance, low cost, easy fabrication, low tolerance, and environmentally benign have become stronger and stronger since the turn of the new century (Fig. 8).

Concerning with high performance, the properties related with load (strength, ductility, toughness), environment (corrosion), time (duration), fabrication (welding, drawing), etc. will be taken into considerations to improve the safety and reliability of components made of steels.

The performance of steel products is closely related to the constitutes and morphology of microstructures. The characterization and effective control of microstructure are now from micron scale to nano scale steadily (to be in nano order). The properties have been raised from the order of 10^6 to 10^9 unit (to be in Giga order). The strength of hot rolled HSLA steel and auto sheet steel has been raised from MPa order to GPa order. The fatigue strength limit of ultrahigh strength steel has been also improved from MPa order to GPa order. The fatigue cycles for steels to undertake have been demanded from Mega cycles to Giga cycles. The rupture time for steel at elevated temperature has been extended from Mega seconds to Giga seconds. The performances of steels in Giga scale are related to precisely controlling of microstructure in nano scale, and closely associated with microstructure characterized with Multi-phase, Meta-stability, and Multi-scale (so-called as M^3 microstructure) (Fig. 9).

Steel makes up approximately 70% of an automobile's overall mass. Advanced steels are no doubt the basis for automobiles to be of high performance, light weight

Fig. 7 The "Bird Nest" for 2008 Olympic Game was made of HSLA steels, Q420 and Q460

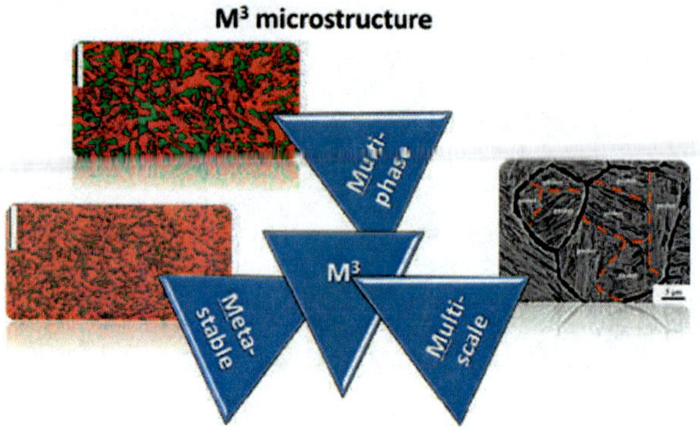

Fig. 9 The performances of steels are associated with microstructure characterized by Multi-phase, Meta-stability, and Multi-scale

Fig. 8 Advanced steels to be of high performance, low cost, easy fabrication, low tolerance, and environmentally benign

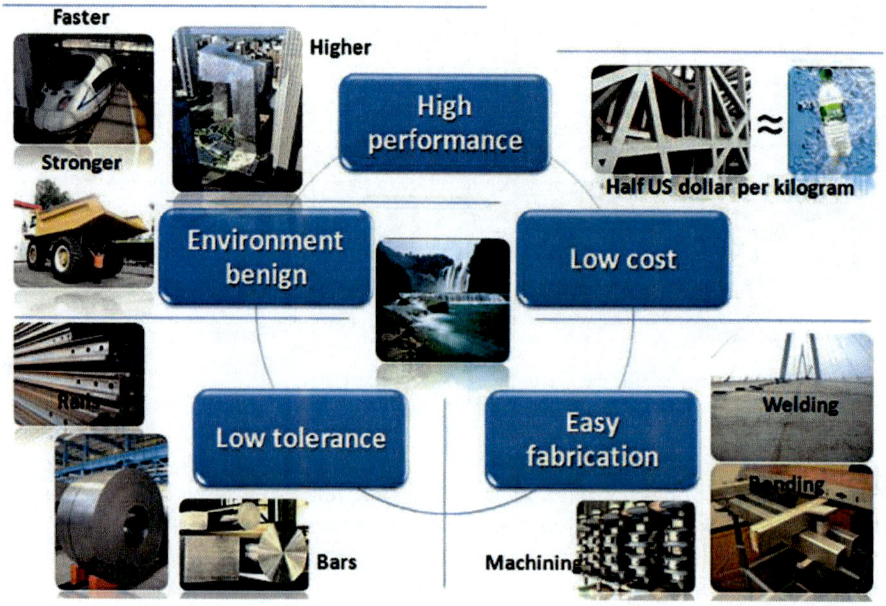

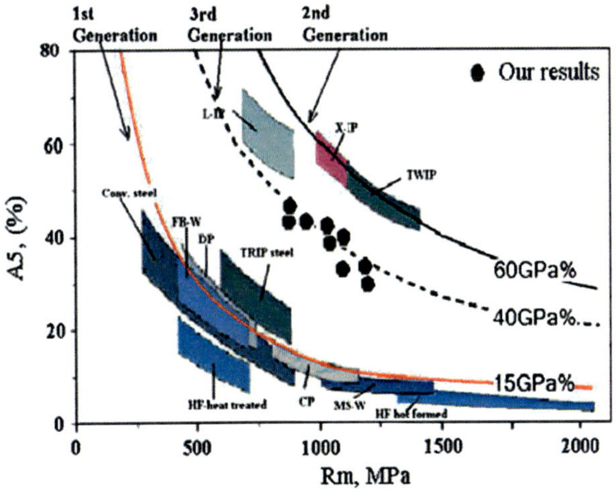

Fig. 10 High strength with high ductility and low cost will be demanded for auto sheet steel

Fig. 11 Luxemburg Pavilion made of weathering steel in Expo 2010, Shanghai

and safety. There are about 30 categories of steel grades produced and used in automobiles or in fabrication today: Al-killed steel, IF steel, BH steel, IS steel, CMn steel, HSLA steel, DP steel, CP steel, martensitic steel, TRIP steel, TWIP steel, austenitic stainless steel, hot stamping martensitic steel, engineering steel, ferritic stainless steel, heat resistance alloy, etc. They are used to manufacture car body and enclosure, engine, transmission system, chassis, suspension parts. Almost every kind of steels could find its way in the manufacture of automobiles, which means that the automobile steels are also very important to the development of all steel products in steel industry.

Nowadays, there are increasing demands for cold sheet steel and coated sheet steel to be in high strength to reduce weight, better ductility to improve formability and safety, low alloy addition and easy fabrication to reduce cost. The development of automobile steels is so fast that nobody could image the future progress precisely. In the last 1990s, people focused their efforts in IF steel and BH steel. And now, DP steel, TRIP steel and hot stamping martensitic steel are being widely used in automobiles, and even to begin with the research of the third-generation sheet steel (Fig. 10).

One of the main disadvantages of steels with low alloying elements is easy to be corrosive in the atmosphere. Stainless steel is one of the ways to overcome this problem, but cost a lot. Another way is to adapt weathering steels for infrastructures and buildings to be of longer duration (Fig. 11). Longer duration will need to pay more attention, not only to resist corrosion, but also to resist heat, cycling load, hydrogen embrittlement, wearing, etc. As a result, the components made of steel will be more effective, and the steel consumption will be reduced.

It is confidently believed that steel will become much better, and eventually leads to a much better world for human beings in the future.

Innovative Steels for Low Carbon Economy

Lejiang Xu

Abstract

As one of the vital structural materials, steel has played an important role in national economic development. Under the background of global warming, holding back carbon footprint has become the main task of our mankind. As a giant source of CO_2 emission, it is rather a severe challenge for steel industry to develop further under *Energy Saving and Emission Reduction* Policy (ESER). This article has reviewed and envisioned such practice on steel production, and analyzed how to make innovation on steel material based on Baosteel's own practice so as to provide material solution for down-stream sectors. High strength, high toughness, long service life and versatility of steel material are the trend for material innovation.

Keywords

Steel material • Innovation • Low carbon economy

1 Introduction

With the sustainable development of China's economy, steel industry, as main raw material source for national economy, has sharply taken off. Especially over the past 10 years, substantial breakthroughs have been made in scale, and output has grown 5.4 times, which makes China the largest steel producing country. Its crude steel output share has been shifted from 15% approx. in 2000 to nearly 50% in 2009.

As an economic development engine, steel industry is also one of the main CO_2 producers. According to the statistics from International Energy Agency, the carbon emission of steel industry accounts for 4–5% [1] of global total amount. While within China, that value is 15.6%, accounting for 43.3% [2] of the steel industry the world over. Therefore, it has become a social issue. Chinese government has solemnly committed that till 2020, CO_2 emission per GDP will be reduced by 40–50% than 2005. As the main carbon emitter, steel industry should take such social responsibility and historical mission.

Energy Saving and Emission Reduction (ESER) of steel industry shall focus on two aspects, one is ESER of steel industry itself; the other is the contribution made by innovative steel material for down-stream sectors. This article has briefly reviewed and analyzed the first scenario. Taking automobile fuel economy, power station boiler, energy transmission, oil–gas transportation and corrosion resistant materials as examples and based on Baosteel's own practices, we focus on the discussion about how to provide material solution through technology innovation for down-stream sectors.

2 ESER of Steel Industry Itself

The basic principle of steelmaking is to reduce ferrous oxide by carbon, than produce carbon saturated hot metal, which is the source to produce liquid steel with different carbon content through oxidation refining. After solidification and rolling,

L. Xu (✉)
Baosteel Group Co., Ltd., Shanghai 201900, China
e-mail: oujh@baosteel.com

Y. Weng et al. (eds.), *Advanced Steels*, DOI: 10.1007/978-3-642-17665-4_2,

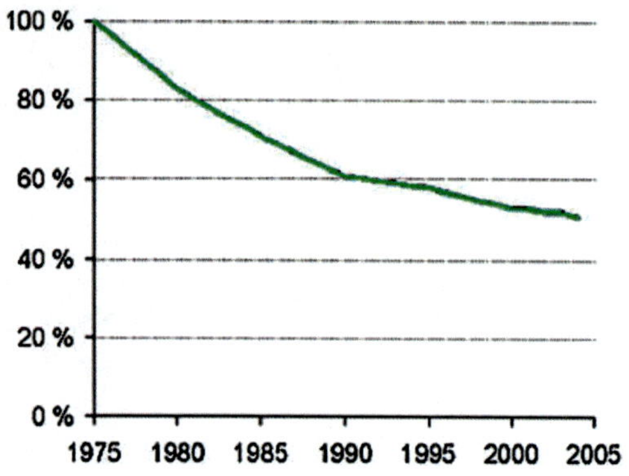

Fig. 1 Energy consumption change per ton of steel in North America, Japan, Europe(1975 =100%)

final product is come out for different clients. Therefore, the main emission of steel industry is CO_2.

In traditional steel production, over 90% CO_2 emission is resulted from energy consumption [3], because carbon energy is dominant in all consumptions. According to China Energy Statistical Yearbook, in 2007, coal has taken up 80% of the total primary energy, thus, energy saving is the priority for CO_2 emission reduction.

Over the past 30 years, global steelmakers have made remarkable improvement in reducing energy consumption through technology upgrade. In such developed areas as North America, Japan, Europe, etc., within the 30 years from 1975 to 2005, average energy consumption per ton of steel has decreased by about 50% [4], see Fig. 1.

According to China Iron & Steel Statistics, standard coal consumption per ton of China's steel industry is reduced from 2.04 ton in 1980 to 0.619 ton in 2009, while CO_2 emission per ton of steel is reduced from 3.22 ton in 1991 to 1.87 ton in 2007, by 42%.

Although great achievements have been made in ESER of China's steel industry, compared with those advanced steelmakers in the developed countries such as Japan, Korea, Germany, the steelmakers in China still have a long way to go in consumption control. Our unit consumption is yet higher than international advanced level by 10–20% [5], for this reason, we still have great space to improve.

In steel production, process renovation is decisive to ESER. For instance, compared to mould casting, continuous casting has saved ingot heating and blooming processes so that consumption can be sharply reduced. Likewise, compared to continuous casting, strip casting enjoys even lower consumption. The process schemes of conventional continuous casting and hot rolling, and strip casting process were shown in Fig. 2. As for the former, two thermal cycles are needed when hot metal is altered into steel plate, while the latter needs only one cycle. Therefore, the consumption for the latter is much lower.

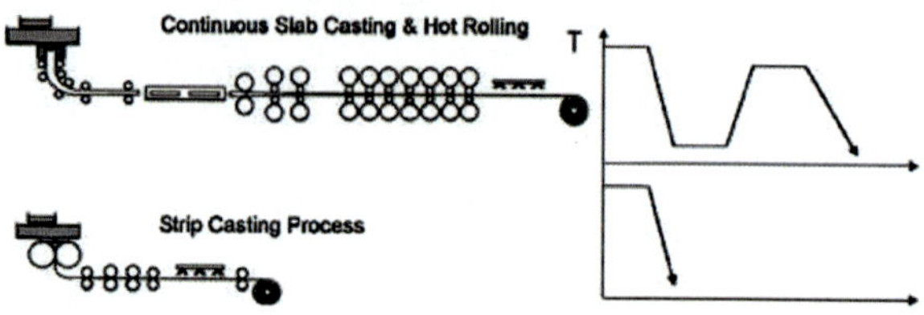

Fig. 2 Schemes of conventional continuous casting and hot rolling and strip casting process

With 10 years' research, Baosteel has successfully developed a brand new low carbon production process in 1,200 mm strip casting pilot line. This year, Baosteel has announced to build an industrial strip casting model line in Ningbo Iron & Steel with annual capacity of 500,000 ton.

3 Innovative Steel Materials Provide Solution for Low Carbon Economy

As a fundamental raw material of national economy, steel is obliged to provide necessary material for technological improvement of other sectors as transportation, energy power, and infrastructure in particular. In the mean time, the development of these sectors raise higher demand for steel materials, this becomes a motive power for its innovation. Then, in the light of fast developed industries such as automobile, oil–gas transmission, power transmission, and power plant boiler, combined with Baosteel's own practices, we will come to the topic of steel material innovation.

3.1 Fuel Economy of Automobile and High Strength of Steel Plate

Among such measures as oil consumption reduction and emission cut in automobile industry, more attention has been attached on lightweight of car body. Statistically speaking, each 10% weight losing could save 3–7% fuel and 13% CO_2. Figure 3 has shown the relations between car weight and fuel efficiency [6].

According China Automotive Lightweight Union, our own-brand passenger cars are 10% heavier than overseas ones of its kind, while larger gap exists in commercial vehicles. In 2010, sales volume of China automobile market expects to exceed 17,000,000, and continues to maintain World No. 1. Rapid increase of automobile ownership results in boosting demand on petroleum. Now, automobile oil consumption accounts for one-third of total oil consumption in China, and is estimated to rise to 57% in 2020 [7]. Therefore, promotion of lightweight research is significant to low carbon society.

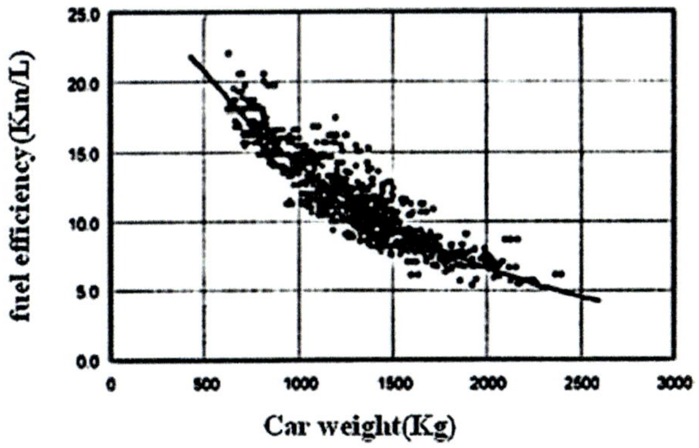

Fig. 3 The correlation between car weight and fuel efficiency

The research result of IISI Automotive Lightweight Project ULSAB-AVC shows [8], massive application of high strength steel and advanced manufacturing technologies (mainly including tailor welded blank, hydraulic forming and hot stamping) are the shortcuts to reduce weight for automobile. Compared with conventional steel, the application of high strength steel could reduce the car weight by 20–25%. In 2009, the application proportion of high strength plate in Chinese automobile industry was only about 25%, while that value abroad was over 50%, while even larger gap exists in application of advanced manufacturing technology. Why these effective solutions are not widely applied in China? On the one hand, China automobile industry needs stronger design capability; on the other hand, domestic steel makers need to capture more core technologies in stable production, application and advanced manufacturing technologies of high strength steels.

In order to promote Weight Reducing & Energy Saving in China automobile industry, Baosteel always stresses R & D in high strength steel and advanced manufacturing technology. Based on lab research, Baosteel has successively built dedicated production lines for ultra high strength plate, tailor-welded blank, hydraulic forming, and hot stamping. Till 2009, Baosteel had owned annual capacity of 200,000 ton ultra high strength plate, 20,000,000 tailor welded blanks, 460,000 hydraulic forming parts and 1,000,000 hot stamping parts. In particular, the dedicated line for ultra high strength steel, started construction in early 2009, has appliedfast cooling technology and multi-functional production process which are jointly developed by Baosteel and MITSUBISHI-Hitachi. After that, the available strength level for cold rolled plate is upgraded from 800 to 1,500 MPa, while for galvanized plate is from 800 to 1,200 MPa. Recently, Baosteel has trial-produced third generation high strength steel—Q&P steel [9], which enjoys higher plasticability than first generation.

High strength material solution is not only fit for automobile industry, but also for other sectors as construction, machinery, container, etc. For instance, screw threaded steel is shifted from Level II (345 MPa) to Level III (400 MPa) and Level IV (500 MPa), which container steel from 345 to 600 MPa and 700 MPa, etc. High strength has become the main trend for innovative steel materials.

Greater efforts shall be made in the field of material science to produce stronger steel to reduce the material consumption, which will not only reduce energy consumption in production, but also make due contributions to ESER for down-stream sectors.

3.2 Oil–Gas Transportation and Pipeline Steels with High Strength and High Toughness

Oil and gas is the crucial energy in modern society. Since the oil and gas fields are usually located in remote areas, long distance pipelines are the most economic, safe and environmental friendly delivery system to transport the oil and gas from field to consumers. In order to save construction investment of pipe line project, enhance transmission efficiency and reduce transportation cost, the operating pressures and diameters of pipeline continue to increase, which requires higher reliability of the pipelines. To handle the demand, the pipeline steels with high strength, high crack propagation arrest toughness at low temperature, excellent weldability are necessary. As for those applied in special areas, H_2S resistance is required. The increasing demand on comprehensive properties of pipeline steel has tremendously promoted the development of modern pipeline steels.

Baosteel's pipe line steel develops at the same pace with the construction of oil and gas pipelines in China. Table 1 has shown the main characteristics of Baosteel's pipe line steel developed over the past 20 years. It indicates that high strength and toughness has become the main theme of Baosteel's pipe line steel development. Steel grade shifted from X42 and X52 20 years ago to X80; impact toughness upgraded from 90 J impact energy to over 240 J at −20°C; maximum thickness increased to 33 mm from 10 mm; available products diversified to coil, heavy plate and welding pipe so as to provide enough material for West–east Natural Gas Transmission Project in China.

Figure 4 shows Baosteel's pipeline steel output in history and grade distribution in 2009. It indicates that the output increases year by year and up to 1,000,000 ton in 2009. Meanwhile, high grade pipeline steel is the main demand in the market. In 2009, the share of the steels of X70 grade and higher is 60%.

Table 1 Baosteel pipe line steel development in variety, grade and product form

Year	90	91	92	93	94	95	96	97	98	99	00	01	02	03	04	05	06	07	08	09
Steel grade	X42					X52														
							X60													
								X65												
											X70									
												X52(HIC), X60(HIC), X65(HIC)								
														X80						
																	X100, X120			
CVN-20°C	≥30 J		≥90 J								≥190 J			≥240						
Thickness	≤10 mm								Up to 17.5 mm								Up to 33 mm			
Variety	Coil																Coil, heavy plate, welding pipe			

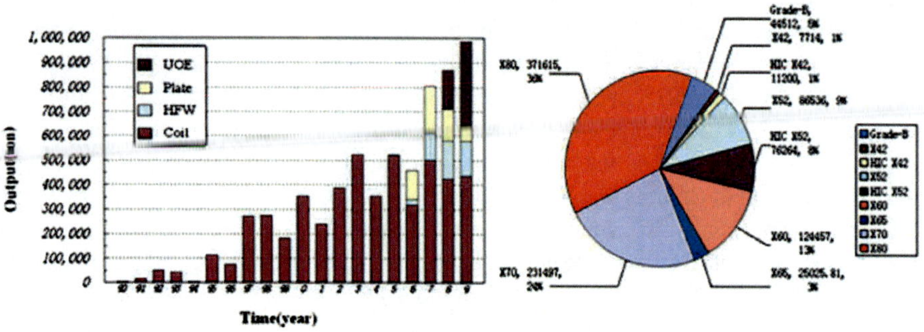

Fig. 4 Baosteel pipe line steel output in history and grade distribution

Same as automotive materials, the developing trend for pipe line steel is the high strength. However, besides high strength, the customers require excellent ductility and toughness also. Therefore, further efforts shall be made by experts in steel material to develop high strength material with excellent ductility and toughness.

3.3 Corrosion Resistance and Long Service Life Design of Steels

Steel is naturally subject to corrosion in service environment. According to the statistic data from Chinese Corrosion Survey Report, the direct economic loss caused by corrosion accounted for about 2–4% of GDP in developed countries, while accounted for about 5% of GDP in China. Meanwhile, the indirect economic losses were immeasurable caused by corrosion-induced equipments damage, mechanical downtime, product quality decline, pollution, and accidents such as explosion and fire. According to the statistic data, about 70% of the failures of oil/gas pipeline were due to corrosion. If the corrosion-resistant steels and appropriate protective measures were adopted, 30–40% of losses caused by corrosion could be retrieved. Therefore, the investigation on the mechanisms of corrosion and the development of long service life steels become the important issues which are urgently needed to be solved.

Demand for corrosion-resistance of steels varies according to service conditions. After a 20-year development, Baosteel has established a product line of corrosion-resistant steel, including weather-resistant steels, H_2S-resistant pipeline and well tube steels, CO_2-resistant 13Cr steels and Ni-based alloys for well tube. These products have been widely applied in various industries, such as containers, railway rolling stocks, automobiles, buildings, off-shore structures and oil and gas field equipments.

There is still a lot of work to do on long service life steels, which contributes a great deal to the construction of the low-carbon society. For instance, the new type corrosion-resistant steel plate recently developed in some country for oil tanker and VLCC exhibits five time higher corrosion resistance than the former product, which not only can dispense with the coating process, but also can promote the safety and the environment conservation of the ships. In order to reach the goal, the material researchers must innovate continuously to develop the steels suitable for various service conditions with longer service life.

Table 2 Relations among vapour parameters, power plant efficiency and coal consumption

Unit type	Vapour pressure (MPa)	Vapour temperature (°C)	Power plant efficiency (%)	Coal consumption for power supply (g/kW h)
Medium pressure unit	3.5	435	27	460
High pressure unit	9	510	33	390
Super-high pressure unit	13	535/535	35	360
Subcritical unit	17	535/535	38	324
Supercritical unit	25.5	566/566	41	300
Ultra-supercritical unit	27	600/600	44	278
Ultra-supercritical unit	30	600/600/600	48	265
Ultra-supercritical unit	30	700	57	215
Ultra-supercritical unit		>700	60	205

3.4 Energy Efficiency of Power Plant Boiler and Ultra-Supercritical Boiler Tube

Electricity is a safe, efficient and clean secondary energy sources, which plays a decisive role in the national economy. It is estimated that until 2020, the total installed capacity of power generators in China will reach 1.186 billion KW. Since the primary energy source in our country is mainly dominated by coal, coal-fired power generation will certainly cause enormous pressure on the environment. Table 2 reveals the relations among vapor pressure, temperature, power efficiency and coal consumption of different units. It is evident that along with the increase of vapour pressure and temperature, the thermal efficiency of power plant boiler improves, while the coal consumption decreases. For instance, the efficiency of the supercritical unit with a vapor pressure of 25.5 MPa and a vapor temperature of 566°C is 41%, and the coal consumption is 300 g/kW h. While the efficiency of the ultra-supercritical unit with a vapour pressure of 30 MPa and a vapour temperature of 700°C can reach 57%, and the coal consumption is reduced to 215 g/kW h. Therefore, the ultra-supercritical unit with high-capacity, high vapor pressure and high vapor temperature represents the direction of future development of power plant boiler.

With many years' unremitting efforts, China has increased the main vapor temperature of power unit to 600°C, and pressure to 26.5 MPa. In the next 10 years, it is estimated that the vapour parameters of coal-fired power generation in China will increase to 700°C and 30 MPa or higher. This will raise a severe demand on the high-temperature strength and oxidation resistance of boiler tubes. Therefore, whether the steel plant is able to produce such boiler tubes becomes one of the restrictions on the development of ultra-supercritical coal-fire power units.

From 1999, Baosteel has been engaged in the study of key materials for supercritical and ultra-supercritical coal-fired power unit with high parameters. The high-pressure boiler steels such as T91, T23, T92, S30432 have been developed one after another. Furthermore, through the demonstration on the application of T91, T92 made by Baosteel on Baosteel's 350,000 kW subcritical unit, and performance tests and assessments in the industry of boiler, the high-pressure boiler tube and the inside screw tube of Baosteel's T91, T23, T92 have been widely used in supercritical and ultra-supercritical coal-fire power unit in China. Up to 2009, accumulated production of high-pressure boiler tube has reached 184,000 tons. Generally speaking, Baosteel is able to supply materials for the whole unit heated surface of boiler at 600°C main vapour in the ultra-supercritical power plant. The following Fig. 5 shows the development of the power plant boiler and the development of Baosteel's boiler tube products.

Steel industry should make greater efforts to develop boiler tubes with higher high-temperature strength and higher oxidation resistance, and supply more competitive steel material for power plant boiler sector.

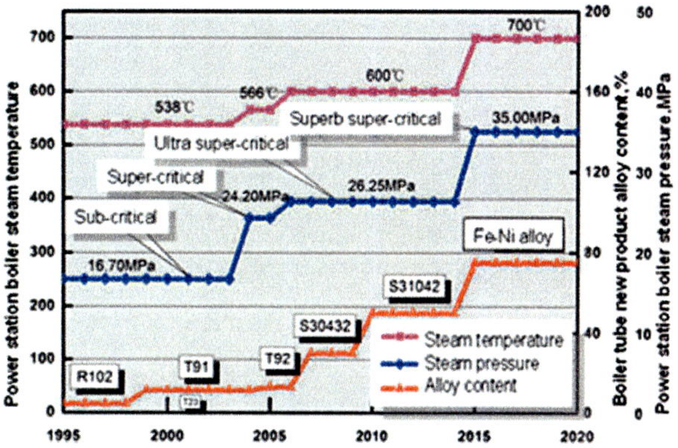

Fig. 5 Developments of the plant boiler and Baosteel's boiler tube products

3.5 Energy-Saving of Transmission and Distribution and Oriented Silicon Steel with High Magnetic Induction

Transformer is one of the key equipments in power sector. Silicon steel is the indispensable material in making the transformers. The transformer made of ordinary oriented silicon steel can cause a power loss of about 1% of the total transmission and distribution capacity. With the transformers made of oriented silicon steel with high magnetic induction, the power loss can be reduced by 40%. If we make a calculation based on the national total power generation capacity of 3 650.6 billion kW in 2009, it means 14.6 billion kW power is saved, which accounts for one-fifth of the national nuclear power generating capacity in 2007 and which is close to one year's power generation volume of Gezhouba Hydropower Station. Besides its energy efficiency, the oriented silicon steel with high magnetic induction can save more than 15% steel consumption for making a same transformer comparing to the ordinary material. Meanwhile, it can cut down copper consumption.

Baosteel has spent 10 years in self-development of the oriented silicon steel production technology. Finally, the production technology of high magnetic induction (HiB) grain-oriented silicon steel with low reheating temperature has been captured. The commercial production of grain-oriented silicon steel has been started from 2008, and 42,000 tons of HiB have been produced in 2009. It is estimated that around 70,000 tons will be produced in this year, thereinto, laser-notched products will fill the domestic gap.

4 Conclusions

The ESER can be implemented in two aspects: the steel industry itself and the contribution made by innovative steel material for down-stream sectors. As an irreplaceable material for the current and foreseeable future human society, there is a great potential for material innovation. We hope great efforts shall be taken by the steel industry staff to make continuous innovations on the material of high strength, high toughness, long service life, and functionalization, etc. so as to provide competitive steel material solutions for downstream users, and make due contributions to the low-carbon society.

References

1. IEA Energy Technology Perspectives 2008; total greenhouse gas emissions from human activities in 2004 were 49 billion tones (IPCC Working Group III, Climate Change 2007)
2. Y. Gan, Modern Steel and Steel Eco-Products Process in Low-Carbon Economy, China Development Forum 2010 of Strip Continuous Galvanizing, Iron and Steel Research Institute (2010)
3. T. Su, Prospect of iron and steel industry under the guidance of a low-carbon economy. Shangdong Metall. **32**(2) (2010)
4. IISI, AISI, JISF, and JFE, Global Steel Sectoral Approach, presentation Washington (2008)
5. K. Xu, Low-carbon economy and steel industry, Iron Steel **45**(3), 1 (2010)
6. S. Takehide, Physical metallurgy of modern high strength steel sheets. ISIJ Int. **41**(6), 520–532 (2001)
7. X. Zhang, Automotive Engineering **31**(1), 1–5 (2009)
8. IISI (2002) ULSAB-AVC Overview report. http://www.worldautosteel.org/uploaded/ULSAB_Overview_Report.pdf
9. W. Li, W. Li, W. Feng (2010) Industry trials of C–Si–Mn steel treated by Q&P Concept in Baosteel. SAE International

Development and Outlook of Advanced High Strength Steel in Ansteel

Xiaogang Zhang

Abstract

The structure steel industry has experienced a revolution during 4 decades. Faced the challenge of global change in climate and environment, the higher strength ductility steels and the environment friendly steels are needed. The R&D for high strength steel production and application in Ansteel has made impressive progress. However, more attention had been paid on the development of new-type high strength steels with higher strength and better properties. The multi-phase microstructure, lower y/s, and corrosion resistance performance structure steel result in new generation high strength steel, which have properties that are often much superior to those exhibited by the older steels. This chapter presents a general review of new generation high strength steel research and development in Ansteel and predicts the development for advanced high strength steel in the foreseeable future.

Keywords

High strength steel • Energy saving • Emission reduction

1 Foreword

In the field of material in twenty-first century, the steel production technology keeps developing at high speed after finance crisis. In 2010, the steel yield will reach 600 million tons in China which means China have become biggest steel production country in the world. Faced the challenge of global change in climate and environment, developing advanced high strength steel is one of the most important method for promoting energy saving and emission reduction. If the steel strength were increased from 400 to more than 800 MPa, the steel consumption would be reduced greatly. So research and development for advanced high strength steel are very meaningful for building a steel great power. The development of high strength steel in Ansteel began in end of last century. As being a strategic target, the work focused on two aspects: firstly the high level production lines of Bayuquan steel project and Ansteel western section project have been built, which greatly promoted the capability of advanced high strength steel production; secondly the development of production technology aimed at advance high strength steel such as high strength hull plate, nuclear power station steel, high strength container steel, power reserve tank steel and high class line pipe steel.

2 The Construction of High Strength Steel Production Line in Ansteel

2.1 Bayuquan Steel Project—the Model of High Strength Production Line

In order to increase capability of high strength steel production, a new production line has been built in Bayuquan. Bayuquan Steel Project was approved by National Development & Reform Commission on 17 May 2006 and was

X. Zhang (✉)
Anshan Iron and Steel Group Corporation, 114021 Anshan, People's Republic of China
e-mail: zxg@ansteel.com.cn

Y. Weng et al. (eds.), *Advanced Steels*, DOI: 10.1007/978-3-642-17665-4_3,

completed and put into operation on 10 September 2008, which possesses an annual capacity of 6.5 million tons of pig iron, 6.5 million tons of crude steel and 6.2 million tons of rolled steel. Its leading product mix focuses on the high value-added products with high technology content, including the container steel plate, pipeline plate, ship plate, mechanical structure steel, boiler plate, vessel plate, bridge plate and building steel. From the beginning of design, Ansteel Bayuquan Steel Project has implemented the energy saving and emission reduction concept to provide "hardware" for production of high strength steel. The product orientation is to substitute low-strength steel by high-strength steel for which many advanced technologies have been adopted in its working procedure to meet different demands.

2.2 Advanced Steelmaking Technology

The steelmaking and continuous casting process consists of three hot metal desulfurization and skimming plants, three 360 t top and bottom blowing converters, a ladle refining furnace (LF), an ANS-OB ladle refining furnace, two RH-TB vacuum degassing devices, two 1,450 mm continuous casting machines, a thick slab continuous casting machine. More than 30 advanced technologies have been adopted such as hot metal pretreatment with compound blowing, high-efficient converter steelmaking automation, new high-quality refining, dynamic soft reduction segment, strand roller electromagnetic stirring, etc.

2.3 Advanced Rolling Technology of Heavy Plate

Heavy Plate Mill uses the 5,500 mm + 5 m 2-stand solution, with the maximum roll force of 10,000 t, in conformity with the technical requirement for a modern medium and heavy plate mill, e.g., the maximum roll force (more than 20 kN/mm), high power (2 kW/mm) and high rigidity (more than 2 kN/mm). The product dimensions are 900–5,300 mm in width, 5–150 mm in thickness, max. 450 mm, 3–25 m in length and is the only heavy plate mill being able to produce 4,800–5,300 mm plates in nation.

2.4 Advanced Hot Rolling Technology

More than 20 advanced technologies have been adopted in hot rolling line by which mill can realize the temperature holding of thin strip in the whole process, high-strength steel controlled rolling, multi production models and flexible production. The hot charging rate is up to 50%.

3 Introduction of Key High Strength Steel in Ansteel

3.1 Development of High Strength Ship Plate

Ansteel has developed shipbuilding steel for a long term and now has the largest yield in producing the most variant dimension and the highest grade of shipbuilding steels in China. Ansteel is also the pioneering steelmaker to produce the ultra-high strength steel for shipbuilding and offshore in the world. So far the ship plate produced by Ansteel adds up to 12 million tons in which the percentage of the high strength steel is up to 50%, and they totally have passed 17 times certificating by various Classification Societies. A number of research projects on high strength steel for navy have been carried out and 3 patents were granted. In May 2006, the steel grade developed by Ansteel from AH32 to FH550, with the maximum thickness of 100 mm were certificated by 9 Classification Societies in the world with the granted 36 patents and 32 proprietary technologies. In 2008, Nickel alloy steels containing 3.5, 5, and 9% Ni for cryogenic service were successfully produced at Ansteel and were also certificated by DNV, LR, and CCS Societies. In 2009, the steel plates for high heat input welding were successfully developed with the maximum thickness of 100 mm and the weld heat input of 100 kJ/cm and were certificated by ABS, CCS, DNV, GL, NK classification societies. Also in 2009, the project of "manufacturing technology innovation and integration of high-performance shipbuilding steel" won the second award of China's State Science and Technology Awards.

3.2 Development of Nuclear Power Steel

In 2006, 15 MnNi steel plates that thicknesses range from 30 to 105 mm were produced in Ansteel and were used in the construction of Qinshan Phase II project (Fig. 1). So far Ansteel has established cooperation with a number of makers of nuclear power equipment and the products deal with API1000, CPR1000 and EPR technology. At meantime these products have been extensively used in SanMen 1# and 2#, HaiYang 1# and 2#, YangJiang, TaiShan, FuQing million-kilowatt class nuclear power station project etc.

In August 2010, AP1000 main conduit used in San Men nuclear power station was successfully forged in Ansteel (Fig. 2). It represents a breakthrough on construction of nuclear power station in China and brings a far-reaching influence on stepping up third generation AP1000 equipment nationalization in China.

Fig. 1 QinShan Phase II project

Fig. 2 San Men AP1000 main conduit

3.3 Development of Pipeline Steel

From beginning of 2001, Ansteel has strengthened the research for high grade pipeline steel. Through hard work of years, a series of achievements have been obtained and put into use in project of natural gas transportation from west to east. In June 2007, X80 (18.4 mm thick and 1,550 mm wide) pipeline coil sheet passed evaluation of making tube for the fist time in HuaBei Petroleum Tube Making Plant and in December 2007 it passed the authentication sponsored by Petroleum of China and Iron and Steel Societies. Until now 150,000 tons of X80 coil plate have been produced and used in project.

The development of X80 flat all the while keeps the leading position in metallurgy industry. Early in 2005, the X80 flat was used in JiNing pipeline project of which made Ansteel the only X80 flat maker and dealer at that time. In recent years X100, X120, high-strain resistant X70 etc. have been developed in succession. Ansteel now has possessed capability to produce variant dimension and the highest grade of pipeline steels. Ansteel is also the pioneering steelmaker to produce the ultra-high strength pipeline steel.

3.4 Development of High Strength Automotive Steel

By means of technological introduction and self-innovation, Ansteel has built a number of automotive production lines with international leading level, formed key production technology of high quality automotive steel plate. By now, Ansteel has developed deep-drawing steel series, high strength deep-drawing steel series, advanced high strength steel series and high quality surface steel series. Meanwhile, Ansteel developed cold rolled steel plate with characteristic to meet user's individual requirements. Ansteel has formed integrated automotive steel series including hot rolled, cold-rolled and hot-dip galvanized steel.

On ASP automotive steel production lines with self-owned intellectual property rights, Ansteel has successfully developed advanced high strength steel represented by DP and TRIP steel with tensile strength grade of 780 N/mm^2, it has been provided to customer commercially.

The low-Carbon low-Silicon no-Aluminum (low-Aluminum) TRIP590 and TRIP780 steel have been developed in Ansteel. By means of breaking the traditional alloy design concept, replacing Silicon and Aluminum with P or P + V, combining the laboratory test and thermodynamics and kinetics calculation, this new type of steel is characterized by low cost, high welding and galvanizing performance, low-temperature toughness and easily production. The performance of TRIP steel is equivalent or superior to the level of similar products abroad. At present, Ansteel has developed TRIP steel sheet with strength grade of 590 and 780 and provided it to customer in batch. The TRIP steel with strength grade of 980 has been developed in laboratory.

3.5 Development of High Strength Container Steel

As an important steelmaker, Ansteel provides directly high strength container steel for more than 40 users, and occupies 22% market share in domestic markets. The main products are Q550NQR1, Q550 J, AS600MC, AS700MC etc. and the trial-producing started in 2006. Among them the output of AS700MC has reached 8000 t whose characters are favorable to welding ability, excellent cold bend and good toughness properties. Ansteel has already completed experimental research work of 700 MPa grade high strength cold formation steel plate which has the atmospheric corrosion resistance performance.

3.6 Development of Water Power Station Steel

In the SanXia construction, to meet the project emergent requirement, Ansteel successfully developed turbine shell steel and pressure tube steel (Fig. 3). In 2004, totally 12,000 tons shell steel and pressure steel of ADB610D were used in the 12 turbine sets located on right bank of San Xia, which symbolized that Ansteel has held the process technology for producing high strength heavy plate.

Fig. 3 Turbine shell made by ADB610D

4 Development and Outlook of High Strength Steel

The economy and society development call for the new generation steel material. Large scale economy construction has been undergoing in China, which requires high strength, high performance and long service life steel in various fields such as high speed railway, over loading bridge, high building, transportation of oil and natural gas, light energy saving car, engineering machinery, big shipping etc. Theoretically, the strength of steel could be more than 8,000 MPa, but the strength of steel lot of used at present still is less than 800 MPa. On the foundation of scientific research already achieved, increasing of strength and service life of steel and developing of advanced high strength steel with the properties of corrosion-resistant, delay rupture-resistant and tired rupture-resistant are possible in technology.

In future, Ansteel will focus on the R&D of advanced high strength steel, in which the core of research still is to study the relationship between microstructure and property. Then by means of reasonable process combined with advanced equipment, the valuable steel materials would be produced. Most of high strength steels with high performance are alloy structure steels and used in QT state. It is not enough to increase the strength from 800 to 1,500 MPa only by grain fining strengthen. The new theories and new technologies are needed to be developed as well. Combined with micros-alloy, rolling control and super cooling, the methods including increasing the clean degree, improving uniformity of steel, selecting reasonable alloy composition and heat treatment etc. should be used for developing new generation high strength steels. One of the developing strategy of 'Ansteel' is to develop more than 1,500 MPa advanced high strength steel.

In twenty-first century, the steel is surely the selected material, and steel industry is still strong, full of vigor. Ansteel wishes with every enterprise to pay great effort for making a steel great power.

Technical Progress and Product Development of TISCO Stainless Steel

Xiao Bo Li

Abstract

TISCO is the earliest and the largest enterprise of stainless steel production in China. After continuous technical reconstruction, especially 500,000t stainless steel enlarging capacity reconstruction and 1.5 million tons of new stainless steel revamp project, stainless steel output reached 3 million tons. In recent years, a group of proprietary technology with independent intellectual property rights was formed through the independent innovation. In aspects of the stainless steel development, TISCO has optimized product structure and the products are widely applied in the high-end market.

Keywords

Stainless steel • Revamp • Proprietary technology • Product structure

Since the reform and opening to the world, especially after entering into twenty-first century, stainless steel production in China has been rapidly developed and the apparent consumption of stainless steel in 2001 amounted to 2.28 million tons, exceeding US and becoming first consuming big country of stainless steel in the world, and in 2009, stainless steel consumption in China by 8.22 million tons, covering 34.25% of total stainless steel consumption in the world, and per capita consumption of stainless steel changed by 0.08 kg from 1998 to 6.32 kg in 2009, higher than the world's average consumption. With the swift economic development in China and the powerful requirement of stainless steel as well as policy support given by the government, state owned enterprises, joint ventures and private enterprises were strongly eager to have put more investments in stainless steel domains during "the Ninth Five-year Plan" and "the Tenth Five-year Plan" and stainless steel output increased very fast, and in 2006, the stainless steel output in China overtook Japan, becoming the largest stainless steel producer in the world. In 2009, stainless steel capacity in China reached 8.8447 million tons, 36% of world's total stainless steel. Nowadays, the capacity of stainless steel in China is able to reach 13 million tons, the one-third of world's total stainless steel. Through ten years quick development, the position of stainless steel of China in the world changes a lot and the whole world focuses its attention upon the development of stainless steel in China. Figures 1 and 2, respectively, give the latest apparent consumption of stainless steel and change of stainless steel output in China.

TISCO is the earliest and the largest enterprise of stainless steel production in China. For years, after continuous technical reconstruction, especially 500,000 tons stainless steel enlarging capacity reconstruction and 1.5 million tons of new stainless steel project, stainless steel output increased rapidly, reaching 3 million tons, and TISCO has developed and changed into a stainless steel enterprise with the international largest size in capacity and installation scale, state-of-art process technology and equipment, the shortest process flow and friendly environment protection.

In addition to the greater assistance helped by the state and strong driving of consumption market of domestic stainless steel, TISCO stainless steel fast development originates from the combination of international first-class full process stainless steel installation, talented person and technical resources, etc., relying on self-innovation

X. B. Li (✉)
Taiyuan Iron and Steel (Group) Co., Ltd., Taiyuan, 030003, Shanxi, China
e-mail: qinly@tisco.cn

Y. Weng et al. (eds.), *Advanced Steels*, DOI: 10.1007/978-3-642-17665-4_4,

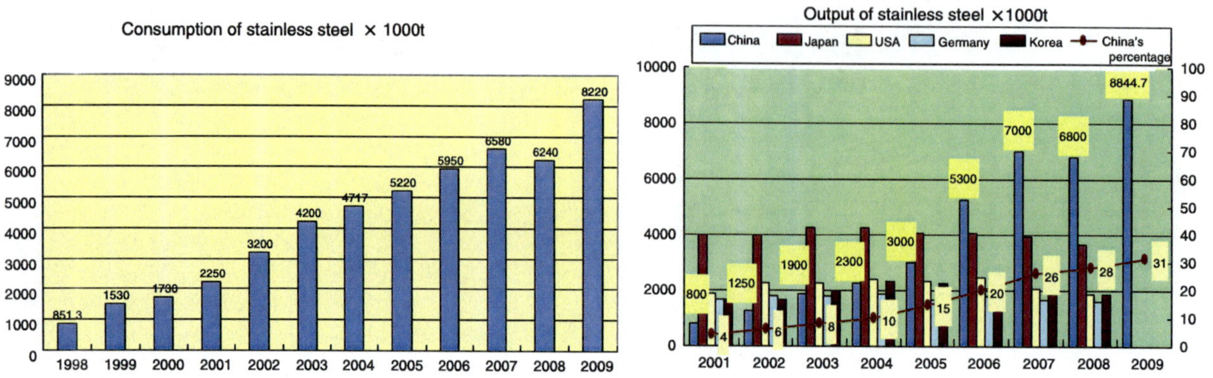

Fig. 1 Change of stainless steel apparent consumption in China

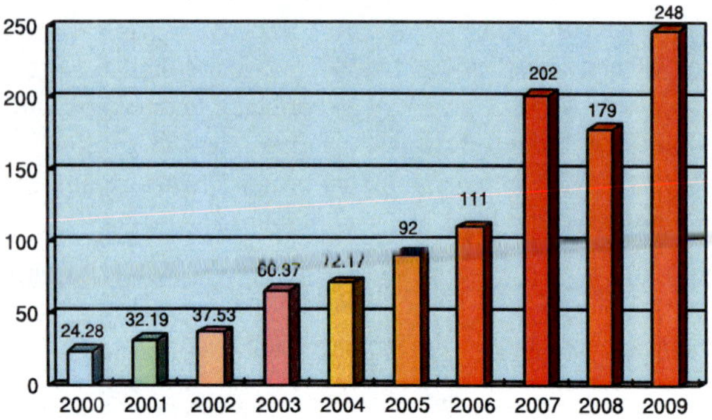

Fig. 2 Change of stainless steel output in China

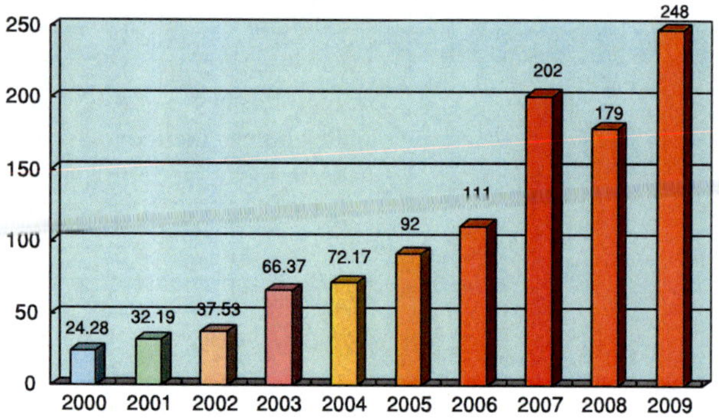

Fig. 3 Latest change of TISCO stainless steel output

and making further technical progress of stainless steel production process, which is also an important reason. Technical progress of stainless steel production made by TISCO recently shows below:

1 Great-Leap-Forward Development of Stainless Steel Output

The development trend of world stainless steel is becoming bigger and bigger in size. In 2001, the average capacity of ten largest stainless steel enterprises in the world were 796,000 tons/year, but till 2006, it was 1,07 million tons/year, increasing by 34%. It is extremely efficient for larger enterprise of stainless steel to cut down the raw material procurement cost and smelting cost, improving production efficiency and product quality.

TISCO began to import the package equipment of 300,00 tons/year stainless steel cold rolling sheet in 1960s of 20 century from Germany, afterwards, in 1980s, TISCO itself developed AOD refining furnace and stainless steel continuous casting technology, establishing the advanced stainless steel production line of "EAF—AOD—continuous casting—Steckel Hot Rolling Mill—four High Cold Rolling Mill—BA Furnace", with annual capacity of stainless steel production by 100,000 tons, and TISCO was called at that time the largest stainless steel enterprise equipped with whole process flow in China. Since entering into twenty-first century, according to the market demand and the support given by the government, we started from August 2000 to the end of 2003 to put 7 billion Yuan RMB for 500,000 tons stainless steel system revamping project with the purpose to build up the global competitive stainless steel enterprise, making the scale and process technical installation level of stainless steel production a new step, and 1 million tons of stainless steel production capacity was formed per year, stepping forward eight powers of stainless steel enterprises in the world. From September 2004 to September 2006, we invested 16.578 billion Yuan RMB to reconstruct 1.5 million tons of stainless steel project ratified by the government, enabling the technical equipment level of TISCO stainless steel production process an international first-class level, and 3 million tons of stainless steel capacity being available per year, and now TISCO has been turned into the largest capacity enterprise of stainless steel production in the world. Figure 3 gives the latest change

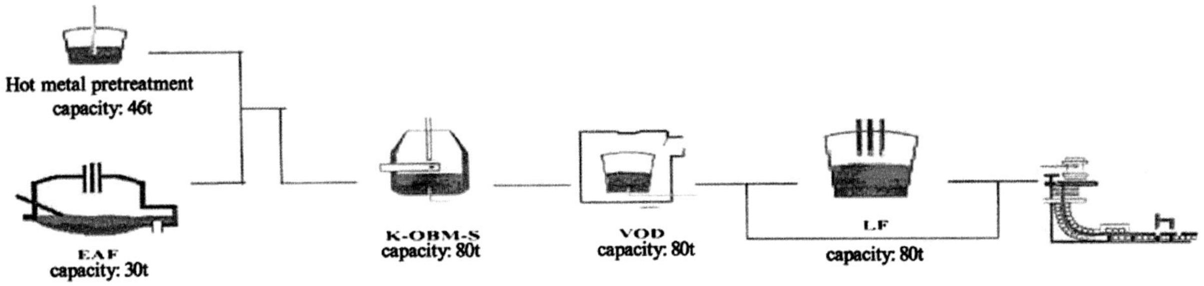

Fig. 4 TISCO triple way for stainless steel production

of TISCO stainless steel output. It can be seen that since TISCO has realized a great-leap-forward development of TISCO stainless steel output since 2000, especially we successfully carried out the revamps of 500,000 tons stainless steel project and newly build up 1.5 million tons of stainless steel item, enabling TISCO to realize a great-leap-forward development of stainless steel output. In 2009, stainless steel output reached 2.48 million tons, the first one in the world, 10.2 times higher than 2000, average increased by 21.76% per year.

Table 1 Latest main technical and economic figures of smelting system

Refining ratio/%	100
Casting ratio/%	94
Slab weight/tons	20
(O)/ppm	≤50
(S)/ppm	≤40
(Cr) comprehensive yield/%	91.3
Slab yield/%	97
Average argon consumption/m^3 $tons^{-1}$	AOD 8.6
	VOD 2.5

2 Relying on Technical Reconstruction and Realization of Optimization and Upgrading of Process Technical Equipment

2.1 Revamp Item of 500,000 tons of Stainless Steel System

In order swiftly to improve the competitiveness of TISCO stainless steel, TISCO applied for ratification of 500,000 tons stainless steel system revamp project in 1999 and this project was approved by the State Council, and its reviewing and approval were completed in 2000 and it was put into operation by the end of 2002. The complete set of revamp included three parts.

2.1.1 Revamping of Smelting System and Its Efficiency

Revamping of smelting system included following two parts: one is that AOD furnace and vertical slab continuous casting machine at No. 3 Steelmaking Plant were revamped, of which, AOD revamp included: furnace volume was enlarged from 18 into 45 tons, AOD blowing changed from conventional AOD blowing into AOD-L, and Austrian VAI expert automatic system was imported for AOD system; and the revamp of vertical slab continuous casting machine included: mold vibration changed from mechanical way into hydraulic way; secondary cooling water system changed from water spraying cooling into steam and water combined dynamic cooling, slab cutting changed from off-line cutting into on-line cutting, at the same time, slag protection powder added into automatic charging system and slab printing machine installed; two is that at No. 2 Steelmaking Plant a new production line of stainless steel based upon hot metal triple way of stainless steel smelting was built up and it was started in August 2000 and completed in December 2002, it is the first one in China and fourth one in the world based upon hot metal as raw material to produce stainless steel by triple way. Its process flow sheet consists: hot metal pretreatment + EAF preheating alloy → K-OBM-S → VOD + LF → continuous casting as shown at Fig. 4. It is characterized by: flexible raw material usage, one part or complete part of dephosphorization hot metal can be applied to produce stainless steel; gunning control is used to achieve satisfied de-carbonization speed; through CO blowing, requirement of heat energy can be ensured; continuous measuring temperature system and pneumatic slag skimming (IRIS slag measuring system); advanced automatic system and software model added.

Through reconstruction of smelting system and several years' production practice as well as further process optimization, now all kinds of technical economic figures reach or approach the international advanced level as shown at Table 1.

Table 2 Latest main technical and economic figure

Item	Actual control level	International level
Longitudinal thickness accuracy (mm)	0.027–0.014	0.014
Transverse thickness accuracy (mm)	0.10	0.08
Width accuracy (mm)	6.7	8.0
Plate convexity	≤6 mm ± 13 μm 95.4%	±15 μm
	>6 mm ± 22 μm 95.4%	
Flatness	25 IU	25 IU

2.1.2 Revamp of Hot Rolling Mill

TISCO 1,549 mm hot rolling mill is second-hand equipment introduced from Nishing Steel, Japan in 1989 with original three roughing mills, six finishing mills, three down coilers and it was in operation in August 1994. There were some problems in this mill such as small profile of stand housing, weak main driven power and serious aged defects, making thinner specification product production more difficult and product quality and size impossible to meet the requirements of downstream cold rolling production and market. By the end of 2000, 1 billion Yuan RMB was invested to revamp this mill from the entire aspects.

Revamping included: (1) No. 3 reheating furnace was added newly to increase 800,000 tons capacity per year; (2) dismantle the original three roughing mills and install one strong vertical roll reversing mill to gain AWC control in order to improve the rolling force, shorten rolling time and ensure the rolling width accuracy; (3) newly added another big reduction F0 finishing mill in order to improve finishing rolling force and widen rolling specifications and ranges; (4) AGC was changed from electrical motivation to hydraulic AGC, electrical loop changed to hydraulic loop to heighten the size precision of product thickness; (5) finishing mill F0–F6 installed with work roll bending device, F4–F6 installed with work roll shifting device to achieve profile shape close-loop control of finishing mill; (6) newly added one fully hydraulic down coiler to enhance the quality of coiling shape.

Through revamping above mentioned, the quality in kind of hot rolling strip of stainless steel can be equivalent to the international level as shown below at Table 2.

2.1.3 Revamp of Cold Rolling System

Revamping of cold rolling included: (1) newly added an APL of 1.1 million tons of hot rolled strip characterized by: newly installed a set of dry, drawing and bending de-scaling device before blasting and pickling treatment to strengthen the pretreatment before pickling; newly installed on-line flattening device at the rear to improve the shape quality of hot rolled strip; (2) newly installed five Sendzmil Rolling Mills to increase the capacity of cold rolling stainless steel strip from 400,000 tons before revamping to 900,000 tons after revamping; (3) newly installed another APL of 500,000 tons cold rolling strip/year, maximum allowable coil weight by 34 tons, process speed maximum by 140 m/min. Main equipment included after being revamped: two hot APL lines, eight Sendzmil cold rolling mills, three cold lines, two flattening units and six 6 slitting shear units, etc.

Table 3 Comparison between TISCO stainless steel product quality and international advanced level's

Index	Existing level	International advanced level
Specifications/mm	0.3–3.0	0.2–3.0
Surface class	2B 2D BA HL	2B 2D BA HL
Varieties	A 70%	A 50–80%
Yield/%	91.5	≥92
Surface roughness/μm	2B ≤ 0.13	2B ≤ 0.13
Plate shape/mm	≤8	<8
Trans. thick differential/mm	±0.01–0.02	±0.01–0.02
Longi. thick differential/mm	±0.01–0.02	±0.01–0.02
Side quality	No burr	No burr

After revamping above mentioned, the quality in kind of cold rolling strip has been further more improved and the accuracy of product size and surface quality have been equaled to the international level (Table 3).

After implementation of "500,000 tons stainless steel system revamping", a great-leap-forward development of stainless steel production has been obtained and until 2004, TISCO has the ability to produce 1 million tons of stainless steel/year and final finished stainless steel product has been amounted to 726,000 tons, ranked the eighth among the world stainless steel main enterprises.

2.2 Newly Installed 1.5 Million Tons of Stainless Steel Project

Based upon the careful analysis of domestic and international stainless steel development tendency, focusing on building up the strategic goal of global most competitive stainless steel enterprise and relying upon original 1 million tons of stainless steel capacity, TISCO began to execute newly 1.5 million tons of stainless steel project in September 2004 and all the new items were put into operation completely in September 2006, and now 3 million tons of stainless steel capacity has been realized as shown below at Table 4.

Table 4 New equipment scale and quality design index of stainless steel project

Name of index			Stainless steel production		International level
			Existing level	New level	
Scale of steelmaking production (10,000 tons)			100	200	–
Coil size (mm)	Hot coil	Thickness	2.5–5.0	2.0–20	2.0–20
		Maximum width	1,320	2,050	2,050
	Cold coil	Thickness	0.3–3.0	0.2–4.75	0.2–4.75
		Maximum width	1,250	2,000	2,000
Yield (%)	Hot coil		97	≥98	≥98
	Cold coil		91	≥92	≥92
Main specifications			300 series	All series	All series
Quality level			Domestic advanced level	International advanced level	Nishing, Japan

2.2.1 Main Construction and Process Technical Equipment Characteristics of New Project

Main construction of new project consists of three parts:

1. *Smelting part* Newly two special stainless steel smelting production lines are installed with annual stainless steel capacity by 2 million tons including two 160 tons EAF, two 180 tons AOD, one 180 tons LF and two 2,150-mm slab continuous casting machines. They are characterized by following three aspects: (a) duplex way of smelting is applied and it is first created by TISCO. Non-treatment common hot metal will be used to be blown in LD converter into low carbon steel liquid, pouring into EAF and adding some scrap for roughing smelting, again refining it in AOD-L furnace, which is characterized by following advantages: relaxing the shortage of scrap resources, reducing harmful element contents, improving usage of carbon chromium ferrous and the beginning blowing temperature of AOD-L furnace, and lowering the production cost; (b) 2,150-mm slab continuous casting machine and hot slab grinding machine are installed in order to realize the hot charging and hot rolling; (c) in order to obtain intensive production, duplex stainless steel production lines are available in one building and it is the first one in the world.
2. *Hot rolling procedure* Newly installed a 2,250-mm hot rolling production line used for 2 million tons of carbon steel and 2 million tons of stainless steel capacity, producing stainless steel coil of 2.0–20 × 1,000–2,100 mm. The international advanced control technology of plate shape is applied in this new unit, including the technology of surface quality automatic inspection, hot coiling process technology of non-core axle thermal baffle and fully hydraulic walking coilers, being the widest hot rolling mill unit just used for stainless steel production in the world.
3. *Cold rolling procedure* Newly installed a APL of No. 2 hot rolled stainless steel strip by 1.15 million tons, used for 2.0–14 × 1,000–2,100 mm stainless steel hot rolled plate, including on-line rolling mill and on-line tension straightening machine, and it is maximum capacity, state-of-art equipment and widest APL of hot rolling stainless steel production in the world.

Newly installed three 700,000 tons wide stainless steel cold rolling mills, of which, two mills are four upright column and 20 high with product specifications by 0.15–6.0 × 1,200–1,625 mm; one is S6 rolling mill with specifications by 0.8–8.0 × 1,200–2,100 mm. Two kinds of rolling mills are characterized by accommodating more greater coiling weight respectively by 30 and 40 tons, producing 6 and 8-mm cold rolling strip, the most thickness stainless steel cold rolled products in the world.

Newly installed a No. 4 cold rolling stainless steel plate APL of annual 700,000 tons capacity, used for 0.4–8 × 1,000–2,100 mm stainless steel cold rolling plate, and it is an APL of the maximum capacity and the widest cold rolling stainless steel in the world.

Relative slitting and transverse cutting lines of cold rolling stainless steel are incorporated.

2.2.2 Efficiency of New Project

By the execution of 1.5 million tons of new stainless steel project, TISCO is able to obtain 3 million tons of stainless steel capacity, and TISCO has become a fully flowing stainless steel enterprise characterized by maximum capacity and equipment, most advanced technology, top level of installation and shortest process flow sheet as well as friendly environment protection internationally.

1. Three million tons of stainless steel capacity shows TISCO being a unique maximum production enterprise in size globally, which is the development trend to heighten centering of stainless steel industry, and superiority of scale economy is priority.

2. More first-class new technology from the world is applied to realize large-scale and modernization of stainless steel equipment. 160 tons EAF, 180 tons AOD converter, 2,150-mm slab continuous casting machine, 2,250-mm hot rolling mill and annually 1.15 million tons of hot rolling strip APL, etc. are belonged to international maximum ones.
3. Through process flow creation, complete flow and compact and high efficient production have been achieved. By using the technology of slab hot grinding and table hot charging, the entire flow speed has been greatly facilitated.
4. Considering the idea of circulation economy, hot cutting slag, scale and dedusting ash, etc. waste material are all recycled to implement full flow cleaning production.
5. More specifications and more series to produce extra wide, thick and thinner specialized products in order to meet the requirement of market, especially greater than 1,600 mm extra wide stainless steel hot rolling coil and cold rolling plate which depended upon the importation for a long time has been produced by ourselves.
6. Special division of work is more properly and explicit to utilize the features of different heat and production line to realize the specialized production based upon steel grade and specifications.
7. The material quality of products has been improved and reached the international most advanced level. The comparison of levels between stainless steel products of TISCO and that of the international top level is as following Tables 5 and 6 after new projects were completed.
8. The whole line costs are less than average level of global stainless steel producers and reached the international advanced level because cheap hot-metal is raw material, and costs of coal, electricity and manpower and investment per ton steel are relatively lower.

Table 5 Material quality of hot rolled coils

Item	2,250-mm hot rolled coils after new projects were complete	The international most advanced level
Thickness tolerance	±(20–50) μm	Hot rolled coil with thickness <5 mm: ±25 μm
Width tolerance/mm	+0–3.5	+0–4
Straightness/IU	±15	±15

Table 6 Material quality of cold rolled products

Item	TISCO cold rolled products after new projects were complete	The international most advanced level
Thickness accuracy	±(0.01–0.015) μm	±(0.01–0.015) μm
Un-flatness/mm m^{-1}	≤2	≤2
2B surface roughness/μm	0.05–0.10	0.05–0.10
BA surface roughness/μm	0.02–0.04	0.02–0.04

3 A Group of Proprietary Technology with Independent Intellectual Property Rights was Formed Through Independent Innovation

In recent years, a group of proprietary technology with independent intellectual property rights was formed through the independent innovation, for example the three-stage stainless steel smelting technology and nitriding stainless steel production technology with molten iron as raw material. Among which relying on the international most advanced "electric furnace + converter furnace + VOD furnace" three-stage stainless steel production line with molten iron as main raw material, the completely new three-stage smelting process with molten iron as raw material was integrated by ourselves forming a complete "three-stage method" smelting stainless steel process technology with independent intellectual property rights and as molten iron as raw materials. This achieved several innovations, whose main technical and economic indexes and the overall technology has reached the leading international advanced level. Four invention patents were declared. The research results played an important role in improving Chinese stainless steel industry concentration, promoting stainless steel production technology progress and product structure adjustment, newly establishing Chinese long process steel enterprises, and renewing stainless steel factories as well. This project was granted with Second Prize of 2005 National Science and Technology Progress, which was the earliest one to use AOD with nitrogen gas for nitrogen alloying stainless steel production and development. Through the development of this proprietary technology, production of nitriding stainless steel without chromium nitride was realized, and has the advantages of low production cost and high control precision of nitrogen content. This technology development drove the TISCO stainless steel product development. We have produced more than 50,000 tons austenitic stainless steel with nitrogen content of 10–3,500 ppm such as 304N, 304NbN, 301LN, A940,

253MA, etc. and duplex stainless steel such as 18-05, 2101 2304, 2205 2507, etc. All these products are the domestic origination, and can completely replace imports. They were used in the Three Gorges Project, CNPC pipeline; petrochemical, chemical and nuclear national key projects, achieving good social and economic benefits. The project won the National Grade II Prize of Science and Technology Progress in 2005. In addition, proportion of non-grinding 304 stainless steel slab is improved greatly, and the surface quality of cold plate has reached the international advanced level through the development of inclusion control technology. Progresses have been also made in Continuous stainless steel casting technology with high efficiency, which makes 304 stainless steel casting velocity up to 1.3 m/min. In 2006, TISCO integrated the first hot strip DRAP line independently with which 2E and No. 1 products of high quality surface has been successfully produced. At the basis of comprehensive recovery and utilization technology of stainless steel resources, a large number of basic research has been carried out in utilization of dust ash and slag and rough raw material (low chromium and low nickel cast iron, iron oxide scale, molybdenum oxide), and developed ferrochrome power injection technology, molybdenum oxide melting reduction technology, and corrosion resistant ceramic preparation technology by use of stainless steel slag, some of these technologies have realized industrialization.

In recent five years, TISCO won successively 29 sciences and technology achievement award at provincial level and above, 4 national award, 3 National Key New Product of stainless steel. It owned 135 patent licenses and led the formulation of 12 stainless steel national standards. At the same time, it presided over some special subjects of national 973-Plan and 863-Plan such as stainless steel used as cab plate of high-speed train and of corrosion failure behavior of key nuclear power material. At present, it owned 700 core technologies mainly in stainless steel among which 100 items has reached international advanced level.

4 To Optimize Product Structure and the Products are Widely Applied in the High-End Market

In aspects of the stainless steel development, in recent years, TISCO has made full use of the whole process stainless steel production equipment with international leading level, talents and technical resources advantage, focused on the market-much need, and made great achievements in development of resource-saving stainless steel such as 400 series ferritic stainless steel, duplex stainless steel, nitrogen-contained stainless steel, stainless steel compound materials as well as high performance stainless steel material applied in special fields of nuclear power, petrochemical, hydropower, environmental protection and high-speed train, etc.

Among them, in the resource-efficiently stainless steel, key production technologies has made a breakthrough in aspects of 400 series ferritic stainless steel, super-low carbon and nitrogen control, non-bonding hot strip rolling, high-efficient continuous annealing of hot strip, production of cold plate with good high brightness, and ferritic stainless steel production with good moldability and wrinkle resistance in recent years. A complete 400 series stainless steel production technologies with independent intellectual property rights and suitable for the enterprise equipment has been formed gradually, and a batch of more than 20 varieties have been developed such as stainless steel T4003 used for new type railway freight cars, ferritic stainless steel 429, 439M and 441 used for car exhaust system which can alternate 304 ferritic stainless steel TTS443/TTS443M, coinage steel CTSZB, super ferritic stainless steel 446 used for seawater condenser, etc. These products have been widely applied in electrical appliances, electronics, railway freight cars, car exhaust system, container, coinage, etc., and its product quality has reached the international advanced level.

The output and proportion of stainless steel Series 400 are significantly increased from 35,500 tons in 2002 to 1.01 million tons in 2009 and 9.46% in 2002 to 42.1% in 2009, respectively, exceeding the world average level of 23%. Emphasized on the duplex stainless steel, the edge crack problems on 2,205-mm duplex stainless steel strip coil has successfully been tackled, making TISCO the only enterprise which can produce 2,205-mm duplex stainless steel cold rolled coil in China. The successful development of S2507 super duplex stainless steel plate with N content of 2,800 ppm and economical duplex stainless steel of 2101 and 2304 has filled up blank of domestic market, replacing importation. TISCO has also made a historic breakthrough in application of duplex stainless steel to shipbuilding, Bridges, petrochemical industry, paper-making, seawater desalination industries, playing a positive role in promoting technological progress and product upgrading in equipment manufacturing and metallurgy industry in China; nitrogen can be substitution of nickel in austenitic stainless steel, high N stainless steel has advantages in saving resources, low cost and high strength, toughness, good resistance, stress corrosion resistance, as well as high adaptability, etc. Nitrogen becomes the top choice material of sustainable development. TISCO has made basic research on high nitrogen stainless steel such as carbon and nitrogen alloying, N addition technology, heat process simulation and thermal processing, welding and welding materials, dissolution and precipitation and tough mechanism, successful smelting high nitrogen stainless steel $10Cr_{21}Mn_{16}NiN$ of N contents above 6,000 ppm under normal pressure with AOD furnace for the first time in the world, and successfully rolling 6-mm steel plate. This material has unique advantages in high strength bolt, wear-resisting screen, coastal bridge deck steel, military deck, etc. Concerning the stainless steel composite material, TISCO has developed more than ten specifications for $00Cr_{22}Ni_5Mo_3N$ dual-phase steel composite, widely used in the fields of petroleum, chemical, salt-producing, water conservancy, food manufacturing equipment, building decoration and so on as well as in many national key project and industries as the Yangtze three gorges project, Qilu, Daqing petrochemical engineering, and vacuum evaporation million tons salt producing project in Chongqing. Al composite steel has been developed successfully for the first time in China and applied to urban subway. The development of stainless steel composite materials and related technology has many fills up the domestic blank.

In respect of development of high performance and high functional stainless steel material in special fields, focusing on high-performance stainless steel and high functional stainless steel material in special fields as new energy and efficiency, environmental protection, high-speed trains, defense in order to realize key materials localization, and as a goal to meet the national economic development, major engineering construction and national defense development demand. In recent years TISCO has successfully developed super austenitic stainless steel pipe and SUPER304H and TP310HCbN seamless steel tube for ultra supercritical boilers, completely meeting design requirements of creep rupture strength, high temperature corrosion resistance, and cold and hot process ability, which has currently been successfully applied to domestic coal-fired power station boiler manufacturing. For the first time in China, TISCO has developed four nuclear steel grades of Z2CN18-10, Z2CND17-12, Z2CN19-10 nitrogen controlled, Z2CND18-12 nitrogen controlled in accordance with the French nuclear RCC nuclear power system and specification. Its quality reaches the level of similar products abroad. The product has passed the nuclear design, production, and nuclear owner authentication, applied to nuclear power projects such as Lingao phaseII, Qinshan phase, Red river, Yangjiang and Ningde, and exported to the foreign countries and applied to nuclear power project in Russia.

In addition, we took tasks of the new project CPR1000 in which core structure material inside reactor, accumulator and boron injection box will be made of nuclear stainless steel plate. Aiming at the corrosion property of oil–gas pipeline in the petrochemical industry, we successfully developed super martensitic stainless steel pipe blank of SUP13Cr-95 and SUP13Cr-110 a with strong ductility and corrosion resistance. The steel can be used for Sinopec and PetroChina, and substitute of the imported materials, which fills the domestic blank. We developed varied strength level of LT, DLT, ST, MT and HT, and cold-rolled steel strip pf 301L which meet EU standard of C700 and C850 coach, which can be applied for light railway, the subway and carriage in line of 200 km long. The successful development of antioxidant levels, heat-resistant stainless steel 253MA meet the domestic needs, realize heat-resistant stainless steel grade serialization, fill up the domestic blank, and replace the imported products. The product has been applied in the TOP 3 power group such as HBC, DBC and SBP. Super austenitic stainless steel plate and coils of 904L has been developed, which can be widely applied in wastewater treatment, chemical industry, paper-making industry, etc. by the steel grade of the properties of uniform corrosion, pitting corrosion, crevice corrosion, inter-crystalline corrosion and stress corrosion cracking. Due to High Eddy-current loss and heat value and low current-carrying existed in the submarine cable with armored galvanized steel wire, submarine cables with non-magnetic stainless steel wire was developed, which have already been used in Guangdong Shantou project of State Grid. Chill-hard hot-press mould plate with high hardness, good surface and shape was developed, which will be used to produce mould for laminate floor, furniture wood and fire-proof plate. Antimicrobial stainless steel of TKJF-1, 2, TKJA TKJF-1 series has been developed. After tested by antibacterial materials testing center of CAS, it has excellent antibacterial properties. Now, it applied for the field of the washing machine, tableware and kitchen ware.

5 Conclusion

In order to make TISCO stronger and perfect, TISCO will speed up a group of product restructuring projects, such as the implementation of stainless steel radial forging machine, 50,000 tons of stainless steel seamless steel tube and 2 million tons of high strength steel precision products in the period of the 12th of five-year. By the emerging and restructuring, TISCO will set up stainless steel sheet project cooperated with Tianjin Steel Tube Group in Tianjin, which will form the capacity of 400,000 tons of cold-rolled strip.

TISCO has become a biggest and most advanced stainless enterprise in the world. However, in order to be a most competitive stainless steel enterprise, and entering the TOP 500, TISCO still have tough jobs to do. TISCO has the obligation to impellent Chinese stainless steel industries development in sound and rapid style with its peers, accelerating the transformation from large to strong.

The State of Steel Industry in India and its Future Prospects

Sanak Mishra

Abstract

India is the fifth largest steel-producing nation in the world. The Indian steel industry accounts for over 7% of the world's total steel production. The domestic crude steel production grew at a compounded annual growth rate of 8.6% during 2004–2005 to 2008–2009. The National Steel Policy of the Government of India has a target for taking steel production up to 110 MT by 2019–2020. While 2007 was an exciting period in the history of Indian steel industry, corresponding to 7% growth over 2006; 2008 witnessed an unprecedented global economic meltdown with only a marginal growth of 3.7%. Consumption declined, in fact, from July 2008 onwards. However, 2009 was a year of great resilience and recovery for the Indian steel industry. For April–December 2009, the provisional data released by Joint Plant Committee indicates a 7.8% rise in consumption of total finished steel. World Steel Association forecasts India's Apparent Steel Usage (ASU) to increase at 13.9% during 2009–2010, over 2008–2009, to reach 63 MT compared to the forecast of 10.7% for The world and 6.7% for China for the same period. Similar figure for 2010–2011 over 2009–2010 stand at 13.7% for India, 2.8% for China and 5.3% for World. The prospective increase in the ASU figures are substantiated through a scrutiny of the consumption patterns in India. While the per capita demand in India, at around 50 kg, is nowhere near to the world average of around 150 kg, or, about 400 kg for developed countries; the rural India, at around 5 kg per person, lags even farther behind in comparison to urban India. However, in consideration of the extensive infrastructure development planned by the government in both rural and urban areas, these consumption figures have a strong scope to increase. Accordingly, a number of major Indian and Global steel players are into a massive capacity expansion mode in India, either through Brownfield or Greenfield route. On a conservative estimate, the steel demand in India is expected to touch around 90 MTPA by 2015 and around 150 MTPA by 2020. Steel supply is, however, expected to reach only around 88 MTPA by 2015 and around 145 MTPA by 2020.While the demand for steel will continue to grow in traditional sectors, specialized steel is also increasingly being employed in various hi-tech engineering industries. Globally, a relation can be observed between steel consumption and the GDP growth rate. Overall, India, being in a high growth phase with huge planned infrastructure development, is bound to witness sustained growth in the steel requirement in the years to come.

Keywords

Indian steel industry • Future prospects • Growth

S. Mishra (✉)
CEO, Greenfield Projects India, ArcelorMittal, 6th Floor,
Plaza M-6, Jasola District Centre, New Delhi 110025, India
e-mail: Sanak.Mishra@arcelormittal.com

Y. Weng et al. (eds.), *Advanced Steels*, DOI: 10.1007/978-3-642-17665-4_5,

1 Introduction to the Indian Steel Industry

The past in the Indian steel landscape can be divided into three major segments (Fig. 1).

1.1 Private Sectors Purely

At the time of independence in 1947, India had only three steel plants—the Tata Iron and Steel Company, the Indian Iron and Steel Company and The Mysore Iron and Steel Limited (Later renamed as Visvesvaraya Iron and Steel Ltd) and a few electric arc furnace-based plants. The period till 1947 is, thus, characterized by a hatching steel industry in the country. This industry had a capacity of about 1 million tonne and was completely in the private sector [1].

1.2 Advent of Public Sector

The first signs of major changes started appearing during the first three Five-Year Plans (1952–1970). During this period, the iron and steel industry was being prepared to be taken under the state control. From the mid-1950s to the early 1970s, the Government of India set up large integrated steel plants in the public sector at Bhilai, Durgapur, Rourkela and Bokaro. Several major policy changes were initiated, including:

- Capacity control measures: Licensing of capacity, reservation of large-scale capacity creation for the public sector units.
- Dual-pricing system: Price and distribution control for the integrated, large-scale producers in both the private and public sectors, while the rest of the industry operated in a free market.
- Quantitative restrictions and high tariff barriers.
- Railway freight equalization policy: To ensure balanced regional industrial growth.
- Controls on imports of inputs, including technology, capital goods and mobilization of finances and exports.

Public sector was a big driving force for the Indian steel industry during this phase. The government created a lot of capacity via the public sector. This large-scale capacity creation contributed to making India the tenth largest steel producer in the world as crude steel production grew markedly to nearly 15 MT in the span of a decade from a mere 1 MT in 1947.

This trend, however, could not be sustained beyond 1970s because the economic slowdown adversely affected the pace of growth of the Indian Steel Industry [1].

1.3 Liberalization

1991–1992 was a remarkable year for India because of its association with the economic liberalization. In 1991–1992 the country replaced the control regime with liberalization and deregulation with a broad focus of globalization in the long term.

The Indian steel industry was impacted by the provisions of the New Economic Policy in the following ways:

- The provision of restricting large scale projects only for public sector was withdrawn, with the effect that private sector could create large projects as well.

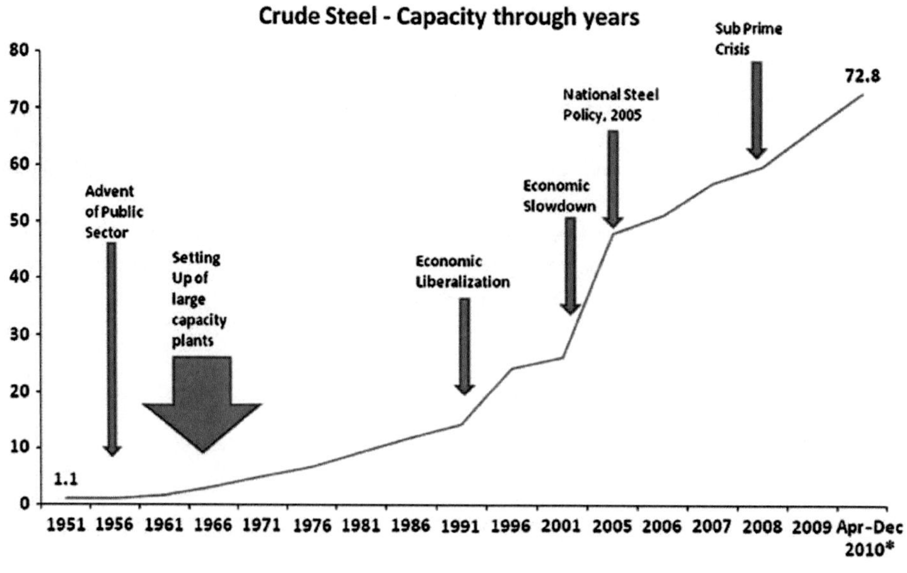

Fig. 1 Crude Steel capacity in India through years

- Pricing and distribution control mechanisms were discontinued.
- Freight equalization scheme was replaced by a system of freight ceiling.
- Quantitative import restrictions were largely removed. Export restrictions were withdrawn.

Post liberalization, the economy, and the steel industry, underwent marked changes. For steel industry, liberalization of economy opened up newer channels for procurement of inputs which were till now being domestically procured, and could now be obtained from overseas markets at competitive rates. Local players now had access to the manufacturing techniques being adopted overseas. Additionally, they also had access to newer markets for their products. On the other hand, the competition from global players, with better technology and operations, also came into picture. Factors such as these increased the need to enhance efficiency levels so as to become internationally competitive.

Indian consumer, on the other hand, had a wide range of products to choose from, whether domestic or imported. This freedom of choice established the superiority of consumer. This, in turn, instigated the steel makers to provide products/service levels in tune with the needs of the consumers.

During this period, large ISPs were set up in the private sector. Some of the notable milestones in the period were:

- Emergence of the private sector with the creation of around 9 million tonne of steel capacity based on state-of-the art technology.
- Reduction/dismantling of tariff barriers, partial float of the rupee on trade account, access to best-practice of global technologies and consequent reduction in costs—all these enhanced the international competitiveness of Indian steel in the world export market.

After 1996–1997, domestic economy witnessed a decline in the growth rate. As a result, the Indian steel industry's pace of growth also slowed down and the performance of the industry fell below average. However, 2002 onwards, global and Indian industry witnessed a turnaround, partly fueled by growth in China. The Indian steel industry was now also playing emphasis on R&D activities, adoption of measures to increase domestic per capita steel consumption and other market development projects, import substitution measures, thrust on export promotion and exploring global avenues to fulfill input requirements.

This rapid pace of growth of the industry and the observed market trends required certain guidelines and frameworks to be put into place to guide the growth and avoid the pitfalls. Thus, National Steel Policy was introduced in November, 2005 to provide a roadmap of growth and development for the Indian steel industry.

The long-term objective of the National Steel Policy was to ensure that India developed a modern and efficient steel industry of world standards, catering to diversified steel demand. The focus of the policy was to attain levels of global competitiveness in terms of global benchmarks of efficiency and productivity [1].

2 Indian Macro Economics—Past, Present and Future (Figs. 2, 3 and 4)

An Edelweiss report states that India's GDP is set to quadruple over the next 10 years and India is likely to be a USD 4.5 trillion (Trn) economy by the year 2020. All sectors (Banking, Broking, Asset Management, Life insurance sector, Domestic Pharma and Healthcare, Media and Entertainment, Education, Premium Urban Housing, and Organized Retail) are expected to witness an average of five to six factor growth [2, 3].

India could move into third place in share of global GDP and may grow even faster than China in 2020–2030 predominantly due to younger demographic composition. By 2020, while India might face problem of over population of

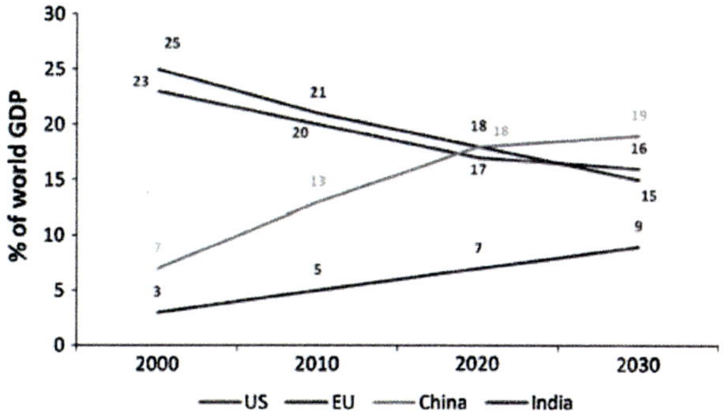

Fig. 2 Projections of India's share in world GDP

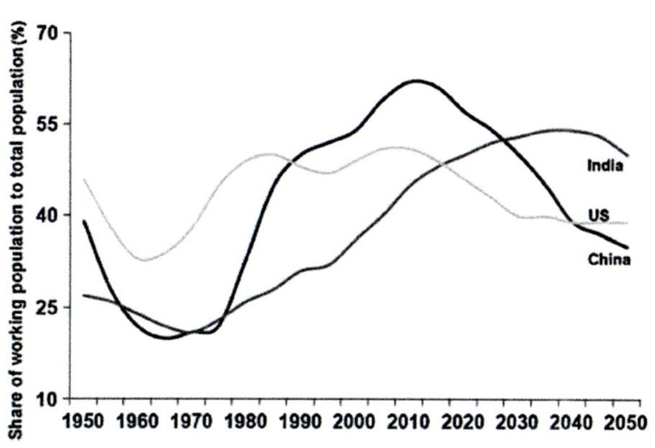

Fig. 3 Projections of share of working population to total population. *Source* UN

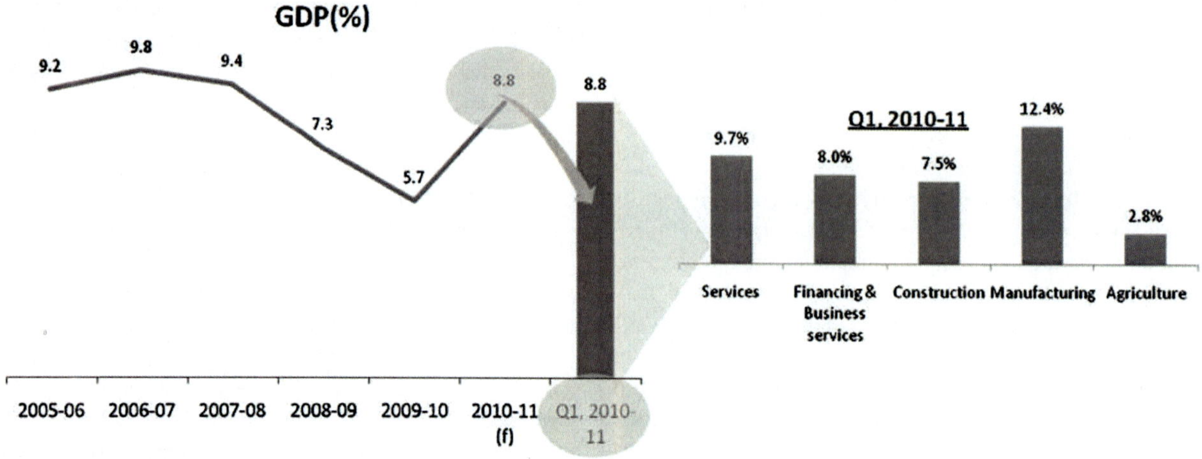

Fig. 4 Sustained positive GDP growth through years, Q1 figures supporting growth

educated youth, developed countries would face a shortage of working age people because of lower birth rates [4].

Even in recent years, when Global economies faced decline in GDP, India still had positive GDP growth rate. The recent Q1, 2011 figures substantiate the expected growth in the Indian economy [5, 3].

3 Existing Indian Steel Landscape

In 2009, the global crude steel production was 1220 MT, a decline of 8% over 2008. During 2009, China was the largest crude steel producer in the world with production reaching 567.8 million tonne, a growth of 13.5% over 2008. India, in 2009, recorded a growth of 2.7% as compared to 2008. India was the fifth largest producer in 2009 and, after China, only other country in the top 10 bracket to register a positive growth during 2009. India also emerged as the largest sponge iron producing country in the world in 2009, a rank it has held on since 2002 [6].

As per the August 2010 figures of World Steel Association, India is globally fourth in crude steel production, at 44.6 MT, after China (425.8 MT), Japan (72.7 MT) and USA (54.3 MT) [7].

India has also established its presence, globally, in the production of sponge iron/direct reduced iron (DRI). The domestic production of sponge iron has been increasing in the country, primarily due to growth of coal-based sponge iron units in the various key mineral-rich pockets of the country. This has enabled the country to achieve and maintain the number one position in the global market [8].

Furthermore, a series of mega projects have been planned in the country. These project, once operational, will pave way for re-structuring of steel industry and its dynamics [1].

3.1 Key Players in the Indian Landscape—Technologies and Geographies

There are four major players in the steel market in India:

1. *Steel Authority of India Limited (SAIL)* SAIL is a leading steel-making company in India. It is a fully integrated iron and steel maker, producing both basic and special steels for domestic construction, engineering, power, railway, automotive and defense industries and for sale in export markets. The product portfolio for SAIL comprises hot and cold rolled sheets and coils, galvanized sheets, electrical sheets, structurals, railway products, plates, bars and rods, stainless steel and other alloy steels. In 2009–2010, SAIL witnessed a crude steel production of 13.51 MT with the saleable steel of 12.6 MT. It is the second largest producer of iron ore in India and possess of having the country's second largest mines network. This gives SAIL a competitive edge in terms of captive availability of iron ore, limestone, and dolomite which are primary inputs for steel making. Iron and Steel production at SAIL is carried out at its five integrated plants and three special steel plants. These plant are located principally in the eastern and central regions of India and situated close to domestic sources of raw materials [9, 10]
2. *Tata Steel* Tata steel, post Corus acquisition in 2007, is the tenth largest steel producer in the world with employee strength of above 81,000 across five continents. During the financial year 2009–2010, the Group recorded deliveries of 24 MT. Tata Steel's overseas ventures and investments in global companies have helped the Company create a manufacturing and marketing network in Europe, South East Asia and the Pacific-rim countries. The Group's South East Asian operations comprise Tata Steel Thailand, in which it has

67.1% equity and Nat Steel Holdings, which is one of the largest steel producers in the Asia Pacific with presence across seven countries. The NatSteel group produces construction grade steel such as rebars, 'cut-and-bend' cages for construction, mesh, precage bore pile, PC wires and PC strand. Tata Steel's Jamshedpur (India) Works has a crude steel production capacity of 6.8 MTPA, which is slated to increase to 10 MTPA by 2011. The Company also has proposed three Greenfield steel projects in the states of Jharkhand, Orissa and Chhattisgarh in India with additional capacity of 23 MTPA and a Greenfield project in Vietnam. The products include hot and cold rolled coils and sheets, galvanised sheets, tubes, wire rods, construction rebars and bearings. Tata Steel has also had made extensive effort toward branding with brands like Tata Steelium, Tata Tiscon, Tata Wiron, Tata Structura, etc. Apart from these product brands, the company also has in its folds a service brand called "steeljunction"—the world's largest retail marketplace for steel [11, 12].

3. *JSW Steel* JSW Steel, the flagship company of the JSW Group, can be traced back to 1982, when the Jindal Group acquired Piramal Steel Limited, which operated a mini steel mill at Tarapur in Maharashtra and renamed it as Jindal Iron and Steel Company (JISCO). It is been ranked 2nd among top 32 "World Class" steelmakers by World Steel Dynamic (June 2010). JSW Steel is a large manufacturer and exporter of galvanized steel in India with its products exported to over 100 countries. JSW has a plant at Salem with an annual capacity of 1 MT. It is on the threshold of a major expansion plan of adding 3.2 MT per annum to its Vijayanagar Plant to achieve 11 MTPA by 2011. By 2020 the company hopes to produce 32 MT of steel annually with Greenfield integrated steel plants coming up in West Bengal and Jharkhand. It has established a strong presence in the global value-added steel segment with the acquisition of a steel mill in US and a Service Center in UK. JSW Steel has also formed a joint venture for setting up a steel plant in Georgia. The Company has further acquired iron ore mines in Chile and coal mines in USA and Mozambique. The product portfolio for JSW comprises pellets, slabs, HR coils/sheets, HR plates, CR coils, Galvanized coils/sheets, Colour coated coils/sheets. It is the leading manufacturer of cold rolled, galvanized and colour coated steel with manufacturing facilities at Vasind & Tarapur in Maharashtra [13, 14].
4. *Essar Steel* Essar Steel is a fully integrated flat carbon steel manufacturer—from iron ore to ready-to-market products with a 2009 capacity of 8.6 million tonnes per annum (MTPA). The company plans to achieve a capacity of 14 MTPA by 2011–2012 with expansion plans in India, as well as Asia and the Americas. An additional advantage to Essar Steel is their high level of forward and backward integration. We are totally integrated—from raw material to finished products, adding value at every stage of the manufacturing process. Essar Steel exports to US and European markets, and to the growing markets of South East Asia and the Middle East. The products find market in consumer sectors, such as automotive, white goods, construction, engineering and shipbuilding. Their clientele include a number of major clients, including Caterpillar, Hyundai, Swaraj Mazda, the Konkan Railway, and Maruti Suzuki [15, 16].

In addition, there are several more primary and secondary players in the market like JSPL, Bhushan Steel, Bhushan Power and Steel Ltd., ISPAT industries, Rashtriya Ispat Nigam Ltd (RINL) etc. In addition, there are several global players planning their foray in the Indian Landscape such as ArcelorMittal, POSCO, JFE etc.

3.2 Government Facilitation/Regulatory Framework

In the Union Budget 2010–2011, the government has allocated US$ 37.4 billion to the infrastructure sector and has increased the allocation for road transport by 13% to US$ 4.3 billion which will further promote the steel industry [8].

The present EXIM policy permits export of iron ore from Goa and Redi sector to all destinations by the iron ore producers; irrespective of the iron content [17].

KIOCL is the canalizing agency for its own products (iron ore concentrates and iron ore pellets) since it is a 100% E.O.U. (export oriented unit) [17].

The export of iron ore with Fe content above 64% is canalized through MMTC [17].

Export of Iron of Goa origin to China, Europe, Japan, South Korea and Taiwan (irrespective of Fe content) and Export of ore from Redi region to all markets (irrespective of Fe content) is not canalized [17].

However, some types of high-grade iron ore (Fe content above 64%) from specific areas like Bailadila in Chattisgarh are allowed to be exported with restrictions on quantity imposed primarily, with a view to meet domestic demand on priority [17].

Off lately, Government has been discouraging the export of raw Iron Ore from the country. In December 2009, the export duty on iron ore lumps was raised from 5 to 10% and export duty of 5% imposed on fines. In May 2010, Export duty on iron ore lumps increased from 10 to 15%. These duties are anticipated to rise further in future [18, 19].

Additionally, the Indian Railways charge additional fare for export of Iron Ore to dissuade it [20].

3.3 Upcoming Capacity/Technology

A host of steel companies have lined up major investment proposals. This includes, in addition to domestic players, several global players like ArcelorMittal, POSCO etc. Furthermore, with an expanding consumer market, the Indian steel industry is likely to receive huge domestic and foreign investments [8].

The domestic steel sector has attracted a staggering investment of about US$ 238 billion, according to government sources [8].

This consists of nearly 222 MoUs signed between the investors and various state governments mostly in the states of Orissa, Jharkhand, Chhattisgarh and West Bengal [8].

- SAIL is planning to set up a 12-million tonne plant in Jharkhand.
- In December, India's largest engineering conglomerate Larsen and Toubro (L&T) and state-owned Nuclear Power Corporation of India Limited (NPCIL) formed a US$ 373 million joint venture for specialized steel and forging products.
- Stainless steel manufacturer and exporter, Varun Industries, is setting up a US$ 172 million stainless steel-cum-alloy steel plant at Rohat, Jodhpur.
- Tata Steel has entered into a joint venture with Japan's Nippon Steel for production and sales of automotive cold-rolled flat products at Jamshedpur. The JV is expected to invest US$ 400 million to set up an automobile venture in India.
- Steel major, JSW Steel has earmarked a capex of US$ 1.6 billion for 2010–2011 and plans to increase capacity of its Bellary plant in Karnataka from 7 to 10 MT by end of 2010–2011 [8].

4 Steeling for the Future

4.1 SWOT Analysis (Fig. 5)

4.1.1 Growth Drivers for the Future

India is amongst the economies which is, and has been, continuously displaying significant growth prospects. The growth drivers for the Indian steel industry depend on a various number of factors.

Government of India has been making substantial investments in infrastructure creation. This is also supported with young demographics of India which indicates more working population by 2020 in comparison to current contemporaries. Also, a latent demand in the rural India will provide a substantial base to drive the growth of Indian steel industry.

Some of the other growth drivers include:

- Investments and growth in steel consuming sectors such as construction, automobiles, infrastructure, consumer durable, oil and gas.
- Merger and Acquisition has evolved to be one of the major growth drivers in the industry leading to economic of scale.
- Young demographics and, subsequently, increased disposable income request more investment in infrastructure; which in turn promotes steel consumption.

4.1.2 Challenges for the Future

There are several challenges to the growth of Indian steel story. The steel industry, by its nature, is a resource consuming sector. Hence, if the Indian steel story has to go to all the way to the targets of 100 + MT, there are several issues to be considered:

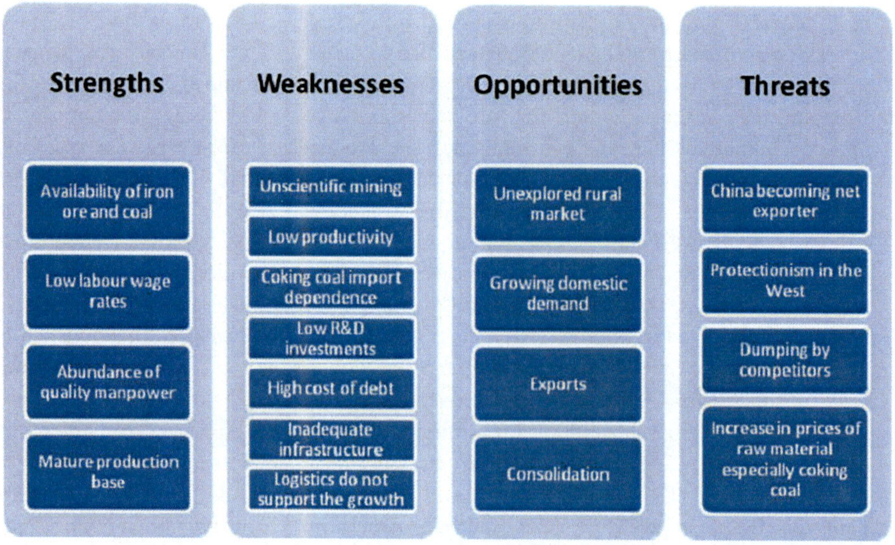

Fig. 5 SWOT analysis of the Indian Steel Industry

- *Land Acquisition* With the land owners demanding higher compensation for their land and the state provision for comprehensive resettlement and rehabilitation policy, Land Acquisition is becoming increasingly difficult with each passing day.
- *Insufficient Infrastructure* While the industries are planning capacities for future production, the infrastructure; in terms of rails, roads and ports, is not increasing in the corresponding proportion. This would, in the long term, result in stagnation of capacities due to raw material and/or finished product stagnating due to lack to support from logistics.
- *Need for R&D* The conventional Blast Furnace route is highly damaging to the environment due to high CO_2 emissions. Steel players need to conduct R&D to produce low CO_2 footprint mechanisms.
- *Lack of indigenous coking coal reserves* Indian steel industry is highly susceptible to international coking coal price volatility due to the lack of domestic reserves. Hence, efforts need to be made to develop processes to utilize domestic resources for production.
- *Unscientific Mining* Old Mining techniques/technology leading to manpower intensive operations. Also associated with unscientific mining methods are problems related to low equipment and manpower productivity. This, in turn, has restricted productivity in India to 4 mt per man shift vis-à-vis Australia's 57 mt and USA 45 mt.
- *Delays due to various Land and Environmental clearances* Due to various hierarchies and bureaucratic levels in India and the various departments whose clearance is required for a steel project, whether or not including mining, an average of 3–4 years are required.
- *Stringent environmental policing* With growing environmental concerns and increasing awareness in public about the environmental impact of mining and steel manufacturing, getting the environmental clearance for the various projects is becoming increasingly difficult.

5 Summary

Indian steel industry, in its modern format, has been existence for around 60 odd years. These years witnessed the installed capacity increase from around 1 MT in 1950s to forecasted 70 + MT in 2010.

India has also grown in its relative position in the world market. As per the latest World Steel Association reports, Indian steel industry now stands at fourth position globally. Supporting this growth in the Indian steel are the youth centered population demographics, government plans of substantial infrastructure investment, latent demand of steel in rural India and accelerating growth in industries such as consumer durables, white good, automobiles, infrastructure and construction etc.

A lot of Domestic and Global players have plans to expand in the Indian steel, either through Brownfield or Greenfield expansion.

On the other hand, Indian steel is also set to face a number of challenges such as progressively stricter environmental norms and more and more emphasis on providing a greened and cleaner working environment, especially in the cases where mining in involved. Similarly, land acquisition is increasingly becoming difficult with troubles in identifying a contiguous patch of land with logistically favorable resources and tribal population which is willing to part from the land. Several states now have a mandate for companies to have comprehensive resettlement and rehabilitation policies and plans for land owners to be provided with alternative dwelling for their lands and existing residences, even before companies can begin work on the land.

While such issues need to be tackled, the Indian steel future looks pretty lucrative with a sufficient Demand–Supply gap to ensure sustained, is not too high, demand for steel.

Acknowledgments The author wishes to acknowledge the assistance and support of Aashish Sood, Executive (Strategy), ArcelorMittal India Ltd. in the preparation of this manuscript.

References

1. Ministry of Steel, Government of India, Annual Report (2009)
2. Convergence, Catch-up and overtaking: How the balance of world economic power is shifting, PricewaterhouseCoopers LLP (2010)
3. International Monetary Fund—World Economic Outlook Database List. http://www.imf.org
4. Edelweiss India 2020 Report
5. Central Statistics Office, Government of India. http://www.mospi.gov.in/
6. Indian in Business. Ministry of External Affairs, Government of India. http://www.indiainbusiness.nic.in/industry-infrastructure/industrial-sectors/steel.htm
7. World Steel Association. http://www.worldsteel.org/pictures/newsfiles/0810%20Production%20figures.pdf
8. India Brand Equity Foundation. http://www.ibef.org/industry/steel.aspx
9. Steel Authority of India Limited website. http://www.sail.co.in/aboutus.php?tag=company-aboutus
10. SAIL Annual Report (2009)
11. Tata Steel website. http://www.tatasteel.com/corporate/company-profile.asp
12. Tata Steel Annual Report (2009)
13. JSW Steel website. http://www.jsw.in/companies/company_JSWSteel.shtml
14. JSW Steel Annual Report (2009)
15. Essar Steel website. http://www.essar.com/section_level1aspx?cont_id=eLiVfqUiZks=
16. Essar Steel Annual Report (2009)
17. Export policy for iron ore, Ministry of Steel, Government of India. http://steel.nic.in/policy.htm#pol2

18. Ministry of Commerce, Government of India. http://commerce.nic.in/eidb/ecomq.asp
19. "Assocham seeks hike in iron ore export duty to 20%", Press Trust of India/New Delhi 17 May 2010. http://www.business-standard.com/india/news/assocham-seeks-hike-in-iron-ore-export-duty-to-20/94617/on
20. Indian Railways. http://indianrailways.gov.in/indianrai-lways/directorate/traffic_comm/freight_rate_circulars.jsp

On the Performance Improvement of Steels through M^3 Structure Control

Han Dong, Xingjun Sun, Wenquan Cao, Zhengdong Liu, Maoqiu Wang, and Yuqing Weng

Abstract

Steels are now considered to be a category of new materials, which have progressed year by year to meet the market requirements for high performance. Although it is well known that the performance of steel is dependent upon microstructure, the potential of microstructure evolution is really unknown by people at present due to the steadily developing steel metallurgy. The study is needed to investigate the potential microstructure evolution phenomena in steels and to develop new technology to improve the properties through microstructure control, by which the safety and reliability of steels in service could be eventually improved remarkably. The issues such as steel metallurgy for ultra cleanliness and ultra homogeneity, phase transformation mechanism of meta-stable austenite subject to temperature and load changes, carbon diffusion and partitioning during transformation, multi-scale characterization of structures, and microstructure stability subject to temperature and load changes are thought to be the key points to clarify the microstructure evolution at the existing circumferences. As a result, it can be expected that the fundamentals of microstructure control featured by multi-phase, meta-stable and multi-scale (M^3) could be established, and then the technologies to process high performance steels could be developed: the third generation HSLA steels with improved toughness and/or ductility ($A_{KV(-40°C)} \geq 200$ J and/or $A \geq 20\%$ as $Rp0.2$ in 800–1,000 MPa), the third generation advanced high strength steels (Rm × $A \geq 30$ GPa% as Rm from 1,000 to 1,500 MPa) for automobiles with improved ductility and low cost, and the third generation heat resistant martensitic steels with improved creep strength ($\sigma^{650}_{10,000} \geq 90$ MPa). It can be expected that the new technology developed will improve the safety and reliability of steel products in service remarkably for infrastructures, automobiles and fossil power station in the future.

Keywords

High performance steels • Microstructure evolution • Phenomenon and mechanisms • Processing

1 Introduction

Steel is one of the most common materials widely used in the world, both for structural and functional applications. Steel has been the basis for economy and society over 2,000 years, and now it is still playing very important roles in the world. Steel is generally believed to be a kind of advanced materials due to its characteristics during processing, fabrication, applications, and also recycling. People cannot image what the world would be if there be no steel around us.

Now, the steel is a kind of basic materials characterized by *massive production* (the crude steel output being around 1.3 billion tons annually in the world), *extended properties* (strength being from 100 to 5,000 MPa, service temperature

H. Dong (✉) · X. Sun · W. Cao · Z. Liu · M. Wang · Y. Weng
Central Iron and Steel Research Institute, Beijing 100081, China
e-mail: donghan@public.bta.net.cn

Y. Weng et al. (eds.), *Advanced Steels*, DOI: 10.1007/978-3-642-17665-4_6,

from −198°C to 650°C, subject to corrosion circumstances from atmosphere to acid, alkali, and salt), *wide applications* (from construction to transportation, machinery, energy, marine, environment protection, resources, weaponry and daily life), *low cost* (price equivalent to mineral water), and *easy recycling* (steel scraps being used to produce high quality steels). It is no doubt that steel is not only the basic material, but also a kind of new materials, to meet the requirements from the development of our economy and society ever since.

Since the late of 1970s, the big change has begun to take place in China steel industry. Over the past ten years especially, the crude steel output of China steel industry has been of approximate four times from 124 Mt in 1999 to 567 Mt in 2009. Great progresses have been taken place not only in quantity of steels, but also in grades and quality of steel products. It is really not simple for one to describe the current situations and predict what will be for China steel industry in the future precisely, due to continually changing in steel products both in quantity and quality, changing requirements from the markets, advanced production facilities introduced, integration tendency for steel enterprises, restriction for environment protection, etc. Nevertheless, it is obvious that China steel industry will still keep to be improved, more new steel processing will be adapted and more advanced steel products will be developed.

Nowadays, the situation which China steel industry has to face is quite different from what the developed countries have met: more and more high strength steel products are needed for long duration buildings and infrastructures in the urbanization; high quality special steel products are required to meet the requirements from equipment manufacture in the industrialization; the steel products resistant to elevated temperature, load, and corrosion is the necking point for the facilities to generate energy; high performance specialty steel products are demanded by weapon manufacture for defense; climate change forces steel processing to be of reduction in energy consumption and waste emission, and of high efficiency in the production of clean steels; there will be certainly a rising requirement for steels to be of high performance from the newly developing industry. Therefore, the driving forces for development of steel processing and products are originated from urbanization, industrialization, climate change, nation safety, and new industry, etc.

Due to the strong demands from the rapid development of urbanization and industrialization, a large amount of different kinds of steel products should be produced for construction, transportation system, energy production and machinery. It is clear as the result that almost one-third of total steel products have been used to building construction and about one-fifth of steel products to infrastructure. The massive production of steel products could result in the enormous consumption of mineral ores, energy, water, and immense emission of waste gases and solids, and shortage in transportations. The steel products being of high performance would reduce the steel consumption and bring down the side effects of massive production. For example, it is proved that 15% in reduction of steel consumption could be obtained if Grade II steel rebar is replaced by Grade III steel rebar. It is well known that the application of high strength steel sheets can reduce the weight of auto bodies. The service efficiency and duration could be improved remarkably through the adoption of high performance steels into the structures and equipments. Heat resistance steel subject to higher temperature may be applied to increase the efficiency of fossil power station. High strength steel can be used to increase the fuel efficiency of vehicles. Corrosion resistance steel could be applied to extend the life duration of structures. As a result, energy will be saved and materials consumption will be reduced. In short, steels to be of high performance is one of the main tendencies for steel industry and end users.

At present, the high performance steels are demanded by almost every sectors, large span and high rise buildings and infrastructures require steels to be of high strength and low yield strength to tensile strength ratio in order to improve the efficiency and aseismic performance. Both high strength and high ductility are needed for sheet steels to raise the fuel efficiency and crashworthiness of vehicles. High rupture strength and low creep rate are desirable for fossil power stations to be at high efficiency.

In brief, there are definitely demands for the development of high performance steels with characteristics of high strength, high toughness, high ductility, low yield ratio, high rupture strength and low creep rate, etc.

In 2009, Ministry of Science and Technology granted a National Basic Research Program of China (973 Program) on high performance steels (2010CB630800) as a succession of previous two phase research programs over last ten years. It aims at the improvement on safety and reliability of structures and equipments in service through the applications of high performance steels, of which will be base on the understanding of microstructure evolution and newly developed processing technologies. The paper will review the target of the program and the newly developments in research on HSLA steel, auto sheet steel, and heat resistance steel.

2 Multi-Phase, Meta-Stable and Multi-Scale Structure

The performance of steel products is closely related to the constitutes and morphology of microstructures. The characterization and effective control of microstructure are now from micron scale to nano scale steadily (to be in nano order). The properties have been raised from the order of 10^6 to 10^9 unit (to be in Giga order). The strength of hot rolled HSLA steel and auto sheet steel has been raised from MPa order to GPa order. The fatigue strength limit of ultrahigh strength steel has been also improved from MPa order to GPa order. The fatigue cycles for steels to undertake have been demanded from Mega cycles to Giga cycles. The rupture time for steel at elevated temperature has been extended from Mega seconds to Giga seconds. The performances of steels in Giga scale are related to precisely controlling of microstructure in nano scale, and closely associated with microstructure characterized with Multi-phase, Meta-stability, and Multi-scale (so called as M^3 microstructure).

There exist certainly some aspects in ambiguous statues in steel science and technology, in which the phenomena and laws are far from totally understanding by human beings.

The ultimate tensile strength of perfect iron crystal was theoretically calculated to be of 21 GPa. In 1956, ultimate tensile strength of 13 GPa was achieved by Brenner in whiskers of iron. But it should be pointed out that the actual steel products are strengthened by defects to resist the movement of dislocations within matrix. The steel with highest tensile strength was obtained in cold drawn pearlitic steel wire developed by Nippon Steel in 1990s. High strain resulted in tensile strength of 5.7 GPa in wire in diameter of 0.04 mm through high dislocation density and dislocation strengthening. Meanwhile, the commercial cold drawn pearlitic steel for tire presents the highest tensile strength of 4.0 GPa in the wire in diameter of 0.2 mm. The data of tensile strength of most martensitic steels are normally within the range of 600–1,800 MPa. The upper limit for martensitic steels to be applied is around 2.2 GPa. Yield strengths of commercial hot rolled HLSA steels are generally in the scale of 400–700 MPa. The highest yield strength being achieved in HSLA steel by TMCP is about 1 GPa. Obviously, the steel products are far from the strength of perfect iron crystal, the binding force of steel matrix. It is indicated that there is a potential for people to improve the strength of steels.

It is certainly true that steels to be strengthened furthermore is always one of the main targets for steel researchers. Generally, the single strengthening mechanism has been well understood, even in atom scale. It is almost impossible for commercial steels to be strengthened by single strengthening mechanism, but by combined strengthening mechanisms. The understanding on the co-existence of combined strengthening mechanisms are still not clear than we wish to. Martensitic steels are strengthened by transformation, during which solution strengthening, precipitation strengthening, grain refinement strengthening and dislocation strengthening are presented. It is still ambiguous for the contributions of different strengthening mechanisms and the co-existence among them during quenching and tempering. Transformation strengthening is composed of several strengthening mechanisms, and widely adopted to strengthen steels. Especially in recent years, strong interests have been paid to the transformations at elevated temperature and lower temperature, which have resulted many new low carbon steels and ultra-low carbon steels. Correspondingly, the microstructure in steel is not simple ferrite plus pearlite, nor bainite, nor martensite, but multi phases. It could be the microstructure composed of ferrite/bainite/martensite, or bainite/martensite, or ferrite/bainite/martensite/austenite, or austenite/bainite/martensite, etc. It is likely that the combination of meta-stable austenite with conventional ferrite and hard bainite/martensite may result in the expected properties. The typical examples are DP steel, TRIP steel, CP steel in commercial applications. The microstructure characterized with multi-phases is the target for steels to be improved in performances.

It can be accepted that strength is raised only when there is enough energy to be absorbed during deformation and fracture. Total elongation and notch impact energy (although not a constitutive property) are commonly used to evaluate ductility and toughness of steels, respectively. In steels, the understanding for structures to control ductility and toughness is conventionally connected with inclusions, precipitates, grains, etc. For multi-phase steel, the coordination of deformation among phases becomes important, but still need to be investigated. For clean steels, the there may be a transition of cracks originated from inclusions to the defects in matrix or boundaries, which is certainly still unknown. Steels can be classified as materials composed of different components in multi-scale. It is necessary to identify the microstructural factors to control ductility and toughness in different scales.

Over past 40 years, TMCP, one of thermomechanical treatments, has been believed to be the most remarkable progress in both physical metallurgy and processing. It is mainly related to the austenite state at high temperature and the transformations during cooling. In recent years, the attractions have been drawn to the stability of austenite and relative transformation in the elevated and lower temperature range: the transformation of meta-stable austenite, the stability of transformed phases, and the behavior of carbon during transformation, etc. The key point of new thermomechanical treatment in the elevated and lower temperature range is microstructure stability and carbon behavior. For martensitic heat resistance steel, the stability of matrix and grain boundaries are supposed to be controlled to improve creep strength. The understanding of the microstructure stability and carbon behavior could certainly help us to achieve the expected multi-phase microstructure, and lead to the target of enhancement in strength, ductility and toughness.

Around the turn of century, researchers in Japan, Korea, Australia, EU and China devoted their efforts on the refinement of ferrite in steels. Grains of 1 μ in size were believed to be the limitation in the bulk specimens processed by heavy hot deformation, which resulted in yield strength of 800 MPa. The state of austenite was thought to be controlled for the beneficiation to grain nucleation and retardation to grain growth. From the point view of refinement, lath bainite or lath martensite are obviously the finest microstructure we have. Meanwhile, the scales to identify bainite or martensite are from austenite in dozens of microns to packet and block in microns, to lath in submicron, to precipitates in nano meters. The multi-scale characterization provides more choices to control microstructure, and then properties. It is still unclear in the transformation mechanism of ultra-low carbon steel. The transformed phases are quite complex, but are beneficial to toughness. It is worth to pay attention to the transformation and the microstructure in ultra-low carbon steel.

As stated above, M^3 microstructure control (multi-phase, meta-stability, multi-scale) is the means to obtain steels with high performance. We will adopt the idea to develop new generation steels for infrastructure, automobile, and boiler pipe. The mechanical properties will be improved significantly, and leads to the improvements in safety and reliability of structures and equipments. It is illustrate briefly in Fig. 1 the target and means of the Program.

3 HSLA Steels with Toughness and/ or Ductility Enhancement

HSLA steel is one of the most rapidly developed and most featuring steel grades in recent years. During the last several decades, there has been a growing trend towards the use of higher strength steel in order to reduce the structural weight and fabrication cost. At present, the HSLA steels produced through on-line TMCP have the yield strength up to 800 MPa, which includes the fine grained ferritic steel with maximum yield strength of about 600 MPa and the ultra-low carbon bainitic steel with the maximum yield strength of about 800 MPa. The off-line quenching and tempering (Q&T) process has been frequently used to produce the steels with yield strength over 800 MPa presently. However, in order to ensure the hardenability during quenching, the Q&T steels have to contain more carbon and expensive alloying elements like Mo, Cr, and Ni, which inevitably deteriorates the weldability and toughness and increases the production cost of steels. Therefore, it is of great significance to produce over 800 MPa grade ultra-high strength steel with lower carbon content and less expensive alloying element through on-line process.

In general, the steel is strengthened at the expense of the sacrifice of some toughness or ductility. In some cases, however, the strength improvement becomes meaningful only if the toughness or ductility is simultaneously improved. For instance, in order to satisfy the strict requirement on self-arrest of propagating ductile fracture in high grade pipeline steels, the minimum requirement on the CVN toughness value is gradually improved with increasing the steel grade: 125 J for X70, 177 J for X80 and 250 J for X100 [1]. Therefore, it is a major concern for structural steels to improve the trade-off balance between strength and toughness or ductility.

At present, the toughness value of high strength steel over 800 MPa in yield strength is ordinarily lower than 100 J, and the total elongation (A_5) is normally <15%. In the third phase 973 program, the concept of the third generation HLSA steel has been proposed with an improved balance between strength and toughness or ductility ($A_{KV(-40°C)} \geq 200$ J and/or $A \geq 20\%$ as $R_{p0.2}$ in 800–1,000 MPa). Furthermore, the third generation HSLA steels should be produced on-line in the existing mills and with a relatively low alloying cost. How to reach this objective? In this section, the desired microstructural features, processing method and the mechanical properties for the third generation HSLA steels are introduced, with emphasis on the improvement in toughness.

3.1 Development of ULCM Steel with Enhanced Toughness

Excellent toughness value can be obtained in ultra-low carbon steels if the sample is fractured in a ductile mode. On the other hand, the effective grain refinement is necessary for the improvement of both DBTT and strength in

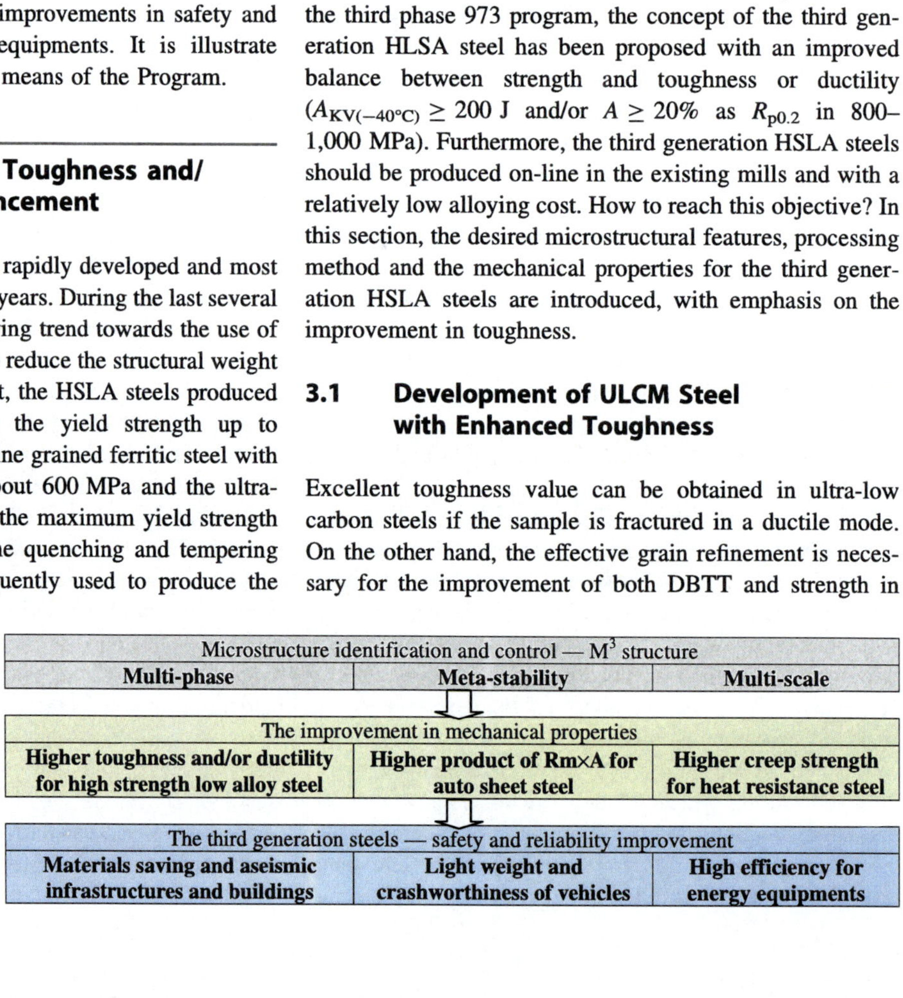

Fig. 1 The schematic illustration of research targets and means of the Program

ultra-low carbon martensitic steel. Thus, the combination of ultra-low carbon design and refinement of the effective grain size in martensite is expected to be a promising route to develop new HSLA steel with enhanced toughness.

Conventionally, the ultra-low carbon martensitic steel is highly alloyed with the expensive elements like Cr, Ni and Mo in order to improve its hardenability, which increases the alloying cost remarkably. In this study, we have made an attempt to use a relatively high Mn content combined with trace boron to improve the hardenablity of ultra-low carbon steel, and thus the alloying cost can be decreased. The Mn content was raised from the level of 1.5wt% in the traditional HSLA steel to the level over 2.0wt% but <3.5wt% in the ULCM steel.

The refinement of the effective grain size in ULCM steel is accomplished by the careful control of rolling process parameters namely the reheating and rolling temperatures as well as the deformation ratios and the conditions for the accelerated cooling. An optimum reheating temperature leads to the best possible initial austenite grain size which is the starting point for the further control of microstructure during the subsequent rolling. The reheating temperature should be as low as possible with the prerequisite that the microalloying element like Nb is ensured to be fully dissolved into the austenite. Furthermore, the maximum deformation ratio with a relatively low deformation temperature during the rough rolling is of great significance for a first grain refining by austenite recrystallization. The finish rolling has to be done close to Ar3 temperature for a proper control of pancaking of non-recrystallized austenite grain (ausforming). The extremely pan-caked austenite is rapidly cooled to the temperature below Ms temperature and then air cooled or coiled, which will create a very fine effective grain size and thus improve both the strength and toughness.

An ultra-low carbon steel was controlled rolled to produce pan-caked austenite grains (ausforming) and then rapidly cooled to produce the martensitic structure according to the principle mentioned above. The main chemical compositions of the tested steel are as follows (in mass%): 0.03%C–2.6%Mn–0.20%Si–0.0013%B. In addition, other alloying or microalloying element, such as Cr, Mo, Nb and Ti are also added properly in order to improve the hardenability and refine the austenite grains. Figure 2 illustrates the yield strength of the tested steel produced by ausforming and by the austenitization treatment (denoted as non-ausformed in this figure). In addition, the relationship of the yield strength of martensite structure with the carbon content, as reported by the Speich [2], is also plotted in this figure. This figure indicates that the strength of ultra-low carbon martensite is relatively low under the condition of traditional austenitization and quenching treatment, and the ausforming can increase the strength significantly. The yield strength of the tested steel is increased from about 800 MPa

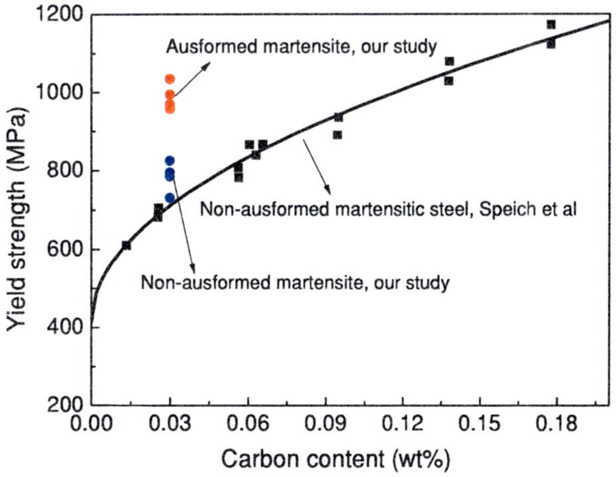

Fig. 2 Yield strength of the ausformed martensite and the non-ausformed martensite and the dependency of yield strength of martensite on the carbon content reported by Speich [2]

to about 1,000 MPa by ausforming. This strengthening is mainly resulted from the refinement of block size. In addition, the higher density of dislocations in the ausformed martensite than in the non-ausformed martensite, which is resulted from the transfer of dislocation from austenite to martensite for the displacive transformation, could be another important reason for the strength increase.

Figure 3a shows the balance of yield strength and Charpy V-notch impact absorbed energy at −30°C for the ausformed martensite. The impact absorbed energy reaches up to 180 J at a yield strength level of 1,000 MPa, which is much higher than that achievable in the conventional low carbon martensitic steels. Figure 3b and c illustrate the impact absorbed energy and the percent of shear area as a function of test temperature, respectively. It is seen that the ductile to brittle transition temperature is lowered remarkably by the ausforming. For instance, the 50%FATT is decreased by 60 K through the ausforming compared with the non-ausformed martensite with the prior-austenite grain size of 12.6 μm.

3.2 HSLA Steels with High Ductility

It is well known that the plastic instability condition in tensile tests is expressed by the formula $\sigma > d\sigma/d\varepsilon$, where σ is the flow stress and ε the true strain. It is seen that a larger strain-hardening rate is required for higher strength steels to avoid the plastic instability and to improve the uniform ductility. Conventional HSLA steels with yield strength over 800 MPa is usually characterized by martensitic or bainitic microstructure with a very high density of dislocations, which leads to a low strain hardening rate and a small uniform ductility. In recent years, multi-phase steels containing some fraction of meta-stable austenite, namely TRIP-assisted steels, have been extensively studied and it is found that the ductility of this kind of steels can be improved significantly through the enhanced strain hardening resulted from strain induced austenite-to-martensite transformation. However, relatively high contents of carbon and Si or Al are essential to acquire the retained austenite constitute in the final microstructure. Furthermore, complex heat treatments like intercritical annealing and stepped cooling are ordinarily necessary for the formation of austenite. All the features mentioned above seem not to be suitable for the production of modern HSLA steels for structural use, because the modern HSLA steels usually require a lower carbon equivalent for better weldability and a simple on-line processing for large scale production.

Ashby et al. [3] indicated that the strain hardening rate of metals depends on the dispersion of hard second phase particles and is proportional to a dispersion parameter $(f/d)^{1/2}$, where f is the volume fraction of the second phase and d the mean diameter of the particles. This presents an important guideline for the design of high strength steel with enhanced ductility. For example, cementite has been used to improve the strength–ductility balance of the ultrafine-grained steels which have a limited strain hardening rate and a small uniform elongation [4]. However, the modern HLSA steel usually has a relatively low carbon content (<0.1%) and thus a small volume fraction of cementite, which will result in small values of f/d and strain hardening rate. Microalloying precipitates could be a candidate for the improvement of strain hardening rate. The diameter of microalloying precipitates is usually as fine as up to several nanometers, so although its volume fraction is also very small (typically 0.1%), the value of f/d should be relative large. Assuming the size of microalloying precipitates is 10 nm, and the volume fraction is 0.1%, then the value of f/d is obtained as 0.1 μm^{-1}, which is comparable to the value of cementite obtained in the ultrafine grained steel with 0.15%C [4]. Furthermore, since the nanometer-sized

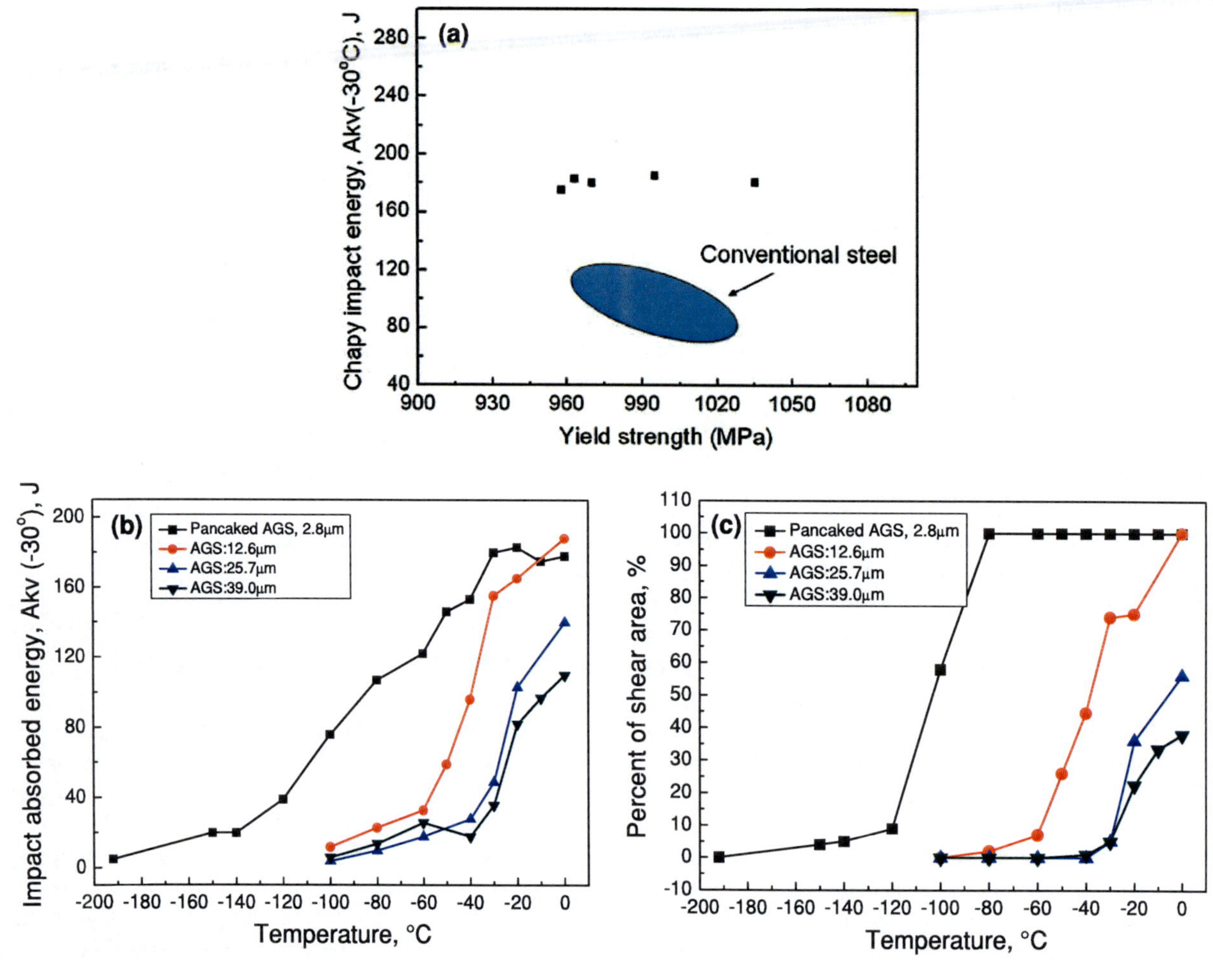

Fig. 3 **a** The balance of yield strength and Charpy V-notch impact absorbed energy at −30°C for the ausformed martensite. **b**, **c** Impact absorbed energy and percent of shear area as a function of test temperature

microalloying precipitates can strengthen the steel significantly by the Orowan mechanism, the density of dislocation in the matrix microstructure could be lower than that of martensitic or bainitic steel at the same strength level. This decrease in density of dislocation can further improve the strain hardening rate.

Therefore, the desired microstructure for the HLSA steel with enhanced ductility is proposed as follows. The matrix microstructure should be fine grained and have a higher density of dislocation than ferrite but a lower one than martensite to ensure a good combination of high strength and high ductility. The acicular ferrite microstructure, which is now widely used in the high grade pipeline steel, could be a good candidate for the desired matrix microstructure. Furthermore, the matrix microstructure should be strengthened by the nanometer sized microalloying precipitates.

The main chemical compositions of the experimental steel are 0.07%C, 1.70%Mn, 0.25%Si, 0.045%Nb and 0.12%Ti. The steel was smelted in a 120 ton converter and a LF furnace. The casting slabs were firstly soaked at 1,250–1,280, followed by rough rolling to reduce the slab thickness from 170 to 32 mm through five passes. Then the slab was finish rolled to a final thickness of 3–5 mm through a six-stand continuous rolling mill. After rolling, the strip was laminar cooled in runout table and coiled finally. The finish rolling temperature was 880–920°C, and the coiling temperature range is 550–680°C.

Figure 4 shows the microstructure of 3 mm thickness strip finish rolled at 880°C and coiled at 550°C. The microstructure exhibits the feature of acicular ferrite microstructure, consisting of quasi-polygonal ferrite, bainite ferrite and a little amount of martensite/austenite (M/A) constituents, Fig. 4a. The TEM image shows a relatively high density of dislocation in the acicular ferrite, Fig. 4b. The results of EBSD analysis show that the microstructure contains a large fraction of small misorientation boundaries up to 44%, Fig. 4d, which is an important feature of acicular ferrite distinguished from that of the polygonal ferrite. In addition, it is found that higher coiling temperature leads to more quasi-polygonal ferrite and less bainite ferrite. The microstructure is refined with decreasing the strip thickness and coiling temperature.

Numerous microalloying precipitates with diameter smaller than 10 nm are observed by TEM, as shown in Fig. 5. It is speculated that these nanometer-sized particles

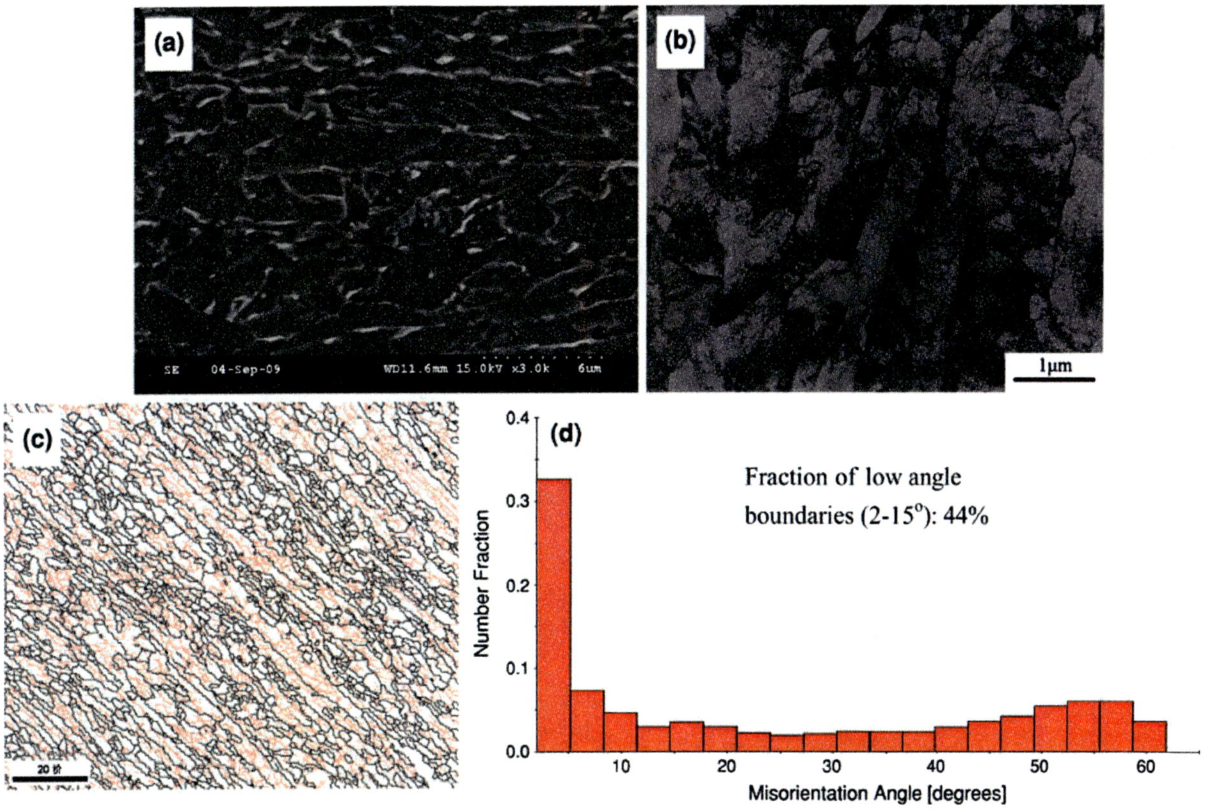

Fig. 4 Microstructure of the tested strip. **a** SEM micrograph showing the feature of acicular ferrite. **b** TEM image showing a relatively high density of dislocations. **c** EBSD boundary map. **d** Misorientation distribution map showing a large fraction of low angle boundaries

Fig. 5 TEM image showing numerous nanometer-sized precipitates containing Ti and Nb

are mainly precipitated from ferrite, which will lead to a significant strengthening effect.

Figure 6a shows the total elongation (A_5) of the developed precipitation hardening steel and the commercial low carbon bainitic steel. It is seen that the low carbon bainitic steel has a total elongation of about 12% on average in the strength range from 700 to 850 MPa, but the precipitation hardening AF steel has a total elongation over 18% at the same strength level. The best balance of strength and total elongation is obtained in the 3 mm thickness strip coiled at 550°C as 850 MPa for yield strength ($R_{p0.2}$) and 24% for total elongation (A_5). The strip has a very fine grained and bainite ferrite dominant AF microstructure, as shown in Fig. 4. Figure 6b shows the change of strain hardening rate with true strain for the low carbon bainitic steel and the precipitation hardening AF strip. It is seen that the strain hardening rate of low carbon bainitic steel decreases rapidly with strain, resulting in low uniform elongation. However, the strain hardening rate of the newly developed strip decreases slowly with strain. Therefore, those curves of strain hardening rate-true strain in the developed strips intersect at larger strains with the true stress–strain curves, i.e., a greater uniform elongation. The reason for the strain hardening rate increases is mainly resulted from the interaction of nanometer-sized precipitates and dislocations. However, further studies are needed to clarify this point.

1. The combination of ultra-low carbon design and a substantial refinement of effective grain size in lath martensite is demonstrated to be a promising route to produce 800–1,000 MPa grade HLSA steels with enhanced toughness. The excellent balance of strength and toughness values is obtained as 950–1,060 MPa for $R_{p0.2}$ and 180 J for $A_{KV(-30°C)}$, which is much superior to that of traditional martensitic steel.
2. Austenite deformation in the non-recrystallization temperature region (ausforming) is capable to refine the effective grain size of ultra-low carbon martensite significantly. Two mechanisms are proposed to explain this. One is by the austenite grain refinement in the direction of thickness, and the other is by the reduction in the fraction of sub-block boundaries with small misorientation and the increase in the fraction of block boundaries with large misorientation.
3. The microstructure with the combination of acicular ferrite and numerous dispersed nanometer-sized particles is demonstrated to be a promising candidate to produce HSLA steels with enhanced ductility. The best balance of yield strength and total elongation is obtained as 850 MPa for yield strength ($R_{p0.2}$) and 24% for total elongation (A_5).

4 Ultrahigh Strength Steel Sheets with Improved Ductility for Automobile

Apart from the weight-lightening, the safety standard of the automobiles was also being paid more and more attention. It was reviewed by American researcher as shown in Table 1 that in 1980s the strengthening of IF steel and the galvanized steel sheets was paid more attention for their cost and erosion properties, in 1990s the strengthening of HSLA, C–Mn, BH steels were widely applied in auto industries due

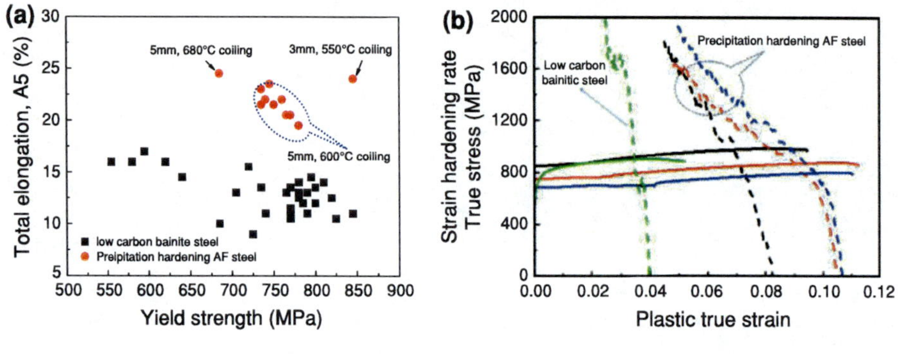

Fig. 6 **a** Balance of yield strength and total elongation for commercial low carbon bainitic steel and the precipitation hardening AF steel. **b** Comparison of strain hardening rate between the two steels

Table 1 The key design, key properties and corresponding steels used in automobiles [5]

Years	Key design	Key properties	Steels
1980	Cost/formidability/erosion	Strengthening	IF, Galvanized sheets
1990	Cost/crashworthiness	Strengthening	HSLA/C–Mn/BH
2000	Cost/energy saving/crashworthiness	Crashworthiness	DP/TRIP/hot-stamped
2020	Crashworthiness/energy saving	Crashworthiness	Third generation steel

to their low cost and higher crashworthiness, and in 2000s the crash properties of DP steel, TRIP steel and Hot stamped martensitic steel were mostly selected in auto industries concerning their low cost, weight-lightening and safety improvement in crash. The developments of automobile steels in the near future would be mainly focused on the crash properties and energy saving, which results in the third generation steel with not only high strength but also the high ductility.

Over the last two decades a significant research effort has been put in the development of Advanced High Strength Steel grades with high strength and high ductility. As an index of formidability and absorbed energy of materials, the product of ultimate tensile strength (Rm) and total tensile elongation (*A*), Rm × *A*, has been applied to tailor the steels for automobile application. For the conventional steels possessing primarily ferrite-based microstructures, such as interstitial free (IF), dual phase (DP), transformation induced plasticity (TRIP), complex-phase (CP), and martensitic (MART) steels, their Rm × *A* is only about 15 ± 10 GPa% named as the first generation steel in Fig. 7a [6–8]. Contrast with this low value, the second generation austenitic steels including twinning induced plasticity (TWIP) steels, Al-added lightweight steels with induced plasticity, and other fully austenitic steel, their Rm × *A* is remarkably high up to 50 ± 10 GPa% named as the second generation steel in Fig. 7a, which was thought to be the best material for cold forming and energy reservoir in crash but much more expensive and difficulty in fabrication [8, 9]. It is clear that the advantage of the first generation steel is their cost-effective but the disadvantage is their low ductility, comparing with that of the second generation steel. The background demonstrated above promoted worldwide study on development of the third generation steel with high Rm × *A* in between these two groups but at relative low cost. In China in 2009, the government launched out the third phase 973 program on the high performance automobile steel, i.e., the third generation automobile steel offering high strength and high ductility, with a target of Rm × *A* no less than 30 GPa% at strength levels of 1.0–1.5 GPa, which just in between the first generation automobile steel and the second generation automobile steels.

In Fig. 7b, the product of Rm × *A* of different kinds of steels, such as conventional steel (IF steel, DP steel, Martensitic steel) [8], conventional TRIP steel [8], Nano-Bainitic steel [10], and TWIP steel and austenitic steel [8], is briefly summarized as a function of austenite volume fraction. It is interesting to be found from Fig. 7b that no matter TRIP effect or TWIP effect in the steel, generally the Rm × *A* linearly increases with increasing of austenite volume fraction from conventional steels to TWIP steel. The slope between Rm × *A* and the austenite volume fraction is ~0.65 GPa%/(1%), indicating the strong dependence of Rm × *A* on austenite volume fraction. It could be expected roughly that 20–40% metastable austenite is essential to obtain steel with Rm × *A* of 30–40 GPa%. Thus the duplex structure with metastable retained austenite in BCC matrix may be a promising way to design the new type automobile steel with both ultrahigh strength and high ductility by means of TRIP/TWIP effects.

How to control the steel microstructure with FCC and BCC phase to get steel with high strength and high ductility? Recently, the conception of microstructure control of

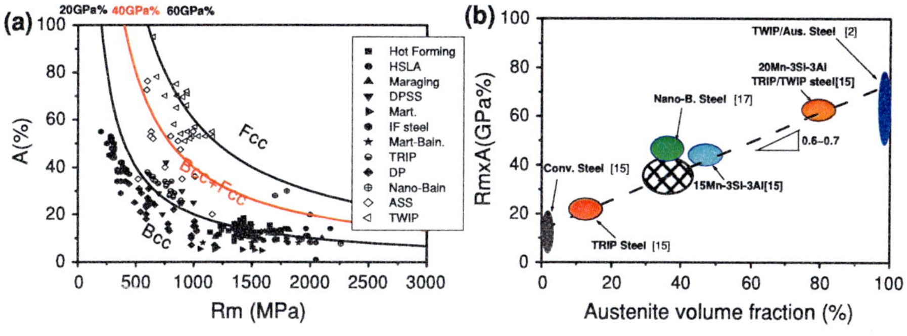

Fig. 7 Dependence of elongation to failure on tensile strength and residue austenite fraction of different steel (**a**) total elongation and ultimate tensile strength and (**b**) product of ultimate tensile strength to total elongation and austenite volume fraction

Multi-phase, Meta-stability and Multi-scale, i.e., the M^3 structure control, was proposed by us. In this study, the M^3 microstructure is not only the characters of the desired microstructure but also the controlling approaches to get this structure. For example, to get the high strength, the microstructure should be controlled in different scales, such as prior austenite grain size scale, the packet, block and lath width scale control. This kinds of controlling of the microstructure were clearly demonstrated in ART-annealed medium manganese steel and the Q&P processed steels, which in turn gives microstructure of processed steels with M^3 features [11–14]. As it was demonstrated very clearly by conventional TRIP steel and Quenching & Partitioning processed steel, the so called "Q&P" steel as firstly proposed by Speer, multi-phase and meta-stability is crucial for the high strength and high ductility [15, 16]. Apart from this Multi-phase and Meta-stability characters, the scale of the microstructure units, i.e., the Multi-scale, is also one of the important characters controlling strength, ductility and toughness of steel by the coupling behavior among phases.

How to develop the M^3 structure in steel by heat treatments? In light of the processing of conventional TRIP steel, isothermal holding of the austenite in the intercritical region (partial phase transformation from austenite to ferrite to develop a FCC + BCC structure) following by bainite holding of the retained austenite (partial phase transformation from the retained austenite to bainite to form a triple phase structure with ferrite, bainite and retained austenite) is the most straightforward manner to obtain BCC–FCC duplex structure. But this process gives a rather soft large grained ferrite matrix, thus cannot be applied to get ultrahigh strength. The hard ferrite matrix and large amount of retained austenite could be obtained by austenite reverted transformation from annealing (simply called ART-annealing in this study), which gives ultrafine grained matrix and large fractioned austenite phase (>20%) [17], or the quenching and partitioning (Q&P) process, which results in a strong hard martensitic matrix and relative large fractioned austenite (5–15%) [13].

In this research, based on the conception of M^3 structure conception, the ART-annealing process was applied to produce the third generation automobile sheet steel aiming at getting automobile steel with ultrahigh strength and high ductility, i.e., with excellent combination of strength and ductility. In order to obtain duplex structure with ultrafine grained hard ferritic matix and metastable austenite phase by ART-annealing, the chemical composition has to be specially designed to control the coarsening of both martensite lath and the newly developed austenite. It is well known that difference from the fast diffusion of interstitial elements, such as carbon, substitutional elements can suppress the microstructure coarsening during ART-annealing at relative low temperature. Amongst the elements applied in the duplex steel and austenite steel and TWIP steel, such as the Si, Mn, Cr, Ni, Mo and Al, Mn is not only a cheap element to enlarge the austenite region and stabilize the austenite phase, but a strong element for substitutional solute hardening.

In our Lab in CISRI of China, Fe–Mn-C steels with 3–9% Mn and varied carbon concentration were designed to develop BCC–FCC duplex steels with TRIP/TWIP effects by ART-annealing. Fe–Mn–C alloys composition design was carried out based economy and solute strengthening principle and its microstructure design methodologies was carried out based on fundamental strengthening mechanisms and metastable austenite deformation stability theory. The cooling rate effect on the hardenability, the quenching temperature effects on microstructure and mechanical behavior, the ART annealing resulted microstructure and mechanical properties and the Mn replacing carbon alloying were studied in detail. Especially, ART-annealing behaviors, the microstructure evolutions and the element (Mn–C) partitioning behaviors were presented with theoretical consideration. The strengthening and ductility-enhancing mechanisms were studied based on the detailed microstructure characterization, mechanical properties examination and the analysis of the relationship between microstructure and mechanical properties. Furthermore, consideration on the strengthening by precipitation of TiC was also tried to increase the strength of Fe–Mn–C alloys.

In this section, the development of the third generation automobile steel at strength level of 1,000 GPa with Rm × *A* of 30–40 GPa% processed by ART-annealing was reported. The ultrafine lamellar ferrite–austenite duplex structure will be demonstrated, the substantially enhanced mechanical properties of the newly developed third generation automobile steel will be presented. At last the strengthening and ductility-enhancing mechanisms of the third generation automobile steel will be proposed tentatively based on the detailed microstructure characterization, mechanical properties examination and the analysis of the relationship between microstructure and mechanical properties.

4.1 Ultrafine Lamellar Ferrite–Austenite Duplex Microstructure in Medium Manganese Steel Processed by ART-Annealing

As an example of the C–Mn steel processed by ART-annealing, the ultrafine lamellar ferrite/martensite structure of Fe-0.2%C–5%Mn steel characterized by TEM was shown in Fig. 4. After 1 h annealing at 650°C, the dark/bright lamellar structure in the original martensite packets was clear with austenite lath parallel to the slightly coarsened martensite lath, Fig. 8a. The equiaxed austenite grains could also be found in the packet boundary or original austenite grain boundary. After 1 h annealing, the width of austenite laths is about 0.2–0.3 μm. With increasing of annealing time, the BCC–FCC duplex microstructure with parallel ferrite laths and austenite laths still remains in Fe-0.2%C–5%Mn steel after 6,12 and 48 h ART-annealing at 650°C as shown in Fig. 8b–d. In addition, it can be seen from Fig. 4 that increasing annealing time improves the uniformity of the microstructure and decreases the precipitation of carbides, suggesting the concurrence of austenite growth by thickening, precipitation dissolve in the interface region between precipitation and the coalescence martensite laths. After 6 and 12 h annealing, the austenite width is about 0.3 μm, slightly thicker than that of 1 h annealed specimen. Even after 48 h annealing at 650°C, the width of both austenite lath and ferrite lath is only about 0.4 μm, indicating the high thermal stability at this high temperature. This may need further study on the mechanism of this kind of thermal stability.

From Fig. 8, it is very interesting that the width of both austenite lath and ferrite lath is still very fine after long time annealing. It was measured the average width is about 0.5 μm for even after 144 h annealing, which still belongs to the submicron grain sized materials, indicating the high thermal stability of Fe-0.2%C–5%Mn steel during ART-annealing process. Also with increasing of annealing time, no precipitation could be found in the studied regions.

The ultrafine lamellar microstructure was also observed through EBSD in FEG/SEM as presented in Fig. 9 as a function of annealing time. In this figure, the austenite phase was revealed by green color but the ferrite phase was presented by the grey color. The blue colored lines depict the low angle boundaries with misorientation lying in between 3 and 15° and the black lines present the high angle boundaries with misorientation larger than 15°.

It can be seen that the austenite grains gradually developed in the initial martensite. In the short ATR-annealing time, for example annealing time lower than 30 min at 650°C the nearly equiaxed austenite grains mainly could be found in the grain boundary or packet boundary as shown in Fig. 9a–c. Due to the low misorientation between martensitic lath, no clear martnsitic lath structure could be identified in these shorted time annealed microstructure. When the annealing time was larger than 30 min at 650°C, the nucleation and growth of lath type austenite grains occur in between the martensitic lath as shown in Fig. 9c and d. Also it can be seen that the parallel lath typed structure becomes clear in these annealed microstructure, because the austenite nucleated in between the martensitic lath highlights the lath structure. After 6 h ART-annealing, the austenite grains nucleated and grew both in the boundary regions and in between the martensitic lath slightly coarsened with increasing annealing time as revealed in Fig. 5e and f.

Deformation microstructure of Fe-0.2%C–5%Mn steel after ART-annealing at 650°C with 6 h was characterized by TEM as shown in Fig. 10a and b. After about 37% uniform uniaxle tension deformation, twined martensite could be distinguished from the annealed martensite.

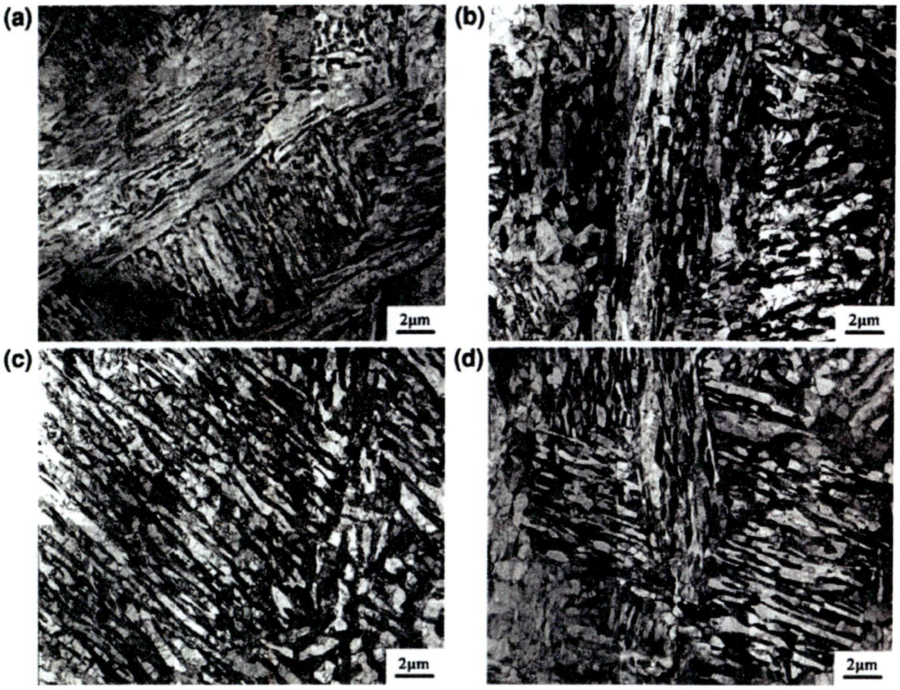

Fig. 8 Microstructures of Fe-0.2%C–5%Mn steel ART-annealed at 650°C with **a** 1 h, **b** 6 h, **c**12 h and **d** 48 h

Fig. 9 Microstructural observation of medium Mn steel ART-annealed at 650°C with different time **a** 1 min, **b** 5 min, **c** 30 min, **d** 1 h, **e** 6 h and **f** 12 h

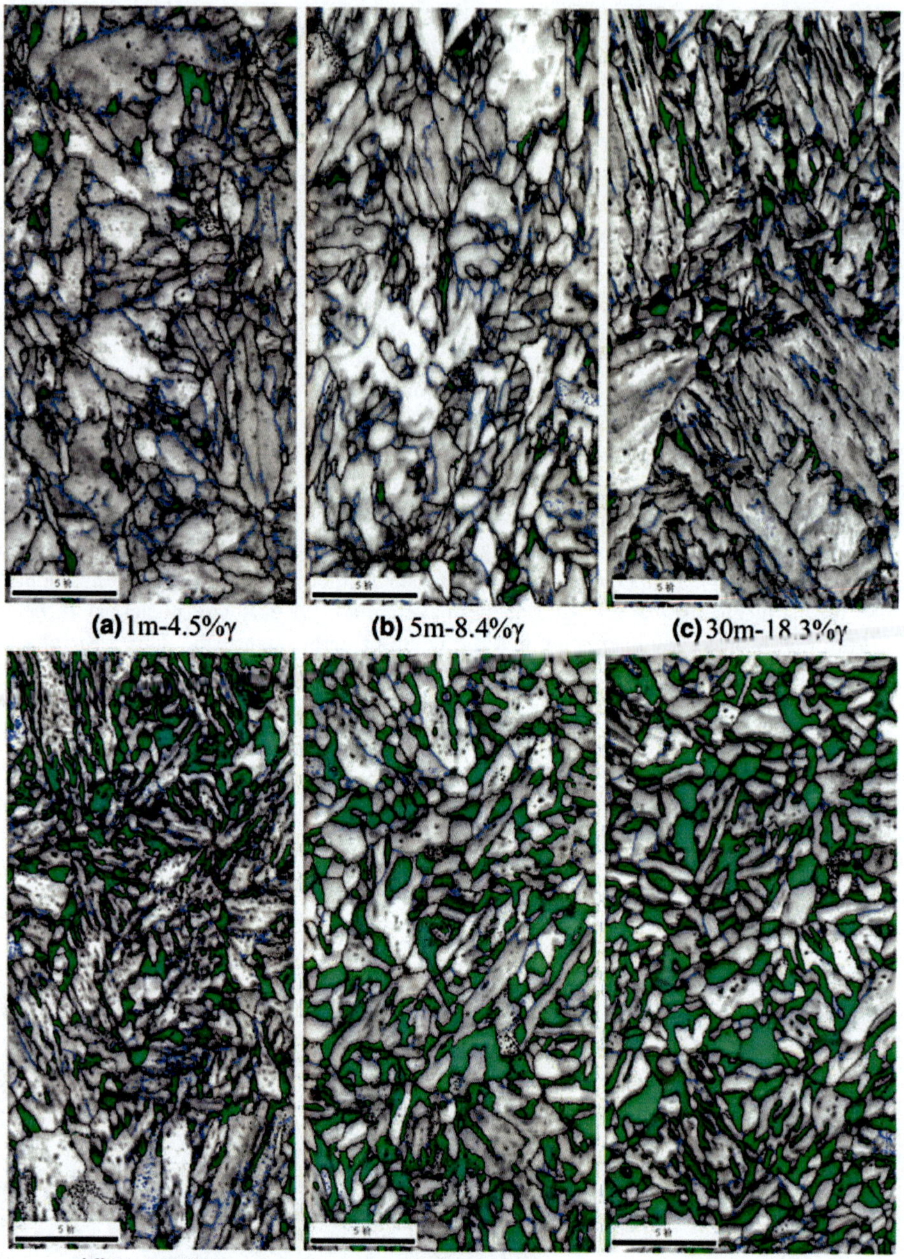

In Fig. 6, one type of martensite with twin plates direction perpendicular to the original austenite plates and another type martensite with twin plates direction parallel to original austenite lath were presented. It can be seen that in one austenite only one packet could be found, which is different from the thermal stability controlled phase transformation. High dislocation density generated in the austenite neighbored martensite was evidenced by XRD measured results of 2.5×10^{15} m^{-2}. It may be expostulated that the high dislocation density in the martensite lath (ferrite) may be attributed to the volume expansion resulted from phase transformation from austenite to martensite.

Based on the above results and discussion, the annealing behaviors of the quenched Fe-0.2%C–5%Mn steel were studied at 650°C with annealing time for up to 144 h. The austenite reverted transformation during annealing in the intercritical region (ART-annealing) were examined by TEM, STEM and XRD. The conclusions could be summarized that (1) ART-annealing of Fe-0.2%C–5%Mn steel resulted in the microstructure evolution from fully

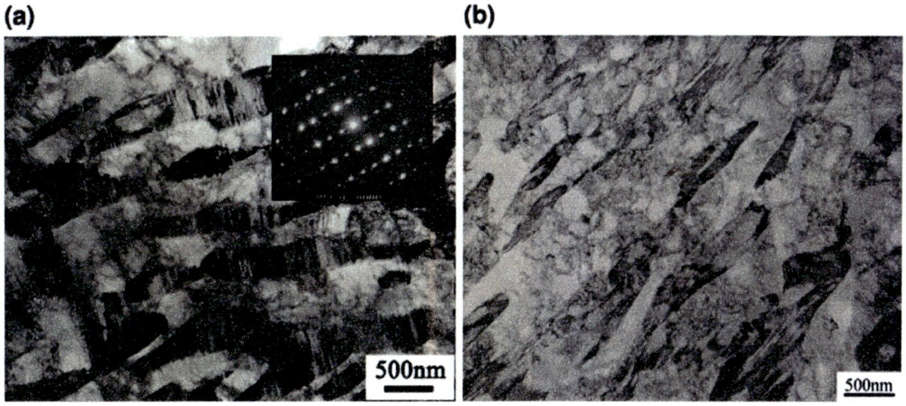

Fig. 10 Room temperature deformation structure at maximum uniform elongation (37%) of Fe-0.2%C–5%Mn steel after 6 h annealing at 650°C (**a**) one type of martensite with twin plates direction perpendicular to the original austenite plates with inserted diffraction paten and (**b**) another type martensite with twin plates direction parallel to original austenite lath

martensite structure to an ultrafine grained duplex microstructure with austenite fraction of ~34%. The final ultrafine lamellar ferrite/austenite duplex structure was obtained by lath scale microstructure, i.e., the ART-annealing. (2) Even after 144 h annealing at 650°C, the width of both austenite lath and ferrite lath remains smaller than 0.5 m, indicating the high thermal stability of the ultrafine grained duplex microstructure of Fe-0.2%C–5%Mn steel, which may be attributed to the slow diffusion rate of Mn in austenite. (3) During deformation process, the austenite gradually transformed into martensite, which suggests that the deformation behavior of ART-annealed medium manganese steel were controlled by the phase transformation induced plasticity (TRIP effects).

4.2 Mechanical Properties of Medium Manganese Steel

The engineering tension stress–strain curves of Fe-0.2%C–5%Mn after air cooling and annealing with different time were given in Fig. 11a. It can be seen from the engineering stress–strain curves that the that increasing annealing time from 1 min to 12 h increases the total elongation (A_T) significantly from 20 to 45%, decreases the yield strength from 830 to 600 MPa, whereas the ultimate tensile strength remains about 960 ± 30 MPa. The true stress–strain curves of studied steel were given in Fig. 11b. It can be seen from the true stress–strain curves that the ultimate tensile stress increases from ~1,000 to 1,350 MPa with increasing annealing time from 1 min to 12 h, suggesting the increased work hardening with increasing of annealing. Interestingly, it can be seen that serrated or jerk flow feature can be observed from the stress strain curves, indicating dynamic strain aging and strong localized deformation during tension test in the steel after 1 h or long time annealing at ambient temperature, Fig. 11b. Furthermore, the true stress–strain curves of the long time annealed specimens assume "S" typed shape, which may indicate the phase transformation during tension deformation process.

The mechanical properties of the medium manganese steels applied in this study were summarized in Fig. 12. It can be seen from Fig. 12a that within the range of ultimate tensile strength from 800 to 1,500 MPa the total elongation can be high up to 45–30%. The product of ultimate tensile strength to total elongation in the Rm range of

Fig. 11 The stress–strain curves of Fe-0.2%°C-5%Mn processed by ART-annealing with different annealing time **a** engineering stress–strain curves and **b** true tensile stress–strain curves

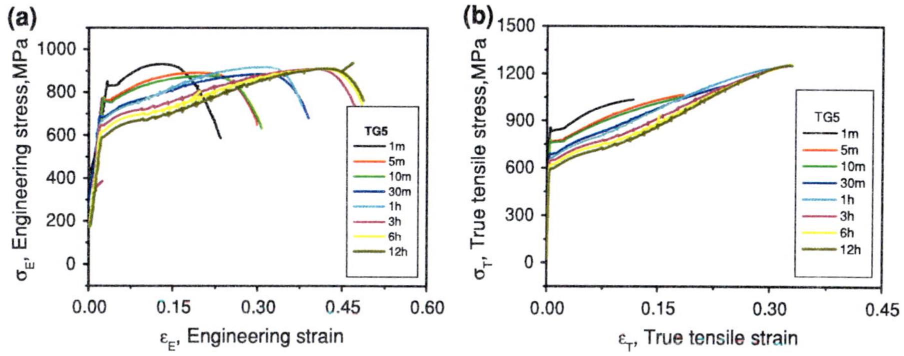

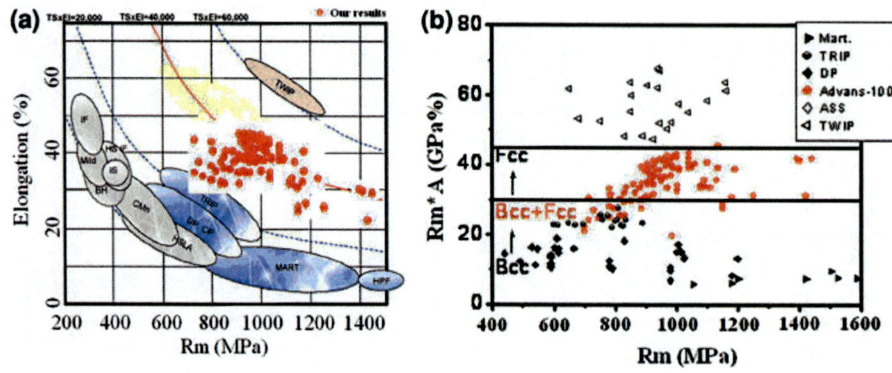

Fig. 12 Summary of the mechanical properties of different medium manganese steels applied in this study **a** total elongation as a function of ultimate tensile strength and **b** product of ultimate tensile strength to total elongation (Rm × *A*) was shown as a function of ultimate tensile strength

800–1,500 MPa is about 40 GPa, which means that the objective of the third generation auto sheet steel can be realized and would give important role in the production in steel industry.

4.3 The Proposed Mechanisms to Improve Ductility

It was well known that the classic strengthening mechanisms are the solute hardening, the precipitation hardening, the grain refinement hardening and the phase transformation hardening, et al. In this study the medium manganese steel after processing by long time annealing assumes ultrafine grained duplex structure with no precipitation can be found. Based on the TEM observation and EBSD characterization, the grain size of these is about 0.5 μm. After tension deformation, no significant grain refinement for ferrite phase was observed and after 37% uniform elongation, the grain size is still about 0.4 μm.

As mentioned above, the true ultimate tensile stress and total elongation are about 1,350 MPa and 45%, respectively, both of which are close to the high alloyed TWIP steel but significantly higher than conventional TRIP steel. Similar to the ultrafine grained IF steel fabricated by equal channel angular extrusion (ECAE), the high yield and tensile strength mainly stem from the grain refinement and high density dislocation [18]. The tensile stress of Fe-0.2%C–5%Mn could also be predicted based on the Hall–Petch law and dislocation strengthening mechanism according to Eq. 2,

$$\sigma = \sigma_0 + M\alpha G b \rho^{\frac{1}{2}} + k_{H-P} D^{-\frac{1}{2}} \qquad (2)$$

where σ and σ_0 are the flow stress and fractional stress, M, α, G, b, ρ, k_{H-P} and D are Taylor factor, dislocation hardening factor, shear modulus, Burgers vector, dislocation density, Hall–Petch slope and grain size, respectively.

Taking $\sigma_0 = 100$ MPa, $M = 3.06$, $\alpha = 0.24$, $G = 83$ GPa, $b = 0.248$ nm, $k_{H-P} = 300\,\mathrm{MPa}\,\mu\mathrm{m}^{\frac{1}{2}}$ and $D = 0.55$ μm for the 6 h annealed steel and $D = 0.4$ μm for the 40% tension deformed steel, the yield stress of Fe-0.2%C–5%Mn steel after 6 h annealing, was calculated to be 595 MPa and the true ultimate tensile stress to be 1,330 MPa, which are in good agreement with experimental results. However, it should be pointed out here that both nano-thickness stacking faults in the γ-phase in as annealed specimens and the nano-thickness martensite twin in as deformed specimens were neglected in grain size measurement, indicating no significant strengthening contribution from stacking faults and martensite twin.

However, if the linear contribution from the different phase is considered based on the strength mixed law, as demonstration by Davies [19, 20] that both yield stress and tensile stress linearly increase with increasing martensite volume fraction, the flow stress of multiphase steel (σ_F) could be formulated by following equation,

$$\sigma_F = f_\alpha \sigma_\alpha + f_\gamma \sigma_\gamma + (1 - f_\alpha - f_\gamma)\sigma_M \qquad (3)$$

Here (3), σ_α, σ_γ and σ_M are the flow stress of ferrite, austenite and martensite, and f_α, f_γ and f_M are the volume fraction of ferrite, austenite and martensite.

It is reasonable to calculate the strength of ferrite phase and austenite phase based on the Hall–Petch law, i.e., $\sigma_\alpha = \sigma_0 + k_{H-P}D^{-\frac{1}{2}}$, which is about $100 + 300 \times 0.5^{-\frac{1}{2}} = 524$ MPa, and the martensite strength could be estimate to be about 2,000 MPa. Then the yield stress is about $\sigma_{0.2} = 524$ MPa and tensile stress is about 790 MPa after 37% elongation but voce equation fitting and the tensile true stress is about $\sigma_{\mathrm{Uitimate}} = 790 + 0.33 \times 1{,}500 = 1{,}285$ MPa when $\sigma_M = 1{,}500$ MPa based on Eq. 3. It can be seen that the calculated stress is also in a good agreement with experimental results when the martensite strength is assumed to be of 1,500 MPa. For the Fe-0.2%C–5%Mn steel, its ultimate stress is over 2,000 MPa as measured by experimental, which means that the martensite is not getting to its ultimate stress. Based on the agreement between the calculated results from Eqs. 2 to 3, it can be concluded that the strength contribution from martensite phase is the same to the dislocation strength contribution. Thus the martensite strength contribution is realized by the dislocation generation during deformation process.

Different from the low ductility of the ultrafine grained IF steel processed by ECAE (normally lower than 10%) [18], the total elongation can be up to 45% at ambient temperature for our processed ultrafine grained Fe-0.2%C–5%Mn steel, implying that the extra-high total elongation cannot be related to the ultrafine grain size. It may be noticed that the total elongation of Fe-0.2%C–5%Mn steel increases from ~22 to ~45% with annealing time increasing from 1 min to 12 h, which assumes a similar trend of austenite fraction with time increasing, suggesting the improved ductility results from the increased austenite fraction. It is well known that the role of austenite phase in TRIP steel is to delay the necking by enhanced the local work hardening due to austenite to martensite phase transformation, which is evidenced by the work hardening rate calculation. Thus the high total elongation of Fe-0.2%C–5%Mn steel is mainly contributed from the highly enhanced TRIP effect because of its high volume fraction of the metastable austenite phase.

In order to understand the mechanism of the enhanced mechanical properties, the uniform tensile elongation and the total elongation were shown as a function of austenite volume fraction in Fig. 9a. It can be seen that both uniform and total tensile elongations increase with austenite volume fraction. Furthermore, the product of ultimate tensile strength to total elongation was shown as a function of the austenite fraction in Fig. 9b as well, from which it can be seen that the product almost linearly increases with austenite volume fraction, which assumes a similar increasing trend of the product increasing with austenite fraction. For comparison, the Rm × *A* values of the typical first generation automobile steel [8], second generation automobile steel [8], nano-bainite steel [10], 15Mn-3Si-3Al steel and 20Mn-3Si-3Al steel [9] were given in Fig. 9b. It can be seen that all the Rm × *A* values nearly linearly increase with increasing austenite fraction no matter TRIP or TWIP effects. Thus, it can be seen that the large austenite fraction is the main reason to the excellent mechanical properties as demonstrated in our study.

Why the mechanical properties increases with increasing austenite fraction as analyzed in Fig. 13? It was found by TEM study that during tension deformation process, the austenite phase transformed into martensite, which results in a high work hardening rate and thus delayed the neck starting and finally gives relative high uniform elongation [21–23]. The higher austenite volume fraction, the larger the uniform elongation and the total elongation could be obtained. However, for a steel with given chemical compositions, the increasing of austenite volume would decrease the austenite stability and thus impair its mechanical stability and finally deteriorates the ability of phase transformation induced plasticity, i.e., the TRIP effects [24, 25], which can interpret the decreased mechanical properties after 6 h annealing as shown in Fig. 13.

The high value of Rm × *A*, i.e., 43 GPa% of Fe-0.2%C–5%Mn steel after 6 h annealing, may be explained by two main reasons. One is the enhanced TRIP effect as discussed above and the other may be the matrix characters, such as its ultrafine grained structure and its low stored energy as well. In order to understand the contribution of TRIP effect, the relationship between product of Rm × *A* of different kinds of steel, and the austenite fraction is re-analyzed in Fig. 13b. It can be seen, no matter TRIP effect or TWIP effect, generally the slope between Rm × *A* and the austenite volume fraction is ~0.6–0.7 GPa%/(1%), indicating the strong dependence of Rm × *A* on austenite volume fraction. Thus the contribution of austenite phase to Rm × *A* in Fe-0.2%C–5%Mn steel could be calculated to be ~18–21 GPa%, almost half of the measured Rm × *A*. In addition, the ultralow stored energy of the

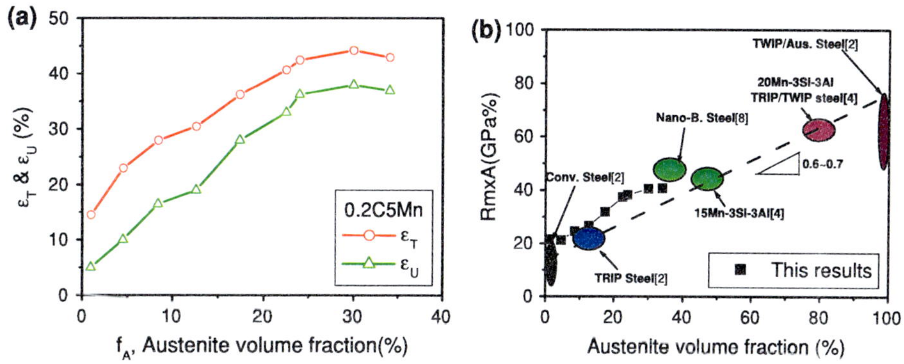

Fig. 13 Mechanical properties dependence on austenite volume fraction **a** relationship between elongation(ε_T and ε_U) and austenite fraction and **b** dependence of the product of Rm × *A* on austenite volume fraction

matrix due to the high temperature annealing may be another reason to the high value of Rm × *A*. During annealing process at 650°C (higher than Acl of 630°C) most of the dislocations were depleted from the materials and thus providing a reservoir for high density dislocations produced by strain induced phase transformation and tension deformation itself during further deformation process. As measured by XRD, the dislocation density before and after 37% uniform tension deformation is about $\sim 3.6 \times 10^{13}$ and $\sim 2.5 \times 10^{15}$ m^{-2}, respectively, which gives dislocation density increment about 2.46×10^{15} m^{-2}, and a stored energy increment (ΔE_S)about 0.63 MJm^{-3} according to Eq. 4 [26]. The expended energy (E_{EXP}) in the matrix during tension deformation process could calculated to be ~22 GPa% according to Eq. 5 [26] with energy storage ratio $\chi = 0.02$ Thus Rm × *A* is roughly the sum of 18–21 GPa% from austenite phase and ~22 GPa% from the matrix, i.e., 39–43 GPa%, which meets very well with the measured one.

$$\Delta E_S = \alpha_1 \beta G b^2 \Delta \rho \tag{4}$$

$$E_{EXP} = \Delta E_S / \chi \tag{5}$$

where α_1 is a constant about 0.5, β the volume fraction of matrix, $\Delta\rho$ the dislocation increment before and after tensile deformation, and χ the energy storage ratio between the stored energy and the expended energy in the matrix.

5 Heat Resistance Steels for Advanced USC Boiler

Boiler steel tube and pipe for power station serve in the environment of elevated temperature and high pressure as well as various corrosions for a long time. The technology of ultra super-critical fossil fuel power plant plays a key role to improve Chinese electricity structure in the near future and is also the one of most important national strategies to achieve emission reduction goal. Advanced heat resistant steels, such as boiler steels, blade steels and rotor steels, have been necking points which prohibit Chinese enterprises to form competitive ability to manufacture ultra super-critical fossil fuel power plants. Although Chinese researchers [27] pioneered "multi-element compound strengthening theory" and contributed much in the fundamentals and product development of boiler steels in the 1960s, technical progress of boiler steels was basically in the in-active in the about quarter century from 1975 to 2000, mainly because of the weak and leak driving force of effective requirements.

As shown in Fig. 14, the steam parameters of Chinese fossil fired power stations had been in very low level in the very long term. However, the situation has been completely changed since 2006, when the very first ultra super critical (USC) fossil fired power plant was launched in Yuhuan, Zhejiang Province. To the end of 2009, 44 sets of 600 MW and 73 sets of 1,000 MW USC power plants have been constructed or under constructing in the mainland of China, which seriously challenges the technical capabilities of Chinese boiler steel sector. By the end of 2008, the requirement of P92, S30432 and S31042 steel tube and pipe used for Chinese 600°C USC power plants depended on import.

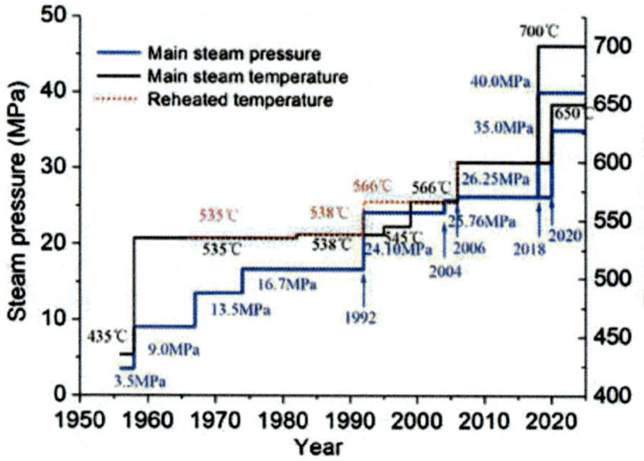

Fig. 14 Steam parameter evolution of Chinese fossil power stations

In 2007, the MOST of China launched a national program with the grant number of 2007BAE51B02 to develop T/P92, S30432 and S31042 steel tubes used for 600°C USC power plants in China. The prime goal of the national project is to realize the localization of above steel grades by the end of 2010. In 2009, the MOST of China launched another national program with the grant number of 2010CB630804 to develop fundamental knowledge of ferritic boiler steels used for 650°C USC power plants. The aim of the national project is to develop a prototype steel pipe used for 650°C USC power plants by the end of 2014. Meanwhile, CISRI and Baosteel had signed a strategic and long-term collaborative program in the R&D of boiler steel products since 2006. The investigation and development of boiler steels and alloys used for 700°C USC power plants is an important part of the technical collaboration. The prototype steel tube and pipe used for 700°C USC power plants is possibly ready by the end of 2014. A national plan to build a demo 700°C USC power plant may be launched by Chinese government in the near future.

It briefly summarized the investigation in laboratories, pilot trials and industrial implementation of P92, S30432 and S31042 boiler steel tube/pipes used for 600°C steam parameter fossil fired ultra super critical (USC) power plants in China in the past decade.

5.1 P92 Steel Pipe

Chengdu Iron and Steel Co. Ltd. of PanSteel Croup (Thereafter as No. 65 plant) melted a heat of P92 steel before 2009 and manufactured a steel pipe with the dimension of Φ298.5 × 33 mm. The metallurgical process was ESR ingot → piercing → pilger rolling process. The mechanical properties of the specimens cut from the manufactured pipe met the requirements of ASTM A213 M-07 both at room temperature and at elevated temperature. However, the δ-ferrite was found in the specimens with the amount of being up to 10% (Fig. 15).

In order to eliminate and reduce the amount of δ-ferrite in P92 steel pipe and to optimize the chemical compositions of P92 steel within the scope of ASME/ASTM standard, CISRI melted 12 heats P92 steel [28]. The nickel and chrome equivalent of 9–12%Cr boiler steels can be expressed as follows:

$$Cr_{eq} = Cr + 0.75W + 1.5Mo + 2Si + 5V + 1.75Nb + 1.5Ti + 5.5Al$$

$$Ni_{eq} = 30C + Ni + Co + 0.5Mn + 0.3Cu + 25N$$

The predicted microstructure of the 12 heats of CISRI P92 steel was plotted in Fig. 14. One of the 12 heats was specially designed to be duplex phase of martensite and δ-ferrite. The experimental measurement agrees well with the above prediction, which implies the fact that it is possible to effectively control the amount of δ-ferrite through the optimization of nickel and chrome equivalents. Meanwhile, experimental trials verified that the hot deformation temperature should not be very high, so as to avoid the increment of δ-ferrite amount. During industrial operation, it is imperative to smoothly and uniformly control thermal evolution of steel pipes to avoid local temperature turbulent. At the same time, CISRI also experimentally investigated the quantitative relationship among the dimension, cooling rate, microstructure and property of P92 steel pipe with

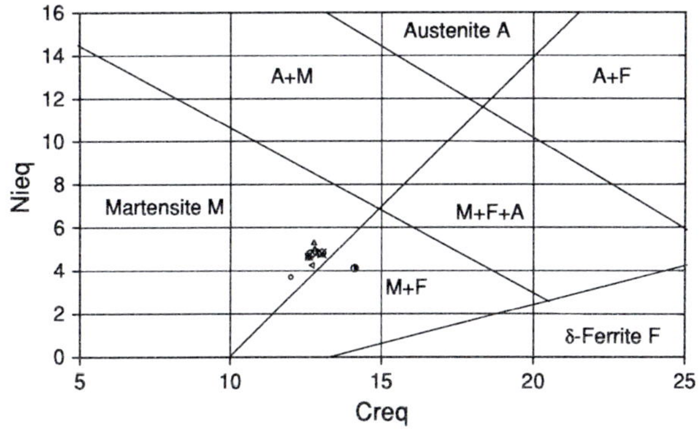

Fig. 15 Schaeffler chart of P92 steels designed by CISRI

Table 2 Mechanical properties of P92 steel melt by No. 65 Plant in 2009

Test place	Rm/MPa	Rp0.2/MPa	A/%	Z/%	HB
At No. 65 Plant	705	515	23.5		209
At CISRI	695	540	23.5	71.5	220
ASME value	≥620	≥440	≥20		≤250

large diameter and thick wall. Fortunately, the adventure has been well rewarded.

Based on aforementioned results, No. 65 Plant industrially melted another heat of P92 steel in the summer of 2009 to make pipes with the diameter of Φ298 and Φ508 mm. CISRI characterized and measured the δ-ferrite amount of the new ingot and found the amount of δ-ferrite is less than 1.0% (about 0.6%). CISRI and Pancheng Steel tested separately the mechanical properties of the P92 steel pipes. The measured data were listed in Table 2. Clearly, the properties met related ASME specifications. The long time creep test of the steel pipe is undergoing at CISRI.

In the past years, some Chinese enterprises, such as Jiangsu Chengde Pipe Co., Inner Mongolia Northern Heavy Industries Co., Hebei Hongrun Heavy Industries Co., industrially and independently manufactured P92 steel pipes in different processes. Their creep data, together with those from CISRI, were plotted in Fig. 16. The extrapolated creep data of those Chinese P92 steel pipes at 625°C for 10^5 h is about 99 MPa, which is higher than 87.6 MPa as claimed by ASME for the same situation (Fig. 16).

5.2 S30432 Steel Tube

S30432 steel was developed by Sumitomo metals industries and Mitsubishi heavy industries of Japan on the basis of TP304H, used for super-heater and re-heater in super critical and/or ultra-super critical power plants, which had been included in ASTM A213/A213M and ASME Code Case 2328-1 since March of 2000.

Comparing to TP304H, strengthening elements such as Cu, Nb, N, B were added into S30432 steel. Meanwhile, the contents of Mn and Si were reduced. CISRI systematically investigated the effect of copper on microstructure and properties of S30432 steel since 1998 [29]. The creep rupture strength of the steel increases with the increase of Cu addition when the Cu content is <4.0%. The creep rupture strength reaches its peak when the Cu content is about 4.0%. And the creep rupture strength decreases when the Cu content is over 4% for the current case. Likewise, the variation of elongation is similar. According to the result of chemical phase analysis, under the same aging condition, the total amount of precipitated carbides including $M_{23}C_6$,

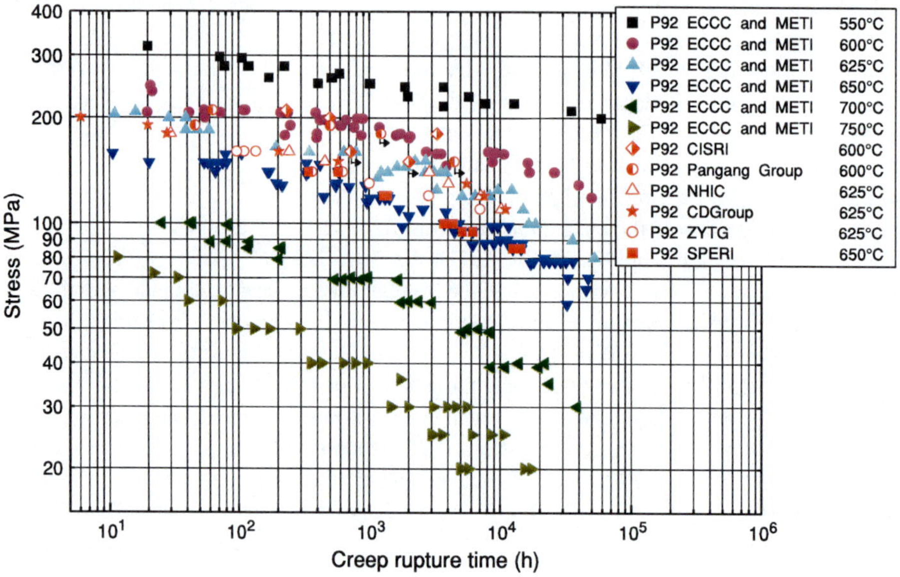

Fig. 16 Current creep data of Chinese P92 steel pipes

Nb(C,N), and M_6C keeps stable with the increase of Cu. Meanwhile, the individual amount of each carbide precipitate also keeps stable. Therefore, this logically indicates that carbides have almost no effect on the change of properties of the steel when the content of Cu increases. The variation of the properties of the steel may be resulting from the Cu-rich precipitates during aging. The precipitated amount, particle size and distribution of Cu-rich phase may mainly contribute to the variation of the properties of the steel. The quantitative relation of the copper content and Cu-rich precipitates was drawn in Fig. 17. When the Cu addition is about 4%, Cu-rich phase has an ideal combination on its precipitated particle numbers, size and spacing, which consequently leads to the highest creep rupture strength in this experimental scope. However, in the specification of Super304H of Sumitomo Metal Industries, the recommended scope of Cu addition ranges from 2.5 to 3.5% and the best control point is 3.0%. When Cu addition is more than 3.5%, it may undermine the hot workability of the steel.

Inter-granular corrosion (IGC) in S30432 steel may occur before service, which challenges the application of the steel tube. In order to better understand the issue, CISRI experimentally investigated the effect of carbon and niobium as well as heat treatment on IGC of the steel during 2003–2006 [30, 31]. As shown in Fig. 18, at the situation of solid solution temperature is higher than 1,100°C, when carbon content is lower than 0.08%, no IGC occurs. When carbon content is higher than 0.08% and niobium content is higher than 0.70%, no IGC occurs too. However, the upper limitation of niobium of S30432 steel in the ASME specification is 0.60%, therefore, it is imperative to control the carbon content in the very narrow scope from 0.07 to 0.08% to avoid the occurrence of IGC of the steel.

Although Mo was not included in ASME/ASTM specifications, it had been realized that the addition of Mo with the scope of 0.20 to 0.40% is very helpful to ensure the sound properties of S30432 steel tubes and therefore CISRI recommended Chinese steel makers to add suitable Mo when they produce S30432 steel tubes.

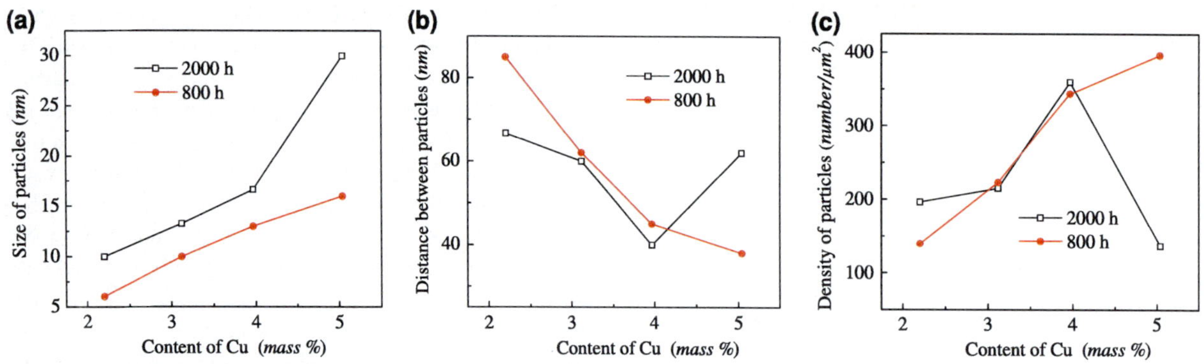

Fig. 17 Effect of copper content on Cu-rich precipitates **a** precipitated particle size, **b** spacing of precipitated particles, **c** particle density

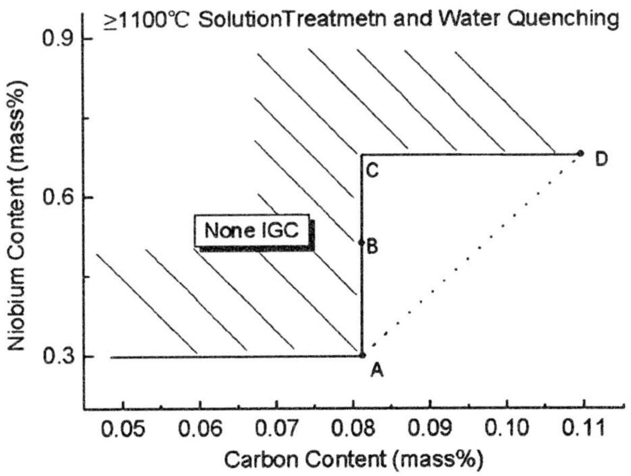

Fig. 18 Effect of carbon and niobium on IGC of S30432 steel tubes

Creep rupture strength and oxidation resistance are two important concerns to verify the successfulness of development of S30432 steel tubes. Both properties directly and closely relate to heat treatment processes. In general, upon the determination of chemical compositions of S30432 steel tube, the higher the solid solution temperature, the higher the creep rupture strength of the steel. However, with the increase of the solid solution temperature, the grains of the steel tubes grow, which undermines the oxidation resistance of the tubes. To solve the problem, high temperature softening treatment would be an effective solution, similar to that of TP347HFG. The temperature of high temperature softening treatment should be 70°C higher than that of final solid solution temperature and its aim is to sufficiently dissolve Nb(C,N) precipitates in austenite during heating and to sufficiently precipitate during subsequent cooling process, to ensure more finer precipitated particles available to pin the grain boundary movement to enhance the creep strength during the service at the temperature of 650–700°C. Larger reduction was carried out during the final cold rolling to obtain finer grains after heat treatment, so as to improve oxidation resistance.

CISRI, BaoSteel Co. Ltd. and Dongfang Boiler Co. Ltd., had carried out systematical investigation on the heat treatment processes of S30432 steel tubes in the recent years. After high temperature softening treatment (>1,230°C) and cold deformation, the temperature of final solid solution of S30432 steel tube should range in the scope of 1,130–1,150°C. It have been experimentally verified that the grains of S30432 steel tube could be in the range of ASTM 7–9.

Holding time is a critical processing parameter when solid solution temperature is high. When solid solution temperature is lower than 1,060°C, the grain grows slowly during holding. However, when solid solution temperature is too low, it may lead to the occurrence of inter-granular corrosion. Therefore, it was indicated that solid solution temperature should be higher than 1,100°C in the ASME specification. The experimental results of CISRI supported the indication. When solid solution temperature is higher than 1,150°C, the grain size showed three continuous regimes during holding. They are fine grain regime, mixed grain regime and coarse grain regime. Thus, the optimized holding time locates in the transition point of fine grain regime and mixed grain regime. The initial fast growth of individual grains can serve as the hint and judge to determine the optimized holding time. In addition, the optimized

Table 3 Dimensions and properties of industrially produced S30432 steel tubes

Producer	Dimension (mm)	Rm (MPa)	$R_{p0.2}$ (MPa)	*A* (%)	HRB, HV
Standard	ASME CC2328-1	>590	>235	>35	<95, <230
BaoSteel	Φ51 × 9.5	610	285	44	HRW156
BaoSteel	Φ47.6 × 6	615	310	41	HRW160
BaoSteel	Φ43.1 × 7	640	320	44.5	HRW156
YiXing steel tube	Φ50.8 × 4.5	655	405	41	HV177
YiXing steel tube	Φ45 × 7	690	450	45	HV177
YiXing steel tube	Φ60 × 7	665	380	45	HRB91
YiXing steel tube	Φ45 × 9.2	625	360	48	HRB87
WuJin Stainless Steel	Φ45 × 9.2	650	335	46	
Great wall special Steel	Φ50.8 × 6.2	660	420	45	HRB87
TISCO	Φ38 × 6.6	610	340	44	HRB78
TISCO	Φ54 × 7	650	370	45	HRB88
JiuLi group	Φ57 × 6.5	660	430	43	
Changshu Huaxin	Φ57 × 6.5	660	450	48	HRB90
Sumitomo metal	Φ51 × 9.5	640	355	59	HRB83

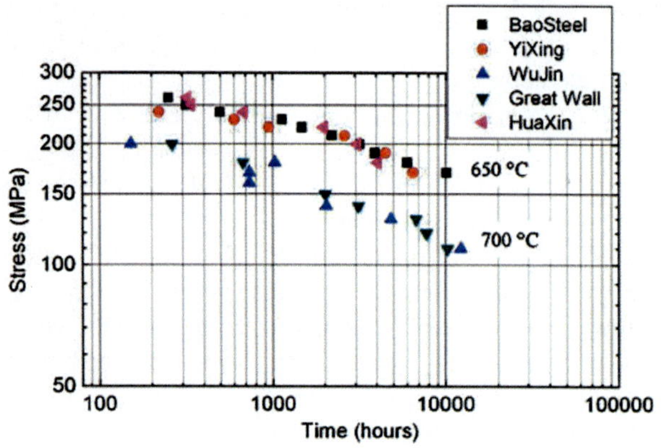

Fig. 19 Creep rupture strength of S30432 tubes made by Chinese manufacturers

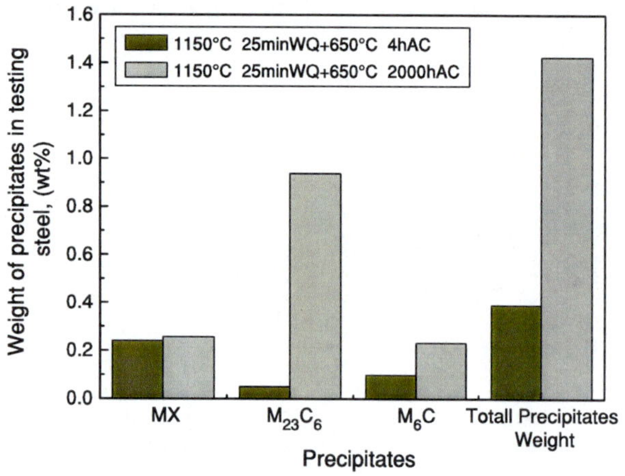

Fig. 20 Mass percentages of precipitates of S30432 steel after aging

holding time also relates to the wall thickness of S30432 steel tubes. Basically, the thicker the steel tube, the longer the holding time.

The dimensions and mechanical properties of S30432 steel tubes manufactured by some Chinese steel makers were listed in Table 3. For comparison, the related data of S30432 tube imported from Sumitomo Metal Company was also listed in the last row. It was found that the elongation of the tube from Sumitomo Metal Company is as high as 60%, when its strength keeps at a reasonable level. From this point of view, there is a big space for Chinese steel makers to further improve the quality and properties of S30432 steel tubes in the future.

According to the modified GB5310 standard, the extrapolated creep rupture strength of S30432 tube at 650°C for 10^5 h is about 117 MPa. And by ASME CC2328-1, the extrapolated creep rupture strength of S30432 tube at 700°C for 10^5 h is about 70.4 MPa. The tested creep rupture strength of S30432 steel tubes made by Chinese steel makers was plotted in Fig. 19. Clearly, the creep rupture strength of S30432 steel tubes made by Chinese steel makers meets the requirements of GB5310 and ASME CC2328-1 [32].

The strengthening mechanism at elevated temperature of S30432 steel tubes was also experimentally exploited in the past years. In addition to solid solution strengthening, precipitation strengthening plays a key role. The major precipitates of S30432 steel during aging and creep testing include MX, Cu-rich particle, $M_{23}C_6$ and M_6C. The evolution of MX, $M_{23}C_6$ and M_6C was plotted in Fig. 20. Cu-rich particle is too small to be measured by chemical phase analysis applied. Fine MX and Cu-rich phases are the dominant strengthening factors. MX precipitates pin dislocations to enhance steel's strength, Fig. 21.

5.3 S31042 Steel Tube

CISRI melted ten heats of S31042 steel to investigate the effect of the variation of nickel, niobium, vanadium, boron

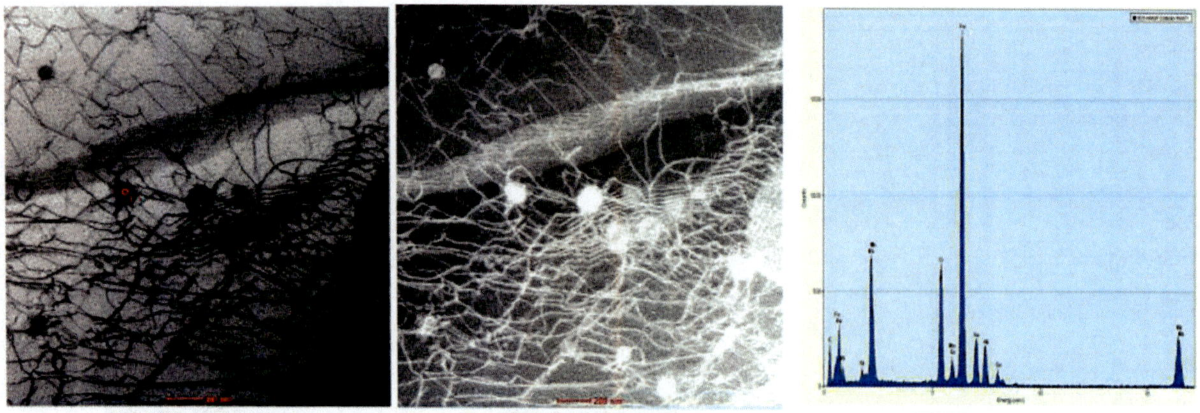

(a) Birght image (b) Dark image (c) Energy spectrum

Fig. 21 HR-TEM image of MX precipitates and dislocations of S30432 creep specimen (650°C, 140 MPa, 10,392 h)

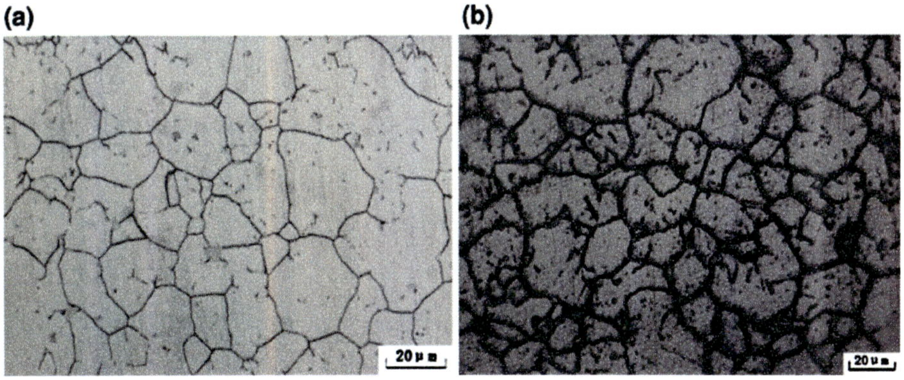

Fig. 22 Coarseness of grain boundaries of S31042 specimens after aging (**a** *left*)1,200°C 30 minWC + 700°C4hAC; (**b** *right*)1,200°C 30 minWC + 700°C300hAC

and rare earth element on microstructure and property of the steel. The effect of solid solution treatment on mechanical properties of the steel was also studied. When the solid solution temperature ranges from 1,150°C to 1,250°C, with the increase of solid solution temperature, the yield strength slightly decreases and the plasticity and strength at elevated temperature keep stable [33]. The plasticity at elevated temperature of the steel is very low due to the agglomeration and growth of $M_{23}C_6$ and Nb(CN)precipitates along the grain boundaries as shown in Fig. 22. It was observed that the interaction among $M_{23}C_6$, MX and dislocations exist, Fig. 23. Not only do the dislocations exist, but also they form dislocation walls along MX precipitates, which results in the fact that the creep strength of the steel decreases slowly at 700°C after long term service.

Before 2008, Baosteel had industrially melted four heats S31042 steel. Based on the experimental results of CISRI, Baosteel melted the fifth heat S31042 steel (40 ton AOD) in the year of 2009. To date, mechanical properties of the steel tube have been inspected and meet the requirement of purchaser's specifications. The long term creep test is undergoing, which is expected to be completed by the end

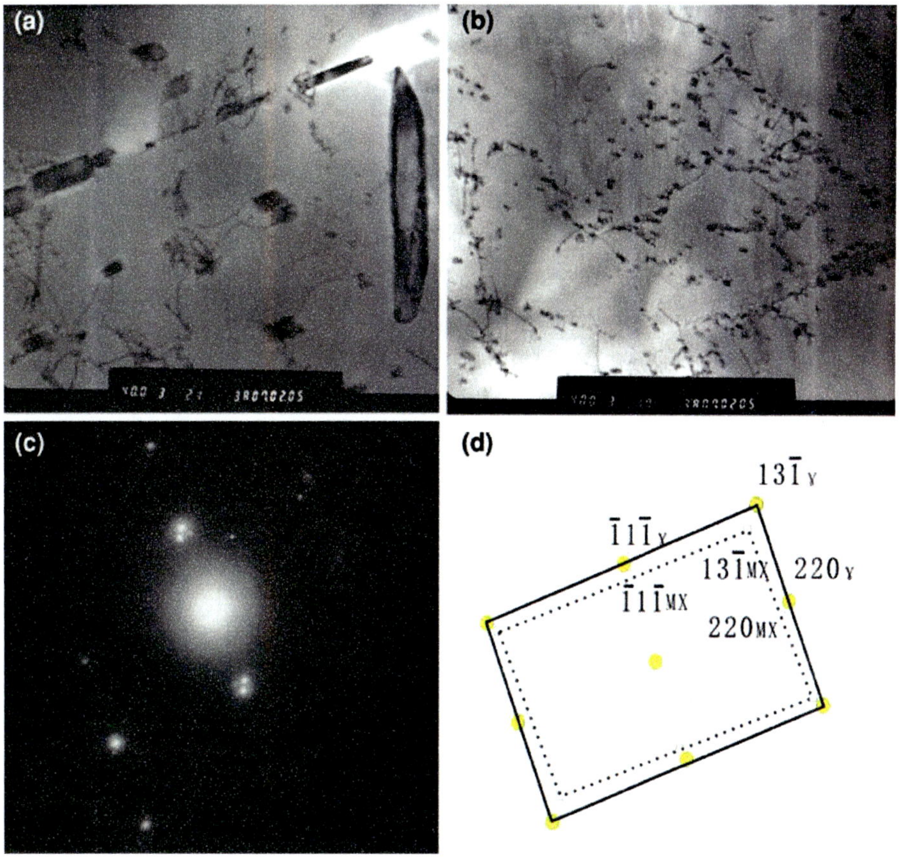

Fig. 23 TEM image of S30432 specimen at 700°C, 135 MPa, 2,226 h
a $M_{23}C_6$ + MX + dislocation wall, **b** dislocation walls, **c** diffraction [010]MX°C[010]γ

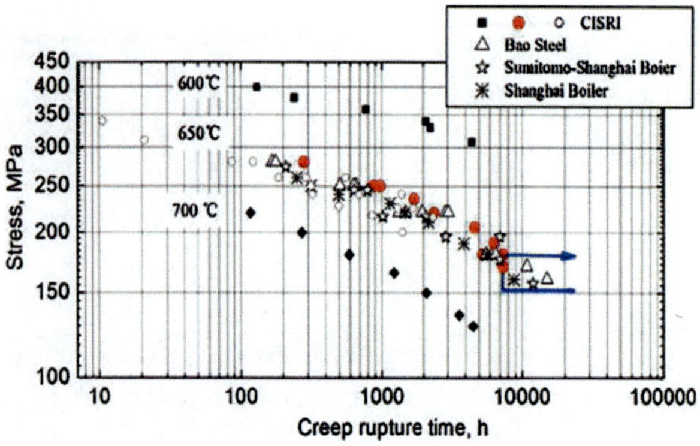

Fig. 24 Creep rupture strength of Chinese S31042 tubes

of 2010. The S31042 steel tube made by Chinese steel makers may enter market in the middle of 2011. The available creep chart of Chinese S31042 steel tubes was drawn in Fig. 24.

5.4 Fundamental Research and Product Innovation of Heat Resistance Steel for Boiler

According to the current progress, Chinese steel makers can provide commercial P92, S30432 and S31042 steel tubes and pipes by the end of 2010. The capacity of delivery and competitiveness of these products will gradually increase. However, the techniques to manufacture P92, S30432 and S31042 in China are far from mature, comparing with that in Japan. Therefore, it is necessary for Chinese researchers to strengthen fundamental knowledge, which takes time. Only when you master the know–how, and further, understand the know–why, an innovation is possible.

Liu and his colleagues at CISRI proposed a novel philosophy on the fundamentals of boiler steel tube/pipe research and development as shown in Fig. 25. Namely, the comprehensive properties of boiler steels can be improved and enhanced through the integration of chemistry, processing and fabrication through the adjustment of stability of microstructures in multi-scales, i.e., the stability of matrix and grain boundary, solution and precipitation of alloying elements during long term service at elevated temperature. The novel philosophy to develop boiler steels can be summarized as: multi-alloying element addition, failure free microstructure, and preferential strengthening.

6 Conclusion Remarks

From point of view of steel grades and quality, the steels used today are definitely different from what we have decades ago. It is worthy to pay more attentions to the circumstance and development of steels in China than ever before, especially due to the rapid progress of industrialization and urbanization. It should be aware of the situation and co-existence of both market and steel industry at this special moment. Matching with the market demands and steel science and engineering, it is believed that more and more new steel grades with higher performances will appear. In order to meet the development of society and economy, we should apply new technologies, new processing, and new equipments to develop high performance steel products with characteristics of high strength and high ductility for auto sheet steel, high strength and high toughness and/or high ductility for HSLA steel, high creep strength for heat resistance steel. Certainly, steel is a kind of new materials changing day by day. In the future, an affordable system of steel production and applications should be constructed to meet the requirements.

The third phase of 973 Program (2010–2014) on steels has been granted by MOST to promote the occurrence of new steel technologies since 2010. As a result, the fundamentals of microstructure control featured by multi-phase,

Fig. 25 Research philosophies to the fundamentals of boiler steels

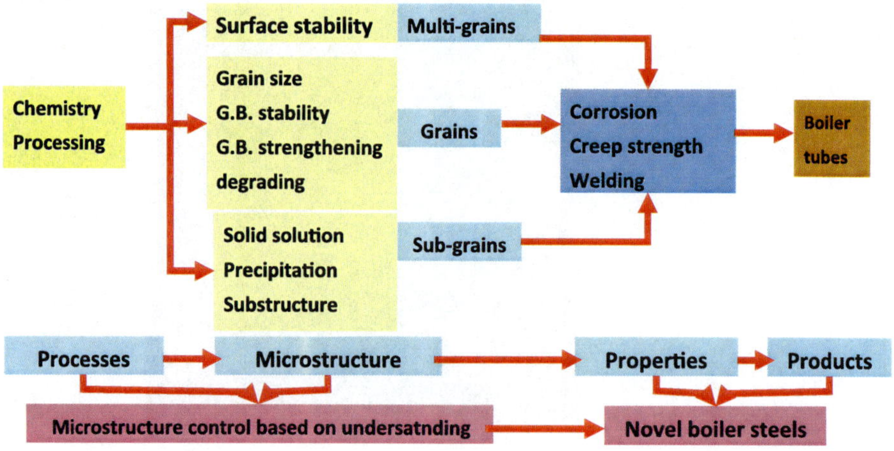

meta-stable and multi-scale (M^3) have been established, and then the targets of high performance steels have been setting up: the third generation HSLA steels with improved toughness and/or ductility ($A_{KV(-40°C)} \geq 200$ J and/or $A \geq 20\%$ as Rp0.2 in 800–1,000 MPa), the third generation advanced high strength sheet steels (Rm $\times$ $A \geq 30$ GPa% as Rm from 1,000 to 1,500 MPa) for automobiles with improved ductility and low cost, and the third generation heat resistant martensitic steels with improved creep strength ($\sigma_{10,000}^{650} \geq 90$ MPa). At present, the product of Rm $\times$ A in auto sheet steels has been doubled to be over 30 GPa%, comparing with conventional auto sheet steels. Charpy impact energy of HSLA steel reached to 180 J at $-30°$ at strength level of 900–1,000 MPa. A novel philosophy on the fundamentals of boiler steel tube/pipe has been proposed based on the understanding advanced heat resistance steels. It can be expected that the new generation steels developed will improve the safety and reliability of steel components in service remarkably for infrastructures, automobiles and fossil power station in the future.

Acknowledgments The authors would like to present sincere thanks to our colleagues, Dr. Jie Shi, Prof. Weijun Hui, Dr. Cunyu Wang, Dr. Le Xu, Dr. Yingjian Zhang, and Prof. Qilong Yong for their great contributions to the research on the third generation steels and the paper preparation. Special thanks to Prof. Yong Gan, the vice president of Chinese Academy of Engineering and the president of CISRI, for his compassionate encourage and thoughtful support to the research program. Ministry of Science and Technology are acknowledged for the financial funding of 973 program of 2010CB630800.

References

1. M. Pontremoli, Metallurgical and technological challenges for the development of high-performance X100-X120 linepipe steels, in *Proceeding of the Second International Conference on Advanced Structural Steels* (ICASS 2004) (Shanghai, China, 2004), p. 39.
2. G. Krauss, Martensite in steel: strength and structure. Mater. Sci. Eng. A **A273–A275**, 40 (1999)
3. M.F. Ashby, The deformation of plastically nonhomogeneous materials. Phil. Mag. **21**, 399 (1970)
4. S.T. Ohmori, K. Nagai, Strain-hardening due to dispersed cementite for low carbon ultrafine-grained steels, ISIJ Int. **44**, 1063 (2004)
5. http://www.Autosteel.org
6. K. Sugimoto, B. Yu, Y. Mukai, S. Ikeda, ISIJ **9**, 1194 (2005)
7. P. Jacques, Q. Furnemont, A. Mertens, F. Delannayz, Phil. Mag. **A81**, 1789–1812 (2001)
8. R. Heimbuch, Overview:Auto/Steel partnership, http://www.a-sp.org
9. G. Frommeyer, U. Brux, P. Neumann, ISIJ **43**, 438 (2003)
10. C. Garcia-Mateo, F.G. Caballero, ISIJ **45**, 1736 (2005)
11. J. Shi, X.J. Sun, M.Q. Wang, W.J. Hui, H. Dong, W.Q. Cao, Scripta Mater. **63**, 815 (2010)
12. J. Shi, W. Q. Cao, H. Dong. Materials Science Forum, **238**, 654–656 (2010)
13. W.C. Yu, Investigation on 30 GPa% Grade Ultrahigh-strength Martensitic–Austenitic Steels. Ph.D. thesis, Central Iron & Steel Research Institute, 2010
14. W.Q. Cao, J. Shi, H. Dong. Materials Science Forum, **29**, 654–656 (2010)
15. J.G. Speer, R.E. Hackenberg, B.C. DeCooman, et al., Phil. Mag. Lett. **87**, 379 (2007)
16. J.G. Speer, D.K. Matlock, B.C. Cooman, et al., Acta Mater. **51**, 2611 (2003)
17. R.L. Miller, Met. Trans. **A3**, 905 (1972)
18. A.A. Gazder, W.Q. Cao, C.H.J. Davies, E.V. Pereloma, Mater. Sci. Eng. **A497**, 341 (2008)
19. R.G. Davies, Metall. Trans. **9A**, 41 (1978)
20. R.G. Davies, Metall Trans. **9A**, 671 (1978)
21. J. Van Slycken, P. Verleysen, J. Degrieck, J. Bouquerel, B.C. De Cooman, . Mater. Sci. Eng. A **460–461**, 516–524 (2007)
22. A. Grajcar, H. Krztoń, J. Achiev. Mater. Manuf. Eng. **35**, 169–176, (2009)
23. A. Grajcar, J. Achiev. Mater. Manuf. Eng. **20**, 111–114 (2007)
24. O. Muránsky, P. Horňak, P. Lukáš, J. Zrník, P. Šittner, J. Achiev. Mater. Manuf. Eng. **14**, 26–29 (2006)
25. V.F. Zackay, E.R. Parker, D. Fahr, Trans. ASM **60**, 252–259 (1967)
26. F.J. Humphreys, M. Hatherly, *Recrystallization and Related Annealing Phenomena*, 2nd edn. (Pergamon, New York, 2004)
27. R.Z. Liu, *Strengthening Mechanism of Low Alloy Heat Resistant Steel* (Metallurgical Industries Press, Beijing, 1981)
28. Z.D. Liu, S.C. Cheng, H.S. Bao, R.X. Shi, Technical Report of P92 steel pipes (Part I) (CISRI Technical Report, Beijing, China, 2008)
29. Y.Yang, The effect of copper on the microstructure and properties of Super304H steel, M.A.Sc. Thesis, CISRI, Beijing, China, 2001
30. S.C. Cheng, Z.D. Liu, H.S. Bao, Effect of carbon content and niobium content on intergranular corrosion of ASME S30432 austenitic heat resistant steel, in *Symposium on Heat Resistant Steels and Alloys for USC Power Plants* 2007, Seoul, Korea, July 3–6, 2007, pp. 201–207
31. H.S. Bao, S.C. Cheng, Z.D. Liu, Effect of heat treatment on properties of ASME S30432 austenitic heat resistant steel, in *Symposium on Heat Resistant Steels and Alloys for USC Power Plants 2007*, Seoul, Korea, July 3–6, 2007, 259–265
32. Z.D. Liu, S.C. Cheng, G. Yang, Y. Gan, S.Q. Xu, S.P. Tan, Research and development of S30432 steel tubes in China used for USC power plants, vol. 45, No. 6, Iron and Steel, 2010, pp. 1–6
33. S.C. Cheng, Z.D. Liu, H.S. Bao, J.Z. Wang, *The investigation of bar-like carbides in HR3C boiler steel, Thermec'2009*, Berlin, Germany, August 23–27, 2009

High-Strength Steels: Control of Structure and Properties

A. S. Oryshchenko and E. I. Khlusova

Abstract

A high level of requirements for structural cold-resistant steels is provided by controlling their structure and properties at every process stage. Development of oil and gas deposits in Siberia, in the Far East, in the arctic shelf of the northern seas and establishment of infrastructure of Northern areas calls for the necessity of applying cold-resistant steels with various strength characteristics levels for fabricated structures, which include: marine oil producing and exploration drilling platforms, arctic service vessels, ice-breakers, tankers, deep-water equipment for platform service, hoisting equipment, reservoirs, cisterns, maritime terminals, pontoons, etc.

Keywords

High-strength economically alloyed steels • Thermomechanical treatment • Heat treatment • Microstructure

Systematic research has been made in CRISM FSUE "Prometey" into the formation of the structure and properties of economically alloyed high-strength steels, technological processes for the production have been developed, as well as a series of new steel grades ensuring the present basic requirements are met, such as competitive ability and cost effectiveness, in combination with high quality for environmentally friendly operation of structures in emergency conditions.

This work gives the ways of controlling the structure at each technological benchmark, and the basic results of research work and testing of flat products made of high-strength steels, including testing of full-scale models. It gives the basic characteristics of high-strength steels developed, applied in ship-building (AB grades) and in oil and gas production (X70–X100 categories).

A. S. Oryshchenko · E. I. Khlusova (✉)
CRISM FSUE "Prometey", St. Petersburg, Russia
e-mail: mail@crism.ru

1 Requirements to the Mechanical Properties and Structure of Steel

Complex requirements are set for high-strength steels:

- wide strength characteristics interval (355–690 N/mm^2) in combination with high plasticity and viscosity for flat steel with the thickness of up to 70 mm (up to 130 mm for individual elements);
- high resistance to brittle fracture, including operation temperatures of up to −50°C;
- high resistance to the effect of static, dynamic and cyclic loads;
- good weldability at ambient temperature;
- resistance to laminary fracture of welded connections, caused by the action of tensile stress in the direction of the plate thickness;
- high crack resistance, including for welded connections;
- corrosion and mechanical strength in sea water;
- even mechanical properties in terms of plate area;
- dense fibrous structure of process sample fracture;
- integrity through the entire plate volume observed by ultrasonic testing;
- stringent plate dimension limits.

Y. Weng et al. (eds.), *Advanced Steels*, DOI: 10.1007/978-3-642-17665-4_7,

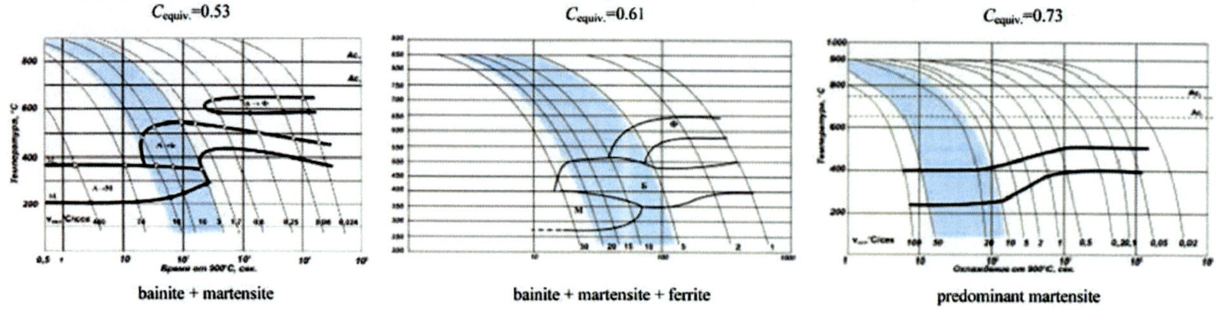

Fig. 1 Thermokinetic diagrams of austenitic transformation of high-strength steels differing in alloying

The complex of requirements for high-strength hull steel, building structures and oil and gas industry determines the necessity of managing the structure and properties of steel through the whole production process:

- For steelmaking, to ensure narrow limits in terms of chemical composition, specified content of carbon and alloying elements, reduce the segregation-related inhomogeneity during casting; increase the steel purity of harmful impurities (sulphur, phosphorus), gases and non-metallic inclusions, prevent non-metallic inclusions from accumulating and their globularization through modifying;
- When selecting alloying, which would provide the formation of a quasiisotropic structure with morphological similarity of structural components (in terms of grain and sub-grain size, dislocation density, dispersion of the carbide phase) in a wide interval of cooling rates;
- For plastic deformation, to refine austenitic grain, prevent the formation of lengthy boundaries at cooling in steel with a mixed structure, and dispargate structural components through creating a sub-structure;
- For tempering, to develop a fine-grained carbide phase of globular morphology.

Pre-dominant influence on steel properties comes from its structure, depending on the level and composition of alloying, thermal and thermo-mechanical treatment technology, and it's the peculiarities of internal structure that cause proneness to brittle fracture. Requirements have been developed to optimal structural condition of low-carbon economically alloyed ferrite, ferrite-bainite and bainite-martensite steels with the yield strength of 355–690 N/mm^2, characterized with an increased resistance to brittle fracture. The most reasonable option is to form a quasiisotropic highly dispersed structure, meeting the requirements to morphological similarity of structural components (firstly by their form—globular or rack-type, dislocation density, grain and sub-grain size, dispersion of the carbide phase.

With the help of high-speed dilatometer DIL 805 Bahr Thermoanalyse, the kinetics of phase transformations is studied, also under the influence of plastic deformation, which allows selecting the alloying composition and developing an optimal chemical composition of steel, Fig. 1.

2 Control of Steel Structure During Smelting

By improving metallurgical smelting processes, applying ladle treatment and modification, controlling the nature and form of non-metallic inclusions, the number and distribution of impurities, grain size, character and type of precipitating carbides, one can essentially increase the cold resistance of many steel grades and create economical technologies for their production.

In continuous cast steel billets, it is most interesting to study segregation of elements within the zone adjacent to the billet's axial line. Enrichment of this zone with elements having influence on the kinetics of austenite disruption at cooling after hot plastic deformation may lead to a structure being formed in the center of the plate which is different from the basic metal by structure or disrupts continuity in ready-made flat steel, as well as to the degradation of welding properties, which is especially important for flat steel with the thickness of more than 40 mm.

The inhomogenous structure caused by precipitations of non-metallic inclusions, especially in the sharp-angle and film form, leads to incontinuities being formed at the boundary between the metallic matrix and inclusions, and may cause localization of the deformation and brittle fracture at low temperatures or under dynamic load. The fracture is initiated first of all by flat pileups of non-metallic inclusions of a round and elongated shape with the diameter of not more than 100 μm or several pileups with a diameter of more than 50 μm, located in one plain or distributed across the plate thickness at a distance of up to 1 mm from one another. Special technological measures (treatment with silicocalcium cored wire, rare-earth metal powder, etc.) allow creating the required steel structure after crystallization, by control of its formation.

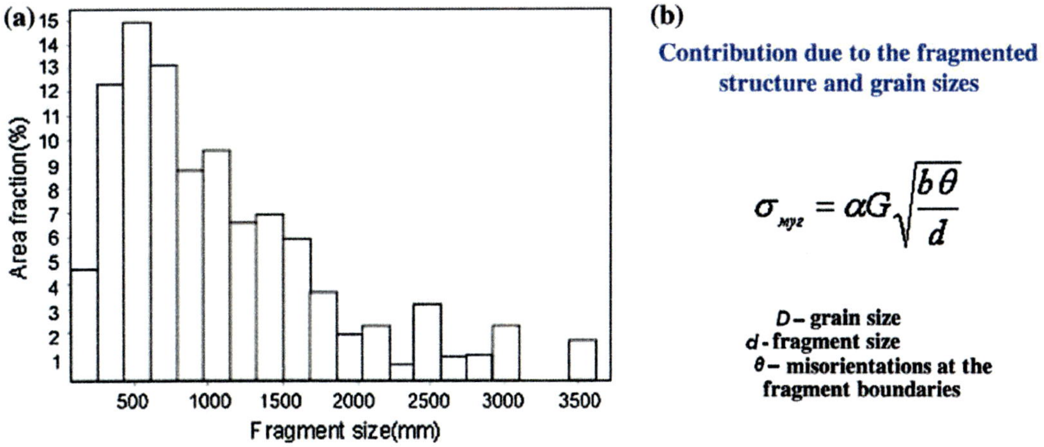

Fig. 2 The share of fragments of nano-, submicro- and micron sizes in high-strength steel after plastic deformation and calculation of contribution into steel strengthening (б)

3 Structure Changes from Thermo-Mechanical Treatment

An essential reserve for making economically alloyed cold-resistant steels is to use the nature of dislocation structures formation after great plastic deformations determining the mechanical properties of deformed steels. Based on the procedure of modeling technological processes with the help of Gleeble 3800 physical modeling complex, principles of selecting technological conditions of plastic deformation are formulated for economically alloyed steels, ensuring the development of an optimal structural condition through gradually increasing of the dispersion of hierarchically conjugated elements of various scale levels: austenitic grain crushing, austenite fragmentation, formation of dispersed structures due to granular bainite inheriting the deformation-induced structure of the fragmented austenite as a result of strike-slip phase transformation at cooling and of the formation of a developed fragmented structure of α-phase, which allows increasing the strength characteristics of economically alloyed steel, with high resistance to brittle fracture being kept.

Phase transformations, being one of the mechanisms of adapting the system to external impact, in this case cause dissipation of energy added by high-temperature plastic deformation. A characteristic feature of phase transformations in low-carbon economically alloyed steels is the fact that mixed structures get formed at continuous cooling: ferrite-perlite, ferrite-bainite or bainite-martensite with various percentages and morphology of structural components. In plate iron, one can ensure formation of structural components with close morphological attributes under the condition that either structures of globular morphology only (ferrite with an insignificant quantity of thin-platy perlite or ferrite and granular bainite), or of rack-type morphology (rack-type bainite and rack-type martensite) will be formed in a sufficiently wide range of cooling rates.

It has been shown that formation of misoriented fragments in austenite, which are of micron and submicron sizes, accelerates transformation in the upper area of the bainite transformation, which, in combination with the fine-grained austenite structure, contributes to granular bainite being formed. It has been established that bainite inherits the deformation-induced structure of hot-rolled austenite, austenite fragmentation is not only inherited but also enhanced as a result of shear phase transformation and thereby, morphological parameters of the fragmented structure of bainite are determined by the parameters of the fragmented structure of deformed austenite. Misorientations at the fragment boundaries may increase through implementing of various options of the effective orientational correlation between the initial (austenite) and final (ferrite) phase, ensuring additional strengthening of steel, Fig. 2.

4 Structural Changes Resulting from Heat Treatment

The main advantage of the traditional strengthening technology with high tempering is the possibility of developing an isotropic structure, which also means, isotropic properties, across the whole thickness of flat steel, also with the thickness of more than 40 mm. In addition, formation of the structure and properties is determined by the steel alloying level, cooling speed after strengthening and modes of subsequent high tempering, influencing the character and kinetics of carbide transformations and structural changes.

Depending on the carbon content, alloying composition and type of the structure obtained after strengthening, the carbide formation process has its own features. This causes explicable changes in steel properties at tempering and determines the range of permissible tempering modes. It has been established that in case steel strength increases, resistance to brittle fracture depends on the morphology and character of location of the carbide phase in the martensite and bainite components. Formation of fine-grained carbide phase or round-shaped carbides of Me_3C and Me_7C_3 types is accompanied by an increase in the impact elasticity, and the precipitation of $Me_{23}C_6$ phase, despite the reduction of strength characteristics, causes no significant changes in impact elasticity in case the tempering temperature increases.

The results of modeling the formation of various type structures at $\gamma \rightarrow \alpha$-transformation (ferrite-bainite, bainite, bainite-martensite) in steel, characterized by the content of chrome, nickel and molybdenum have shown that the greatest danger comes from the formation of ferrite, which causes a low level of yield strength, but, which is most important, a low content of the fibrous component in the cross-sectional view of the samples after dynamic bending and reduction in resistance to brittle fracture, Fig. 3.

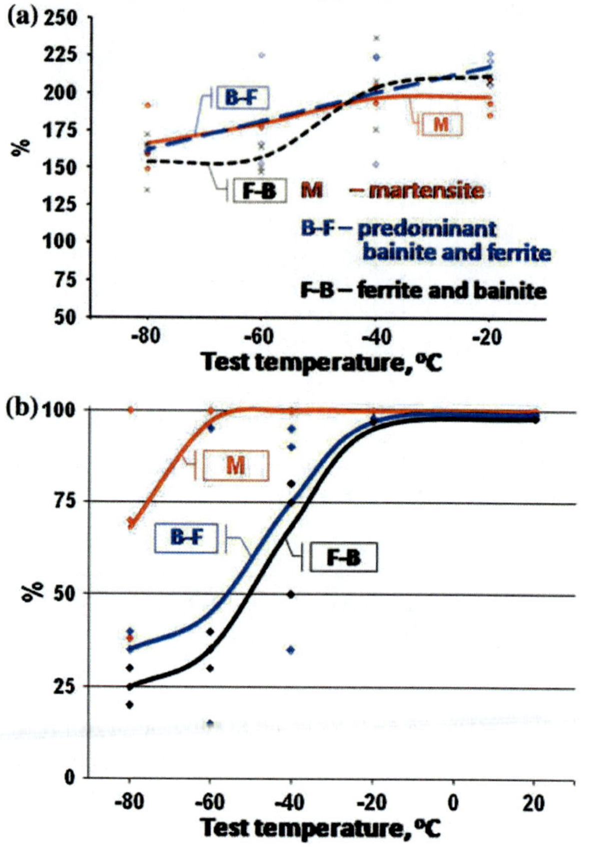

Fig. 3 Impact energy of high-strength steels differing in structure at lowering testing temperatures (**a**) and the share of the fibrous component in the cross-section of samples after the testing (**b**)

5 Quality Control

A complex assessment of working capacity characteristics, maintainability and welding capacity of high-strength steels is carried out in a certified Promtest KM mechanical testing laboratory, having a fleet of special equipment in stock. The quality of ready-made rolled products made of ship-building or building steels in Russia is controlled based on the chemical composition, mechanical properties (tensile and impact-bending tests, including after mechanical ageing), dimensions, surface quality, presence of internal defects shown by ultrasonic testing, austenite grain size, cold-bending testing of technological samples, fracture type and weight (for ship-building steel). Rolled steel can be accepted by batches or on a plate-by-plate basis. The most stringent requirements are set for special high-strength steels for ship-building, the least stringent to building steels. Some types of testing are used in Russia only. In ship-building, the fracture type of technological samples is often determined by the thickness equal to the plate thickness, at room and subzero temperatures (Tкб). The presence of 70–100% of the fibrous component in the cross-section of the sample (for steels of various strength category) points to its high resistance to brittle fracture.

The critical brittleness temperatures **Tкб** and NDT, determined based on the results of testing large-size samples, confirm the high resistance of high-strength steels to brittle fracture, Table 1.

Table 1 Critical brittleness temperatures for high-strength steels

Steel grade	Thickness (mm)	NDT(°C)	Tкб(°C)
F500 W	45	−80 −95	−20
	80	−75 −80	−15
F620	41	−85 −100	−28
	80	−70	−25
	100	−95	
F690	60	−80	−47
	70	−120	−75
	150	−100	−12

Application of steels with fine-grained structure and essential reduction in metal hardness in the HAZ due to the reduction of the carbon component by 20–30% ensures the guaranteed level of metal's fracture strength, including in the HAZ of welded structures during the most infavorable thermal cycles of welding. Economical alloying, in

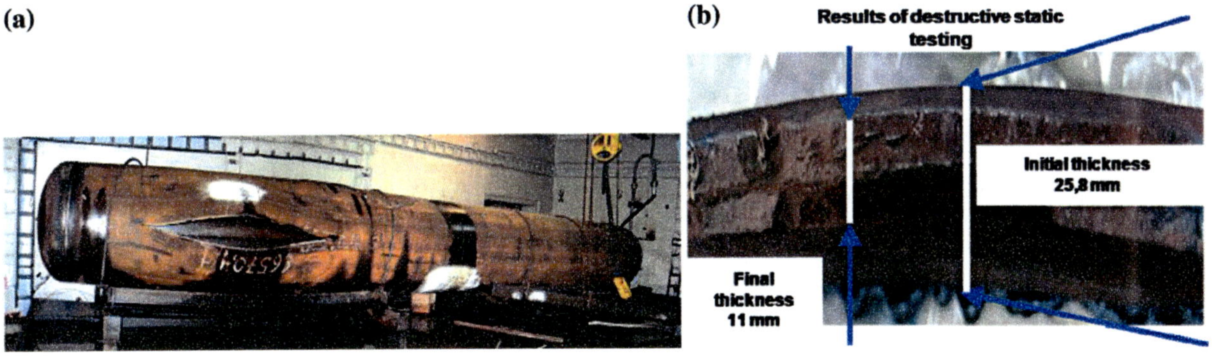

Fig. 4 Pipe type after cyclic tests (**a**) and narrowing of the lateral section in the cross-section before static fracture of a pipe which gets earlier cyclically loaded (**b**)

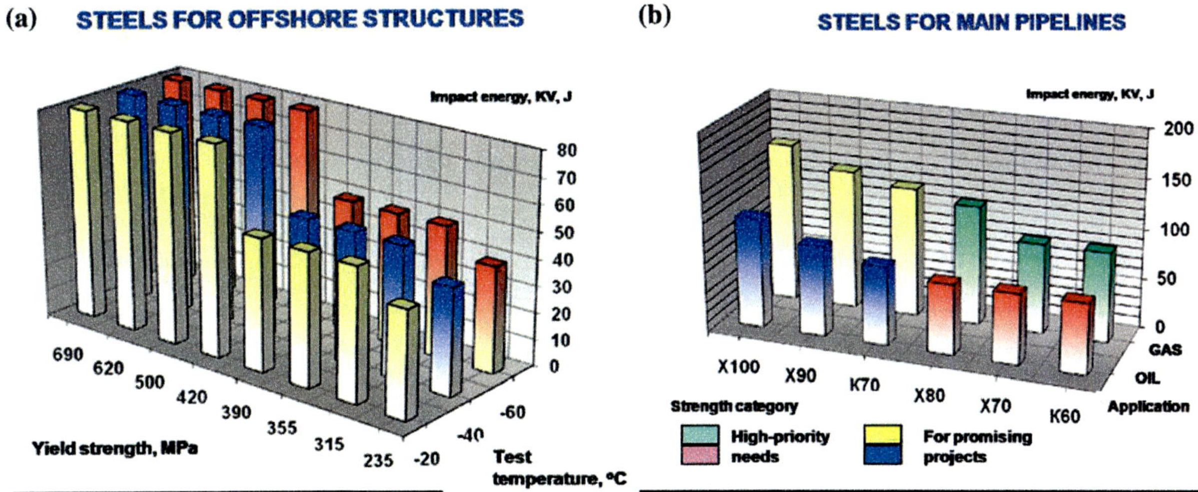

Yield strength	Maximum thickness(mm)
235, 315	70
355, 390	70
420, 460	70
500, 620	70
690	40

Strip category	Maximum thickness(mm)	DWTT, at −20°C
K60 (X70)	40	90
X80 (K65)	35	90
K70	30	90
X90	20	90
X100	20	90

Fig. 5 New steels for ship-building (**a**) and main pipelines (**b**)

combination with restricted content of micro-alloying elements ensures, due to strengthening ability reduction, minimum hardness values in the weld adjacent zone of high-strength steels, which reduces proneness to cold crack formation. In addition, a reduction in the carbon content reduces structural inhomogeneity in case of heating in the intercritical temperature interval, reduction in the content of carbide-forming elements, in combination with a lowered carbon content, contributes to the homogenization of austenite in case of heating for welding, increase in chemical homogeneity in the HAZ of welded structures and homogeneity of phase transformations in the process of cooling after welding.

High resistance to static and cyclic loads is ensured, which is evidenced, for example, by the almost 50% narrowing of the lateral section in the cross-section before static fracture of a pipe which gets earlier cyclically loaded, Fig. 4.

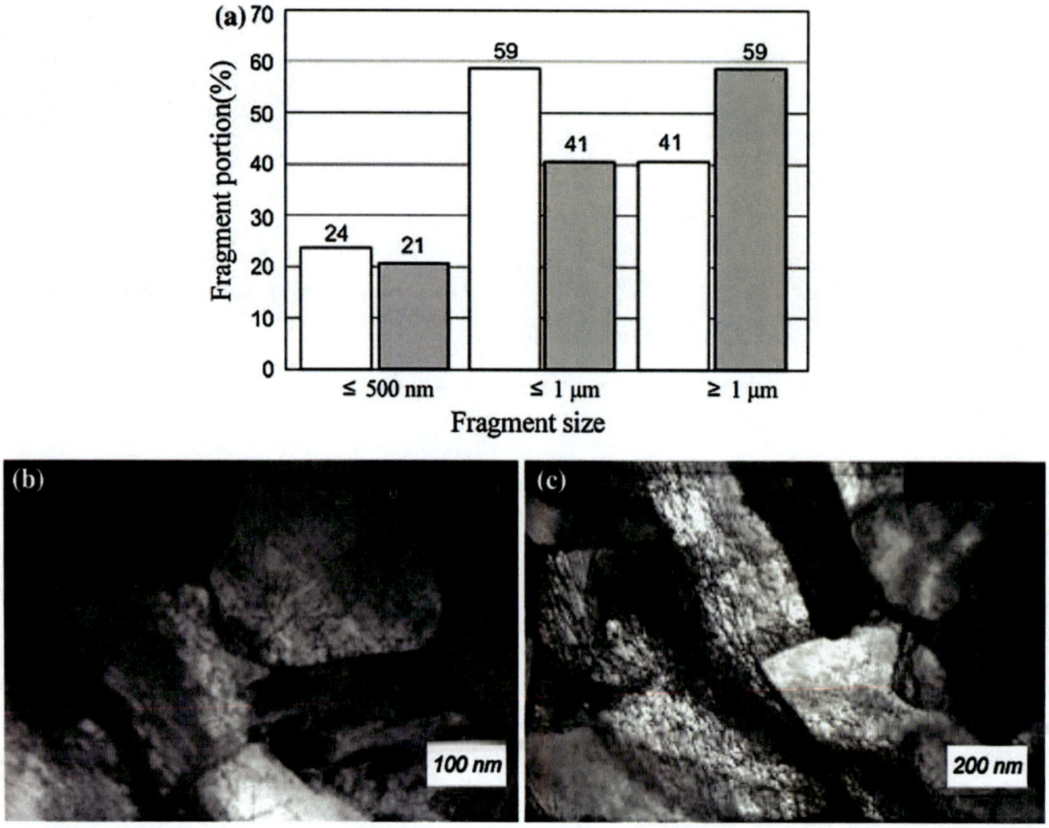

Fig. 6 Results of assessing the share and sizes of fragments (nanostructure elements) by EBSD-analysis in flat X80 industrial pipe steel with the thickness of 28 mm, (**a**) light columns—on the surface of the steel, *grey*—in the middle in terms of thickness and its structure: (**b**) sub-grains in ferrite (300–1000 μm), (**c**) subgrains in bainite (200–500 μm)

6 Introduction of the Developed Steel

The developed concept of influencing the structure and properties of high-strength steels has allowed developing a series of platel economically alloyed steels, which, in terms of the combination of alloying, hardness, cold resistance and crack resistance, meet the ecological reliability requirements in case of operation of marine equipment structures in emergency conditions, Fig. 5a. Those steels are applied for such marine platforms as Prirazlomnaya, Arkticheskaya and Moss Maritime. A series of pipe steels has been developed, Figs. 5 and 6, with a high resistance to brittle and plastic fracture and increased environmental reliability for operation in corrosion environments, in conditions of increased seismicity, low temperatures, for the construction of Bovanenkova–Ukhta gas pipeline, East-Siberia-Pacific Ocean oil pipeline, and prospective projects.

Technogenic catastrophes becoming frequent recently have called for additional requirements to increasing strength characteristics, resistance to brittle fracture in steels for building structures, including materials for hydroelectric power plants. Steels have become necessary that are capable of withstanding extreme loads and keep the initial properties to the most at low temperatures, in case of short-time heating, in addition to all the operational requirements set for steels for metal structures. Even in such a traditional industry as agricultural production, a sharp increase in equipment reliability is also necessary.

It has been shown that significant improvement in the consumer properties complex is possible in case of applying steels with elements of nanostructure, Fig. 6, obtained through controllable crystallization, precision thermal heating, intensive plastic deformation, based on physical modeling of the structural changes process.

Thus, the structure and physical and mechanical properties of high-strength economically alloyed steel are formed at all the stages of technological conversions and may be changed in case of operational loads. From this point of view, stability of steel structure may be provided by technologicval measures during the melting, rolling, heat and thermomechanical treatment of plate steel, ensuring increase in the metallurgical quality of steel and formation of a quasiisotropic structure.

Introduction of new steels in industry will allow reducing the self-cost of plate steel production by 20–25%, including strip, through using a significantly lower quantity of alloying elements during melting, reducing the steel intensity of structures through increasing strength characteristics without reductions in the plasticity and viscosity of steel, reducing labor and energy consumption for installation and construction (welding) of structures, increasing the maintenance-free periods and accident-free operation up to 50 years.

Ultra-high Strength Steel Treated by Using Quenching–Partitioning–Tempering Process

T. Y. Hsu (Zuyao Xu) and Xuejun Jin

Abstract

Quenching–Partitioning–Tempering (Q–P–T) process is developed to yield ultra-high strength and high toughness steel based on Q–P process suggested by Speer et al. The chemical composition of Q–P–T steel is designed as <0.5C, 1.5Si (or Al), 1.5Mn, 0.02Nb, 0.2Mo (mass%) in which complex carbide precipitation during tempering may contribute further hardening; or <0.5C, 1.5Si, 1.5Mn (mass%) in which η (θ) carbide precipitation may offer strengthening beside the martensite formation. Ultimate tensile strength and total elongation of >2,000 MPa and >10%, ~1,500 MPa and ~15%, and ~900 MPa and ~20% are obtained with Q–T–P processed steels containing 0.4, 0.2 and 0.1C, respectively. Steels processed by Q–P–T generally contain ~5% retained austenite. Several examples are given in this article. The thermodynamics and kinetics as well as strengthening and toughing mechanisms for Q–P–T process are briefly discussed.

Keywords

Ultra-high strength steel • Quenching–Partitioning–Tempering process

1 Introduction

Steel, a significant material, acts as major role in the development of the modern construction and industry and will further play its important role in this century. Materials scientists devoted themselves to develop high strength steels in order to save energy and raw materials and realized achievements, e.g., in last century, steels possessing excellent mechanical properties were developed, such as martensitic age hardening steel [1] and martensitic second hardening steel [2–4] giving ultimate strength of 2,000 MPa and toughness of 40 J. However, these steels contain high alloying elements, e.g. AF1410 [5–7] or AerMet100 [8, 9] steel contains 10Ni-2Cr-1Mo-14Co (mass%) that makes their cost too high to be widely utilized. Recently, Bhadeshia et al. [10–13] established very strong steel with ultimate tensile strength 2.5 GPa and fractural toughness 30–40 MPa $m^{1/2}$, yet it contains 0.78–0.98 mass%C possessing poor weld ability and potential brittleness. A modified 4,340 steel with carbon content about 0.5 mass%C developed by Fang et al. [14] brings ultimate tensile strength >2,200 MPa and $K_{IC} > 50$ MPa $m^{1/2}$. Krauss [15] revealed that >0.5% of carbon content in carbon and low alloyed steels would lead temper embrittlement resulted from cementite formation [16].

Today we are facing serious challenge of CO_2 reduction and urgently required to reduce the total scale of steel production and develop ultra-high strength steel with considerable toughness and low cost that may be produced by a novel heat treatment process. It is recognized that steels, after heat treatment, should contain considerable amount of retained austenite so as to ensure toughness [17, 18], including resistance of hydrogen embrittlement [19].

T. Y. Hsu (Zuyao Xu) (✉) · X. Jin
School of Materials Science and Engineering, Shanghai Jiao Tong University, 200240 Shanghai, China
e-mail: zyxu@sjtu.edu.cn

Y. Weng et al. (eds.), *Advanced Steels*, DOI: 10.1007/978-3-642-17665-4_8,

2 Carbon Partitioning from Martensite to Austenite

In 1979, Thomas et al. [20] found and later confirmed [21] that in a quenched 0.27 mass%C steel, the retained austenite enriched substantially with carbon of 0.4–1.04 mass% is trapped between lath martensite. They suggested that during quenching there might occur carbon partitioning from martensite to austenite [22]. By comparison the calculated durations between the formation of lath martensite and carbon partitioning owing to the different solubility of carbon in martensite and austenite, with carbon concentration profile as shown in Fig. 1, Hsu and Li [23] in 1983 showed that the carbon partition may keep pace with, or slightly lag behind, the formation of lath martensite. The time required for equalization of enriched austenite is at least one order of magnitude slower than the formation of lath martensite. Our work also emphasizes that the carbon partitioning is not a main process in lath martensite transformation and the formation mechanism of low carbon martensite differs from that of upper bainite.

Solute partitioning, closely related to phase transformation, is a significant procedure in heat treatment process. It also occurs in two-phase-field annealing, i.e., solute partitioning from ferrite to austenite as in TRIP steels [24] as well as in isothermal holding for bainite formation, from bainite to austenite. In Si-content steel, carbon partitioning may cause the formation of carbide free bainite [25, 26].

Si or Al may play important role for partitioning, since Si or Al does not dissolve in Fe_3C and prevent the Fe_3C precipitation but not the complex carbide formation. Content of Si or Al in steel is beneficial to carbon partitioning.

3 Quenching–Partitioning–Tempering Process

In 2003, Speer et al. [27] suggested Quenching–Partitioning (Q–P) process, i.e., austenitization–quenching to M_s–M_f-partitioning at quenching temperature (QT) called one-step treatment or at above QT or above M_s (two-step treatment)-water quenching to room temperature, giving the treated Si-containing steel rather high strength along with considerable elongation and toughness resulted from the existence of considerable amount of carbon enriched retained austenite through partitioning. Their first tested example showed that a 0.35C-1.3Mn-0.74Si (mass%) steel treated by Q–P process, as: austenitizing at 920°C for 510 s, followed by quenching at 300°C salt bath for 400 s or 320°C for 200 s and water quenching to room temperature, contains 6% retained austenite, with carbon content 0.97–1.47 mass% and 1.23–1.58 mass%. Although in [27], their thermodynamics and kinetics analysis for Q–P process are incomplete, the Q–P novel heat treatment making high strength steel containing certain amount of retained austenite [28] is certainly a great contribution for developing ultra-high strength steels. The combined mechanical properties of Q–P treated steels show superior mechanical properties as comparison with dual-phase, TRIP and general martensitic steels e.g. in [29, 30] as shown in Fig. 2 [31].

Based on previous literatures, especially the Q–P process suggested by Speer et al., the present author designed the structure and chemical compositions of ultra-high strength steel as fine lath martensite; dispersed complex or ε (η) carbide precipitated in martensite and retained austenite

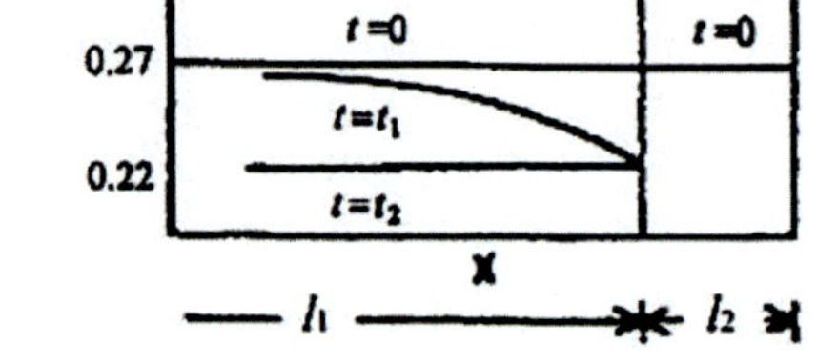

Fig. 1 Sketch of carbon concentration profile in martensite and retained austenite for quenched 0.27%C steel

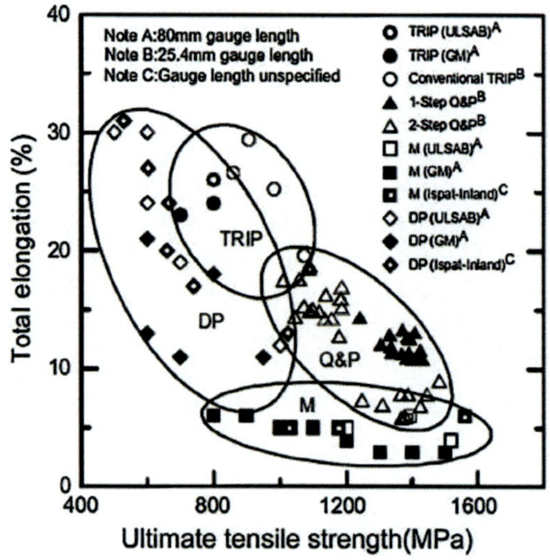

Fig. 2 Total elongation vs. UTS for *DP* (dual phase), TRIP, *M* (martensitic) and Q–P sheet steels

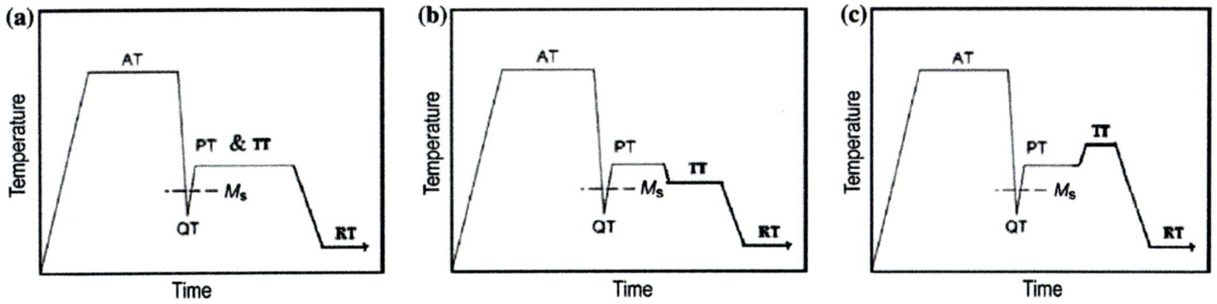

Fig. 3 Schematic Q–P–T heat treatment cycles. Note: *AT, QT, PT, TT* and *RT* are austenitizing, quenching, partitioning, tempering (precipitation) and room temperatures, respectively

with certain carbon content, and considerable thickness as well as fine grain size of original austenite [32], and the chemical composition as <0.5C, 1.5Si (or Al), 1.5Mn with (or without) 0.2Mo and 0.02Nb (mass%). The present author [32, 33] also proposed a Quenching–Partitioning–Tempering (Q–P–T) process in which the tempering process leading the precipitation of complex carbides in steels containing complex carbide forming elements e.g. Mo and Nb, or precipitation of η (θ) transitional carbide in steels without carbide forming elements, for further increment of strength.

In Q–P–T process, the tempering temperature (TT) may be higher or lower than the partitioning temperature (PT), i.e., TT > PT or TT < PT, as shown in Fig. 3 [33]. As one-step Q–P, for steels without carbide forming element, one step Q–P–T can also be done that is not shown in Fig. 3 [33].

The strength and toughness of the Q–P–T treated steel are determined by the amounts of martensite, retained austenite, the carbon contents in martensite and austenite and the dispersion of carbides, in turn, depend upon the QP, TP and TT as well as the durations at QP, PT and TT.

Thus, an optimum Q–P–T heat treatment schedule should be made by analysis of experimental results and thermodynamics and kinetics of Q–P–T for treated steel as discussed briefly in [33]. Based on the thermodynamics studies on martensitic transformation FCC(γ)–BCC(α) [34], the present author [35] and his coworkers calculated Ms in Fe–C [36, 37] and iron-base alloys [38–40]. From solution of the Fick's law [41], or by using the DICTRA software package [42], the carbon partitioning can be easily calculated. By referencing thermodynamics studies on tempering in ferrous alloys, e.g. [43, 44], the temperature of carbide precipitation, that is the temperature at which $\Delta G = 0$ can be obtained. During rather long duration at PT or TT, there may occur bainitic transformation in austenite, such as work shown by Zhong et al. [45] that the interface between austenite and martensite would move during holding, implying that there may probably occur the bainite formation. The prediction of the bainite formation temperature in a Q–P–T steel can be obtained by using modified $\Delta G = 0$ value from $\Delta G = 0$ of Fe–C [46].

Koistinen and Marburger equation [47] has been extensively used to predict the volume fraction of martensite, f, formed at quenching temperature, T_q, as M_s is known, as

$$f = 1 - \exp\left[-\alpha(M_s - T_q)\right] \tag{1}$$

in which $\alpha = 1.10 \times 10^{-2}$ for plain carbon steels with carbon content 0.37–1.10 mass%, obtained from experiment. Equation (1) was derived by Magee [48] and is a well known kinetics equation for an athermal martensitic transformation. In derivation of Magee's equation, the change in free energy per unit volume accompanying the martensitic transformation, ΔG, is considered to be a function of temperature only. However, as stated above, in lath martensite formation, carbon partitioning may occur. Thus, ΔG is a function of not only temperature but also the carbon content in austenite and the Magee's equation, or the Koistinen and Marbarger equation should be modified. The present author and his co-workers [49, 50] derived a kinetics equation in general form, which can be applied even to low-carbon steels as

$$f = 1 - \exp\left[\beta(c_2 - c_1) - \alpha(M_s - T_q)\right] \tag{2}$$

where $\beta = \bar{v}\varphi(\partial\Delta G/\partial c)$, c_2 and c_1 represent the carbon contents in austenite before and after quenching, respectively. The subject was given as a part of my lecture to ICOMAT-1995 [51]. Equation (2) will reduce to Eq. 1 in case of martensitic transformation in high or medium carbon steels, $c_1 = c_2$ for too low M_s that may lead weak carbon partitioning. Equation 2 should be simplified with experimental data for easy calculation of martensite amount formed at T_q. Kinetics of Q–P–T process needs to be systematically studied.

4 Examples of Q–P–T Process Performance

4.1 0.488C-1.195Mn-1.185Si-0.98Ni-0.21Nb (mass%) Steel [52]

A 0.488C-1.195Mn-1.185Si-0.98Ni-0.21Nb (mass%) steel treated as austenitizing at 850°C for 300 s, followed by quenching to a 95°C salt bath for 10 s, isothermal holding at tempering temperature 400°C for 10–1,000 s, respectively, and water quenching to room temperature, gives the mechanical properties, i.e., ultimate tensile strength, R_m and total elongation El as shown in Fig. 4. The microstructure of the sample treated by partitioning and tempering at 400°C for 10 s contains lath martensite, in which a great part of martensite with lath width <100 nm, 4.1 volume% retained austenite of nanoscale in thickness and highly dispersed fine complex carbides of nanoscale in size embedded in martensite. The volume fraction of retained austenite varies with tempering duration, as 4.1, 6.4, 5.2, 4.3 and 4.0 volume% after tempering at 400°C for 10, 30, 60, 300, and 900 s, respectively. The increment of volume fraction of retained austenite from 4.1 to 6.4 by the increase of partitioning duration from 10 to 30 s implying the effect of carbon partitioning and the decrease of volume fraction of retained austenite from 6.4 to 5.2, 4.3 and 4.0 volume% by the long duration at 400°C shows that during holding at 400°C there may occur the carbide precipitation as well as phase transformation of austenite to bainite. This steel possesses ultimate tensile strength over 2,000 MPa with elongation over 10% when the holding duration for partitioning and tempering is less than 300 s (2,160 MPa with 11% elongation after 10 s), showing an excellent performance of Q–P–T process.

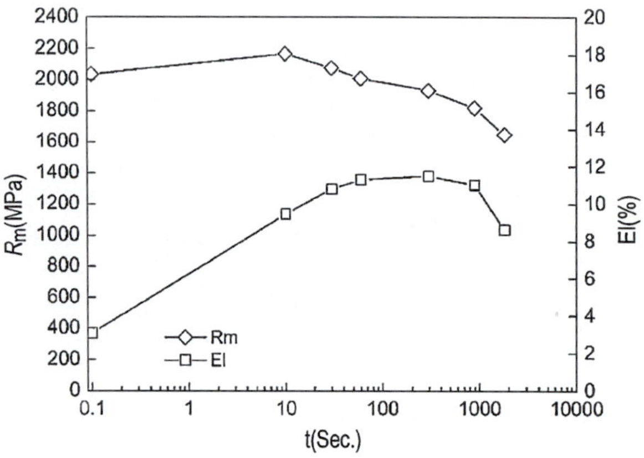

Fig. 4 Tensile strength Rm and elongation El vs. tempering duration (t/s) for Q–P–T treated 0.485C-1.195Mn-1.185Si-0.98Ni-0.21Nb (mass%) steel

4.2 0.2C-1.5Mn-1.5Si-0.05Nb-0.13Mo (mass%) Steel [53]

A 0.2C-1.5Mn-1.5Si-0.05Nb-0.13Mo (mass%) steel treated by Q–P–T process, i.e., austenitizing at 920°C for 300 s followed by quenching at 220°C salt bath, partitioning and tempering at 400°C for 20 s, and water-quenching to room temperature, possesses ultimate tensile strength of 1,500 MPa, with total elongation of 15%. The volume percentages of retained austenite are 4.5, 7.0, 5.4, 5.2 and 5.8 volume% after holding at 400°C for 10, 20, 40, 80 and 180 s. The width of some martensite laths is smaller than 100 nm yet that of some other is greater than 100 nm and the size of complex carbides is in nanoscale.

4.3 0.20C-1.18Si-1.44Mn-0.05Nb (mass%) Hot-rolled Sheet Steel [54]

A 0.20C-1.18Si-1.44Mn-0.05Nb (mass%) hot-rolled sheet steel with thickness of 2 mm treated with Q–P–T process as austenitizing at 900°C for 300 s, followed by quenching at 330 or 300°C salt bath for 10 s partitioning, tempering at 400°C for 30 s, and water quenching to room temperature, gives ultimate tensile strength R_m 1200 MPa with total elongation El 18% and Charpy impact energy of 48 J at room temperature as well as 28 J at −40°C, or $R_m = 1270$ MPa with El = 15.7% as quenched at 300°C for 10 s.

4.4 0.15C-2.1Mn-0.45Si-0.015Nb-0.03Al (mass%) Steel [55]

This steel is a typical dual-phase steel gives ultimate tensile strength of 1,127 MPa, yield/strength ratio 0.872 with total elongation of 7% after treated by traditional heat treatment process for dual-phase steels, and after treated by Q–P–T, it shows only slightly reduced strength, $R_m = 1{,}047$ MPa, very low yield/strength ratio of 0.458, and much high elongation of 15.5%, 1.2 times greater than that of treated as dual-phase steel.

4.5 0.17C-1.5Mn-1.5 Si and 0.17C-1.5Mn-1.5Si-0.04Nb-0.089V (mass%) Steels [56]

The 0.17C-1.5Mn-1.5 Si steel T_1 is a typical TRIP steel, which contains no carbide forming element and may serve as Q–P steel after treated with Q–P process. The 0.17C-1.5Mn-1.5Si-0.04Nb-0.089 V steel T_5 is a typical Q–P–T steel containing small amount of carbide-forming elements

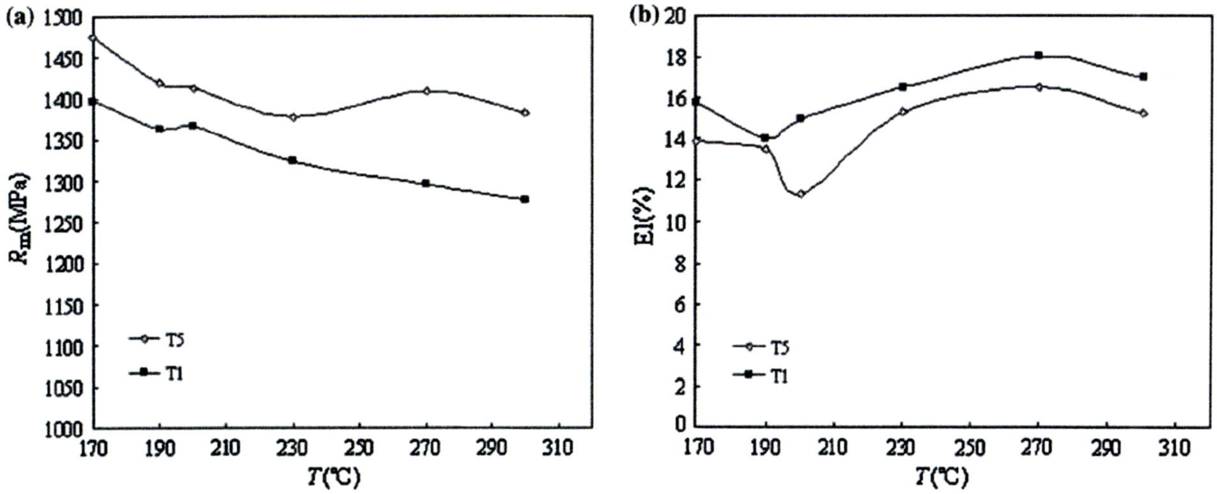

Fig. 5 Ultimate tensile strength Rm (**a**) and total elongation El% (**b**) Vs quenching temperature for 0.17C-1.5Mn-1.5 Si steel (T_1), and 0.17C-1.5Mn-1.5Si-0.04Nb-0.089V steel (T_5)

Nb and V. After general heat treatment, i.e., traditional process for TRIP steel, T_1 possesses ultimate tensile strength of 950 MPa with elongation of 22%. The above mentioned two steels possess ultimate tensile strength R_m and total elongation El% after Q–P for T_1 and after Q–P–T process for T_5, i.e., austenitizing at 900°C for 240 s, followed by quenching at 170–300°C for 10 s, partitioning and tempering at 400°C for 90 s, and water cooling to room temperature as shown in Fig. 5.

From Fig. 5, it is seen that as quenched at 270°C, the T_1 steel gives $R_m \approx$ 1,280 MPa, much higher than that it serving as TRIP steel and El $\approx$ 18%, slightly lower than 22% of TRIP steel, and that R_m of T_5 is much higher than that of T_1, revealing that the tensile strength of Q–P–T steel is remarkably higher than that of Q–P steel, with only slightly reduced elongation.

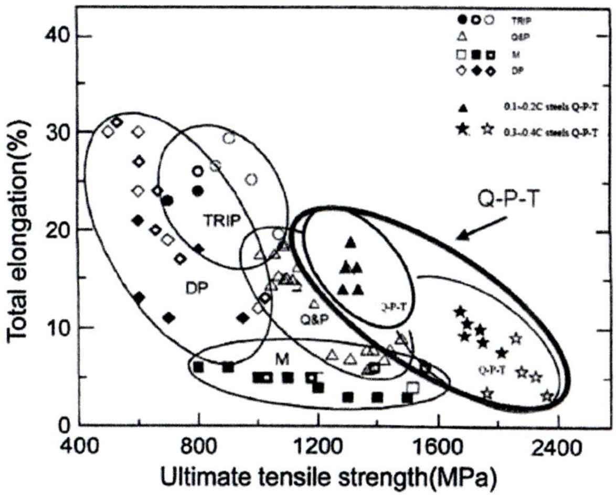

Fig. 6 Utimate tensile strength R_m vs. total elongation El for DP, TRIP, M (martensitic), Q–P treated and Q–P–T treated steels (modified from Fig. 2 [31])

4.6 0.41C-1.27Si-1.30Mn-1.01Ni-0.56Cr (mass%) Steel [57]

A 0.41C-1.27Si-1.30Mn-1.01Ni-0.56Cr (mass%) steel treated by using Q–P–T process, i.e. austenitizing at 820°C for 10 min, followed by quenching in 180°C salt bath holding for 180 s and water quenching to room temperature, (a one step Q–P–T process) possesses ultimate tensile strength 2,468 MPa, yield strength 1,550 MPa and elongation 11.6%. In the microstructure of the treated steel, there exists ε-carbide, revealing ε-carbide precipitation during partitioning process. So in the so-called "one-step Q–P process" actually there exists tempering and this process is certainly a one-step Q–P–T process.

Seung et al. [58] showed that a 0.1C-1.51Mn-1.48Si-0.02Nb-0.04 V-0.015Ti-0.304Mo steel treated by using Q–P process, i.e., austenitizing at 1,000°C followed by quenching to 223°C, holding at 350°C for 120 s and water quenching to room temperature, shows very nice mechanical properties, i.e., yield strength of 700 MPa, ultimate tensile strength of 900 MPa, Charpy energy absorption 100 J at 20°C and 60 J at −200°C. The steel they studied contains carbide forming elements Nb, V, Ti and Mo and some carbide may precipitate during holding at 350°C. So, most probably there occurred tempering (precipitation) and may be actually a Q–P–T steel.

As mentioned above, Q–P–T treated steel possesses pronounced balanced mechanical properties, even superior over Q–P treated steel and Fig. 2 can be modified as shown in Fig. 6.

5 Strengthening and Toughening Mechanisms

Strengthening mechanisms in Q–P–T treated steels are as follows:

1. Lath martensite with high dislocation density, especially with carbon pinning effect. Previous work pointed out that reduction of lath width contributes only small increment of strength. However, the present author considers that although there are small angles between lath martensite, yet as the lath width is in nanoscale, considerable strengthening effect should be shown, as an example, 0.485 C steel mentioned in Sect. 3.
2. Dispersed complex carbide precipitation contributes strengthening, e.g. the Nb-containing carbides, with average size of 5 nm, precipitate at 400°C after 10 s holding that leads strength increment about 130 MPa compared to the average carbide size increasing only to 35 ± 10 nm, after holding for 1800 s in the case of 0.485C-1.195Mn-1.185Si-0.98Ni-0.20Nb (mass%) steel [52].

Recently, by means of SEM and TEM combined with EDS analysis, Wang et al. [59] characterize the carbide in this steel after tempering at 400°C for 10 s and confirm that during the tempering, fine NbC precipitates with size of (5 ± 3) nm in martensite, with orientation relationship $(110)_{NbC}//(100)_\alpha$ and $[\bar{1}10]_{NbC}//[01\bar{1}]_\alpha$ or Baker-Nutting one. ε-carbide can also form in Q–P–T process as the results of Grange et al. [60, 61] showed. During the partitioning process, the depletion of carbon in martensite is an advanced stage as shown by Santofimia et al. [62].

Lath martensite behaviors limited toughness. Toughness of Q–P–T treated steel depends on the amount of retained austenite and their distribution. The ideal feature may be that austenite with thickness >10 nm is trapped between lath martensite. TRIP effect is another category for discussion in austenite-containing steels.

Strengthening and toughening mechanisms for Q–P–T steel should be further studied.

6 Conclusion

Steels with designed compositions as <0.5C-1.5Mn-1.5Si with or without 1Ni-0.02Nb-0.20Mo and processed by using lower temperature austenitization, quenching at QT (between M_s and M_f), and carbon partitioning at PT (generally above M_s), or tempering for complex carbide precipitation at PT (>QT or <QT) or for η (θ) carbide precipitation at lower temperature (~QT), and then water quenching to room temperature, possess good combined mechanical properties. For example, UTS >2,000 MPa and elongation >10%, ~1,500 MPa and ~15%, and ~900 MPa and ~20% are obtained with steels containing 0.4, 0.2, and 0.1 C, respectively. The microstructures of steels treated by using Q–P–T process generally contain ~5% retained austenite of considerable thickness trapped between fine lath martensite. There are also dispersed complex carbides or η (θ) carbides embedded in martensite. Preliminary tests showed that the combined mechanical properties of Q–P–T processed steel are superior over steels of dual-phase, TRIP, general martensitic and Q–P processed. Heat treatment schedule for certain Q–P–T steel should be optimized according to thermodynamics and kinetics of its Q–P–T process. The strengthening and toughening mechanisms of Q–P–T processed steels need further study.

References

1. F. Floreen, Met. Rev. **13**, 115 (1968)
2. V.K. Chandhok, J.P. Hirsh, E.J. Dulis, Trans. AIME **224**, 858 (1962)
3. V.K. Chadhock, J.P. Hirsh, E.J. Dulis, Trans. ASM **56**, 677 (1963)
4. G.P. Speich, D.S. Dabkowski, L.F. Porter, Metal. Trans. **4**, 303 (1973)
5. C.D. Little, P.M. Machmeire, US Patent 4,076,525, 28 Feb 1978
6. P.M. Machmeire, C.D. Little, M.H. Horowity, R.P. Oates, Met. Technol. **6**, 291 (1979)
7. R. Ayer, P.M. Machmeire, Metal. Trans **27A**, 2510 (1996)
8. M. Schmidt, R. Hemphill, 34th Sgamore Army Materials Technology Laboratory, Watertown, MA, p. 239, 1990
9. R. Ayer, P.M. Machmeire, Metall Trans. **24A**, 1943 (1993)
10. F.G. Caballero, H.K.D.H. Badeshia, J.A. Mawella, D.G. Jones, P. Brown, Mater. Sci. Technol. **18**, 279 (2002)
11. F.G. Caballero, H.K.D.H. Badeshia, Curr. Opin. Solid State Mater. Sci. **8**, 251 (2004)
12. H.K.D.H. Badeshia, Mater. Sci. Technol. **21**, 1293 (2005)
13. H.K.D.H. Badeshia, in *Proceedings of the 3rd International Conference on Advanced Structural Steels*, Gyeongju, 33–40, Aug 2006
14. C.G. Fang, H. Dong, J. Shi, Y.L. Lin, Q.L. Yong, W.J. Hui, M.Q. Wang, Y.Q. Weng, Ordnance Mater. Sci. Eng. **29**(2), 31 (2006). (in Chinese)
15. G. Krauss, Metall. Trans. **32A**, 861 (2001)
16. T. Ando, G. Krauss, Metall. Trans. **12A**, 1283 (1981)
17. G.Y. Lai, W.E. Wood, R.A. Clark, V.F. Zackay, E.R. Parker, Metall. Trans. **5**, 1663 (1974)
18. T.Y. Hsu (Xu Zuyao). Shanghai Metals **17**(1), 1–6 (1995) (in Chinese)
19. R. Ritchic, M.H.C. Cedeno, V.F. Zackay, E.R. Parker, Metall. Trans. **9A**, 35 (1978)
20. B.V. Rao, G.Thomas, in *Proceedings of ICOMAT-79*, M.I.T., 12–16, 1979
21. M. Sarikaya, G. Thomas, in *Proceedings of International Conference on Solid to Solid Phase Transformations*, TMS-AIME, 999–1004, 1982
22. M. Sarikaya, G. Thomas, J.W. Steeds, S.I. Barnal, G.D.W. Smith, in *Proceedings of International Conference on Solid to Solid Phase Transformations*, TMS-AIME, 1421–1425, 1982
23. T.Y. Hsu (Xu Zuyao), X. Li, Scripta Metall. **17**, 1285 (1983)
24. B.C. DeCooman, in *Proceedings of International Conference on TRIP-aided High Strength Ferrous Alloys*. Gardracht Redaktions and Industrie Press Service Bad Haryberg. Germany, 2002

25. H.K.D.H. Badeshia, *Bainite in Steels* (The Institute of Materials, Cambridge Press, London, 2001), pp. 373–385
26. D.K. Matlock, G. Krauss, J.G. Speer, J. Mater. Process. Technol. **117**, 324 (2001)
27. J.G. Speer, D.K. Matlock, B.C. Decooman, J.G. Schroch, Acta Mater. **51**, 2611 (2003)
28. D.V. Edmonds, K. He, F.C. Riggo, B.C. DeCooman, D.K. Matlock, J.G. Speer, Plenary Lecture at ICOMAT-05, Mater. Sci. Eng. **438–440**, 25 (2006)
29. A.M. Streicher, J.G. Speer, D.K. Matlock, B.C. Decooman, in *Proceedings of International Conference on Advanced High Strength Sheet Steel for Automotive Applications*, AISI, Warrendable PA, pp. 51–64, 2004
30. B.C. DeCooman, J.G. Speer, in *Proceedings of 3rd International Conference on Advanced Structural Steels*, Gyeongju, Korea, 885–891, Aug 2006
31. J.G. Speer, D.V. Edmonds, F.C. Riggo, D.K. Matlock, Curr. Opin. Solid State Mater. Sci. **8**, 219 (2004)
32. T.Y. Hsu (Xu Zuyao), PRICM-6, Mater. Sci. Forum **561–565**, 2283 (2007)
33. T.Y. Hsu (Xu Zuyao), Int. Heat Treat. Surf. Eng. **2**(2), 64 (2008)
34. T.Y. Hsu (Xu Zuyao), Acta Metall. Sinica **16**, 426–429 (1980). (in Chinese)
35. T.Y. Hsu (Xu Zuyao), J. Mater. Sci. **20**, 23 (1985)
36. T.Y. Hsu (Xu Zuyao), Chang Hongbing, Luo Shoufu, J. Mater. Sci. **18**, 3206 (1983)
37. T.Y. Hsu (Xu Zuyao), Chang Hongbing, Acta Metall. **32**, 343 (1984)
38. H. Chang, T.Y. Hsu (Xu Zuyao), Acta Metall. **34**, 333 (1986)
39. T.Y. Hsu (Xu Zuyao), in *Proceedings of ICOMAT-86*, The Japan Inst Metals, 245 (1987)
40. L. Li, T.Y. Hsu (Xu Zuyao) J. Heat Treating **4**, 140 (1985)
41. J. Crank, *The Mathematics of Diffusion* (Oxford University Press, Oxford 1956) p. 45
42. A. Borgenstam, L. Hoglund, J. Agren, A. Engstron, J. Phase Equilib. **2**, 269 (2000)
43. G. Ghost, G.B. Olson, Acta Mater. **50**, 2099 (2002)
44. T.Y. Hsu (Xu Zuyao), C. Siwei, Mater. Sci. Technol. **1**, 1025 (1985)
45. N. Zhong, X.D. Wang, Y.H. Rong, L. Wang, J. Mater. Sci. Technol. **22**, 751 (2006)
46. T.Y. Hsu (Xu Zuyao), Mou Yiwen, Acta Metall. **32**, 1469 (1984)
47. D.P. Koistinen, R.E. Marburger, Acta Metall. **7**, 59 (1959)
48. C.L. Magee, Am. Soc. Metals, Metals Park, Ohio, 115 (1970)
49. T.Y. Hsu, W. Lu, Y. Wang, Iron and Steel **30**(4), 52 (1995) (in Chinese)
50. T.Y. Hsu (Xu Zuyao), *Martensitic Transformation and Martensite*, (2nd edn.) (Science Press, Beijing, 1999) p. 563
51. T.Y. Hsu (Xu Zuyao), J. de Phys. IV **5**, C8–351 (1995)
52. X.D. Wang, N. Zhong, Y.H. Rong, T.Y. Hsu (Xu Zuyao), L. Wang, J. Mater. Res. **24**, 260 (2009)
53. N. Zhong, X.D. Wang, Y.H. Rong, L. Wang, Mater. Sci. Eng. **506A**, 111 (2009)
54. Y. Wang, S. Zhou, Z.H. Guo, Y.H. Rong, Mater. Sci. Forum **654–656**, 37 (2010)
55. X.S. Liao, Dr. Degree Thesis, Shanghai Jiao Tong University, 2010
56. Z.L. Hu, X.D. Wang, L. Wang, Y.H. Rong, Trans. Mater. Heat Treat. **31**(4), 76 (2010). (in Chinese)
57. H.Y. Li, X.W. Lu, W.J. Li, X.J. Jin, Metall. Mater. Trans. **41A**, 1284 (2010)
58. C.H. Seung, C.A. Jae, Y.N. Saug, S.J. Kim, H.C. Yang, J.G. Speer, K. David, Met. Mater. Inter. **13**(6), 439 (2007)
59. X.D. Wang, W.Z. Xu, Z.H. Guo, L. Wang, Y.H. Rong, Mater. Sci. Eng. **527A**, 3373 (2010)
60. R.A. Grange, C.R. Hribal, L.F. Porter, Metall. Trans. **8A**, 1775 (1977)
61. S.S. Nayak, R. Anumolu, R.D.K. Misra, K.H. Kim, D.L. Lee, Mater. Sci. Eng. **498A**, 442 (2008)
62. M.J. Santofimia, L. Zhao, J. Sietsma, Metall. Trans. **40A**, 46 (2009)

Part II

Physical Metallurgy Frontier

Long-term Stabilization of Steel Availability under Limited Resources

Kotobu Nagai

Abstract
Japan faces severe and complicated resources problems. The population reached its peak and decreases gradually. Now the domestic market has been saturated. The raw material prices have risen suddenly and the product values have been deflated. The present paper discusses the change in new boundary conditions for new paradigm of economic growth based on the large material stock and high material technology of steels.

Keywords
Limited resources Product-to-product • Reliability • High performance • Education

1 Boundary Change

1.1 Population Peak

People dream the eternal growth of society, however, should rationally recognize the change in the boundary conditions. When the population change is forecast under the present trend, each population is not always growing for centuries. Fig. 1 is a case of a century prediction of the population change for China, Korea, Germany, and Japan. In order to understand the variation behavior for a whole and make the comparison among the countries easier, a kind of normalization is done for the populations. Namely, based on the China peak population, each population is multiplied by 30, 18, and 12 for Korea, Germany, and Japan, respectively, so that all the peaks become in the same level.

In the population growth among four, Germany headed, Japan ran the second, and Korea the third. Germany and Japan had already the population peak in 2003 and 2004, respectively. Korea is already saturated and predicted to have the peak in 2020s. China is still growing, but also predicted to have the peak in 2020s. After the peak, the four countries will lose the fundamental base for economic growth in terms of the quantity of products when the people feel full of the modern products.

1.2 Steel Production

Germany and Japan increased the steel production rapidly after the World War II and reached the saturation in 1970s. Germany seems to go ahead of Japan by 6 years as shown in Fig. 2a. Both did not or could not increase the steel production more, probably since the market was saturated domestically and globally. And both did not decrease the production largely, which means they kept their competitiveness in the global market. But Japan has had almost twice saturation level as Germany, probably since Japan has extorted more than Germany. Fig. 2b shows the variation in apparent domestic use in the Japan market. The apparent domestic use still contains some steel that is exported as a shape of final product like cars. Thus, the net domestic need seems to be around (6–7) $\times 10^8$ t in the so-called industrialized countries in the expression of Fig. 1.

About 20 years after the jumps of Germany and Japan, Korea made also a rapid increase in steel production. Korea's case is similar to that of Japan. Namely, Korea also has produced steel both for domestic needs and for overseas

K. Nagai (✉)
National Institute for Materials Science, Ibaraki 3050047, Japan
e-mail: Nagai.Kotobu@nims.go.jp

Y. Weng et al. (eds.), *Advanced Steels*, DOI: 10.1007/978-3-642-17665-4_9,

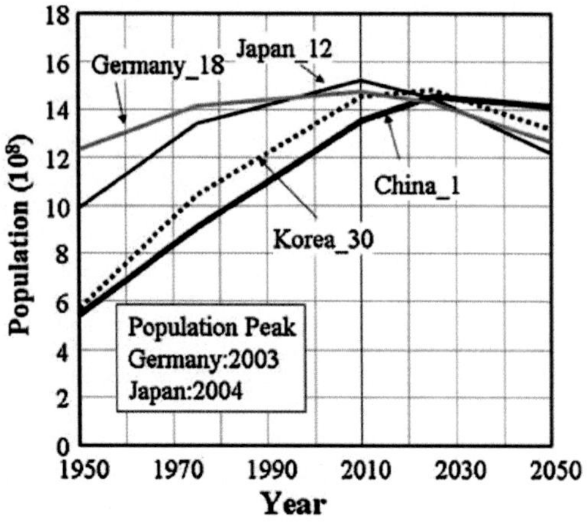

Fig. 1 Population change in 1950–2050

market. And now Korea produces more steel than Japan, which means Korea exports more steel than Japan.

Again about 20 years after the jump of Korea, China is making another rapid increase in steel production. People think about how the steel production changes in China. According to Fig. 2a, the prediction has quite a wide range, and in the maximum case it takes 10 years more to reach the peak of 15 × 10^8 t level. On the other hand, according to Fig. 2b or in the minimum case, China is close to the peak of (7–10) × 10^8 t level. The fluctuation may correspond to the export. One 10^8 t export, roughly equivalent to the need for two 10^8 people in the industrialized countries, is quite huge and beyond common sense as trade.

Anyway, by 2020, the steel production in the East Asia will reach the saturation. Then the East Asian steel people should sincerely consider what the new era could be.

1.3 Steel Stock

Steel once produced would not disappear. Steel stock means the steel amount in registered use. The steel stock increases gradually, since the service life of steel is fairly long compared with that of other material like plastics. After the service life, the steel is discarded as steel scrap. Finally the steel scrap appears as much as the steel produced if the steel production keeps a constant level.

The historical statics is not yet well accumulated as for the steel scrap generation. Japan as well as Korea is poor in natural resources. Both have a good statics about the steel scrap, since the steel scrap is the precious treasure for both poor countries. In Fig. 3, the variation in steel stock estimated in two countries is shown as per capita. According to this statics, Japan and Korea already are reaching the saturation to the level 10 t per capita. Any one nation from baby to eldest person may save 10 t of steel! In Japan the steel scrap generation becomes almost equivalent to the domestic need, which means Japan is ready to be self-sufficient from the viewpoint of the steel amount.

Korea has accumulated the steel stock twice faster than Japan. Surprisingly, as shown in Fig. 3b, the variation of Japan matches perfectly that of Korea when the time scale is shortened by half. It is considered that China is accumulating the steel stock rapidly. In Fig. 3b, a prediction is drawn as the similar variation as Korea would occur after 20 years delay. If true, in 2030, the steel scrap generation might be equivalent to that to the domestic need, in other words, China may become more self-sufficient in steel resources.

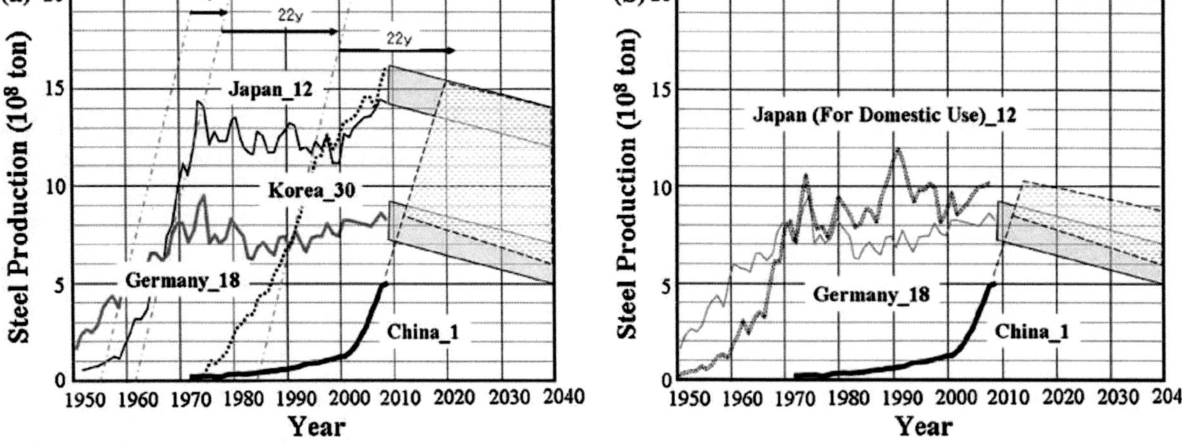

Fig. 2 Variations in rude steel production. **a** Gross production; **b** apparent domestic use

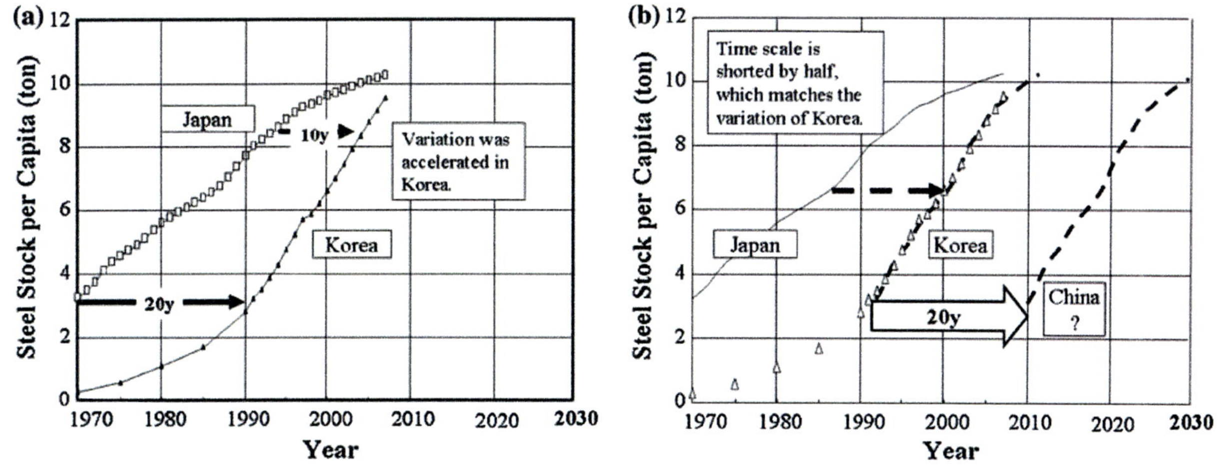

Fig. 3 Variations in steel stock (a) Variations in steel stock in Korea and Japan; (b) a prediction

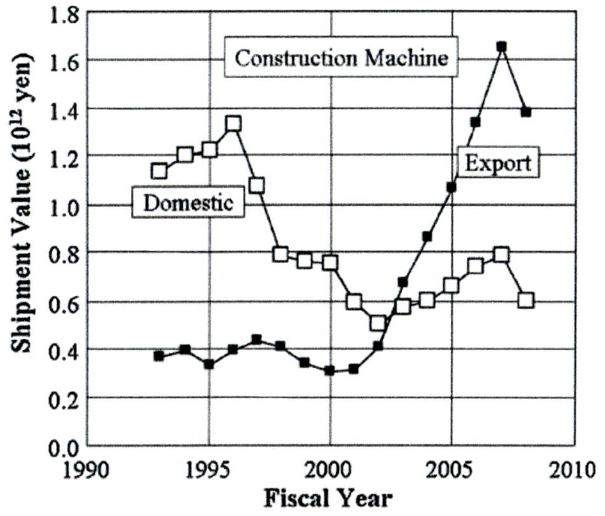

Fig. 4 Change in main market for construction machine industry of Japan

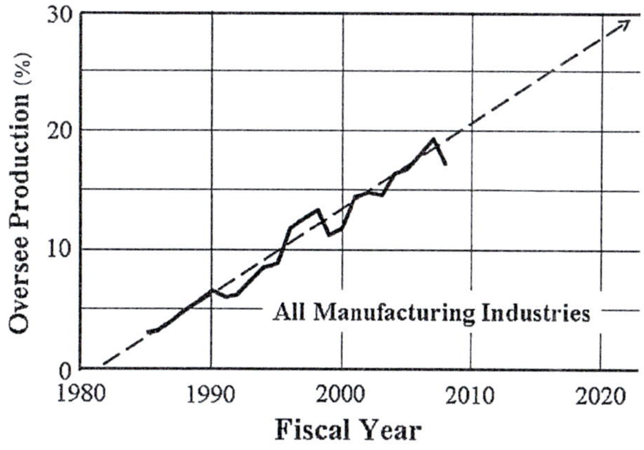

Fig. 5 Increase in overseas production of all manufacturing industries of Japan

2 Globalization of Manufacturing

Steel industry is hard to relocate its large-scale plants quickly in a global scale. On the other hand, other manufacturing industries are not so hard. Those industries move form shrinking market to expanding one. Look at a case in Japan.

Fig. 4 shows the annual variation in the domestic part and the export part of construction machines fabricated in Japan. For some decades the ratio of the domestic need has been around 60–70% for every industrial product. It has been true until 1995 even in Fig. 4. But in this case, the domestic market has shrunk suddenly and never come back. Fortunately the overseas market has increased after 2001 and the Japan manufacturing industry has gained the growing market successfully to keep the total production rate good enough. The phenomena are happening in many manufacturing industries like automobile one in Japan. Exporting a large quantity is not as beneficial to each company as fabricating the same quantity in or near overseas market. Very naturally, they are eager to establish the factories in or near the growing market. Hence, the ratio of overseas production is increasing linearly every year as seen in Fig. 5. About 30% production is predicted to be overseas around 2020. Direct impact of this change could be more decrease in the domestic need for steel in the near future in Japan.

2.1 Chapter Summary

1. The population reaches the peak. Japan had already it in 2004 and Korea and China will have it in 2020s.

2. The world steel society has invited new growing leaders in the turn as Germany, Japan, Korea and now China. Every leader increased the steel production quickly, reached the peak, and kept the saturated level.
3. After the domestic market reached the saturated level, keeping the overseas share becomes plays an essential role on maintaining the production. At the same time, the steel scrap generation becomes almost equivalent to the net domestic need in terms of quantity.
4. After the population peak, the domestic need may shrink in the sense of quantity. Manufacturing industry seeks for the alternative market in oversea.
5. After getting some stable market in oversea, the manufacturing industry would establish overseas plants in or near the most promising market.
6. Finally, both the net and the indirect domestic needs will decrease gradually. The similar phenomena will happen from leader to leader with some decade interval.

3 Paradigm Change

Japan is a poor country, since little natural resource is granted. The author thinks about a paradigm change in order to found sustainable fundamentals for such the poor country. Steel is highly expected to work as one of the main materials in the society. So the steel availability is hoped under the limited resources. Steel must be available for get and use in cleverer ways.

3.1 Product-to-Product Steel

Low cost fundamentals are requisites for sustainability. Air and water must be clean for healthy life. Energy is strongly desired to be as cheap as possible. Raw material is also requested to be as cheap as possible. High efficiency use of these is also essential to survive the life most securely.

In order to get cheaper raw material as for steel, there could be some different options like the utilization of low grade ores, more efficient use of direct reduced iron, and full use of steel scrap. All the options are important to realize. Especially for domestic market in the poor country like Japan, recycling would be most essential. Steel scientists should challenge for "Product-to-product steel". The product-to-product steel is inherently designed to reuse in the same product. In some case, the steel do not have to be re-melted for reuse. Good conversation or collaboration among the related industries would bring about better solution for this idea.

Plant design, plant scale, and plant location would be quite different from the present.

3.2 High Reliability Steel Product

Longer and more secure usage of steel product becomes much more important. The reliability belongs not only to material factors of steel but also many factors related with product. Long term data accumulation is inevitable. Based on the most reliable long term data base, the mechanism must be clarified. Only the highly educated and trained expert can tell the best solution for material selection and problems of accident or wrong application. Global collaboration is a key to pave this way for common benefits. Not only the steel people but also all the related engineering people must commit this duty. Especially governmental institutions should coordinate this approach.

3.3 High Performance Steel Product

Even any good material is used as a part of product. Any good material has no merit as itself. For best efficiency of material, only the best combined idea of functions of material and part can reveal the solution. Especially the combined approach by material and manufacturing is highly respected. Simulation and characterization will work best in this approach. A platform for such challenge would be only the hot laboratory. Not only steel but also other excellent materials could be best candidates for the best solution. Comprehensive knowledge and understanding are required more and more.

3.4 Steel Education

Only the younger generation can support and run the new paradigm. We must develop new education system to foster new type of materials scientists and engineers. Some functions of the new system should be global and open. Steel people are believed to play the leading role on this reform of engineering education.

4 Summary

In 10 years, East Asia would enter in a saturated condition as for steel production. How the steel people act in such the situation will be a key for each nation. I hope all the steel people lead the society to establish a new steel era with highest efficiency in use of material and energy.

Grain Boundary Carbon Segregation Estimated by McLean and Seah-Hondros Models

Setsuo Takaki, Nobuo Nakada, and Toshihiro Tsuchiyama

Abstract

Our recent study has revealed that the Hall–Petch coefficient in ferritic iron has clear carbon concentration dependence in the range up to 60 ppm-C and it has been proposed that such a behavior deeply relates to the segregation of carbon at grain boundary. On the estimation of carbon concentration at grain boundary, two kinds of model could be applied: McLean model and Seah-Hondros model. In this study, the carbon concentration at grain boundary in ferritic iron was estimated by using these models for the temperature of 973 K. Under the presumption that the saturation atomic carbon concentration at grain boundary is 0.25 (equivalent to Fe_3C), 6 ppm-C and 57 ppm-C were obtained from McLean model and Seah-Hondros model, respectively, for the critical matrix concentration that causes the grain boundary saturation. Hence, it was concluded that Seah-Hendros model is reasonable to explain the carbon concentration dependence of the Hall–Petch coefficient in ferritic iron.

Keywords

Ferritic steel • Grain boundary segregation • Carbon • Hall–Petch coefficient • Critical grain boundary strength

1 Introduction

On the yielding of polycrystalline ferritic steels, it has been believed that the appearance of clear yield point is due to the Cottrell locking of dislocations by interstitials such as nitrogen and carbon, and that the yield stress is significantly lowered by reducing the total content of interstitials [1]. However, author et al. has recently found that the Hall–Petch coefficient (k_y) of ferritic steel has marked carbon concentration dependence in the range below 60 ppm: The value of k_y is as small as 100 ∼ 200 MPa·μm$^{1/2}$ in pure iron but enhanced to 600 MPa·μm$^{1/2}$ by adding 60 ppm of carbon [2]. On the other hand, yield stress (σ_y) of ferritic steels can be increased by grain refinement following the Eq. 1 as a function of grain size d [3–5].

$$\sigma_y = 100 + 600\,d^{-1/2} \qquad (1)$$

This means that yielding of ferritic iron is depending on the mechanism of grain refinement strengthening and the decrease of yield stress by purifying is due to the decrease in k_y. According to the dislocation pile-up model [6], k_y is given by the following equation:

$$k_y = M(2Gb\tau^*/\pi k)^{1/2} \qquad (2)$$

where M, G, b, k and τ^* are Taylor factor (2 in bcc metal), shear modulus (80 GPa for bcc iron), Burgers vector (0.25 nm for iron), constant depending on the nature of dislocation (∼1) and critical grain boundary strength required for generating dislocations from grain boundary, respectively. By putting $(2/\pi k)^{1/2} \approx 1$, the above equation is simplified as follows:

S. Takaki (✉) · N. Nakada · T. Tsuchiyama
Department of Materials Science and Engineering, Kyushu University, 744 Motooka, Nishi-ku, Fukuoka, 819-0395, Japan
e-mail: takaki@zaiko.kyushu-u.ac.jp

Y. Weng et al. (eds.), *Advanced Steels*, DOI: 10.1007/978-3-642-17665-4_10,

$$k_y = M(Gb\tau^*)^{1/2} \quad (3)$$

It should be noted that the possible factor affected by solute carbon is τ^* only, if we can accept the dislocation pile-up model for the yielding mechanism of polycrystalline ferritic steel.

In this paper, the value of τ^* was calculated from k_y which has already reported by our research group [2] and then the carbon concentration dependence of τ^* was compared with the results on carbon grain boundary segregation which were estimated from the models of Seah-Hondros and McLean.

2 Experimental

The materials used for investigating the carbon concentration dependence of k_y is ferritic iron with various range of grain size and carbon concentration up to 60 ppm [2]. Grain size was controlled by changing the annealing time at 973 K after the cold rolling of 90% reduction in thickness. Annealed specimens were quenched into water to suppress carbon segregation and precipitation of Fe_3C that occur on cooling from the annealing temperature. Interstitial free (IF) steel containing 0.024%Ti was used in this research to evaluate the true Hall–Petch coefficient in pure iron without interstitials. The amount of carbon, nitrogen and oxygen in the IF steel is 4, 10 and 23 ppm, respectively, thus 0.024%Ti is enough to fix all of these solute interstitials as Ti(C + N) and TiO_2.

3 Results and Discussion

3.1 Evaluation of the Critical Grain Boundary Strength in the Dislocation Pile-up Model

The Hall–Petch plots are shown in Fig. 1 for IF steels. The friction stress (stress at $d^{-1/2} = 0$) is slightly different depending on the amount of Mn, Si, P and the other impurities, but the slope k_y is almost same at 150 MPa·μm$^{1/2}$ in all IF steels [7, 8]. This value is about 1/4 of the k_y in plain low carbon steel; 600 MPa·μm$^{1/2}$ [5]. Figure 2 displays the influence of solute carbon on the k_y value of ferritic iron [2]. It should be noted that small amount of carbon below 20 ppm gives a large effect to the k_y value. D.V.Wilson has reported the increase of k_y by the aging at 363 K and proposed the effect of carbon segregation as to the change of k_y on the basis of the dislocation pile-up model [9]. According to this model, yielding of polycrystalline metals occurs when the applied shear stress at grain boundary exceeds a critical value τ^* which is

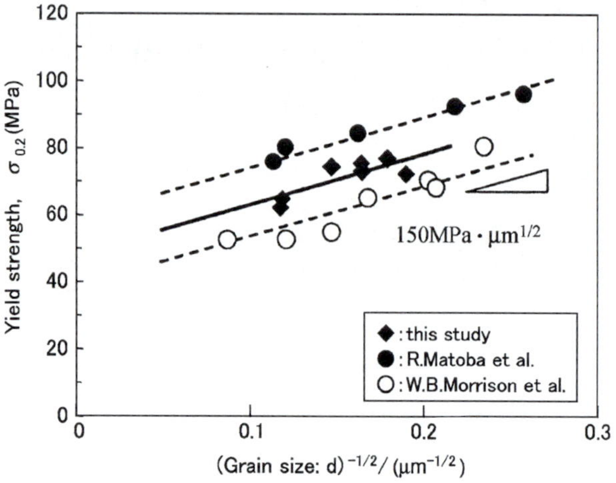

Fig. 1 Hall–Petch relation in IF steel

required for generating dislocations from grain boundary. Author et al. recently succeeded a TEM observation of the dislocations which has been emitted from grain boundary just after yielding, although the material used is not ferritic steel but a high nitrogen austenitic steel [10]. Putting M = 2 for the Eq. 3, the following equation can be constructed to obtained τ^* from k_y.

$$\tau^* = k_y^2/4Gb \quad (4)$$

The calculated τ^* is plotted in Fig. 3 as a function of the amount of solute carbon. This figure clearly shows linearity below 60 ppm and then the tendency of leveling off above that. This result suggests that the carbon concentration at grain boundary has saturated by the addition of 60 ppm carbon and further carbon addition does not give any influence to the behavior of dislocation emission at grain boundary. Auger spectrum analysis is useful technique to investigate the behavior of grain boundary segregation of impurities but this method could not be applied in this investigation because the materials used never caused grain boundary fracture even at cryogenic temperature. Therefore, the grain boundary segregati on behavior of carbon was discussed by applying grain boundary segregation models proposed by Seah-Hondros and McLean.

3.2 Outline of Seah-Hondros Model

Figure 4 displays the concentration profile at grain boundary, X_M: concentration in matrix, X_{M0}: solubility in matrix, X_B: concentration at grain boundary and X_{B0}: saturation concentration at grain boundary. M.P.Seah and E.D.Hondros has proposed the following equation [11] to estimate the

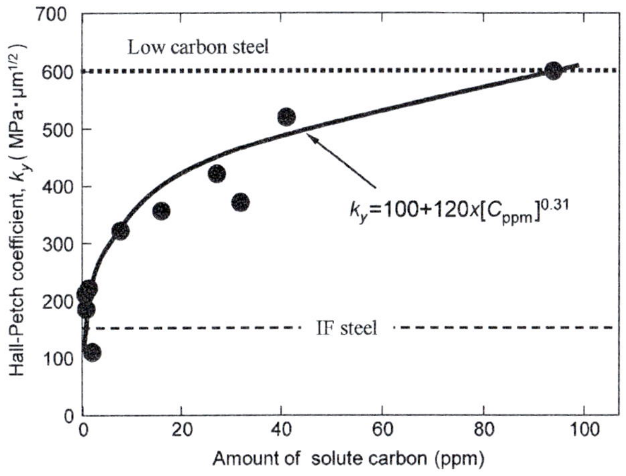

Fig. 2 Influence of solute carbon on the Hall-Petch coefficient in ferritic steel

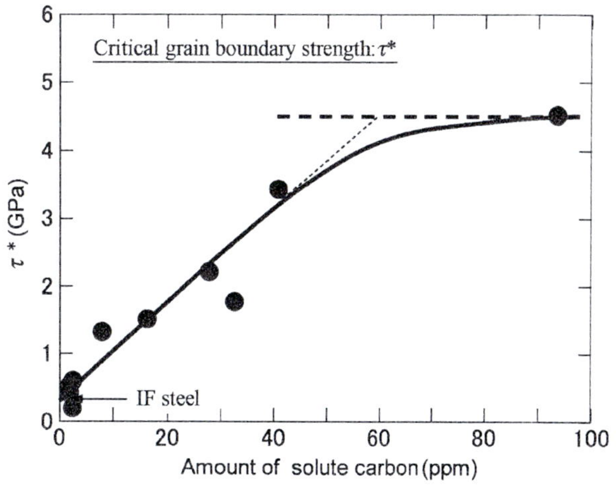

Fig. 3 Relation between solute carbon content and critical grain boundary strength in ferritic steel

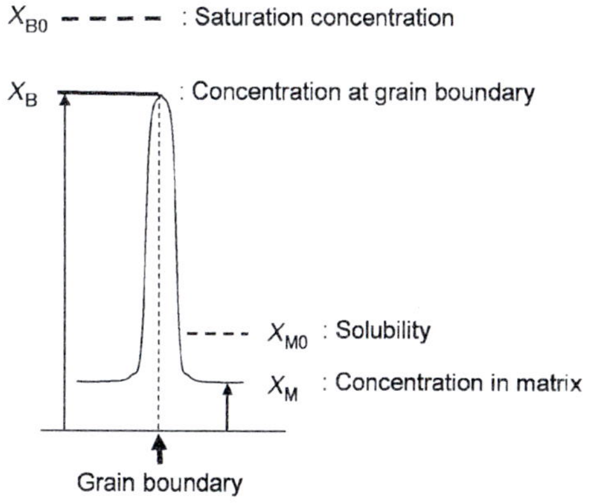

Fig. 4 Illustration showing the concentration profile grain boundary

grain boundary segregation behavior of impurities by modifying an adsorption model.

$$(X_{B0}/X_B)\{X_M/(X_{M0}-X_M)=(1/K)+\{(K-1)/K\}(X_M/X_{M0}) \quad (5)$$

where K is a thermal factor expressed by $K = \ln(Q/RT)$ and Q is a thermal energy representing the difference between matrix and grain boundary, although the physical meaning is not mentioned clearly in the original paper [11]. It is characteristic for this model that the values of K and Q does not depend on alloy systems as shown later and the behavior of grain boundary segregation is directly depends on the solid solubility of alloying elements. Figure 5 displays the saturation balance between matrix (X_M/X_{M0}) and grain boundary (X_B/X_{B0}) for the cases of various K. It is found that (X_B/X_{B0}) increases with (X_M/X_{M0}) and the value of K affects greatly the grain boundary segregation behavior of alloying elements. In practical application of Seah-Hondros model, the most important point is to estimate the value of K precisely.

When the concentration of matrix is very small ($X_M \ll X_{M0}$), the above equation can be rewritten as follows.

$$(X_B/X_{B0})=K(X_M/X_{M0}) \quad (6)$$

This equation means that the grain boundary segregation of alloying elements is increased linearly to the saturation ration in matrix (X_M/X_{M0}). The above equation is also rewritten as follows.

$$(X_B/X_{B0})/X_M=K(1/X_{M0}) \quad (7)$$

The left term is called the enrichment factor (β) and this factor can be experimentally obtained by Auger spectrum analysis. Figure 6 shows the relation between atomic solid solubility and the enrichment factor at 1000 K for many binary alloying systems [12]. It is interesting that the segregation at grain boundary is surely promoted with decreasing the solubility and most of data are on a narrow band regardless of the alloying systems. This result strongly supports the reasonability of Seah-Hondros model. When $X_{M0} = 1$, the value of β gives K, as is obvious from the above equation. Hence, we can estimate at $K = 2.4 \sim 10$ at 1000 K and the value of K gives $Q = 7.32 \sim 19.14$ kJ/mol, respectively. It is characteristic of Seah-Hondros model that the obtained value for Q can apply to any alloying systems and the nature of alloyi ng systems is reflected by the atomic solid solubility. In Fig. 7, temperature dependence of K is shown for the case of $Q = 7.32$ kJ/mol and $Q = 19.14$ kJ/mol. The reasonable value of K should be between the drawn two curves.

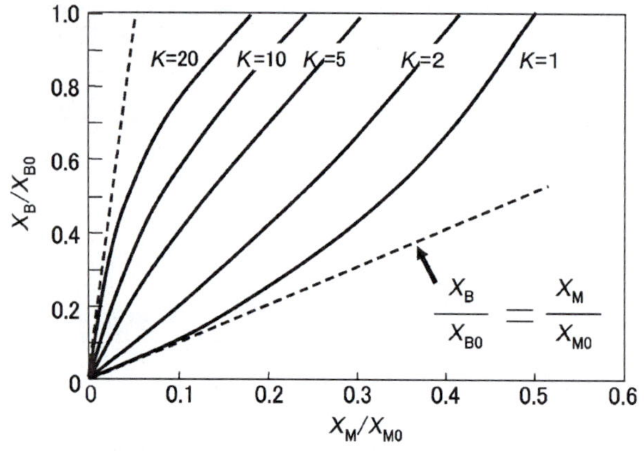

Fig. 5 Effect of K in the relation between (X_M/X_{M0}) and (X_B/X_{B0}) on the Eq. 5

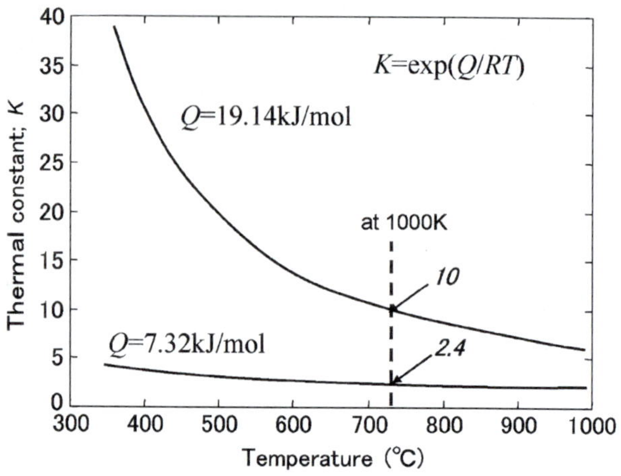

Fig. 7 Temperature dependence of thermal factor K for the case of $Q = 7.32$ kJ/mol and $Q = 19.14$ kJ/mol

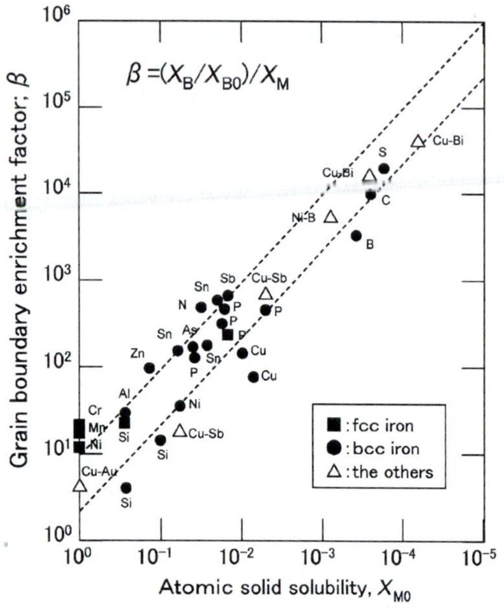

Fig. 6 Relation between atomic solid solubility and grain boundary enrichment factor at 1000 K

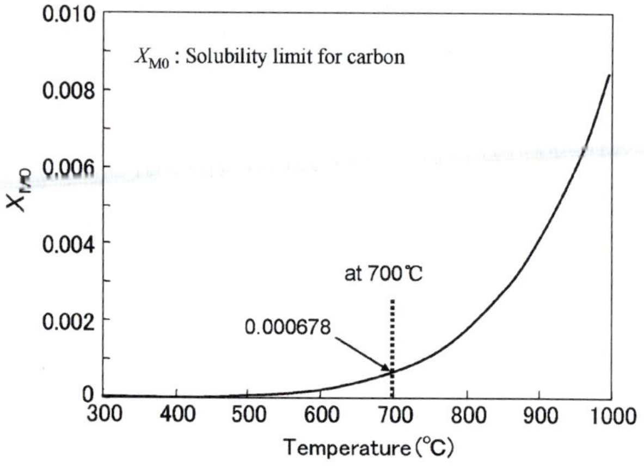

Fig. 8 Temperature dependence of atomic solubility limit of carbon in Fe–C alloy (After the THERMO-CALC)

3.3 Estimation of the Carbon Concentration at Grain Boundary by Seah-Hondros Model

In Seah-Hondros model, the solubility of alloying element gives a large influence to the grain boundary segregation. In Fe–C alloy, the precise solubility of carbon can be obtained as a function of temperature using the THERMO-CALC. The result obtained is displayed in Fig. 8. Since the effect of carbon segregation was discussed for the case of 973 K in the previous Fig. 3, the grain boundary segregation behavior of carbon at this temperature is mentioned from here.

The atomic solid solubility of carbon (X_{M0}) is 0.000678 at 973 K. The saturation concentration (X_{B0}) at grain boundary is not known but it is probably equal to 0.25 which is carbon fraction of Fe_3C. On the other hand, K at this temperature is estimated at a certain value between 2.48 and 10.69 which are correspondent to $Q = 7.32$ kJ/mol and $Q = 19.14$ kJ/mol, respectively. By putting these values into the Eq. 5, the carbon concentration at grain boundary (X_B) can be obtained as a function of the carbon concentration in matrix (X_M) as shown in Fig. 9. This result indicates that carbon of 34 ~ 57 ppm is enough to make grain boundary saturate by carbon atoms. In the previous Fig. 3, it was suggested that 60 ppm carbon is enough to make τ^* enhance to 4.5 GPa which is the maximum value obtained in low carbon steels. If we can accept linear relationship between τ^* and the carbon concentration at grain boundary, a reasonable carbon segregation behavior would be

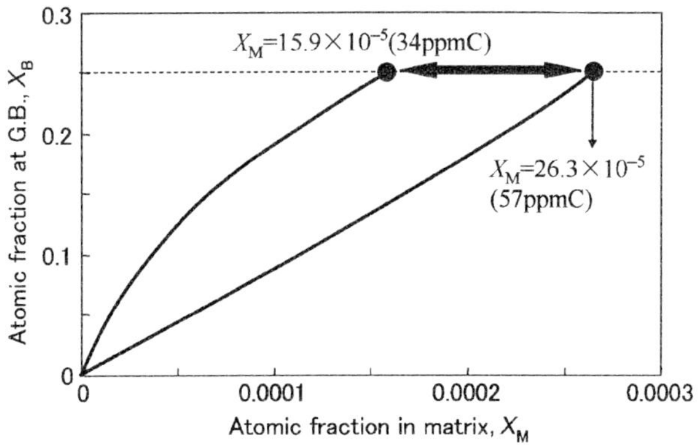

Fig. 9 Relation in the carbon concentration between matrix and grain boundary for the cases of $K = 2.48$ and $K = 10.69$ in Seah-Hondros model

estimated by assuming $Q = 7.32$ kJ/mol in Seah-Hondros model.

3.4 Estimation of the Carbon Concentration at Grain Boundary by McLean Model

D.McLean has proposed the following equation considering the concentration ratio balance between matrix ($X_M/(1 - X_M)$) and grain boundary ($X_B/(1 - X_B)$) [13].

$$(X_B/(1-X_B)) = K * (X_M/(1-X_M)) \quad (8)$$

where $K*$ is a thermal factor represented by $K* = \exp(Q*/RT)$. $Q*$ is the segregation energy which is identical to the alloying element in alloy systems. In the case of carbon in ferritic iron, $Q* = 76$ kJ/mol has already been reported [14]. Therefore, the value of $K*$ can be calculated at 12080 for 973 K. By putting this value into the above equation, the carbon concentration at grain boundary (X_B) can be obtained as a function of the carbon concentration in matrix (X_M) as shown in Fig. 10. This result suggests that 6 ppm carbon is enough to make grain boundary saturate by carbon atoms. In general, the precipitates of Fe_3C are found in low purity iron containing several tens ppm carbon but never found in high purity iron containing carbon less than 10 ppm. From this fact, it may be concluded that $Q* = 76$ kJ/mol is over estimation for the segregation energy of carbon in McLean model. If reasonable segregation energy was obtained for the carbon segregation, McLean model could be applied to the estimation of carbon grain boundary segregation in ferritic steels.

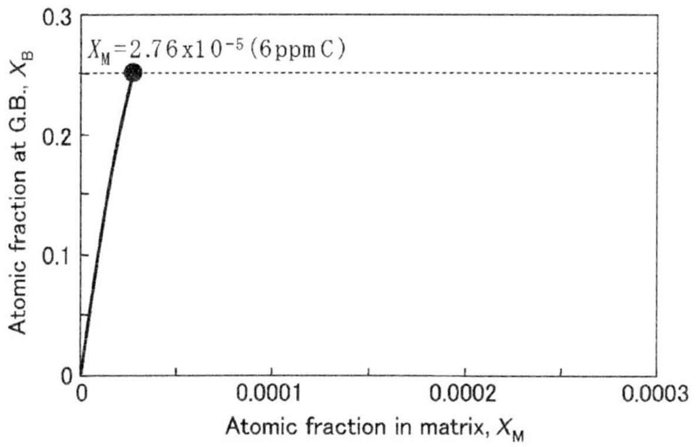

Fig. 10 Relation in the carbon concentration between matrix and grain boundary obtained by using McLean model

4 Conclusion

In the relation between carbon content and the critical grain boundary strength $\tau*$ which is the shear stress required for dislocation emission from grain boundary, clear linearity below 60 ppm carbon and also the tendency of leveling off above that were found. From this result, it was suggested that the carbon concentration at grain boundary has saturated by the addition of 60 ppm carbon and further carbon addition does not give any influence to the behavior of dislocation emission at grain boundary. On the estimation of grain boundary segregation of carbon, it was concluded that Seah-Hondros model can reasonably be applied by assuming $Q = 7.32$ kJ/mol: The carbon concentration in matrix 57 ppm is enough to make grain boundary saturate by carbon atoms at 973 K.

Since a clear relationship between solid solubility and the grain boundary segregation is found in many alloying systems and precise estimation of solid solubility becomes easy due to marked development in thermo-dynamic calculation, it may be recommended to apply Seah-Hondros model for the grain boundary segregation behavior in many alloying systems. It is convenient that the value $Q = 7.32$ kJ/mol is not dependent on alloy systems, thus in any alloy systems, grain boundary segregation behavior might be estimated only by the solid solubility of the alloying element.

References

1. J.R. Low, M. Gensamer, Trans. AIME **158**, 207 (1944)
2. K. Takeda, N. Nakada, T. Tsuchiyama, S. Takaki, ISIJ Int. **48**, 1122 (2008)
3. Y.Kimura, S.Takaki, *Proceedings of 1998 PM World Congress*, EPMA, Granada, p. 573 (1998)
4. Y. Kimura, S. Takaki, J. Stat. Theory Prac. **41**, 13 (2000)

5. M. Etou, S. Fukushima, T. Sakai, Y. Haraguchi, K. Miyata, M. Wakita, T. Tomida, N. Imai, M. Yoshida, Y. Okada, ISIJ Int. **48**, 1142 (2008)
6. N.J. Petch, J. Iron Steel Inst. **174**, 25 (1953)
7. R. Matoba, N. Nakada, Y. Futamura, T. Tsuchiyama, S. Takaki, Tetsu-to Hagane **93**, 513 (2007)
8. W.B. Morrison, W.C. Leslie, Metall. Trans. **4**, 379 (1973)
9. D.V. Wilson, Metal Sci. J. **1**, 40 (1967)
10. T. Tsuchiyama, Y. Fujii, Y. Terazawa, T. Ando, S. Takaki, ISIJ Int. **48**, 861 (2008)
11. M.P. Seah, E.D. Hondros, Scripta Met. **7**, 735 (1973)
12. M.P. Seah, Metal Phys. **10**, 1043 (1980)
13. D.McLean, Grain Boundaries in Metals, Oxford University Press (1957)
14. H. Erhart, H.J. Grabke, Met. Sci. **15**, 401 (1981)

Nano-Preciptates Design with Hydrogen Trapping Character in High Strength Steel

Fu-Gao Wei, Toru Hara, and Kaneaki Tsuzaki

Abstract

Nano-precipitates of alloy carbides TiC, NbC and VC which have the same NaCl-type crystal structure in tempered martensite have been characterized by means of high-resolution transmission electron microscope (HRTEM) and correlated to the hydrogen trapping property. The examination of whether the amount of hydrogen absorbed by the TiC particles depends on their surface area or volume indicate that the coherent and semi-coherent TiC particles trap hydrogen at the precipitate/matrix interface at ambient temperature while the incoherent TiC particles trap hydrogen inside themselves only at high temperatures. Coherent and semi-coherent NbC and VC particles also demonstrate a surface area dependence of hydrogen trapping capacity with NbC > TiC ≫ VC. Contrary to TiC, incoherent NbC and VC particles are unable to trap hydrogen.

Keywords

High strength steel • Hydrogen embrittlement • MX carbide • HRTEM • Thermal desorption spectrometry

1 Introduction

Titanium carbide has been known to have a strong interaction with hydrogen through the electrochemical permeation method [1] and the thermal desorption spectrometry (TDS) analysis [2, 3]. The activation energy for the desorption of hydrogen from TiC precipitates in steels was estimated to be as high as 95 [1] or 87 kJ/mol [2].

Our recent study also showed high activation energy for TiC particles [4]. However, these results only seemed to be valid for incoherent TiC particles. The coherent or semi-coherent TiC particles in the 0.42C-0.30Ti steel demonstrated a different hydrogen trapping characteristic; the activation energy is much lower than that for the incoherent particles [5]. Pressouyre and Bernstein [1, 6] pointed out earlier that the TiC particle becomes more reversible when its coherency with the matrix is increased; however, no experimental evidence has been provided. On the other hand, an autoradiography imaging of hydrogen [7] suggested that hydrogen is trapped at the precipitate/matrix interface. Thus, the first purpose of the present study is to investigate the relationship between the size and coherency of the TiC precipitate and its hydrogen trapping property. The first step is to produce a series of TiC precipitates of various interfacial characters ranging from coherent through semi-coherent to incoherent by heat treatment. The second step is to characterize these precipitates on an atomic scale; however, this was not possible due to the strong magnetism interference of low-alloy ferritic or martensitic steel. The third step is to

This article was originally published in "Effects of Hydrogen on Materials", ASM International, ISBN-13: 978-1-61503-003-3, (2009), pp. 448–455.

F.-G. Wei (✉)
Technology Development Department, Yakin Kawasaki Co., Ltd., 4-2 Kojima-cho, Kawasaki-ku, Kawasaki, Kanagawa 210-8558, Japan
e-mail: fugao.wei@nyk.jp

T. Hara · K. Tsuzaki
Structural Metals Center, National Institute for Materials Science, 1-2-1 Sengen, Tsukuba, Ibaraki 305-0047, Japan

Y. Weng et al. (eds.), *Advanced Steels*, DOI: 10.1007/978-3-642-17665-4_11,

correlate the size and coherency of the TiC precipitate to its hydrogen trapping property with an emphasis of obtaining the answer to the two questions: (1) how does the hydrogen trapping characteristic change with a successive variation of interfacial character from coherent to incoherent, and (2) does the amount of hydrogen trapped by (semi-)coherent and incoherent TiC precipitates depend on their surface area or volume? The answer to the second question determines the trap sites of hydrogen to be the precipitate/matrix interface or to be inside the precipitate.

The second purpose of the study is to investigate the hydrogen trapping properties of NbC and VC precipitates which have the same crystal structure as that of TiC. NbC and VC, along with TiC, are important carbides for microstructural control in microalloyed steels and low-alloy high strength steels. However, little is known regarding their hydrogen trapping properties. Thus, they are examined with an emphasis on comparing them to the TiC precipitate.

2 Experimental Method

Three steels containing equal molar fractions of TiC, NbC and VC, respectively, i.e. 0.05C-0.20Ti-2.0Ni, 0.05C-0.41Nb-2.0Ni and 0.05C-0.24V-2.0Ni in mass% were prepared. They were received in the form of hot-rolled plates that contained carbide precipitates in the ferrite matrix. The conditions of heat treatment, hydrogen charging and TDS analysis of the present steels were the same as those for the 0.05C-0.20Ti-2.0Ni steel in the previous study [8]. The major conditions are: (1) quenching from 1,350 C and tempering at 300 ~ 1,000 °C for 3 h, (2) dimension of the specimen for TDS analysis being ϕ5 mm × 40 mm, (3) cathodic charging of hydrogen in a 3 mass% NaCl + 0.3 mass% NH_4SCN) aqueous solution at a current density of 1 A/m^2 for a short time of 1 h to investigate the deep traps associated with TiC precipitates, (4) hydrogen analysis by TDS at a heating rate of 100 °C/h, (5) estimation of the activation energy for hydrogen desorption from traps by fitting the experimentally measured desorption spectrum to the Kissinger's first-order reaction kinetic formula, and (6) microstructural observation by means of HRTEM. The tempering temperature for NbC- and VC-containing steels was limited to 700 °C and the hydrogen charging time was extended to 48 h until the saturation of hydrogen entry.

3 Results and Discussion

3.1 Microstructural Characterization of TiC Precipitates on the Atomic Scale

Austenitization at 1,350°C dissolved all the TiC particles in austenite. Then a fully martensitic microstructure was produced by the quenching that followed. Tempering at 300–1,100°C after the quenching produced a series of TiC precipitates with interfacial characters changing from coherent to incoherent as expected. No precipitates other than TiC were observed. TiC began to precipitate at 500°C as a coherent precipitate and changed to semi-coherent particles at about 550°C. The semi-coherent character existed to 800°C beyond which it lost its coherency with the matrix. Coherent and semi-coherent TiC precipitates exhibit a disk-like shape with its broad face on the {100} planes of ferrite and satisfies the Baker–Nutting orientation relationship with the ferrite, i.e. $(100)_{TiC}//(100)_{\alpha}$, $[011]_{TiC}//[001]_{\alpha}$. A TiC precipitate in the 0.05C-0.20Ti-2.0Ni steel which was tempered at 600°C is shown in Fig. 1. The imaging process of the HRTEM micrograph revealed that there are misfit dislocations on the broad faces of the precipitate and the side faces as well [9]. The change in the size and shape of the TiC precipitate with tempering temperature is summarized in Fig. 2. Both the diameter and the thickness of the disc-like (semi-)coherent TiC precipitate increase with increasing tempering temperature with an abnormal increase at 800°C. Incoherent TiC precipitates formed above 800°C were observed to be approximately spherical.

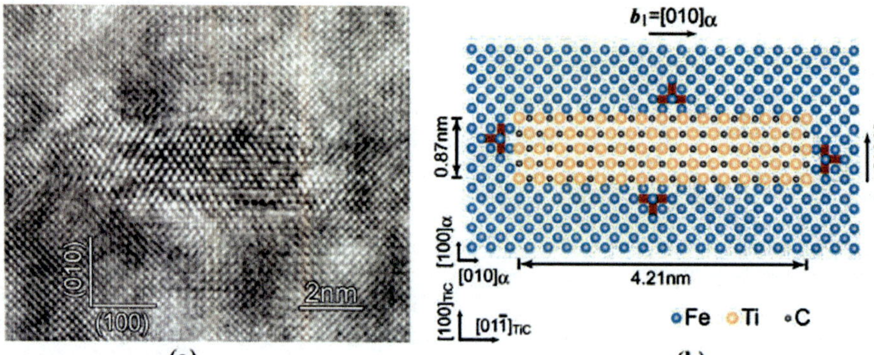

(a) (b)

Fig. 1 Semi-coherent TiC precipitate in the 0.05C-0.20Ti-2.0Ni steel tempered at 600°C. **a** HRTEM micrograph and **b** schematic illustration of the lattice mismatch

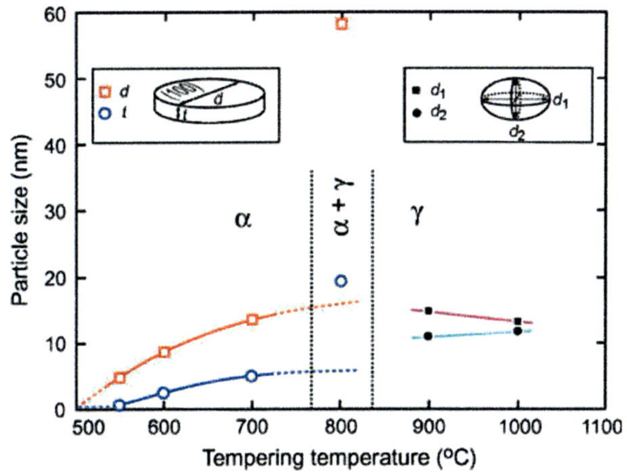

Fig. 2 Change in size and shape of TiC precipitate in the 0.05C-0.20Ti-2.0Ni steel with tempering temperature

3.2 Hydrogen Trapping of TiC Precipitates with Interfacial Characters from Coherent to Incoherent

The results of the TDS analysis on samples tempered at 300–1,100°C are shown in Fig. 3. A short hydrogen charging time of 1 h was selected for detecting the deepest trap associated with the TiC precipitates. Only a small peak at about 120°C is present for the samples that were tempered below 500°C. However, the desorption peak becomes larger and shifts to about 230°C when tempered at 500°C. On tempering above 550°C, the 230°C peak splits into two peaks; one remains at 230°C and the other (the subsidiary peak) shifts towards a high temperature as the tempering temperature increases. The 230°C peak becomes smaller when the tempering temperature increases and disappears in the spectrum of the sample that was tempered at 1,000°C.

The peak located at 120°C supposedly resulted from the hydrogen trapped by the grain boundaries and dislocations in the martensitic matrix [8]. The 230°C desorption peak was contributed from the hydrogen that was trapped by the broad face of the disc-shaped TiC precipitate. The subsidiary peak that split from the 230°C peak is from the side face of the precipitate. The hydrogen desorption activation energy deduced from the 230°C peak was about 55.8 kJ/mol for the samples that were tempered from 600 to 800°C. By dividing the hydrogen content by the total surface area of the broad faces, the hydrogen trapped by the broad faces of the disk-like semi-coherent TiC precipitate was estimated to be constant and about 1.3 atoms/nm^2 (Fig. 4). The facts of the constant hydrogen content per unit surface area and the constant desorption activation energy of relatively high values over the range of the tempering temperature of 600–800°C strongly suggest that hydrogen associated with the 230 °C peak is trapped at the same deep site on the broad face of the semi-coherent TiC precipitate. The deep site is probably the core of misfit dislocations as shown in Fig. 1b. This result implies that the higher the misfit dislocation density, the more the hydrogen will be trapped at the precipitate/matrix interface. The NbC is predicted to trap more hydrogen than TiC while VC will trap less hydrogen than TiC because NbC and VC show a higher and a lower lattice mismatch with the ferrite matrix than TiC, respectively. Another prediction is that hydrogen is likely to occupy the less distorted sites other than the misfit dislocation core on the interface when more hydrogen is charged into the steels.

Fig. 3 Evolution of the TDS spectrum of the 0.05C-0.20Ti-2.0Ni steel quenched and tempered at various temperatures. Hydrogen charging time 1 h

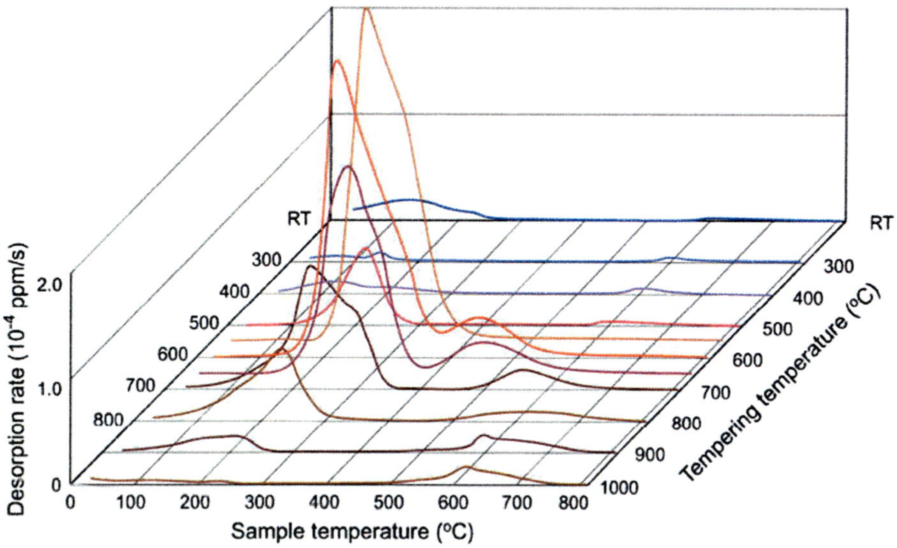

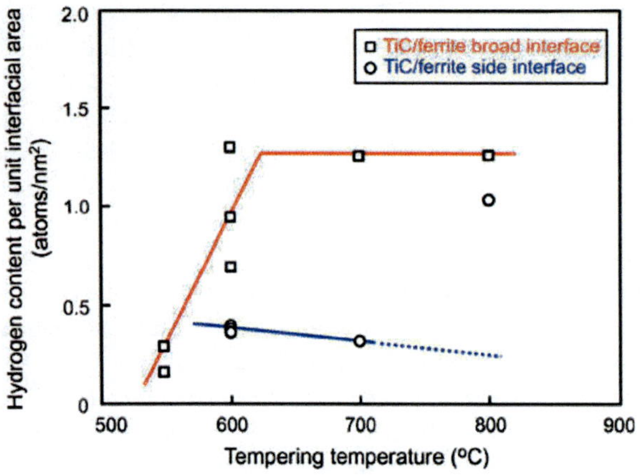

Fig. 4 The hydrogen trapped by the unit area of the broad and the side faces of the semi-coherent TiC precipitate, respectively. Hydrogen charging time 1 h

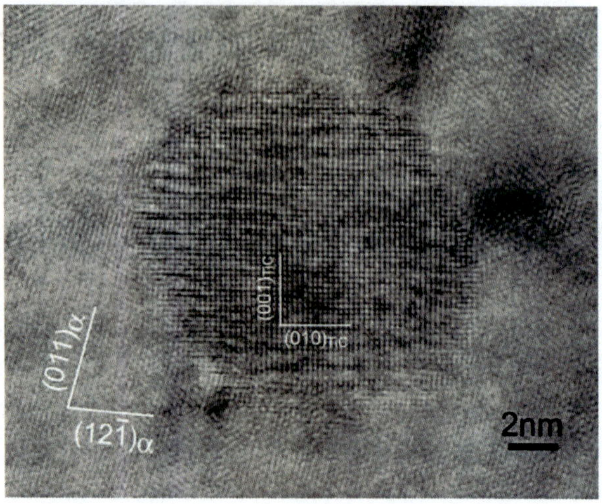

Fig. 5 An incoherent particle in the 0.05C-0.20Ti-2.0Ni steel

In contrast to the 230°C peak, the subsidiary desorption peak that is separated from it cannot be rationalized by the constant hydrogen content per unit surface area as shown in Fig. 4. The subsidiary desorption peak is interpreted to result from the hydrogen trapped at the carbon vacancy sites within the TiC precipitate beneath the side interface. The carbon vacancies are supposed to form during the Ostward ripening process of TiC precipitates during which the smaller particles begin to dissolve from the side face preferentially with a faster dissolution rate of carbon than titanium [10].

3.3 Hydrogen Trapping Characteristics of Incoherent TiC Particles

TiC particles have long been considered as a strong trap which can absorb hydrogen easily at ambient temperature. As described above, (semi-)coherent TiC precipitates are able to trap hydrogen easily during cathodic charging at room temperature. However, our recent study [11] found that incoherent TiC particles do not trap hydrogen at all at room temperature. Incoherent TiC can trap hydrogen but this is only possible at high temperatures. Figure 5 shows an incoherent TiC particle in the as-received 0.05C-0.20Ti-2.0Ni steel. The incoherent TiC particles that remained undissolved during austenitization absorb hydrogen from the environmental atmosphere. The hydrogen trapped by the incoherent TiC, which is characterized by a desorption peak at around 600°C, depends on the amount or more accurately the volume [8] of undissolved TiC particles (Fig. 6a). At room temperature, the incoherent TiC particles cannot be charged with hydrogen even with a long charging time and a

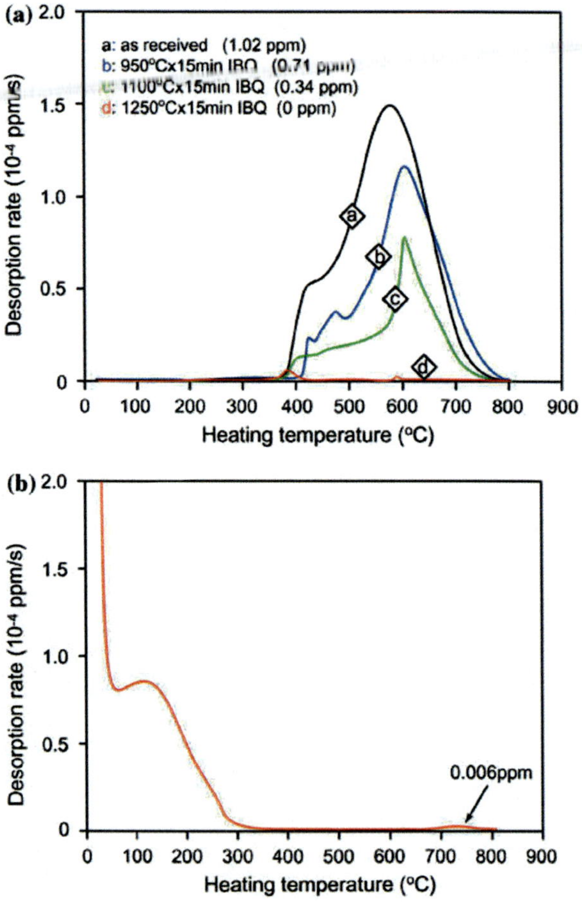

Fig. 6 TDS analysis of hydrogen trapped by incoherent TiC particles in the as-received 0.05C-0.20Ti-2.0Ni steel. **a** Analysis without cathodic charging after austenitization at various temperatures, and **b** analysis after vacuum degassing at 950°C and prolonged cathodic hydrogen charging at a high current density

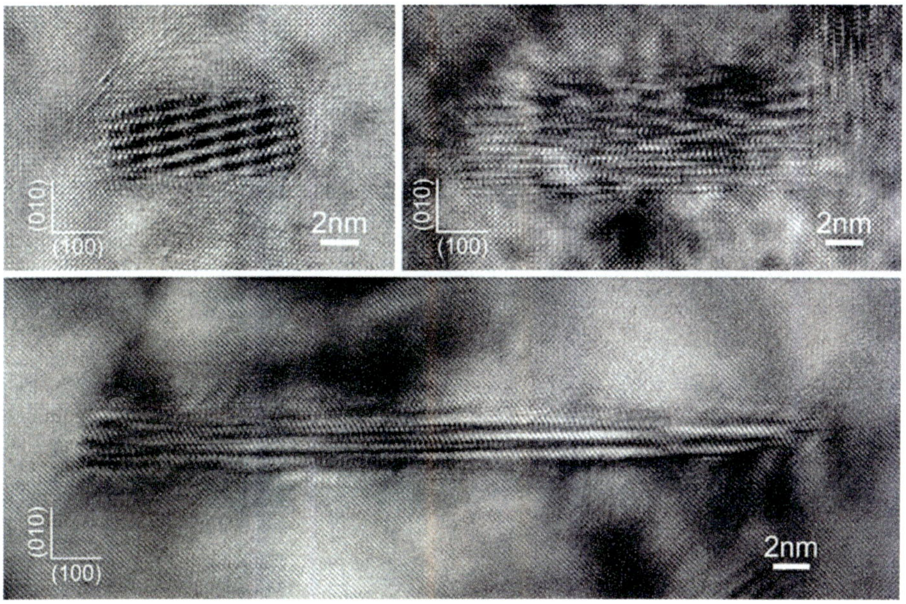

Fig. 7 HRTEM micrographs of NbC, TiC and VC precipitates in the 0.05C-0.41Nb-2.0Ni, 0.05C-0.20Ti-2.0Ni and 0.05C-0.24 V-2.0Ni steels quenched and tempered at 700°C, respectively

high current density (Fig. 6b). The inability of hydrogen trapping of incoherent TiC particles at low temperatures is explained by the trap site of carbon vacancies which requires high activation energy for both trapping and detrapping of hydrogen.

3.4 Hydrogen Trapping of NbC and VC Precipitates

NbC and VC nano-precipitates show a disk-like appearance like that of TiC and maintain the Baker–Nutting orientation relationship with ferrite. However, their diameters differ noticeably compared to that of TiC although they all have almost the same thickness (Fig. 7) [12]. The diameter of the disk-shaped carbide is proportional to the reciprocal of its lattice misfit (Fig. 8a), which means that there is the same number of misfit dislocations on the broad face of each particle if the same misfit dislocation configuration is assumed (Fig. 8b).

Figure 9 shows the TDS spectra obtained after hydrogen charging to saturation for the three steels tempered at 700°C containing the NbC, TiC and VC precipitates, respectively. Contrary to TiC, NbC and VC do not have a desorption peak at around 600°C. The reason for this difference has not been clarified yet. The precipitates contribute significantly to the desportion peaks below 250°C. The hydrogen trapping ability of the alloy carbides varies in the descending order of NbC $>$ TiC $\gg$ VC. This finding shows a good agreement with the above prediction based on the density of misfit dislocations on the broad face of the disk-shaped alloy carbide precipitate.

The results obtained in this study emphasize that the hydrogen trapping property, i.e. hydrogen trapping capacity and the hydrogen–precipitate interaction energy, depends considerably on the type of alloy carbides, precipitate/ferrite interfacial character and the crystal defects within the precipitate. Composition selection and microstructural control of alloy carbides may offer a variety of options for the application of alloy carbides as hydrogen trap sites in the development of hydrogen embrittlement resistant steels.

Conclusions

(1) (Semi-)coherent TiC precipitates trap hydrogen at precipitate/ferrite interfaces. The core of misfit dislocations on the broad face of the disk-shaped precipitate is supposed to act as hydrogen trap sites at a low hydrogen concentration.

(2) Incoherent TiC particle is unable to trap hydrogen at ambient temperature. High temperatures are required for incoherent TiC particle to trap and detrap hydrogen. The carbon vacancies are probably the trap sites for hydrogen in incoherent TiC particles.

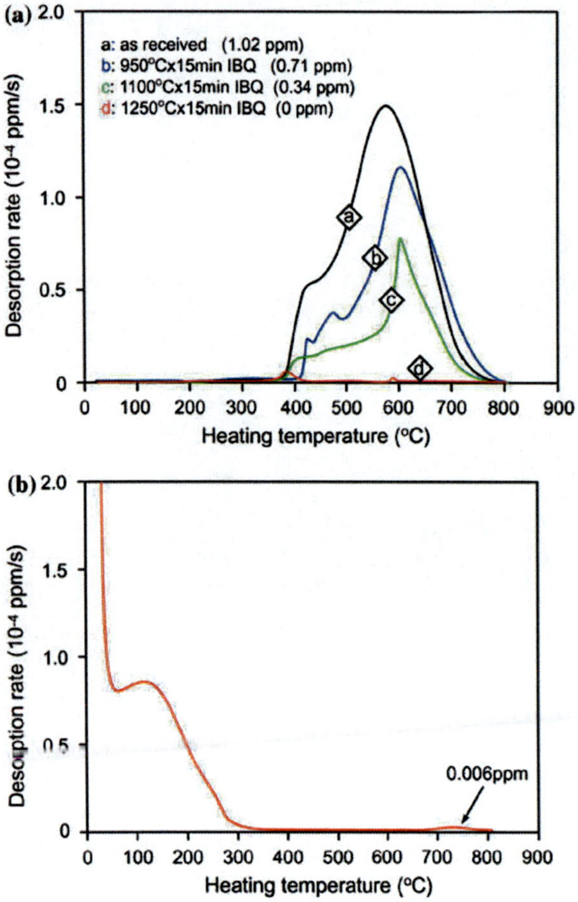

Fig. 8 Size and morphology of disk-like NbC, TiC and VC in the steels tempered at 700°C. **a** Relationship between diameter and lattice misfit, and **b** illustration of misfit dislocations on the broad faces of the precipitates

(3) The hydrogen trapping ability of (semi-)coherent alloy carbides varies in the descending order of NbC > TiC ≫ VC. Unlike TiC, neither NbC nor VC show any desorption peak at high temperatures around 600°C.

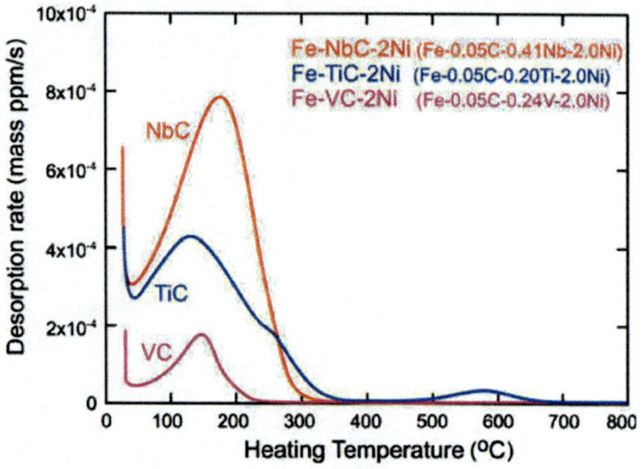

Fig. 9 TDS spectra for 0.05C-0.41Nb-2.0Ni, 0.05C-0.20Ti-2.0Ni and 0.05C-0.24V-2.0Ni steels quenched and tempered at 700°C. Charging time: 48 h

References

1. G.M. Pressouyre, I.M. Bernstein, Metall. Trans. A **9A**, 1571 (1978)
2. H.G. Lee, J.Y. Lee, Acta Metall. **32**, 131 (1984)
3. S.M. Lee, J.Y. Lee, Acta Metall. **35**, 2695 (1987)
4. F.G. Wei, T. Hara, K. Tsuzaki, Metall. Mater. Trans. B **35B**, 587 (2004)
5. F.G. Wei, T. Hara, T. Tsuchida, K. Tsuzaki, Iron Steel Inst. Jpn. Int. **43**, 539 (2003)
6. G.M. Pressouyre, I.M. Bernstein, Metall. Trans. A **10A**, 1571 (1979)
7. A. Asaoka, G. Lapasset, M. Aucouturier, P. Lacombe, Corrosion **34**, 39 (1978)
8. F.G. Wei, K. Tsuzaki, Metall. Mater. Trans. A**37A**, 331 (2006)
9. F.G. Wei, T. Hara, K. Tsuzaki, Phil. Mag. **84**, 1735 (2004)
10. F.G. Wei, T. Hara, K. Tsuzaki, Ferrum (Bull. Iron Steel Inst. Jpn.) **12**, 766 (2007)
11. F.G. Wei, K. Tsuzaki, Metall. Mater. Trans. A**35A**, 3155 (2004)
12. F.G. Wei, K. Tsuzaki, Second Place Award of International Metallographic Contest, Class 3: Electron Microscopy—Transmission and Analytical (2008), http://www.metallography.com/ims/contest.htm

Micro-Mechanical Behavior of Inclusions in Advanced Steels

Xishan Xie, Yanpin Zeng, Miaomiao Wang, and Hongmei Fan

Abstract

Inclusions are unavoidable even in super-clean advanced steels because of the necessary melting process. The effect of two kinds of typical inclusions such as TiN and AlN have been studied by means of specially designed SEM in situ tensile and fatigue tests for two advanced ultra-high-strength steels MA250 and GE1014 respectively. These unique experiments of SEM in situ tensile and fatigue test directly trace the entire process of crack initiation and propagation till fracture from tested specimens. TiN inclusion in MA250 often characterizes with large blocky cubic morphology. Crack easily initiate at the sharp corners of TiN or TiN/matrix interfaces or very often directly initiate inside TiN particles because of its brittleness. These cracks easily propagate to the matrix and to cause early failure. For elimination the harmful effect of TiN on this kind ultra-high-strength steel. An advanced ultra-high-strength steel without Ti designated as GE1014 was developed by GE Power, USA. Very small inclusion AlN particles (only several microns) can exist in GE1014. If these AlN small particles distribute in steel as inclusion chains, cracks often directly initiate at the inclusion chains among AlN small particles and line up to develop voids, which rapidly propagate to the matrix till early failure. These valuable experimental results reveal the harmful effect of inclusions in micro-scale and can be connected with tensile and fatigue loading processes for understanding early failure mechanisms, which are helpful for practical life prediction of these advanced steel components.

Keywords

Inclusion • In situ tensile • In situ fatigue • Crack • Ultra-high-strength steel

1 Introduction

Vanadium, niobium and titanium are important strengthening elements not only for micro-alloying high strength low alloy (HSLA) steels but also for advanced high strength and ultra-high-strength steels by means of precipitation strengthening. However, V, Nb and Ti characterize with strong tendency to combine with carbon especially with nitrogen in formation of MC, MN or M (CN). If these phases precipitate directly form molten metals during solidification, they play a role as inclusions in these advanced steels.

X. Xie (✉) · Y. Zeng · M. Wang · H. Fan
University of Science and Technology Beijing,
Beijing 100083, China
e-mail: xishanxie@mater.ustb.edu.cn

M. Wang
Shanghai Power Equipment Research Institute,
Shanghai 20024, China

H. Fan
Shougang Research Institute of Technology,
Beijing 100041 China

Y. Weng et al. (eds.), *Advanced Steels,* DOI: 10.1007/978-3-642-17665-4_12,

Table 1 Nominal chemical compositions and representative mechanical properties of two investigated steels

No.	Steel	Chemical composition (mass %)							Mechanical properties			
		C	Ni	Cr	Co	Mo	Ti	Al	$\sigma_{0.2}$ (MPa)	σ_b (MPa)	δ (%)	Ψ (%)
1	MA250	0.002	18	0.2	8.5	5	0.45	0.09	1,848	1,917	12–15	50–55
2	GE1014	0.2	14	2.5	10	14	–	0.9	1,889	2,068	12–15	50–55

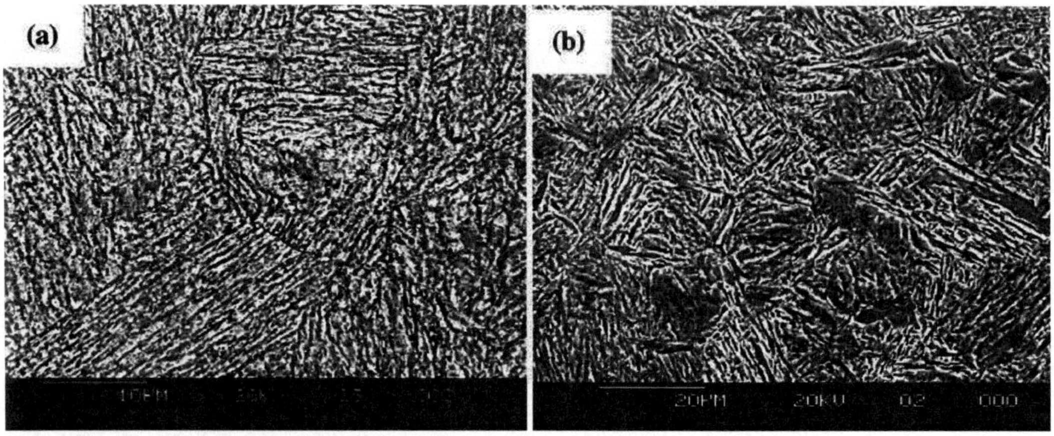

Fig. 1 SEM structure images of **a** MA250 and **b** GE1014 after standard heat-treatments

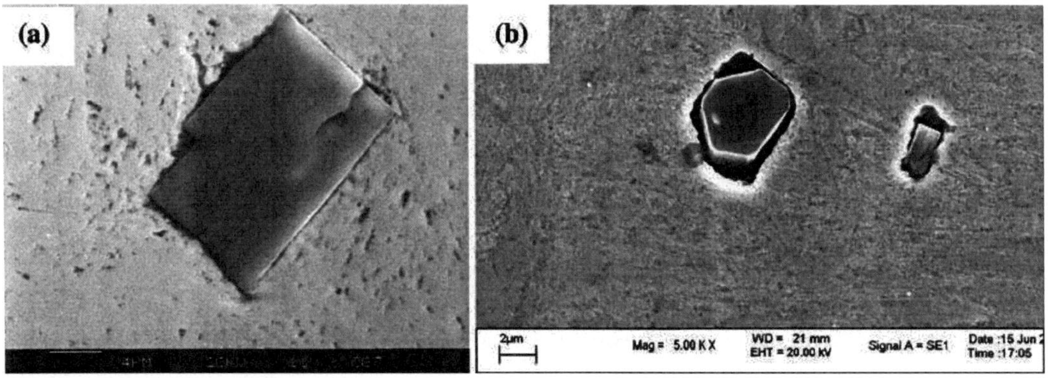

Fig. 2 Typical morphologies of TiN inclusion (12 × 17 μm^2) in **a** MA250 and **b** AlN (3–5 μm) in GE1014

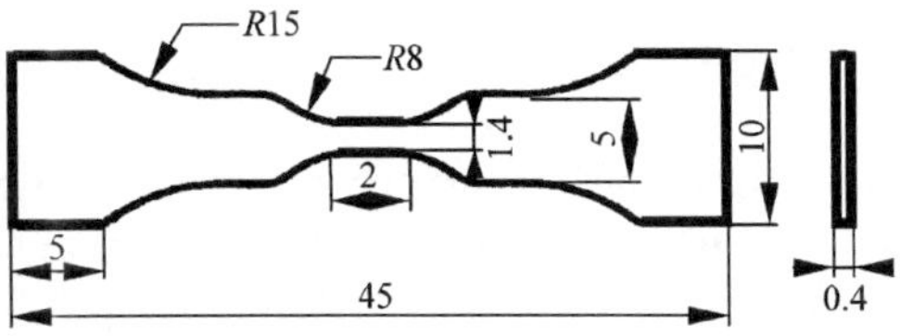

Fig. 3 Schematic of SEM in situ test specimen

Aluminum is a deoxidation element normally used in melting process of steels and sometimes also plays a role as strengthening element for inter-metallic phase precipitation strengthening, such as Ni_3Al, etc. However aluminum can combine with nitrogen or oxygen to form aluminum nitride (AlN) or oxide (Al_2O_3), which can retain in steels as inclusions. Aluminum nitride AlN may combine with other oxides such as MgO and sometimes also with sulfide such as MnS to form complex inclusions.

The mechanical properties such as yield and ultimate tensile strengths, especially low cycle fatigue properties (LCF), of ultra-high-strength steel rotating shaft components have been seriously degraded by the existence of non-metallic inclusions which are introduced into steels during melting process. Inclusions act as origins for crack initiation, then decrease tensile properties and LCF

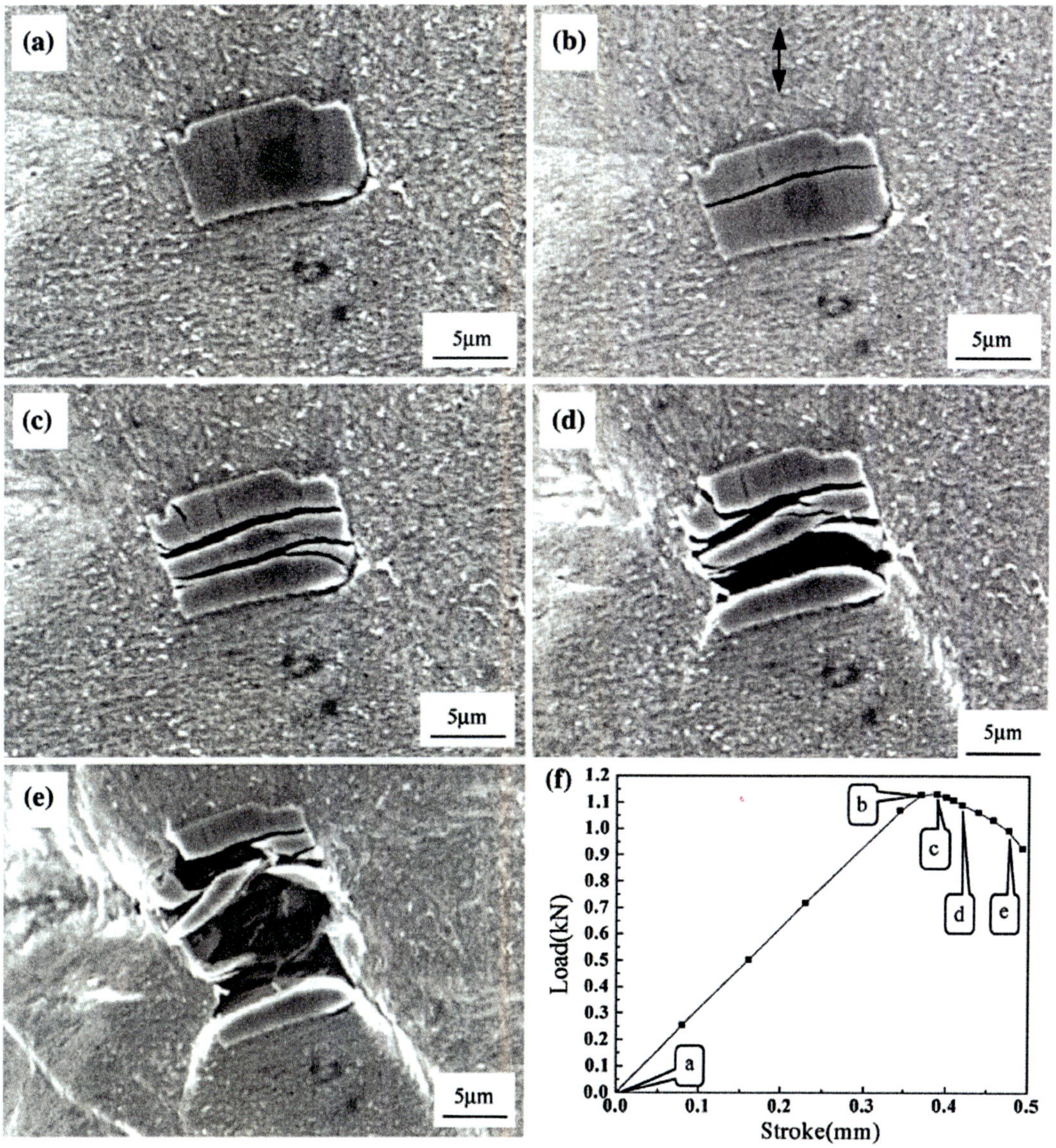

Fig. 4 The micro-mechanical behavior of TiN inclusion and load–stroke diagram during in situ tension process in MA250

capability of ultra-high-strength steels. It is important to decrease inclusion fraction in these steels by advanced metallurgical processing. However, getting completely clean steel without inclusions is currently impossible. In the past time many research papers had been done dealing with the effect of inclusions on mechanical properties of high strength steels [1–4]. However, recently only a few of publications have been published on micro-mechanical behaviour of inclusions in steels or alloy [5–7] and a very few systematical investigations on mechanical property tests have been carried out. Present paper concentrates the effect of inclusions on micro-mechanical behavior, such as crack initiation and propagation at tensile and fatigue tests. The effect of two typical inclusions TiN and AlN in two ultra-high-strength steels MA250 and GE1014 have been investigated in detail.

2 Materials and Experiments

Two kinds of ultra-high strength steels (MA250 and GE1014), which are used for today's large aircraft engine shafts, have been used for this investigation. The nominal chemical compositions and typical mechanical properties after standard heat treatments are listed in Table 1.

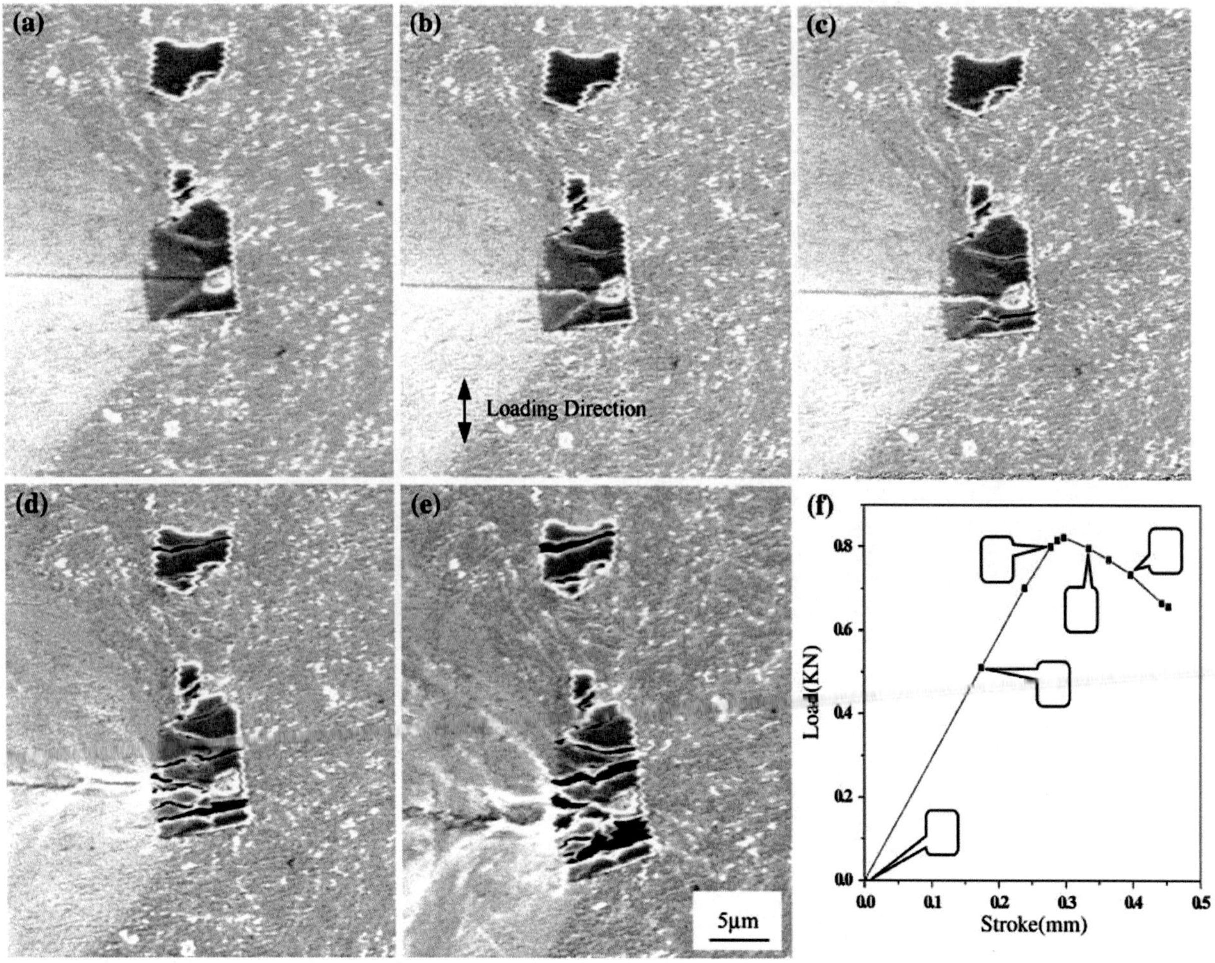

Fig. 5 The micro-mechanical behavior of TiN inclusion and load–stroke diagram during in situ tension process in MA250

It can be seen that both steels characterize with very high strengths and good ductilities. The structures of both steels characterize with tempered martensite as shown in Fig. 1. Two typical inclusions TiN in MA250 steel and AlN in GE1014 steel are shown in Fig. 2, respectively.

A careful specimen preparation was conducted. The flat tension and fatigue specimens with dimension shown in Fig. 3 including the investigated inclusions were selectively cut from the test material. These inclusions must be located at the surface in the gauge length section of specimens. Each sample was mechanically polished. For easy finding the inclusion during SEM in situ tension and fatigue tests, a pyramid indentation was made nearby each observed inclusion by micro-hardness tester.

SEM in situ tension tests were conducted on the tensile loading stage in the chamber of SEM equipment SS 550 which can control the load and displacement at ambient temperature. Similar with in situ tension tests, LCF in situ tests were conducted on the fatigue loading stage in SEM chamber, which can control the cyclic stress and number of cycles at ambient temperature.

3 Results and Discussion

3.1 Micro-Mechanical Behaviour of TiN Inclusion in MA250 Ultra-High-Strength Steel

Figure 4 shows the micro-behavior of a rectangular TiN inclusion with the area of 12.5 × 7 μm^2 at the beginning of tensile test. The load–stroke curve during in situ tension process is simultaneously shown in Fig. 4f. The loading direction is almost perpendicular to the TiN inclusion. Before the applied load increased to the point b (before yielding) as shown in Fig. 4f, the TiN inclusion and matrix have almost not changed apparently. When the load reached point b (nearby yielding), the first crack immediately

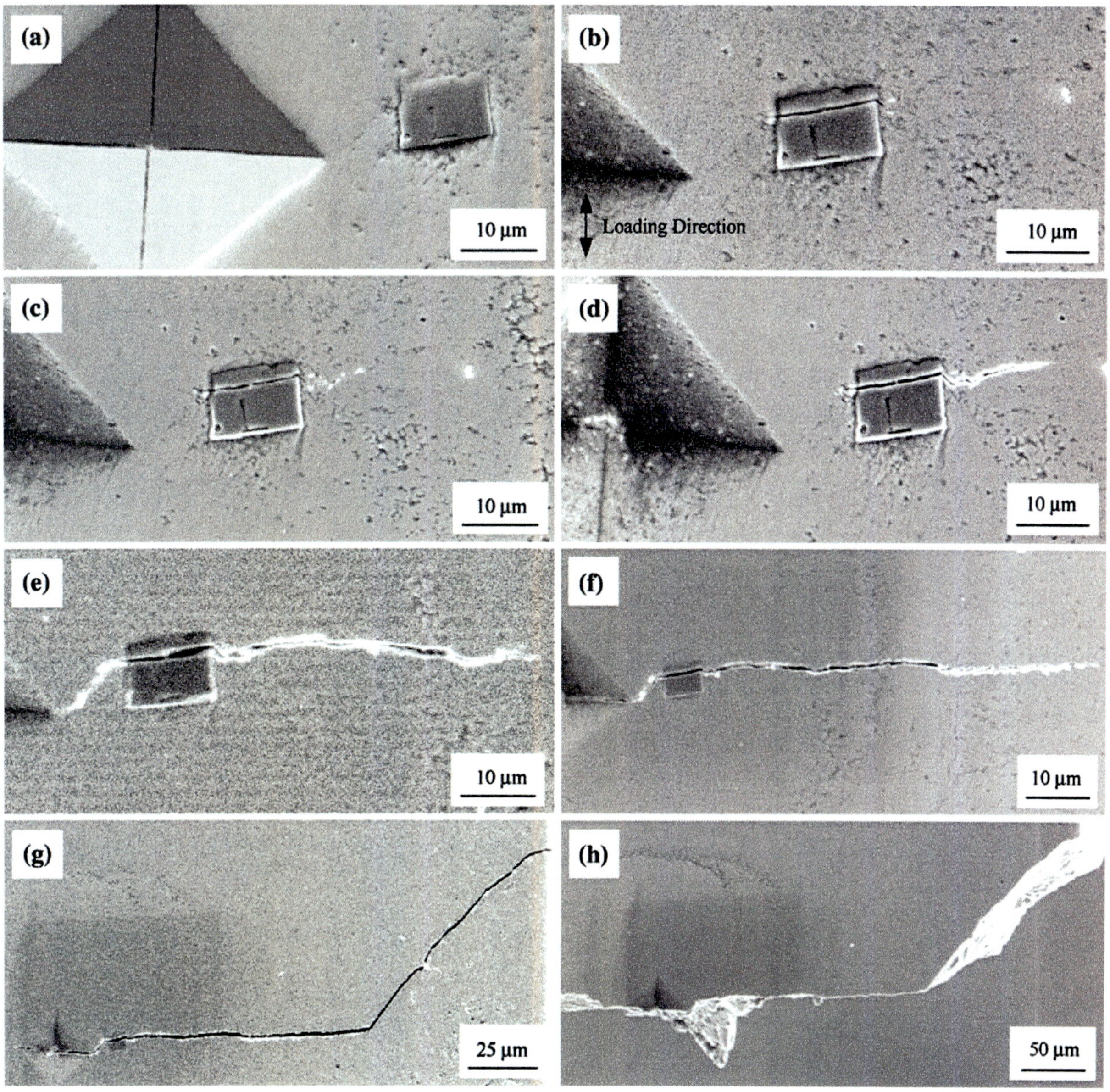

Fig. 6 The micro-mechanical behavior of TiN inclusion during in situ fatigue process. ($\sigma_{max} = 1{,}429$ MPa, $R = 0.1, f = 8$ Hz), fatigue cycles *N* are: **a** 0; **b** 165,445; **c** 214,604; **d** 240,380; **e** 297,476; **f** 383,234; **g** 429,810 and **h** 432,950

occurred in TiN inclusion and developed across whole inclusion (see Fig. 4b). With the increasing of applied load to reach the point c in Fig. 4f, the second crack occurred almost parallel to the first crack as indicated in Fig. 4c. Meanwhile other small cracks occurred which were oriented 45° to the first and second cracks. At this time the whole TiN inclusion were divided into three small pieces (see Fig. 4c). With continuous loading after yielding, the engineering applied stress began to decrease, some obvious slip bands appeared at specimen surface and the cracks became wider than before. Meanwhile the cracks oriented in about 45° to the loading direction (see Fig. 4d) and propagated seriously not only in TiN inclusion but also into matrix (see Fig. 4e). When the engineering applied stress decreased to 1,544 MPa, the specimen failed.

Figure 5 shows the microscopic behavior of several pieces of rectangular TiN inclusions at in situ tension process. The sizes of two major inclusions are about $8 \times 5.5\ \mu m^2$ (upper one) and $10 \times 12\ \mu m^2$ (lower one). The loading direction is indicated as arrows. A very small crack can be observed carefully in the lower TiN inclusion (see Fig. 5a) before loading. It might be formed during the

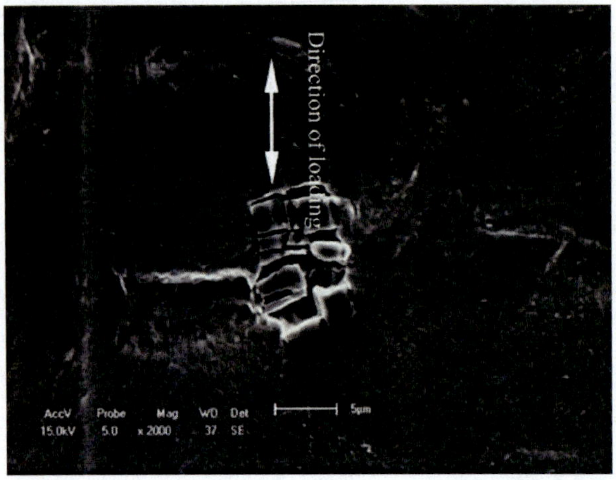

Fig. 7 The micro-mechanical behavior of a TiN inclusion (6 × 10 μm^2) during in situ fatigue process. (σ_{max} = 1,400 MPa, R = 0.1) at N_f = 16,000 cycles

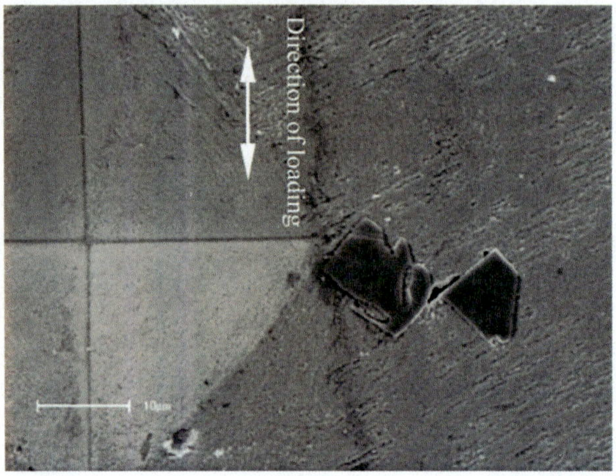

Fig. 8 The micro-mechanical behavior of a couple of TiN inclusions (8 × 10/7 × 9 μm^2) during in situ fatigue process (σ_{max} = 1,411 MPa, R = 0.1) at N_f = 1,411 cycles

deformation of forging process. When the load was below point b (engineering applied stress level at about 0.6σs) as shown in Fig. 5f, the inclusions and matrix did not change apparently. However, the first very short crack (about 2 μm) formed in the lower TiN inclusion even at the engineering applied stress level lower than yielding (see Fig. 5b). This small crack propagated with the increasing of engineering applied stress. When the applied load reached point c (nearby yield stress level) as indicated in Fig. 5f, the first small crack grew to a length of about 6 μm (see Fig. 5c). When the engineering applied stress level was over yielding and reached point d, many cracks formed in the lower TiN inclusion and also two small cracks formed in the upper TiN inclusion (see Fig. 5d), and the deformation behavior in matrix were also clearly observed. When the applied load reached point e before failure, all cracks divided the inclusions into many pieces as shown in Fig. 5e.

Figure 6 shows the micro-mechanical behavior of a typical rectangular TiN inclusion (8 × 19 μm^2) at in situ fatigue process. It can be seen that a small crack in the direction perpendicular to the loading direction had already formed at the forging process (see Fig. 6a). The maximum tensile stress 1,429 MPa is about 0.75 yield stress. The crack firstly propagated into the matrix at 165,445 cycles as shown in Fig. 6b. As the fatigue cycles increased to 214,604 cycles, this crack propagated from both sides into the matrix (see Fig. 6c). Fatigue crack propagated apparently in the matrix at 240,380 cycles, and its length was about 30 μm as shown in Fig. 6d. Fatigue crack length increased with the cycling and its total length reached 68 μm at 297,476 cycles as shown in Fig. 6e. Meanwhile this crack propagated rapidly and its length developed to 180 μm at 383,234 cycles (see Fig. 6f). When fatigue cycles reached 429,810, this long crack continuously propagated to the specimen surface in the direction of 45° (see Fig. 6g). Tested specimen failed at 432,950 cycles with a very long crack initiated directly from inclusion as shown in Fig. 6h.

Figure 7 shows a typical cracking behavior in TiN and at the interface of TiN/matrix. The crack firstly initiated in TiN inclusion and more crack were formed in TiN with increasing of fatigue cycles. The crack could also form at the interface of TiN/matrix. All the cracks are perpendicular to the loading direction as indicated. Figure 8 shows a crack initiated at the interface of TiN/matrix of two small TiN inclusions. This crack is in 45° to the loading direction.

3.2 Micro-Mechanical Behaviour of AlN Inclusion in GE1014 Ultra-High-Strength Steel

From above mentioned experimental results it can be concluded that TiN inclusion in MA250 is very harmful for mechanical properties. For elimination TiN in this kind of ultra-high-strength steel GE Power developed a new ultra-high-strength steel designated as GE1014 without Ti but alloyed with Al (see Table 1). GE1014 characterizes with higher strengths than MA250. From the chemical composition of GE1014 it can imagine that TiN is eliminated but AlN or partial Al_2O_3 inclusions appear.

The main inclusion in GE1014 is AlN (and partial Al_2O_3) with smaller particle size of 3–5 μm in average than TiN with 12–15 μm in MA250 (see Fig. 2). Figure 9 shows the micro-mechanical behaviour of a small size AlN (2 × 5 μm^2) inclusion and load–stroke diagram during in situ tension process. Before yielding no crack initiated in

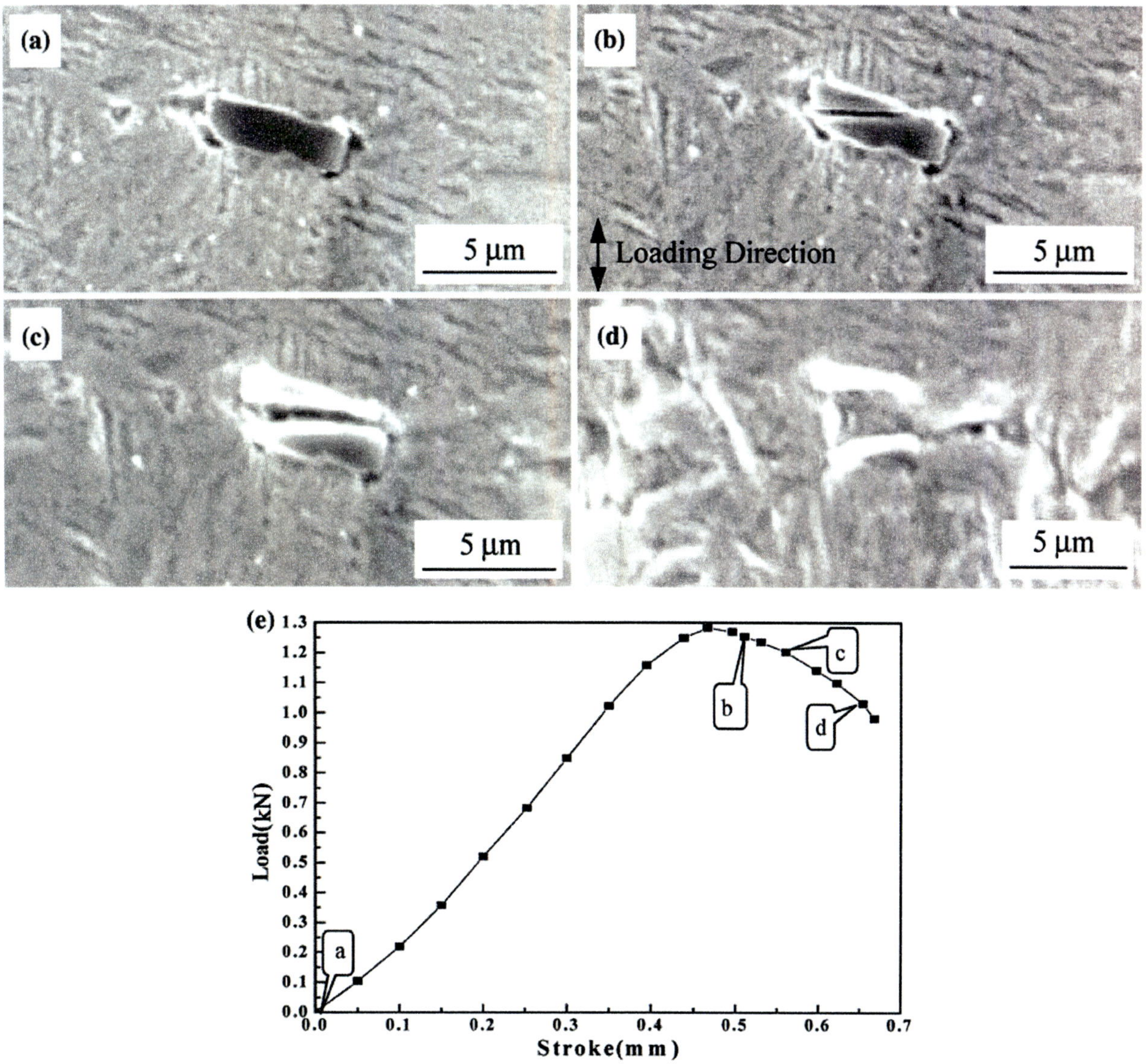

Fig. 9 The micro-mechanical behavior of a small size (2 × 5 μm^2) AlN inclusion and load–stroke diagram during in situ tension process in GE1014

AlN inclusion or at the interfaces of AlN/matrix. However a clear crack perpendicular to the loading direction formed in AlN just after yielding (see Fig. 9b, e at the point of b). This crack quickly developed (see Fig. 9c, d with e at the relevant points c and d). This crack continuously developed till to failure.

Figure 10 shows the micro-mechanical behaviour of a very small size AlN (2 × 4 μm^2) inclusion and load–stroke diagram during in situ tension process. It can be clearly seen that a void already existed before loading at AlN/matrix interface (see Fig. 10a with e) because of the result of hot deformation process. This existed void easily propagated at the interfaces of AlN/matrix (see Fig. 10b, c with e at the point of b and c). In this case existed void from hot deformation process induced further crack propagation till to failure.

In some cases if the small particle size AlN lined up as a chain of inclusions, it will be very harmful for mechanical properties. Figure 11 shows the micro-mechanical behavior of a chain of AlN inclusions with the length of 88 μm during fatigue test ($\sigma_{max} = 1{,}417$ MPa, $R = 0.1$, $f = 5$ Hz). It can be seen that this small crack initiated in the matrix between AlN inclusions as arrow indicated in Fig. 11b ($N = 102{,}556$). With the increasing of cycling ($N = 107{,}645$) this small cracks propagated to the left and right directions among AlN inclusions. This crack can

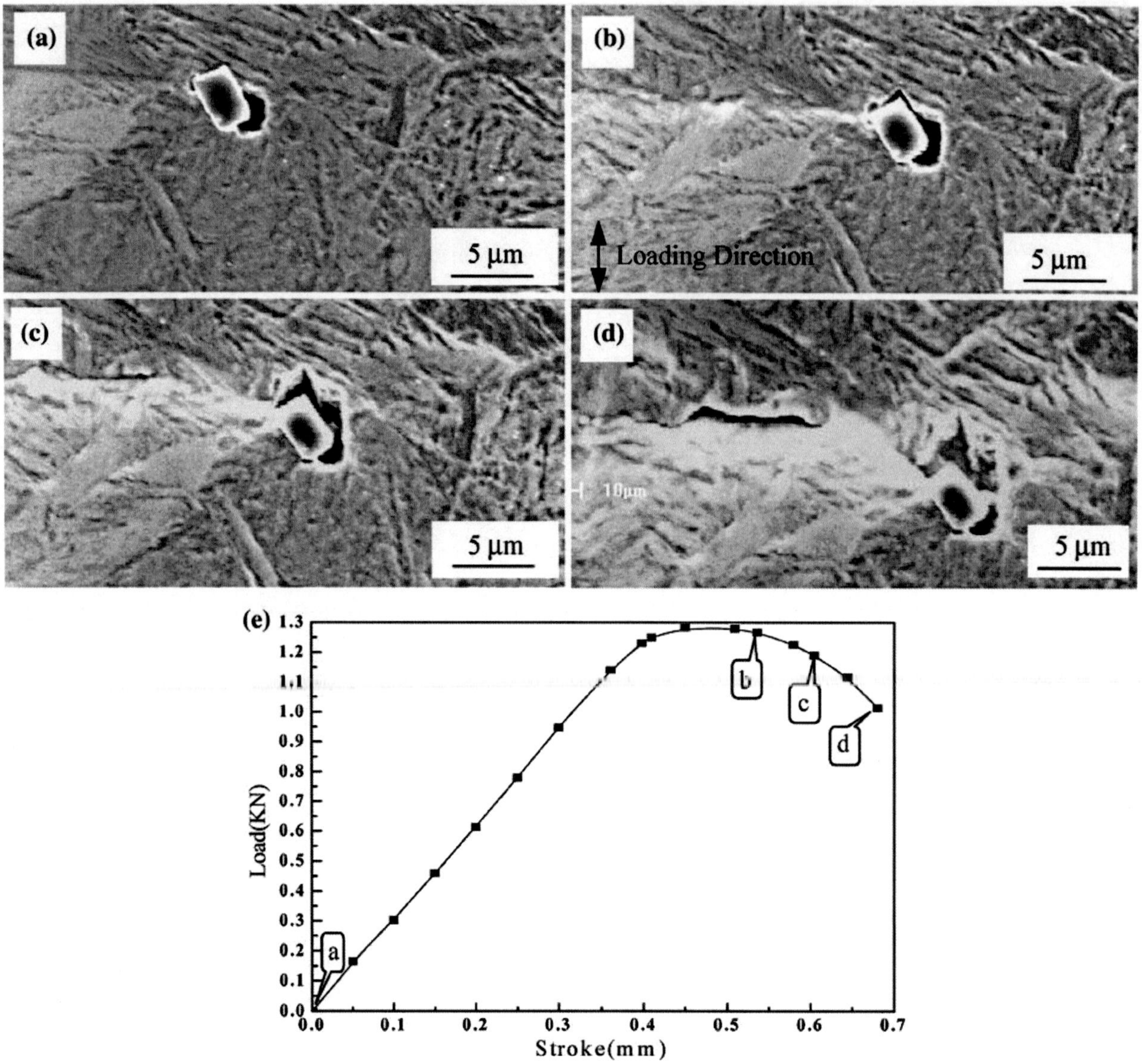

Fig. 10 The micro-mechanical behavior of a small size (2 × 4 μm^2) AlN inclusion and load–stroke diagram during in situ tension process in GE1014

propagate to the matrix and through AlN inclusions as indicated in Fig. 11c. With further increasing of cycling ($N = 112{,}665$) a long crack (~90 μm) formed and propagated to the right side matrix as shown in Fig. 11d and both side matrix as shown in Fig. 11e ($N = 115{,}659$). This long crack continuously propagated to the matrix and crack length reached 165 μm at 126,972 cycles (see Fig. 11f). Furthermore this crack quickly propagated to specimen surface and its length reached 630 μm at 132,996 cycles (see Fig. 11g). Finally this specimen was suddenly failed at 133,006 cycles (see Fig. 11h).

Fractography of failed specimens shows that the fatigue crack initiated as arrow indicated at the inclusions (see Fig. 12a). A chain of inclusions is clearly shown in Fig. 12b at higher magnification.

4 Discussion

Any kind of inclusions are unavoidable in advanced steels because of necessary melting process. Two premium quality super-high-strength steels MA250 and GE1014 both are

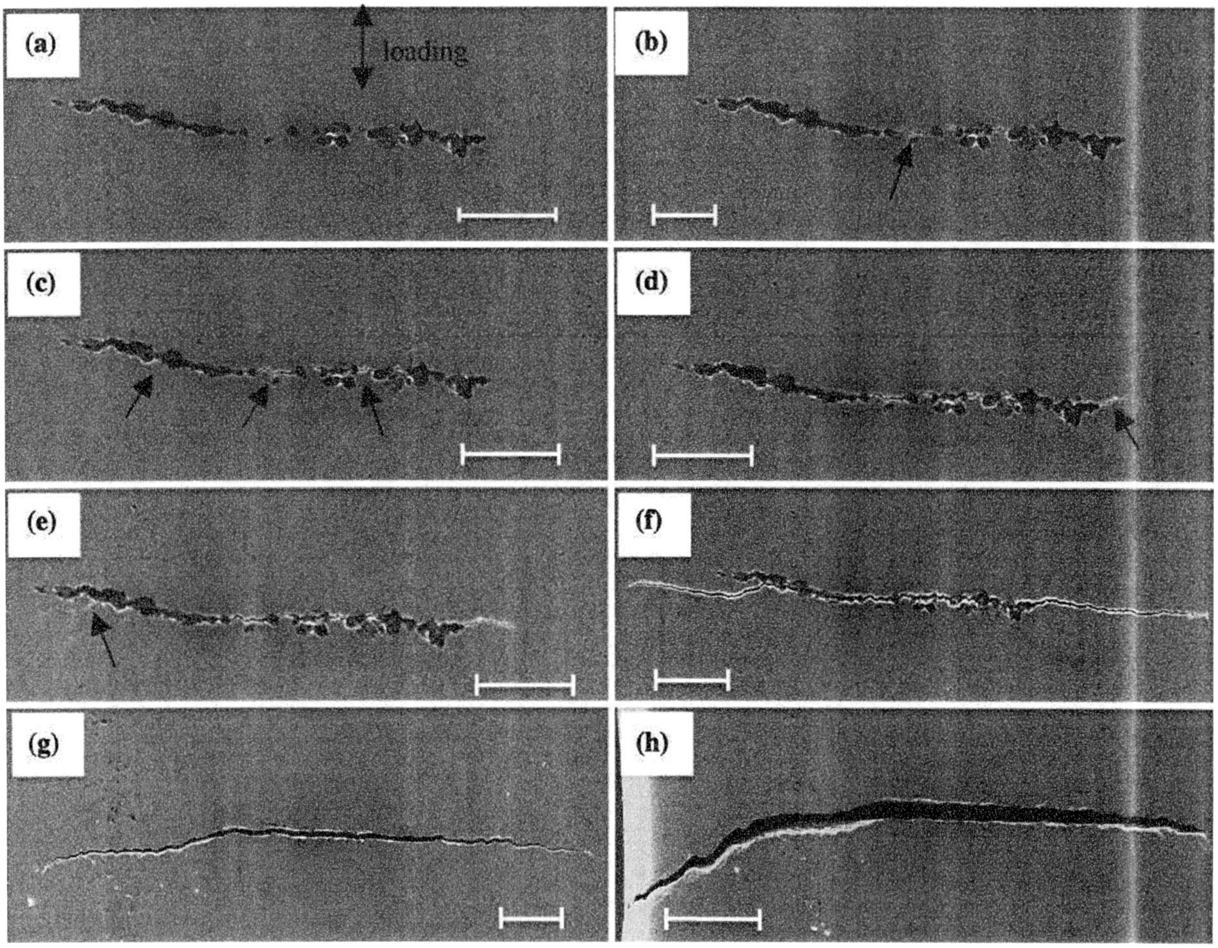

Fig. 11 Initiation and propagation of the crack induced by a chain of AlN inclusions under in situ fatigue process. ($\sigma_{max} = 1{,}417$ MPa, $R = 0.1, f = 5$ Hz), fatigue cycles N are: **a** 0; **b** 102,556; **c** 107,645; **d** 112,665; **e** 115,659; **f** 126,972; **g** 132,996 and **h** 133,006

produced by advanced vacuum melting (VIM + VAR). However, there are still a certain amount of inclusions in these advanced steels. The blocky TiN inclusion characterized with almost cubic morphology is very often in average size large than 10 μm in MA250. For elimination of this big size TiN inclusion in MA250 a new ultra-high-strength steel GE1014 (without Ti, but alloying with Al) with higher strength than MA250 was developed. Instead of large size TiN inclusion in MA250 the small particle AlN inclusion normally is in an average size of 5 μm or even less.

From above mentioned in situ tension and fatigue experimental results it can be concluded that inclusions such as TiN in MA250 and AlN in GE1014 both are very harmful for mechanical properties. At tensile process crack can initiate in TiN before yield and also in AlN after yielding. Cracks in TiN can easily propagate into matrix with the increasing of stress at tensile test or with the stress cycling at fatigue test. Cracks can also initiate at the interface of TiN/matrix. From view point of mechanical properties degradation developed by TiN inclusion, the fraction and size of TiN inclusions should be controlled to lower fraction and smaller size level.

The degradation effect of small size (~5 μm) AlN on mechanical properties in GE1014 is lower than that of large size (≥10 μm) TiN in MA250. However, a chain of small AlN inclusions can be formed in GE1014 because of some reasons from melting and hot deformation. This kind of small size AlN inclusion chain is very harmful on mechanical properties degradation. Cracks are very easily initiated among these small AlN inclusions and line up as a long crack. This long crack quickly propagates to the matrix till failure. It is important that the AlN inclusion chain in GE1014 should be eliminated by advanced metallurgical processing.

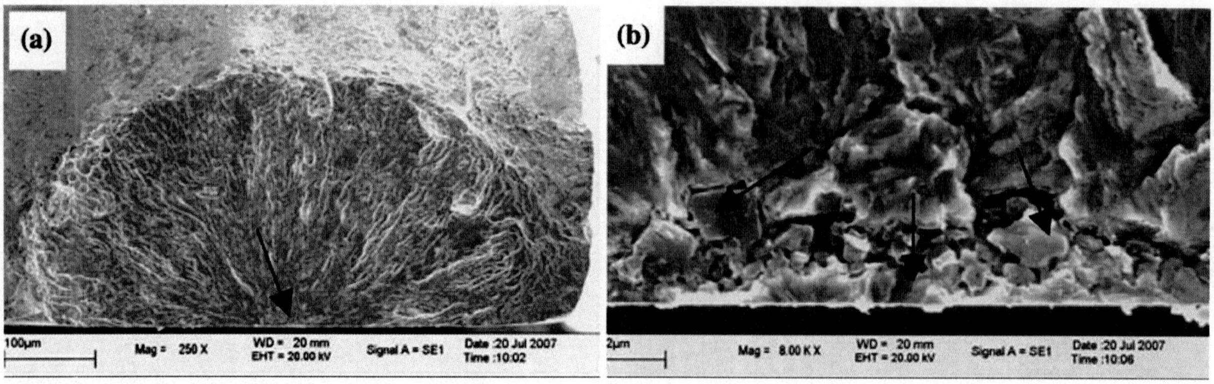

Fig. 12 Fractography of the failed sample shown in Fig. 11

5 Conclusions

(1) TiN and AlN both are typical inclusions in ultra-high-strength steels MA250 and GE1014 respectively. These inclusions are very harmful not only for tensile but also for fatigue properties.

(2) Cracks can initiate in the large size TiN inclusion or at the interfaces of TiN/matrix and also easily initiate at a chain of small AlN inclusions to form a large size crack. These cracks can easily propagate into matrix with the increasing of stress at tensile test or with the stress cycling at fatigue test.

(3) For improvement of ultra-high-strength steel quality the fraction and size of TiN inclusion should be controlled to a lower and smaller size levels in MA250 steel and the AlN inclusion chain should be avoided by advanced metallurgical processing in GE1014.

Acknowledgments Authors would like to thank GE Aircraft Engines, USA for support this project and special thanks to Drs. Mark Rhoads and Jon Groh for materials offering and generous discussions on this project.

References

1. K. Tanaka, T. Mura, Metall. Trans. A **1**, 117 (1982)
2. B. Hong, J. Inner Mongolia Polytechnic Univ. **4**, 14 (1977) (in Chinese)
3. Q.Y. Wang, C. Bathias et al., Int. J. Fatigue **24**, 1269 (2002)
4. S.X. Li, China Basic Sci. **4**, 14 (2005)
5. X.S. Xie, L.N. Zhang et al., in *Superalloys 2004, TMS* (2004) p. 451
6. Y.P. Zeng, J.X. Dong et al., J. Univ. Sci. Technol. Beijing **2**, 202 (2005) (in Chinese)
7. J.M. Zhang, J.F. Zhang et al., Mater. Sci. Eng. A **394**, 126 (2005)

Dislocation Assisted Phase Transformation Observed in Iron Alloys

Yoon-Uk Heo, Masaki Takeuchi, Kazuo Furuya, and Hu-Chul Lee

Abstract

This study examined the phase transformation of ordered intermetallic precipitates to thermodynamically stable austenite in iron alloys assisted by a dislocation glide or climb. In Fe-Mn-Ni alloys, ordered fct θ-MnNi precipitates formed in the martensite grains and at the lath and grain boundaries during aging after quenching. With further aging, stacking faults and twins formed in the θ-MnNi particles by the glide of Shockley-type 1/6<112> partial dislocations. The crystal structure of the twins was similar to the face centered cubic structure, and these twins transformed to austenite as a result of iron diffusion into the twins. In Fe–Ni–Ti alloys, austenite nucleated first at the interface of the ordered hexagonal η-Ni_3Ti intermetallic precipitates and martensite matrix. However, with further aging, an ordered face centered cubic (fcc) γ'-Ni_3Ti phase formed from the η-Ni_3Ti phase by the dislocation climb of 1/4<0001> type edge dislocations. The final transformation to austenite occurred by the diffusion of iron into the γ'-Ni_3Ti phase.

Keywords

Phase transformation • Ordered intermetallic phase • Twinning • Dislocation climb

1 Introduction

In many alloy systems, metastable transition phases appear before the precipitation of the thermodynamically equilibrium phase during aging of supersaturated solid solutions. Precipitation of the GP zone, θ'' and θ' phases before the precipitation of an equilibrium θ phase in aluminum-copper alloys is a well known example [1]. In iron–carbon alloys, hexagonal ε- or orthorhombic η-carbide appears before the precipitation of cementite during tempering at lower temperatures. These transition phases are kinetically favored phases and finally transform to an equilibrium phase with further tempering treatments. The interface between the transition phase and metal matrix is the most favorable nucleation site for the equilibrium phase. One typical example of such a phase transition in iron alloys is the transition of η-Fe_2C carbide to equilibrium cementite Fe_3C carbide. Nakamura and co-workers reported that θ'-carbide (phase mixture of κ- and θ-carbides) nucleates at the interface of η-carbide and -martensite matrix, and a final transformation also occurs by the nucleation of equilibrium θ-carbide at the interface of θ'-carbide and -martensite matrix [2, 3]. In this case, the interface of martensite matrix and carbides are not coherent and the interface can be a favorable heterogeneous nucleation site for other carbide phases. However, when the interface between the transition phase and matrix is coherent or semi-coherent, the interface may not be energetically favorable for nucleation or the coherency strain would perform an active role in the transition behavior of the precipitates.

Y.-U. Heo · H.-C. Lee (✉)
Graduate Institute of Ferrous Technology, Pohang University of Science and Technology, Hyoja-dong San 31, Nam-gu, Pohang, 790-784, Korea
e-mail: huchul@postech.ac.kr

M. Takeuchi · K. Furuya
National Institute of Materials Science, 3-13 Sakura, Tsukuba, 305-0003, Japan

Y. Weng et al. (eds.), *Advanced Steels*, DOI: 10.1007/978-3-642-17665-4_13,

This article presents two examples of non-conventional transformation of transition precipitates to stable austenite assisted by dislocation glide or dislocation climb.

2 Transformation of Ordered Face Centered Tetragonal ($L1_0$) θ-MnNi Phase to Austenite

Fe-Mn-Ni alloys have an excellent age hardening response during aging after quenching. Alloy hardening during aging was attributed to the precipitation of an intermetallic θ-MnNi phase in the martensite matrix [4]. However, the presence of θ-MnNi precipitates at the grain boundaries reduces the grain boundary strength and aged alloys showed no tensile ductility, even after a very short aging treatment [5, 6]. However, after extended aging, the θ-MnNi precipitates transform to an austenite phase and the tensile ductility recovers [7, 8].

At the early stage of the transformation, many fine bands or faults developed in the θ-MnNi particles, as shown in Fig. 1. The high angle annular dark field (HAADF) image of these bands in Fig. 2 show that these bands are stacking faults or twins that developed in θ-MnNi particles by the glide of Shockley-type 1/6<112> partial dislocations. Stacking faults mainly form in double layers. A θ-MnNi intermetallic phase is an anti-ferromagnetic material where the spin moments of nearest manganese are anti-parallel to each other. 1/6<112> partial slips brings manganese atoms with a parallel spin moment to the nearest site. This situation can be avoided by the formation of a double stacking fault. The crystal structure of non-cubic crystal twins is not necessarily identical to that of the matrix. The c/a ratio of the θ-MnNi crystal is 0.94. However, the patterns obtained from the twin were almost identical to those of austenite. The c/a ratio of the twin crystal was determined to be 1.02. The crystal structure of the twin is similar to that of austenite despite the fact that atomic arrangement in the twin is still ordered.

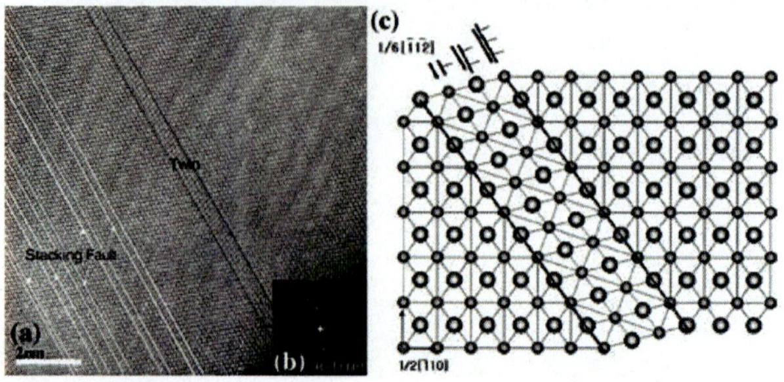

Fig. 2 STEM HAADF image of a twin and stacking faults developed in a θ-MnNi crystal. **a** STEM HAADF image of a twin and stacking faults, **b** FFT pattern of the image, **c** atomic model of a twin crystal

Figure 3 shows the electron energy loss spectra (EELS) of a twin band and surrounding matrix. Iron peaks can be observed in the EELS spectra of the edge of the twin, positions 1 and 2. However, no iron peak was observed in the EELS spectra of the matrix and the center of the twin band. The θ-MnNi phase does not have iron solubility. However, the EELS spectra clearly show that iron diffusion occurs in the twin. As discussed in the previous paragraph, the crystal structure of the twin is similar to that of austenite (fcc) and EELS analysis revealed iron diffusion into the twin. Iron diffusion into the twin crystal will certainly change the twin crystal to austenite. In other words, twin bands transform to austenite bands according to the

Fig. 1 Development of stacking faults and twin bands in θ-MnNi precipitates after aging for 12 h at 440°C

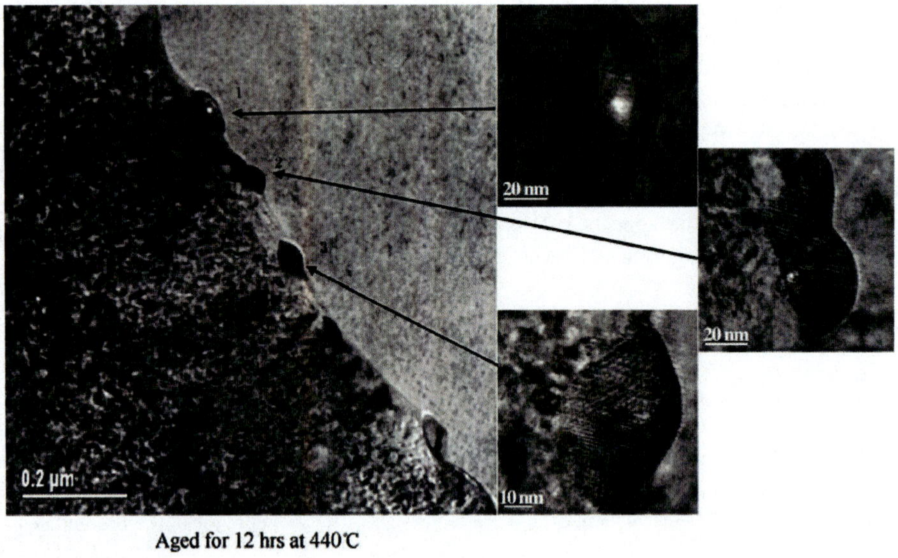

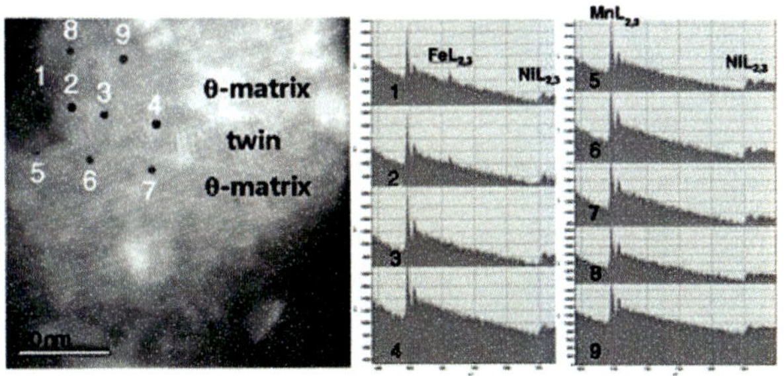

Fig. 3 STEM image of twin bands and EELS spectra of a twin band and matrix. The positions of EELS analysis are shown in the STEM image

diffusion of iron into the twin bands. In the later stages of aging, extensive development of twin bands and austenite bands were observed in θ-MnNi particles and the final transformation of θ-MnNi particles to austenite particles occurs by the coalescence of these austenite bands.

3 Transformation of Ordered Hexagonal (DO_{24}) η-Ni_3Ti to Austenite

Fe-Ni-Ti alloys are another type of age hardening iron alloy. These alloys harden by the precipitation of an ordered hexagonal η-Ni_3Ti intermetallic phase in the matrix. The η-Ni_3Ti phase is not a thermodynamically stable phase, and η-Ni_3Ti particles transform to a more stable austenite phase after extended aging [9, 10]. Austenite nucleates at the interface of η-Ni_3Ti precipitates by a diffusion process [11]. Figure 4a shows the growth of austenite at both sides of a η-Ni_3Ti particle after aging at 530°C for 8 h. The {111} plane of austenite grew parallel to the (0001) plane of the η-Ni_3Ti phase, as shown in Fig. 4b, and the atomic misfit in these two planes was only 0.55%. They form a coherent interface, and coherency stresses is expected to develop at the interface with the growth of austenite at the interface.

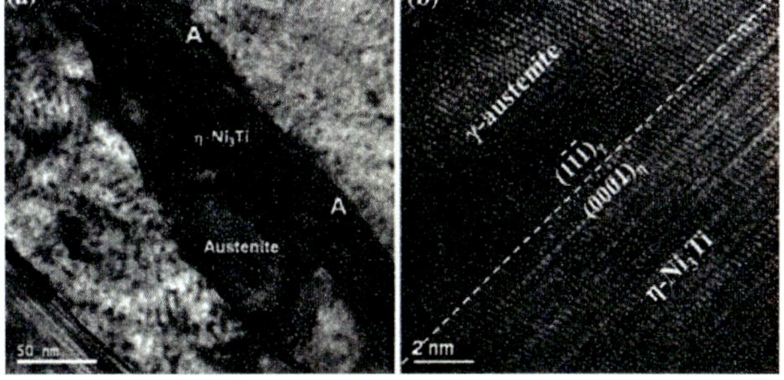

Fig. 4 Growth of austenite at the interface of an η-Ni_3Ti precipitate (530°C, 8 h aging)

Figure 5 shows a high-resolution electron micrograph of the austenite–η-Ni_3Ti interface area after tempering at 530°C for 164 h. The HRTEM image shows a new phase formed between the austenite and η-Ni_3Ti phase. The fast Fourier transformation (FFT) of the images in the white squares in austenite and the new phase are given on the left side of the micrograph. A comparison of the FFT diagram from the new phase with that of austenite showed that the new phase has an ordered cubic ($L1_2$) structure. The lattice parameter of $L1_2$ is similar to that of austenite, which suggests that the new $L1_2$ phase is a γ'-Ni_3Ti phase, another metastable form of η-Ni_3Ti crystal [8]. The HRTEM image in Fig. 5 shows that the η-Ni_3Ti phase at the lower end of the γ'-Ni_3Ti band does not fully transform to a γ'-Ni_3Ti crystal, typical modulation texture of η-Ni_3Ti crystal still remains in this area. Figure 6 presents another HRTEM image of this area. Figure 6a shows numerous defects in this crystal (circled area). An analysis of the reconstructed image using (0004) and (000-4) diffraction beams in Fig. 6b shows that these defects are edge dislocations with a Burgers vector of 1/4 [0001]. The climb of this type of dislocation forms intrinsic stacking faults by removing one layer of the (0001) plane from the DO_{24} η-Ni_3Ti crystal.

The stacking sequence of (0001) planes in a DO_{24} crystal is *ABACABAC* type, as shown in Fig. 7a. By removing one type of (0001) layer (A-layer in Fig. 7b), the stacking sequence of the close packed planes can change to the *ABCABC* type, which is the sequence of a face centered cubic (fcc) crystal. The result of the dislocation climb is the formation of an ordered fcc γ'-Ni_3Ti crystal. By removing another type of A-layer, the packing sequence can be changed to *CBACBA*, which are in a twin relation with each other. The γ'-Ni_3Ti crystal forms in

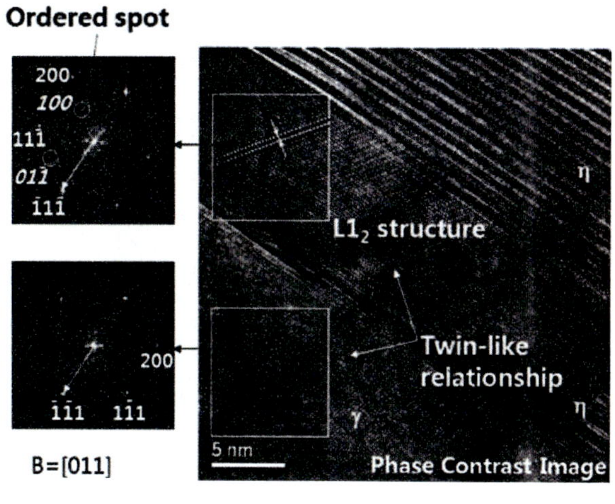

Fig. 5 Formation of a γ'-Ni_3Ti phase at the interface of η-Ni_3Ti and austenite and FFT image from the γ'-Ni_3Ti (*upper diagram on the left*) and austenite phase (*lower diagram*)

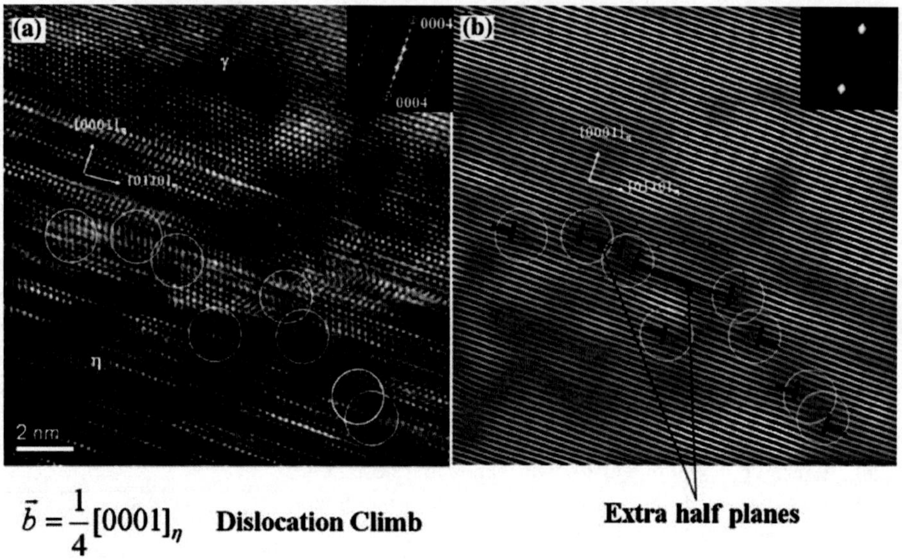

Fig. 6 Dislocation climb in the η-Ni_3Ti phase. **a** Phase contrast image, **b** reconstructed image with (0004) and (000-4) beams

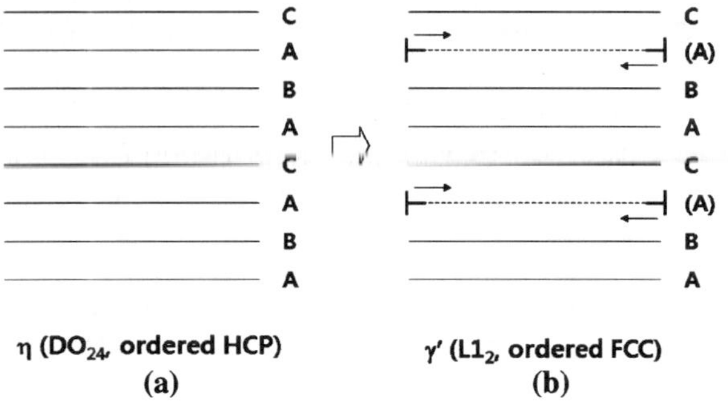

Fig. 7 Formation of a $L1_2$ structure from a DO_{24} crystal by the removal of (0001) planes due to a dislocation climb

either an identical or twin orientation with the neighboring austenite phase. For the final transformation of the γ'-Ni_3Ti crystal to austenite, iron diffusion into the γ'-Ni_3Ti crystal is expected. Figure 8 shows the EELS spectra of the position marked in the η-, γ'-, and γ phase. The EELS spectra from the positions 7, 8 and 12, 13 shows no trace of iron diffusion into the η-Ni_3Ti crystal. The EELS spectra from the γ'-Ni_3Ti crystal showed a difference. The EELS spectra of positions 10 and 11 clearly shows the presence of iron in this crystal. However, the EELS spectrum of position 5 shows no trace of iron and inconclusive results from position 4.

The presence of a η-phase between the austenite and γ'-Ni_3Ti may prevent iron diffusion from the austenite to

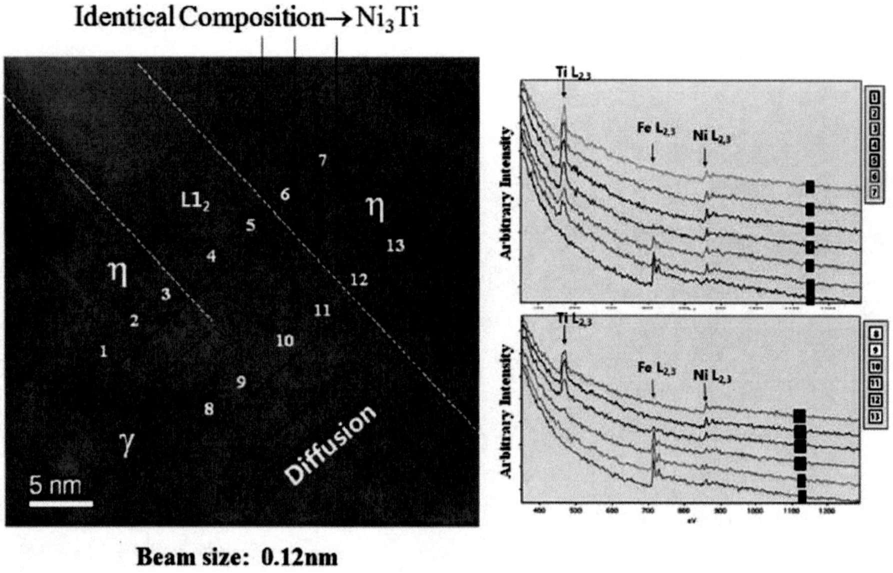

Fig. 8 EELS spectra of the positions marked in the η-, γ'-, and γ-phase

γ'-phase. The iron peaks in the EELS spectra from positions 1, 2 and 3 may have originated from austenite surrounding the η-phase. With the diffusion of iron into the γ'-Ni_3Ti crystal, the characteristic superlattice spots disappear from the diffraction patterns of the γ' crystal, which means that this γ'-Ni_3Ti band can transform to an austenite band.

4 Summary

Two different types of phase transformation assisted by the dislocation are summarized in this report. In the first case of θ-MnNi precipitates, the glide of 1/6<112> type Shockley partial dislocations forms a twin band in the ordered fct θ-MnNi particles and iron diffusion occurred into these twin bands to transform the twin bands to an austenite band. The final transformation of the particles occurs by the coalescence of these twin bands. In the second case, the climb of the 1/4<0001> type edge dislocations transforms the ordered hexagonal η-Ni_3Ti crystal to an ordered fcc γ'-Ni_3Ti crystal. Iron diffusion into the γ'-Ni_3Ti crystal transforms this crystal to fcc austenite.

References

1. D.A. Porter, R.E. Eastering, *Phase Transformations in Metals and Alloys*, 2nd edn. (Chapman & Hall, London, 1993), p. 291
2. Y. Nakamura, S. Nagakura, Trans. JIM **27**, 842 (1986)
3. Y. Nakamura, T. Mikami, S. Nagakura, Trans. JIM **26**, 876 (1985)
4. M. Tanaka, T. Suzuki, M. Yodogawa, J. Jpn. Inst. Metals **31**, 1075 (1967)
5. S. Hussein Nedjad, M. Nili Ahmadabadi, T. Furuhara, Mater. Sci. Eng. A **490**, 105 (2008)
6. H.-C. Lee, S.H. Mun, D. Mckenzie, Metall. Trans. A **34**, 2421 (2003)
7. S. Hussein Nedjad, M. Nili Ahmadabadi, R. Mahmudi, T. Furuhara, T. Maki, Mater. Sci. Eng. A **438–440**, 288 (2006)
8. Y.-U. Heo, M. Kim, H.-C. Lee, Acta Mater. **56**, 1306 (2008)
9. D.M. Vanderwalker, Metall. Trans. A **18A**, 1191 (1987)
10. Y.-U. Heo, M. Takeguchi, K. Furuya, H.-C. Lee, Acta Mater. **57**, 1176 (2009)
11. M.-S. Kim, Y.-U. Heo, H.-C. Lee, Solid State Phenom. **118**, 469 (2006)

Solution and Precipitation of Secondary Phase in Steels: Phenomenon, Theory and Practice

Qilong Yong, Xinjun Sun, Gengwei Yang, and Zhengyan Zhang

Abstract

The secondary phases exert strong influence on the mechanical properties, processing properties and performance of steels. The volume fraction, size, shape and distribution of the secondary phases must be effectively controlled in order to obtain the excellent properties and performance. On the basis of equilibrium solubility product formula, precipitation thermodynamics and kinetics, and classical nucleation and growth theory, the theoretical calculation method and procedure for the PTT (precipitation fraction–time–temperature) curve and NrT (nucleation rate–temperature) curve for main secondary phases precipitated in austenite and in ferrite have been proposed. The calculation results are well in agreement with practical experimental results. By analyzing the calculation results, some important deductions for the practical control of secondary phases in steels have been made.

Keywords

Secondary phase • Control • Precipitation • Theoretical calculation

1 Introduction

Corresponding with the term of primary phase (that is matrix) we preferred to use the term of secondary phase rather than second phase (or else there would be third phase, fourth phase and so on). Moreover, the secondary phases here include all the phases beside matrix phase. No matter the secondary phase itself is of beneficial or not, the effect of secondary phase on the steel properties is determined by the size, fraction, shape and distribution of secondary phase [1–4]. For example, under the adequate controlled conditions, the MnS phase that is traditionally considered as harmful inclusion in steel can be uniformly distributed and refined to about 50 nm thus being converted to beneficial secondary phase.

The matrix microstructure has essential effect to the properties and performance of the steel material and has been drawn the attention of many researchers and engineers and achieved very important progress in recent years. Matrix grains has been refined to 1–5 μm and grain refinement has been considered one of the most important strengthening and toughening mechanisms in engineering structural steels [5]. However, further refining the grain size has faced the challenge of strengthening and toughening efficiency and production technique difficulty. On the other hand, the secondary phases have shown to play important roles to the mechanical property, fracture characteristic, fatigue behavior of the steel. Very fine secondary phases can give birth to very strong strengthening effect while do not impair the uniform plasticity and only induce somewhat lower harmful effect to toughness compared to the other strengthening methods.

The secondary phases stable at high temperature can effectively retard matrix grain growth so that exert important effect to grain refinement. Thus, the secondary phases and their control technique would be the most important development trend for upgrading the properties and performance of the steels [1].

Q. Yong (✉) · X. Sun · G. Yang · Z. Zhang
Central Iron and Steel Research Institute, Beijing 100081, People's Republic of China
e-mail: yongql@126.com; yongqilong@nercast.com

Y. Weng et al. (eds.), *Advanced Steels*, DOI: 10.1007/978-3-642-17665-4_14,

2 The Role of Secondary Phases in Steels

The main beneficial roles played by secondary phases in steels can be summarized as follows.

The secondary phase undissolved in matrix while maintaining enough volume fraction and enough fine size at high temperatures such as soaking or reheating temperatures can effectively inhibit the coarsening of matrix grains so as to provide the basis for grain refinement. The grain size is proportional to the volume fraction of secondary phase while inversely proportional to the size of secondary phase. Therefore, increasing the volume fraction and reducing the size of secondary phase would be benefiting to grain refinement [6].

The secondary phase strain-induced precipitated in hot rolling process can effectively retard the recrystallization of deformed austenite grains thus benefit the non-recrystallization control rolling process. Also, increasing the volume fraction and reducing the size of secondary phase are beneficial [1–3, 5].

The very fine secondary phase strain-induced precipitated in austenite or precipitated in ferrite can produce strong precipitation strengthening. The precipitation strengthening increment is proportional to the square root of the volume fraction of secondary phase while approximately inverse proportional to the size of secondary phase, thus, increasing the volume fraction and reducing the size of secondary phase benefit the precipitation was strengthening. When the size of secondary phase in the range of 2–5 nm, the precipitation strengthening increment would be more than hundreds MPa even if the volume fraction of secondary phase is $\sim$0.1% (the addition amount of secondary phase forming element less than 0.1%). It is obvious that this strengthening method is very effective. It is worth mentioning that the precipitation strengthening of very fine and uniformly distributed secondary phase would not impair the uniform plasticity of the material [1–3].

The formation of carbide or nitride can fix the carbon or nitrogen atoms thus to avoid the harmful effect of interstitial solutes on deep-drawing property (uncontinuous yield) of IF steels or to avoid the harmful effect of intergranular corrosion in stainless steels. For complete fixing all the interstitial solutes, the amount of carbide or nitride forming elements must somewhat be larger than stoichiometric ratio [7]. On the other hand, the re-dissolution of carbides would release the carbon element to produce the bake hardening effect in BH steels [8].

The secondary phases which size is in 0.2–0.8 μm can promote the formation of intragranular ferrite thus to promote the ferrite grain refinement and to reduce the harmful effect of secondary phases on plasticity and toughness [9]. The later role is very much important.

The properly coarse and uniformly distributed hard secondary phases can evidently increase the wear-resistance of the material. The effect is proportional to the volume fraction of secondary phases while the size of secondary phases possesses somewhat complex relation with the wear resistance.

Some secondary phases have special effect to the performance of material, such as the antiseptic role of copper-rich phase and/or silver-rich phase, the lubricate role and chip breaking role of sulfide or sulfur-bearing phase, the antifriction and noise reduction role of graphite, and so on.

On the other hand, the secondary phases have some negative effects especially to the plasticity and toughness of steels.

The coarse secondary phases could be the source of micro crack, and the size of micro crack is determined by the size of the secondary phase particle [10]. Only the micro crack whose size is greater than critical size tends to propagate and finally leads to the rupture. Therefore, to control the size of largest secondary phase particle to less than an adequate value (for 400 MPa grade steels about 100 μm, for 1,000 MPa grade steels about 20 μm, and for 2,000 MPa grade steels about 10 μm) is essential for raising the fracture strength and the fatigue lifetime of steels. Furthermore, in order to restrain the harmful effect of coarse secondary phases, the shape of large secondary phase particles must be also controlled to optimal spherical shape thus to avoid having sharp edges. Moreover, the distribution of large secondary phases must be uniform avoiding banding or concentration on grain boundaries.

Low melting temperature secondary phases which tend to be in fusion state in hot rolling process and in heat treatment process causing hot embrittlement of the steels, thus they are never permitted to exist in steels in a noticeable quantity.

Some kinds of secondary phases such as the sigma phase in stainless steels, iron phosphide eutectic phase in structural steels, which remarkably damage the plasticity and toughness of steels, must be eliminated.

3 The Precipitation Parameters and Precipitation Theory

To obtain adequate volume fraction of secondary phase and to control the average size of precipitates, we must well understand the precipitation thermodynamics and kinetics.

3.1 Solubility or Solubility Product Formulae and Precipitation Free Energy of Secondary Phases

The equilibrium precipitation amount can be calculated from the solubility or solubility product formula of the secondary phase in matrix. The solubility or solubility

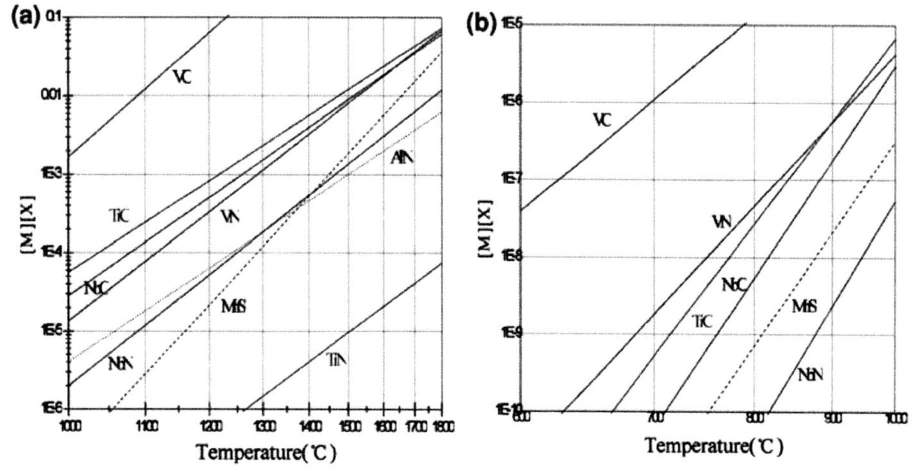

Fig. 1 The comparison for solubility products of common secondary phases in austenite (**a**) and in ferrite (**b**)

product formula of the secondary phase can be represented as follows:

$$\log[C] = A - B/T(\text{for single-element secondary phase}) \tag{1}$$

$$\log\{[M]\cdot[X]\} = A - B/T \quad (\text{for double} - \text{element secondary phase}) \tag{2}$$

where [M] or [X] is the equilibrium solubility in mass percent of element M or X, respectively; T is absolute temperature in K; A and B are specific constants for specific kind of secondary phase in specific matrix (austenite or ferrite).

Large number of the solubility or solubility product formulae for various secondary phases in austenite or in ferrite determined by using phase analysis experiments or by thermodynamically theoretical derivation is available. The commonly applied formulae for frequently used carbides and/or nitrides in austenite and/or in ferrite are plotted in Fig. 1 where the solubility product is in logarithmic coordinate while the temperature is in °C coordinates.

The equilibrium soluble amount at certain temperature of single element (besides Fe) as secondary phase as ε-Cu, Fe_3C, and Fe_4N could be calculated directly from the solubility formula. The equilibrium soluble [M] and [X] of double-element secondary phase MX could be calculated from the solubility formula and stoichiometric ratio equation as follows:

$$\log\{[M]\cdot[X]\} = A - B/T \tag{3}$$

$$\frac{w_M - [M]}{w_X - [X]} = \frac{A_M}{A_X} \tag{4}$$

where w_M and w_X are the mass percent of element M and X in steel, respectively, and A_M and A_X are the atomic weight of element M and X, respectively.

The solubilities of [M], [C], and [N] and the chemical formula coefficient x of triple-element secondary phase MC_xN_{1-x} formed from two completely mutually soluble double-element secondary phases with one common element can be calculated from the simultaneous solubility formulae and the stoichiometric ratio equations as follows:

$$\log\left\{\frac{[M]\cdot[C]}{x}\right\} = A_1 - B_1/T \tag{5}$$

$$\log\left\{\frac{[M]\cdot[N]}{1-x}\right\} = A_2 - B_2/T \tag{6}$$

$$\frac{w_M - [M]}{w_C - [C]} = \frac{A_M}{xA_C} \tag{7}$$

$$\frac{w_M - [M]}{w_N - [N]} = \frac{A_M}{(1-x)A_N} \tag{8}$$

Here it is supposed that the elements M, C, and N only either exist in solid solution or in MX or MC_xN_{1-x} phases without forming another phase.

If the steel was soaked at temperature T_H and then cooled to the temperature T and holding for precipitation, the mole free energy ΔG_M for precipitation of MC or MC_xN_{1-x} phase can be calculated as follows:

$$\begin{aligned}\Delta G_M &= \ln 10\cdot RT(A - B/T) - \ln 10\cdot RT\log([M]_H\cdot[C]_H)\\ &= -19.1446\{B - AT + T\log([M]_H\cdot[C]_H)\}\end{aligned} \tag{9}$$

$$\begin{aligned}\Delta G_M = &-19.1446\{xB_1 + (1-x)B_2\\ &-[xA_1 + (1-x)A_2]T + T\log\left([M]_H\cdot[C]_H^x\cdot[N]_H^{1-x}\right)\}\end{aligned} \tag{10}$$

where R is the ideal gas constant 8.31441 J/mol, $[M]_H$, $[C]_H$ and $[N]_H$ are the solubilities (in mass percent) of M, C and

Fig. 2 The variation of chemical formula coefficient x (**a**) and mole free energy ΔG_M (**b**) with temperature for 0.10%C–0.05%Nb steels for NbC_xN_{1-x} precipitated in austenite

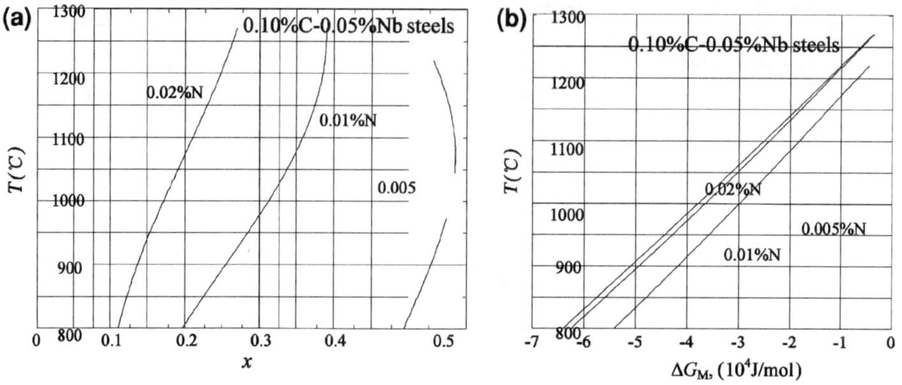

Fig. 3 The variation of chemical formula coefficient x (**a**) and mole free energy ΔG_M (**b**) with temperature for 0.10%C–0.10%V steels for VC_xN_{1-x} precipitated in austenite

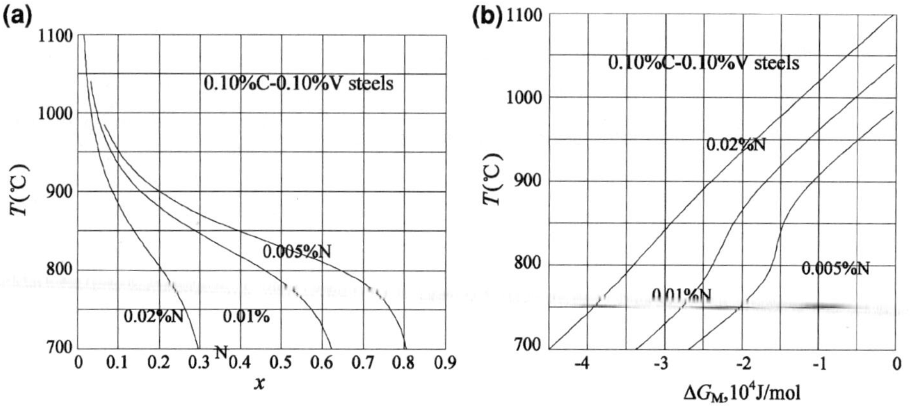

N elements at temperature T_H, respectively. If T_H is higher than the full solution temperature for MC or MC_xN_{1-x} phase, $[M]_H$, $[C]_H$ and $[N]_H$ must be replaced by the contents (in mass percent) of elements M, C and N in the steel.

Respective calculation and experiment results have shown that, the variation of chemical formula coefficient x with temperature for Nb (C,N) precipitation in austenite is comparatively small (commonly not surpassing 0.2, see Fig. 2a) as the solubility values of NbC and NbN are very close (see Fig. 1a). However, the variations of x with temperature for Ti (C,N) and V (C,N) precipitation in austenite are marked. In the higher temperature range, x-values are close to 0 that means the precipitates are approximate to TiN or to VN. When it is in the lower temperature range, x-values are comparatively large that means the precipitates are close to TiC or to VC (see Fig. 3a). The phenomenon is a consequence of that the solubility values of TiC and TiN or those of VC and VN differentiate markedly (see Fig. 1a). On the other hand, because of the effect of mixing entropy, the mole free energy ΔG_M for precipitation of Nb (C,N) in austenite is almost in linear relation with temperature (see Fig. 2b); while in the case Ti (C,N) or V (C,N), the relations are somewhat complex, i.e. in some temperature ranges evidently deviated from strait line (see Fig. 3b).

3.2 The Orientation Relationship between Secondary Phase and Matrix and the Specific Interfacial Energy

In the nucleation and growth process, the orientation relationship between secondary phase and matrix has been maintained, such as the parallel relationship (cube on cube) between M(C,N) and austenite, the Baker–Nutting relationship between M(C,N) and ferrite, the Pitsch–Schrader relationship between Fe_3C and ferrite, and so on. Thus, the interface between secondary phase and matrix would be semi-coherent and the specific interfacial energy of various secondary phases with different iron matrixes could be calculated from the interface mismatch dislocation theory [11, 12]. As the misfits in most cases are different in different orientations, the specific interfacial energy values in different orientations are different. This difference leads to that the precipitated secondary phase has specific shape with certain aspect ratio.

In Table 1, the orientation relationships between common secondary phases and iron matrix, the misfits in various orientations and the calculated aspect ratio values of secondary phases are listed [1].

The calculated results for specific interfacial energy values of various secondary phases with different iron

Table 1 Orientation relationship and aspect ratio for common secondary phases

Secondary phase	Matrix	Orientation relationship	Misfits in different orientation	Shape of secondary phase	Aspect ratio
NbC	Austenite	$(001)_{NbC}\|\|(001)_{\gamma}$, $[010]_{NbC}\|\|[010]_{\gamma}$	0.187 in all orientations	Sphere	1
VC	Austenite	$(001)_{NbC}\|\|(001)_{\gamma}$, $[010]_{NbC}\|\|[010]_{\gamma}$	0.132 in all orientations	Sphere	1
TiC	Austenite	$(001)_{NbC}\|\|(001)_{\gamma}$, $[010]_{NbC}\|\|[010]_{\gamma}$	0.159 in all orientations	Sphere	1
NbN	Austenite	$(001)_{NbC}\|\|(001)_{\gamma}$, $[010]_{NbC}\|\|[010]_{\gamma}$	0.176 in all orientations	Sphere	1
VN	Austenite	$(001)_{NbC}\|\|(001)_{\gamma}$, $[010]_{NbC}\|\|[010]_{\gamma}$	0.122 in all orientations	Sphere	1
TiN	Austenite	$(001)_{NbC}\|\|(001)_{\gamma}$, $[010]_{NbC}\|\|[010]_{\gamma}$	0.145 in all orientations	Sphere or cube	1
NbC	Ferrite	$(001)_{NbC}\|\|(001)_{\alpha}$, $[010]_{NbC}\|\|[110]_{\alpha}$	0.359 in $[001]_{NbC}$, 0.093 in $[010]_{NbC}$ and $[001]_{NbC}$	Round plate	1.62
VC	Ferrite	$(001)_{NbC}\|\|(001)_{\alpha}$, $[010]_{NbC}\|\|[110]_{\alpha}$	0.315 in $[001]_{VC}$, 0.031 in $[010]_{VC}$ and $[001]_{VC}$	Round plate	2.39
TiC	Ferrite	$(001)_{NbC}\|\|(001)_{\alpha}$, $[010]_{NbC}\|\|[110]_{\alpha}$	0.336 in $[001]_{TiC}$, 0.061 in $[010]_{TiC}$ and $[001]_{TiC}$	Round plate	1.82
NbN	Ferrite	$(001)_{NbC}\|\|(001)_{\alpha}$, $[010]_{NbC}\|\|[110]_{\alpha}$	0.348 in $[001]_{NbN}$, 0.077 in $[010]_{NbN}$ and $[001]_{NbN}$	Round plate	1.69
VN	Ferrite	$(001)_{NbC}\|\|(001)_{\alpha}$, $[010]_{NbC}\|\|[110]_{\alpha}$	0.307 in $[001]_{VN}$, 0.020 in $[010]_{VN}$ and $[001]_{VN}$	Round plate	2.98
TiN	Ferrite	$(001)_{NbC}\|\|(001)_{\alpha}$, $[010]_{NbC}\|\|[110]_{\alpha}$	0.324 in $[001]_{TiN}$, 0.044 in $[010]_{TiN}$ and $[001]_{TiN}$	Round plate	2.05
MnS	Austenite	$(001)_{MnS}\|\|(001)_{\alpha}$, $[110]_{MnS}\|\|[100]_{\alpha}$	0.375 in $[001]_{MnS}$, 0.080 in $[110]_{MnS}$ and $[1\bar{1}0]_{MnS}$	Plate	3.84
MnS	Ferrite	$(100)_{MnS}\|\|(111)_{\alpha}$, $[011]_{MnS}\|\|[\bar{1}01]_{\alpha}$	0.052 in $[100]_{MnS}$, 0.089 in $[011]_{MnS}$ and $[01\bar{1}]_{MnS}$	Nearly spherical	1.03
Fe_3C	Ferrite	$(001)_{Fe_3C}\|\|(211)_{\alpha}$, $[100]_{Fe_3C}\|\|[01\bar{1}]_{\alpha}$, $[010]_{Fe_3C}\|\|[\bar{1}11]_{\alpha}$	0.104 in $[001]_{Fe_3C}$, 0.024 in $[100]_{Fe_3C}$, 0.041 in $[010]_{Fe_3C}$	Spheroid	1.70: 1.45: 1

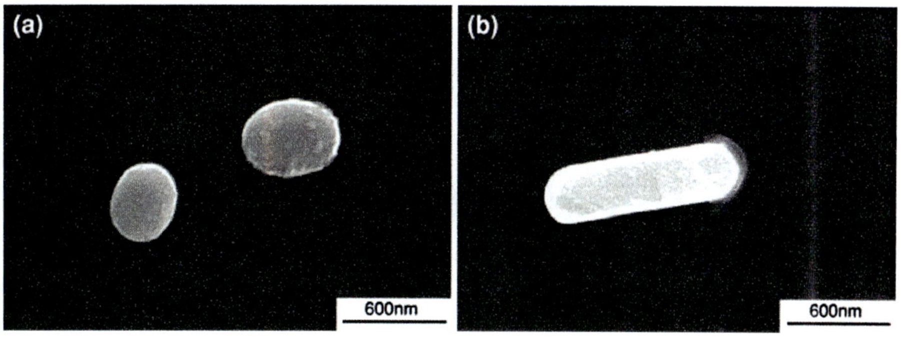

Fig. 4 The shape and aspect ratio of MnS particles precipitated in ferrite (**a**) and in austenite (**b**)

matrixes at 1,000°C are in the range of 0.2–0.6 J/m^2 and consistent with the commonly estimated values, that means the calculating method is rational. The calculated shape and aspect ratio has been observed and substantiated for microalloy nitrocarbides. In recent research work, MnS precipitates in ferrite and in austenite are observed and shown in Fig. 4.

It has been noticed that the interfacial energy is varied with temperature approximately as the elastic modulus with temperature. Because the precipitation temperature range is wide, the variation range of interfacial energy value is large either. This will seriously influence the precipitation dynamics [11, 12]. A lot of relative works have not considered this variation and thus leading their research results becoming dubious.

3.3 The Revision of the Theory of Nucleation on Dislocations

Some secondary phases such as TiN, MnS, BN would precipitate in interdendritic region at solidification process, while others precipitate along matrix grain boundaries in higher temperature range, their sizes are larger than

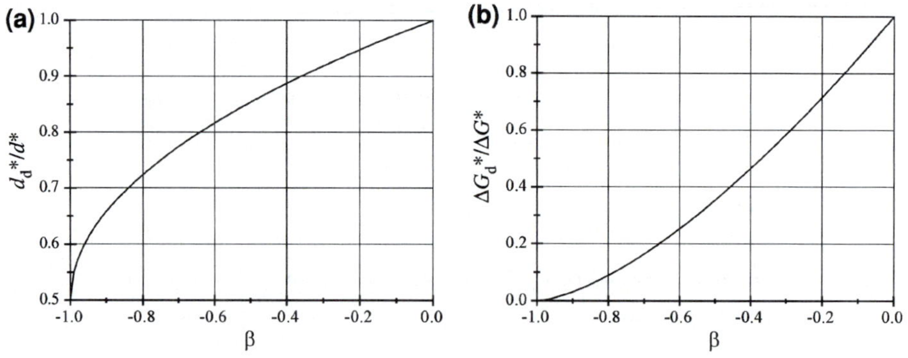

Fig. 5 The variations of relative critical nucleus size (**a**) and relative critical nucleation energy (**b**) with parameter β for nucleation on dislocations versus. homogeneous nucleation

magnitude of μm so that these precipitates should be appropriately restrained. In lower temperature range, nucleation on dislocations has been confirmed as the main precipitation mechanism for most of the secondary phases precipitated in iron matrix [13].

The classical theory of nucleation on dislocations proposed by Cahn [14] has some imperfectness and leads to that the nucleation of secondary phase very easily takes place with the critical nucleation free energy towards zero. We [15] have revised the theory and proposed a new parameter $\beta = A\Delta G_V/2\pi\sigma^2$(where $A = Gb^2/[4\pi(1-\nu)]$ for edge dislocation or $A = Gb^2/(4\pi)$ for screw dislocation, ΔG_V the volume free energy for precipitation, σ the specific interfacial energy, G the shear elastic modulus of matrix and b the Burgers vector of dislocation) to replace the parameter α in Cahn theory, as $\beta = \alpha/4$ which trends to -1 more slowly than α so that the critical nucleation free energy would not easily move towards zero.

The critical nucleus size d_d^* and critical nucleation free energy ΔG_d^* for secondary phase formed on dislocations would be (Fig. 5):

$$d_d^* = -\frac{2\sigma}{\Delta G_V}\left[1+(1+\beta)^{1/2}\right] = \frac{1}{2}\left[1+(1+\beta)^{1/2}\right]d^* \quad (11)$$

$$\Delta G_d^* = \frac{16\pi\sigma^3}{3\Delta G_V^2}(1+\beta)^{3/2} = (1+\beta)^{3/2}\Delta G^* \quad (12)$$

where d^* and ΔG^* is the critical nucleus size and the critical nucleation free energy for homogeneous nucleation, respectively.

4 The Precipitation Process Control

The classical nucleation and growth theory could be applied for the precipitation of secondary phases. Considering above revisions, the theoretical calculation method and procedure for the PTT (precipitation fraction–time–temperature) curve and NrT (nucleation rate–temperature) curve for major secondary phases precipitated in austenite and in ferrite has been proposed. The calculated results are consistent very well with experimental results in almost all the precipitation systems and all the steels. This confirms that the theoretical calculation method is correct [1].

From the calculated results and the experimental results, we come to some important findings, which are very useful for the practical control of the precipitation process of secondary phases.

(1) The PTT curve of Nb (C, N) precipitated in austenite and in ferrite is typical C shape, the nose temperature precipitated in austenite for common HSLA steel chemical composition at 880–930°C (see Fig. 6a), in ferrite at 670–720°C (see Fig. 6b).
(2) The PTT curve of V (C, N) precipitation in austenite is ε shape when the N content lower than 0.01% and C shape when N content higher than 0.015%. The nose temperature (if ε shape, upper nose) at 820–880°C. As the upper nose located in very long time, V (C, N) does not precipitate practically in low N steels (see Fig. 7a).
(3) The PTT curve of V (C, N) precipitation in ferrite is C shape when the N content is lower than 0.005% and a monotonic decreasing curve when N content higher than 0.01%, i.e. the higher the temperature, the faster the precipitation process (see Fig. 7b). This means that the inter-phase precipitation could be very easily take place.
(4) The precipitation of microalloy carbonitride in ferrite would be affected by the formation of pearlite, because the formation of pearlite would consume soluble C thus reduce the precipitation free energy and retard the precipitation process (see Fig. 6b–7b).
(5) The deformation of matrix promotes the precipitation process of secondary phases because of the effect of deformation storage energy, i.e. so called deformation induced precipitation. In austenite area, about 50% deformation could shorten the precipitation beginning time one order of magnitude and make the nose temperature raise 30–50°C (see Fig. 6a).
(6) The NrT curve is of reverse C shape or reverse ε shape. Although the NrT curves having reversed C shape or reverse and shape, they obey the same rule as PTT

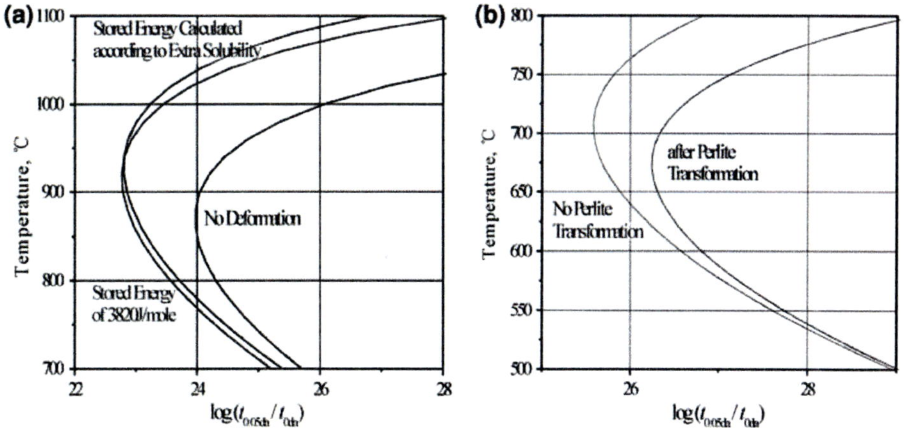

Fig. 6 The PTT curve for Nb (C, N) precipitated in austenite (**a**) and in ferrite (**b**) for 0.10%C–0.005%N–0.06%Nb steel (nucleation on dislocations, nucleation rate attenuated to zero with time)

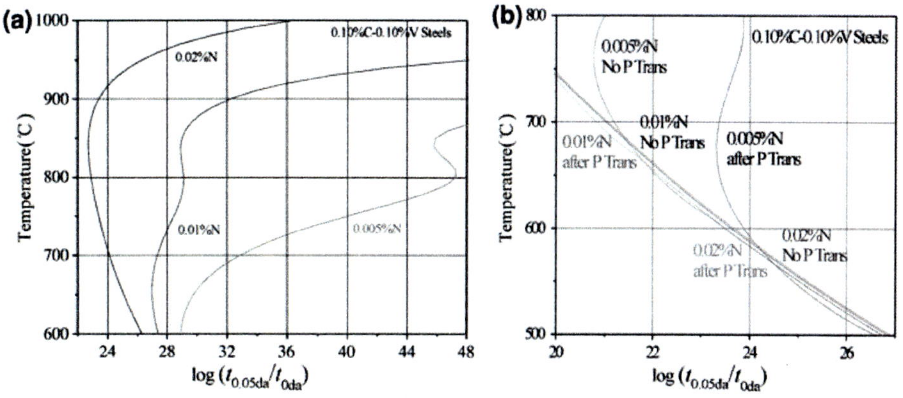

Fig. 7 The PTT curve for V (C, N) precipitated in austenite (**a**) and in ferrite (**b**) for 0.10%C–0.005%N–0.10%V steel (nucleation on dislocations, nucleation rate attenuated to zero with time)

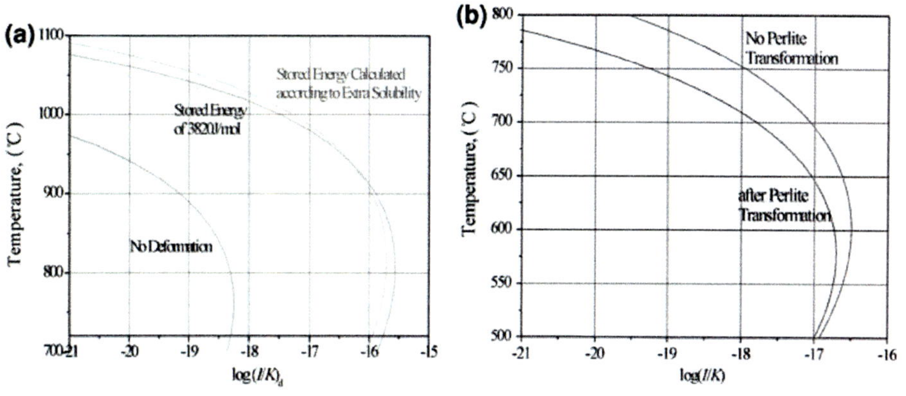

Fig. 8 The NrT curves for Nb (C,N) precipitated in austenite (**a**) and in ferrite (**b**) for 0.10%C–0.005%N–0.06%Nb steel (nucleation on dislocations)

curves mentioned above. The nose temperature of NrT curve would be about 80–100°C lower than PTT curve for the same precipitate system with same nucleation mechanism (compare Fig. 6 with Fig. 8).

(7) Once the precipitation started, the nucleation rate of secondary phase would be rapidly attenuated to zero with time, as the formation of precipitate nucleus would simultaneously reduce the super-saturation degree and leads to sharp decline of the precipitation free energy in the surrounding micro–region. Thus, holding at the nose temperature of NrT curve could obtain the finest precipitates, as at that temperature the nucleation rate is largest. In secondary hardening steels or PH steels, holding at under aging temperature would be obtained finer precipitate. However, the precipitation beginning time and the finish time would be delayed.

(8) The precipitation process of MnS in austenite for common HSLA steels is very rapidly and significantly

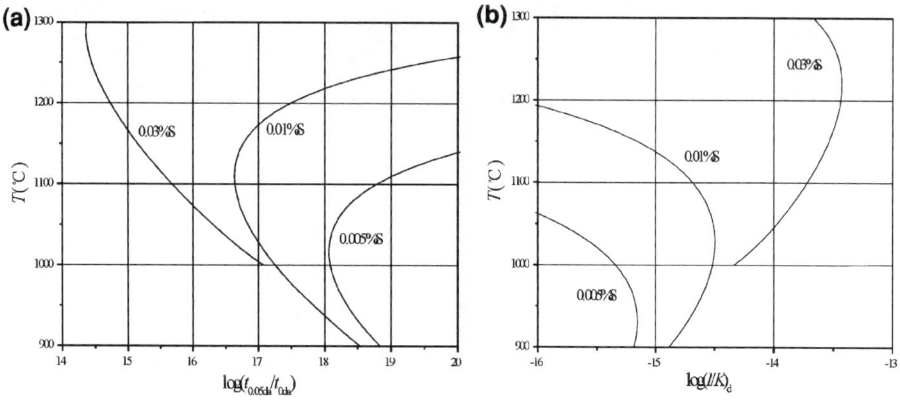

Fig. 9 The PTT curve (**a**) and NrT curve (**b**) for MnS precipitated in austenite for 5%Mn different S content steels (nucleation on dislocations)

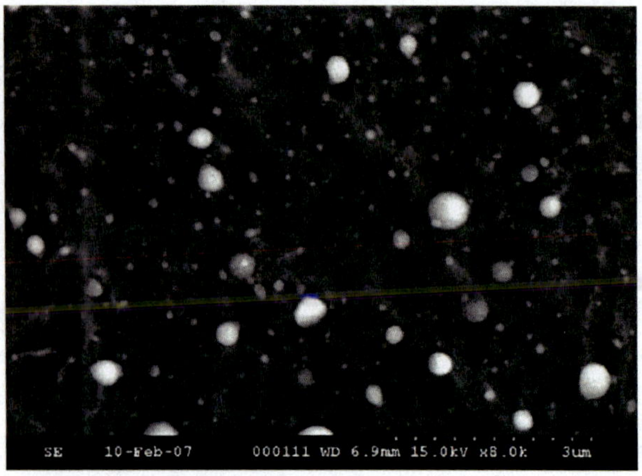

Fig. 10 The Fe_3C particles in 0.9%C–0.3Cr% steel: coarser ones undissolved in austenite, finer ones deformation induced precipitated in martensite

affected by the S content. For the forepart typical S content of 0.03%, the nose temperature of PTT curve would be at 1,289°C, so the MnS particle size would be very large and commonly greater than magnitude of μm. While when the S content reduced to 0.01 or 0.005% in recent clean steels, the nose temperature of PTT curve would be markedly drop to 1,110 or 1,017°C, respectively, (see Fig. 9), and the precipitate size would be in the range of 20–100 nm that is beneficial to properties.

(9) The deformation induced precipitation can also take place in ferrite area but have not drawn enough attention. Recently, we have proposed a special deformation heat treatment process for 0.9%C steel and obtained very fine Fe_3C precipitates, whose size is in the range of 30–50 nm as shown in Fig. 10. Due to the martensite maintains high hardness and the additional Fe_3C precipitate volume fraction is very large, its super-hardness of about HRC68–72 has been achieved.

5 Conclusions

(1) The main roles of secondary phases in steels have been reviewed. The volume fraction, size, shape and distribution of the secondary phases must be effectively controlled in order to obtain the beneficial effects of second phase precipitation while eliminate its harmful effects.

(2) According to the equilibrium solubility product formula, precipitation thermodynamics and kinetics and classical nucleation and growth theory, some revision on the traditional theory has been made, moreover, the theoretical calculation method and procedure for the PTT (precipitation fraction–time–temperature) curves and NrT (nucleation rate–temperature) curves of major secondary phases precipitated in austenite and in ferrite have been proposed.

(3) Some important principles for the control of secondary phases in steels and their practical application have been analyzed and introduced.

Acknowledgments This work is supported by National Basic Research Program of China (973 Program), Project no. 2010CB630800.

References

1. Q. Yong, *Secondary Phases in Steels* (Metallurgical Industry Press, Beijing, 2006)
2. T. Gladman, *The Physical Metallurgy of Microalloyed Steels* (The Institute of Materials, London, 1997)
3. F.B. Pickering, *Physical Metallurgy and the Design of Steels* (Applied Science Publication, London, 1978)
4. J.W. Martin, R.D. Doherty, *Stability of Microstructure in Metallic Systems* (Cambridge University Press, London, 1976)
5. Y. Weng (ed.), *Ultra-Fine Grained Steels* (Metallurgical Industry Press, Beijing, 2006)

6. T. Gladman, On the theory of the effect of precipitate particles on grain growth in metals. Proc. Roy. Soc. **A294**, 298–309 (1966)
7. H. Takechi, Metallurgical aspects on interstitial free sheet steel from industrial viewpoints. ISIJ Int. **34**(1), 1 (1994)
8. A.K. De, S. Vandeputte, B.C. DeCooman, Kinetics of low temperature precipitation in a ULC-bake hardening steel. Scrip. Mater. **44**, 695 (2001)
9. T. Furuhara, J. Yamaguchi, N. Sugita, G. Miyamoto, T. Maki, Nucleation of proeutectoid ferrite on complex precipitates in auetenite. ISIJ Int. **43**, 2028 (2003)
10. Y. Murakami (ed.), *Metal Fatigue: Effect of Small Defects and Nonmetallic Inclusions* (Elesevier Science Ltd., Oxford, 2002)
11. Q. Yong, Y. Li, Z. Sun, B. Wu, Theoretical calculation of specific interfacial energy of semicoherent interface between microally carbonitrides and austenite. Acta Metall. Sin. **2**, 153 (1989)
12. Q. Yong, Y. Li, Z. Sun, B. Wu, Theoretical calculation of specific interfacial energy of semicoherent interface between microally carbonitrides and ferrite. Chin. Sci. Bull. **34**, 1747 (1989)
13. Q. Yong, Y. Li, K. Zhao, The isothermal transformation kinetics time exponent versus the nucleation and growth mechanism. J. Yunnan Polytech. Univ. **6**(3), 7 (1990)
14. J.W. Cahn, Nucleation on dislocations. Acta Metall. **5**, 169 (1957)
15. Q. Yong, Theory of nucleation on dislocations. Chin. J. Met. Sci. Tech. **6**, 239 (1990)

Ways to Manage Both Strength and Ductility in Nanostructured Steels

Nobuhiro Tsuji

Abstract

Nanostructured steels composed of ultrafine grains (UFG) with sizes smaller than 1 μm perform surprisingly high strength but sometimes show limited tensile ductility. In the present paper, systematic experimental results on mechanical properties of the nanostructured steels with ferrite single phase are firstly shown. The limited tensile ductility of the nanostructured ferritic steels was due to very small uniform elongation, which was attributed to the early plastic instability in the UFG microstructures. This basic understanding suggests one of the ways to overcome the low tensile ductility: if the strain-hardening of the matrix is enhanced by any means, such as dispersing fine second phase in the matrix, both high strength and adequate ductility can be managed even in nanostructures. Actual examples of the nanostructured steels that could achieve good strength–ductility balance were also introduced. Dispersing fine carbides within the UFG ferrite matrix was actually effective to manage both strength and ductility. Also ultrafine dual-phase structure composed of ferrite and martensite resulted in both high strength and large uniform elongation. It was also shown that transformation induced plasticity caused by deformation induced martensite transformation of metastable austenite could work in nanostructured steels. The present results clearly indicate that using multi-phases is the promising direction for managing both high strength and adequate ductility in nanostructured steels.

Keywords

Plastic instability • Strain-hardening • Multi-phase • Carbides • Martensite

1 Introduction

It has recently become possible at least in laboratory scale to fabricate nanostructured metals (including steels) composed of ultrafine grains (UFG) with mean grain sizes smaller than 1–2 μm [1–8]. The obtained nanostructured metals perform excellent mechanical properties, such as surprisingly high strength and good low-temperature toughness [2–11]. However, the nanostructured metals have limited tensile elongation in many cases unfortunately [9, 10]. The aim of this review paper is firstly to make it clear why the tensile ductility is limited in the nanostructured metals. Based on this understanding, the second purpose is to show how we can overcome this issue, demonstrating several experimental examples of the UFG steels, in which both high strength and adequate ductility are managed.

2 Ways to Obtain Nanostructured Steels

In case of steels, there are mainly two different ways to fabricate the UFG structures: one is the thermomechanically controlled processing (TMCP) using phase transformation

N. Tsuji (✉)
Department of Materials Science and Engineering,
Kyoto University, Kyoto 606-8501, Japan
e-mail: nobuhiro.tsuji@ky5.ecs.kyoto-u.ac.jp

Y. Weng et al. (eds.), *Advanced Steels*, DOI: 10.1007/978-3-642-17665-4_15,

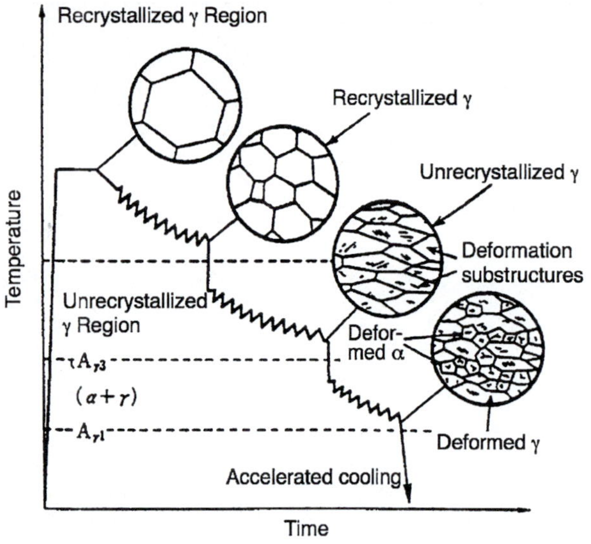

Fig. 1 Schematic illustration showing the microstructure evolution during the controlled rolling [13]

in steels, and the other is the severe plastic deformation (SPD) applying very high plastic strain to the materials [4–6]. These two processes make different kind of nanostructures.

One of the typical examples of TMCP is so-called controlled rolling for plate steels [12, 13]. Figure 1 schematically illustrates the microstructure evolution in the controlled rolling. The purpose of the controlled rolling is to obtain fine-grained ferrite (α) in final plates of low-C steels. When the steels are rolled in high temperature austenite (γ) region, recrystallization of γ quickly happens after plastic deformation (rolling). Grain refinement of γ by recrystallization is effective to obtain fine-grained α eventually, since grain boundaries of γ act as preferential nucleation sites of α. However, refinement of γ by recrystallization is limited, as grain growth quickly occurs at high temperature as well. When the deformation (rolling) is carried out at lower temperature, unrecrystallized γ can be obtained. The unrecrystallized γ including dislocation substructures is much more preferable than recrystallized γ for obtaining fine-grained α, because the deformation structures act as in-grain nucleation site for α. In the controlled rolling, small amount of alloying elements, such as Nb and Ti, are added, in order to disperse fine carbides and nitrides in γ matrix. The fine carbides and nitrides inhibit recovery and recrystallization of γ to reserve unrecrystallizaed structures till $\gamma \rightarrow \alpha$ phase transformation. Cooling the steels at high cooling rate is additionally effective to obtain fine-grained α, since large driving force for phase transformation under large supercooling increases the frequency of nucleation. Fine carbides and nitrides are again effective to pin the grain growth of α. The minimum grain size of α obtained through conventional controlled rolling is about 5 μm [13]. When this type of TMCP is carried out under more severe conditions, i.e., at much lower temperature in larger one-pass reduction, grain size of 1 μm has been achieved [3, 4, 7]. In whichever case, the fine-grained α obtained are equiaxed grains without dislocation substructures inside, as they form through diffusional transformation.

In SPD, bulky metals are deformed up to ultrahigh plastic strain above logarithmic equivalent strain of 4–5 by special deformation processes, such as equal-channel angular extrusion (ECAE), high pressure torsion (HPT) and accumulative roll bonding (ARB) [14]. UFG structures are obtained in the as-deformed state in SPD. The formation mechanism of the UFGs during SPD is understood in terms of *grain subdivision* [15, 16]. The UFG structures obtained by SPD have characteristics significantly different from those obtained through TMCP route. An example of UFG structures obtained through SPD is shown in Fig. 2, which shows TEM microstructure (a) and corresponding grain boundary map (b) of the interstitial free (IF) steel heavily deformed to equivalent strain of 5.6 by ARB process [17]. The microstructure was observed from the transverse direction (TD) of the sheet material. The grain boundary map was constructed from the precise orientation data obtained by Kikuchi-line analysis of the identical area in TEM. The ARB processed IF steel is filled with the UFGs elongated to the rolling direction (RD). The grain sizes are fairly uniform and the average thickness of the UFGs is 200 nm. The boundary map clearly shows that the UFGs are mostly surrounded by high-angle grain boundaries having large misorientation. From a viewpoint of misorientation, therefore, they are certainly 'grains'. However, they have elongated morphology and involve dislocations as well as sub-boundaries (low-angle boundaries) inside. That is, the UFG structure shown in Fig. 2 is essentially a deformation structure. This is reasonable, since the UFG structure is fabricated by deformation (SPD) and the formation mechanism is grain subdivision. The characteristics of the structure would of course affect the mechanical properties.

3 Typical Mechanical Properties of Single-Phased Nanostructured Steels

In this section, typical mechanical properties of a nanostructured steel having α single phase are shown and discussed. The nanostructured IF steel with various fine grain sizes was fabricated by SPD and subsequent annealing. Here, the ARB process was used as SPD. The ARB is a kind of SPD process using rolling deformation, and the principle and processing details have been shown in previous papers [6, 13, 14, 18]. Chemical composition of the IF steel used is

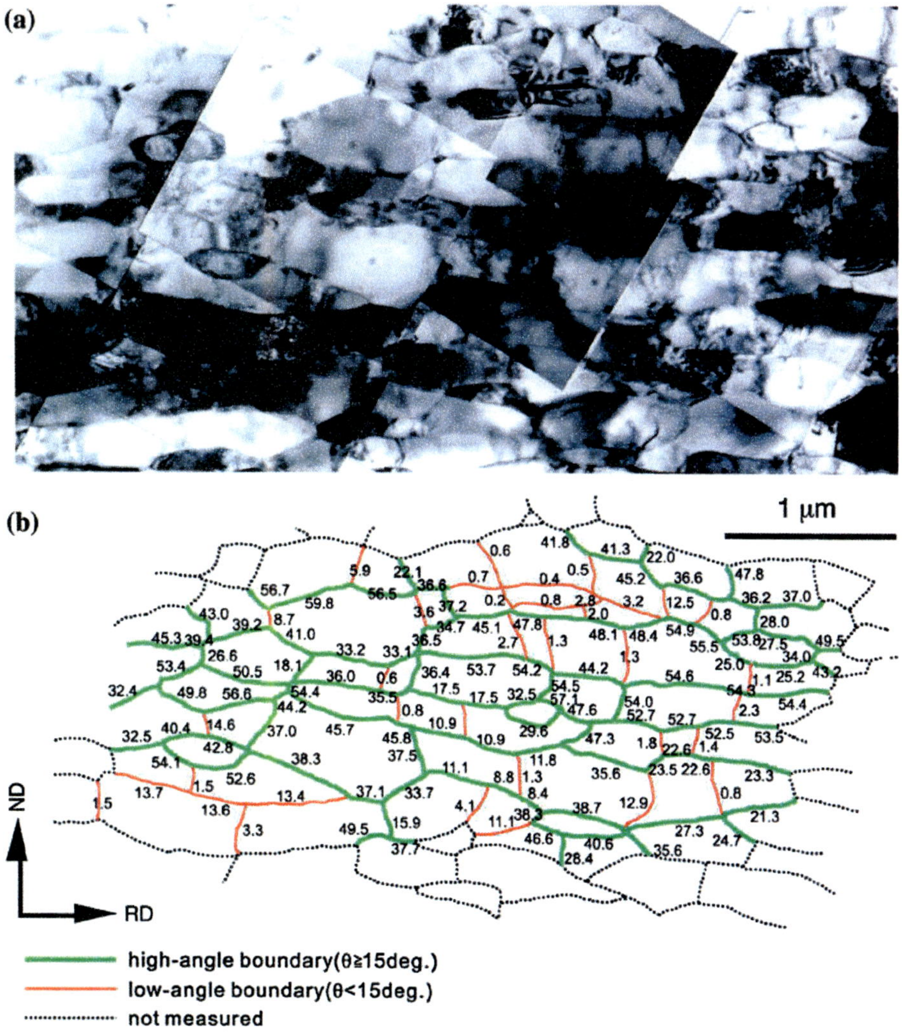

Fig. 2 TEM microstructure (**a**) and corresponding grain boundary map (**b**) of the IF steel ARB processed by seven cycles (equivalent strain of 5.6) at 500°C [17]. Misorientation of the boundaries (in degree) is superimposed in (**b**). Observed from TD

Table 1 Chemical composition of the IF steel studied (mass%)

C	N	Si	Mn	P	Cu	Ni	Ti	Fe
0.002	0.003	0.01	0.17	0.012	0.01	0.02	0.072	Balance

shown in Table 1. The starting sheet had fully recrystallized microstructure with a mean grain size of 20 μm. Two sheets of the IF steel with thickness of 1 mm were stacked, held at 500°C for 600 s in an electronic furnace with Ar atmosphere, and then roll-bonded by 50% reduction in one pass without lubrication. These procedures correspond to the first ARB cycle. The roll-bonded sheet was cut into two, and the above-mentioned procedures were repeated for the second ARB cycle. The ARB was repeated up to seven cycles. Because equivalent strain of 50% rolling is 0.8, the total equivalent strain accumulated in seven cycles of ARB is 5.6. The ARB processed sheets were annealed at various temperatures ranging from 400 to 800°C for 1.8 ks. Tensile specimens with gage width of 5 mm and gage length of 10 mm, which was 1/5 miniaturized of JIS-5 specimen, were cut from the sheets ARB processed and annealed. Tensile test of the specimens was carried out at room temperature at an initial strain rate of 8.3×10^{-4} s^{-1}. Tensile direction was parallel to RD. The microstructures of the specimens were characterized by transmission electron microscopy (TEM) and by electron back-scattering diffraction (EBSD) analysis in a scanning electron microscope equipped with field-emission type gun (FE-SEM).

The microstructure of the as 7-cycle ARB processed specimen was already shown in Fig. 2. Figure 3 shows EBSD maps of the IF steel specimens 7-cycle ARB processed and then annealed at various temperatures for 1.8 ks. TEM microstructures of the ARB processed and annealed

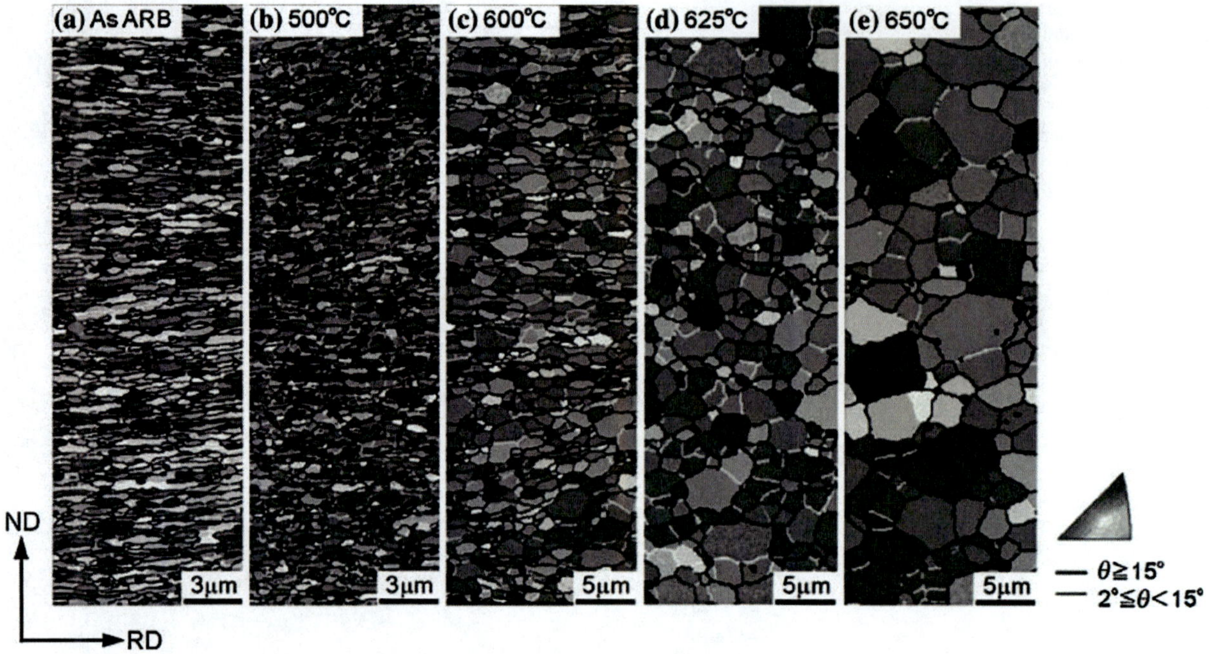

Fig. 3 EBSD map[5] of the IF steel ARB processed by seven cycles and then annealed at various temperatures for 1.8 ks. Observed from TD. **a** as-ARB processed; **b** annealed at 500°C; **c** 600°C; **d** 625°C; **e** 650°C

IF steel are also shown in Fig. 4. The as-ARB processed specimen (Fig. 3a) showed the UFG structure elongated to RD, which corresponded well with Fig. 2. The mean thickness of the elongated grains was about 200 nm. By annealing at relatively low temperature, recovery at grain interior occurred to decrease the dislocation density within the UFGs Figs. 3b, 4b, 4c. After annealing at 600°C (Figs. 3c, 4d), almost all the grains were dislocation free already, though the grain shape was still elongated slightly. At the same time, grain growth gradually happened with increasing annealing temperature, to make the grain size coarse. The specimen annealed at 625°C (Figs. 3d, 4e) showed equiaxed grains free from dislocations, which were difficult to be distinguished from conventional recrystallized microstructures. However, the mean grain size of the 625°C annealed specimen was still very fine (1.6 μm), which cannot be obtained through conventional deformation and recrystallization. The change in the UFG structure (grain growth) was fairly uniform in this material.

The engineering stress–strain curves of the IF steel ARB processed and annealed are shown in Fig. 5. The annealing temperature and resulted mean grain size of each specimen are also indicated in the figure. Strength of the as 7-cycle ARB processed specimen reached over 900 MPa, which was three times higher than that of the starting material. On the other hand, tensile ductility significantly decreased by the ARB. The stress–strain curve showed the maximum strength at early stage of tensile deformation, followed by macroscopic necking. Thus, the uniform elongation was limited within a few percents. This is a typical mechanical property of strain-hardened materials, which corresponds with the characteristic of deformation structures in the UFGs fabricated by SPD (Fig. 2). Thus, annealing process was carried out to remove the features of deformation structures. The strength of the specimens decreased with increasing annealing temperature (or with increasing mean grain size), while tensile ductility (uniform elongation) recovered only after the mean grain size became over 1 μm (Fig. 5). As was mentioned above, the 600°C annealed specimen with a mean grain size of 0.76 μm had small number of dislocations within the grains already. Thus, it can be concluded that the UFG ferritic steel having mean grain size smaller than 1 μm certainly shows limited tensile ductility, especially limited uniform elongation. It has been reported that the UFG ferritic steels fabricated through TMCP route also show similar mechanical properties [3, 7], though they have equiaxed UFG structures without deformation substructures.

It was confirmed that the nanostructured ferritic steel certainly showed limited tensile ductility, especially limited uniform elongation. Though only the experimental results for the ferritic IF steel were shown, similar conclusion has been obtained also in austenitic steels [19] and other metals like Al and Cu [10, 20, 21]. The limited uniform elongation

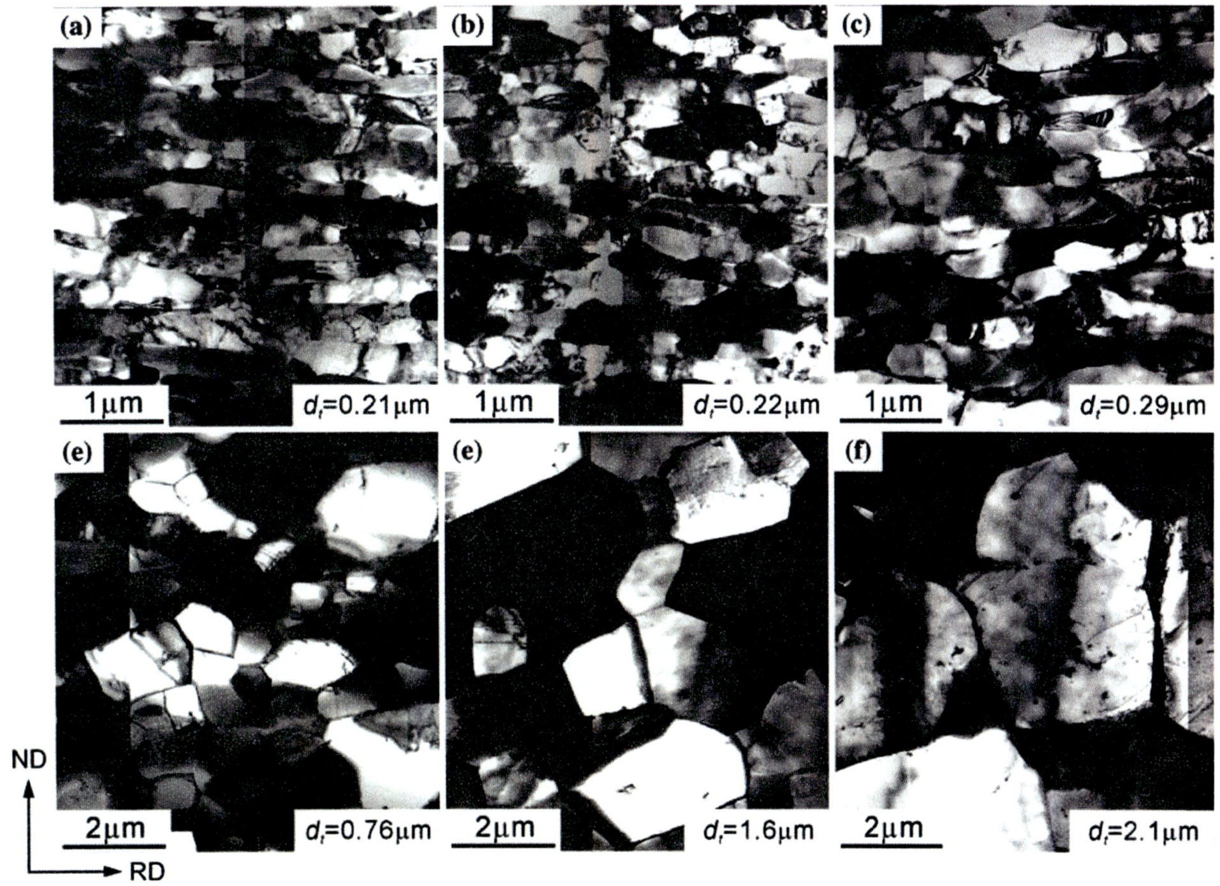

Fig. 4 TEM microstructures of the IF steel ARB processed by seven cycles and subsequently annealed at various temperatures for 1.8 ks. Observed from TD. **a** as-ARB processed; **b** annealed at 400°C; **c** 500°C; **d** 600°C; **e** 625°C; **f** 650°C

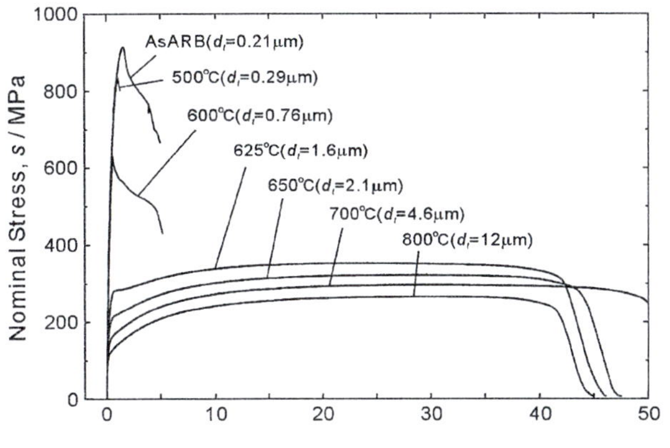

Fig. 5 Engineering stress–strain curves of the IF steel ARB processed by seven cycles and then annealed at various temperatures for 1.8 ks. The annealing temperatures and resulted average grain sizes of the specimens are also indicated in the figures

in the nanostructured metals can be understood in terms of plastic instability [10, 22]. Plastic instability corresponds to necking propagation in tensile test, so that it determines the uniform elongation of the materials. The simplest equation for the plastic instability condition of strain-rate insensitive materials (typical metals) is known as Considère equation,

$$\sigma \geq \left(\frac{d\sigma}{d\varepsilon}\right) \tag{1}$$

where σ is flow stress (true stress) and $d\sigma/d\varepsilon$ is strain-hardening rate [23]. The condition can be schematically illustrated in Fig. 6. In the figure, yield strength of a material increases by a certain strengthening mechanism like grain refinement strengthening. Here it is assumed for simplicity that the strain-hardening rate does not change even if the material is strengthened. According to Eq. 1, the position at which two curves [flow stress (σ) and strain-hardening rate ($d\sigma/d\varepsilon$)] meet is the point of plastic instability. The figure clearly shows that the plastic instability condition is achieved at earlier stage of tensile deformation as the yield strength becomes higher. Grain refinement increases the strength of metallic materials, and especially yield strength is significantly increased by fine grain

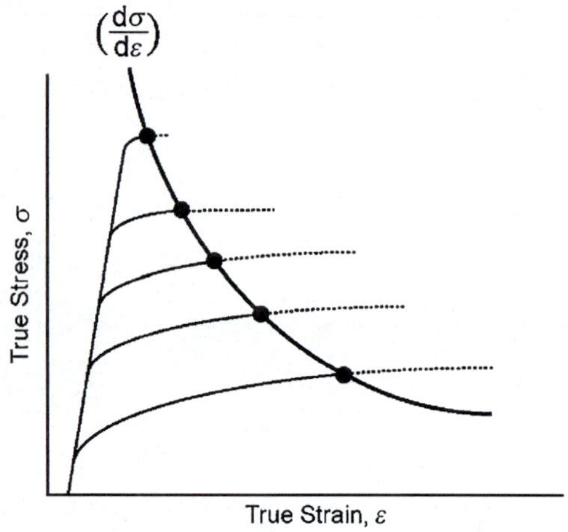

Fig. 6 Schematic illustration showing the change in points of plastic instability as yield strength increases. It is assumed that the strain-hardening rate is constant

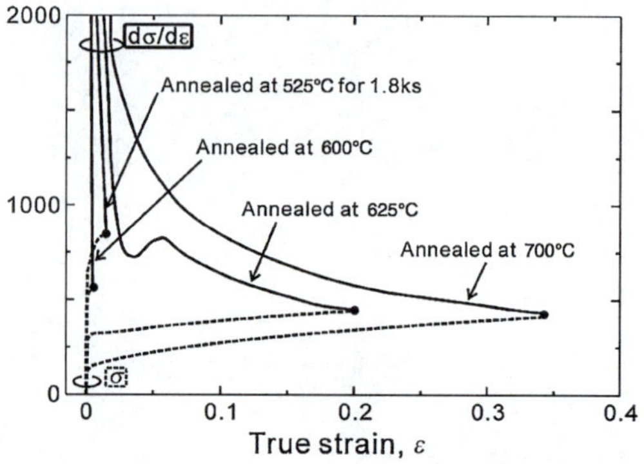

Fig. 7 True stress (*broken lines*) and strain-hardening rate (*solid lines*) as a function of true plastic strain in the IF steel ARB processed and then annealed at various temperatures for 1.8 ks. The positions at which two curves meet are the points of plastic instability, which correspond well with the uniform elongation of the specimens

structure. On the other hand, strain-hardening after macroscopic yielding is not enhanced by grain refinement, as can be seen from Fig. 5. Rather decrease in strain-hardening has been found in the UFG Al [10]. Consequently, early plastic instability occurs in the nanostructured metals, resulting in limited uniform elongation in tensile tests.

The plastic instability condition was verified for the IF steel specimens ARB processed and annealed. Figure 7 shows true stress (broken lines) and strain-hardening rate (solid lines) as a function of true plastic strain in the IF steel ARB processed and then annealed at various temperatures for 1.8 ks [24]. The specimens correspond to those shown in Figs. 3 and 4. As was explained in Fig. 6, the points at which two curves meet correspond to the plastic instability condition. The strains at plastic instability points in Fig. 7 agreed well with the uniform elongation determined from the engineering stress–strain curves of the same specimens. This means that the uniform elongation of the UFG IF steel is certainly determined by plastic instability, as is discussed above.

4 Ways to Manage Both Strength and Ductility in Nanostructured Steels

4.1 Dispersing Fine Carbides in UFG Matrix of Ferrite

It has been confirmed that the limited tensile ductility (limited uniform elongation) in the nanostructured ferritic steel is attributed to the early plastic instability. The early plastic instability seems an essential and inevitable feature of UFG microstructures. However, this understanding also tells us that the issue could be overcome if the strain-hardening rate of the UFG matrix is enhanced by any means. It should be noted that the nanostructured materials with limited tensile ductility reported in previous studies have been mostly single-phased materials like pure metals. By making the nanostructures multi-phased, for example, we may manage both high strength and adequate ductility even in nanostructures. Here, one of the actual examples is shown.

When we fabricate nanostructures in metals, quite heavy plastic strain is necessary in most cases of both TMCP and SPD routes. On the other hand, we have found that nanostructures can be obtained without heavy deformation in low-C steels when as-quenched martensite is used as the starting microstructure [24–28]. When the as-quenched martensite is conventionally rolled by 30–70% reduction in thickness at RT and annealed at appropriate warm temperature, submicrometer grains can be formed. This easy fabrication of nanostructures is thought to be attributed to the characteristics of martensite in carbon steels [26]. Martensite in steels is a metastable phase having high free energy, and it is a kind of fine-grain structure involving high density of lattice defects in the as-transformed state [29]. Additionally, it should be noted that the as-quenched martensite in steels is a supersaturated solid solution of carbon. Consequently, solute carbon uniformly precipitates as fine carbides within the ferrite matrix during warm-temperature annealing following to conventional cold-rolling. That is, the obtained UFG structure through this route (martensite method) is a multi-phased nanostructure composed of UFG ferrite and finely dispersed carbides [25, 26].

A plain low-C steel (JIS SS400) having chemical composition shown in Table 2 was used. Hot-rolled sheet 2 mm thick was austenitized at 1000°C for 900 s and then water-quenched to get martensite structure. Since plain low-C steels have low hardenability, austenite grains were

Table 2 Chemical composition of the JIS SS400 steel studied (mass%)

C	N	Si	Mn	P	S	Fe
0.13	0.004	0.01	0.37	0.020	0.004	Balance

coarsened at relatively high austenitizing temperature. The austenite grains were coarsened to be 270 μm in mean grain size, so that fully martensitic microstructure was obtained in the quenched specimen [26]. The as-quenched martensite was conventionally cold-rolled by 50% reduction in thickness in several passes. The cold-rolled specimen was annealed at various temperatures ranging from 400 to 600°C for 1.8 ks.

Figure 8a shows a TEM microstructure of the SS400 steel started from as-quenched martensite, cold-rolled by 50% and then annealed at 500°C for 1.8 ks. The microstructure was observed from TD. Nearly equiaxed UFGs of which mean grain size was 200 nm were observed. In this TEM observation, Kikuchi-line analysis of the identical area was carried out and misorientation between adjacent grains was calculated from the precise orientation data obtained. The result is represented as a boundary misorientation map in Fig. 8b. The boundary misorientation map indicated that many of ultrafine ferrite grains are surrounded by high-angle boundaries. Within such ultrafine ferrite grains, a number of fine carbides (cementite) with various sizes precipitated uniformly. This is because the specimen before 500°C annealing was a supersaturated solid solution of carbon, as was described before. It is concluded, anyhow, that the multi-phased nanostructure composed of ferrite and cementite is fabricated through the martensite method.

Engineering stress–strain curves of the SS400 steel specimens started from as-quenched martensite, cold-rolled by 50% and then annealed at various temperatures for 1.8 ks are shown in Fig. 9. Tensile tests were carried out under the same conditions described in Sect. 3. The as-rolled specimen showed very high strength over 1.5 GPa, but it had limited uniform elongation similar to the SPD material (Fig. 5). Flow stress decreased with increasing

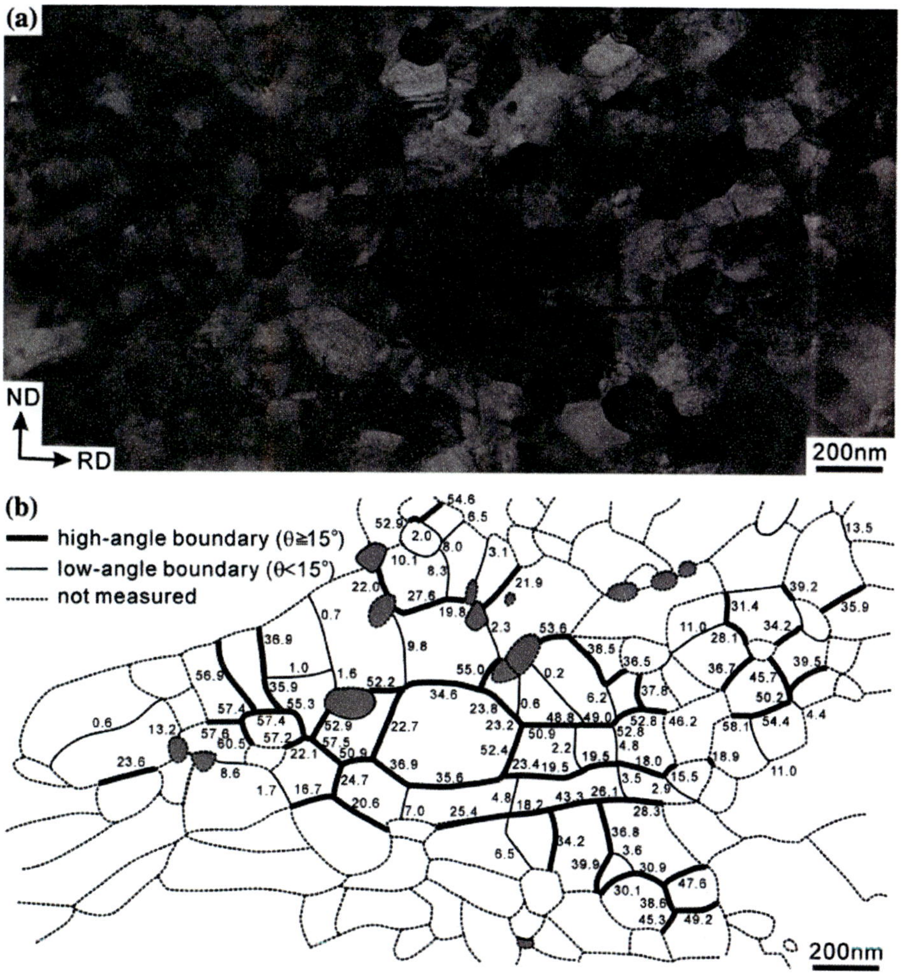

Fig. 8 TEM microstructure (**a**) and corresponding boundary misorientation map (**b**) of the 0.13% C steel started from as-quenched martensite, cold-rolled by 50% and then annealed at 500°C for 1.8 ks [26]. Observed from TD

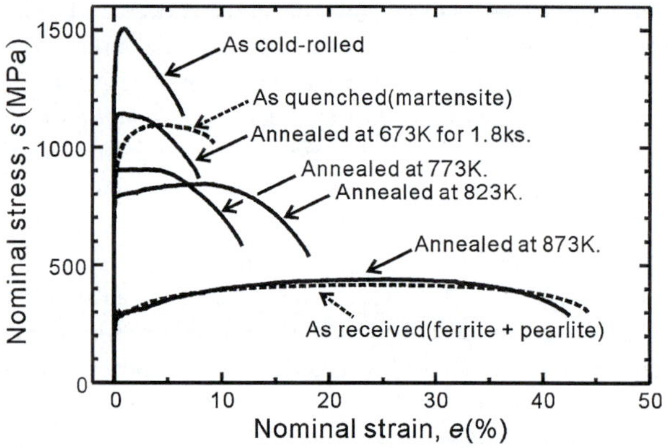

Fig. 9 Engineering stress–strain curves of the SS400 steel having multi-phased UFG structure composed of ferrite and fine carbides [25]

annealing temperature. However, the specimens annealed at 500°C (773 K) or 550°C (823 K), that showed multi-phased UFG structures of ferrite and cementite like Fig. 8, performed obvious strain-hardening after macroscopic yielding. As a result, these specimens showed adequate uniform elongation as well as high strength. Especially the 550°C annealed specimen showed 0.2% proof stress of 710 MPa, tensile strength of 870 MPa, uniform elongation of 8% and total elongation of 20%. Because starting sheet was a 400 MPa class steel (JIS SS400), the multi-phased UFG specimen obtained had strength more than two times higher than that of the starting hot-rolled sheet with ferrite–pearlite structure. It has been confirmed that the uniform elongation of these materials are again understood by plastic instability [24, 28]. This indicates that the finely dispersed carbides in the multi-phased UFG steel enhanced the strain-hardening to delay the plastic instability [26]. It is, therefore, concluded that dispersing fine precipitates (carbides) within the ferrite matrix is certainly effective to manage both high strength and adequate ductility in nanostructured steels.

4.2 UFG Dual-Phase Steels

In commercial steels for automobile applications, dual-phase (DP) structures composed of ferrite + martensite or ferrite + bainite are sometimes used for managing both high strength and ductility. If we can make nanostructured DP steels, they are expected to show excellent mechanical properties. A low-C steel (JIS SM490; 490 MPa class) with the chemical composition shown in Table 3 was used for making UFG-DP structures. The hot-rolled sheets with ferrite + pearlite structure was ARB processed by six cycles (equivalent strain of 4.8) at RT with lubrication. The as-ARB processed specimen showed very fine lamellar structures elongated to RD. The mean lamellar spacing was 60 nm. The as-ARB processed sample having such a microstructure was annealed at various temperatures between A_1 and A_3 (i.e., in $\alpha + \gamma$ two phase region) for various periods. High heating rate of 100 K/s was achieved by the use of induction heating system. After the intercritical annealing for relatively short periods, the specimens were ice-brine quenched. Figure 10 shows one of typical SEM microstructures of the ARB processed and intercritically annealed specimen. In the SEM image, ferrite and martensite reveal in dark and light contrasts, respectively. In ice-brine quenching, austenite (γ) transformed to martensite. The UFG-DP structure composed of UFG ferrite and UFG martensite was obtained. Both ferrite and martensite grains were nearly equiaxed. The mean grain sizes of ferrite and martensite were 1.3 and 0.84 μm, respectively, and the fraction of martensite was 28.6% in this case. These microstructural parameters varied depending on the annealing conditions.

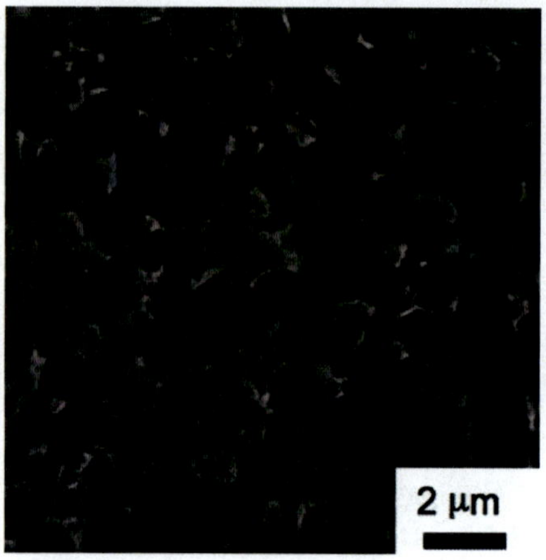

Fig. 10 SEM micrograph of the SM490 steel ARB processed by six cycles at RT, intercritically annealed at 740°C for 60 s, and then ice-brine quenched. Observed from TD

Engineering stress–strain curves of the SM490 steel ARB processed, intercritically annealed and quenched are shown in Fig. 11. The as-ARB processed specimen performed very high strength of about 1.4 GPa but showed limited uniform and total elongation. Intercritical annealing

Table 3 Chemical composition of the JIS SM490 steel studied (mass%)

C	N	Si	Mn	P	S	Nb	Sol. Al	Fe
0.138	0.002	0.01	0.64	0.013	0.002	0.02	0.031	Balance

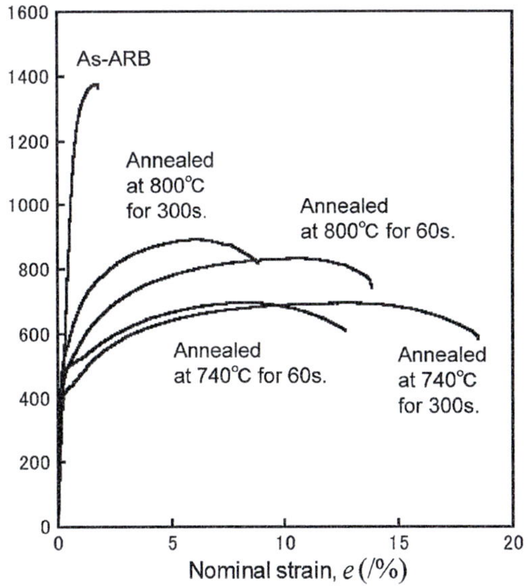

Fig. 11 Engineering stress–strain curves of the SM490 steel having ultrafine grained dual-phase structures

greatly decreased the flow stress of the material, but all UFG-DP steels showed enhanced strain-hardening. As a result, fairly high strength and adequate tensile ductility were both managed in these specimens. It can be concluded, therefore, that dual-phase structure can provide good strength–ductility balance even in UFG structures. Because the used steel had plain compositions, grain growth of nano-lamellae was not significantly suppressed even when a high heating rate was used. Addition of some kinds of alloying elements, such as strong carbide former (V, Ti, Nb, etc.), would inhibit grain growth during intercritical heat treatment to result in much finer DP structure. Such nano-DP steels are expected to perform more excellent mechanical properties.

4.3 UFG TRIP Steels

When metastable austenite is deformed, deformation induced martensite can form depending on the chemical composition and deformation conditions. When the deformation induced martensite appears at necked regions in tensile specimens, it can stop the propagation of necking as martensite in steels is much harder than austenite. Consequently, large tensile ductility can be obtained in such a kind of steels. This phenomenon is called transformation induced plasticity (TRIP) [30]. In this last section, it is assessed whether TRIP is useful in nano-structured steels [31].

A 15%Cr-10%Ni steel was used as the starting material in the present study. The chemical composition of the alloy, that was designed to make its martensite transformation temperature around room temperature, is shown in Table 4. The 15%Cr-10%Ni steel was provided to the ARB process. Two pieces of the sheets 1 mm in thickness, 30 mm in width and 150 mm in length were stacked to be 2 mm thick after surface treatment (degreasing and wire-brushing), kept in an electrical furnace at 600°C, which was above the austenite transformation finishing temperature (A_f) of the alloy (440°C), for 600 s. The heated sheets were roll-bonded by 35% reduction in the first pass and reheated under the same condition before being rolled by 50% total reduction in the second pass. These two rolling passes are considered as the one ARB cycle. The roll-bonded sheet was cooled not in water but in air in order to prevent transformation into martensite, and then cut into two pieces having nearly initial dimensions. Such a procedure was repeated up to six cycles ($\varepsilon = 4.8$). Tensile test of the ARB processed samples was carried out at a nominal strain rate of 8.3×10^{-4} s^{-1} at room temperature. The specimen for tensile test was 10 mm in gage length and 5 mm in gage width, which was the 1/5 miniaturized size of the JIS-5 specimen. Microstructural observations by optical microscopy (OM) and TEM were conducted for the ARB processed specimens before and after tensile test. Thin foils parallel to TD of the ARB processed sheet were prepared by twin-jet electropolishing in a 100 ml $HClO_4$ + 900 ml CH_3COOH solution. The mean grain sizes were measured by mean interception method. The volume fractions of phases (austenite or martensite) were determined by EBSD in FE-SEM.

Figure 12 shows engineering stress–strain curves of the 15%Cr-10%Ni steel ARB processed by various cycles at 600°C. The tensile strength of the ARB processed specimen reached to 900 MPa and furthermore the elongation were higher than 30%. Especially the elongation of the specimens ARB processed by four and five cycles were beyond 40%. These curves are very characteristics, unusual and deferent from those of the typical ARB processed materials with stable single phase. In the previously studied ARB materials, the strength became higher with increasing the number of ARB cycles, and the elongation dropped down below 10% by one ARB cycle. However, both the strength and the elongation of this alloy were increased by ARB. Yield drop

Table 4 Chemical composition of the 15%Cr-10%Ni steel studied (mass%)

C	Si	Mn	P	S	Cr	Ni	O	N	Fe
0.0024	<0.01	0.02	<0.005	0.0010	14.99	10.21	0.0068	0.0008	Balance

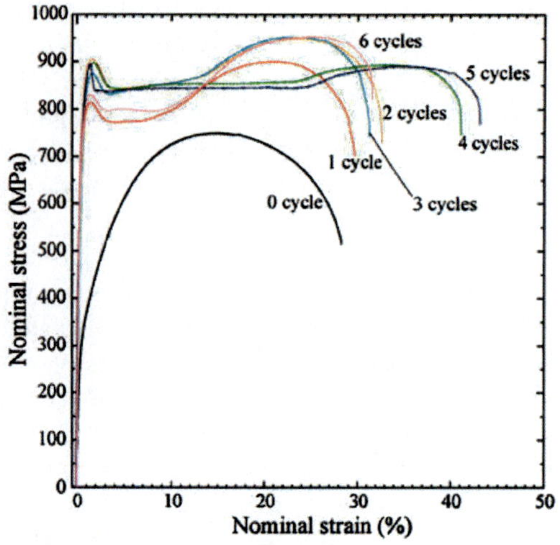

Fig. 12 Engineering stress–strain curves of the 15%Cr-10%Ni steel ARB processed by various cycles at 600°C

phenomena followed by unchanged stress periods were also appeared in the stress–strain curves. These unusual mechanical properties have not been observed in other SPD materials before.

The microstructures and phase ratios of the specimens before or after tensile tests were investigated in order to understand the unique mechanical properties, especially the large tensile ductility in the steel. Figure 13 shows TEM microstructures and selected area electron diffraction (SAED) patterns of the 15%Cr-10%Ni steel ARB processed by four cycles at 600°C (a) and subsequently deformed to 41% nominal strain by tensile test (b). The TEM image (a) shows that the steel ARB processed exhibits an elongated nanostructure, which is a typical structure of the materials produced by ARB. The mean thickness of the elongated grains is 210 nm. The result of the SAED pattern analysis indicates that the microstructure is austenite. In other words, the UFG austenite steel was produced by four cycles of the ARB process. The TEM image (b) of the specimen deformed to 41% strain by tensile test shows a similar microstructure to the as-ARB processed specimen. The mean grain thickness is 170 nm. Though both microstructures showed a similarity, the SAED analysis clearly indicated that the microstructure of the tensile-deformed specimen is martensite (bcc phase). Typical lath martensite microstructure composed of packets, blocks and laths [29] was not observed in this martensite transformed from the ultrafine grained austenite during tensile test. It is reasonable to consider that martensitic transformation occurs during deformation, and that the large elongation appeared in the nanostructured austenitic steels (Fig. 12) is attributed to the deformation-induced transformation. It can be concluded that TRIP phenomenon is effective to manage both strength and ductility also in nanostructured steels.

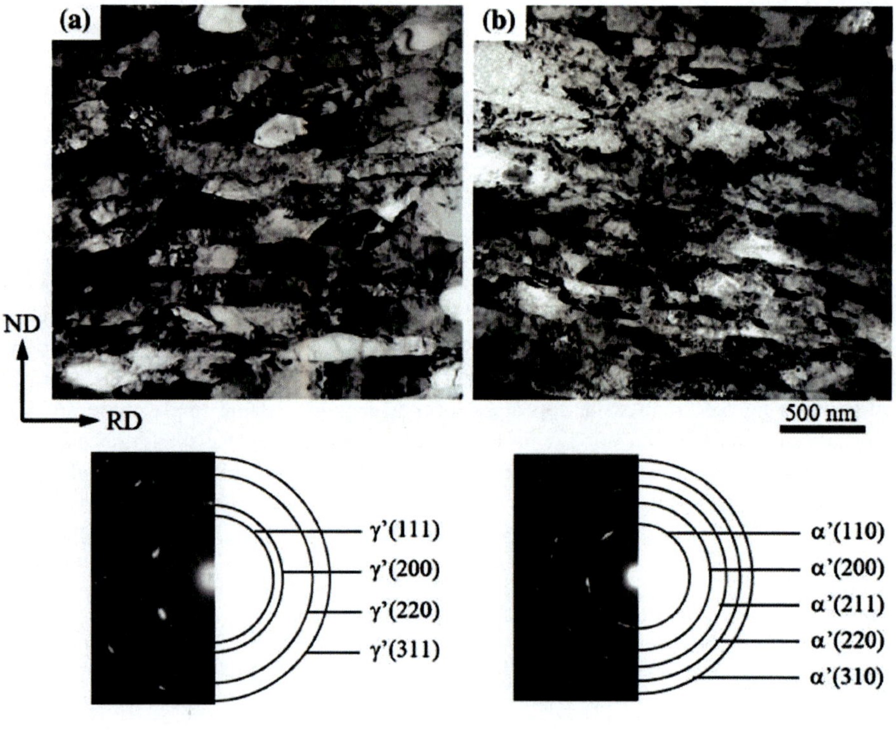

Fig. 13 TEM images and SAED patterns of the 15%Cr-10%Ni steel ARB processed by four cycles at 600°C. **a** Before the tensile test (as-ARB processed), and **b** after the tensile test (deformed to 41% strain by tensile test)

5 Summary

In the present paper, it was clearly shown that nanostructured steels with single phase certainly perform limited uniform elongation. The small uniform elongation is understood in terms of early plastic instability in the UFG microstructures. This understanding, however, also suggests a possibility to manage both high strength and adequate ductility in the nanostructured steels. Actually several experimental examples of the multi-phased UFG steels that could have good strength–ductility balance were represented in this paper.

From a practical viewpoint, tensile ductility is certainly important. In this sense, the present paper indicates that the future studies on the nanostructured steels should be directed to the materials having "multi-phased" structures. By the way, it should be also noted that the UFG metals show such instability only for tensile stress. Though tensile ductility is certainly limited, the UFG metals definitely do not lose plasticity and are not brittle. Actually, they show rather large post-uniform elongation in many cases, and it has been known that they can be heavily deformed in compression, rolling, bending, and so on. This should be emphasized even from a practical point of view. Additionally, it has been recently found that the nanostructured metals sometimes show surprising properties that cannot be understood in terms of conventional commonsense of metallurgy and materials science [22, 32–34], which suggests that further fundamental studies are desired for the nanostructured metals also from academic viewpoint, in order to pioneer a new frontier of metallic materials.

Acknowledgments This study was financially supported by the Grant-in-Aid for Scientific Research on Innovative Area, 'Bulk Nanostructured Metals', through MEXT, Japan, and the support is gratefully appreciated.

References

1. R.Z. Valiev, R.K. Islamgaliev, I.V. Alexandrov, Prog. Mater. Sci. **45**, 103 (2000)
2. S. Torizuka, K. Nagai, A. Sato, J. Jpn. Soc. Tech. Plast. **45**, 287 (2001)
3. Y. Hagiwara, M. Niikura, M. Shimotomai, Y. Abe, Y. Shirota, J. Jpn. Soc. Tech. Plast. **45**, 402 (2001)
4. N. Tsuji, Tetsu-to-Hagane **88**, 359 (2002)
5. M.J. Zehetbauer, R.Z. Valiev (eds.), *Nanomaterials by Severe Plastic Deformation* (Wiley-VCH, Weinheim, 2004)
6. B.S. Altan (ed.), *Severe Plastic Deformation Toward Bulk Production of Nanostructured Materials* (NOVA Science Publishers, Inc., New York, 2006)
7. R. Song, D. Ponge, D. Raabe, J.G. Speer, D.K. Matlock, Mater. Sci. Eng. A **441**, 1 (2006)
8. H.K.D.H. Bhadeshia, Mater. Sci. Eng. A **481–482**, 36 (2008)
9. A.A. Howe, Mater. Sci. Tech. **16**, 1264 (2000)
10. N. Tsuji, Y. Ito, Y. Saito, Y. Minamino, Scripta Mater. **47**, 893 (2002)
11. N. Tsuji, S. Okuno, Y. Koizumi, Y. Minamino, Mater. Trans. **45**, 2272 (2004)
12. I. Kozasu, *Controlled Rolling and Controlled Cooling* (ISIJ, Tokyo, 1997)
13. T. Maki, Tetsu-to-Hagane **81**, N547 (1995)
14. N. Tsuji, Y. Saito, S.H. Lee, Y. Minamino, Adv. Eng. Mater. **5**, 338 (2003)
15. X. Huang, M. Tsuji, N. Hansen, Y. Minamino, Mater. Sci. Eng. A **340**, 265 (2003)
16. N. Hansen, Metall. Mater. Trans. A **32A**, 2917 (2001)
17. N. Tsuji, R. Ueji, Y. Saito, Mater. Jpn. **39**, 961 (2000)
18. Y. Saito, N. Tsuji, H. Utsunomiya, T. Sakai, R.G. Hong, Scripta Mater. **39**, 1221 (1998)
19. H. Kitahara, N. Tsuji, Y. Minamino, Mater. Sci. Forum **503–504**, 913 (2006)
20. N. Kamikawa, Ph.D. thesis, Osaka University, 2006
21. N. Takata, S.H. Lee, C.Y. Lim, S.S. Kim, N. Tsuji, J. Nanosci. Nanotechnol. **7**, 3985 (2007)
22. Y. Wang, M. Chen, F. Zhou, E. Ma, Nature **419**, 912 (2002)
23. R.H. Wagoner, J.L. Chenot, *Fundamentals of Metal Forming* (Wiley, New York, 1997)
24. R. Ueji, Ph.D. thesis, Osaka University, 2004
25. N. Tsuji, R. Ueji, Y. Minamino, Y. Saito, Scripta Mater. **46**, 305 (2002)
26. R. Ueji, N. Tsuji, Y. Minamino, Y. Koizumi, Acta Mater. **50**, 4177 (2002)
27. R. Ueji, N. Tsuji, Y. Minamino, Y. Koizumi, Sci. Technol. Adv. Mater. **5**, 153 (2004)
28. N. Tsuji, R. Ueji, Y. Minamino, Trans. Mater. Res. Soc. Jpn. **29**, 3529 (2004)
29. H. Kitahara, R. Ueji, N. Tsuji, Y. Minamino, Acta Mater. **54**, 1279 (2006)
30. V.F. Zackay, E.R. Parker, D. Fahr, R. Busch, Trans. ASM **60**, 252 (1967)
31. T. Maekawa, H. Kitahara, N. Tsuji, Adv. Mater. Res. **26–28**, 413 (2007)
32. W.P. Tong, N.R. Tao, Z.B. Wang, J. Lu, K. Lu, Science **299**, 686 (2003)
33. X. Huang, N. Hansen, N. Tsuji, Science **312**, 249 (2006)
34. X. Huang, N. Kamikawa, N. Tsuji, N. Hansen, ISIJ Int. **48**, 1080 (2008)

Steels: Data Exploration for Discovery and Data-Sharing

Guoquan Liu

Abstract
We are coming into the data exploration age of science, so many things cannot be done without support from databases and data analysis. The same is true to steels. With some examples of trend analysis and material property chart analysis, this article attempts to discuss the importance and necessity of data exploration and data-sharing.

Keywords
Steel • Databases • Data-sharing • Data exploration • Future of steel

1 Introduction

Materials Science and Engineering discipline seeks to explain and control one or more of the four basic elements: structure and composition of a material; synthesis and processing of the material; properties of the material; performance of the material [1, 2]. As one of the most important engineering materials, steel has its own science and technology, as demonstrated by its development history and scientific literature.

Recently, the people start to consider the science evolution in the way focusing on data, as shown in Fig. 1. According to such kind of approaches, science was originally empirical, like Leonardo (da Vinci), making wonderful drawings of nature; next came the theorists who tried to write down the equations that explained the observed behaviors, like Kepler or Einstein; then when we got to complex enough systems, there came the computer simulations, the computational branch of science [3]. Now, we should be prepared to get into the data exploration part of steel science and engineering. That is, we are rich in information and data, including experimental, theoretical, computational, and simulated data related to the structure/property/processes/function/performance linkage of steels. Are the data well treated and fully used? How to explore and make full use of the data? To answer the second question may be extremely difficult; however, it should be helpful to do some primary work by discussing related examples and situations of steel scientific data sharing and exploration.

2 Trend Analysis: Evolution of Engineering Materials

Trend analysis involved historical data, current data, and predicted data for future development. One should collect these data and evaluate them, then construct databases, and do the data exploration to discovery the trend.

In this way, Ashby from Cambridge University (UK) analyzed the evolution of engineering materials with time [4], as shown in Fig. 2. It is very easily to see from the figure that, though metals once dominated among engineering materials in last century, they have been in serious competition with other newly developed engineering materials for several decades. The relative importance of

G. Liu (✉)
School of Materials Science and Engineering, University of Science and Technology Beijing, Beijing 100083, China
e-mail: g.liu@ustb.edu.cn

G. Liu
State Key Laboratory for Advanced Metals and Materials, University of Science and Technology Beijing, Beijing 100083, China

Y. Weng et al. (eds.), *Advanced Steels*, DOI: 10.1007/978-3-642-17665-4_16,

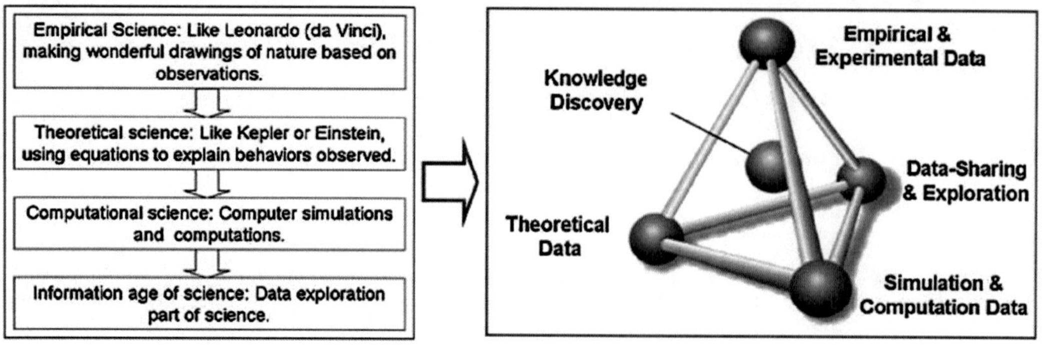

Fig. 1 A new way to view the evolution of science focused on data (design based on the idea from [3])

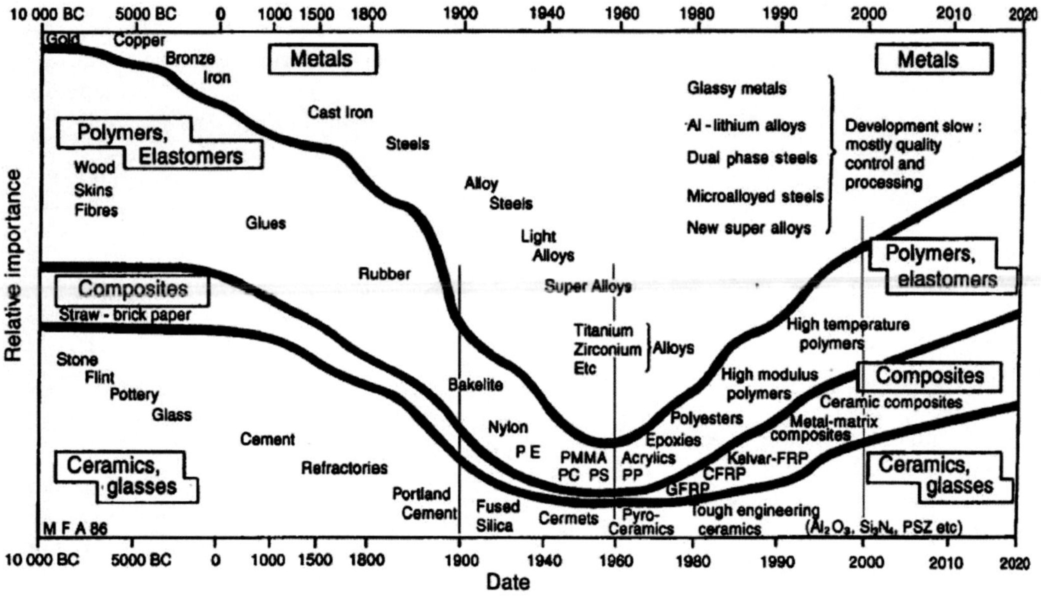

Fig. 2 The evolution of engineering materials with time [4] (The time scale is non-linear)

metals is lowering gradually after 1960, when advanced polymers, advanced ceramics, composites, and semiconductors, etc., have been becoming available and significantly changing our lives.

The evolution of alloy steels with time during the year of 1900–1960 in the United States is illustrated in Fig. 3. It is shown that the wars, especially World War II, greatly promoted the development of alloy steels which were badly needed for producing weapons, and had intense impact on the alloy steel's category [5]. A general picture can be easily drawn from such kind of extensively data based illustrations, and tells us that alloying of steels was influenced heavily by the availability of the alloying elements to the steel producer (the country). By the 1960s, "engineering materials" means "metals". Engineers were given courses in metallurgy; other materials were barely mentioned [4].

In 1991, Eagar from MIT [6] pointed out that steel is still the most common metal, having an overwhelming dominance in estimated world markets, comprising over 95% by volume, due to its relatively low cost, excellent properties and ease of fabrication, see Fig. 4. According to Eagar's estimation it should be pointed out that, although the trend shown in Fig. 2 is correct as a whole, the share of metals in engineering materials has not been diminishing such fast as shown in Fig. 2. One of reasons may be that from 1960 onwards, the "relative importance" of materials shown in Fig. 2 is based on data for the teaching hours allocated to each materials family in UK and US universities [4], without considering developing countries China, India, and the like; while the projection to 2020 relies on estimates of material usage in automobiles and aircrafts by manufactures [4] without considering the great demand of materials by

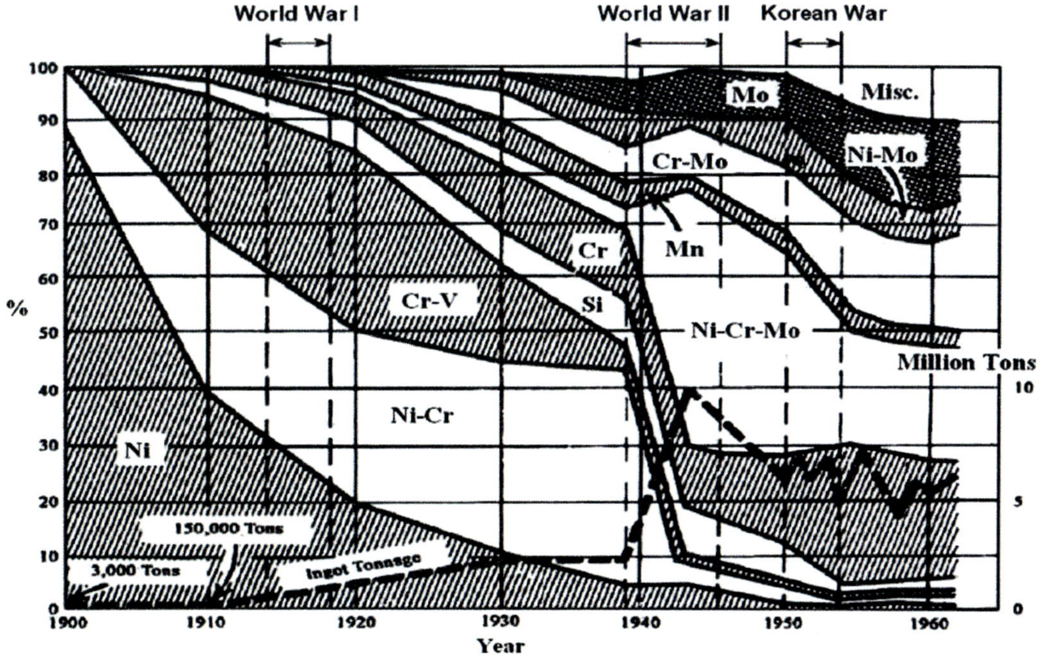

Fig. 3 Ingot tonnage and leading category of alloy steels in USA during the year of 1900–1960 [5]

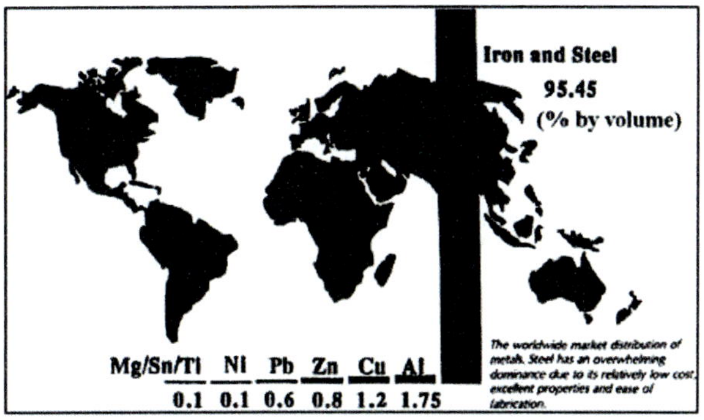

Estimated World Markets for Advanced Materials In the Year 2000

	Market Volume	Estimated Annual Growth
Fine Ceramics	$5-billion	20%
Composites	$15-billion	10%
Semiconductors	$100-billion	5%
Steel	$500-billion	2%

Fig. 4 Steel still has overwhelming dominance in estimated world markets for advanced engineering materials [6]

world's infrastructure, especially in developing or less-developed countries. Thus, the "relative importance" of metals especially steels has been underestimated in Fig. 2 after 1960.

Let us see the case in developing country such as China. In a recent excellent presentation based on extensive data analysis, Li [7] discussed the issue on "Iron & Steel in China in Post-Economic Crisis Era". Even the crisis always has a negative effect on steel production, steel industry remains strong and keeps growing after 1960, especially from 2000 onward after a 26 year slow increase (Fig. 5a). On the other hand, Fig. 5b shows that, as a developing country China's steel production reached 46.56% of the world production when China produced about 5.68 hundred million tons in 2009. Thus, nobody says again "Steel industry is sunset industry" in China today, and it explains why China universities currently offer more lectures on steels than western developed countries. Accordingly, the part of Fig. 2 from 2000 onwards needs to be revised if it is used for the whole world.

Li focused on the impact of the crisis on both the world's steel production and the steel technology development [7]. As shown in Fig. 5, world steel production increased slowly after the oil crisis in 1974 (increased only 100 million tons for the following 26 years). However, a variety of new steel technology developed rapidly in that period which means that the crisis provided not only challenges but also opportunities for steel development. Again in 2009, the recent economic crisis led to a marked reduction of world's steel production. Now we are also facing "climate crisis" and should pay great effects on creating green economy and a sustainable low-carbon world, which may offer new opportunities for development and breakthrough of steel technology [7].

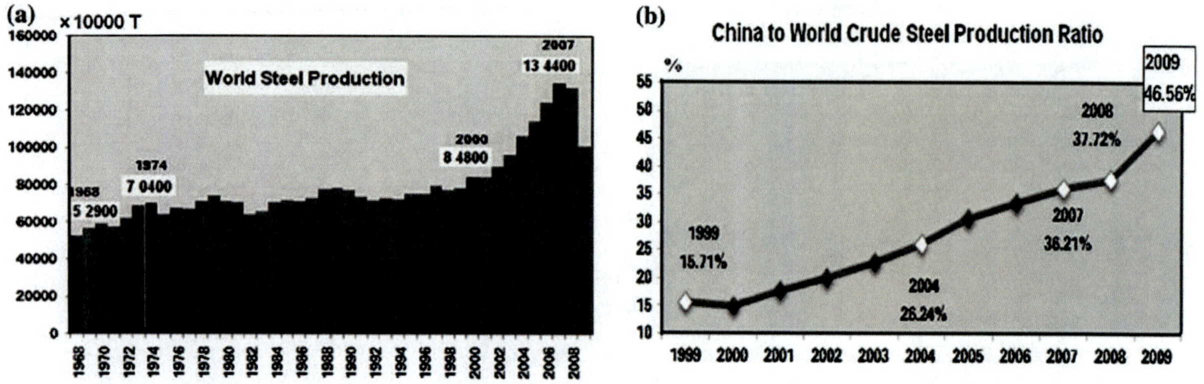

Fig. 5 a Variation of world's steel production with the year of 1968–2009; b The percentage occupied by China's steel industry [7]

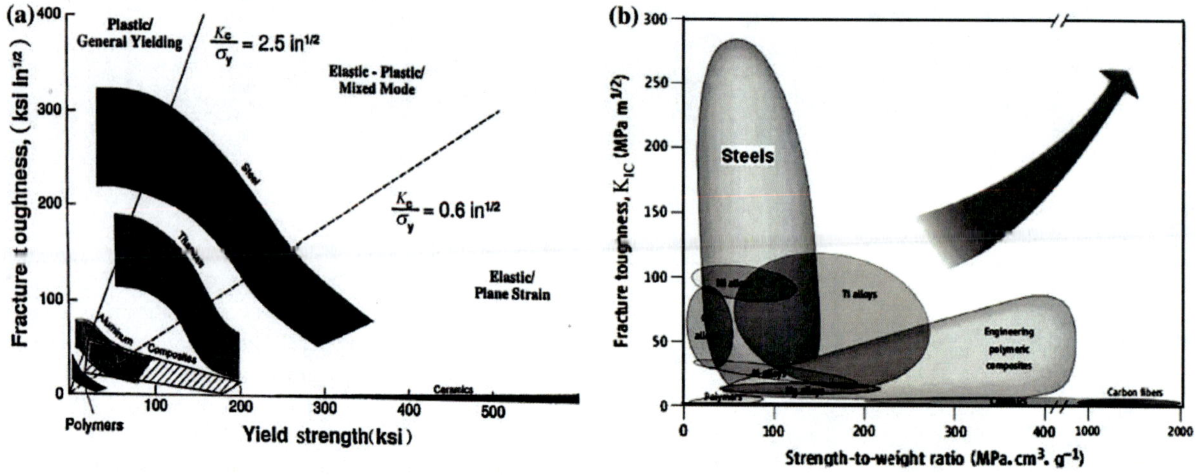

Fig. 6 a Steel has advantages of higher fracture toughness among advanced engineering materials (quoted from [6]), the three regions defining elastic and plastic behavior are described for materials of approximate 1 in. in thickness. b Steel has advantages of higher fracture toughness among advanced engineering materials (quoted from [8])

3 Property Chart Analysis: Exploring Material Properties in Pair

Material properties limit performance. So Ashby proposed to use so-called "Material Property Chart" to survey the materials and their properties that is useful especially for selection of engineering application [4]. Actually, many material scientists and material developers have been using similar charts for long plotting one property against another and mapping out the fields in property-space occupied by each materials class even by individual materials.

As illustrated in Fig. 2, the relative importance of metals started to lower after 1960 when other advanced engineering materials have been becoming available, i.e. the age of advanced materials started. Then, what is the future of metals? This question led to two important and inspiring articles with exactly the same title "The Future of Metals", one in 1991 from Eagar (a MIT professor) [6] and the other in 2010 on *Science* from Lu (a Chinese scientist) [8]. Either from economic point of view or from scientific point of view, they discussed advantages and limitations, opportunities and challenges of metals, especially steels, in the future. They all used strength (or specific strength)—toughness charts (Fig. 6a, b) to identify the advantages of steels as if by prior agreement.

Using a material property chart (called as "modified ratio-analysis diagram" in Ref. [6]), Eagar showed clearly that metals especially steels, have the most favorable combination of strength and toughness for most modern constructions, Fig. 6a.

Lu also used a material property chart (Fig. 6b), vividly demonstrating that steels have very high fracture toughness (a measure of the energy required for propagating cracks) but with limitations of low specific strength (the strength/weight ratio). It answers why the other advanced materials

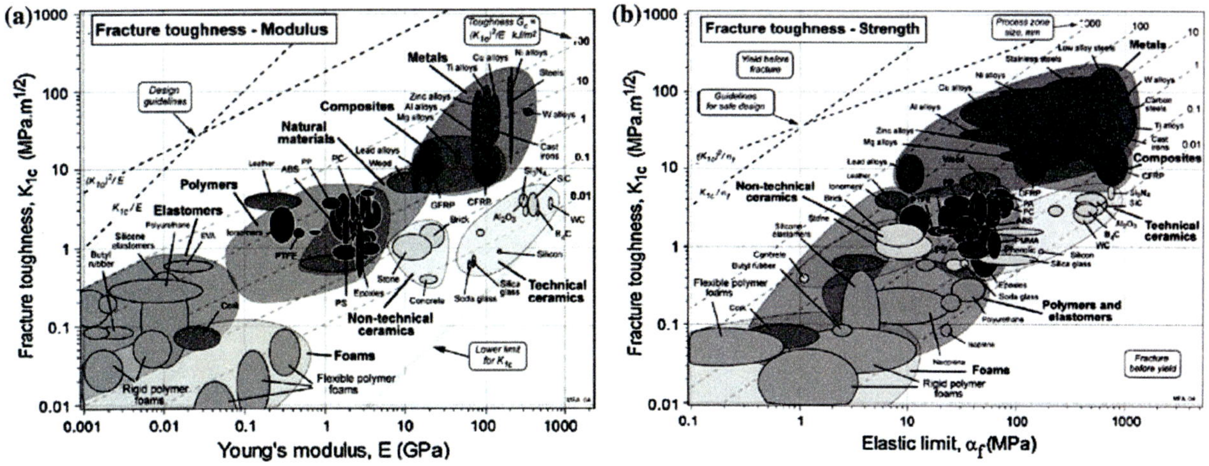

Fig. 7 Examples of combinational property charts [3]: fracture toughness-modulus (**a**) and fracture toughness-strength (**b**)

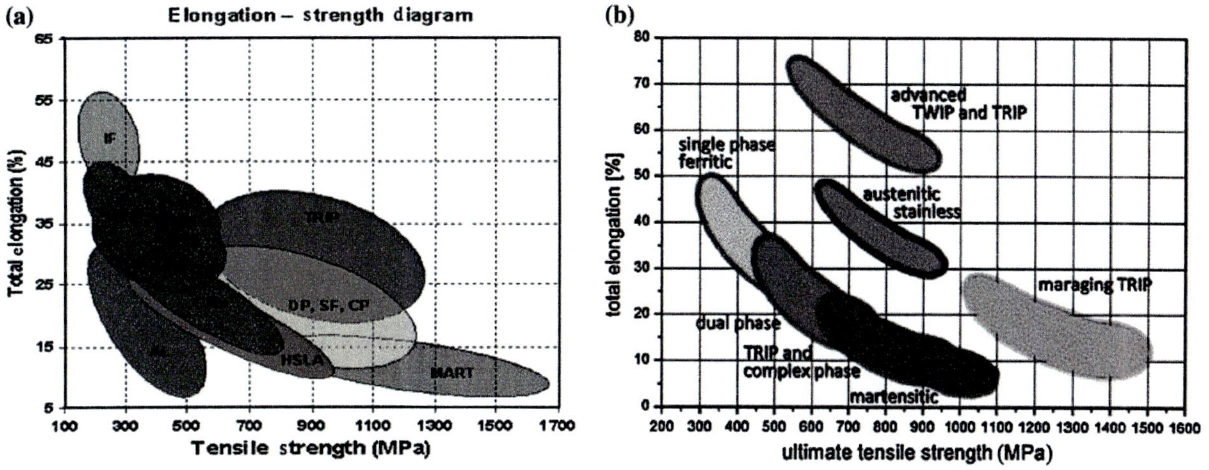

Fig. 8 Other combinational property charts, showing that new mechanisms and new steel variety are leading to higher strength–higher ductility steel products. **a** A chart for US Steel's products [9], **b** quoted from very recent research literature [10]

with higher strength/weight ratios used in the world's most fuel-efficient commercial jetliner—the Boeing 787 Dreamliner are substantially replacing traditional steels [8].

Thanks to these material property charts such as Fig. 6 as well as Fig. 7, we know that the steels are the toughest known materials, i.e. they have excellent fracture toughness and exceptional combination of strength and toughness. That is one of main reasons why steels own a bright future and remain to be the most common metal continually used in extremely large quantities for bridges, buildings, ships, transportation vehicles, and the like.

On the other hand, the low strength/weight ratios of steel would affect the application cases where weight is a primary concern [8]. If not the case, then as Eagar proposed, the growth of the world's infrastructure, especially in less-developed countries, as well as the maintenance of a high standard of living in developed countries, will create a steadily growing demand for metal (especially steel) products, and the metals market will continue to thrive into the twenty-first century [7]. His prediction has been verified by the growth of steel industry especially after the year of 2000 (Fig. 4).

The steel scientists and engineers also like to use material property charts to distinguish material classes and to find the key characteristics of steels. As illustrated in Fig. 8, they are keeping to try their best to further enhance the steel properties, including R/D of new steel products such as dual phase steels, advanced TRIP (Transformation-Induced Plasticity) steels, advanced TWIP (Twinning-Induced Plasticity) steels, and maraging (age hardenable) TRIP steels. For this type of applications, here the strength–elongation charts are used instead.

4 Further Discussion on Data-Sharing and Data Exploration

As demonstrated in previous two sections by only few examples, it is already seen vividly the importance and the necessity of collection, preservation, management (databases), exploration of data for discovery, or in general, of an emerging discipline called *materials informatics* [11]. However, when people are busying with their own work on research or steel production or product manufacture, huge volume of invaluable data generated are wasted day by day due to ignoring data-sharing, due to the data are kept unknown or unavailable to others. Actually, the data can be copied and repeatedly used, integrated and further explored, while the original data producer or owner often loses nothing.

In the "PROJECT PROPOSAL: World Materials—An open-access online materials database for everyone" he drafted, Pierre Villars from Switzerland describe the current situation in following words: "At present in the area of telecommunication and software development there are overall revolutionary developments (Internet, powerful search engines, relational database management systems, etc.). Here science fiction became reality. This in full contrast to the area of materials databases development where overall there exist many small materials databases, but with little compatibility, no overall concept, limited continuity, limited financial stability, etc. *Here we are still in Stone Age*. There is a big lack of the existence of a comprehensive fully relational, fully compatible materials database system. Without drastic changes within the next few years this will ultimately lead to the bottleneck in materials research and development." He appealed the public for attentions paying to worldwide data-sharing, and appealed for the creation of *World Materials*, an open-access online material database for everyone.

Governments and enterprises of different countries are paying for materials databases and data-sharing projects at different levels, for public benefit, commercial or other purposes. According to "The Outline Plan for The National Science and Technology Infrastructure Program during 2004–2010", China's government started the National Scientific Data Sharing Program in National Science and Technology Base Platforms in 2005. A *Ferrous Materials Data-Sharing Network* is now under construction by cooperation of University of Science and Technology Beijing and Central Iron and Steel Research Institute, as part of the *National Materials Science Data Sharing Network of China*, to meet the great demand from the society, industry, education, and the development of science and technology.

5 Conclusions

(1) It seems that the relative importance of steels and metals has often been underestimated, especially when the factors of developing or less developed countries are not correctly included.

(2) Various property chart analyses show that steel may be the toughest engineering material, owns exceptional toughness–strength combinations, and the properties of steel are continually improving by applying new technology. This helps steel to keep or even increase its share in materials market.

(3) Data exploration and data-sharing are invaluable approaches to steel science and technology, as demonstrated by above trend analysis and material property chart analysis.

Acknowledgments This work was mainly financially supported by the National Materials Science Data Sharing Network Project of China (No. 2005DKA32800), and partially from the National 973 Project (No. 2007CB209800), the High Technology Research and Development Program of China (No. 2007AA03A223), and the Science and Technology Project of Beijing, China (No. D08050303450802).

References

1. Materials Science and Engineering for the 1990s, Report of the Committee on Materials Science and Engineering, National Research Council (National Academy Press, Washington, 1989)
2. M.C. Flemings, R.W. Cahn, Acta Mater. **48**, 371 (2000)
3. A. Szalay, quoted in *New York Times*, May 20, 2003. Also see: Combinatorial Sciences and Materials Informatics Collaboratory (CoSMIC), http://mse.iastate.edu/cosmic/what2.html. Accessed July 13, 2010
4. M.F. Ashby, *Materials Selection in Mechanical Design*, 3rd edn. (Elsevier, Amsterdam, 2005)
5. K.-X. Guo (K.H. Kuo), Mater. Sci. Eng. **19**(3), 2–9 (2001) (in Chinese)
6. T.W. Eagar, Welding J. **70**(6), 69 (1991)
7. S. Li, Iron & steel in China in post-financial crisis era. Presented at the Low-Alloy Steel Society Meeting, CSM, held in Central Garden Hotel, Beijing, on June 26, 2010 (in Chinese)
8. K. Lu, Science **328**, 319 (2010)
9. Automotive Steel Technology, Formability chart: material based on strength and elongation, http://www.ussteel.com/corp/auto/tech/index.asp. Accessed July 13, 2010
10. D. Raabe, D. Ponge, O. Dmitrieva, B. Sander, A new group of 1.5 GPa steels with unexpected high ductility based on nano-precipitates (2010), http://www.mpie.de/index.php?id=2821. Accessed July 13, 2010 (also see: D. Raabe, D. Ponge, O. Dmitrieva, B. Sander, Nanoprecipitate-hardened 1.5-GPa steels with unexpected high ductility. Script Mater. **60**(12), 1141–1144 (2009))
11. G. Liu, J. Wu, H. Wang, C. Li, in *Proceedings of the 2nd Asian Materials Database Symposium*, March 10–14, 2010. Sanya, China (2010), p. 62

Long Life High Strength Steels to Resist Fatigue Failure and Delayed Fracture

Weijun Hui, Han Dong, Yuqing Weng, Jie Shi, and Maoqiu Wang

Abstract

In recent years, economic and environmental considerations have increased the need to safely extend the service life of components and structures beyond their original design life. It well known that fatigue and delayed fracture are the two important mechanisms for the failure of steel components and structures in service. Based on our systematic studies of the fatigue failure and delayed fracture behaviours of high strength steels in the last 10 years, we proposed a new kind of concept to improve both fatigue failure resistance and delayed fracture resistance of high strength steels, which comprises the combination of inclusion modification, structure controlling, hydrogen trap controlling and grain boundary strengthening, and it is called IST & GST technology in this paper. Firstly, basic considerations for the improvement of service life of high strength steels are introduced. Secondly, methods such as inclusion modification, structure controlling and hydrogen trapping controlling are discussed for improve the fatigue failure and delayed fracture resistance of high strength steels. Finally, based on IST & GST technology, examples of development of long life high strength steels such as a newly developed 2,000 MPa grade high strength spring steel with excellent fatigue failure resistance and 1,500 MPa grade high strength bolt steel with superior delayed fracture resistance are introduced.

Keywords

High strength steel • Fatigue failure • Delayed fracture • Non-metallic inclusion • Microstructure • Hydrogen trap

1 Introduction

Higher-strength materials are the main objectives for which material scientists have strived for many decades. With the development of modern industry, there is an increasing demand for higher strength steels with tensile strength exceeding about 1,200 MPa to be used for high strength components and structures, mainly for the purposes of higher performance, weight reduction and cost saving, such as the use of higher strength bolts could decrease the number, size and weight of bolts [1]. The ideal structural steel combines high strength, to resist failure by plastic deformation, with high toughness, to resist failure by crack propagation, and with high performance, to possess longer service life. However, these objectives are often contradictory and frustrating, such as steels with higher strength are, ordinary, less torrent of internal flaws and liable to fracture at smaller applied loads [2].

In recent years, economic and environmental considerations have increased the need to safely extend the service life of components and structures beyond their original design

W. Hui (✉) · H. Dong · Y. Weng · J. Shi · M. Wang
Central Iron and Steel Research Institute, Beijing 100081, China
e-mail: huiweijun@nercast.com

Y. Weng
The Chinese Society for Metals, Beijing 100711, China

Y. Weng et al. (eds.), *Advanced Steels*, DOI: 10.1007/978-3-642-17665-4_17,

life. It well known that fatigue and delayed fracture are the two important mechanisms for the failure of steel components and structures in service. With increasing the strength of steel, its toughness generally decreases, and its fatigue failure susceptibility increases, particularly its delayed fracture susceptibility increases remarkably. Therefore, the application of higher strength steel requires the improvement of its toughness, fatigue failure and delayed fracture resistance simultaneously. Toughening of high strength steels has been an important subject of extensive studies over past decades and remarkable achievement has been obtained. Therefore, the improvement of fatigue failure and delayed fracture resistance is mainly concerned in this paper.

Actually, in the last over 10 years, two consecutive national programs, namely "Important Basic Research of New Generation Steel Materials"(October 1998–October 2003) and "Metallurgy Basic Research on Iron and Steel Quality Improvement and Service Lifetime Increase" (October 2004–October 2009) have been conducted in China and one of its objects is to significantly improve the service life of high strength steels for structural components [3]. Similarly national programs have also been implemented in Japan and Korea [4, 5].

2 Basic Considerations for the Improvement of Service Life of High Strength Steels

2.1 Fatigue Failure

Fatigue is a common failure mechanism of structural components in service, and has been extensively studied for more than one century. Unfortunately, accidents or even disasters, resulting from fatigue failure, still happen almost every day. In practical engineering situations, high-cycle fatigue (HCF) failures (life normally lower than 10^7 cycles) are very common. While in many industries, the required life time of many components such as in engine, automobile, railway, air plane, offshore structures, bridge and specific medical facilities, often exceeds 10^8 cycles. In fact, it has been found that components or structures, made of high strength steels, still suffer fatigue failure at higher number of cycles, though these steels was considered to display a fatigue limit at a high number of cycles (typically $>10^6$) [6–9]. Therefore, solutions to prevent such a failure are a strong demand with the development of modern industries such as automobile, machines, construction and railway. Therefore, there is global interest in, and need for, understanding of the fatigue behaviour of structural materials in the HCF regime, especially in the very high cycle fatigue (VHCF) regime.

It is well known that there is a good correlation between rotating bending fatigue strength, σ_w, of smooth specimens and tensile strength, R_m, for low or medium strength steels as follows:

$$\sigma_w = 0.5R_m(R_m \leq 1200\,\text{MPa}) \tag{1}$$

In this case, fatigue cracks tend to initiate from the specimen surface and therefore are termed as surface fracture. However, this linear correlation does not hold and there is more scatter or even declination in fatigue strength values for high strength steels with strength level over about 1,200 MPa. In this case, the origins of fatigue fracture are not always at the specimen surface but often some distance away, particularly for HCF especially VHCF causing the so-called internal fracture or fish-eye fracture. As a result, σ_w is generally much lower than the predicted value by Eq. 1, and could be calculated by the following equation [10]:

$$\sigma_w = \frac{C(HV + 120)}{(\sqrt{area})^{1/6}} \times [(1 - R)/2]^{\alpha} \tag{2}$$

where $\sqrt{area}$: square root of the projected area of small defects or inclusions, μm; HV: Vickers hardness, kgf/mm^2; C is the location constant and surface, subsurface, and interior inclusions equal to 1.43, 1.41 and 1.56, respectively; $\alpha = 0.266 + \text{HV} \times 10^{-4}$.

Internal fractures originate from internal defects, which are mostly inclusions, but in some cases are microstructural defects [6–12]. The defect especially inclusion size and property are thought to be the main factors that control the fatigue properties of high strength steels. Therefore, many attentions have been paid to minimize both the size and number of inclusions and the prevention of fish-eye fracture would certainly improve the fatigue property of high strength steels.

Our and other researcher's further experimental results show that there might exist a critical size of inclusions, below which the fatigue fracture origins will not initiate from the inclusion but from specimen surface or internal microstructural defects [7, 11, 13–15]. For example, our recent work shows that for clean 54SiCrV6 and clean 50CrV4 steels, in which the inclusion size (≤3.0 μm) is smaller than the critical size (about 3.5 μm for the steels investigated), the fatigue failure originated from the inclusion clusters, whereas for clean 54SiCr6 steel, in which the inclusion size is smaller than 1 μm, the fatigue failure hardly occurs in the 10^6–10^9 cycles regime [15].

However, in practice, it is a great challenge to control inclusion size to such extreme low level. Therefore, it is suggested that other methods such as controlling microstructure besides controlling inclusions should be explored to minimize the detrimental effect of inclusions and to improve the fatigue properties of high strength steels. Wei et al.'s study about the fatigue behavior of 1,500 MPa bainite/martensite duplex-phase high strength steel showed that the steel has higher fatigue strength and fatigue crack threshold and lower crack propagation rate compared to fully martensitic steel [16]. Our recent study of the VHCF and fatigue crack growth (FCG) behaviors of newly developed 2,000 MPa ultra high strength spring steel with Bainite/Martensite duplex microstructure obtained through isothermal quenching also confirmed this suggestion [17].

A very important characteristic of SEM observation of fracture surface is that the vicinity of the inclusion at the fracture origin in the case of long life has a particular rough and granular morphology compared with the area surrounding it inside the fish-eye. Murakami et al. firstly reported this phenomena and called it the Optically Dark Area (ODA), for it looked optically dark when observed by an optical microscope [18]. This particular region is named GBF in this paper. The relative size of GBF to the size of the inclusion at the fracture origin increases with increase in fatigue life. Thus, the GBF is considered to have a crucial role in the process of VHCF failure and is considered that the fatigue life in the VHCF regime is mostly spent in the crack initiation and early crack growth inside the GBF. Figure 1 shows SEM fractographs of a typical "fish-eye" with an inclusion in the centre and a GBF surrounding the inclusion [19].

Many investigators have tried to explain the formation mechanism of GBF and obtained some interesting results. The most possible mechanism is that proposed by Murakami etc. who considered that the formation of GBF might be related to hydrogen trapped by inclusion [9]. Our recent experimental results provide a strong implication that hydrogen trapped by inclusion plays an important role in the VHCF regime for high strength steels [20, 21]. Other results [22, 23] also confirmed that the formation of GBF is closely related to hydrogen trapped by inclusions. Therefore, it is suggested that stronger hydrogen trap other than inclusion be introduced into high strength steel to alleviate the harmful effect of hydrogen trapped by inclusion. As discussed later, the hydrogen trapping effect of finely precipitated carbides or carbonitrides of microalloying elements such as V and Ti could improve the DF resistance of high strength steels. Thus, improvement in fatigue properties of high strength steels owing to these fine carbides or carbonitrides is also expected from the viewpoint of suppressing or delaying GBF formation duo to the hydrogen trapping effect.

Based on above considerations and our systematic studies of the fatigue failure behaviour of high strength steels, we proposed a new kind of concept to improve fatigue failure resistance which comprises the combination inclusion modification, structure controlling and hydrogen trap controlling, and it is called IST technology.

2.2 Delayed Fracture

Delayed fracture (DF) is a phenomenon in which components such as bolt or pre-stressed concrete steel bar

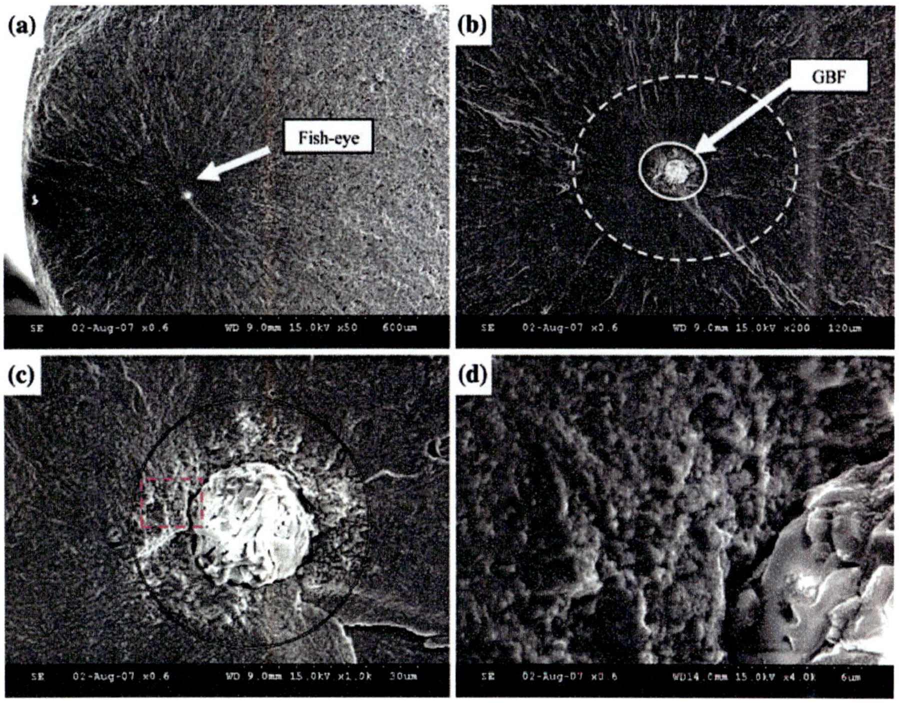

Fig. 1 SEM fractographs showing a "fish-eye" (**a**) with an inclusion in the center (**b**) and a GBF surrounding the inclusion (**c**) and its high-reservation magnification (**d**) of crack initiation site of 60SiCrVA steel ($\sigma_a = 725$ MPa, $N_f = 3.87 \times 10^8$ cyc) [19]

suddenly fail after a certain period of duration after loading. It is a kind of environmentally enhanced embitterment, resulting from the interaction among materials, environment and stress [24, 25]. As stress corrosion cracking (SCC) of high strength martensitic steel in aqueous environment is actually a hydrogen-induced cracking phenomenon, therefore, the DF discussed in this paper is referred to either hydrogen-induced DF or SCC in aqueous environment of high strength martensitic steel. DF susceptibility has a strong relationship with strength, particularly when tensile strength exceeds about 1,200 MPa, the fracture stress decreases significantly, which severely limits the applications of higher strength martensitic steels [25]. Therefore, DF is a major challenge for applications and receives continuous attention and efforts for decades.

Since DF is induced by hydrogen, how hydrogen facilitates crack initiation and propagation will be a key point. As the solute hydrogen content is rather small, therefore, the experimentally measured quantity of hydrogen to induce DF is not the hydrogen in real solute condition, but the one stored in various traps in the steel. As DF often occurs at ambient temperature, it is considered that such fracture results from diffusible hydrogen rather than non-diffusible hydrogen, which has been supported by some experimental evidences [26, 27].

From the aforementioned recognition, in the development of DF resistant high strength steels, a rational approach can be taken in response to the DF process composed of the entry of hydrogen, the transportation and accumulation of hydrogen to stress concentrated region such as prior austenite grain boundaries, and the formation of cracks. The proposed approach comprises: (1) preventing the entry of hydrogen; (2) rendering the absorbed hydrogen harmless; and (3) improving microstructural homogeneity such as grain boundary properties.

Furthermore, it has been realized by a number of studies in laboratory specimens and failure bolts in service that delayed fracture in quenched and tempered steels is usually initiated and propagated from prior austenite grain boundary [1, 28]. The interface between carbide and matrix has been verified as one of the strongest traps for hydrogen [29, 30]. Carbide in the form of films precipitates at prior austenite grain boundaries causes failure along prior austenite grain boundaries and deteriorates the fracture strength of high strength steels in the presence of hydrogen [31]. Therefore, prior austenite grain boundaries should be strengthened by several methods such as reduction of grain boundary segregation, grain refinement, prevention of film-like carbides, controlling the characteristics and distribution of hydrogen traps to improve the delayed fracture resistance of high strength steels [32].

Based on aforementioned considerations and our systematic studies of the delayed fracture behaviour of high strength steels [32], we proposed a new kind of concept to improve delayed fracture resistance which comprises the combination grain boundary strengthening, structure controlling and hydrogen trap controlling, and it is called GST technology.

3 Methods to Improve the Fatigue Failure Resistance of High Strength Steels

3.1 Inclusion Modification

For decades, inclusions have been associated with the general problem of fatigue failure in steels. It well known, as mentioned above, that inclusions have a strong effect on the fatigue properties of high strength steels, which is more significant in the HCF especially VHCF regime. Inclusion modification including size refining and composition controlling is one of the main efforts to improve the fatigue properties, especially in the VHCF regime of high strength steels.

Figure 2 shows the statistical results of 100 fatigue crack surfaces for commercial spring steels 60Si2CrVA and 60Si2MnA [33]. Obviously, fatigue fractures initiated from inclusions is only 39% at 10^7 cycles; however, it increased to 71% at 10^9 cycles, this is to say, the initiation from

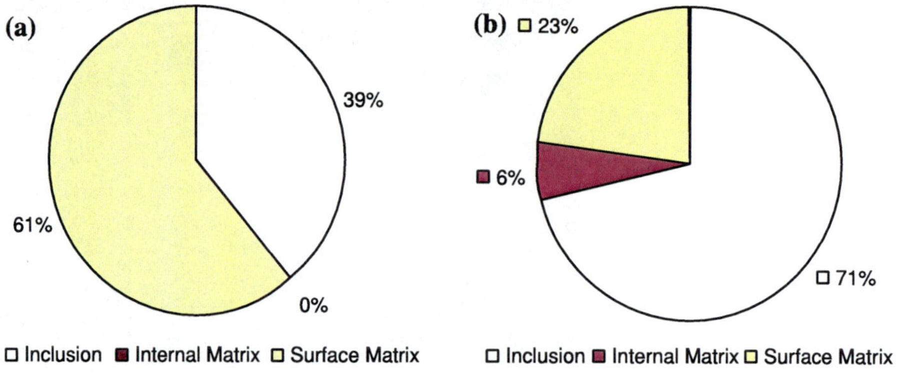

Fig. 2 Statistics of crack initiation site features at fatigue fracture surfaces under high cycle (**a**) and very high cycle fatigue (**b**) of 100 fatigue crack surfaces for commercial spring steels 60Si2CrVA and 60Si2MnA

internal inclusions significantly increased in the VHCF regime. Hong et al.'s analysis of 58 group literatures data also found that fractures preferentially initiate from internal inclusions when tensile strength exceeds about 1,600 MPa, while fractures initiate from either surface or internal, especially in the lower strength level [34].

3.1.1 Inclusion Size

The properties of clean steels are highly affected by the few large inclusions. The most deleterious inclusions are the largest ones, which are almost invariably collections of small particles rather than one single precipitate. Thus, the occurrence of large inclusions is of special importance, as could be seen from Fig. 3. It is clear that most fractures are initiated from subsurface large inclusions and their sizes tend to decrease for the further interior ones. Although the average inclusion size is about 25 μm, the occurrence of much larger inclusions is still frequent.

Murakami etc. conducted intensive studies on the effects of inclusions on fatigue behaviors of high strength steels, and proposed the $\sqrt{area}$ parameter model (Eq. 2), which quantitatively pointed the significant effect of inclusion size on fatigue strength. From Eq. 2, it could be seen that if the size of inclusion decreases to one half of it original size, fatigue strength will increase about 15%. Figure 4 shows the variations of fatigue strength with inclusion size

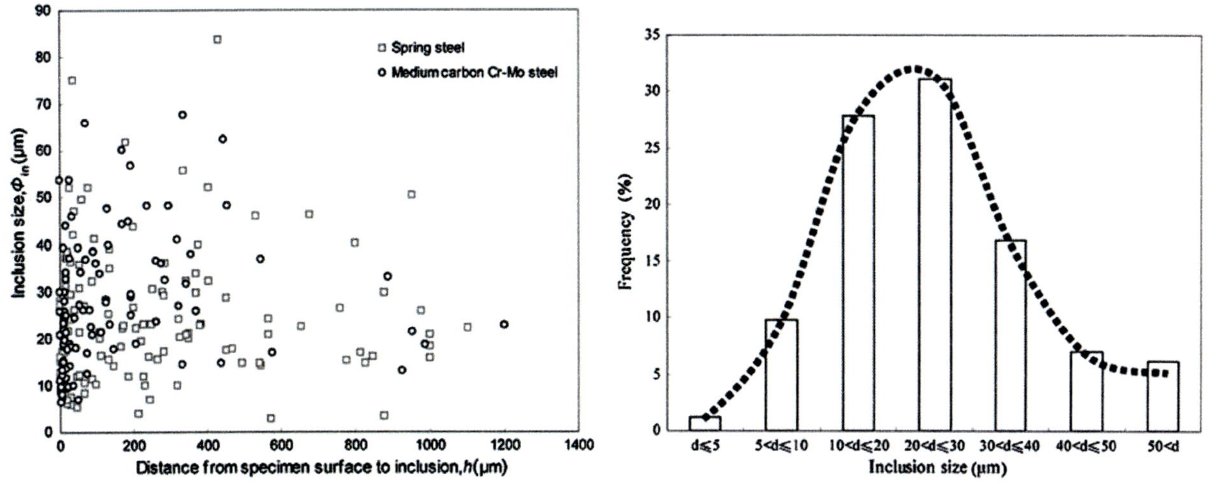

Fig. 3 Statistical results of 244 data obtained from fracture origins of high strength spring steels and medium-carbon Cr-Mo steels in the HCF and VHCF regime

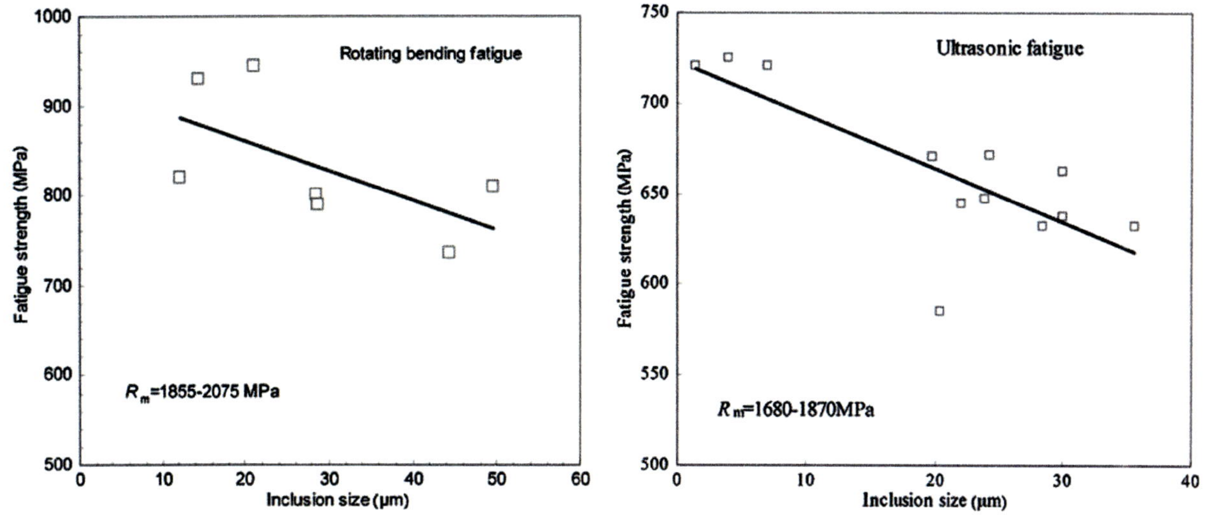

Fig. 4 Variations of fatigue strength with inclusion size (average size of inclusions at fracture origins for failed specimens) of high strength steels, showing a strong dependence of fatigue strength on inclusion size

(average size of inclusions at fracture origins of failed specimens) and Figs. 5 and 6 show our results of the S-N curves of two different spring steels 60Si2CrV and 50CrV4 with different inclusion sizes [15, 19, 35]. It is clearly that decreasing the size of inclusion could considerably increase both fatigue strength and fatigue life. Therefore, reducing the size of inclusions especially the size of detrimental large inclusions is one of the major and effective methods to improve the fatigue properties of high strength steels.

Large oxide inclusions are dangerous and much more harmful than small inclusions. A critical inclusion size is usually defined, above which inclusions are dangerous and can cause the failure of steel products [36]. Based on Murakami et al.'s the $\sqrt{area}$ parameter model and assumed the inclusion to be spherical, we deduced expressions to estimate the critical inclusion size [14]:

$$\phi_{in,c} = C\left(1 + \frac{120}{HV}\right)^6 \qquad (3)$$

where C is a constant of 0.813, 0.528, 0.969 for surface(here the inclusion is assumed as hemispherical), subsurface and interior inclusions, respectively. Obviously, the critical inclusion size decreases with increasing the hardness of steel and the minimum critical inclusion size is of subsurface inclusion. Further experimental results indicate that this method of estimating the critical inclusion size is reasonable [14, 33]. For two groups of specimens of 42CrMo steel with extremely small inclusions(within 1 μm), all the fatigue crack initiated from surface matrix; and for two commercial 42CrMo steels, the fatigue cracks initiating at inclusions is just because the inclusion sizes in these steels overrunning their estimated critical inclusion sizes.

Finally, it should be noted that as it is impossible to produce steels without inclusions, control over the size distribution becomes more and more important since it largely determines the probability of the presence of a large inclusion at a critical location.

3.1.2 Inclusion Composition

Stress localization at the interface between the inclusion and matrix is the origin of fatigue failure initiation. The effect of inclusions on the fatigue crack initiation depends on the chemistry, size, density, location relative to the surface, and morphology. In terms of the effect of inclusion chemistry on fatigue crack initiation, the most detrimental inclusions, in decreasing order of harm, are calcium aluminates, Al_2O_3, and spinel. This is precisely the order in which the average thermal expansion coefficients increase [37]. Large exogenous inclusions of slag or refractory origin are always detrimental to fatigue properties of high strength steels because of their large size and irregular shape.

Different inclusion types also have different geometries, affecting the stress concentration at the steel/inclusion interface during fatigue loading. Smooth and elongated inclusions in the direction of major principle stress, such as manganese sulphide, have lower stress concentrations, besides smaller $\sqrt{area}$, than globular inclusions such as calcium aluminate or sharp edged inclusions such as titanium nitride. A higher stress concentration facilities both initiation and early propagation of microcracks and, thus, affects fatigue life adversely. However, the same inclusion can have different effects on fatigue strength depending on the direction of loading. For example, Abe et al.'s study on two spring steels shows that the transverse direction fatigue

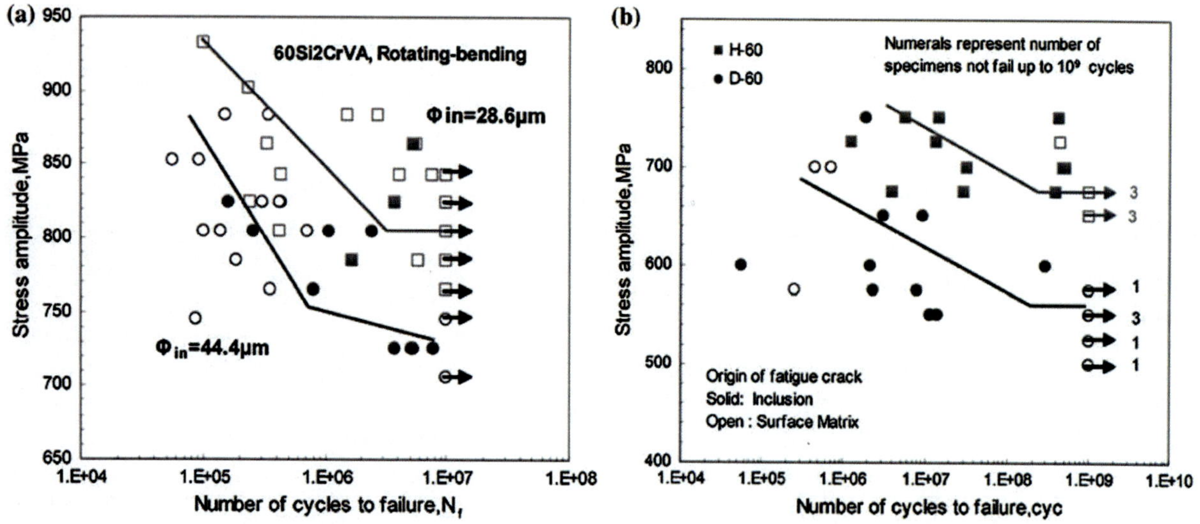

Fig. 5 S-N curves of two spring steels 60Si2CrVA with different inclusion sizes in HCF regime (**a**) and VHCF regime (**b**), showing that decreasing the size of inclusion could considerably increase both fatigue strength and fatigue life (oxygen content of both spring steels D-60 and H-60 is about 10 ppm)

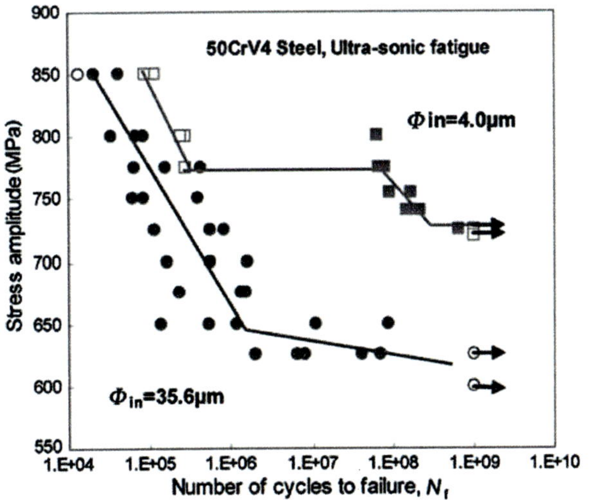

Fig. 6 S-N curves of spring steel 50CrV4 with different inclusion sizes, showing that decreasing the size of inclusion could considerably increase both fatigue strength and fatigue life

tests showed fish-eye fractures with the origins of large and lengthened MnS inclusions and the fatigue strengths were about a half of those in the longitudinal(rolling) direction, which origins were mainly small Al_2O_3, TiN and matrix [38].

The control of deformable oxide inclusions is critical to product performance. Extensive work has been carried out over the years, both theoretically and experimentally, to engineer deformable oxide inclusions [39–41]. Through readjusting the content of the easily oxidizable elements in molten steels, it is possible to research the objective of controlling composition of secondary deoxidization products and making them deformable and then changing to long fine stringers during hot rolling. The square root of projection area of inclusion $\sqrt{area}$ of these long fine stringers of inclusion in the longitudinal direction is so small that is difficult to initiate crack, this is the main purpose of oxide inclusion composition and morphology controlling.

It should be noticed that, as mentioned above, a firm opinion that hard inclusions are more detrimental than soft inclusions for fatigue strength has prevailed for long years. However, an thorough studies by Murakami et al. [42] shows that this opinion was proved incorrect, i.e., the chemical composition of inclusions is not crucial factor controlling fatigue limit, even if the chemical composition influences the rigidity of inclusions and the residual stresses around inclusions. On the contrary, it is verified that two crucial factors which control the fatigue strength are the Vickers hardness of matrix HV and the square root of projection area of inclusion $\sqrt{area}$, as could be seen from Eq. 2.

3.1.3 Cleanliness

The volume fraction of inclusions depends directly on the oxygen and sulphur content. Advances in steelmaking technology have resulted in remarkable decreases of the contents of oxygen and sulphur, and hence a significant decrease of the inclusion content. For the improvement of fatigue properties of steel, the number of the large seized inclusion should be reduced and since the control of individual inclusions is difficult, the most practical way to achieve this is to lower the oxygen content. Iikubo et al.'s result shows that the fatigue limits of ultra-clean spring steel do fall on the straight line extrapolated from the data for low and medium strength steels [43]. In fact, the cleanliness was improved by decreasing the oxygen content, and this led to a decrease in the size of inclusions.

It should be noticed, however, the number of the large seized inclusion should be reduced while reducing oxygen content. Our investigation of the fatigue fracture behaviors of two kinds of commercial 42CrMo steel with deferent cleanliness by using rotating bending fatigue test showed that the fatigue strength of the two steels is almost the same when tensile strength level is 1,500 MPa; however, the fatigue strength of high cleanliness 42CrMo steel (H-42CrMo) is obviously higher than that of low cleanliness 42CrMo steel (L-42CrMo) when tensile strength level is 2,000 MPa (Fig. 7) [44, 45]. SEM analysis of the fractured surface of the specimens with tensile strength level of 2,000 MPa showed that fatigue crack initiates mainly from non-metallic inclusions, whereas from matrix surface for specimens with tensile strength level of 1,500 MPa. Although impurity content such as oxygen of H-42CrMo steel (13 ppm) is much lower than that of L-42CrMo steel (42 ppm), the sizes of inclusion at the origins of the fractures are much bigger. Granulometric analysis of inclusion determining its size distribution also shows that there are a certain amount of large spinel inclusions for the H-42CrMo steel (Fig. 8). This kind of large spinel inclusion deteriorates the fatigue strength of H-42CrMo steel, however, which is still higher than that of L-42CrMo steel at the same strength level. Our further investigation of the fatigue fracture behaviors of two spring steels.

60Si2CrVA shows that there was significant difference of the fatigue properties of the 60Si2CrVA steels, although their oxygen content is equal (Fig. 5) [19, 35, 44]. SEM examination of fatigue fracture surface reveals that these differences were mainly caused by the difference of inclusion size. That is to say, both fatigue life and fatigue strength increase with decreasing inclusion size. The measurements on the inclusion particles from the commercial 38Si7 spring steel with extremely low oxygen content (<6 ppm) has also proved this finding. As shown in Fig. 9, almost all fatigue failures initiate from the large oxide inclusion.

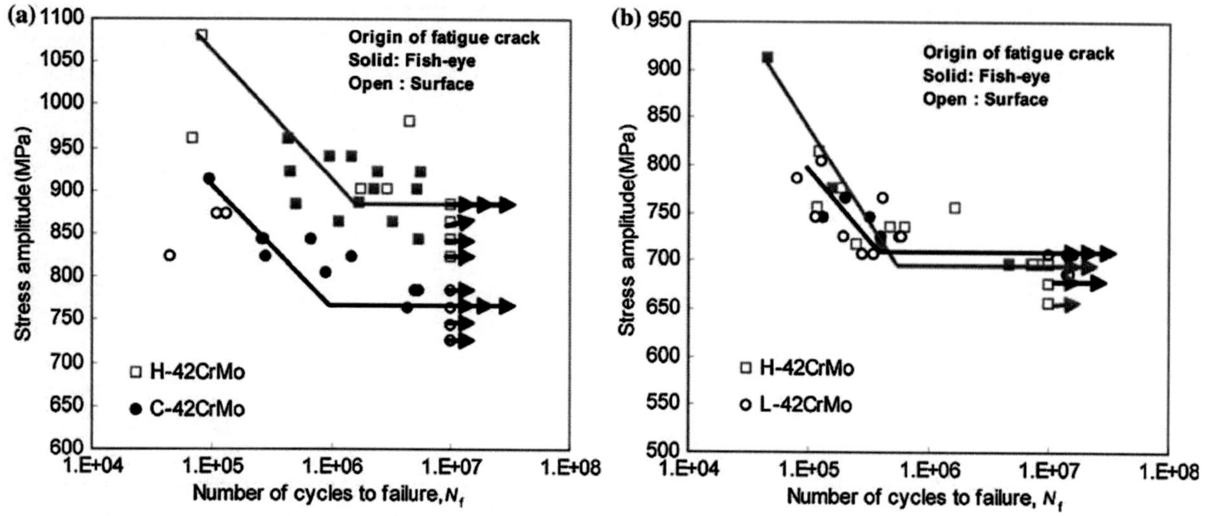

Fig. 7 S-N curves of rotating bar two-point bending fatigue of 42CrMo steel (H-42CrMo refers to high cleanliness 42CrMo steel with [O] = 13 ppm, L-42CrMo to low cleanliness 42CrMo steel with [O] = 42 ppm) (**a**) R_m = 1,960–2,020 MPa; (**b**) R_m = 1,510–1,560 MPa

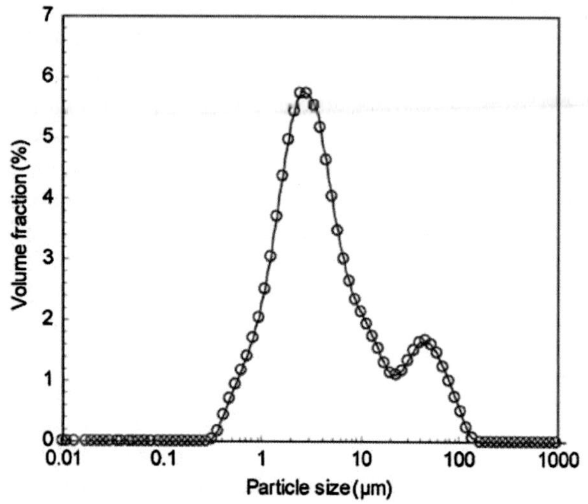

Fig. 8 Granulometric distribution of inclusion of high cleanliness H-42CrMo steel

Therefore, besides controlling steel cleanliness such as oxygen content, controlling of the number of large seized inclusion is of special important, otherwise, incorrect conclusion may be obtained.

Fukumoto and Mitchell [46] proposed that the so called "zero-inclusion steel" in which the inclusion size is smaller than 1 μm could improve the fatigue performance of high strength steel substantially. Our investigation of the fatigue behaviors of zero-inclusion and commercial 42CrMo steels in the VHCF regime showed that fatigue cracks in most samples of the commercial ones initiated at inclusions, while the fatigue cracks of all samples initiated from the surface matrix in the case of the zero-inclusion ones [13]. When the applied stress is around the fatigue strengths, the specimens of the two commercial steels may facture in 10^6–10^9 cycles regime. However, if the applied stress is around the fatigue limits, the specimens of the two zero-inclusion steels are hardly fractured when the number of cycles is greater than 2 × 10^6. Therefore, compared with the fatigue strength of the commercial counterpart, the fatigue limits if zero-inclusion steel is increased moderately, but the improved reliability of the VHCF life is really encouraging.

3.2 Structure Control

As mentioned above, it is often observed that the aforementioned fish-eye fracture in high strength steels originates from internal defects, which are mostly non-metallic inclusions, but in some cases are microstructural defects. A very important characteristic of SEM observation of fracture surface is the formation of GBF area. GBF is considered to have a crucial role in the process of VHCF failure and is considered that the fatigue life in the VHCF regime is mostly spent in the crack initiation and early crack growth inside the GBF. Therefore, microstructural condition also plays an important role in the process of VHCF failure and is expected to improve the fatigue property of high strength steels through structure controlling.

Furthermore, experimental results show that there might exist a critical size of inclusions, below which the fatigue fracture origins will not initiate from the inclusions but from specimen surface or internal microstructural defects. In practice, it is a great challenge to control inclusion size to such extreme low level. Therefore, other methods such as controlling microstructure besides controlling inclusions should be explored to minimize the detrimental effect of

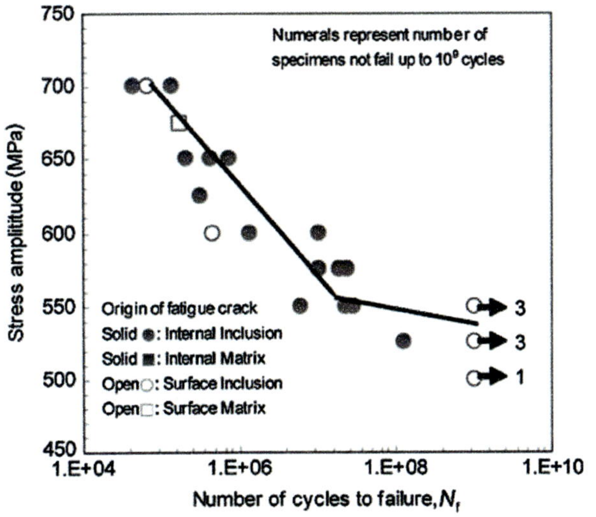

Fig. 9 S-N curve of very high-cycle fatigue for 38Si7 spring steel with extremely low oxygen content [O] < 6 ppm

inclusions and to improve the fatigue properties of high strength steels.

From the above considerations, we recently studied the very high cycle fatigue and fatigue crack growth (FCG) behaviors of both newly developed 2,000 MPa medium-carbon spring steel, namely NHS, and AISI 4340 steel with Bainite/Martensite duplex microstructure (designed as B/M) obtained through isothermal quenching and fully martensite (designed as M) for comparison [17].

NHS steel was melted in a 150 kg vacuum induction furnace and its chemical composition was 0.45C-2.0Si-0.75Mn-0.9Cr. In order to obtain different microstructures, three heat treatment procedures were used in this investigation. The austenitizing and isothermal temperatures were varied for obtaining different bainite/martensite duplex microstructures, which were designated as B/M1 and B/M2, respectively. Table 1 gives the detail heat treatment procedures.

Figure 10 shows the microstructures of B/M1, B/M2 and M. For the isothermal heat treated samples B/M1 and B/M2, the microstructure consists of mainly martensite and lower bainite and bainite in B/M1 is thinner and more uniformly distributed than that in B/M2. For sample M, typical tempered martensite was observed. The total retained austenite amount of B/M1 and B/M2 samples are about 7.1 and 3.4 vol%, respectively, while the total retained austenite amount of the M sample is far less than 1.5 vol%. The kinds of retained austenite thin films not only eliminate a harmful effect of carbide, but may also make the tip of fatigue crack blunt. This blunting is able to abate stress concentrate on the fatigue crack tip. Therefore, they could decrease the fatigue crack growth rate.

The mechanical properties of the B/M1, B/M2 and M samples are listed in Table 2. Though having nearly identical tensile strength of about around 2,000 MPa, the fracture toughness of B/M1 and B/M2 samples are obviously higher than that of M sample. Furthermore, B/M1 sample possess the best combination of strength and toughness.

The S-N curves of the three samples are shown in Fig. 11. Unlike that of the conventional S-N curve, which often has a horizontal line representing the existing of fatigue limit, it can be seen that all of the three samples display a continuously decreasing stress-life response and no obvious horizontal line exists, even at a great number of cycles (10^6–10^9), which is more correctly described by the fatigue strength at 10^9 cycles. Furthermore, the data of martensite microstructure is rather scattered than those of bainite/martensite duplex microstructures. Such as for the martensite microstructure, the range for failure cycles span from 10^7 to 10^9 cycles at stress amplitude of about 725 MPa (close to the fatigue strength). The fatigue strengths are 715 and 635 MPa for B/M1 and B/M2 samples, respectively, while is 715 MPa for M sample. The fatigue strength of B/M1 is equal to that of M, however, the former has evidently longer life than that of the latter (Fig. 11d).

As seen from Fig. 11, for the B/M1 sample, all the fracture origins were internal inclusions both at the higher and lower stress amplitude, whereas for the B/M2 sample, all the fracture origins except one which was internal matrix, were internal inclusions. For the M sample, most

Table 1 Heat treatment procedures for the obtaining of different microstructures

Sample	Quenching or isothermal quenching	Tempering
B/M1	875°C (30 min)+325°C(120 s)+oil quenching	300°C(120 min)+air cooling
B/M2	900°C (30 min)+350°C(120 s)+oil quenching	300°C(120 min)+air cooling
M	925°C (30 min)+oil quenching	350°C(90 min)+air cooling

Table 2 Summary of mechanical properties of the three kinds of microstructures

Sample	R_m (MPa)	$R_{p0.2}$ (MPa)	K_{IC} (MPa·m$^{1/2}$)	ΔK_{GBF} (MPa·m^1)	ΔK_{th} (MPa·m$^{1/2}$)
B/M1	1,975	1,770	83	4.0	4.1
B/M2	2,065	1,775	64	3.6	3.6
M	2,025	1,760	53	3.6	3.7

Fig. 10 Microstructures of NHS steel with different microstructures **a** B/M1, **b** B/M2, **c** M, **d** TEM of B/M1

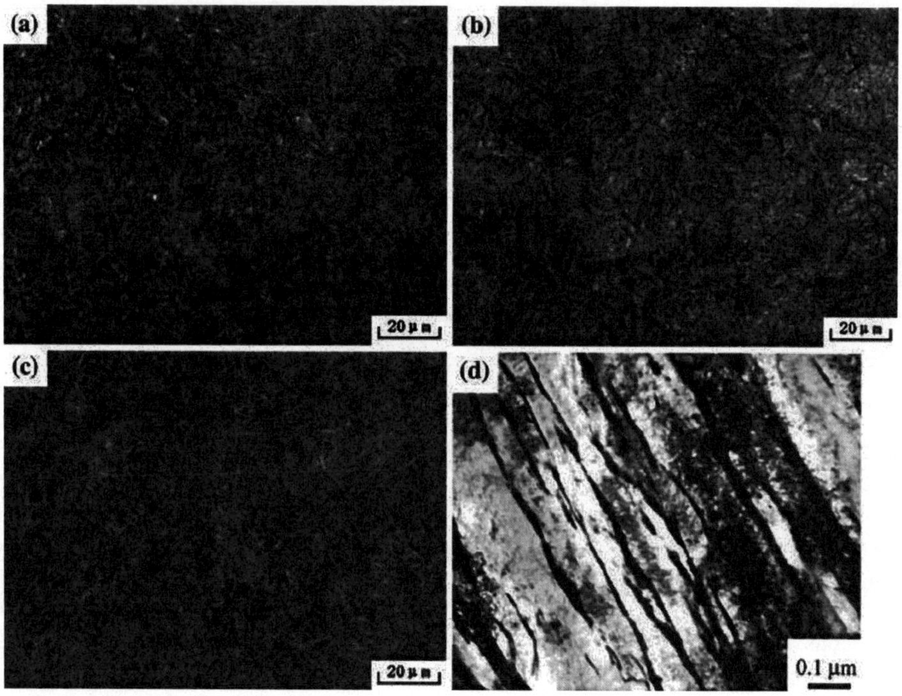

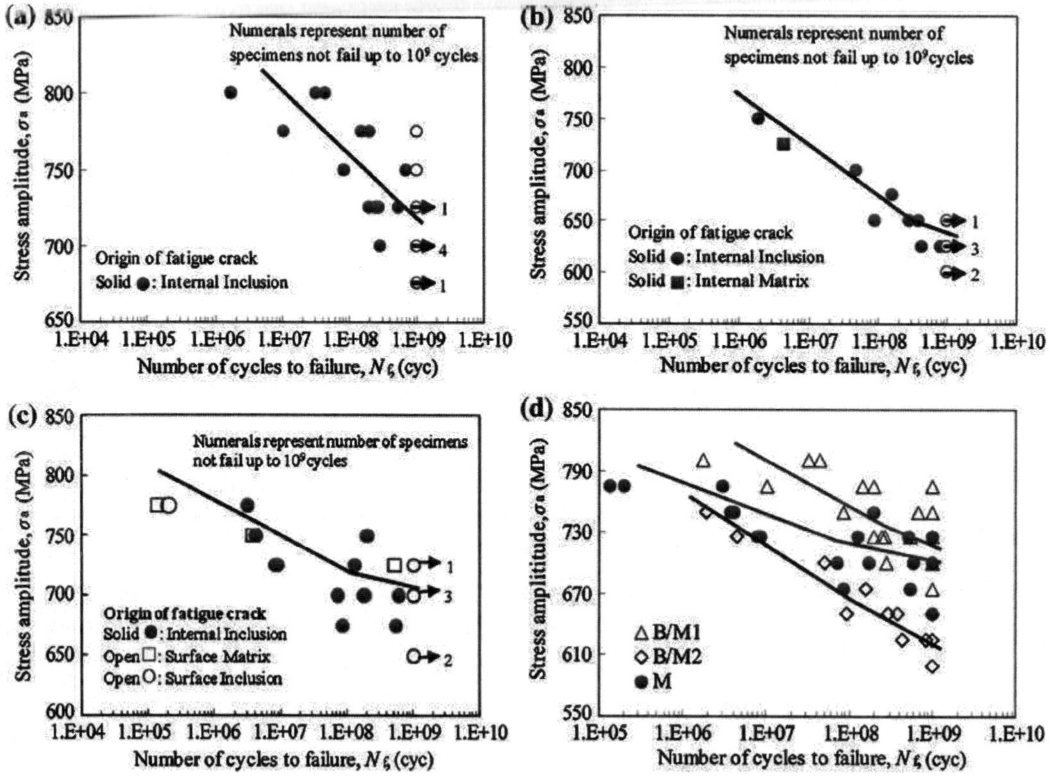

Fig. 11 S-N curves of the three samples **a** B/M1; **b** B/M2; **c** M; **d** combined S-N curve

fracture origins were internal inclusions at the lower stress amplitude, the others were surface matrix or surface inclusion at the higher stress amplitude. This implies that fatigue crack initiation is mainly controlled by inclusion, while in some cases, microstructure also has effect on fatigue crack initiation. Fish-eye fractures in most samples originate from internal inclusions, but in one case from microstructural defect. There are often rough regions (GBF) surrounding inclusions in the fish-eye at cycles beyond about 1×10^6. Example of the fracture surface in fish-eye fracture originated from internal matrix is shown in Fig. 12.

The relationship of da/dN and $\triangle K$ is shown in Fig. 13. The threshold values of $\triangle K_{th}$ are also listed in Table 2. The values of $\triangle K_{th}$ correspond well with these of $\triangle K_{GBF}$. This fact confirms the hypothesis, that is to say, all the values of $\triangle K_{GBF}$ of the same steel with different microstructures almost equal to these of $\triangle K_{th}$, which indicates that the value of $\triangle K_{GBF}$ is the threshold value governing the beginning of stable fatigue crack propagation.

From the fatigue crack propagation curves of the three kind of microstructures, it also could be seen that the fatigue crack growth rate da/dN of the B/M1 sample is obviously lower than that of the B/M2 sample, and also lower than that of the M sample when $\triangle K$ is less than about 10 $MPam^{1/2}$. This indicates that bainite/martensite duplex microstructure with thin and uniformly distributed bainite has the lowest da/dN. The large bainite existing in the B/M1 sample accelerates the fatigue crack propagation.

Similar results were also obtained for ultrahigh strength steel AISI 4340. Therefore, it is suggested that properly controlled microstructure of Bainite/Martensite duplex microstructure instead of full martensite is a promising way to minimize the detrimental effect of inclusions and to improve the fatigue properties of high strength steels.

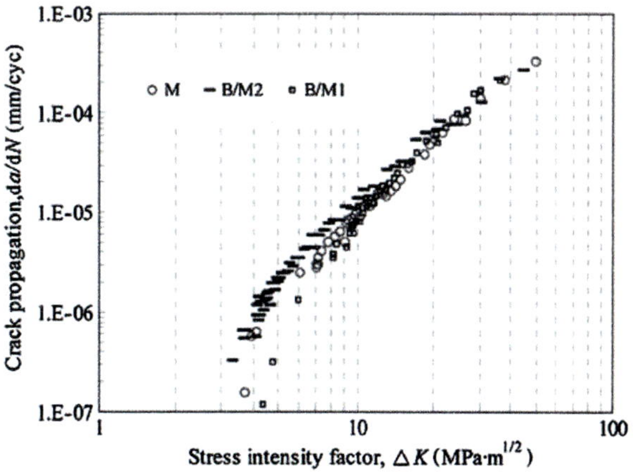

Fig. 13 Relationship between stress intensity factor range $\triangle K$ and fatigue crack growth rate da/dN of the three kinds of microstructures

4 Methods to Improve the Delayed Fracture Resistance of High Strength Steels

4.1 Grain Boundary Strengthening

DF susceptibility of the tempered martensitic microstructure is closely related to the prior austenite grain boundary strength [47, 48]. Therefore, strengthening of grain boundary is an effective way to inhibit crack initiation and propagation along grain boundaries so that DF resistance could be improved. Methods for strengthening grain boundaries include: (1) to reduce the segregated amount of embrittling elements at grain boundaries; (2) to control grain boundary carbides; (3) to refine prior austenite grains. The last will be presented in next section, so the former two methods will be discussed below.

4.1.1 Reducing Segregation of Impurities at Grain Boundaries

To lower the contents of P and S impurities in steels could reduce their segregation at grain boundaries, which would lead to cleaner and stronger boundaries to inhibit crack initiation [49, 50]. Consequently, DF resistance of high strength steels could be improved. An example is shown in Fig. 14. Results of SCC test and sustained load tensile test indicate that both K_{ISCC} and DFSR of the high clean 42CrMo steel are significantly higher than those of the commercial 42CrMo steel [32].

Nevertheless, the possible lowest contents of P and S are constrained by steelmaking technology and the production

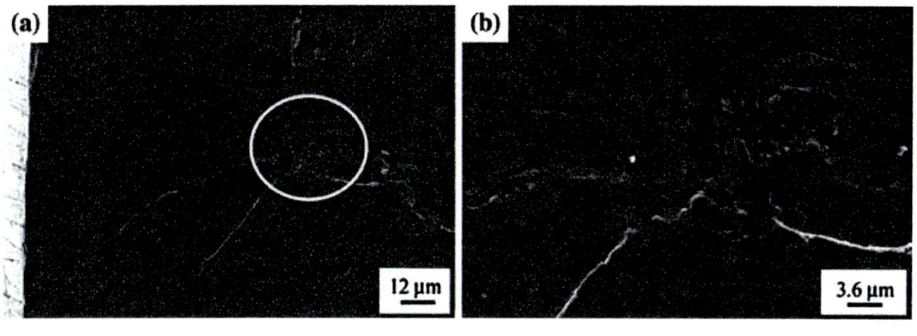

Fig. 12 SEM fractographs showing internal matrix at fracture origin of B/M2 sample **a** low magnification, **b** high magnification of the fatigue initiation area shown by white circle in **a** ($\sigma_a = 725$ MPa, $N_f = 4.37 \times 10^6$ cyc)

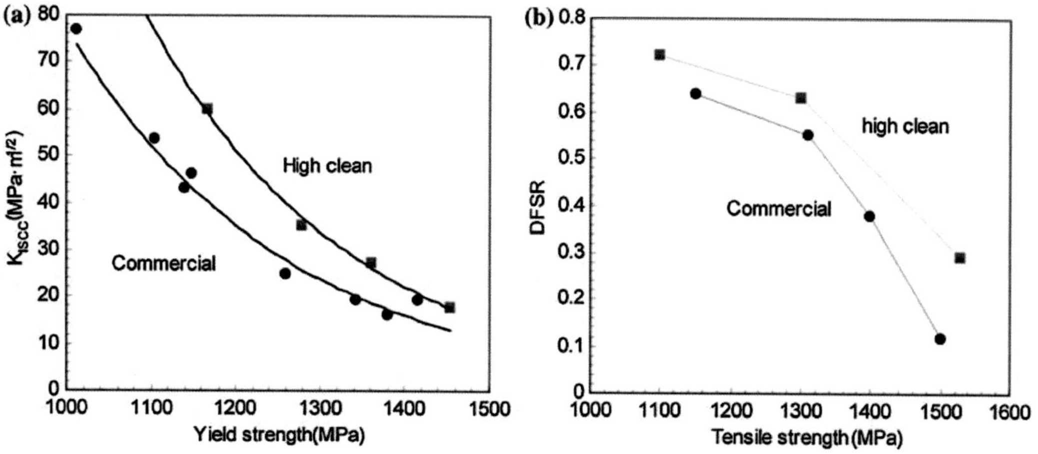

Fig. 14 Variation of K_{ISCC} (**a**) and DFSR (**b**) with strength of 42CrMo steel (commercial: 0.023P-0.018S-0.006O-0.007N-0.65Mn-0.33Si; high cleanliness: 0.001P-0.002S-0.002O-0.003N-0.06Mn-0.06Si)

cost. Another alternative is to add some elements into steels which can reduce or even diminish the segregation of S and P at boundaries. It can be done by two ways: (1) add elements to capture impurities and then reduce their segregation at grain boundaries, such as use Mo to scavenge P; (2) reduce the concentration of alloying elements that promote P segregation in austenite, such as Mn, Si and Cr. In Cr-Mo steels, it has been proved that Mn and Si could promote the segregation of P in austenite, so that removal of these two elements could lower the segregation of P. The aforementioned high clean 42CrMo steel contains very low contents of Mn and Si besides low contents of P and S. Therefore, the excellent DF resistance of high clean 42CrMo steel is related to its very low contents of Mn and Si.

4.1.2 Controlling Grain Boundary Carbide

Interface between carbides and the matrix are strong hydrogen traps in steels [24]. If carbides are mainly situated at grain boundaries, hydrogen will be enriched at boundaries as well, which would lead to boundary embrittlement and thus DF crack may nucleate at boundaries and propagate along them. Therefore, controlling grain boundary carbides such as changing the morphology of boundary carbides, reducing the quantity of carbides at boundaries or even obtaining carbide-free boundaries would improve the DF resistance of martensitic steels.

It is suggested that tempering temperature has a significant influence on DF behavior of martensitic steels [32]. At low-temperature tempering, film-type carbides often form at grain boundaries, leading to more hydrogen trapped at boundaries and thus poor DF resistance. At high-temperature tempering, film-type grain boundary carbides are broken and spheroidized, leading to less hydrogen trapped at boundaries and retardation of intergranular crack propagation. On the other hand, lots of carbides precipitate within grains, thus leading to more hydrogen trapped within grains. Both could strengthen grain boundaries and then improve DF resistance. Figure 15 shows the change of SCC fracture

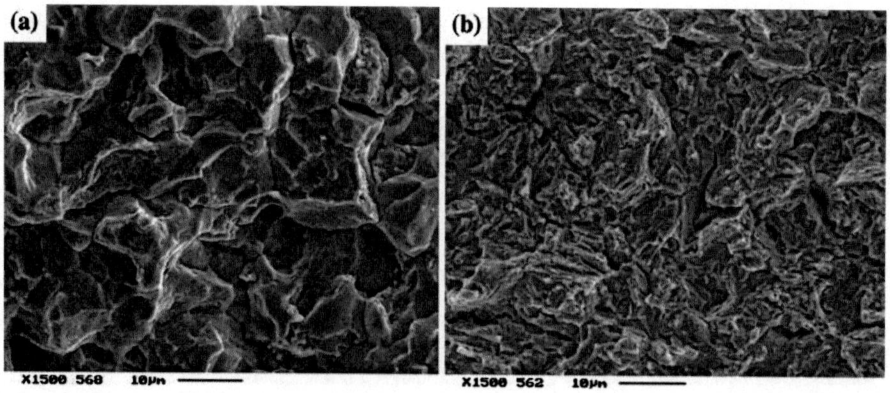

Fig. 15 Fracture surfaces of SCC specimens of 42CrMoVNb steel tempered at different temperature, showing the change of fracture from intergranular dominant mode (**a**) to transgranular dominant mode (**b**). **a** 410°C, K_{ISCC} = 10.5 MPa·m$^{1/2}$; **b** 600°C, K_{ISCC} = 37.1 MPa·m$^{1/2}$

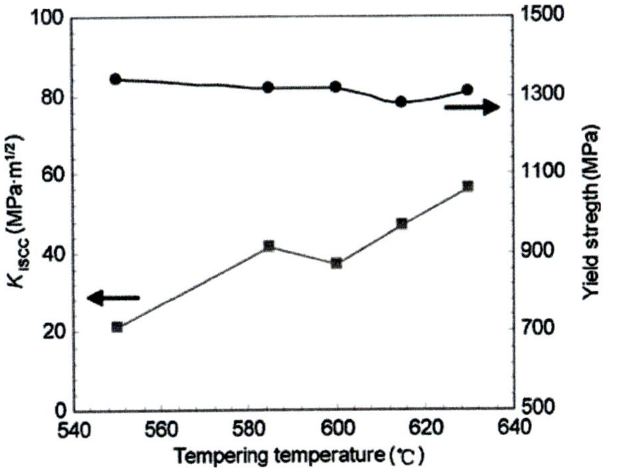

Fig. 16 Variations of K_{ISCC} and yield strength with tempering temperature of 42CrMoVNb steel

mode from intergranular to transgranular for 42CrMoVNb steel with increasing tempering temperature [51].

Nevertheless, higher tempering temperature generally means lower strength for low alloyed steels. One feasible solution to the purpose of obtaining a better balance between strength and DF resistance is the addition of secondary hardening elements Mo and V. Both elements could form carbides at high tempering temperature, which not only contribute to higher strength but also act as hydrogen traps. Figure 16 shows the variations of K_{ISCC} and yield strength with tempering temperature. DF resistance rises with increasing tempering temperature at the same level of strength for the 42CrMoVNb steel. This indicates that high tempering temperature is a very effective way to improve DF resistance for high strength steels.

4.1.3 Influence of Mo Alloying

Mo as a common alloying element is added into the structural steels generally for the two objectives: one is to improve hardenability; the other is to prevent temper embrittlement. Besides the addition of Mo could increase the strength of martensitic steels due to the precipitation of Mo containing carbides, Mo could also strengthen grain boundaries, as proved by Weng in 1980s [52]. Our work again shows that Mo can segregate at prior austenite grain boundaries and thus strengthen them, leading to improved DF resistance [53, 54].

When various contents of Mo were added into 40CrV steel, it was found that the strength-tempering temperature curve shifts up when the content of Mo increases, particularly at tempering temperature higher than 500°C due to the precipitation hardening of Mo/V alloy carbides, as shown in Fig. 17. Results from the sustained load tensile test and SCC test are given in Fig. 18. DFSR and K_{ISCC} of the tested steels both decrease obviously with increasing strength. In

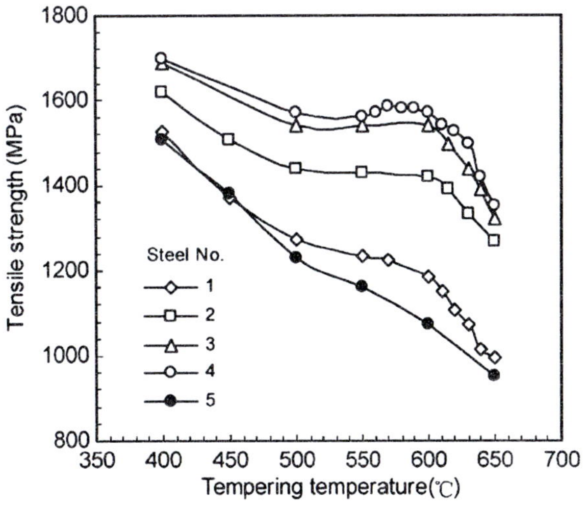

Fig. 17 Variations of tensile strength with tempering temperature of 40CrV steels with various molybdenum contents (No. 1: 0.40C-1.1Cr-V, base steel; No. 2: 0.50Mo; No. 3: 1.15Mo; No. 4: 1.54Mo; No. 5: 42CrMo)

general, increasing content of Mo leads to higher DFSR and K_{ISCC} until 1.15%Mo, beyond which DFSR and K_{ISCC} decrease a bit but are still higher than those with low Mo content (Fig. 19). Therefore, DF resistance does not have a monotonous dependence on the content of Mo in the investigated range of strength variation.

The beneficial effect of Mo on improving the DF resistance of martensitic steel is suggested to be the results of one or more of the following factors [54]: (1) the retardation of softening and secondary hardening of Mo raise the tempering temperature for a given strength; (2) Mo-carbides precipitated during high-temperature tempering act as hydrogen traps; (3) segregation of Mo at grain boundaries to control impurities and strengthen grain boundaries.

By the EDS microanalysis of Nos. 2–4 steels at different positions, it was found that the solute Mo could easily segregate at grain boundaries besides the rest Mo forms carbides. Such Mo segregation does not change with tempering temperature. Figure 20 gives the EDS microanalysis results of Mo concentration across one boundary in No. 2 steel tempered at 585°C. Mo concentration reaches the maximum of $1.7 \pm 0.4\%$ at grain boundary, while it decreases sharply with the increasing distance h from the boundary; for example, $0.7 \pm 0.1\%$ at $h = 2$ nm and close to the bulk concentration at $h = 10$ nm. Microanalysis in the No. 2 steel tempered at 500 and 650°C both gave very similar results too.

Zhang et al. [55] has successfully used EELS and EDS mapping techniques to study the grain boundary of 42CrMo1VNb and 42CrMo steels. Figure 21a shows the Fe edge of EELS recorded at bulk and grain boundary (GB) of 42CrMo steel. The grain boundary has a tall, narrow peak at

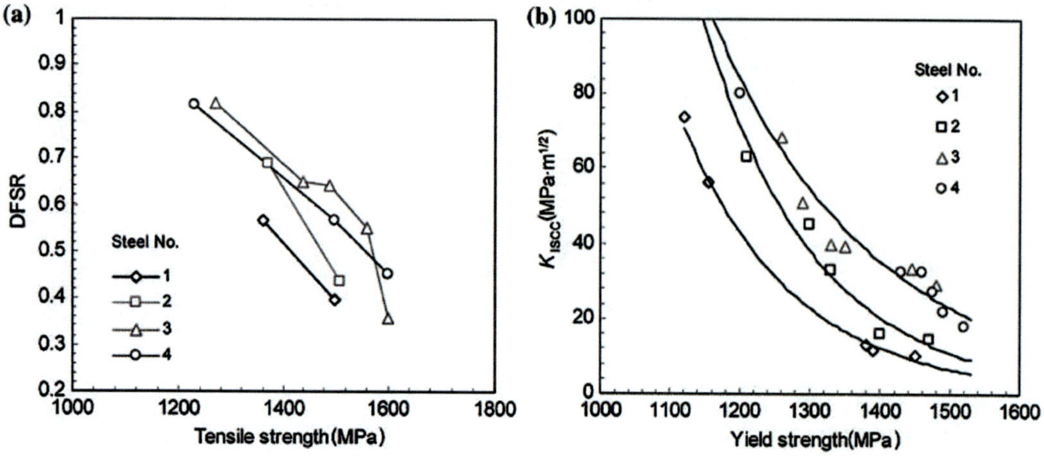

Fig. 18 Variations of delayed fracture strength ratio (DFSR) (**a**) and threshold stress intensity factor of SCC, K_{ISCC} (**b**) with strength of 40CrV steel with various molybdenum contents (No. 1: 0.40C-1.1Cr-V, base steel; No. 2: 0.50Mo; No. 3: 1.15Mo; No. 4: 1.54Mo; No. 5: 42CrMo)

the onset of Fe edge compared with the short and broad Fe edge of the bulk. This indicates the grain boundary is weaker than the bulk so that intergranular fracture occurred in this steel as the experiment observed. Figure 21b shows the result of EELS measurement on a typical boundary of 42CrMo1VNb steel (EDS elemental mapping has showed that there was some Mo and Cr segregation in that grain boundary). The result shows that the grain boundary has a short and broad peak at the onset of Fe edge compared with the peak of the bulk, indicating that the grain boundary is stronger than the bulk. This is in agreement with the experimental result that transgranular fracture occurred in the sample of 42CrMoVNb steel. Since 42CrMo1VNb steel contains more Mo than 42CrMo steel, it can be concluded that the segregation of Mo at grain boundary could increase grain boundary strength.

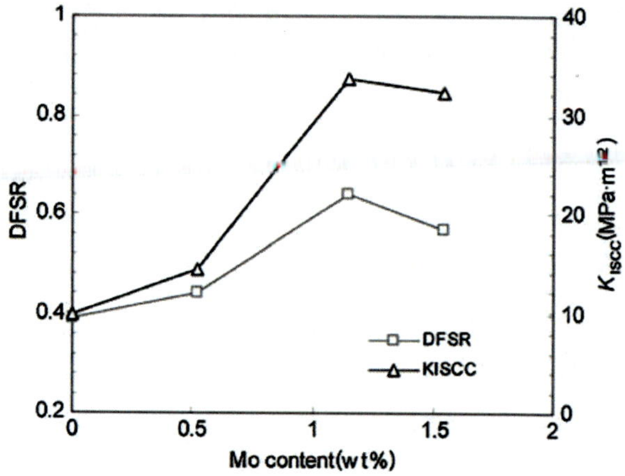

Fig. 19 Variations of delayed fracture strength ratio (DFSR) and threshold stress intensity factor of SCC, K_{ISCC} with molybdenum content of 40CrV steel

4.2 Structure Controlling

DF in high strength steels generally initiates from prior austenite grain boundaries and then propagates along them. Therefore, besides carbides and segregation of impurities at grain boundaries, prior austenite grain size is another major factor affecting DF.

Austenite grain size in the range of 4.7–120 μm were obtained in a commercial 42CrMo steel by varying austenitizing temperature and applying thermal cycling and the SCC behaviors in 3.5 pct NaCl aqueous solution were investigated by using bolt-loaded modified WOL type specimens. It was found that threshold stress intensity factor, K_{ISCC}, decreases gradually with increasing prior austenite grain size, d, but K_{ISCC} could be improved when d is larger than 65 μm, as shown in Fig. 22. Further investigation shows that the dependence of K_{ISCC} on d has a strong relationship with the relevant size of plastic zone ahead of the crack tip, R, and d [32]. That is to say, when d is 65–120 μm and larger than R, K_{ISCC} increases with increasing d and when d is 4.7–22 μm and smaller than R, K_{ISCC} decreases with increasing d.

K_{ISCC} is the threshold stress intensity factor for crack non-propagation and measured under the plain strain condition. The real DF process in components such as bolts, however, consists of crack initiation and propagation, and crack initiation often plays particularly important role [56]. Therefore, the sustained load tensile test was carried out by using round notched tensile specimens in Walpole solution to investigate the influence of microstructure refinement on DF behavior of martensitic steels. As strength has a significant influence on DF resistance for the martensitic steels. In the studied range of strength, delayed fracture strength

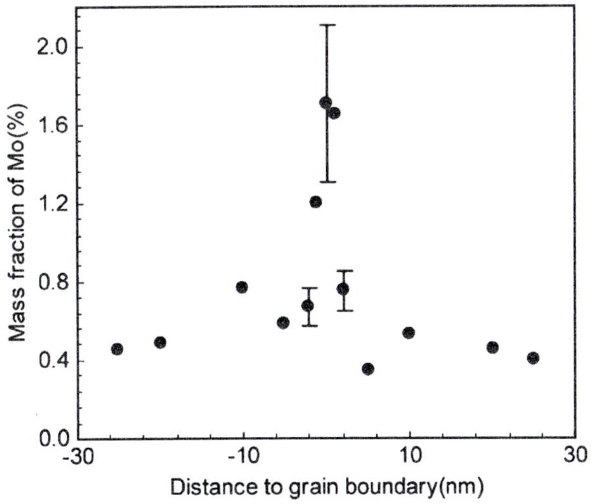

Fig. 20 Result of energy dispersive X-ray spectroscopy (EDS) analysis showing the distribution of Mo around grain boundary of No. 2 steel (0.50 %Mo) tempered at 585°C for 2 h

ratio (DFSR) of 42CrMoVNb steel decreases with increasing strength (Fig. 23). At the same level of tensile strength, DFSR could be improved remarkably when the prior austenite grain size d decreased from 20 to 8 μm; but improved a little slower when *d* decreased from 8 to 4 μm; and there was no noticeable increase when *d* decreased from 4 to 2 μm (Fig. 23a). Due to the refinement of austenite grain, the increment of yield strength is higher than that of tensile strength, therefore, the variation of DFSR with yield strength is more significant (Fig. 23b) and DFSR could be improved with decreasing *d* until it reached 4 μm, beyond that DFSR changed very little. This implies that further austenite grain refinement would be difficult to improve DF resistance of high strength martensitic steels. The reason for this kind of delayed fracture behavior could be discussed mainly from factors of the stress concentration and segregation at grain boundary.

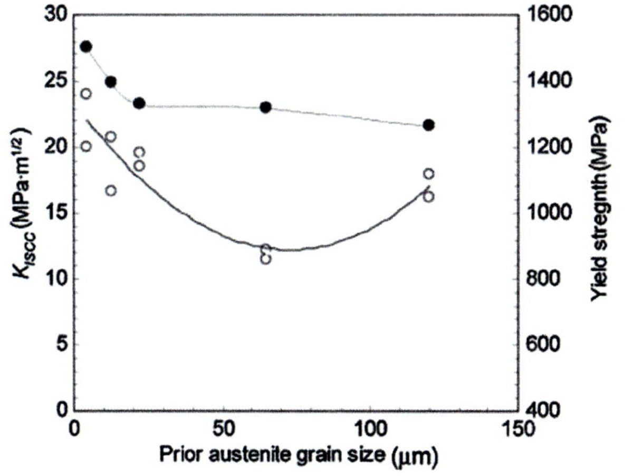

Fig. 22 Variations of yield strength and K_{ISCC} with prior austenite grain size *d* of commercial 42CrMo steel tempered at 400°C

4.3 Hydrogen Trap Controlling

There are many kinds of defects or microstructural heterogeneities in steel can act as hydrogen trapping sites, such as dislocation, microvoid, grain boundary, inclusion and precipitate [24]. The function of a trap is related to the diffusivity of hydrogen that controls the kinetics of entry or accumulation at stress concentration, thus influencing DF resistance. Microalloying elements, such as V, Ti and Nb, have strong affinity to carbon and nitrogen and can form stable carbides or carbonitrides, which could act as hydrogen traps.

Figure 24 shows the results of the sustained load tensile tests of the dependence of DF resistance on the tensile strength and the V content in 42CrMo steel [57]. An increasing content of V leads to more dispersed VC precipitates during high temperature tempering and also more undissolved VC particles during austenitizing (Fig. 25). As a result, more hydrogen are captured by VC, leading to the

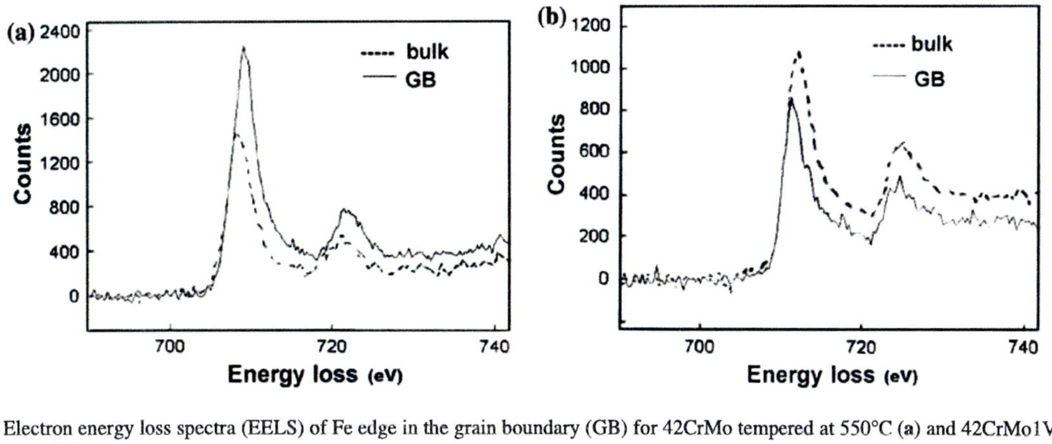

Fig. 21 Electron energy loss spectra (EELS) of Fe edge in the grain boundary (GB) for 42CrMo tempered at 550°C (**a**) and 42CrMo1VNb steel tempered at 600°C (**b**)

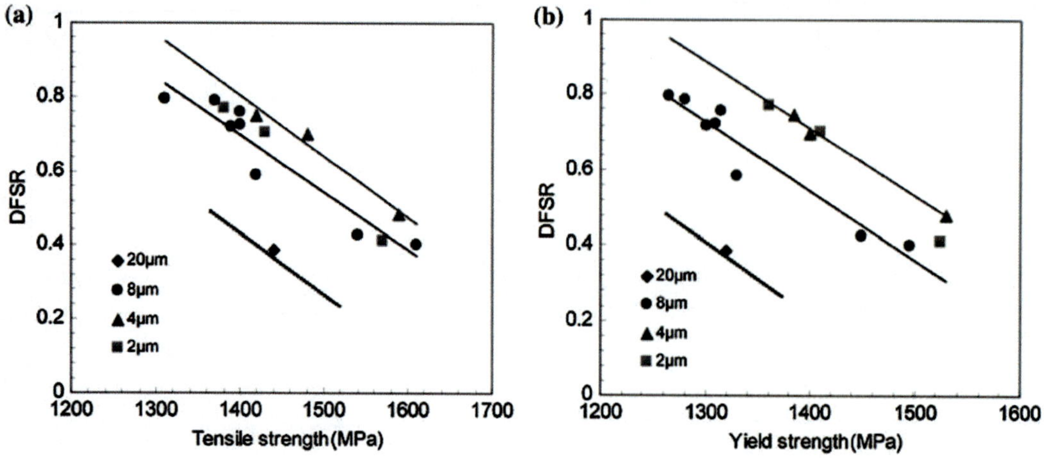

Fig. 23 Variations of delayed fracture strength ratio (DFSR) with tensile strength (**a**) and yield strength (**b**) of 42CrMoVNb steel with different prior austenite grain sizes

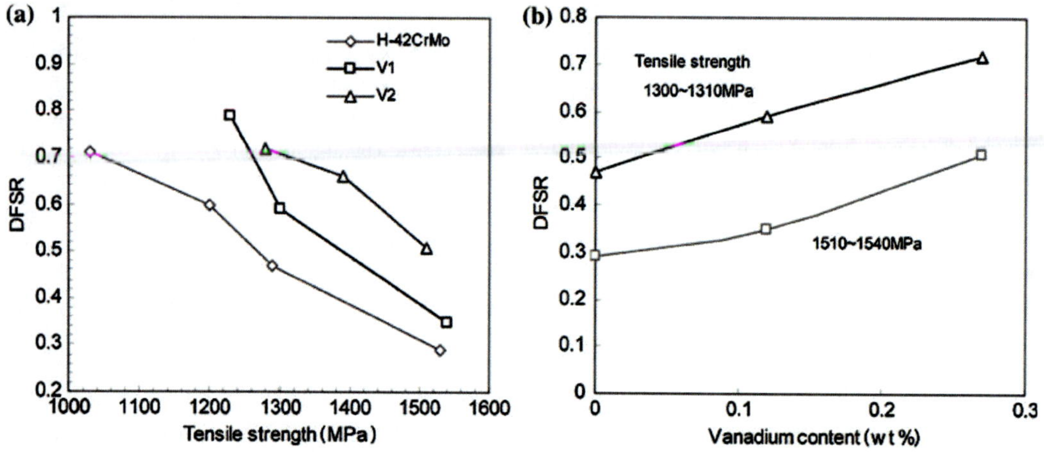

Fig. 24 Variations of delayed fracture strength ratio (DFSR) with tensile strength (**a**) and vanadium content (**b**) of 42CrMo steel with different vanadium content (H-42CrMo: base steel with high purity; V1: base steel + 0.12 %V; V2: base steel + 0.27 %V)

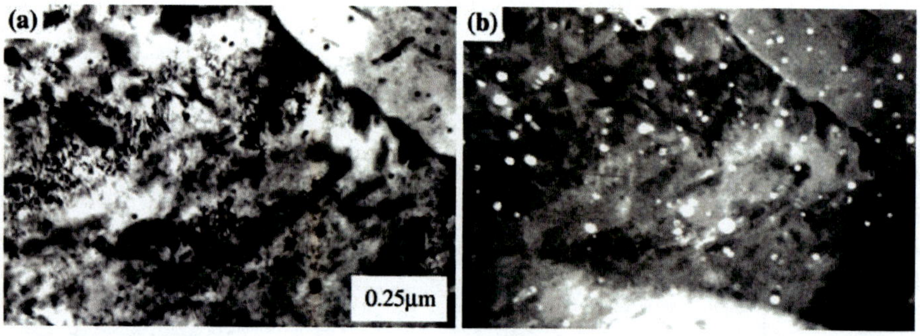

Fig. 25 VC carbide morphology of 42CrMo + 0.27 %V steel tempered at 600°C for 2 h (TEM, bright field (**a**) and dark field (**b**))

curve of DFSR vs. tensile strength shift to up-right corner and DFSR increases with V content. Figure 26 shows fractographs in the DF crack initiation area of specimens fractured in Walpole solution of 42CrMoV steel and 42CrMo steel. At similar tensile strength, i.e. 1,280 and 1300 MPa respectively, the failure mode is transgranular for 42CrMoV steel, whilst mixed mechanism of intergranular and transgranular for 42CrMo steel.

Thermo-Calc calculation and TEM observation show that a considerable amount of VC precipitates does not

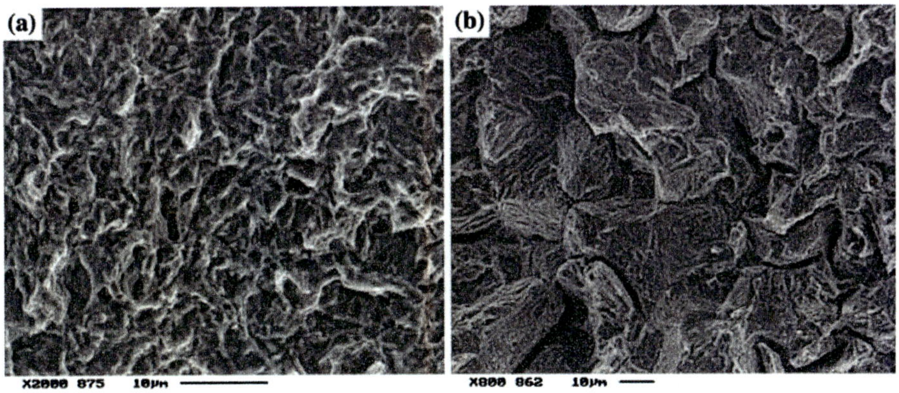

Fig. 26 Fracture surfaces in crack initiation area of sustained load tensile DF specimens of 42CrMo + 0.27 %V steel (**a**) and high clean 42CrMo steel (**b**). **a** 600°C, R_m = 1,280 MPa, σ_a = 1,660 MPa, t = 140 h; **b** 470°C, R_m = 1,300 MPa, σ_a = 1,390 MPa, t = 88 h; (R_m: tensile strength; σ_a: applied stress; t: failure time)

resolve and still remain in the solution when V-microalloyed steel is soaked at normal austenization temperature of 920–940°C [32]. Therefore, the absorbed hydrogen content in the V-microalloyed steel is obviously higher than that in the steel without V after cathodic hydrogen charging (Fig. 27). When the steel was tempered at around 600°C, the absorbed hydrogen content was markedly increased for the precipitation of nano-scale VC coherent particles [58]. When the steel was as tempered at 700°C, the absorbed hydrogen content was decreased but was still as high as that of the as-quenched condition duo the spherical VC incoherent precipitates.

Furthermore, when tempering temperature is higher than 500°C for V-microalloying steel, the dispersed precipitation of VC not only acts as strong hydrogen trap, but also leads to precipitation hardening. Therefore, the tempering temperature for V-microalloyed 42CrMo steel is higher than that of non-microalloyed 42CrMo steel at the same strength

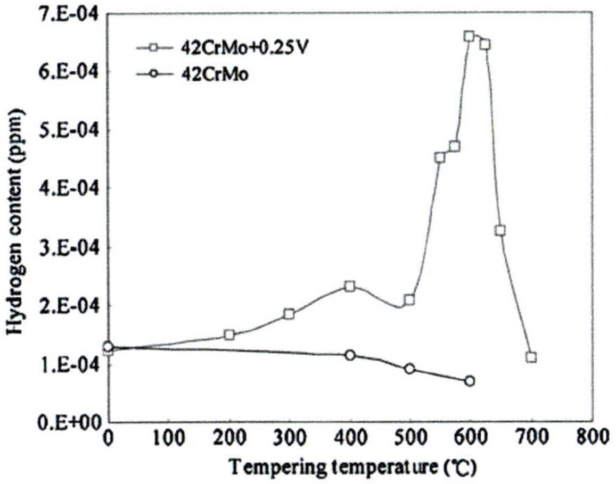

Fig. 27 Variations of absorbed hydrogen content with tempering temperature for the V-bearing and 0.40 carbon steels

level. This is also one of the main reasons for improved DF resistance of V-microalloying steel.

5 Development of Long Life High Strength Steels

5.1 2,000 MPa Grade High Strength Spring Steel with Excellent Fatigue Failure Resistance

Recently, considerable efforts have been made in the development of high performance spring steels with higher design stress to meet the need for weight and cost savings. Since the increase of design stress requires higher fatigue strength and sag resistance, the hardness and strength of spring is usually increased. For example, when the design stress is 1,200 MPa grade, the required hardness is to be between 52 and 54 HRC [11, 59]. However, as for the conventional spring steels such as 60Si2MnA (equivalent to SAE9260) and 50CrVA (equivalent to SAE6150), which has been the most popular spring steels in Chinese automotive industries, its application is generally limited up to 1,000 MPa design stress for their dramatically reduced corrosion fatigue strength and delayed fracture strength. Furthermore, with the increase of the hardness of spring steel, its fatigue strength is more scatter or even declination [43]. Therefore, for the development of new spring steels, the emphasis has been focused on increasing the strength while improving ductility, toughness, delayed fracture resistance, fatigue strength and corrosion fatigue strength [11, 43, 59–61].

As one of the applications of IST technology, a new type of 2,000 MPa grade high strength spring steel NHS (medium carbon Si–Cr-Ni-V–B) with excellent fatigue resistance as well as sag resistance was developed. The following describes the main characteristics of the new high strength spring steel.

5.1.1 Material Design and Mechanical Property

Compared to that of conventional spring steels, carbon content of the new spring steel is reduced to improve the toughness. However, a certain content of carbon should be set to obtain the required high strength. For suspension springs with tensile strength over 1,900 MPa, significant attention should be paid to corrosion fatigue strength. Therefore, suitable nickel is added to restrain corrosive pit forming, thus improving its corrosion fatigue property. Vanadium, titanium and boron elements are also added to improve delayed fracture strength and sag resistance. Furthermore, silicon content is a slightly higher than that of conventional spring steels to compensate the adverse effect on sag resistance by lowering carbon content. Optimized alloy design enables that both high tensile strength over 2,000 MPa and high yield strength could be obtained when the new steel is tempered at 350°C.

Figure 28 shows the relationship between mechanical properties and tempering temperature. Table 3 compares the mechanical properties of steel NHS and 60Si2MnA tempered at different temperatures. It can be realized that the ductility and toughness of steel NHS are higher than those of 60Si2MnA when tempered to same temperature below 400°C. Therefore, steel NHS has a tensile strength of 2,000 MPa grade when tempered at 350°C and also good ductility and toughness at the temperature. Low carbon content and the addition of boron are supposed to be one of the main reasons for the improvement [62].

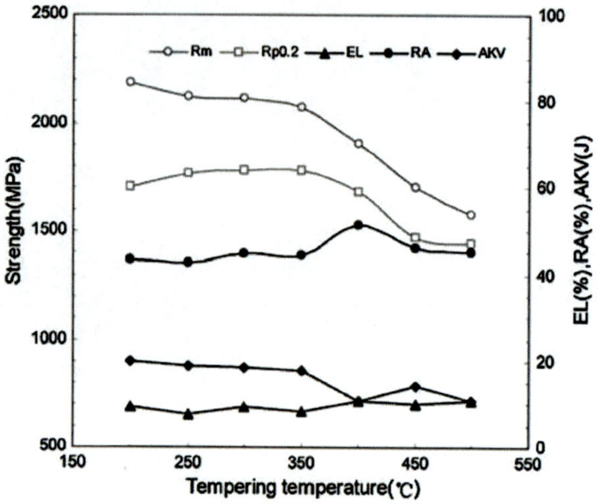

Fig. 28 Variations of mechanical properties with tempering temperature of steel NHS

5.1.2 Fatigue Property

Figure 29 shows the S-N curves of the experimental steels. Here NHS-1, NHS-2, NHS-3 and NHS-4 represent steel NHS with different melting process. As for rotating bar two-point bending fatigue test, the data in the S-N curves of 60Si2MnA and NHS-2 could be interpreted with two straight lines, like that of the conventional S-N curve. The horizontal line represents the fatigue strength at 10^7 cycles. However, for NHS-1, its fatigue life continues to drop and no obvious horizontal line exists, like that of 60Si2CrVA, SUP12 and SWOSC-V steels [63, 64]. As for tension–compression ultrasonic fatigue test, similarly, all the fatigue lives continue to drop and no obvious horizontal line exist.

Fractographic examination of all the fatigue fractured specimens at 10^7 cycles by SEM revealed, as seen from Fig. 29a, for NHS-2, which has the highest fatigue strength, fracture origins were specimen surface at the high stress amplitudes, the others were internal inclusions at the lower stress amplitudes, whereas for NHS-1 and 60Si2MnA, all the fracture origins were internal inclusions or specimen surface both at the higher and lower stress amplitudes, respectively. However, almost all the fracture origins of specimens at 10^9 cycles were internal inclusions as seen from Fig. 29b.

The analysis results of the inclusions in the fatigue crack initiation site are given in Tables 4 and 5 for HCF and VHCF, respectively. As for HCF, the inclusions as fracture origins were TiN and Al_2O_3 composite inclusions for steel NHS-1, while were Al_2O_3 and CaO composite for steel NHS-2. The difference in inclusion sizes as the results of different steel melting process is the main reasons for the

Table 3 Mechanical properties of the steels used

Steel	Steel UHS				60Si2MnA			
Tempering temperature (°C)	300	350	400	450	300	350	400	450
Hardness (HRC)	56.0	54.7	53.3	48.2	57.5	55.4	50.3	46.2
Tensile strength (MPa)	2,115	2,075	1,905	1,705	2,285	2,265	1,925	1,655
Yield strength (MPa)	1,780	1,780	1,680	1,475	2,025	2,040	1,820	1,525
Elongation (%)	9.3	8.3	10.8	10.0	6.5	8.3	9.3	10.9
Reduction of area (%)	44.8	44.3	51.5	46.3	29.5	35.5	48.4	45.4
Charpy impact value (J)	18.5	18.0	14.0	14.5	5.2	5.3	12.4	17.0

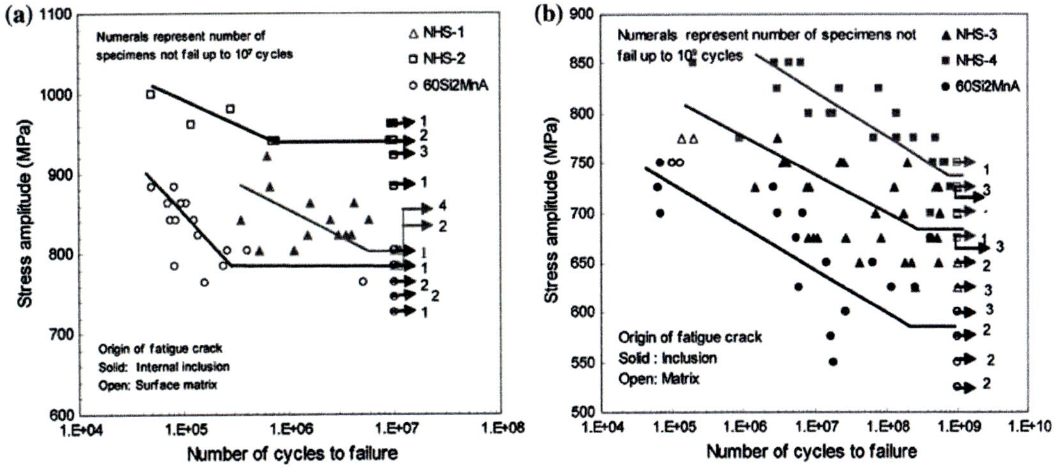

Fig. 29 S-N curves obtained by the rotating bar two-point bending fatigue test (**a**) and tension–compression ultrasonic fatigue test (**b**)

different fatigue behaviour of the two NHS steels. Obviously, the difference of fatigue behaviour of steels NHS and 60Si2MnA can be due to the difference of strength level. It is known that the fatigue strength ratio of rotary bending fatigue is around 0.50. It can be seen from Table 4, both NHS-2 and 60Si2MnA have comparatively high fatigue strength ratio, though the former has much higher tensile strength than that of the latter. It should be noted that for VHCF, the influence of inclusion on fatigue behaviour is more significant. Therefore, more attention should be paid to minimize both the size and number of inclusions to further improve the fatigue property of steel NHS.

Figure 30 shows the S-N curves of corrosion fatigue tests. Steel NHS shows much higher corrosion fatigue characteristic than that of 60Si2MnA. Observation of the fracture surfaces of all failed specimens by SEM revealed that all fractures occur from the bottom of corrosive pit. This means that the pit formation could be the focal point to evaluate the results of corrosion fatigue test and has been confirmed by others [59, 60]. Figure 31 shows the relationship between the corrosion pit depth and number of corrosive cycles. It can be seen that the difference of pit depth is rather small and it tends to increase rapidly over 7 cycles. After 10 cycles, the pit depth of NHS is around 35 μm, while is over 80 μm for 60Si2MnA.

5.1.3 Sag Resistance

Figure 32 shows the variations of Bauschinger loop area with tempering temperature. It can be seen that, with increasing tempering temperature, the loop area slightly increases and then decreases after reaching its maximum value. This is consistent with the results of other spring steels [61]. The occurrence of a maximum in loop area at around 350°C is strongly related to the characteristics of

Table 4 Summary of tensile strength and fatigue strength at 10^7 cycles of the steels used

Steel	Tensile strength R_m (MPa)	Fatigue strength σ_{-1} (10^7) (MPa)	Fatigue strength ratio σ_{-1} (10^7)/R_m	Average inclusion size at the origin (μm)	Inclusion type
NHS-1	2,010	810	0.40	49	TiN, $Al_2O_3{\cdot}MgO{\cdot}SiO_2$
NHS-2	2,065	945	0.46	21	$Al_2O_3{\cdot}(CaO)_x$
60Si2MnA	1,640	780	0.48	–	–

Table 5 Summary of tensile strength and fatigue strength at 10^9 cycles of the steels used

Steel	Tensile strength R_m (MPa)	Fatigue strength σ_{-1p} (10^9) (MPa)	Fatigue strength ratio σ_{-1p} (10^9)/R_m	Average inclusion size at the origin (μm)	Inclusion type
NHS-3	2,025	700	0.35	12	TiN, $Al_2O_3{\cdot}(CaO)_x$
NHS-4	2,080	730	0.35	15	TiN, Al_2O_3
60Si2MnA	1,690	585	0.35	20	$Al_2O_3{\cdot}CgO{\cdot}SiO_2$

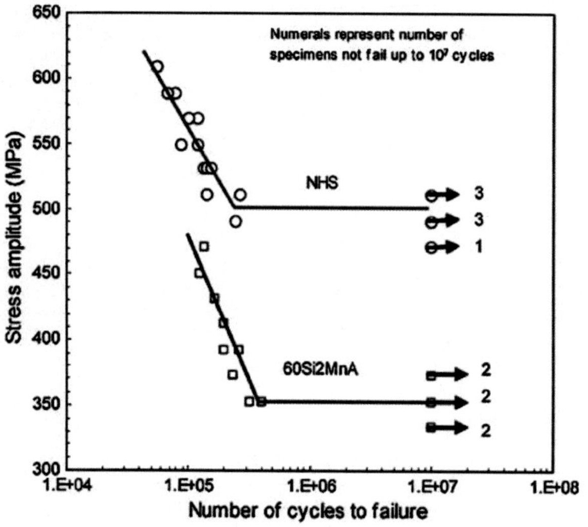

Fig. 30 S-N curves obtained by the rotating bar two-point bending corrosion fatigue tests

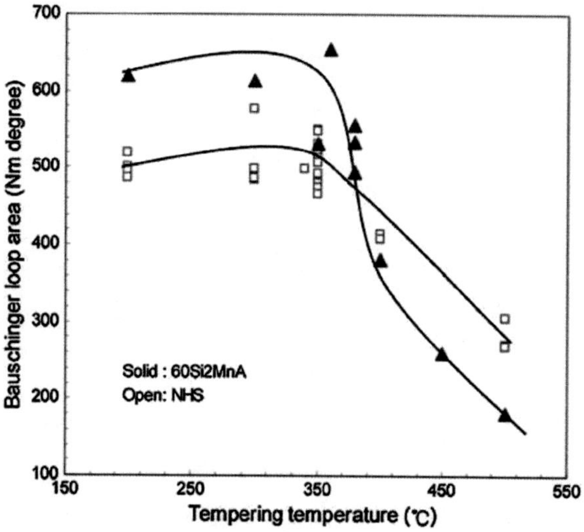

Fig. 32 Variations of Bauschinger loop area with tempering temperature

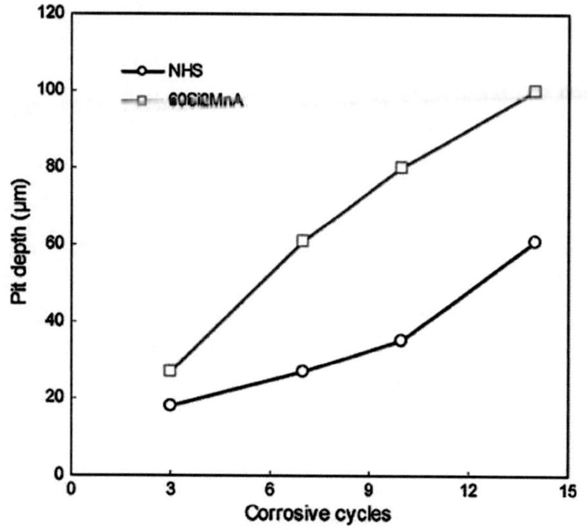

Fig. 31 Relationship between corrosion pit depth and number of corrosive cycles

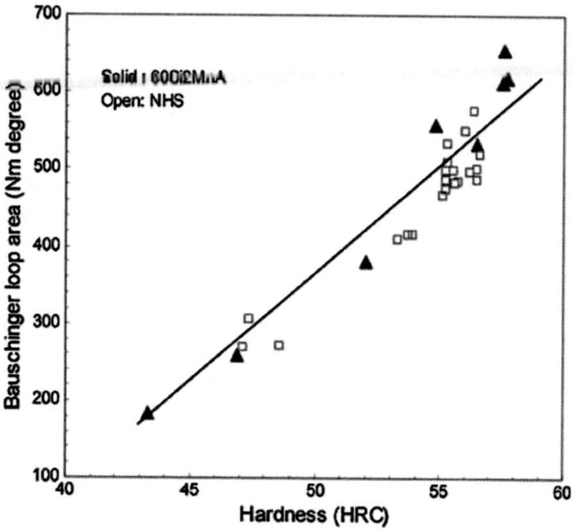

Fig. 33 Relationship between Bauschinger loop area and hardness

carbides existing at each tempering temperature. At temperatures below around 400°C, the loop area of 60Si2MnA is higher than that of steel NHS owing to its higher carbon content and thus higher hardness (see Table 3). At temperatures above around 400°C, the loop area of NHS is higher than that of 60Si2MnA, this is mainly owing to the higher hardness of the former owing to the additions of Cr and V, where Cr affects the behavior of tempered carbides, V causing the precipitation of vanadium carbonitrides.

The relationship between loop area and hardness as shown in Fig. 33, shows that the loop area increases with increasing hardness and there are little difference between steels NHS and 60Si2MnA at same hardness. Therefore, the sag resistance of steel NHS with hardness of about 54HRC when tempered at 350°C is much higher than that of 60Si2MnA with its practical hardness up to 50HRC.

5.2 1,500 MPa Grade High Strength Bolt Steel with Excellent Delayed Fracture Resistance

Based on the aforementioned understanding of the fundamentals of the DF behavior of high strength martensitic steels, the proposed concept to improve DF resistance comprises the combination of grain boundary strengthening, structure controlling and hydrogen trap controlling, and it is

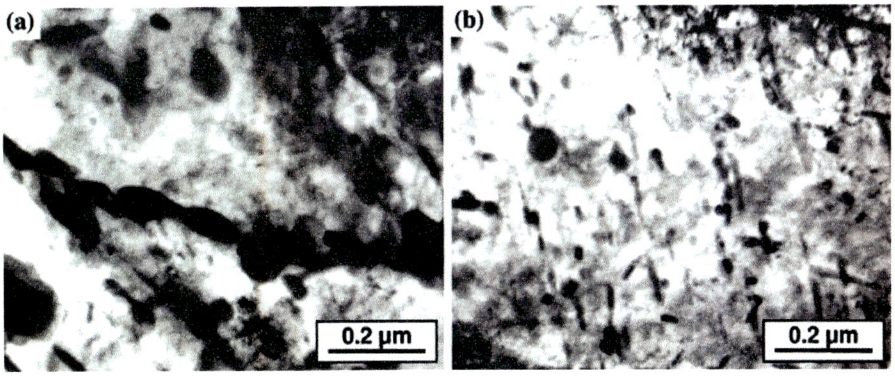

Fig. 34 TEM microstructures of ADF1 steel tempered at 600°C, showing the broken and spheroidized carbide at prior austenite grain boundary (a) and the dispersed alloy carbide precipitates (b)

called GST technology. ADF series high strength structural steels have been developed with excellent DF resistance, which are characterized by: (1) the tensile strength increases to 1,500 MPa grade by Mo/V precipitation hardening and grain refinement; (2) grain boundary bonding is strengthened by Mo segregation, and hydrogen diffusivity is delayed and its enrichment at grain boundary is weakened through the introduction of hydrogen traps of carbides/carbonitrides; (3) grain refining through microalloying and heat treatment to improve strength, toughness and DF resistance at the same time.

Grain growth in ADF1 steel is much slower than that in 42CrMo steel, the active energy for migration of austenite grain boundary in ADF1 steel is higher than that in 42CrMo steel by 158.8 kJ/mol, and consequently, the austenite grain is as fine as less than 10 μm, when the steel is austenitised at 950°C. ADF1 steel possesses rather fine microstructure, in which V/Mo alloy carbides are dispersed extensively and the carbides at grain boundaries are not continuous and spheroidized after tempering at high temperature (Fig. 34).

Variations of tensile properties of ADF1 steel with tempering temperatures are shown in Fig. 35. With the rise of tempering temperature from 400 to 650°C, the strength of 42CrMo steel decreases monotonously largely due to the rapid coarsening of cementite. There are a significant secondary hardening of ADF1 steel when the tempering temperature is higher than 500°C and this secondary hardening reaches a peak at temperatures around 600°C. The addition of strong carbide forming vanadium and higher molybdenum content not only significantly retards softening but also forms fine alloy carbides which produce a hardness and strength increase at high tempering temperatures.

At the same strength level, both DFSR and K_{ISCC} of the ADF1 steel are noticeably higher than that of 42CrMo steel (Fig. 36). SEM observation of the fracture surfaces of DF specimens reveals that the failure is mainly transgranular for ADF1 steel whilst mainly intergranular for 42CrMo steel (Fig. 37).

Owing to the good combination of strength, toughness and DF resistance, the ADF series high strength steels may be used to produce high strength bolts, crafts and connectors et al. Some Chinese automobile makers are now producing these components made from the ADF steels, and some of them have been industrialized for massive production.

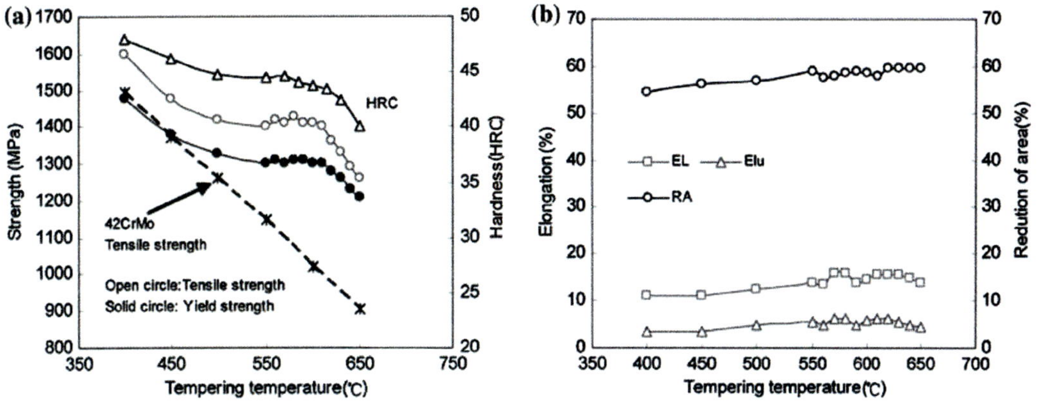

Fig. 35 Variations of hardness, strength (a) and elongation, reduction of area (b) with tempering temperature of ADF1 steel (El: elongation; E_{LU}: uniform elongation; RA: reduction of area)

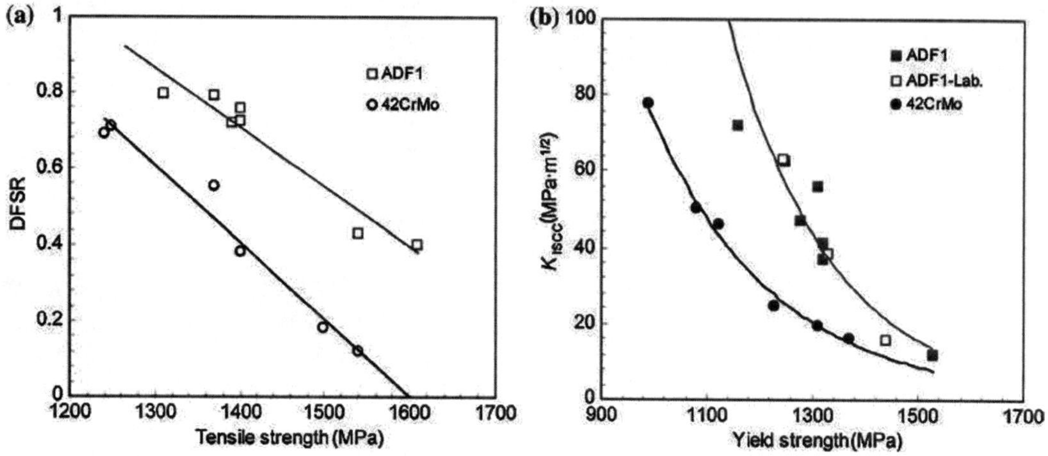

Fig. 36 Variations of DFSR (**a**) and K_{ISCC} (**b**) with strength of ADF1 steel

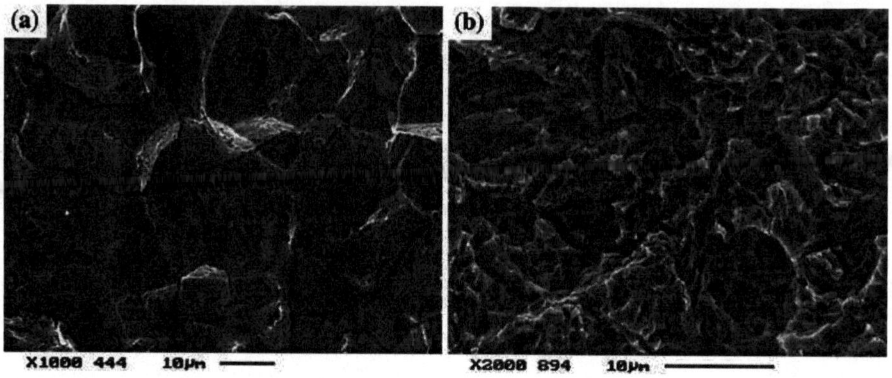

Fig. 37 Fracture surfaces in crack initiation area of sustained load tensile DF specimens of 42CrMo steel (**a**) and ADF1 steel (**b**), showing intergranular fracture for 42CrMo steel and transgranular fracture for ADF1 steel (tensile strength of both steels is 1,500 MPa)

6 Summary

The demand of the fuel economy improvement has been heightened from a point of resources saving and environment protection. Therefore, there are an increasing need to safely extend the service life of components and structures beyond their original design life. It well known that fatigue and delayed fracture are the two important mechanisms for the failure of steel components and structures in service. Based on our systematic studies of the fatigue failure and delayed fracture behaviours of high strength steels in the last 10 years, we proposed a new kind of concept to improve both fatigue failure resistance and delayed fracture resistance of high strength steels, which comprises the combination of inclusion modification, structure controlling, hydrogen trap controlling and grain boundary strengthening, and it is called IST & GST technology in this paper. Some basic considerations and proposal methods are introduced and discussed in this paper. Based on IST & GST technology, examples of development of long life high strength steels such as a newly developed 2,000 MPa grade high strength spring steel with excellent fatigue failure resistance and 1,500 MPa grade high strength bolt steel with superior delayed fracture resistance are introduced. Anyway, much more efforts should be paid to perfect this technology and to develop more long life high strength steels.

References

1. W.J. Hui, H. Dong, Y.Q. Weng, Iron and Steel **3**, 69 (2001)
2. J.W. Morris Jr, Z. Guo, C.R. Krenn, Y.H. Kim, ISIJ Inter. **6**, 599 (2001)
3. Y.Q. Weng (ed.), *Ultra-Fine Grained Steels* (Metallurgical Industry Press, Beijing, 2009)
4. A Sato, Ultra Steel 2000, in *Proceedings of the International Workshop on the Innovative Structural Materials for Infrastructure in 21st Century*, 2000
5. W P Lee, Ultra Steel 2000, in *Proceedings of the International Workshop on the Innovative Structural Materials for Infrastructure in 21st Century*, no. 33, 2000

6. J.M. Zhang, Z.G. Yang, S.X. Li et al., Acta. Metall. Sinica. **3**, 259 (2006)
7. Y. Furuya, S. Matsuoka, T. Kimura et al., Tetsu-to-Hagane **8**, 630 (2005)
8. Q.Y. Wang, J.Y. Berad, A. Dubarre, G. Baudry, S. Rathery, C. Bathias, Fatigue Fract. Eng. Mater. Struct. **22**, 667 (1999)
9. Y. Murakami, N.N. Yokoyama, J. Nagata, Fatigue Fract. Eng. Mater. Struct. **25**, 735 (2002)
10. Y. Murakami, H. Usuki, Int. J. Fatigue **5**, 299 (1989)
11. C.S. Lee, K.A. Lee, D.M. Li, S.J. Yoo, W.J. Nam, Mater. Sci. Eng. **A241**, 30 (1998)
12. J.M. Zhang, J.F. Zhang, Z.G. Yang, G.Y. Li, G. Yao, S.X. Li, W.J. Hui, Y.Q. Weng, Mater. Sci. Eng. **A 394**, 126 (2005)
13. Z.G. Yang, S.X. Li, J.M. Zhang, J.F. Zhang, G.Y. Li, Z.B. Li, W.J. Hui, Q.Y. Weng, Acta. Mater. **52**, 5235 (2004)
14. Z.G. Yang, J.M. Zhang, S.X. Li, Q.Y. Wang, W.J. Hui, Y.Q. Weng, Mater. Sci. Eng. **A 427**, 167 (2006)
15. J.M. Zhang, S.X. Li, Z.G. Yang, Y. Li, W.J. Hui, Y.Q. Weng, Int. J. Fatigue **4**, 765 (2007)
16. D.Y. Wei, J.L. Gu, H.S. Fang, B.Z. Bai, Z.G. Yang, Int. J. Fatigue **26**, 437 (2004)
17. Y.H. Nie, W.T. Fu, W.J. Hui, H. Dong, Y.Q. Weng, Fatigue Fract. Eng. Mater. Struct. **32**, 189 (2009)
18. Y. Murakami, T. Nomoto, T. Ueda, Fatigue Fract. Eng. Mater. Struct. **22**, 581 (1999)
19. H.M. Zhao, W.J. Hui, Y.H. Nie, Y.Q. Weng, H. Dong, Chinese J Mater. Res. **5**, 526 (2008)
20. Y.D. Li, Z.G. Yang, Y.B. Liu, S.X. Li, G.Y. Li, W.J. Hui, Y.Q. Weng, Mater Sic. Eng. A. **A489**, 373 (2008)
21. Y.D. Li, Z.G. Yang, S.X. Li, Y.B. Liu, S.M. Chen, W.J. Hui, Y.Q. Weng, Adv. Eng. Mater. **7**, 561 (2009)
22. Y. Murakami, H. Konishi, K. Takai, Y. Murakami, Tetsu-to-Hagane **86**, 777 (2000)
23. Y. Murakami, N.N. Yokoyama, K. Takai, J. Soc. Mater. Sci. Japan **10**, 1068 (2001)
24. W.Y. Chu (ed.), *Hydrogen Damage and Delayed Fracture* (Chinese Metallurgical Industry Press, Beijing, 1998)
25. S. Matsuyama (ed.), *Delayed Fracture* (Nikkan-Kogyo Press, Tokyo, 1989)
26. H. Matsumoto, F. Nakasato, T. Kushida, Sumimoto Met. **4**, 128 (1996)
27. N. Suzuki, N. Ishii, T. Miyagawa et al., Tetsu-to-Hagane **2**, 227 (1993)
28. F. Nakasato, M. Takahashi, Met. Technol. **11**, 449 (1980)
29. G.M. Pressouyre, I.M. Bernstein, Metall. Trans. **A 11**, 1571 (1978)
30. J.P. Hirth, Metall. Trans. A **11A 6**, 861 (1980)
31. G.F. Li, R.G. Wu, T.C. Lei, Metall. Trans. A **23A10**, 2879 (1992)
32. W.J. Hui, Y.Q. Weng, H. Dong (eds.), *High Strength Bolt Steels* (Metallurgical Industry Press, Beijing, 2009)
33. S.X. Li, Y.Q. Weng, W.J. Hui, Z.G. Yang (eds.), *Very High Cycle Fatigue Properties of High Strength Steels—Effects of Nonmetallic Inclusions* (Metallurgical Industry Press, Beijing, 2010)
34. Y.S. Hong, A.G. Zhao, G.A. Qian, Acta. Metall. Sinica **7**, 769 (2010)
35. H.M. Zhao, W.J. Hui, Y.H. Nie, H. Dong, Y.Q. Weng, Iron and Steel **5**, 66 (2008)
36. R. Kiessling, *Proc 2nd Int Conf "Clean Steel"* (Institute of Metals, London, 1983)
37. M. Larsson, A. Melander, A. Nordgren, Mater. Sci. Tech. **3**, 235 (1993)
38. T. Abe, Y. Furuya, S. Matsuoka, Tetsu-to-Hagane **6**, 711 (2003)
39. H. Suito, R. Inoue, ISIJ Int. **36**, 528 (1996)
40. W.J. Hui, H. Dong, X.G. Zeng, X.Q. Xing, Q.Li, Iron and Steel **9**, 68 (1999)
41. X. Zhang, S.V. Subramanian, Wire J. Int. **12**, 102 (1999)
42. Y. Murakami, T. Toriyama, Y. Koyasu, S. Nishida, Tetsu-to-Hagane **6**, 678 (1993)
43. T. Iikubo, Y. Ito, H. Hayashi, T. Saito, N. Takagi, Denki-Seiko (Electric Furnace Steel) **1**, 23 (1986)
44. W.J. Hui, H.M. Zhao, Y.H. Nie, Y.Q. Weng, H. Dong, in *CMS*2007 *Annual Meeting Proceedings* (Metallurgical Industry Press, Beijing, 2007)
45. Y.H. Nie, W.J. Hui, W.T. Fu, Y.Q. Weng, Chinese J. Mater. Res. **21**, 123 (2007)
46. S. Fukumoto, A. Mitchell, in *Proceedings of the* 1991 *Vacuum Metallurgy Conference on the Melting and Processing of Specialty Materials*, 1991
47. C.L. Briant, H.C. Feng, C.J. McMahon Jr, Metall. Trans. A **9A 5**, 625 (1978)
48. Y.Q. Weng, C.J. McMahon Jr, Mater. Sci. Eng **3**, 207 (1987)
49. S. Komazazki, S. Watanabe, T. Misawa, ISIJ Inter. **11**, 1851 (2003)
50. J. Yu, C.J. McMahon Jr, Metall. Trans. **A 11A 2**, 291 (1980)
51. W.J. Hui, H. Dong, Y.Q. Weng et al., Trans. Mater. Heat Treat. **6**, 37 (2006)
52. Y.Q. Weng, Dissertation, University of Pennsylvania, 1984
53. W.J. Hui, H. Dong, Y.Q. Weng, in *Proceedings of First International Conference on Advanced Structural Steels (ICASS* 2002), no. 243, 2002
54. W.J. Hui, H. Dong, Y.Q. Weng et al., Acta. Metall. Sinica **12**, 1274 (2004)
55. X.Z. Zhang, Y. Ma, H.H. Zhou, Workshop on New Generation Steel **13–16**, 185 (2001)
56. S. Yamasaki, T. Takahashi, Tetsu-to-Hagane **7**, 460 (1997)
57. W.J. Hui, H. Dong, Y.Q. Weng, J. Iron and Steel Res. Inter. **4**, 63 (2003)
58. T. Tsuchida, T. Hara, K. Tsuzaki, Tetsu-to-Hagane **11**, 771 (2002)
59. A. Yoneguchi, J. Schaad, Y. Kurebayashi, Y. Ito, SAE Technical Paper Series 2000-01-0564
60. M. Shimotsusa, N. Ibaraki, T. Ikeda, T. Nakayama, Wire J. Inter. **31**, 78 (1998)
61. W.J. Nam, C.S. Lee, D.Y. Ban, Mater. Sci. Eng. **A289**, 8 (2000)
62. Y.H. Nie, W.J. Hui, H. Dong, J. Shi, W.T. Fu, Iron and Steel **39**, 51 (2004)
63. W.J. Hui, Y.H. Nie, H. Dong, Y.Q. Weng, C.X. Wang, J. Mater. Sci. Tech. **24**, 787 (2008)
64. T. Abe, Y. Furuya, S. Matsuoka, Fatigue Fract. Eng. Mater. Struct. **27**, 159 (2004)

Part III

Auto Sheet Steels

Metallurgical Perspectives on Advanced Sheet Steels for Automotive Applications

Debanshu Bhattacharya

Abstract

Two major drivers for the use of newer steels in the automotive industry are fuel efficiency and increased safety performance. Fuel efficiency is mainly a function of weight of steel parts, which in turn, is controlled by gauge and design. Safety is determined by the energy absorbing capacity of the steel used to make the part. All of these factors are incentives for the US automakers to use both Highly Formable and Advanced High Strength Steels (AHSS) to replace the conventional steels used to manufacture automotive parts in the past. Highly Formable Steels are generally ultra-low carbon steels fully or partially stabilized by alloying elements such as Ti or Nb. AHSS is a general term used to describe various families of steels. The most common AHSS is dual-phase steel which consists of a ferrite–martensite microstructure. These steels are characterized by high strength, good ductility, low yield to tensile strength ratio and high bake-hardenability. Another class of AHSS is the multi-phase steel which has a complex microstructure consisting of various phase constituents and a high yield to tensile strength ratio. Transformation Induced Plasticity (TRIP) steels is the latest class of AHSS steels finding interest among the US automakers. These steels consist of a ferrite–bainite microstructure with significant amount of retained austenite phase and exhibit the highest combination of strength and elongation, so far, among the AHSS in use. High level of energy absorbing capacity combined with a sustained level of high n value up to the limit of uniform elongation as well as high bake-hardenability make these steels particularly attractive for safety critical parts and parts requiring complex forming. Finally, martensitic steels with very high strengths are also in use for certain parts. All of the above kinds of steels will be discussed in this paper.

Keywords

Automotive • Formable steels • Advanced high strength steel

1 Introduction

The use of advanced steels in the automotive industry has grown explosively over the last decade. Their applications can be broadly classified into two major categories: outer body panels and body-in-white parts. In the case of outer body panels, the major driver has been the continuing quest for fuel economy and hence, weight reduction. This translates to more formable steels for manufacture of complex-design parts as well as part consolidation. In addition, reduction of gauge for weight savings has led to the use of high strength IF and IF-bake hardenable steels. However, in addition to dent resistance, stiffness is another requirement of outer body panels and even with higher steel strength, thickness can not be reduced below a certain minimum from the stiffness point of view. So, while some efforts are still

D. Bhattacharya (✉)
ArcelorMittal, Global R&D, 3001 E. Columbus Drive,
East Chicago, IN 46312, USA
e-mail: Debanshu.Bhattacharya@arcelormittal.com

Y. Weng et al. (eds.), *Advanced Steels*, DOI: 10.1007/978-3-642-17665-4_18,

ongoing in this field, the majority of activity in the gauge reduction arena has been the development of advanced high strength steels (AHSS) for structural applications For the body-in-white parts, major drivers have been reductions of weight as well as an increased emphasis on safety performance. In these applications, the focus, therefore, has been on first, the development of higher and higher strength steels with good formability for gauge reduction, and second, on steels with higher energy absorption capacity for safety enhancement parts. Steels for the first type of application include multi-phase (complex phase) steels and dual phase steels. The second type of applications include dual phase but particularly the Transformation Induced Plasticity (TRIP) steels. Finally, martensitic steels with very high strengths are also in use for certain parts.

2 Highly Formable Steels

2.1 Interstitial-Free (IF) Mild Steels

In the classic IF Mild Steel, the carbon content is at the ultra low carbon level (typically less than 50 ppm). All the C and N are combined with strong nitride and carbide forming elements such as Ti and Nb using standard stoichometric equations [1]. The resulting steel thus has low yield strength, high elongation, n-value and r-value. For these reasons, IF steels have been applied on mild steel parts where improved formability was desired.

The SAE specification J2329, [2] covers the mechanical property requirements of mild steels. The IF mild steels are mostly used for grades 4 and 5 products in this specification, and maybe grade 3 in some cases, particularly for hot dip coated steels.

In the USA, these steels are termed CQ (Commercial Quality), DQ (Drawing Quality), DDQ (Deep Drawing Quality) and EDDQ (Extra Deep Drawing Quality) with the EDDQ steel as the product with the most formability being used currently. In Japan, however, steels with higher formability termed "Super Formable" (Super-EDDQ) are being developed [3, 4]. The hierarchy of these formable steels is shown in Fig. 1 [5] where the r-value and n-value of these steels are depicted. The basic metallurgical principles used to produce these steels have been well outlined by Ushioda et al. [5].

There are mainly two metallurgical techniques being used to produce these steels: refinement of hot rolled grains by high reduction during hot rolling just above the Ar_3 and ferritic hot rolling using lubrication [6]. Using the first technique, r-value of 2.5 and n-value of 0.27 has been obtained [7]. Using ferritic, lubricated hot rolling, r-value of as high as 3.0 have been obtained [8]. In South Korea, steels with high formability are also being developed: SEDDQ (Super EDDQ) and HEDDQ (Hyper EDDQ). SEDDQ steels have a minimum r-value of 2.5 while HEDDQ steels have r-values of greater than 2.7 [9]. Techniques used to produce these steels are reduction of alloying elements such as C, N, sulfur (S) and phosphorus (P), and increased hot rolling and cold rolling reductions to achieve the desired crystallographic textures [10].

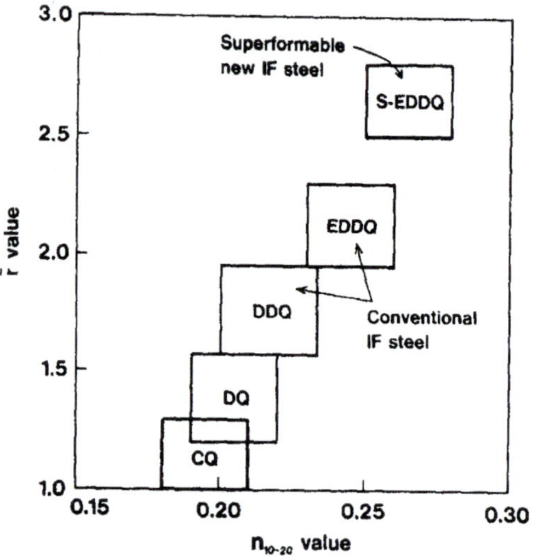

Fig. 1 Super formable steel sheets

There has been no demand or request for such super formable mild steels in the USA. It is generally felt that such steels would invariably have lower strength and would actually be too "soft" for most applications as well as for in-plant handling.

Since IF mild steel is a mature product, little new research has gone into the fundamental metallurgical aspects of these steels in recent times. However, two aspects have received attention: precipitation characteristics of stabilizing elements and the role of Nb in (Nb containing) IF steels. In the classic model, Ti forms TiN first, then TiS and finally TiC to fully stabilize the steel using standard stoichometric formulae. However, work by the University of Pittsburgh has shown that the precipitation kinetics is much more complicated, forming the H-phase, $Ti_4C_2S_2$ by intercalation. The two precipitation models are shown in Fig. 2 from Hua et al. [11].

The role of Nb in IF steels, (Nb-Ti or Nb only) is quite complex and still not completely understood. One effect of Nb, which has been well established, is its effect on crystallographic texture [12–17]. Nb improves the r_{45} value and as a result, there is a decrease in r, when compared to Ti-IF steels. The other known effect of Nb is its effect on galvannealing behavior. This was shown by the work of Tokunaga and Kato [18], where the powdering tendency of Nb-Ti steel was found to be superior to Ti-IF steels (see Fig. 3). Recent work by Cheng [19] has shown that while under certain conditions, the galvannealing rates are comparable, it is more consistent in Nb-Ti-IF steel than in Ti-IF steel (Fig. 4).

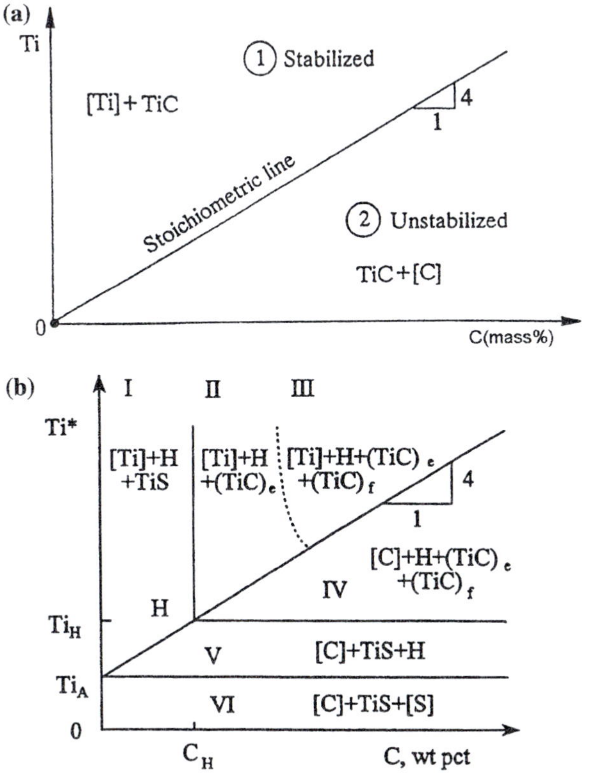

Fig. 2 Precipitation map in a Ti-stabilized ULC steel, **a** traditional, **b** proposed

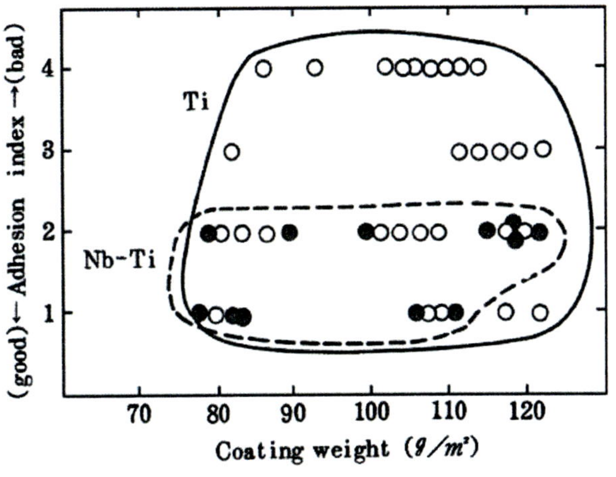

Fig. 3 Coating adherence of Ti-IF and Nb + Ti-IF steels

The mechanism of this behavior is believed to be differences in the outburst phenomenon observed in these two kinds of IF steels. However, the reason behind such differences in outburst reaction is not completely clear. It has been suggested [20] that since the affinity of Nb and P is lower than Ti and P; Nb does not scavenge P, as does Ti through the formation of FeTiP phosphides. As a result, P is left at

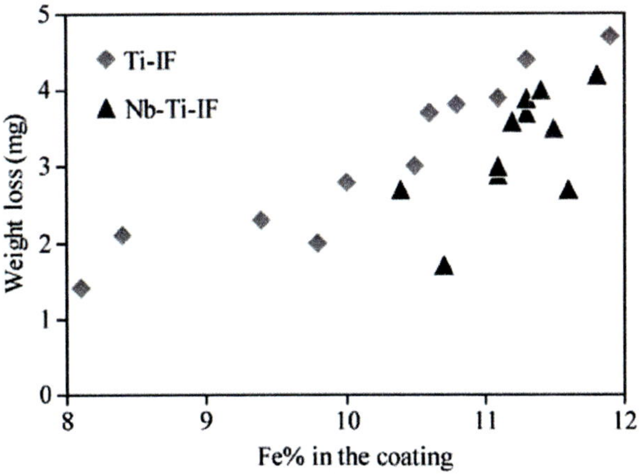

Fig. 4 Effect of Nb on the powdering characteristics of IF steels

the grain boundaries, thus, reducing the outburst phenomenon. In addition, work at the University of Pittsburgh by DeArdo et al. [21] has shown segregation of Nb to grain boundaries, again reducing the outburst reaction. Some recent work by Bhattacharya and Cheng at ArcelorMittal [22] has yielded some interesting results. Analysis by Auger Electron Spectroscopy (AES) and Glow Discharge Optical Emission Spectroscopy (GDOES) of the surface of samples quenched from the temperature at which the steel would enter the galvanizing pot, show that there is significant diffusion of Nb to the surface in Nb-Ti steels. It appears, therefore, that the control of outburst reaction in Nb-Ti-IF steels needs to be explained by Nb diffusion to both the surface and the grain boundaries.

3 High Strength IF Steels with No Bake-Hardening (HSIF)

The most commonly used HSIF steel in the USA is a fully stabilized 340 MPa minimum tensile strength product designated variously as 35E, HFT 340^{+}, etc. This has been used for outer body panel applications for some time now, and has been very successful in improving the denting performance.

In addition, two other HSIF steels have been developed—a 390 MPa minimum tensile strength and a 440 MPa minimum tensile strength product. While there has been little use of the 390 product, the 440 product has been in limited use in the USA. All HSIF products are essentially Ti + Nb (in some cases Ti) stabilized IF steels strengthened with P and Mn.

As indicated above, the high strength in these steels is obtained through the addition of P and Mn. The addition of P to IF steels, however, creates two unique issues. One is a phenomenon called secondary work embrittlement (SWE)

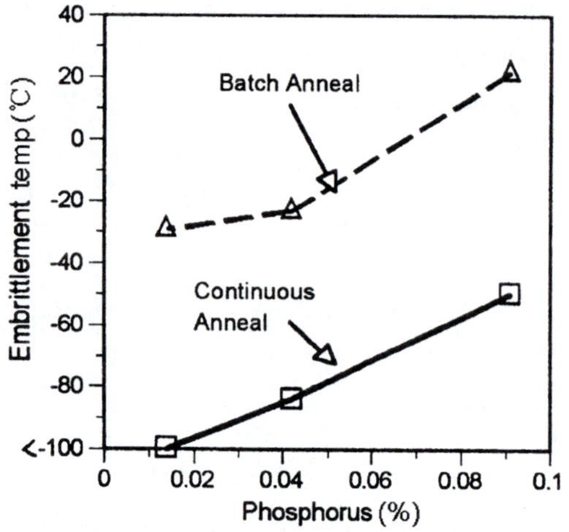

Fig. 5 Effect of P addition on SWE in a Ti + Nb steel

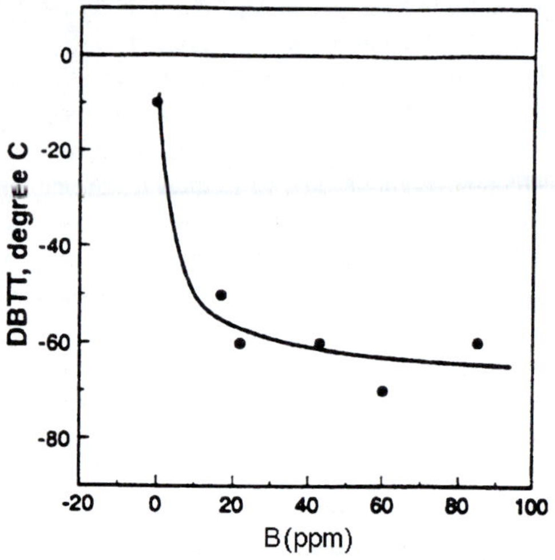

Fig. 6 Effect of B addition on SWE

or cold work embrittlement (CWE), in which the steel fractures in a brittle, intergranular mode during a secondary forming process after an initial large drawing strain. The other issue with P addition is the effect of P on the galvannealing behavior of IF steels.

SWE in IF steels has some interesting features.

First, it is observed even in non-rephosphorized IF steels, in particular, when batch-annealed [12, 23].

Second, SWE in IF steels increases with increasing P content [24, 25].

These are clearly shown in Fig. 5 [24] where the effect of P on SWE in batch annealed and continuous annealed steels is depicted.

Third, addition of boron (B) greatly reduces propensity for SWE in IF steels [23, 26], as shown in Fig. 6, from the work of Takahashi et al. [26].

Finally, SWE in IF steels depends on the agent used for stabilization. Steels stabilized with Ti generally have higher transformation temperature (SWE) than steels with Nb or Nb + Ti [26].

The mechanism of SWE has been well researched [27]. It is agreed that in IF steels, grain boundaries are depleted of carbon. As a result, P segregates to the grain boundaries and weakens them thus causing SWE. The segregation of P to grain boundaries and the effect of B have been elegantly shown by Yamada et al. [27] For a bulk concentration of 0.1% P, the grain boundary concentration is as much as 0.6%. B addition causes B segregation to the grain boundaries, displacing P and thus, reducing SWE.

However, the addition of B affects mechanical properties, texture and bake-hardenability [28]. One interesting effect of B is its influence on the texture memory effect. As the work of Ushioda et al. [5] and others have shown, annealing of IF steels, above the Ac_3 temperature, causes a deterioration of texture and r-value, especially, in low Mn steels. Recent unpublished work by Yakubovsky, Fonstein and Bhattacharya at ArcelorMittal has shown that addition of B causes the texture memory to be retained.

The reasons behind lower SWE with Nb addition are more complicated. One theory is that NbC is less stable than TiC and, hence, dissolution during annealing causes some C to segregate back to the grain boundaries. However, other work [29] has shown that Nb itself segregates to the grain boundary causing reduced P segregation and, hence, SWE. Overall, SWE is an important aspect of HSIF steels and should be considered as important while selecting applications for these steels.

With respect to the effect of P on galvannealing, P affects both the galvannealing rate as well as the phase structure of the coatings [30, 31] which in turn can affect the adherence of the coating. [32]. This phenomenon can be both advantageous and problematic. Galvannealing process parameters have to be modified in order to accommodate the galvannealing rate of P added steels, but also, P addition can be used to control and obtain the desired phase structure of the coating.

3.1 High Strength IF Steels with Bake Hardening (HSIF-BH)

As mentioned earlier, bake hardening steels need to contain C in solution, therefore they are not "strictly IF" steels. However, for the purposes of this paper, they would still be termed "IF".

The desire to improve dent resistance while maintaining the superior formability of "IF" steels has led to the development of high strength, IF bake hardenable (HSIF) steels. This has been driven by the very interesting observation that bake hardenable grades, of equivalent or even slightly lower in-panel yield strengths, impart superior dent resistance performance [33, 34].

To obtain reasonable bake hardenability in IF steels, it is generally believed that 5–20 ppm carbon in solution is needed [4, 35]. Lesser amounts result in insufficient bake hardening while higher amounts of carbon lead to unacceptable aging response at room temperature.

Now, the optimum amount of carbon in solution can be achieved by several methods:

Method 1: Leave all the carbon in solution at steelmaking while tying up nitrogen with the requisite amount of titanium as titanium nitride [5].

Method 2: Leave an optimum amount of carbon in solution. This is accomplished by controlling the precipitation of carbide by the addition of Ti or Nb [36, 37].

Method 3: Ensure carbon precipitation initially by using a suitable carbide former. In the subsequent annealing process, sufficient precipitates are dissolved to obtain the required amount of carbon in solution. Niobium has been used in the past in this approach [38]. However, high annealing temperatures are needed to redissolve NbC precipitates. Another idea, using the same concept, claimed to use vanadium as a carbide former and dissolution of vanadium carbide during annealing to obtain the necessary amount of carbon in solution [39, 40]. However, work by Girina and Bhattacharya at ArcelorMittal [41] has clearly shown that vanadium (V) plays no role since vanadium carbide precipitation does not occur in normally processed steel and the bake hardening is imported by the carbon left in solution from steelmaking (similar to method 1).

In North America, HSIF-BH steels, which are available, use most of these methods to obtain bake hardening in IF steels; although the Ti-V steels should be viewed as using method 1 and not method 3, as discussed above.

For a review of the current HSIF and HSIF-BH steels in the USA, the reader is referred to [42].

In general, three HSIF-BH grades are of major use in North America: 180Y BH, 210Y BH, and 340T BH. Note that while the first two steels are specified by minimum yield strength, the last one is specified by minimum tensile strength. A 340 MPa min. tensile strength steel may also meet a min. 210 MPa yield strength; however, it is not required to do so.

As pointed out earlier, the bake hardenability in these steels arises from the intentional solute carbon in the steel. Because of the presence of solute carbon, these steels should also be prone to aging during storage. The effect of such aging on subsequent mechanical behavior is, therefore, of interest. In particular, since both aging and bake-hardenability arise from the presence of carbon in solution, the interrelationship between aging and bake-hardenability need to be studied.

The effect of prior aging on subsequent mechanical behavior has been studied in low carbon steels [43]. However, in that work, the samples were aged after pre-straining. While this represents the case where formed panels await point treatment, it does not reflect the common situation where coils are stored before stamping. In ultra low carbon (ULC) steels, the return of Yield Point Elongation (YPE) after aging and its relationship to carbon in solution have been studied in detail [44–47]. However, to the best of the author's knowledge, no work on the effect of prior aging on subsequent bake-hardenability in ULC steels had been reported until the recent work at ArcelorMittal [48]. In this work, an HSIF-BH steel with ~350 MPa tensile strength (~230 MPa yield strength) was aged naturally (35°C) for 0 to 36 weeks (9 months) and artificially at 100°C for 0 to 4 h. Mechanical properties and bake-hardenability (2% strain, 170°C, 20 min) were then evaluated to establish the effect of aging on subsequent mechanical behavior including bake hardening.

Since the samples were aged, both naturally and artificially, on unstrained condition, this represents the more common situation of storage of coils before stamping and subsequent paint curing. In addition, samples were artificially aged at 100°C for 1 h, then naturally aged up to 3 months and tested to evaluate if any further natural aging occurs after the initial artificial aging.

The results obtained in this work were very interesting. Figure 7 shows the effect of storage at 35°C on mechanical properties and clearly shows the effect of natural aging. There was little change in tensile strength, yield strength and r-value with storage. Both ductility (% elongation) and n-value decreased with aging, but at a decreasing rate with time. The majority of reductions occurred within the first 6 weeks of storage and there was no further deterioration of properties. Return of YPE of <0.5% occurred after 9 months of storage, in the worst case scenario. The maximum level of YPE acceptable for a stamped part to be free from stretcher strains (St-St) remains a point of conjecture and is likely to vary from part to part. It has been reported that up to 0.4% YPE causes no stretcher strains [49]. However, based upon the experience of supplying BH steels for many years by ArcelorMittal and other steel companies, it is safe to say the even 0.5% YPE should be free from St-St during forming.

Another interesting aspect of this study was the effect of anisotropy. There was no effect of direction (L vs. T) on strength or elongation after aging; however, n-value is predominately affected in the T direction and YPE is affected more in the L direction by aging. While the exact reason has not been established for this effect, one possibility is the differential diffusivity of C atoms in the L vs. T direction leading to easier formation of Cottrell atmospheres in one direction as well as different directional dislocation density.

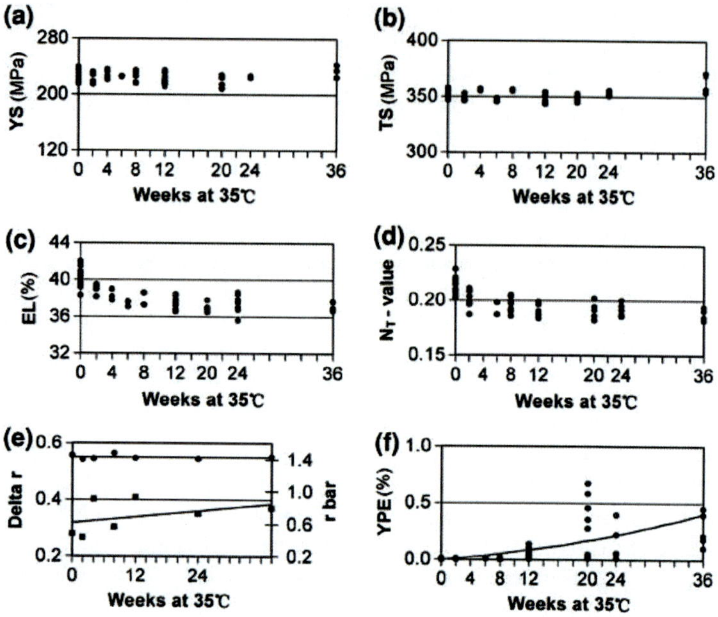

Fig. 7 Effect of aging time at 35°C on **a** yield strength, **b** tensile strength, **c** total elongation, **d** n-value, **e** r-value, **f** YPE

The effect of artificial aging was also studied. These results again show that there was no significant change in tensile strength, yield strength and r-bar; total elongation and n-value decreased; the maximum decrease occurring after 1 h at 100°C. YPE showed no significant change until 4 h of holding, increasing to 0.14%, a small effect.

Results of the combined effect of artificial and natural aging show no significant effect of subsequent natural aging after the initial artificial aging. This suggests that 1 h aging at 100°C is a good test to predict the long-term behavior of these steels.

The effect of natural aging (prior to the 2% prestrain for BH testing) on the subsequent bake hardenability is shown in Fig. 8. It is clear that there is no significant effect of natural aging on bake hardenability.

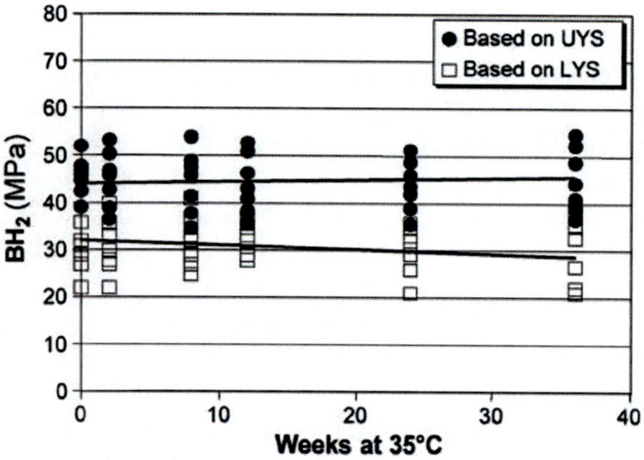

Fig. 8 Effect of natural aging on BH_2

Effect of artificial aging (100°C, 1 h) on BH_2 (BH after 2% strain) was also evaluated and showed that there is little effect of aging on bake hardening.

Finally, the excess carbon was calculated using stoichometric relationships and shows that BH_2 increases with excess carbon and that at least ~10 ppm carbon is needed to obtain significant bake hardening. This is similar to results reported earlier [47, 50].

The results of the study at ArcelorMittal [48] also clearly show that there is no effect of aging on bake-hardenability response (BH_2) in this steel. This is interesting as it is also seen that BH_2 is clearly dependent on the excess carbon in solution. Presumably, therefore, as carbon atoms in solution migrate to form Cottrell Atmospheres during aging, solute carbon available for bake hardening should be decreased. In work published earlier on samples aged after pre-straining [43], there was such a decrease in bake hardening with aging as expected from a decrease in solute carbon. However, this does not seem to be the case in this work for samples aged before prestraining. There are two possible explanations for this observation. First, there has been some thought that there may be other mechanisms which contribute to strain aging in addition to Cottrell Atmospheres. Such mechanisms may not reduce the solute carbon available for bake-hardending and this will have no significant effect on BH_2. The other possibility is that in pre-strained samples, there is a high density of dislocations, which, during subsequent aging, lead to significant Cottrell Atmospheres using carbon atoms. In unstrained and subsequent aged samples, either the dislocation density is simply not high enough to reduce solute carbon significantly through Cottrell Atmospheres, or the 2% straining

Table 1 Dual Phase steels and their mechanical property requirements

Product	TS_{min} (MPa)	YS (MPa)	TE_{min} (%)
Cold Rolled 590 MPa Dual Phase (CR 590 DP)	590	305–450	24
Cold Rolled 780 MPa Dual Phase (CR 780 DP)	780	420–550	14
Cold Rolled 980 MPa Dual Phase (CR 980 DP)	980	600–720	10
Galvanized 600 MPa Dual Phase (GI 600 DP)	600	340–410	23
Galvanized 780 MPa Dual Phase (GI 780 DP)	780	420–550	14
Galvannealed 590 MPa Dual Phase (GA 590 DP)	590	300–410	23
Galvannealed 780 MPa Dual Phase (GA 780 DP)	780	440–560	12
Galvannealed 980 MPa Dual Phase (GA 980 DP)	980	600–720	10

after aging essentially unlocks the Cottrell Atmospheres to provide sufficient solute atoms for "secondary" pinning. Thus, in pre-strained and aged samples, there is an effect of aging on BH_2 while in samples pre-strained after aging there is no effect of aging on bake-hardenability.

4 Advanced High Strength Steels (AHSS)

4.1 Dual Phase Steels

Among AHSS, dual phase steels are gaining the widest usage among automakers. This is because they provide an excellent combination of strength and ductility while at the same time they are widely available due to the relative ease of manufacture. As a result, a large number of dual phase products have been developed and used in the USA.

Table 1 is a summary of the dual phase product property requirements. Requirements for the same product sometimes vary widely; hence only representative property targets are listed. Also, requirements of total elongation depend on steel gauge; values shown here are for 1.6–2.0 mm range. Finally, the test results are for JIS-T samples except for GI 600 DP where it is ASTM-L.

4.1.1 Cold Rolled Dual Phase Steels

The cold rolled dual phase steels described here have been developed using the advantages of the water quench continuous annealing line.

All the steels developed are based on annealing in the two-phase (inter-critical) temperature region and the consequent increase in carbon content in austenite in comparison with the average carbon content in the steel. Thus, as shown in Fig. 9, carbon in austenite at a lower inter-critical temperature C_{γ_2}, is higher than that at a higher temperature, C_{γ_1}, at the same total steel carbon content.

As is clear from Fig. 9, the closer to A_{c1} the annealing temperatures are, the higher the C_γ (carbon content in austenite) and higher its hardenability. Thus, effects of annealing temperature (T_{an}) and cooling rate are

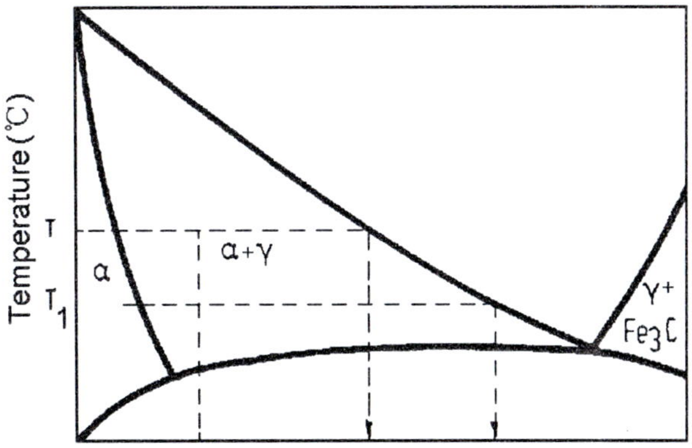

Fig. 9 Pseudo binary Fe (Me)-C diagram, illustrating concentration of carbon in austenite as a function of heating in two-phase region

interrelated. The lower the T_{an} in the $\alpha + \gamma$ region and therefore the higher C_γ, the lower the permissible cooling rate that allows martensite transformation while avoiding pearlite and/or bainite transformation.

Direct quenching from inter-critical temperature range allows achieving very high strength of steels without expensive alloying. By water quenching directly from the inter-critical region but without initial slow cooling, any desired volume fraction of martensite, which will be equal to the amount of formed austenite, can be obtained [51]. However, the final properties of directly water-quenched steels are sensitive to fluctuations of annealing parameters, which affect the amount of formed austenite.

Interrupted cooling cycle is a combination of beneficial features of direct quenching and relatively slow initial cooling. The lowest temperature of water quenching is important for the shape of steel sheet, its YS/TS ratio and partly its elongation. Figure 10 presents effects of quenching temperature, T_q, (beginning of water quenching) at various annealing temperatures on properties of 0.1C-1.5Mn-0.3Si steel. Similar results were presented for various amounts of C and Mn in an earlier work [52]. As shown, the lower the annealing temperature (higher stability of austenite), the larger the temperature plateau of quenching temperature

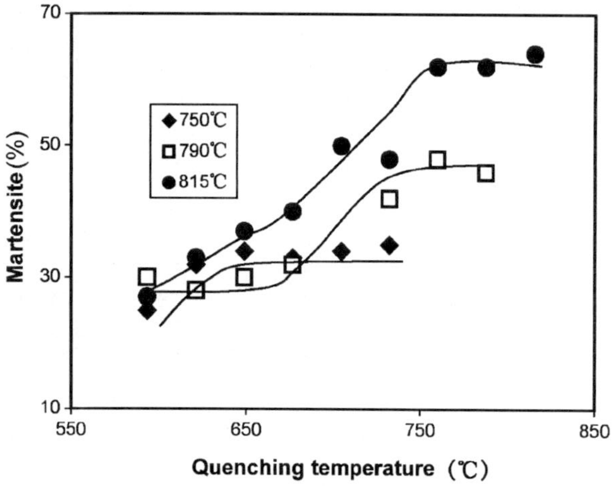

Fig. 10 Volume fraction of martensite as a function of quenching temperature after various annealing temperatures for a 0.1C-1.42Mn steel

Table 2 Mechanical properties of cold rolled dual phase steels

Product	TS (MPa)	YS (MPa)	TE (%)
CR 590 DP	625	370	26
CR 780 DP	820	470	18
CR 980 DP	1030	650	14

where no changes in volume fraction of martensite and therefore TS occur.

As was shown in the earlier work [52], after over-aging at 260°C it is possible to meet the necessary requirements of CR 590DP grade (see Table 1) at a rather wide range of C and Mn content, T_{an}, and T_q. The optimal chemistry was determined based on the best combination of flexibility of annealing (hardenability) and weldability. Typical properties of a 0.10C-1.0Mn-0.3Si commercially produced CR 590DP grade are presented in Table 2.

The same principles can also be used to produce CR 780 DP and CR 980 DP. Higher volume fractions of martensite are obviously needed for the higher strength steels and the processing parameters have to be controlled accordingly in order to achieve the required properties. Typical properties for these products are also given in Table 2.

Representative microstructures for two of the products CR 590 DP and CR 980 DP are shown in Fig. 11.

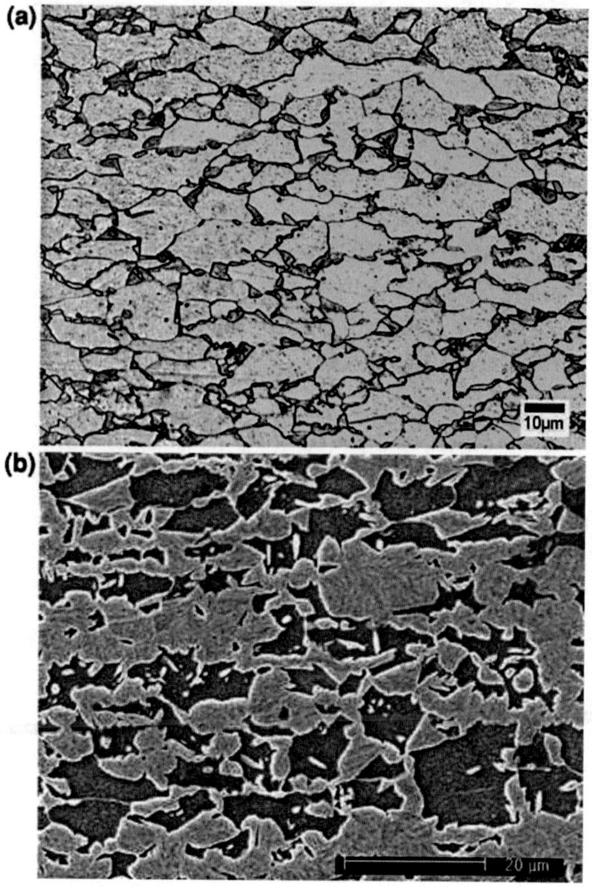

Fig. 11 Typical microstructure of **a** CR 590 DP and **b** CR 980 DP, SEM 3000×

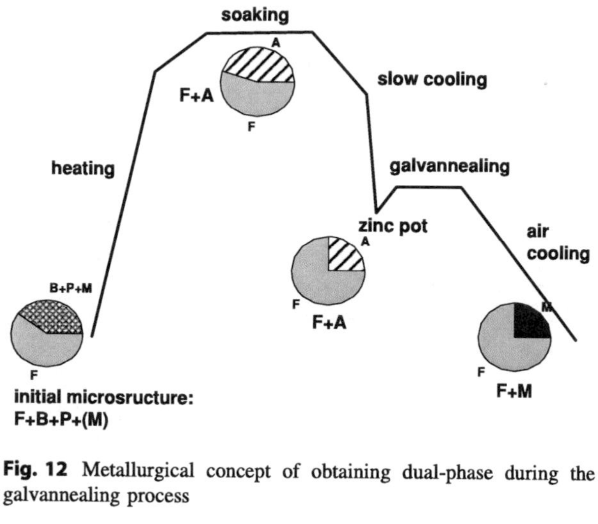

Fig. 12 Metallurgical concept of obtaining dual-phase during the galvannealing process

4.1.2 Galvannealed Dual Phase Steels

Property requirements of Galvannealed Dual Phase Steels have been presented in Table 1. Low YS and therefore low YS/TS ratio can be achieved only by obtaining ferrite–martensite dual-phase structure.

A schematic of the metallurgical concept to obtain dual phase structure after galvannealing is presented in Fig. 12. Inter-critical annealing, as was described above, can be used to obtain austenite enriched by carbon. The basic idea is to have such a combination of carbon and manganese content that it ensures a very high stability of gamma-phase, sufficient to prevent, as much as possible, any decomposition of austenite during galvanizing/galvannealing. The final austenite to martensite transformation should take place during final air-cooling.

Additional contribution to enrichment of austenite by carbon takes place during the initial, relatively slow, cooling typical for all galvannealing lines. As a result of low cooling rate there is enough time for "new ferrite" formation from austenite at sufficiently high temperatures when a near-equilibrium carbon redistribution from ferrite to remaining austenite can be achieved. This phenomenon has some important practical consequences such as significantly decreasing the sensitivity of the final structure and properties to annealing temperature. This feature has been observed in several studies and is sometimes called the effect of "self-stabilization" of dual-phase structure [53, 54]. In fact, the higher the annealing temperature and higher the amount of initial austenite, the lower its stability due to its lower carbon content and the greater the portion of "new ferrite" formation. As a result of variable amount of "new" ferrite, a roughly constant amount of martensite (and tensile strength) can be produced in dual-phase steels over a reasonably wide "window" of annealing temperatures, sometimes up to 40–80°C.

The GA 590DP steel has been developed using the cycle presented above. The chemical composition is a steel of ~0.06–0.10% carbon alloyed only by manganese, Its high Mn content combined with a high carbon content in the final portion of austenite, which can reach 0.4–0.6%, ensures both high hardenability of austenite and low martensite start temperature, M_s.

Zinc pot temperature is practically constant (~460°C), and cooling rate is determined by line speed, which cannot be changed without sacrificing productivity. Thus critical variable parameters of processing GA 590DP steel include annealing and galvannealing temperatures, T_{an} and T_{ga}, respectively.

Figure 13a, showing mechanical properties as a function of annealing temperature of this steel, confirms the low sensitivity of tensile strength (volume fraction of martensite) to annealing temperature, as discussed above, inherent to slow cooling from two-phase temperature region. On the other hand, total elongation increases with higher annealing temperatures due to a larger volume of ductile "new ferrite" formed by decomposition of a larger volume of lower carbon austenite. Yield strength decreases in parallel to TS decrease due to a lower martensite volume fraction. So the highest T_{an} should be limited only by the need to achieve the necessary minimum tensile strength.

Figure 13b presents effects of T_{ga}. The higher the galvannealing temperature, the more likely that the remaining austenite would decompose partly by bainite reaction before

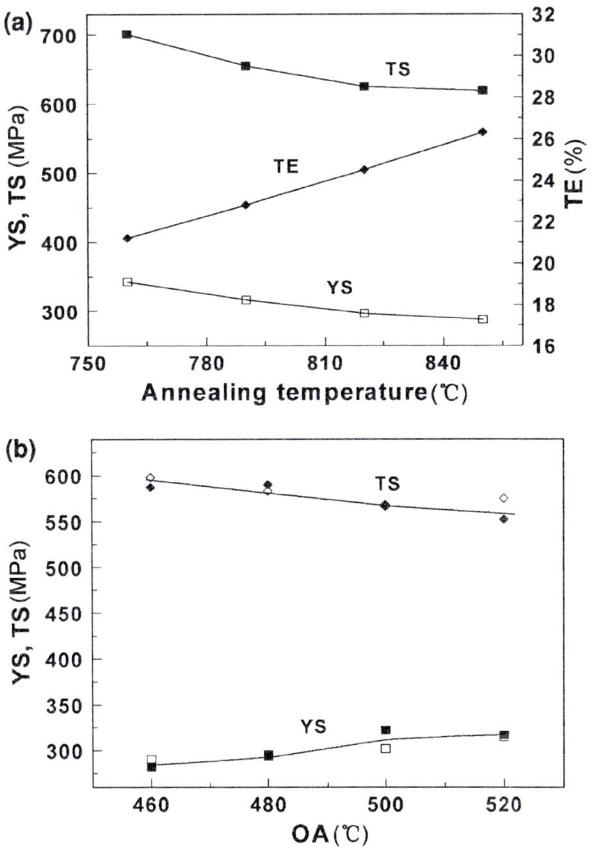

Fig. 13 Effect of **a** annealing and **b** overaging (galvannealing) temperature on tensile properties of GA 590 DP

martensite transformation during final air-cooling. This results in lower final strength of steel. Thus, T_{ga} should be kept as low as possible while still being sufficient to ensure quality of coating.

While the basic metallurgical principles remain the same for higher strength steels such as 780 DP and 980 DP, increased alloying is, of course, necessary.

4.1.3 Galvanized Dual Phase Steels

Requirements of Galvanized dual phase products have been presented in Table 1. Obtaining dual-phase structure using common HDG lines can be achieved using the same concept shown in Fig. 12. The key factor of this approach is sufficient, rather high alloying (>2% Mn and/or additions of Cr, Mo, V and so on). However, this approach can result in welding problems on the part of both the manufacturer and the customer.

ArcelorMittal has a HDG facility that is equipped with the Zinquench (ZQ) technology [55] whereby a strip can enter the zinc pot at temperatures as high as 600–630°C. Very high heat conductivity and a very large volume of zinc in the pot equipped with mixing devices result in very high

Table 3 Mechanical properties of galvannealed and galvanized dual phase steels

Product	TS (MPa)	YS (MPa)	TE (%)
GA 590 DP	620	355	26
GI 600 DP	640	350	27
GA 780 DP	795	490	15
GA 980 DP	1045	675	14

cooling rate (close to 100°C/s) in the range of snout temperature (T_{sn}) to zinc pot temperature. This technology provides a unique opportunity to keep a higher amount of remaining austenite at leaner alloying. This is further enhanced by the other critical advantage of ZQ, which is to improve the wettability of the steel strip so that it is possible to have good quality coating at Si content as high as 0.65%.

At the same time, the contribution to strengthening by silicon provides the same product strength at less martensite volume fraction and therefore gives an additional option to decrease the content of carbon or alloying elements that could negatively affect carbon equivalent and, therefore, weldability of steels. Typical microstructure of GI contains the dominant martensite type with a very small portion of bainite as strengthening phases in ferrite matrix. Regardless of the combination of annealing and snout temperatures the final strength displays practically linear dependence on volume fraction of obtained martensite that is typical for dual-phase steels [56].

Again, higher strengths such as 780 or 980 MPa need higher amounts of alloying to increase hardenability of martensite under the relatively slow cooling rate of standard hot dip galvanizing lines. Typical properties of these steels are given in Table 3.

Typical microstructure of coated dual phase steels is given in Fig. 14.

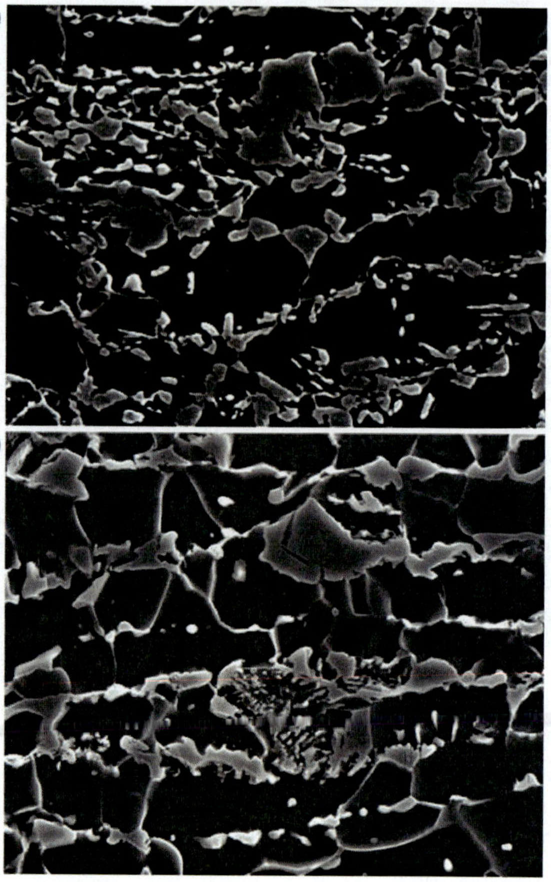

Fig. 14 Microstructure of **a** GA 590 DP and **b** GI 600 DP steels, SEM, 3000×

Table 4 Mechanical properties of multiphase steels

Product	TS (MPa)	YS (MPa)	TE (%)
CR 590 MP	690	515	23
CR 980 MP	1040	795	12
GA 590 MP	620	505	26

4.2 Multiphase Steels

Multiphase steels, also referred to as complex phase steels in Europe, are steels with a higher level of yield strength at the same comparable tensile strength levels of dual phase steels. Such steels are gaining popularity in the USA in recent times.

To obtain the high YS/TS ratio, different metallurgical principles need to be used for cold rolled and for galvannealed products. For cold rolled steels, achievement of higher YS/TS ratio of >0.7 is possible with a higher overage temperature on an appropriate dual phase structure. For galvannealed steels, however, higher YS cannot be obtained from an initial dual phase structure since in the galvannealing process, martensite is formed only in the final step of air-cooling and no further overaging is possible. The only way to gain yield strength in galvannealed multiphase structure is to obtain appropriate mixture of pearlite, bainite as well as ferrite straightened by grain refinement and precipitation strengthening by Nb.

Typical mechanical properties for these steels are given in Table 4.

Typical microstructures for a cold rolled and galvannealed multiphase steels are shown in Fig. 15.

4.3 TRIP Steels

TRIP steels, based on Transformation Induced Plasticity effect offer the highest combination of strength and elongation [57], which is a measure of high level of energy

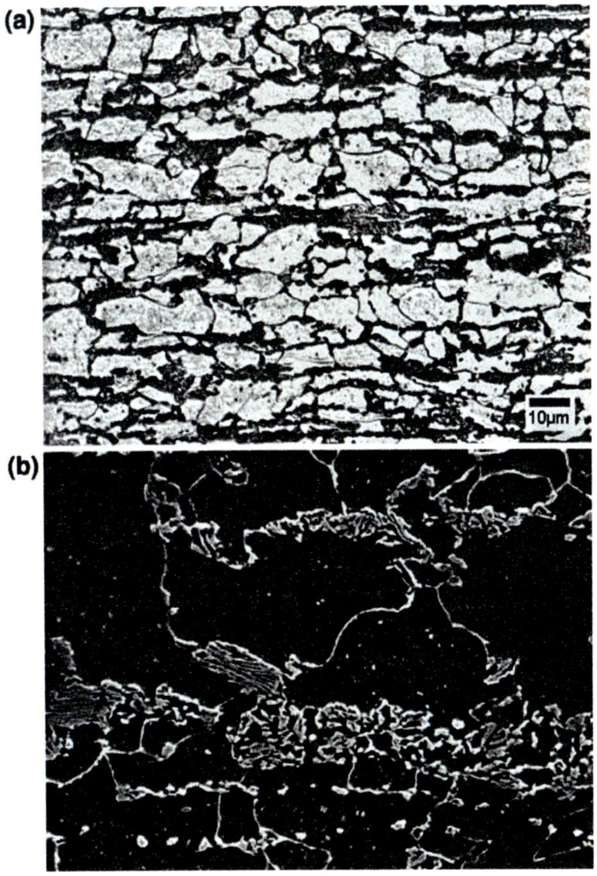

Fig. 15 Microstructure of a CR 590 HY and b GA 590 HY, 3000×

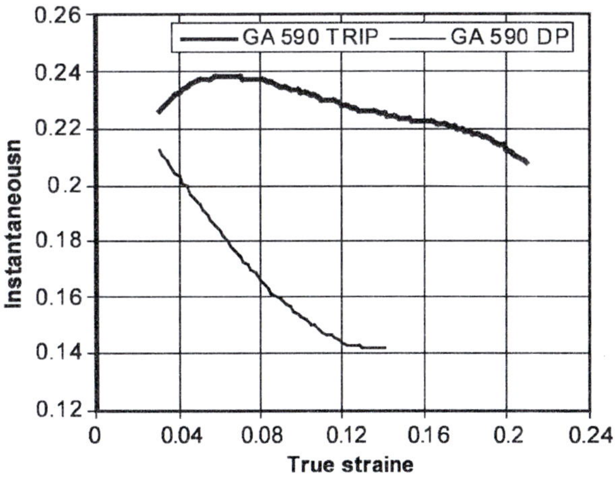

Fig. 16 Comparison of n vs. e curves for TRIP and dual phase steels of the same strength

Table 5 Target mechanical properties of TRIP steels

Product	TS $_{min}$ (MPa)	YS (MPa)	TE $_{min}$ (%)
Cold Rolled 590 TRIP	590	350–495	31
Cold Rolled 780 TRIP	780	410–500	21
Galvannealed 590 TRIP	590	360–510	26
Galvanized 590 TRIP	590	380–480	27
Galvannealed 780 TRIP	780	410–560	19
Galvanized 780 TRIP	780	440–500	21

absorption. Simultaneously, TRIP steels display high n-value up to the limit of uniform elongation as shown in Fig. 16 [58]. In addition, they also show high bake hardening compared to dual phase steels [59].

Initial enrichment of austenite by carbon takes place during inter-critical annealing as shown in Fig. 9. The relatively slow initial cooling and the rather rapid cooling down to the temperature of isothermal holding result in further enrichment of the remaining austenite by carbon, enhancing austenite stability. Further growth in its stability occurs during the austenite to bainite transformation in the presence of strong ferrite forming elements. This significantly retards carbide formation, part of the bainite reaction, and helps to keep all carbon in the remaining austenite. The elements commonly used are Si, Al and P [60]. As a result of such a high carbon content of more than 1–1.3% (author's data) or even 1.6% [61] in the final portion of austenite, the martensite start temperature becomes lower than room temperature. This stable retained austenite (RA) transforms to martensite under subsequent mechanical stress/strain resulting in Transformation Induced Plasticity as discovered by Zackay [57].

Implementation of this phenomenon to commercial production requires the development of a chemistry/cycle combination that ensures the necessary rate of bainite reaction to match with available holding time in the bainite region, inherent to a given facility. In other words, one should ensure overlapping the bainite temperature–time region with the real Bainite Isothermal Temperature Time (BITT) restrictions. For relatively long holding times, this bainite reaction should not occur at a high rate so as to prevent carbide formation and to retain a significant amount of high carbon austenite. In contrast, in HDG lines a composition, which is characterized by a fast bainite reaction at temperatures of zinc pot and galvannealing, should be selected so that even these short times would be sufficient for significant transformation to bainite.

Based on these principles, the following TRIP steels have been developed in the USA (Table 5).

Typical microstructure for TRIP steels using LaPera tint etching [62] are shown in Fig. 17.

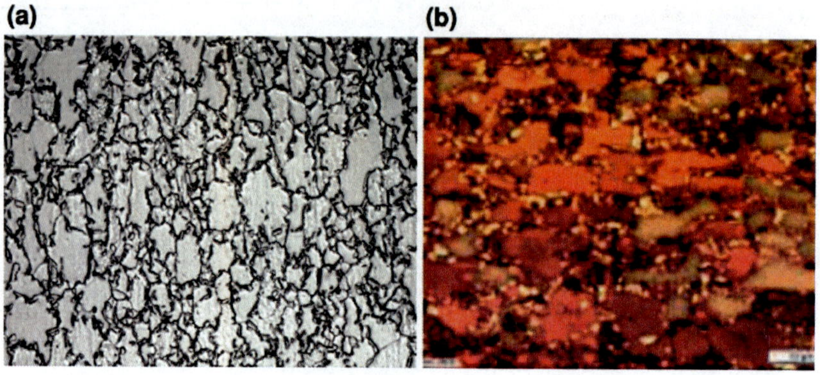

Fig. 17 Microstructures of a typical TRIP steel using **a** Nital (1000×) and **b** LePera (2000×)

Table 6 Typical mechanical properties of martensitic steels

Product	TS (MPa)	YS (MPa)	TE (%)
M 130	1054	923	5.4
M 160	1178	1020	5.1
M 190	1420	1213	5.1
M 220	1585	1350	4.7

4.4 Martensitic Steels

Using water quenching in a continuous annealing line, steels with 100% martensite are produced. These steels offer very high strength although ductility is lower than other AHS steels. The strength of the steel is controlled by the carbon content and a complete austenitizing temperature is used to obtain a fully martensitic structure. The martensitic steels in production in the USA is given in Table 6.

5 Future

Currently dual phase, multiphase steels and TRIP steels are already used in production vehicles in the US Martensitic steels have also been in use for bumpers and door beams for some time now. The immediate developments are 1180 MPa TS DP and 980 MPa TS TRIP steels. The next generation of AHSS in limited use is a new class of steels based on Twin Induced Plasticity, called TWIP steels. These offer very high elongations of 60–80% at comparable strength levels. However, since these steels are heavily alloyed, their manufacturability and hence their eventual use remains to be seen. The other development focus in North America is on a family of steels with properties between the current DP/TRIP steels and the new TWIP steels, called the "Third Generation AHSS" or "3G AHSS". Looking further ahead, the new generation of AHSS is likely to be in the areas of high formability AHSS, AHSS with high modulus, low density steels and very ultra high strength AHSS. These are depicted schematically in Fig. 18 [63].

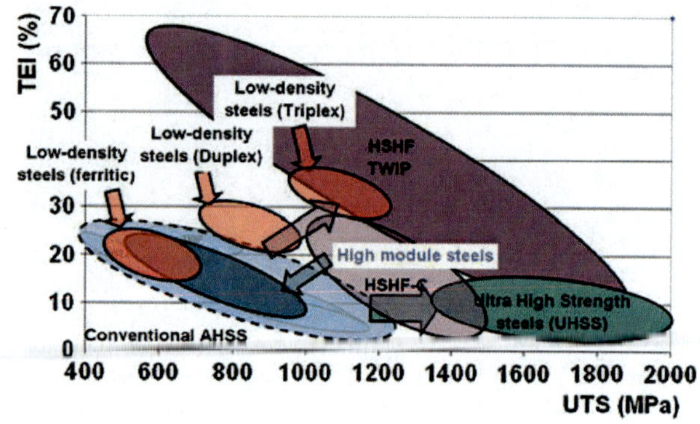

Fig. 18 Future developments in AHSS

References

1. I. Gupta, D. Bhattacharya, *Metallurgy of Vacuum Degassed Steel Products* (TMS, Warrendale, 1989), p. 43
2. SAE Specification J2329, *Categorization and Properties of Low Carbon Automotive Steel Sheets* (SAE, Warrendale, 1997)
3. H. Takechi, *International Forum for Physical Metallurgy of IF Steels* (ISIJ, Tokyo, 1994), p. 1
4. H. Takechi, in *Microalloying '95 Conference Proceedings* (ISS, Warrendale, 1995), p. 71
5. K. Ushioda et al., *International Forum for Physical Metallurgy of IF Steels* (ISIJ, Tokyo, 1994), p. 227
6. T. Asamura, in *International Symposium on Modern LC and ULC Sheet Steels for Cold Forming; Process and Properties*, Aachen (1998), p. 1
7. I. Itami et al., SAE Paper No. 930783 (1993)
8. T. Senuma, K. Kawasaki, ISIJ Int. **34**, 51 (1994)
9. O. Kwan, C. Yim, G. Kim, in *Proceedings-IF Steels 2000* (ISS, Warrendale, 2000), p. 111
10. K. Chin, *Posco Research Reports* (1999)
11. M. Hua, C.I. Garcia, A.J. DeArdo, *Modern LC and ULC Sheet Steels for Cold Forming-Processing and Properties.* (Aachen, 1998), p. 145
12. M. Hua, C.I. Garcia, A.J. DeArdo, Met. Trans. A, TMS **9**, 1769 (1997)
13. M. Hua, C.I. Garcia, A.J. DeArdo, *Modern LC and ULC Sheet Steels for Cold Forming-Processing and Properties* (Institute of Ferrous Metallurgy, Aachen, 1998), p. 145

14. R. Hook, *Metallurgy of Vacuum Degassed Steel Products* (TMS, Warrendale, 1990), p. 263
15. O. Hashimoto et al., *Proceedings-Advances in the Physical Metallurgy and Applications of Steels* (The Metals Society, London, 1982), p. 95
16. S. Satoh et al., Trans. ISIJ **24**, 838 (1984)
17. L. Ruiz-Aparicio, C.I. Garcia, A.J. DeArdo, Met. Trans. A **32A**(2), 417 (2001)
18. Y. Tokunaga, H. Kato, *Metallurgy of Vacuum-Degassed Steel Products* (TMS, Warrendale, 1989), p. 91
19. C. Cheng, in *Proceedings-42nd Mechanical Working and Steel Processing Conference* (ISS, Warrendale, 2000), p. 225
20. M. Gullinan et al., in *Conference Proceedings-Galvatech 95* (ISS, Warrendale, 1995), 295
21. A.J. DeArdo, in *International Symposium-Niobium 2001* (AIME-TMS, Orlando 2001)
22. D. Bhattacharya, C. Cheng, in *Proceedings-Galvatech 2004* (AIST, Warrendale, 2004), p. 509
23. S. Bhat, B. Yan, J. Chintamani, T. Bloom, *Iron and Steelmaker*, 22, September, 33 (Iron and Steel Society, USA, 1995)
24. R. Pradhan, *International Forum for Physical Metallurgy of IF Steels* (ISIJ, Tokyo, 1994), p. 165
25. E. Yasuhara et al., in *Proceedings-38th Mechanical Working and Steel Processing Conference* (ISS, Warrendale, 1996), p. 409
26. N. Takahashi et al., *Metallurgy of Continuous Annealed Sheet Steel* (TMS, Warrendale, 1982), p. 133
27. M. Yamada et al., *Tensu-Tu-Hagane*, ISIJ **8**, 1049 (1987)
28. O. Girina, D. Bhattacharya, in *Proceedings—IF Steels 2000* (ISS, Warrendale, 2000), p. 35
29. J. Rege, C.I. Garcia, A.J. DeArdo, in *Proceedings-39th Mechanical Working and Steel Processing Conference* (ISS, Warrendale, 1997), p. 149
30. Y. Hisamatsu, in *International Conference on Zinc and Zinc Alloy Coated Steel Sheet, Galvatech 89*, Japan (1989), p. 3
31. C. Cheng et al., *2nd International Conference on Zinc and Zinc Coated Steel Sheet, Galvatech 92*, Germany (1992), p. 122
32. C. Cheng, L. Franks, *The Use and Manufacture of Zinc and Zinc Alloy Coated Sheet Steel Products into the 21st Century, Galvatech 95*, Chicago (1995), p. 723
33. M.F. Shi et al., SAE Technical Paper No. 970158 (1998)
34. P. Belanger et al., in *Proceedings—IF 2000* (ISS, Warrendale, 2000), p. 13
35. W. Bleck, R. Bode, O. Maid, L. Meyer, in *Proceedings-Symposium on High Strength Sheet Steels for the Automotive Industry* (ISS, Baltimore, 1994), p. 141
36. A. VanShick, D. Vanderschueren, S. Vandeputte, J. Dilewijns, in *Proceedings-39th Mechanical Working and Steel Processing Conference* (ISS, Warrendale, 1997), p. 225
37. A. Pichler et al., *Proceedings-39th Mechanical Working and Steel Processing Conference* (ISS, Warrendale, 1997), p. 63
38. T. Irie et al., *Metallurgy of Continuously Annealed Sheet Steel* (TMS, Warrendale, 1982), p. 155
39. P. Mitchell, T. Gladman, in *Proceedings-39th Mechanical Working and Steel Processing Conference* (ISS, Warrendale, 1997), p. 37
40. K. Taylor, J. Speer, in *Proceedings-39th Mechanical Working and Steel Processing Conference* (ISS, Warrendale, 1997), p. 49
41. O. Girina, D. Bhattacharya, in *Proceedings-41st Mechanical Working and Steel Processing Conference* (ISS, Warrendale, 1999)
42. D. Bhattacharya, (2004) in *Proceedings—Modern Developments in Metallurgy and Technologies for Automotive Industries* (CBMM, Moscow, 2004), p. 76
43. R. Pradhan, SAE Technical Paper No. 910290 (1991)
44. N. Kozima et al., Sumitomo Metals **95**(5), 12 (1993)
45. A. DeArdo, in *Proceedings-IF Steels 2000* (ISS, Warrendale, 2000), p. 131
46. K. Tanikawa et al., NKK Tech. Rev. **72**, (1995)
47. T. Tanioku et al., SAE Technical Paper No. 910293 (1991)
48. Z. Niemczura, I. Gupta, N. Hake, D. Bhattacharya, in *Proceedings-43rd Mechanical Working and Steel Processing Conference* (ISS, Warrendale, 2001), p. 185
49. O. Akisue et al., Tetsu-Tu-Hagane **67**, 462 (1981)
50. A. Okomoto et al., Sumitomo Search **39**, 183 (1989)
51. A. Nishimoto et al., ISIJ **21**(11), 778–782 (1981)
52. I. Gupta, P.H. Chang, *Technology of Continuously Annealed Cold Rolled Sheet Steel, Conference Proceedings* (TMS-AMIE, Detroit, 1984), pp. 263–276
53. G. Eldis, in *Structure and Properties of Dual Phase Steels, Conference Proceedings*, TME-AMIE (1979), pp. 202–220
54. R. Lagneborg, in *Dual Phase and Cold Pressing Vanadium Steels in the Automobile Industry, Conference Proceedings* (1979). pp. 43–59
55. R. Patil, P. Sippola, U.S. Patent 6177140, Jan 2001
56. R. Patil, O. Girina, D. Bhattacharya, *Development of a Dual Phase High Strength Galvanized Steel Using Zinquench Technology, Galvatech* (ISS, Warrendale, 2004), pp. 439–447
57. V. Zackey et al., Trans. ASM **60**, 252–259 (1967)
58. O. Moriau et al., *Proceedings of International Conference on TRIP-Aided High Strength Ferrous Alloys*, Ghent (2002), pp. 247–251
59. O. Yakubovsky, N. Fonstein, D. Bhattacharya, in *Proceedings of International Conference on TRIP-Aided High Strength Ferrous Alloys*, Ghent (2002), pp. 263–270
60. S. Traint et al., in *Proceedings of 44th MWSP Conference* (2002), pp. 139–152
61. J. Mahieu et al., Met. Mat. Trans. **33A**(8), 2573 (2002)
62. F. LePera, J. Metals **32**(3), 38 (1980)
63. H. Guyon et al., *Materials in Car Body Engineering*, Automotive Circle International (2010)

Recent Development of Nb-Containing DP590, DP780 and DP980 Steels for Production on Continuous Galvanizing Lines

K. Cho, K. V. Redkin, M. Hua, C. I. Garcia, and A. J. DeArdo

Abstract

Dual phase steels for production on CGL, having tensile strength range from 590 to 1200 MPa, was developed in Nb-bearing Cr–Mo steels with carbon contents 0.06 and 0.15 mass%. The role of Nb in these steels, as well as the formation and transformation characteristics of austenite as a function of intercritical annealing temperature, cooling rate from intercritical annealing temperature to 460°C were investigated. The mechanical property tests of the steels treated by CGL were performed. The OIM and EBSD image quality analyses were employed in this investigation for analyzing microstructures and the phase transformation products of austenite. The results show each strength grade dual phase steel has the excellent combination of tensile strength and ductility properties. The addition of Nb can further improve the tensile strength of the Cr–Mo dual phase steel without ductility reduction. This relates to microstructure refined by the Nb in steels and the proper combination of volume fraction of recrystallized ferrite and martensite.

Keywords

Austenite formation • Cooling rate • Dual-phase • Image quality • Martensite • Niobium • Recrystallization

1 Introduction

There is no question that the automobile industry has been under constant pressure to develop better performing vehicles in the areas of safety, cost and ecological responsibility. This demand has resulted in a series of pioneering efforts between the global steel and automobile industries which led to the successful development of the Ultra Light Steel Auto Body (ULSAB) in 1998 and continues with the ULSAB-Advanced Vehicle Concepts (AVC) program. The Advanced High Strength Steels (AHSS) used in the ULSAB-AVC program have enabled automakers to reduce fuel consumption, improve passenger safety, apply mature and proven technologies, manufacture vehicles at reasonable prices and increase recyclability. It has been estimated that by the year 2006 the use of AHSS (Dual-Phase and TRIP) in the range of 590–780 MPa Tensile Strength by the Japanese and European automobile industries will increase approximately 35% from the amounts used in the year 2001. The North American automotive industry has been slower than its competitors to adapt the usage of AHSS in their vehicles. This trend is expected to change in the near future [1].

The particular advantage in mechanical properties and overall features offered by the Dual-Phase steels over other Advanced High Strength Steels are making these steels a

K. Cho · K. V. Redkin · M. Hua · C. I. Garcia · A. J. DeArdo (✉)
The Basic Metals Processing Research Institute, Department of Mechanical Engineering and Materials Science, University of Pittsburgh, Pittsburgh, PA 15261, USA
e-mail: deardo@pitt.edu

K. Cho · C. I. Garcia · A. J. DeArdo
Department of Mechanical Engineering, University of Oulu, Oulu, Finland

Y. Weng et al. (eds.), *Advanced Steels*, DOI: 10.1007/978-3-642-17665-4_19,

Table 1 Advantages and disadvantages of AHSS over conventional steels

Steel type	Advantage	Disadvantage
AKDQ – DQ- DDQ	Low cost and excellent formability	Not applicable in structural components
IF Steels	Excellent combination of strength and formability. Good consistency of mechanical properties	Limited weldability of higher strength grades
HS Solid Solution IF		
Bake Hardenable		
HSLA	Familiarity, excellent weldability and widely available	Relatively low formability, variability in properties and springback issues as the yield strength increases
Dual-Phase	High strength, formability acceptable, crash-worthiness similar to TRIP steels	High alloy content, some springback issues, potential for edge cracking, formability remains a challenge
	No major welding issues	
TRIP	High crash-worthiness, high strength with small forming strains	Require expensive annealing cycles, carbon equivalent can be high, limited weldability
Martensitic	Ultra high strength	Low formability, geometrical variability between slit coils

Table 2 Chemical composition and rolling conditions of experimental steels

Steel	Element(mass%)								Rolling conditions		
	C	Mn	Si	P	Al	$Cr^{+}Mo$	Nb	N	T_{Finish}	T_{coil}	%CR
BDP-590	0.06	1.50	0.40	0.010	0.06	0.48	0.02	0.006	900°C	550°C	60
BDP-780	0.06	1.48	0.40	0.010	0.03	0.48	0.04	0.006	900°C	550°C	60
BDP-980	0.151	1.76	0.42	0.010	0.064	0.83	0.024	0.52	900°C	550°C	60
BDP-1180	0.150	1.74	0.41	0.010	0.072	0.83	0.043	0.52	900°C	550°C	60
BDP-590 N	0.051	1.51	0.10	0.011	0.06	0.50	0.02	0.0063	900°C	550°C	60

premier candidate in the production of components such as: bumper beams, motor compartment-longitudinal rails, rocker inners, pillar reinforcements and door inners. These are few of the applications that are being considered by the global automobile industry regarding the future use of DP-AHSS [1, 2]. Table 1 shows a comparison of the advantages and disadvantages of AHSS over the steels traditionally used by the automobile industry.

The major objective of this paper is to present the development of a series of strength grades with low-carbon BDP-590, BDP-780 BDP-980 and BDP-1180 Nb-bearing steels that is being conducted at the Basic Metals Processing Research Institute, University of Pittsburgh. In addition, of low-carbon BDP-590 and BDP-590 N with low silicon for improved weldability were also developed. The detailed microstructural characterization and the resulting mechanical properties will be presented and discussed in this paper.

2 Experimental

The chemical composition of the steels and the rolling conditions used in this investigation are presented in Table 2.

The laboratory heats were melted in a vacuum induction-melting furnace. The ingot sizes were 550 mm in length × 210 mm in width × 210 mm in thickness. After melting, ingots from each steel composition were sectioned and machined into 210 mm × 210 mm × 50 mm samples for subsequent processing. Prior to hot rolling these samples were reheated at 1200°C for 2 h and then hot rolled to a final thickness of 6 mm. The finish rolling and coiling temperatures are shown in Table 2. The surface of the as-hot rolled samples was cleaned prior to being cold rolled 60%.

2.1 Intercritical Annealing

Samples from the steels in the cold rolled condition were used to study the recrystallization behavior of ferrite and the formation of austenite during intercritical annealing. The intercritical annealing treatments were conducted with the aid of a computer controlled Research Incorporated high-speed furnace. The furnace was operated using similar processing conditions of a typical continuous galvanizing line (CGL) line, i.e. a heating rate of 3°C/sec and 60 s holding time at the given intercritical temperature. After intercritical annealing the specimens were immediately water quenched

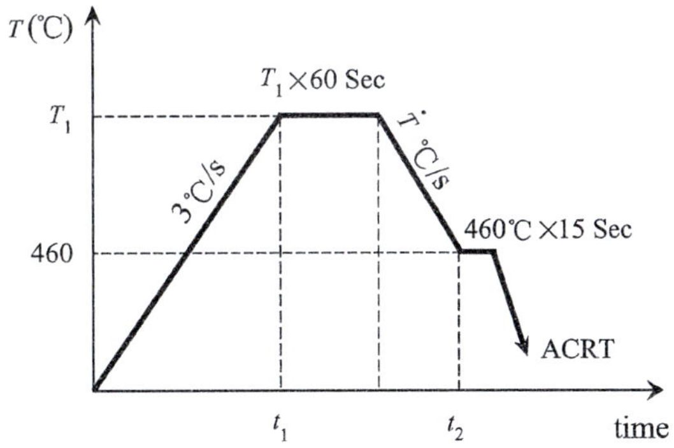

Material	T_1 (°C)	$\dot{T}$ (°C/s)
BDP-590 BDP-590N	740	3 8 15
BDP-780	800	8 15
BDP-980	735 740	15
BDP-1180	740 760	15 20

Fig. 1 Gleeble CGL simulation process of BDP-590, BDP-780, BDP-980 and BDP-1180

to room temperature. The range of temperatures explored in the intercritical region was between 725 and 810°C.

2.2 CGL Simulation Annealing Studies

The information provided from the intercritical annealing experiments was used to design the CGL simulation annealing studies. After holding at the desired intercritical annealing temperature, the specimens were cooled at three cooling rates. The cooling rates were based on the prediction of a critical cooling rate (CCR) equation used for the steel compositions being investigated. The simulation studies were conducted in a Gleeble 3500 unit. Samples from each steel in the cold rolled condition approximately 25 cm long and 5 cm wide were subjected to the CGL simulation treatment described in Fig. 1.

2.3 Microstructural Analysis

The microstructural analysis was conducted using standard electron optics techniques (OM, SEM and TEM). The SEM examination was conducted using a Philips XL-30 FEG SEM microscope. Thin foils were prepared and analyzed using a JEM-200CX TEM operated at 200 kV. The recrystallization behavior of ferrite and the formation of austenite during intercritical annealing were assessed using the automated TSL-SEM orientation imaging microscopy (OIM) system attached to the XL-30 FEG SEM microscope. A detailed discussion of this technique and its application to the characterization and quantification of microstructures can be found in the literature [3, 4]. In addition to OIM, the volume fraction of austenite (observed as martensite) and ferrite were measured using an automated image analysis BioQuant IV system.

2.4 Mechanical Testing

The mechanical properties of the specimens subjected to the CGL simulation were evaluated using a computer controlled MTS 880 system. Two ASTM sub-size tensile specimens (25.4 mm nominal gage and 6.4 mm nominal width) per condition were tested at a cross-head speed of 2 mm/min. In addition, the VHN values of both the intercritically annealed and CGL simulated process samples was measured and recorded.

3 Results and Discussion

3.1 Formation of Austenite

The formation of austenite during intercritical annealing was predicted using the commercial software package JMat-Pro. The desired phase balance of ferrite and austenite (martensite/lower bainite) to obtain the BDP-590, BDP-780, BDP-980 and BDP-1180 tensile strength was based on the work by Bucher and Hamburg shown in Fig. 2 [5]. Their work suggests that in order to obtain the tensile strength required for BDP-590 and BDP-780, BDP-980 and BDP-1180 properties, the volume fraction of martensite/lower bainite required is approximately 15%, 32%, 50% and

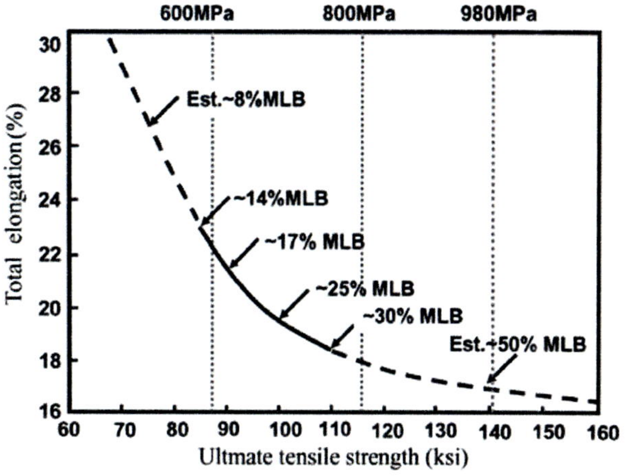

Fig. 2 Percent of dual-phase microstructure (martensite/low bainite) (MLB) for various strength-ductility combinations [5]

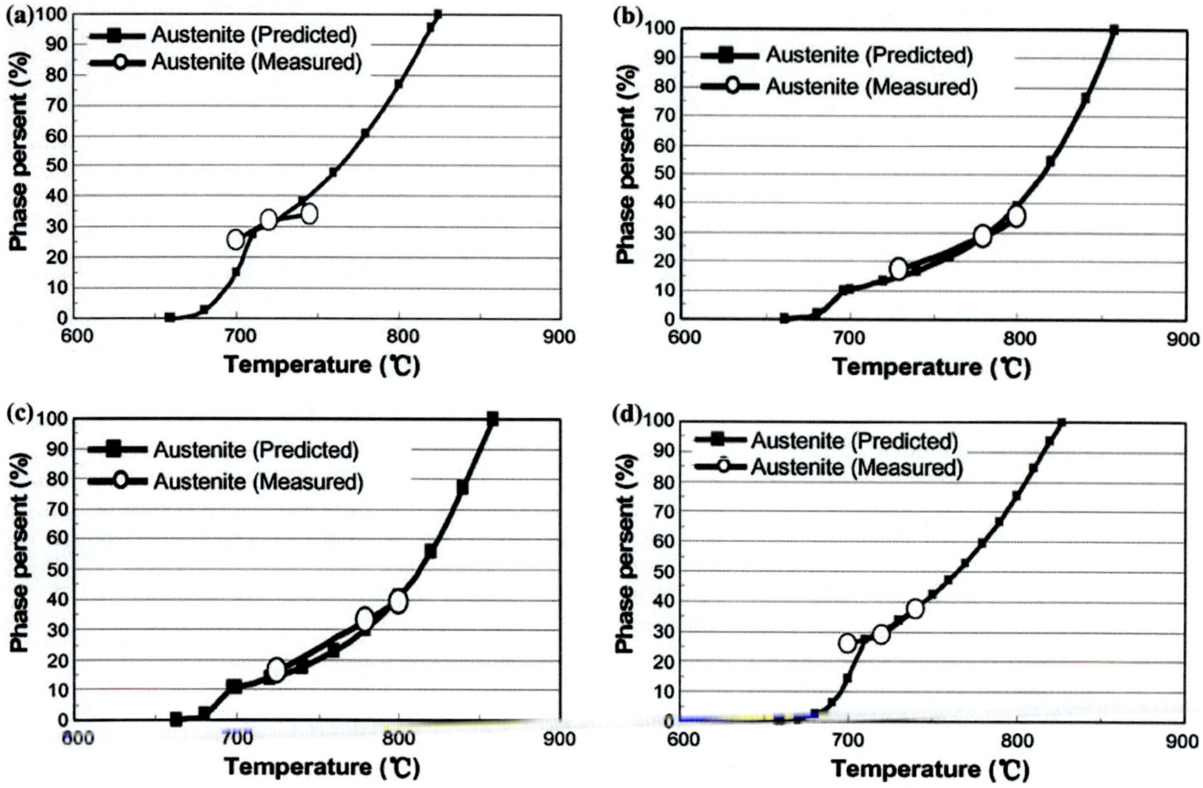

Fig. 3 Comparison of austenite formation between the theoretical prediction and the measured values (**a**) BDP-590, (**b**) BDP-780, (**c**) BDP-980, (**d**) BDP-1180

Table 3 Formation of austenite during intercritical annealing and predicted tensile strength of the resulting dual-phase microstructure after water quenching at room temperature

Material	$T_{intercritical}$ (°C)	Grain size of α (μm)	Vol.% of γ (measured)	VHN	Predicted tensile strength (MPa)
BDP-590	725	3.7 ± 1.3	16.2	200 ± 2.2	641
	780	4.6 ± 1.7	33.1	260 ± 7.2	823
	800	4.8 ± 1.9	39.1	272 ± 5.2	861
BDP-780	730	3.8 ± 1.5	17.1	246 ± 5.7	780
	780	4.8 ± 1.6	29.0	274 ± 2.1	867
	800	5.0 ± 1.4	39.1	285 ± 2.6	902

larger than 50% respectively. The JMat-Pro software package provided the predictions of the intercritical annealing temperatures to attain the desired phase balance. Figure 3 shows a comparison of the formation of austenite during intercritical annealing between the theoretical prediction and the measured volume fraction. Table 3 shows the measured volume fraction of austenite, the VHN hardness number of the resulting dual-phase microstructure and its predicted tensile strength (MPa) of BDP-590 and BDP-780.

The predicted Tensile Strength was calculated using a modified relationship to the one proposed by Davis [6]:

$$\text{UTS (MPa)} = 0.0017(\text{VHN})^2 + 2.2494(\text{VHN}) + 123.31$$

The predicted tensile strength values are shown in Table 3. These results clearly suggest that the low-C Nb-bearing steels used in this investigation should be able to attain the desired tensile strength required for BDP-590 and BDP-780. The typical optical micrographs of the dual-phase (ferrite and martensite) microstructures observed in both steels after the intercritical annealing treatments are shown in Figs. 4 and 5. A detailed examination of the microstructure shows that the size and distribution of the ferrite grains and the islands of the second phase martensite seem to be similar in both steels at the lower intercritical temperatures. However, at higher intercritical annealing temperatures the distribution of the ferrite grains in the 0.02 Nb steel appears to become less uniform compared to that of the 0.04 Nb steel. The effect of Nb present as a solute or as precipitate on the recrystallization and grain growth behavior of either austenite or ferrite during thermal processing has been examined in other studies [7–10]. The general understanding suggests that the pinning force exerted by the presence of Nb either as a solute or precipitate or as a combination of both, seems to retard the recrystallization of austenite or ferrite and to prevent grain coarsening.

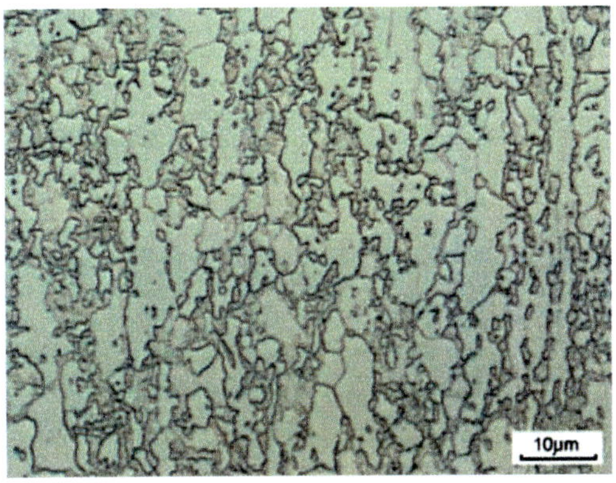

Fig. 4 OM of BDP-590 steel after low intercritical annealing temperatures

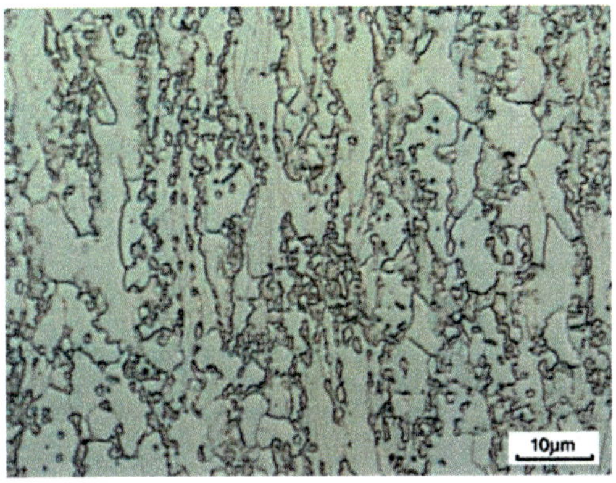

Fig. 5 OM of BDP-780 steel after low intercritical annealing temperatures

3.2 CGL Simulation Treatments

The typical SEM microstructure after the CGL simulation studies conducted in a Gleeble 3500 unit are illustrated in Fig. 6. The quantitative analysis of the ferrite grains and the second phase microstructure (martensite) reveals an average ferrite grain size of about 2.4 μm, while the average size of the martensite island is approximately 700 nm, see Figs. 7 and 8. The average size of the ferrite grains decreases slightly with increasing the cooling rate from the intercritical annealing temperature, while the size of the martensite islands seems to remain unchanged. The quantitative analysis of the microstructure and the tensile properties resulting from the CGL simulation treatments for BDP-590 and BDP-780 are presented in Table 4. The results illustrated in this table show a very good correlation between the intercritical annealing (Table 3) results and the CGL simulation experiments. The low Nb (0.02) steel seems to be more immune, "less sensitive", to variations in the amount of martensite formed with changes in the cooling rate from the intercritical temperature than the high Nb (0.04) steel. This observation might be strongly related to the higher synergistic effect of Nb + Cr + Mo on the overall hardenability of the steel. That is, the overall combined solute content of Nb + Cr + Mo prior to cooling from the intercritical annealing temperature is expected to be higher in the

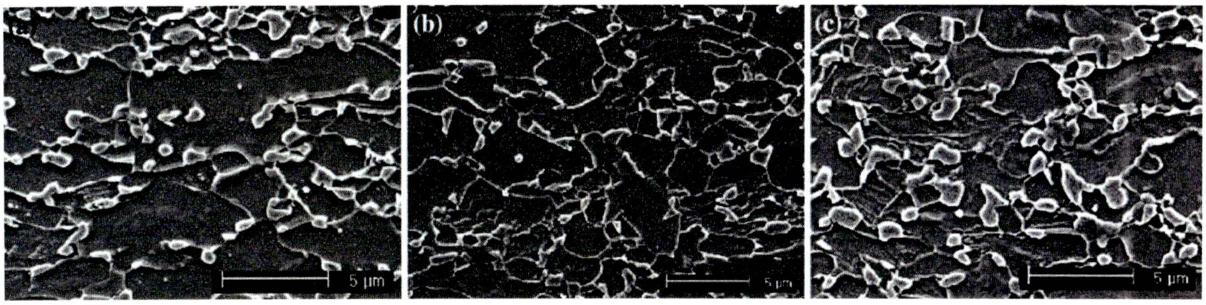

Fig. 6 Typical SEM micrographs observed in both steels after the CGL simulation treatment: (**a**) BDP-590 3°C/s (**b**) BDP-780 8°C/s (**c**) BDP-780 15°C/s

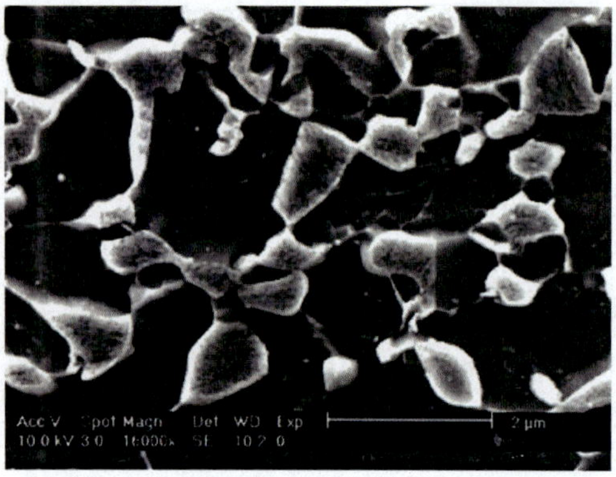

Figs. 7 and 8 SEM micrographs showing the typical size of the martensite islands in BDP-590 steel after the CGL simulation treatments

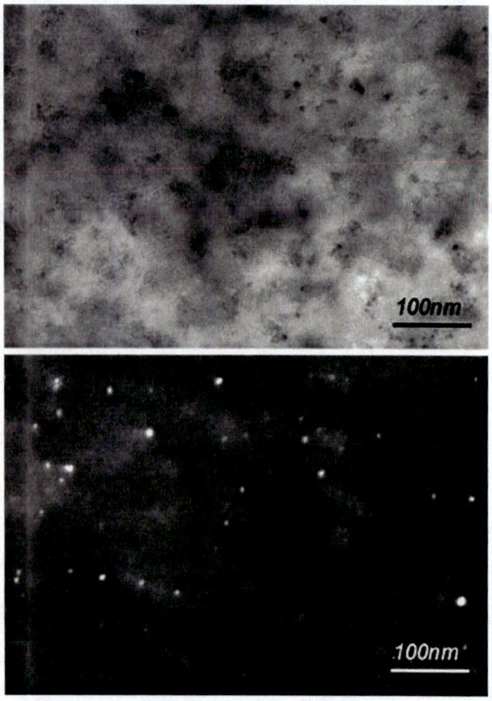

Fig. 9 Bright and dark field TEM micrographs showing fine (approximately 5 nm in size) NbC precipitates observed in steel BDP-780

BDP-590 steel than in the BDP-780 steel. The TEM analysis of the microstructures in both steels after cooling from the intercritical temperature to 460°C and then WQ to RT revealed a fine dispersion of NbC precipitates in the 0.04 Nb steel, see Fig. 9. These precipitates have the typical B–N crystallographic relationship with respect to the matrix, indicating that they were formed in ferrite. Their formation most likely took place during cooling to the coiling temperature and during the coiling simulation prior to cold rolling. The presence of these precipitates in the 0.04 Nb steel indicates that they do not undergo complete dissolution during the CGL simulation treatment. While in the 0.02Nb steel the dissolution of the NbC precipitates was nearly complete.

3.3 Mechanical Properties After CGL Simulation Treatments

The results of the mechanical properties after the CGL simulation treatments in the Gleeble 3500 unit are presented in Fig. 10 and Tables 4 and 5. The results illustrated in Fig. 10 show the variation of yield strength, tensile strength and % total ductility with increasing the % martensite, independent of the steel composition. The linear relation between the tensile properties and the volume % of the

Table 4 Quantitative analysis of the microstructure and mechanical properties of BDP 590 and BDP 780 resulting from the CGL simulation treatments

Material	Cooling rate (°C/s)	Grain size of α (μm)	Vol.% of γ (measured)	YS (MPa)	UTS (MPa)	EI (%)
BDP-590	3	2.4 ± 1.0	17 ± 2.7	410	719	28
	8	2.4 ± 1.1	18 ± 3.7	408	725	25
	15	2.4 ± 1.0	21 ± 2.3	407	730	25
BDP-780	8	2.1 ± 0.8	24 ± 4.0	466	810	25
	15	2.0 ± 0.8	30 ± 3.1	489	830	24

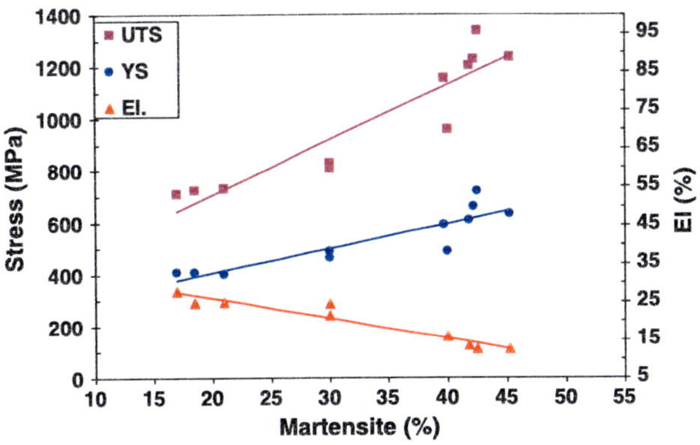

Fig. 10 Relation between the tensile properties and the volume % of martensite independent of the steel composition

second phase (martensite) is in good agreement with previous studies [11–14]. Table 4 illustrates the effect of cooling rate on the resulting volume fraction of martensite and the corresponding tensile properties. The results from this table clearly indicate that BDP-590 and BDP-780 tensile properties with excellent ductility can easily be achieved using CGL simulation treatments at a relatively wide range of moderate cooling rate conditions from the intercritical annealing temperature. Table 5 shows the effect of the Nb content, intercritical annealing temperature on martensite volume fraction and the tensile properties of BDP- 980 and BDP-1180. The result of the mechanical properties of low silicon BDP 580 N, which composition is same as the BDP-580 except the silicon content, is shown in Table 6. It reveals that the good combination of the tensile strength and ductility, which is same as the BDP-590 steel, can be achieved in the low silicon BDP-580 N steel using moderate cooling rate from the intercritical annealing temperature.

The important features of the final dual-phase microstructure obtained in this work are: (1) a fine and uniform ferrite grain size, (2) at least 50% of recrystallized ferrite prior to cooling from the intercritical temperature, (3) a nano-size second phase martensite (approximately in the range 600–700 nm in size), and (4) a relative low "sensitivity" of the final microstructure to cooling rates. The sensitivity of the microstructure to cooling rate seems to decrease as the Nb + Cr + Mo content in solid solution increases during intercritical annealing.

3.4 General Discussion

The effect of Nb on both the recrystallization of ferrite and the formation of austenite during intercritical annealing or CGL processing of cold rolled dual-phase steels is fairly well understood [7–9, 15]. The major effects of Nb can be summarized as follows: (a) the formation of fine NbC precipitates during coiling leads to the refinement of the hot band microstructure, (b) the presence of fine NbC precipitates delays the recrystallization behavior of ferrite during intercritical annealing, (c) the undissolved NbC controls the recrystallized ferrite grain size and inhibits grain growth, (d) the fine austenite grain size formation in the intercritical region, (e) the lowering of the M_s start temperature due to the fine austenite, and (f) the enhancement of the stability of the retained austenite. The microstructural results obtained in this investigation are in good agreement with the above general understanding of the effect of Nb. For example, in this investigation the average grain size of the ferrite was less than 3 μm and the average size of the second phase

Table 5 Quantitative analysis of the microstructure and mechanical properties of BDP-980 and BDP-1180 resulting from the CGL simulation treatments

Material	Cooling rate(°C/s)	Nb (%)	ICAT (°C)	Vol.% of γ (measured)	YS (MPa)	UTS (MPa)	EI (%)
BDP-980	15	0.043	740	39.7	590	1154	–
	15	0.024	735	40.4	489	960	16.0
BDP-1180	15	0.024	760	41.8	608	1204	13.6
	20	0.024	760	45.2	634	1237	12.5
	15	0.043	740	42.2	660	1230	–
	20	0.043	760	45.2	722	1338	12.6

Table 6 Quantitative analysis of the microstructure and mechanical properties of BDP-590 N resulting from the CGL simulation treatments

Material	T(°C)	Cooling rate (°C/s)	Vol.% of α′ (measured)	YS (MPa)	EI (%)	UTS (MPa)
BDP-590 N	740	8	16.8	385	>23	649
		15	17.4	391	23	669

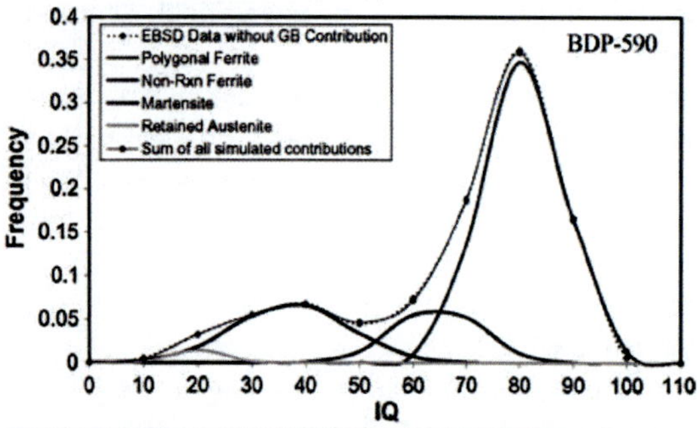

Peak (IQ)	Vol. %	Phase
80.6	67.1	Polygonal Ferrite
84.7	13.2	Non-Rxn Ferrite
37.4	18.1	Martensite
20.2	1.6	Retained Austenite

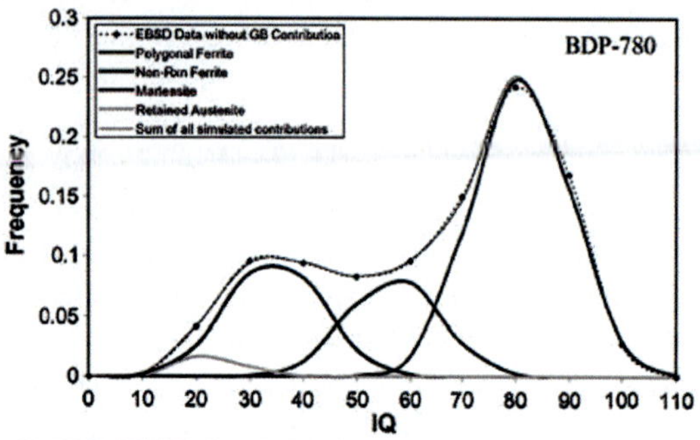

Peak (IQ)	Vol. %	Phase
81.2	57.3	Polygonal Ferrite
57.1	18.2	Non-Rxn Ferrite
34.6	22.1	Martensite
23.2	2.4	Retained Austenite

Figs. 11 and 12 IQ analysis of the dual-phase microstructure after CGL processing

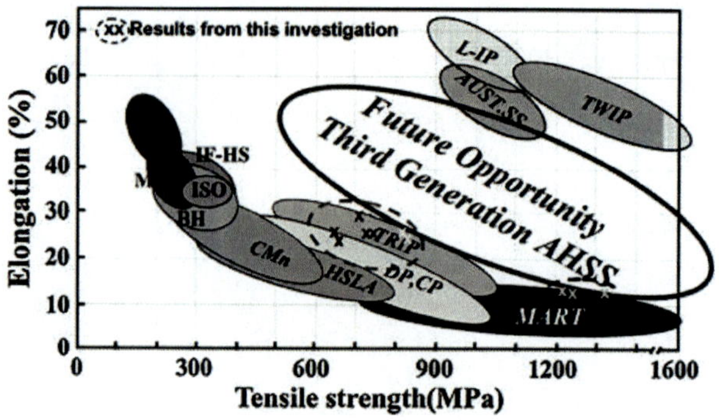

Fig. 13 Comparison of the strength and ductility of BDP-590, BDP-590 N, BDP-780, BDP-980 and BDP-1180 with other complex HSLA steels [17]

martensite was approximately 700 nm. These results certainly support the evidence that the presence of Nb either as solute or as fine NbC precipitate or a combination of both, leads to the overall refinement of the microstructure during CGL processing. In addition, the results of this study also showed that the presence of Nb retards the recrystallization behavior of ferrite, see Figs. 11 and 12. These figures show the results from the OIM and image quality (IQ) analysis of the dual-phase microstructure after CGL processing for steels BDP-590(3) and BDP-780(15). The numbers in parenthesis refer to the cooling rate from the intercritical temperature to the holding temperature of 460°C. The results from these figures clearly show that the final microstructure is composed by recrystallized polygonal ferrite, non-recrystallized ferrite and martensite. No evidence of retained austenite or bainite was detected using the IQ analysis. These results were collaborated by TEM analysis. The phase balance determined by the IQ analysis is in excellent agreement with the quantitative assessment of the microstructures using the computer controlled image analysis BioQuant IV system, see Table 4. The use of OIM and IQ techniques to identify and quantify the microstructural components in complex HSLA and multi-phase (TRIP) steels has been successfully used in other studies [4, 16].

In summary, based on the results of this investigation and others reported in the literature, the role of Nb on microstructural refinement and its effect on the kinetics of ferrite recrystallization were confirmed. However, it appears that it is very unlikely that full recrystallization of ferrite will take place under typical CGL or CAL processing conditions independent on the steel composition. IQ analysis on the recrystallization behavior of several simple and complex steel systems with and without Nb additions has clearly shown that during cold rolling not all the ferrite grains undergo the same amount of deformation [4, 16]. That is, not all the ferrite grains have the same driving force (statistical and geometrical necessary dislocations, i.e. store energy) to undergo recrystallization during intercritical annealing. Hence, the overall driving force for the recrystallization of ferrite during intercritical annealing will be dictated by both the stored energy of the matrix and by the pinning effects on the ferrite grain boundaries exerted by the presence of precipitates and solutes. The microstructural state of the ferrite is important since it is generally agreed that the ductility of dual-phase steels is controlled by the lattice perfection of the ferrite. The understanding of the recrystallization behavior of ferrite and the formation and decomposition of austenite during CGL processing provided the guidelines to the development of the excellent package of mechanical properties exhibited by the steels investigated. A comparison of the tensile properties of the low-C, Nb-bearing steels used in this investigation to those reported in the literature is shown in Fig. 13.

4 Conclusions

The results of this investigation showed that a series of strength grades DP steels tensile properties can be obtained in a low carbon Nb-bearing dual-phase steels using standard CGL processing conditions. The excellent combination of strength and ductility was obtained using moderate cooling rates from the intercritical annealing temperatures. The central features of the dual-phase microstructure were: (a) at least 50% of the ferrite was recrystallized, (b) fine and uniform ferrite grain sizes (<3 μm), (c) nano-size martensite islands (700 nm), and (d) excellent hardenability due to the synergistic effect of Nb + Mo + Cr.

References

1. C.D. Horvath, The Future Revolution in Automotive High Strength Steel Usage. www.autosteel.org (2004)
2. S.G. Fountoulakis, Recent Trends in Automotive Applications of AHSS, ISG Private Communication (2007)
3. F.J. Humpreys, H. Jazaeri, J. Microsc. **3**, 241 (2004)
4. J. Wu et al., ISIJ Int. **2**, 254 (2005)
5. J.H. Bucher, E.G. Hamburg, *Structure-Property Relationships for VAN-QN Dual-Phase Steels, Formable HSLA and Dual-Phase Steels*, pp. 142–150
6. J.R. Davis (ed.), *Metals Handbook: Desk Edition*. (ASM International, Materials Park, 2006) p. 72
7. E. Buddy Damm, M.J. Merwin, (eds.), *Austenite Formation and Decomposition* (The Iron and Steel Society (ISS) and the Minerals, Metals, and Materials Society (TMS), Warrendale, 2003)
8. W. Bleck, A. Frehn, J. Ohlert, in *Proceedings of the Niobium 2001, Niobium Science and Technology*. (TMS, Warrendale 2002) p. 727
9. S.R. Goodman, in *Conference Proceedings of International Conference on Technology and Applications of HSLA Steels*. (1984) p. 239
10. C.I. Garcia, A.K. Lis, A.J. DeArdo, SAE International Congress and Exposition, paper 910143 (1991)
11. R.A. Kot, B.L. Bramfitt (eds.), *TMS-AIME*. (Warrendale, PA 1981) p. 3
12. R.G. Davies, Metall. Trans. **9A**, 41 (1978)
13. R.G. Davies, Metall. Trans. **9A**, 671 (1978)
14. T. Okita et al., Nippon Kokan Technical Report (Overseas) (1985) p. 25
15. K. Hulka, The Role of Nb in Multi-Phase Steel, Niobium Products Company Report, GmbH Steinstrasse 28, D-40210 Dusseldorf, Germany (2002)
16. J.E. Garcia-Gonzalez, Ph.D Thesis, Department of Materials Science and Engineering, University of Pittsburgh, 2005
17. AISI-DOE RFP (2006)

Lightweight Car Body and Application of High Strength Steels

Mingtu Ma and Hongliang Yi

Abstract

Improvement of safety, reduction of energy consumption, and reduction of emission become one of the most highlighted issues for automotive industry in recent years. One of the most significant solutions, i.e. lightweight car body has been described in this paper. Design, implementation, and characterization of parameters of lightweight car body are reviewed as well. In the later part of this paper, the development of high strength steels (HSS) and advanced high strength steels (AHSS) and their typical properties and microstructures are introduced. The application of these high strength steels for lightweight car body can largely improve safety and performance of vehicles.

Keywords

Lightweight car body • High strength steel • Advanced high strength steel

1 Introduction

The automotive industry is developing fast worldwide, especially in China. The annual output of automobiles increases more than 10% in China since 2000. In China, 13.64 million units had been produced in 2009 and more than 15 million units are expected to be produced in 2010. Automotive industry promotes the progress of human civilization. However, it leads also to the problems of safety, fuel consumption and emission of pollutants [1]. Due to increasing number of the retaining and annual output of automobiles, emissions of CO_2 and other environment pollutants become ever worse. Energy consumption and environment pollutants due to automobiles have been serious problems. At the same time, because the price of international oil climbs fast and the requirements of the whole performance by customer are much higher, automotive lightweight becomes more and more significant. This paper illustrates the signification, conception, and comprehension of automotive lightweight. Characterization of parameters, and execution methods of realization of lightweight car bodies are also clarified.

M. Ma (✉)
China Automotive Engineering Research Institute,
Chongqing 400039, People's Republic of China
e-mail: mingtuma@126.com

H. Yi
Graduate Institute of Ferrous Technology, Pohang University of Science and Technology, Pohang 790-784, Republic of Korea

2 Tendency of Automotive Industry: Lightweight

Seventy-five percent of the energy consumption is related to the mass of whole car body [2] and the lightweight car body therefore can reduce the energy consumption and accordingly CO_2 emission effectively. For instance, under the condition of Europe IV emission regulation, gasoline consumption in 100 km (Y) and mass of whole car body (X) follows Eq. 1 [1] in Ford 'TRANSIT' car:

$$Y = 0.003X + 3.343\,4 \tag{1}$$

In the case of commercial vehicle, reduction in mass of 1,000 kg can result in decreasing 6–7% energy consumption.

Y. Weng et al. (eds.), *Advanced Steels*, DOI: 10.1007/978-3-642-17665-4_20,

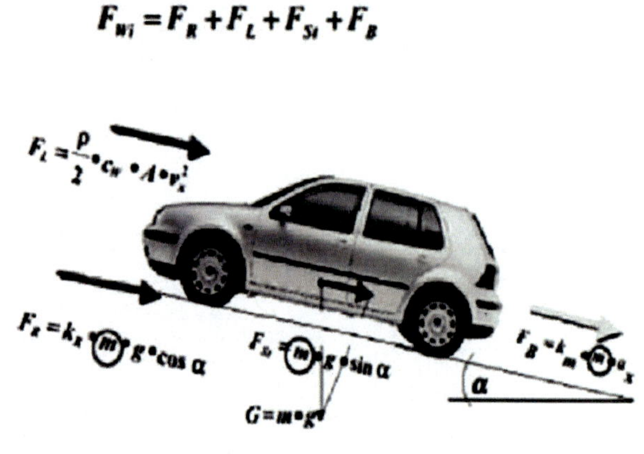

Fig. 1 Influence of mass on oil consumption [3]

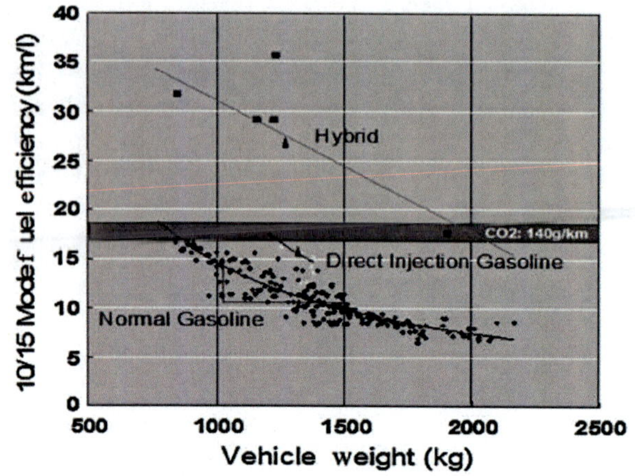

Fig. 2 Relation of 10/15 mode fuel efficiency with vehicle weight [1]

Except the air dynamic resistance, all the other resistance is related to the mass of car body, such as rolling resistance of wheel, resistance of gravity during slope climbing, resistance for acceleration, etc. [3] (Fig. 1). That is proven by experimental results (Fig. 2) [1].

3 Characteristic Parameters of Automotive Lightweight (Fig. 3)

Lightweight design should include weight reduction of the body-in-white (BIW) and performance improvements. The functional requirements for stiffness, crash and car lifetime have to be considered in selection of the materials for a body in white. The stiffness of a structure can be influenced by the E-modulus and panel thickness and construction. Crash and vehicle lifetime are controlled by the strength of the material and thus affected by high strength steel grades. In order to express lightweight effect, a specific lightweight coefficient L is suggested in Ref. [4]. The specific lightweight

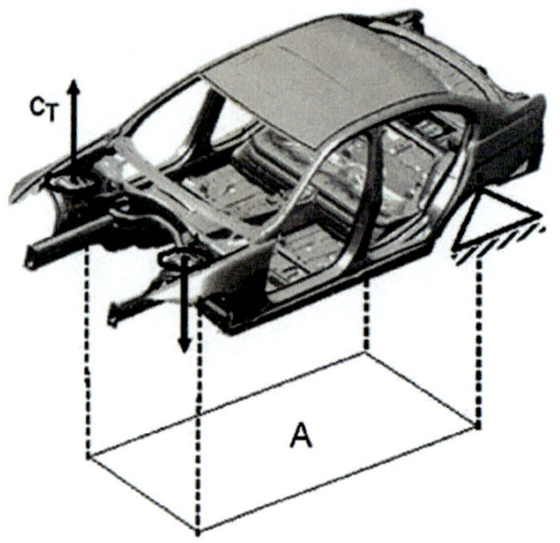

Fig. 3 The specific lightweight coefficient

coefficient L is suggested by Bruno Lüdke from BMW and is expressed by the formula (2). L is the ratio of structure weight m_{Ger} (excluding glass) to static torsion stiffness C_t (with glass) and the associated area A (footprint track × wheelbase), In Eq. 2, area A is an important parameter for the design of automobiles. The class of vehicles and the performance C_t of BIW are determined by this parameter. L can be also as a customer benefits evaluating parameter. Increased customer benefits are reflected in a decreased lightweight coefficient [1, 4].

$$L = \frac{m_{Ger}}{C_t \times A}\left[\frac{\text{kg}}{\text{Nm}/^{\circ} \times \text{m}^2} \times 10^3\right] \tag{2}$$

In order to reflect more directly the lightweight effect, a lightweight exponent L^i is presented and suggested based on Eq. 1. Lightweight exponent L^i is expressed in Eq. 2,

$$L^i = \frac{L_1 - L_2}{L_1} = 1 - \frac{m_{Ger2}}{m_{Ger1}} \cdot \frac{C_{t1} \times A_1}{C_{t2} \times A_2} \tag{3}$$

L^i is non-dimension factor.

L_1 represents the lightweight coefficient of the original vehicle, and L_2 represents the lightweight coefficient of the lightweight vehicle. When the properties of BIW are invariable, Eq. 3 can be simplified as Eq. 4,

$$L^i = 1 - \frac{m_{Ger2}}{m_{Ger1}} \tag{4}$$

Similarly, when the weight of the vehicle is invariable, but the properties of BIW are enhanced, a vehicle is also lightweight, and then Eq. 3 can be expressed as Eq. 5,

$$L^i = 1 - \frac{C_{t1} \times A_1}{C_{t2} \times A_2} \tag{5}$$

In order to reflect comprehensively the influence of lightweight on the whole performance, the ratio of loads for front axes to loads of rear axes must be kept to be 1:1, and

Table 1 The targets of ULSUB and ULSUB-AVC

	2000BIW	ULSAB	ULSAB-AVC
Crash requirements	2000 Year's law	2000 Year's law	2004 Year's law
BIW mass (kg)	270	203	218
BIW costs (US$)	979	947	972

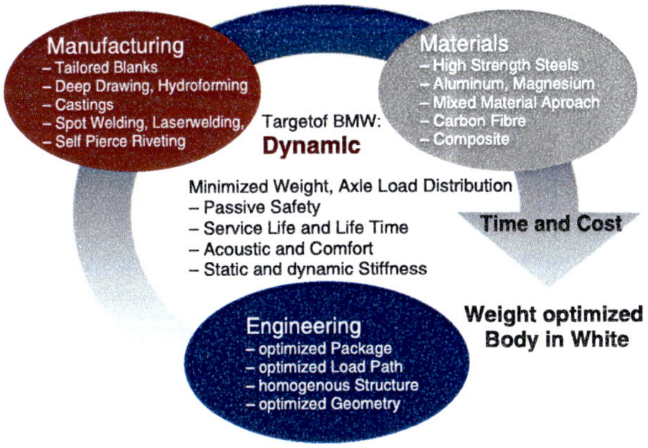

Fig. 4 Lightweight design approaches

the functions, cost and weight reduction all should be considered. Take a ULSAB-AVC for instance, three factors were required, i.e. (1) lightweight of BIW must satisfy crash requirements, (2) mass relative to original vehicles should reduce 20%, (3) no additional cost, as seen in Table 1.

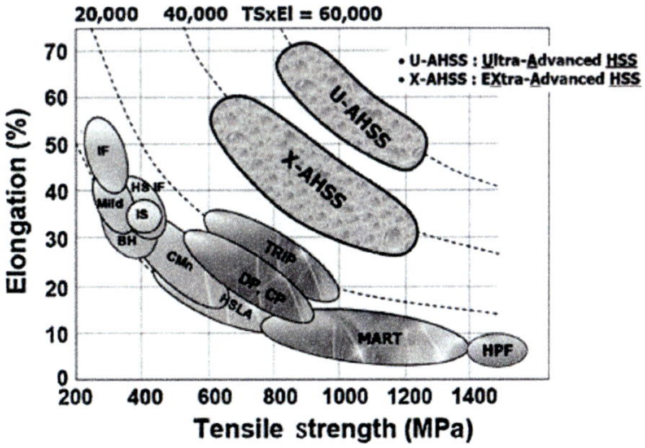

Fig. 6 Typical mechanical properties of steels for automotive application

4 Lightweight Design Approaches

Lightweight design approaches are shown in Fig. 4. From Fig. 4, a lightweight design should be a superior integration of different materials, a superior integration of different specialties, a superior integration of design, materials and advanced manufacturing technologies. A lightweight design can be performed by optimizing geometry. But the better

Fig. 5 Standardized crash simulations

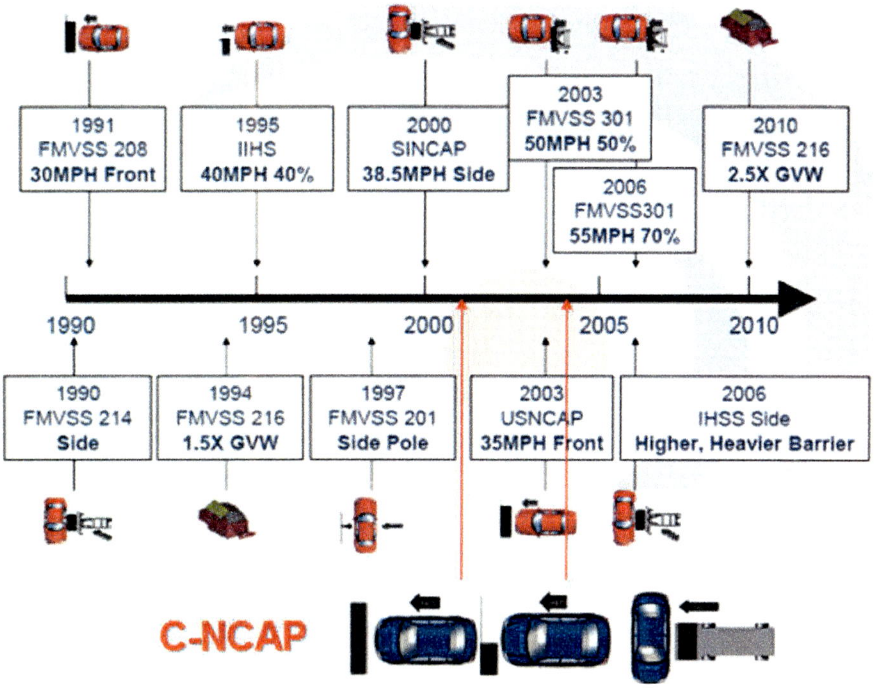

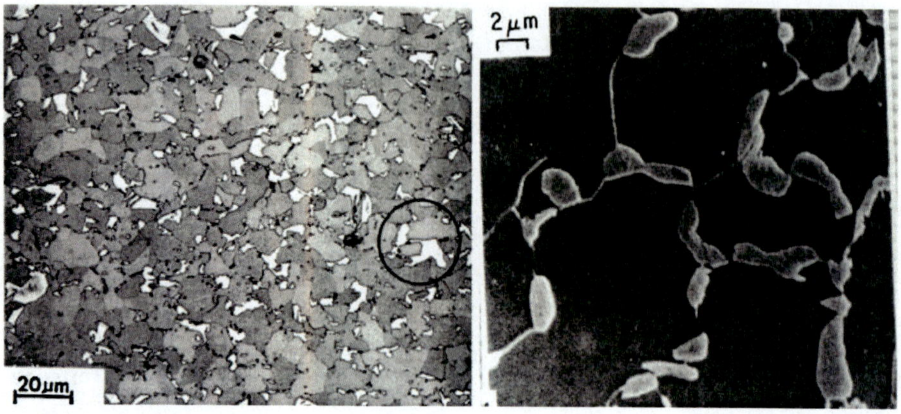

Fig. 7 Optical micrographs (**a**) and SEM photo of dual phase steel (**b**)

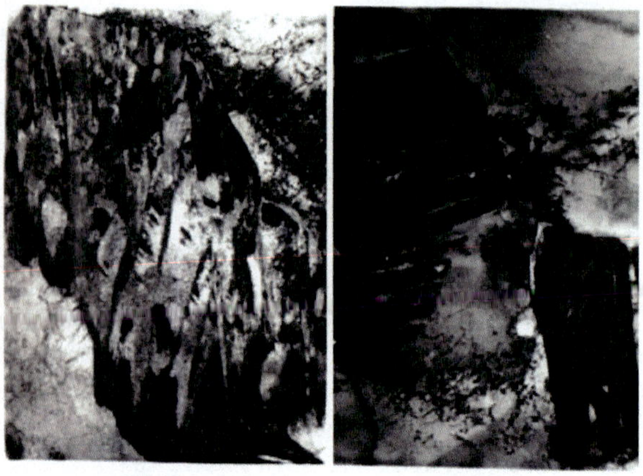

×15000

(a) Lath island of martensite and dislocations in ferrite

×2100

(b) Twin island of martensite and dislocations in ferrite

Fig. 8 TEM micrographs of dual phase steel

effect of lightweight can be able to perfect only by the dominant integration of design, materials and manufacture.

The parameter of mass is related to the strength of materials, specific strength, ductility, modulus, geometry and forming technologies of materials. The choice of advanced forming technology is important, for the application of SHSS (super high strength steel) to improve the safety performance. Due to the poor formability of SHSS, the hot forming must be used in order to attain the strength more than 1,500 MPa. The high strength floor panels of vehicle body can be manufactured by roll forming. TWB (tailored welding blanks) technology can reduce the numbers of parts and weld points, resulting in improvement of the fatigue performance and sealing performance as well as reduction of weight. Hydro-forming can reduce the numbers of parts and improve the performance of parts (for example, sub-frame of engine). These form the foundation of the design [5, 6].

The computer simulation is an effective method in execution of approaches of automotive lightweight. The load paths for crash and formability for a part can be simulated, and thus reasonable technological parameters can be attained for lightweight design and selection of materials and advanced forming technologies.

5 Lightweight and Passenger Safety

Standardized crash simulations such as the EURO NCAP or the North American NHST tests are for enhancing the requirements for passenger safety in case of collision, which including frontal impact, side impact, etc. (Fig. 5). USA

Fig. 9 Twin island of martensite (**a**) and crystal lattice fringe (**b**)

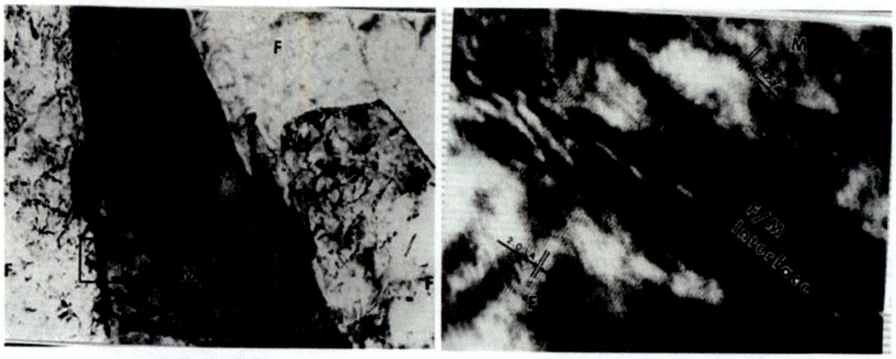

(a) ×22000

(b) (110) crystal lattice

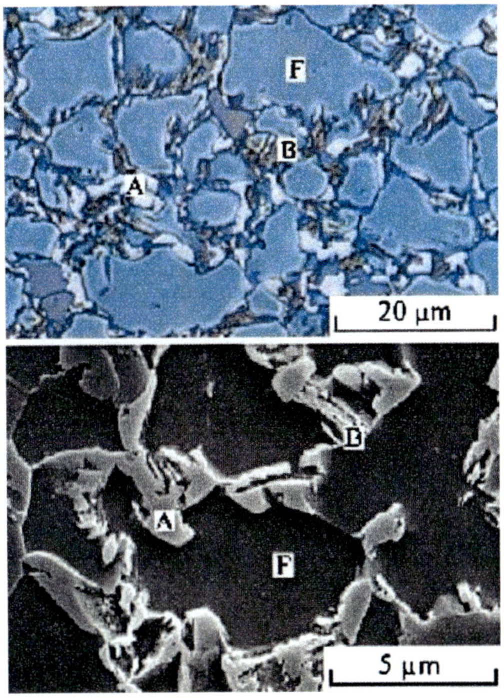

Fig. 10 SEM micrographs of TRIP steel (ferrite, bainite and remind austenite) [15]

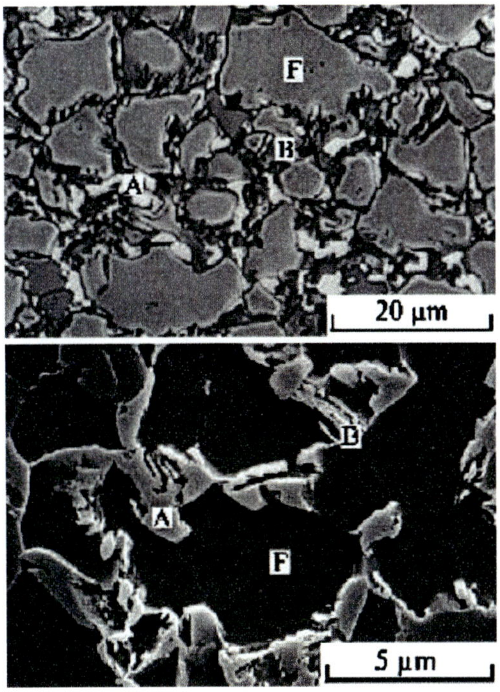

Fig. 11 TEM micrographs of TWIP steel: **a** before deformation and **b** after deformation [12]

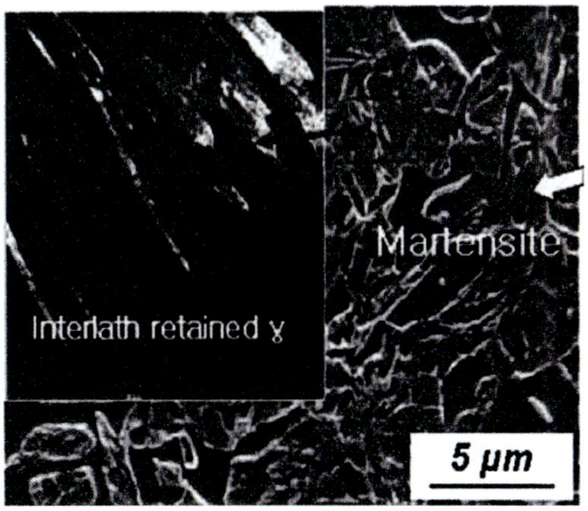

Fig. 12 TEM and SEM micrograph of Q&P or Q&PT steel [12]

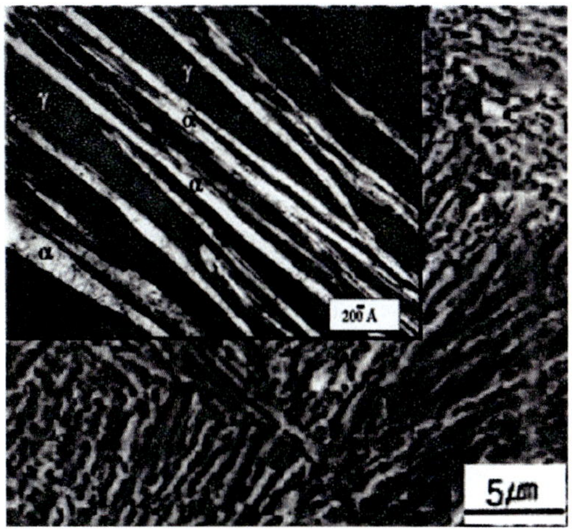

Fig. 13 TEM and SEM micrograph of nano-bainite steel with nano-size austenite film

government even increased the velocity from 64 to 80 kph in frontal test recently and enhanced the loading from 1.5 times of mass of car body to 2.5 times of that for roof crush test. Lightweight by applying high strength steels in BIW is one of the most effective solutions for the combination of energy reduction and safety improvement.

6 High Strength Steel (HSS) and Advanced High Strength Steel (AHSS)

6.1 Concept of HSS and AHSS

Many attentions have been focused onto the development of HSS and AHSS (Fig. 6) in recent years, whose application on vehicles plays important roles in lightweight of BIW and enhancement of safety. Conventional HSS includes C–Mn,

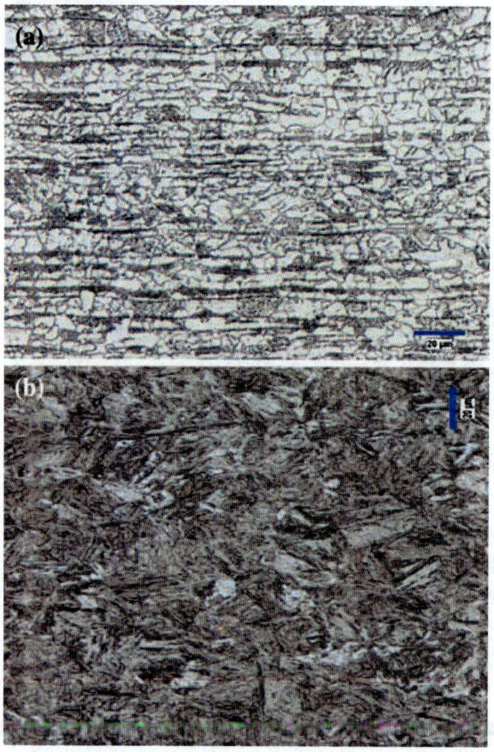

Fig. 14 Optical micrographs of hot press forming steel. **a** Before hot press forming; **b** after hot press forming

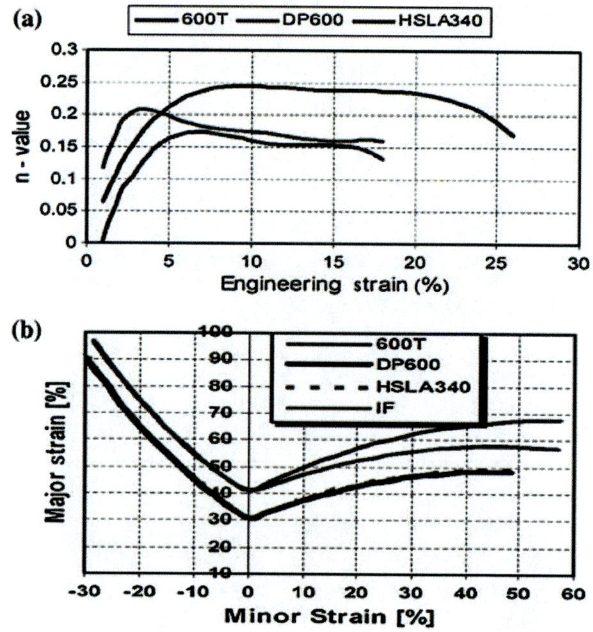

Fig. 15 Formability of high strength steels. **a** Instant *n* value and **b** forming limited curves measured from sheet with 1.2 mm thickness

bake hardening (BH), high strength interstitial free (IF) and high strength low alloying (HSLA) steels [6]; AHSS includes dual phase (DP) [7], transformation-induced plasticity (TRIP) assisted [8, 9], complex phase (CP) and martensitic (M) steels [10]; next generation of AHSS represents ultra-advanced HSS [11], twining-induced plasticity (TWIP) steels [12], and extra-advanced HSS [11] including quenching and partitioning (Q&P) [13], quenching and partitioning tempering (Q&PT) [14] and super bainitic steels [9, 11].

Table 2 Typical mechanical properties of high strength steels for automotive application [13, 16]

Code	YS (MPa)	TS (MPa)	YPE (%)	*n* (10%-UE)	TE (%)
IF	150	300		0.24	46
DQSK	170	300		0.22	43
BH340	220	345		0.19	37
IF-rephos	220	345		0.22	38
HSLA340	350	445	2.6	0.17	28
DP600	340	600		0.17	27
DP800	450	840		0.11	17
DP1000	720	100		0.06	11
TRIP600	380	631		0.23	34
TRIP800	470	820	0.6	0.23	28
Q&P	950	1,080		0.12	18
Q&PT	900	1,600		0.18	27
HPF	1,200	1,500			10

Dual phase steel has been developed since the end of 1970s and widely applied for commercial auto parts formed by cold pressing. Its microstructure consists of soft ferrite as continuous matrix and hard martensite island (Fig. 7), which performs many advantages for application such as: high initial work hardening rate, high ductility and high strength, low ratio of ultimate tensile strength to yield strength, and high bake hardening ability. TEM micrographs of dual phase steel are shown in Fig. 8. Twin island of martensite and crystal lattice fringe are shown in Fig. 9.

Table 3 Static and dynamic mechanical properties of high strength steels

Steel, Gizrfe	YS	UTS	A	B	n	C
BH300	312	419	363	701.22	0.725	0.020
400w	325	460	403	599.06	0.593	0.0214
HSLA350	352	443	370	684.6	0.594	0.0188
HSS590	409	589	455	1006.8	0.663	0.0146
DP600	448	660	478	741.5	0.399	0.0141
TKIP590	428	329	472	1024.7	0.676	0.0119
DP800	547	827	583	900.17	0.357	0.0095

Table 4 Fatigue behavior of advance high strength steels

	BH300 GI	440W GA	HSLA350 GI	HSS590 CR	DP600 GI	TRIP590 EG	DP800 GA
σ'_f	549	841	806	886	983	813	1205
ε'_f	0.969	0.468	1.920	0.480	0.211	0.496	0.104
B	−0.063	−0.105	−0.098	−0.095	−0.101	−0.063	−0.101
C	−0.614	−0.523	−0.668	−0.538	−0.457	−0.572	−0.394
K′ (MPa)	530	966	671	983	1363	871	2104
n′	0.097	0.198	0.133	0.173	0.219	0.109	0.253
Endurance limit, σ'_e	193	2.09	203	230	228	336	307
Notch endurance limit, σ'_n	120	130	125	144	142	178	147
$K_f = \sigma_{-1}/\sigma_n$ at 10^7 reversals	1.61	1.61	1.62	1.60	1.61	1.89	2.09

TRIP-assisted steels can be considered as a modified dual phase steel, which consists of allotriomorphic ferrite as matrix, bainitic ferrite and retained austenite, in some case containing small amount of martensite. That, like dual phase steel, achieves high work hardening rate and combination of strength and ductility, in which the retained austenite transforms into martensite during deformation and therefore supplies extra plasticity compared with dual phase steels, which is the so-called TRIP effect standing for transformation induced plasticity. The enriched carbon can increase

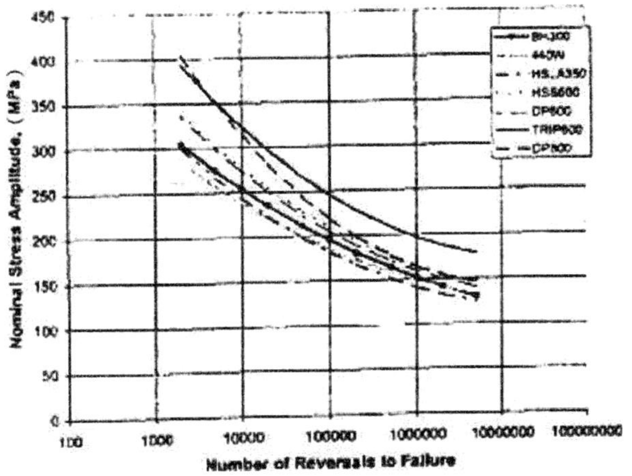

Fig. 16 Notch fatigue behavior of advance high strength steels

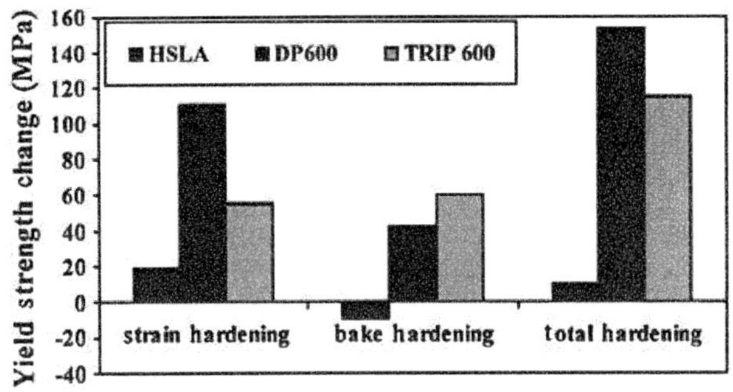

Fig. 17 Strain hardening and bake hardening properties of HSLA, DP600, and TRIP600, which are baked at 170°C for 30 min after 2% uniaxial prestrain

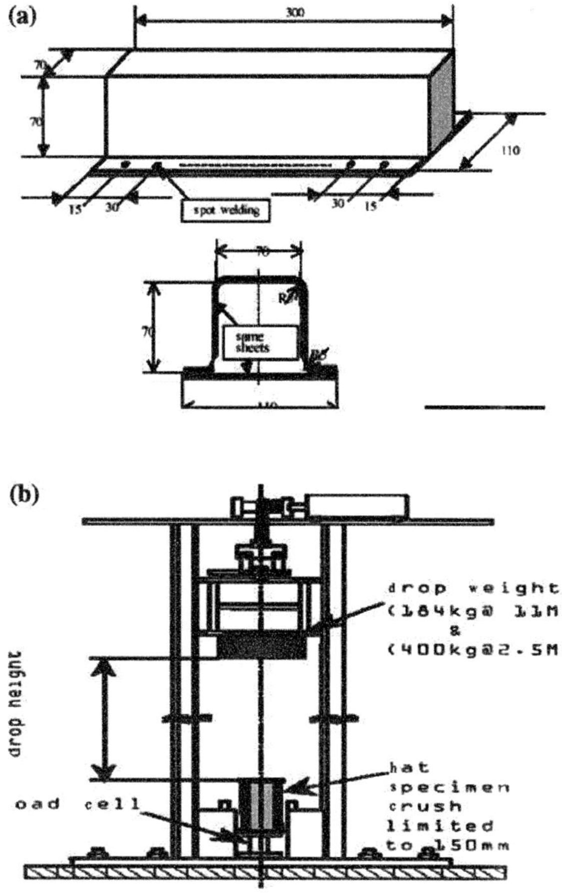

Fig. 18 Drop weight system measuring crash energy absorption **a** test sample **b** drop weight system

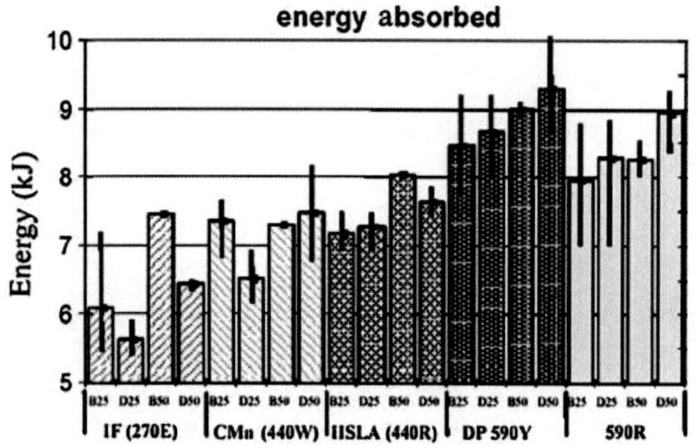

Fig. 19 Crash energy absorption of HSS and AHSS

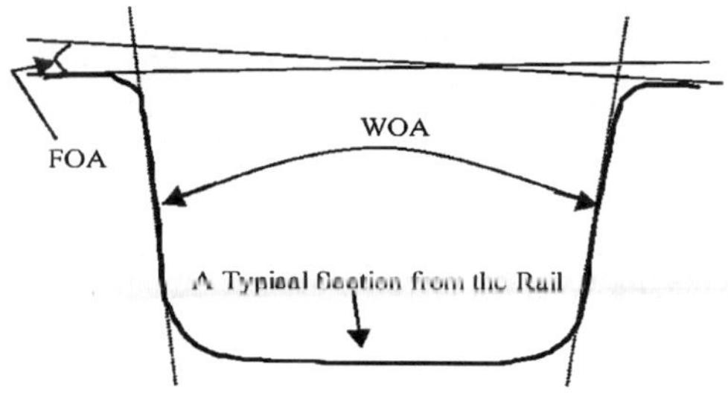

Fig. 20 Parameter and sample measured spring back of HSS sheet

the stability of austenite against martensite transformation, which partitioning into austenite during intercritical annealing and isothermal holding for bainitic transformation where the cementite precipitation is retarded by addition of silicon and aluminium. The retained austenite has to be sufficient stable to sustain until the large strain. Hot press forming steel is of full martensite (Fig. 10) and therefore extremely strong, which is suitable for the application on the anti-intrusion parts.

The U-AHSS represents TWIP/TRIP steel, which conventionally contains more than 20 mass% of manganese and some other solutes such as silicon, aluminium, etc. This kind of steel performs extraordinary ductility combined with toughness, strength and delayed fracture resistance, which is achieved by the transformation into ε' martensite from austenite during deformation (Fig. 11).

The Q&P steel consists of martensite as matrix and retained austenite with lath morphology and therefore performs strength as high as martensitic steels and combines suitable ductility due to the TRIP effect (Fig. 12). The carbide is permitted to precipitate and therefore the toughness of martensite matrix can be improved in the Q&PT steels.

Nano-size austenite film (Fig. 13) is retained between bainitic ferrite by low temperature bainite transformation in the super bainite steels, which behaves excellent combination of strength and ductility.

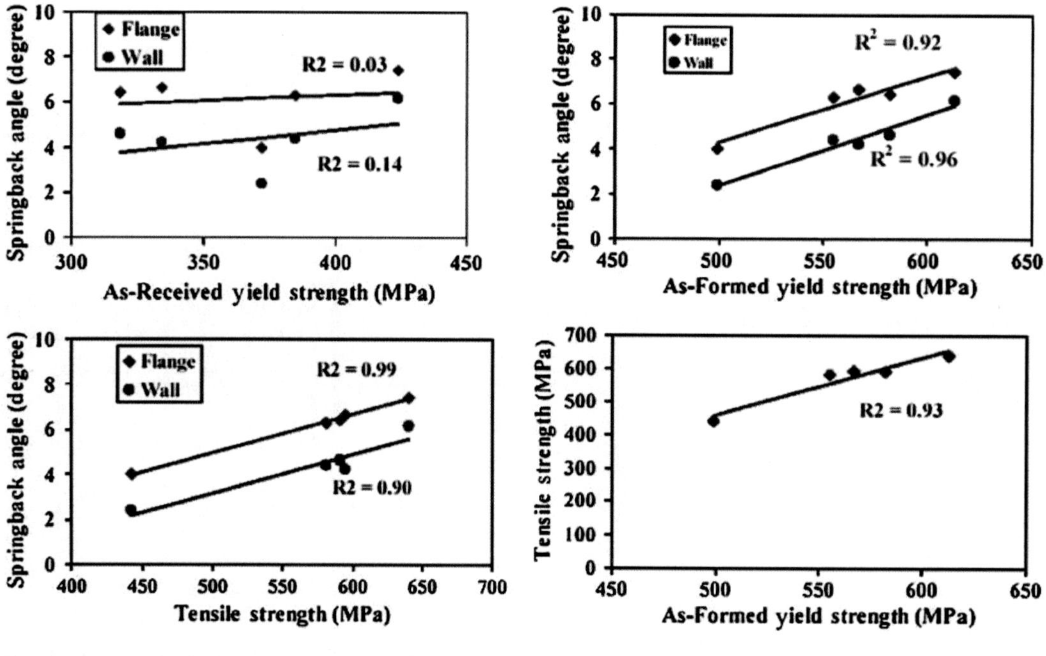

Fig. 21 The effects of mechanical properties on spring back angle

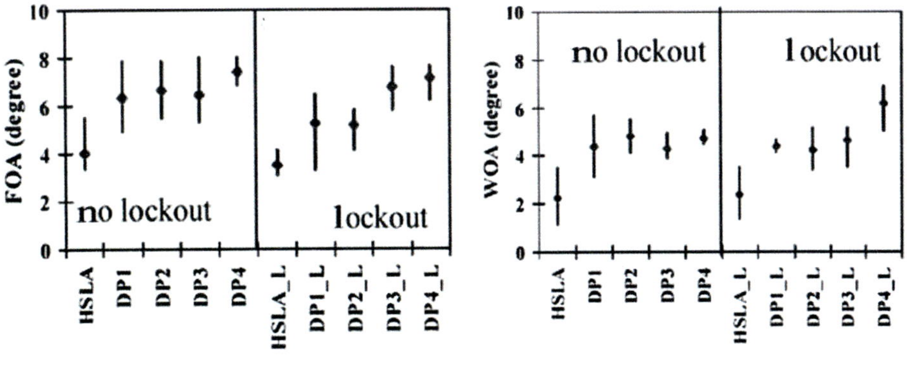

Fig. 22 FOA of advanced high strength steel

Hot press forming martensite is of martensite microstructure after hot press forming (Fig. 14) and behaves super strength and some elongation.

6.2 Mechanical Properties and Formability of HSS and AHSS Steels

Advanced high strength steels achieve better ductility and high work hardening exponent (Table 2), and accordingly better formability (Fig. 15), compared with the conventional high strength steels [13, 16].

6.3 Special Properties of High Strength Steels

6.3.1 Mechanical Properties under the High Strain Rate

Crash worthiness is one of the crucial properties of high strength steels due to its conventional application on

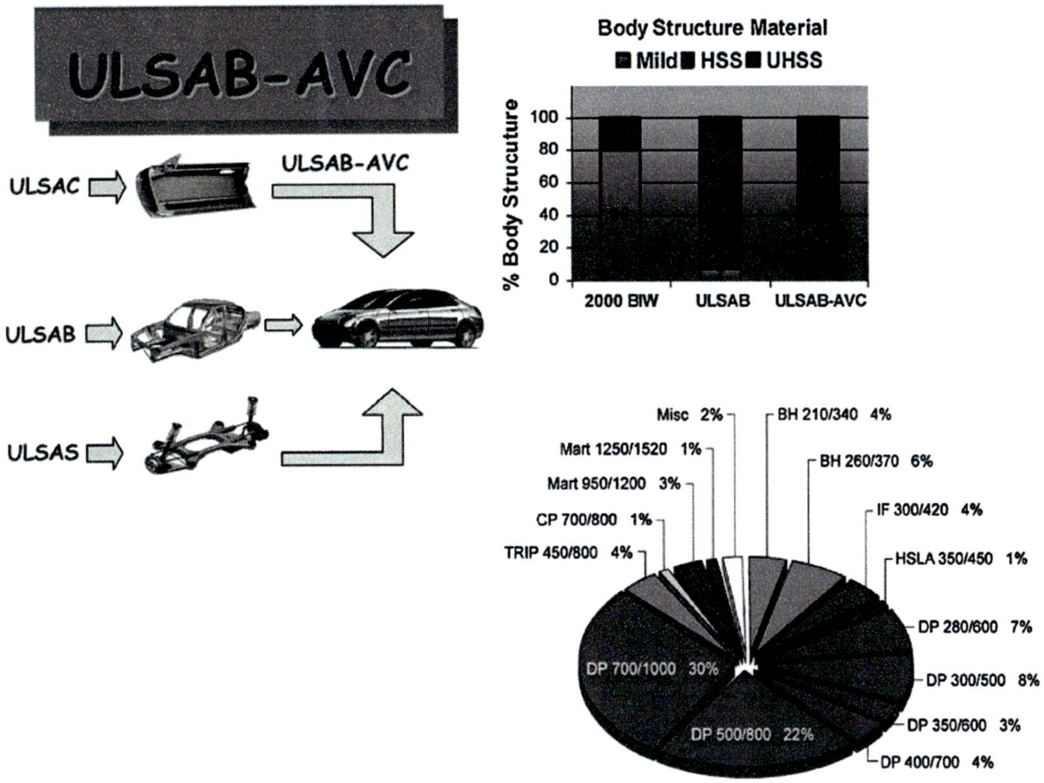

Fig. 23 Concept of ULSAB-AVC and application of HSS steels

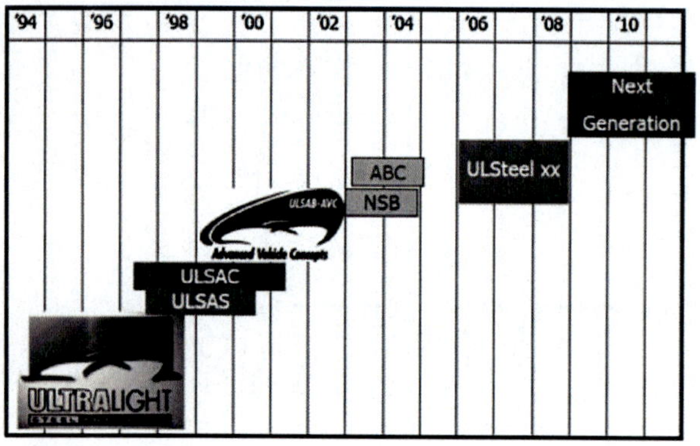

Fig. 24 Development and project of application of AHSS after ULSAB-AVC

Fig. 25 SKODA concept design and the car

structural auto parts, which can be characterized properly by high speed tensile property measured on a servo electro-hydraulic universal testing machine or split Hopkinson bar apparatus (HBST). The data is analyzed using the Johnson–Cook equation [18] as following:

$$\sigma = (A + B\varepsilon^n) * [1 + C * \ln(\varepsilon)]$$

In the range of strain rate from quasi-static to 1,000 s^{-1}, the elastic module is almost constant and the strength and energy absorption before necking increase with increasing strain rate (Table 3) [16, 17, 19, 20, 21].

6.3.2 Fatigue Behavior

The fatigue strength of high strength steels usually increases with their ultimate tensile strength (Table 4 and Fig. 16) [7, 20]. The TRIP-assisted steels perform higher fatigue strength and notch fatigue strength compared the other steels at similar strength level.

6.3.3 Strain Hardening and Bake Hardening

Dual phase steels behave much better overall hardening properties including both strain hardening and bake hardening, compared with TRIP-assisted and HSLA steels (Fig. 17) [16].

6.3.4 Energy Absorption of Auto Parts Made of AHSS Advanced High Strength Steels

Energy absorption is the crucial property for the anti-crash auto parts, which can be measured using a drop weight system (Fig. 18). The parts made from AHSS (TRIP600 and

Table 5 After ULSAB: steelmakers' developments

	ATLAS ('01)	NSB ('02)	ABC ('03)
Target vehicle	Cabrio (Karmann)	Zafira (Opel)	European 5-door, C–D class
Body type	Space frame (convertible)	Space frame	Monocoqne
Forming technologies	Hydroforming Roll forming	Hydroforming Tailored tube Conical tube Double sheet Roll forming Davex profile	TWB Patch Roll forming Hot press forming
Materials	No special comments on materials	Dual phase: 42% Complex phase: 15% Microalloyed: 12% Retained austenite: 8% Partial martensitic: 7% Deep drawing: 16%	TS > 1,000 MPa: 46% 600 < TS < 1,000: 10% 350 < TS < 600: 37% Mild: 7% Sandwich steel Usi-light: spare wheel tub Quiet steel: dash panel TWIP: B-pillar reinf. Usibor1500: B-pillar outer
Results	Weight: 19% Stiffness: 9% Cost: 6%	Weight: 24% Stiffness: torsional 12% Bending: 44% Cost: 2%	Weight: 21% Stiffness: 9% (torsional) Cost: 9%

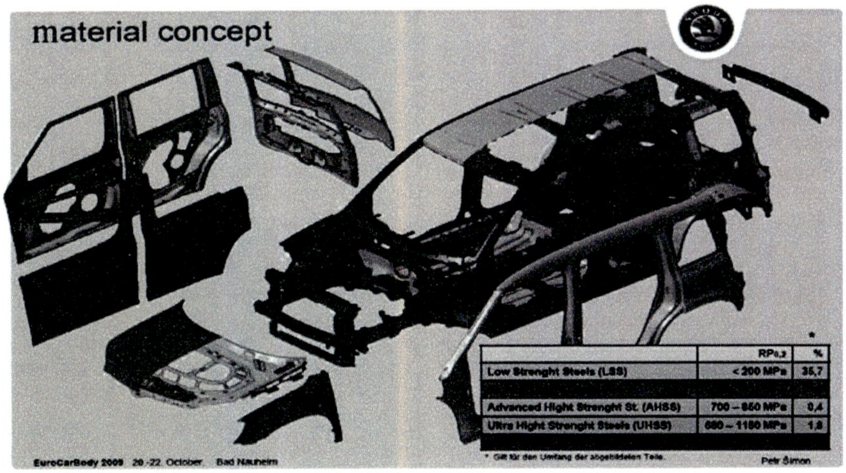

Fig. 26 Material concept for new model SKODA

DP600 steels) perform much better energy absorption property compared with the conventional HSS (C–Mn and HSLA steels) (Fig. 19).

6.3.5 Spring Back

The high strength steels usually behave high yield strength and high initial work hardening rate, therefore meet trouble during pressing due to spring back, which can be measured using a rail specimens shown in Fig. 20. The effect of mechanical properties on spring back is shown in Fig. 21. The FOA of advanced high strength steels is shown in Fig. 22.

7 The Application of Advanced High Strength Steel in Lightweight Automobiles

7.1 Application of AHSS and HSS in ULSAB-AVC

The typical example of application of advanced high strength steel in automotive lightweight is ULSAB-AVC (Fig. 23) [16]. After ULSAB-AVC, in order to achieve automotive lightweight the development situation and project of application of AHSS are listed in Fig. 24 and Table 5.

7.2 Application of HSS and AHSS in Developing of New Model SKODA [22]

New model SKODA are shown in Fig. 25 which have the major parameters: Weight of BIW, including the door up covers et all: 373.6 kg; Uncap impact, 5star; static torsion stiffness 21.5 kNm/deg; project area 3.957 m^2; the material concept for SKODA is shown in Fig. 26.

8 Summary

The rapid development of automotive industry is bringing three serious problems for our life including safety, consumption of energy, and emissions of CO_2. Lightweight car body is one of the significant approaches to solve those problems. Lightweight design includes not only weight reduction of the body-in-white (BIW) but also performance improvements. It can be characterized with either lightweight coefficient or lightweight index. Lightweight design must integrate application of different materials, advanced forming technology, and optimized geometry design.

The development of high strength steel (HSS) and advanced high strength steel (AHSS) provides opportunity for the lightweight materials for the development of automotive industry. Their application effectively improves the performance and safety of lightweight car body.

References

1. M. Ma, H. Yi et al., Eng. Sci. **11**(9), 20 (2009)
2. Boech (ed.), *Automotive Handbook* (Academic and Book Periodicals, Beijing, 1986) (in Chinese)
3. M. Bertram, Improving sustainability in the transport sector—through weight reduction and the application of aluminum, in *The Symposium on the International Aluminum Conference of Application in Automobile Industry* (keynote paper), Dalian, July (2007)
4. B. Lüdke, M. Pfestorf, Functional design of a "Lightweight body in white"—how to determine body in white materials according to structural requirements, Niobium Microalloyed Sheet Steels For

Automotive Applications, TMS (The Minerals, Metals & Materials Society, Warrendale, 2006)
5. M. Ma, H. Lu, Z. Li, Mater. Mech. Eng. **7**, 5 (2008)
6. M. Ma, *Advanced Steel of Application in Automobile* (Chemical Industry Press, Beijing, 2008)
7. M. Ma, B. Wu, *Dual Phase Steel-Physical and Mechanical* (Metallurgy Press, Beijing, 2009)
8. H. Yi, δ-TRIP Steel. Thesis for Doctor of Philosophy. Pohang University of Science and Technology, 2010
9. H.K.D.H. Bhadeshia, *Bainite in Steels*, 2nd edn. (Printed and bound in the UK at the University Press, Cambridge, 2001)
10. International Iron and Steel Institute, Advanced high strength steel (AHSS) application guidelines. www.worlddutosteel.org (2005)
11. O. Kwon, Next generation automotive steels at POSCO. POSCO global EVI forum (2008)
12. Y. Kang, *Modern Automobile Steel Sheet—Technology Forming Theory and Technique* (Metallurgy Press, Beijing, 2009)
13. Z. Xu, Heat Treatment **22**(1), 1 (2007)
14. Z. Xu, Heat Treatment **23**(2), 1 (2008)
15. W. Li, in *Symposium of Automobile Material of SAE-CHINA* (2005), Jiang Yin
16. M.F. Shi, China America Automotive Materials Seminar, Detroit (2003)
17. M. Ma, M.F. Shi, Iron and Steel (7), 68 (2004)
18. K. Xu, C. Wong, B. Yan, H.A. Zhu, SAE SP-1765, 19 (2003)
19. Y. Chunxia, S. Wen et al., Iron and Steel Suppl. **40**(11), 749 (2005)
20. B. Yan, K. Xu, in *44th MWSP Conference Proceeding*, vol. XL (2002) p. 493
21. B. Yan, in *44th MWSP Conference Proceedings*, vol. XL (2002) p. 509
22. Skoda Yeti, Eurocar Body, Bad Nauhei (2009)

Design of Lean Maraging TRIP Steels

Dirk Ponge, Julio Millán, and Dierk Raabe

Abstract

We present a design strategy for a new type of age hardenable ultrahigh strength TRIP-assisted steels with a good tensile elongation. The alloys have a low carbon content (below 0.02 mass% C), 9–15 mass% Mn and, in order to reduce costs, only minor additions of Ni, Al, Ti, and Mo (of the order of 1–2 mass%). After quenching the microstructure of these steels comprises martensite and different amounts of retained austenite. During age hardening the strength and ductility can be increased simultaneously: The strength is increased by the formation of nanosized intermetallic precipitations in the martensite. At the same time new austenite is forming by a partitioning of Mn and Ni from martensite to austenite. The increase of the uniform elongation during aging is a result of a tempering of the quenched martensite and a TRIP effect due to the austenite. The precipitation state is also of importance for the ductility. Atom probe results reveal high contents of Ni, Mn and Al in the homogeneously distributed particles. The nano-sized precipitates have a high dispersion owing to the good nucleation conditions in the heavily strained martensite matrix in which they form. The approach of combining a moderate precipitation hardening with a TRIP effect enables to produce steels with good combinations of strength, ductility and low costs.

Keywords

Precipitation hardening • TRIP effect • Martensite • Tempering • Ductility

1 Introduction

Steels with high ultimate tensile strength (UTS) above 1 GPa and good ductility (total elongation (TE) of 15–20% in a tensile test) are of paramount relevance for lightweight engineering design strategies and corresponding CO_2 savings, Fig. 1 [1, 2].

In this work we report about a design approach for precipitation hardened ductile high strength martensitic and austenitic-martensitic steels (up to 1.5 GPa strength). The alloys are characterized by a low carbon content (0.01% C), 9–15% Mn to obtain different levels of austenite stability, and minor additions of Ni, Ti, Al and Mo (up to 2%) (all compositions in wt%, unless otherwise stated). Mn, Ni, Ti and Al are required for creating precipitates during aging heat treatment.

Hardening in these materials is realized by combining the TRIP effect with a maraging treatment (TRIP: transformation-induced plasticity; maraging: martensite aging through thermally stimulated precipitation of particles). The TRIP mechanism is based on the deformation-stimulated athermal transformation of metastable austenite into martensite and the resulting matrix and martensite plasticity required to accommodate the transformation misfit [3–11]. The maraging treatment is based on hardening the heavily strained martensite through the formation of small intermetallic precipitates (of the order of several nanometers). These particles act as highly efficient obstacles against

D. Ponge (✉) · J. Millán · D. Raabe
Max-Planck-Institut für Eisenforschung GmbH,
40237 Düsseldorf, Germany
e-mail: d.ponge@mpie.de

Y. Weng et al. (eds.), *Advanced Steels*, DOI: 10.1007/978-3-642-17665-4_21,

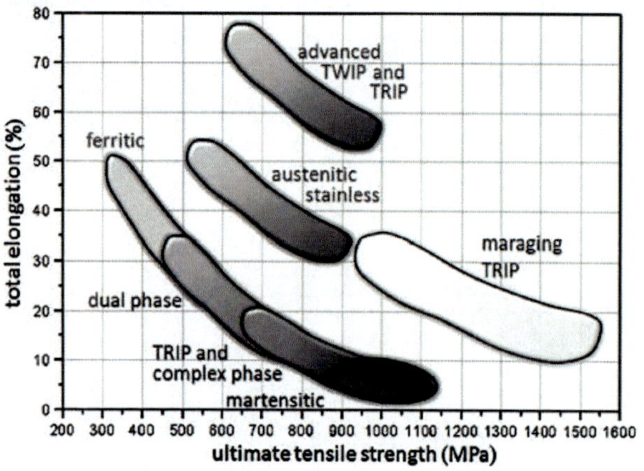

Fig. 1 Overview of typical strength-ductility profiles of different types of steels. The strength is expressed in terms of the ultimate tensile strength measured during tensile testing and the ductility is expressed in terms of the total elongation. The data represent regimes such as published in the references given below. TRIP: transformation-induced plasticity; TWIP: twinning-induced plasticity; Complex phase: multiphase steels (e.g. austenitic-martensitic steels which may contain also bainite); maraging TRIP: new steel concept that includes strain hardening mechanisms based on transformation induced plasticity and formation of intermetallic nanoparticles in martensite during aging. This new approach leads to a simultaneous increase in both strength and total elongation enhancing the regime of formable ultrahigh strength steels by 0.5 GPa

dislocation motion through the Orowan and Fine-Kelly mechanisms enhancing the strength of the material [12–17]. While both types of alloys, i.e. TRIP steels [3–11] and maraging steels [12–17] have been well investigated, the combination of the two mechanisms in the form of a set of simple Fe–Mn alloys as suggested in this work, namely, the precipitation hardening of transformation-induced martensite by intermetallic nanoparticles, opens a novel and lean alloy path to the development of ultrahigh strength steels that has not been much explored in the past [18, 19]. We refer to these alloys as lean maraging TRIP steels.

Related steel design trends that are based on small-scaled second phase precipitates are also pursued by using oxide [20–23], nitride [24, 25], or Cu nano-sized particles [26]. Other pathways to the design of ultrahigh strength steels have been realized in ultra fine grained materials obtained by advanced thermomechanical processing [27–34] or by accumulative roll bonding [35].

The joint maraging TRIP approach justifies a more detailed study since it opens up a new approach to increase the strength of conventional steels at relatively lean alloying costs, i.e. without large quantities of expensive alloying elements, owing to the small volume fraction of precipitates required in maraging steels [12–17]. The negative side of such strategies lies often in the fact that an increase in tensile strength is typically accompanied by a corresponding drop in ductility, Fig. 1.

In the current case, however, the opposite trend is found, that is, we observe for two of the alloys under investigation (9% Mn, 12% Mn) the surprising simultaneous increase of both, strength and ductility upon aging. This most unexpected effect was not explained in the literature before.

The materials investigated combine in part different hardening mechanisms. The first one is the formation of mechanically induced martensite for the alloys with 0.01% C and 12% or respectively 15% Mn since they have retained austenite fractions between 15 vol.% and 50 vol.%. This part of the approach follows, hence, the same hardening principles as conventional TRIP steels [3–11]. The second mechanism is the strain hardening of the ductile low carbon α'- and ε-martensite phases and of the remaining retained austenite in those cases where it is present. The third mechanism is the formation of nano-sized particles in the as-quenched and also in the deformation-induced martensite during final aging. The latter heat treatment is consequently also referred to as martensite age hardening or maraging [12–15].

The nano-sized precipitates have a high dispersion owing to the good nucleation conditions in the heavily strained martensite matrix in which they form. Another aspect which might promote the formation of fine precipitates during cooling is the fact that the austenite, which contains the solute atoms required for the formation of precipitates in the martensite (Ni, Ti, Al), is stable until relatively low temperatures owing to the elevated Mn content. When during quenching the austenite becomes finally transformed into martensite the diffusion length of the atoms involved in precipitation is limited by the low temperature. While in conventional Ni-Co maraging steels the nanosized precipitates are usually assumed to be ordered (intermetallic), the nature and composition of the precipitates encountered in the present alloys are not so clear yet.

Besides the optimization of yield strength, ultimate tensile strength (UTS), and total elongation (TE) we also aim to design steels that are characterized by a homogeneous deformation behavior, hence the unexpected increase in TE upon aging is of high relevance. Most high strength steels (except for TWIP steels; TWIP: twinning induced plasticity) have a deficit with respect to the homogeneity of strain hardening since they reveal a drastic increase in strength at the beginning of deformation where it is not required but fail to provide sufficient strain hardening for the compensation of strain localization at later stages of deformation.

Here we report about a 9% Mn, a 12% Mn, and a 15% Mn alloy. All three materials have a very low carbon content and minor additions of Ni, Ti, Al, and Mo to form nano-precipitates. The main difference among the three alloys consists in the Mn content and hence, in the amount of retained austenite they contain (an increase in the Mn content lowers the equilibrium transformation temperature between ferrite and austenite). The results are compared to the mechanical properties of a conventional Ni-Co based maraging steel [12, 13].

2 Experimental

Four alloys were investigated, namely, one standard 17% Ni-11% Co maraging steel and the three maraging TRIP Fe–Mn steels (9% Mn, 12% Mn, 15% Mn), Table 1 [12, 13]. The three Fe–Mn alloys have a low carbon content and minor additions of Ni, Ti, Al, and Mo to form precipitates in the martensite [36]. The main difference among the three materials consists in their Mn content (~9%, ~12%, ~15%) and hence, in the volume fraction and stability of the retained austenite they contain after quenching.

The alloys were melted and cast to round billets of 1 kg each in a vacuum induction furnace. Annealing and swaging of the as-cast alloys was conducted to ensure homogenization of the microstructure and removal of segregation effects. After annealing at 1,150°C for 1 h swaging was conducted in 8 passes between 1,000°C and 1,150°C. The billets were swaged from a diameter of 27.0 mm to 13.5 mm which corresponds to a logarithmic strain of 1.39. This was followed by air cooling to 800°C and a water quench to room temperature. The rods were reheated to 1,100°C for 0.5 h, hot rolled in 6 passes to a total logarithmic strain of 1.9 into strips with a thickness of 4 mm and water quenched. These strips were cold rolled to a thickness of 1.5 mm corresponding to a logarithmic strain of 1. The subsequent solution heat treatment was performed at 1,050°C for 0.5 h followed by a final water quench. For heat treatments above 1,000°C argon gas atmosphere was used to prevent oxidation. Final aging heat treatments were conducted at 450°C at times between 1 min and 3,000 h. After aging the samples were quenched in water.

Flat tensile specimens were machined in the as quenched and in the aged state with a thickness of 1 mm, width of 4 mm and a gage length of 10 mm. A strain gage extensometer was used for precise determination of the strain. Tensile testing was conducted on a Zwick ZH 100 tensile testing machine at a constant cross head velocity corresponding to an initial strain rate of 10^{-3} s^{-1}.

Hardness testing according to EN ISO 6507-1 was conducted using a Zwick 3212 hardness tester to determine the Vickers hardness with a load of 49.05 N (HV5).

Characterization of the chemical and microstructural homogeneity of the cast, formed, and heat treated samples was conducted by using optical and scanning electron microscopy (SEM) in conjunction with EDX (energy dispersive x-ray spectrometry) and high resolution EBSD (electron back scatter diffraction). The SEM was a JEOL JSM-6500F field emission scanning electron microscope (FE-SEM) operated at 15 kV. The EBSD scans were carried out in areas of about 100 × 270 μm^2 in cross sections in the middle of the samples at a step size of 100 nm. Additional high resolution EBSD maps were taken at 50 nm step size on 25 × 25 μm^2 areas. Samples were ground using SiC paper (8 µm). Subsequently, the samples were mechanically polished using diamond suspensions of 3 µm and 1 µm. Final polishing was done using a SiO_2 suspension (0.1 µm). In order to study the possible influence of mechanical polishing on premature transformation of retained austenite (which is metastable against shear loads) samples were also prepared by electropolishing using 500 ml methanol, 500 mL 2-butoxyethanol and 60 ml 70% perchloric acid for 20 s. Both preparation methods provided similar microstructure results.

Transmission electron microscopy (TEM) images were taken on the solution-treated, quenched plus finally age hardened sample with 12% Mn in order to study the size and spatial distribution of the nanoparticles which are formed during aging. For TEM sample preparation the material was first thinned to a thickness below 100 µm by mechanical polishing. Standard 3-mm TEM discs were then punched and electropolished into TEM thin foils using a Struers Tenupol twin-jet electropolishing device. The electrolyte consisted of 5% perchloric acid ($HClO_4$) in 95% ethanol cooled to −30°C. The thinned specimens were then investigated in the field emission transmission electron microscope JEOL JEM 2200 FS operated at 200 kV. The analysis was carried out in scanning TEM mode (STEM) using a bright field (BF) detector.

The chemical composition of some of the nanoparticles observed in the material was studied at atomic scale resolution using atom probe tomography (APT) with a local electrode technique (IMAGO LEAP 3000X HR

Table 1 Chemical composition of the steels used (mass%)

Steel	C	Ni	Mo	Co	Ti	Si	Al	S	P	O	N	Mn	Fe
17 wt% Ni-11 wt% Co	0.024	17.20	3.89	11.4	1.55	0.078	0.176	0.003	<0.001	<0.003	0.0010	0.23	bal.
9 wt% Mn	0.007	2.00	1.07	–	1.04	0.047	0.086	0.003	<0.001	<0.002	0.0006	8.86	bal.
12 wt% Mn	0.010	2.06	1.12	–	1.09	0.057	0.116	0.010	<0.001	<0.002	0.0012	11.9	bal.
15 wt% Mn	0.006	2.06	1.10	–	1.03	0.078	0.095	0.005	<0.001	<0.002	0.0009	14.7	bal.

metrology device). This system provides an excellent mass resolution which is essential for the analysis of multi-component steels. The configuration includes a laser based atom probe capability in addition to voltage pulsing. The measurements presented in this work were conducted in laser mode where the LEAP electrode applies a static field to the specimen while an ultra-fast laser pulse triggers the removal of the atoms. The use of the reflectron method used in the current configuration is important for an optimal time of flight (TOF) precision. Sample tip preparation for APT analysis was conducted using perchloric acid chemical etching.

3 Results

Figure 2 shows the mechanical data of the four alloys. Figure 3 presents the corresponding microstructure results (phase distribution of α'-martensite and of retained austenite in the as-quenched state). Figure 2a shows the engineering stress–strain curves of the conventional Ni-Co maraging steel in as-quenched, age hardened, and 15% cold rolled plus age hardened state. The latter state was studied to investigate the effect of retained austenite on the aging response of the alloys. The corresponding results obtained from phase and texture determination via EBSD are presented in Fig. 3a, confirming that the Ni-Co maraging steel contains no or little retained austenite in the as-quenched state.

Figure 2b–d present the corresponding results for the three Mn-based steels (Fig. 2b: 9% Mn; Fig. 2c: 12% Mn; Fig. 2d: 15% Mn). The yield strength (YS) of the material with 9% Mn is about 350 MPa, its ultimate tensile strength (UTS) about 810 MPa, and the total elongation (TE) about 6% in the as-quenched state. The properties after aging heat treatment (48 h at 450°C) are surprising. The UTS lies above 1 GPa (as expected) while the TE does not drop upon precipitation strengthening as observed for conventional Ni-

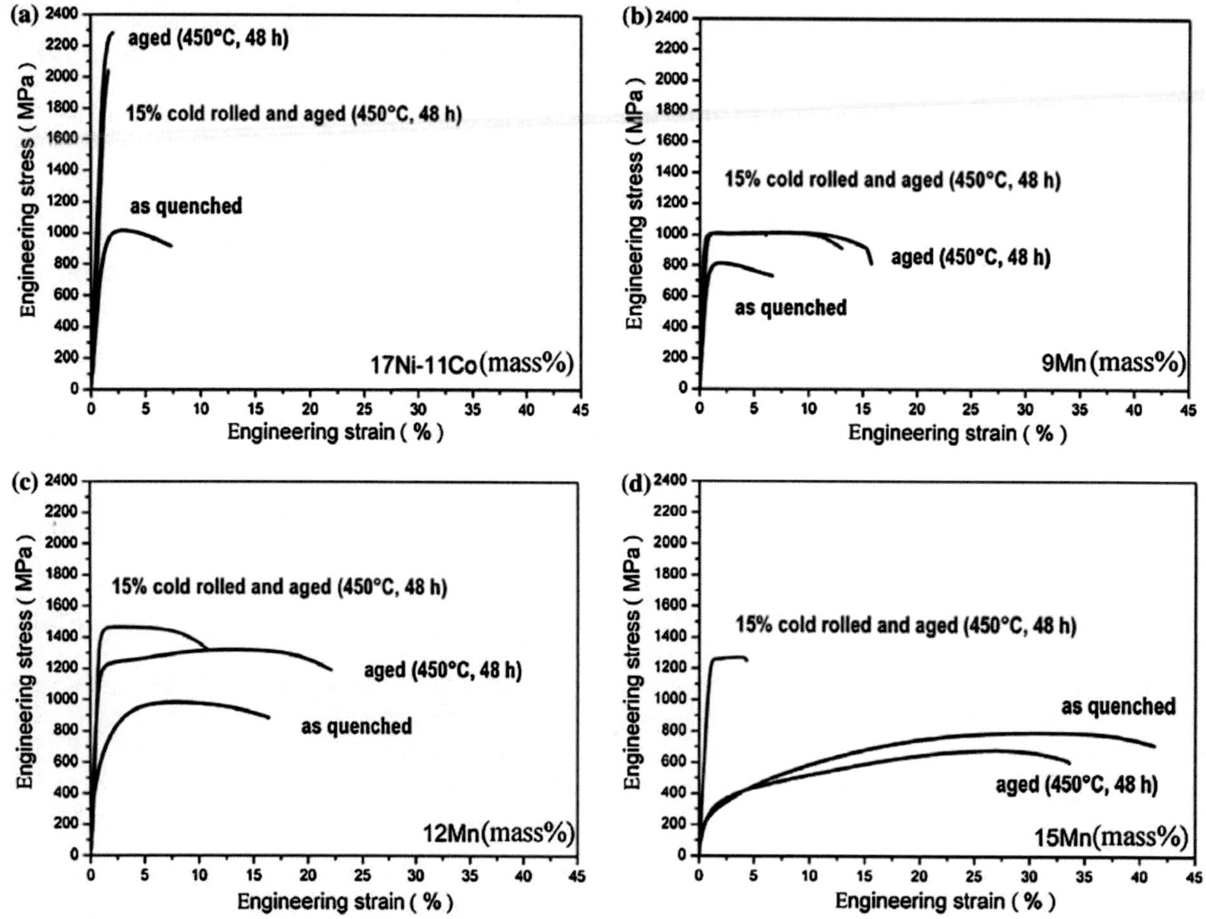

Fig. 2 Engineering stress–strain curves for the four steels, Table 1. **a** Standard 17% Ni-11% Co maraging steel; **b** Fe–Mn steel with 9% Mn; **c** Fe–Mn steel with 12% Mn; **d** Fe–Mn steel with 15.% Mn. All alloys have a very low carbon content and minor additions of Ni, Ti, Al and Mo to form precipitates. The main difference among the Fe–Mn alloys consists in their Mn content and hence, in the amount of retained austenite they contain. Three sets of data are shown, namely, in the as-quenched, age hardened, and 15% cold rolled plus age hardened state

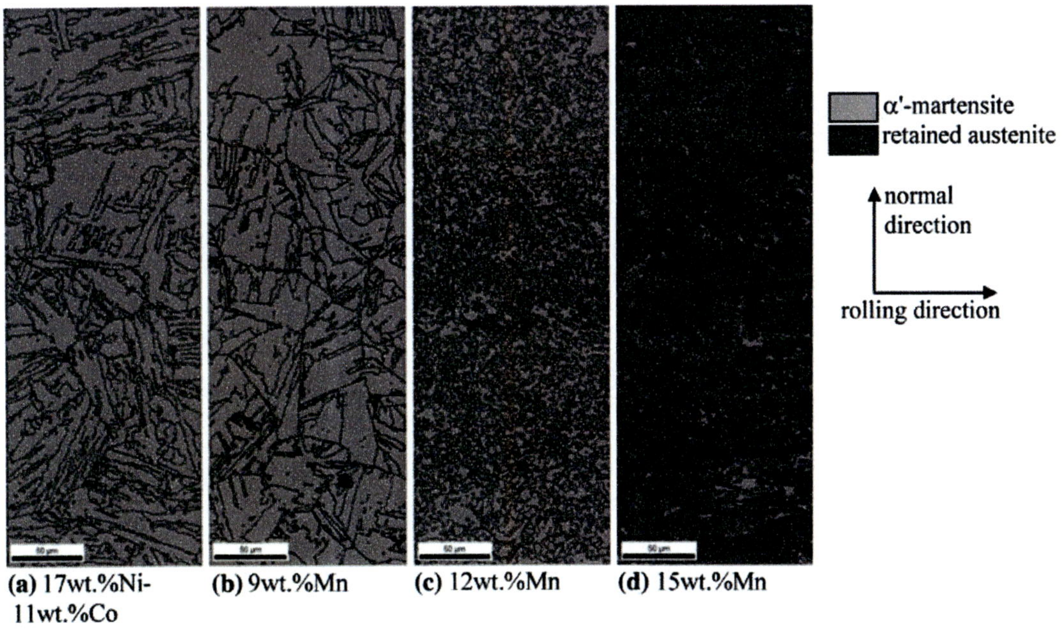

Fig. 3 Microstructure results for the four alloys after quenching obtained via EBSD analysis. Phase distribution of α'-martensite (*light gray* areas) and retained austenite (*dark gray* areas). **a** Conventional 17%Ni-11%Co maraging steel; **b** Fe-9%Mn steel; **c** Fe-12%Mn steel; **d** Fe-15%Mn steel, Table 1

Co based maraging steels [12–16] but it increases from 6% to more than 15%. This means that the aging treatment of this martensitic alloy simultaneously increases both strength and ductility. Fifteen percent pre-deformation (by cold rolling) of the same sample prior to aging yields similar properties as without pre-deformation. This result indicates that no retained martensite is involved. This is confirmed by the EBSD maps. The EBSD analysis for the 9% Mn alloy (Fig. 3b) reveals coarse α'-martensite lamellae with longitudinal dimensions of up to 100 μm, but no retained austenite appears in the solution annealed and quenched state. This means that it is not an age hardenable TRIP steel (referred here to as maraging TRIP steel) but a Mn-based maraging steel [18, 19]. Similar observations are made for the microstructure of the 12% Mn alloy, Figs. 2c and 3c. This sample also consists of an α'-martensite matrix but it contains up to 15 vol.% retained austenite and some ε-martensite. During aging additional austenite is formed.

Hence, this material represents an age hardenable TRIP steel as it can undergo hardening both, via mechanically induced martensite formation and also through precipitation hardening of the as-quenched and of the mechanically induced martensite. The EBSD map for the 12% Mn alloy shows a considerably finer α'-martensite microstructure than that observed in the 9% Mn sample, Fig. 3b, c. Dilatometry and ferromagnetic data suggest in part a higher austenite fraction of up to 20 vol.%. The differences between dilatometry, ferromagnetic characterization, and EBSD analysis can be attributed to the limited statistics provided by EBSD.

Also EBSD yields surface information only. The size of the retained austenite islands lies between 1 μm and 20 μm. The ε-martensite lamellae are below 2 μm and occupy an overall fraction of about 1–2 vol.%. The 12% Mn alloy has a YS of about 325 MPa, UTS of nearly 1 GPa, and TE of about 16% in the as-quenched state, Fig. 2c. After aging (450°C for 48 h) the UTS increases to more than 1.3 GPa and the TE to 21%. Fifteen percent cold rolling prior to aging leads to a strong increase in strength (nearly 1.5 GPa UTS) but also to a drop in TE (about 10%).

Regarding the ductility it is most remarkable that both steels (9% Mn and 12% Mn) show, irrespective of their retained austenite content, the surprising feature of a simultaneous increase in both, UTS and total elongation upon aging. While the UTS increases by 25–30% the total elongation increases by more than 150% (from 6 to 15%) for the 9% Mn sample and by 31% (from 16 to 21%) for the 12% Mn alloy. This increase of both properties represents a very unusual feature of these ultrahigh strength materials.

Figure 2d shows the engineering stress–strain curves for the 15% Mn sample steel in as-quenched, age hardened, and 15% cold rolled plus age hardened state. The corresponding phase map is given in Fig. 3d. The alloy contains a high volume fraction of retained austenite in the as-quenched state. A high resolution EBSD analysis shows that in addition to α'-martensite the sample also contains some ε-martensite [37].

The as-quenched 15% Mn alloy has a YS of about 160 MPa, a UTS of about 800 MPa, and a TE of about 40%, Fig. 2d. After aging (450°C for 48 h) the UTS drops to about 700 MPa and the TE to 32%. Fifteen percent cold rolling prior to aging leads to a strong increase in strength (above 1.2 GPa UTS) but also to a drop in TE (about 3%). This means that the steel with the most stable austenite and highest Mn content (15%) does not reveal the same unusual feature of an increase in ductility upon aging heat treatment as the two steels with smaller Mn content (9 and 12%).

The BF-STEM micrographs that were taken exemplarily on the age hardened maraging TRIP steel (12% Mn) show that the nanoscaled precipitates reveal a narrow size distribution with an average diameter of 6–8 nm, Fig. 4a, b.

Figure 4b also reveals some of the retained austenite which is free of precipitates. Local EDX analysis conducted in convergent beam mode show an increased content in Ni, Mn, Ti, and Al in these particles when compared to the surrounding matrix [13, 14, 38]. The area density of the nanoparticles was about 300 μm^{-2}. The corresponding volume density is estimated as 5×10^3 μm^{-3} (about 2–3 vol. %), Fig. 4. Slightly elongated nanoparticles were observed at the interfaces between the martensite lamellae.

In addition to the EDX-TEM analysis we also conducted APT measurements on the alloy with 9% Mn. Figure 5 shows the microstructure after aging at 450°C for 48 h (Fig. 5a) and after 192 h (Fig. 5b). Table 2 provides an overview of the compositions of the probe, matrix and particles of the alloy with 9% Mn, aged for 48 h.

The particles are enriched in Ni, Mn, Fe, Al and Ti. These observations indicate that the particles encountered do as a rule not assume a simple binary and ternary composition but seem to be more complex. To compare the composition of the particles with the total probe concentration, for each element i an enrichment factor k_i was calculated by the following equation:

$$k_i = [i_{\text{particles}}]/[i_{\text{probe}}] \quad (1)$$

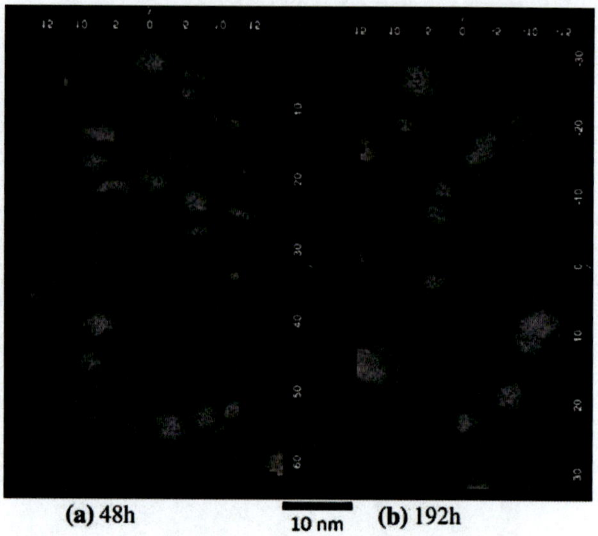

Fig. 5 Atom probe tomography (ATP) measurements conducted on the maraging alloy with 9% Mn after aging at 450°C for (**a**) 48 h and (**b**) 192 h; The *gray* surfaces highlight regions with an atomic Ni concentration above 14 at.%

with $[i_{\text{particles}}]$ as the concentration of element i in the particle and $[i_{\text{probe}}]$ as the concentration of element i in the total probe volume. A value of k_i larger than 1 signifies a tendency of element i to enrich in the particles and a value smaller than 1 a tendency of a depletion in the particle. Al and Ni show the highest values of k_i. The small values of k_i of Fe and Mo show the tendency of a depletion of these elements in the particles.

The chemical composition of the particles is not changing significantly when the aging time is increased from 48 to 192 h. But the ATP results show an increase of the volume fraction of the particles from 1.5 to 4.3% and an

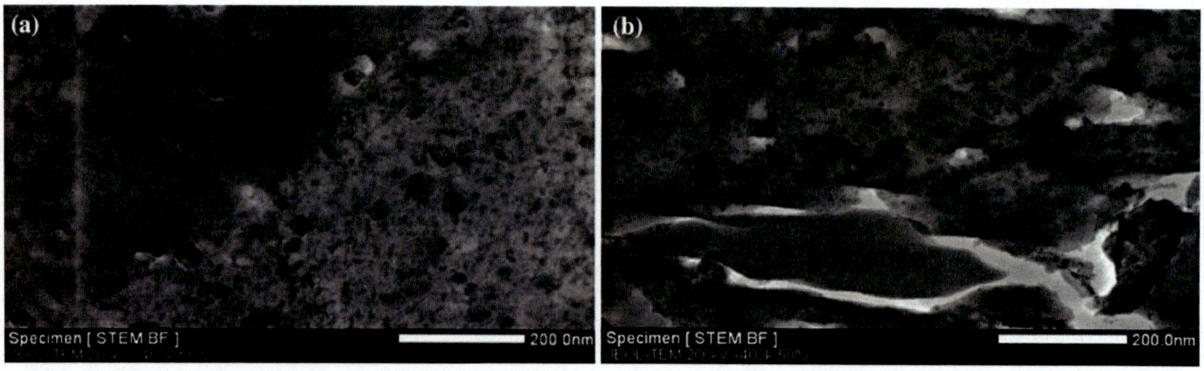

Fig. 4 TEM images of nanoparticles (and some larger particles) formed in the aged 12% Mn alloy (450°C, 48 h). The nanoparticles have an average diameter of 8–12 nm. Local EDX analysis shows an increased content in Ni, Mn, Ti, and Al in the particles relative to the matrix. **a** BF-STEM images of particles in two neighboring martensite lamellae. **b** BF-STEM images of particles in the martensite and particle-free austenite regions

Table 2 Composition of probe, matrix and particles and enrichment factor of ATP measurement shown in Fig. 5a

	at.% (probe)	at.% in matrix	at.% in particles	Enrichment factor k_i
Mn	8.38	8.10	24.70	2.95
Ni	2.08	1.44	39.99	19.23
Mo	0.62	0.64	0.12	0.19
Ti	1.20	1.16	3.91	3.26
Al	0.26	0.15	7.02	27.00
Si	0.13	0.12	0.28	2.15
Fe	87.33	88.40	23.97	0.27

Table 3 Volume fraction and mean diameter of particles for conditions shown in Fig. 5

Aging time at 450°C	48 h	192 h
Volume fraction (%)	1.47	4.34
Mean diameter (nm)	4.69 ± 0.71	6.13 ± 2.17

increase of the mean particle diameter from 4.7 nm to 6.1 nm with increasing the aging time (Table 3).

4 Discussion

The stress–strain data for the different specimens show in principal two types of behavior before aging. The first two samples, i.e. the conventional highly alloyed 17% Ni-11% Co maraging steel and the maraging steel with 9% Mn both reveal an α'-martensitic microstructure in the as-quenched state, Fig. 3. This means that both materials do not contain metastable retained austenite, hence no TRIP effect occurs. Their mechanical performance, observed in tensile tests, Fig. 2a, b, reflects this fact. Both materials reveal the same strength-elongation profile in the as-quenched and in the 15% pre-rolled state, i.e. no TRIP-related hardening was observed.

In contrast, the 12%Mn and the 15%Mn samples, Table 1, both show a pronounced increase in strength upon pre-deformation (plus subsequent aging), Figs. 2 and 3. This observation is attributed to the mechanically-stimulated transformation of the retained austenite leading to a TRIP effect. Corresponding EBSD and X-ray measurements show that the austenite present in both specimens in the as-quenched or aged state is gradually transformed into martensite during deformation, Fig. 6.

The motivation for attributing the additional strain hardening capacity of the 12% and 15%Mn alloys essentially to the TRIP effect becomes clear when comparing the data for the 15% pre-rolled specimens among the four samples, Fig. 2. While the samples with 12% and 15%Mn contain retained austenite, Fig. 5b, and, hence, show a strong increase in strength upon pre-deformation, the conventional Ni-Co maraging steel and the 9% Mn samples both do not contain austenite and, therefore, do not reveal any change in strength when 15% cold rolled prior to tensile testing.

The second approach to classify the mechanical test result lies in the analysis of the unexpected ductilization upon aging observed for some of the specimens: Two alloys reveal a surprising increase both in strength and in the total elongation after aging, namely, the lean maraging steel with 9% Mn (no retained austenite), Figs. 2b and 3b and the maraging TRIP steel with 12%Mn (containing retained austenite), Figs. 2c, 3c and 6.

For the 9% Mn maraging steel we observe that the UTS increases by 25–30% to more than 1 GPa and the TE by more than 150% (from 6% TE to more than 15% TE) due to the aging heat treatment (450°C for 48 h). For the 12% Mn maraging TRIP alloy we observe after aging an increase in UTS to more than 1.3 GPa and of the TE from 16% to more than 21%. This increase in both mechanical properties (UTS, TE) represents a very unusual feature of these ultrahigh strength materials, Fig. 1. All other strengthening methods explored so far in the field of ultra high strength steels lead to a decrease in the ductility rather than to its enhancement [34].

The difference in retained austenite content between the two alloys (9% Mn, 12% Mn) means that the unexpected ductilization effect seems not to coincide with the occurrence of the TRIP effect, as one of the materials does by practical standards not seem to contain retained austenite (9% Mn).

The increase in strength upon aging heat treatment is attributed to precipitation hardening. The TEM and APT results, Figs. 4 and 5, reveal that the steels contain nano-sized particles such as also found in conventional maraging steels [12–17] and that the particle dispersion is very high, even after the long aging treatment (450°C, 48 h) used in this study (shown here exemplarily for the 12% Mn steel). The observation of a high maintained strength and high reluctance of the nanoscaled precipitates to coarsen was for conventional maraging steels reported before [15].

Besides this strong effect of the nano-precipitates formed during aging also the high retained dislocation content observed in the martensite matrix is important for the high strength. The TEM data revealed that the dislocation density in the martensite is very high, about 10^{15}–10^{16} m^{-2}. It is noteworthy that the dense dislocation arrangement prevailed even after the aging heat treatment. The dislocation density could also play an important role for the nucleation of the precipitates and their very high dispersion since we observed that many precipitates were located at dislocations.

It is also important for the plastic properties that the (nearly) carbon-free martensite matrix is rather ductile. Conventional carbon-based martensitic steels typically reveal very poor ductility, Fig. 1.

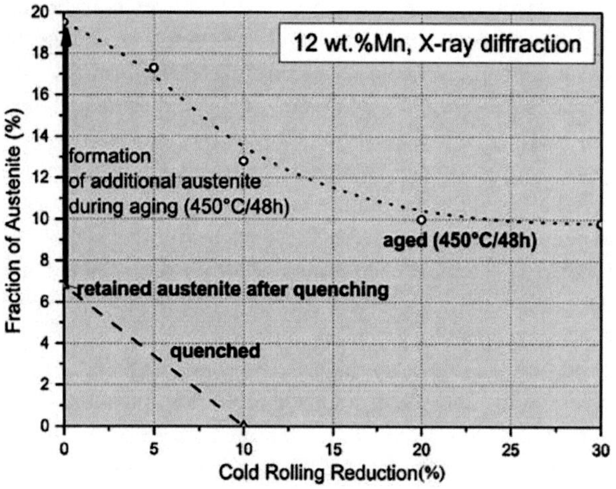

Fig. 6 X-ray diffraction results of 12%Mn steel; In the as-quenched condition 7% of austenite was detected. After cold rolling of the quenched steel with a reduction of 10% practically all the austenite transformed to martensite. During the aging of the steel immediately after quenching the volume fraction of austenite is increasing to nearly 20% due to the formation of new austenite. During rolling of the aged steel a part of the austenite transformed to martensite

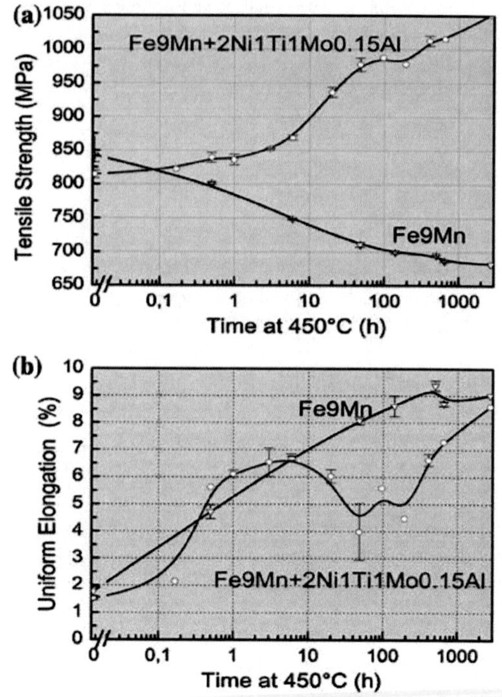

Fig. 7 Results of room temperature tensile tests as a function of time at 450°C for the age hardenable steel Fe9Mn + 2Ni1Ti1Mo0.15Al (9% Mn and further additions of 2% Ni, 1% Ti, 1% Mo and 0.15% Al) and the reference steel Fe9Mn with only an addition of 9% Mn, but no further additions for age hardening

However, irrespective of this plausible relationship between the nano-precipitates and the increase in flow stress, a central point of our observations is the simultaneous strong increase in the total elongation after aging, Fig. 2b, c. Three effects are conceivable to explain this phenomenon, namely, first, the kinetics associated with delayed austenitization, second, the kinetics of precipitation and third, the effect of tempering the as-quenched martensite during aging.

Our X-ray results (Fig. 6) indicate that partial re-transformation into austenite during aging might play a role for ductilization at least for the specimen with 12% Mn. On the other hand the alloy with only 9% Mn shows the same ductilization effect although partial re-transformation at 450°C is thermodynamically not so likely for this alloy owing to its low austenite-stabilizing Mn content, Fig. 2b. After 48 h aging at 450°C no austenite was detectable by X-ray diffraction for the alloy with 9% Mn.

A second explanation for the ductilization effect can be seen in a possible Orowan hardening mechanism. Orowan hardening would increase the strain hardening rate. According to the Considère criterion the necking of the tensile test sample can be delayed up to the strain where the strain hardening rate equals the true flow stress.

A third explanation for the ductilization effect might be the tempering of the as-quenched martensitic microstructure of the steels. For the investigated steels it is not possible to separate the effects of tempering the as-quenched martensite and precipitation hardening because both mechanisms depend on diffusion of Fe and the alloying elements. At the aging temperature of 450°C both tempering and precipitation hardening will take place simultaneously. Therefore, another steel was investigated as a reference material: This steel (in the following designated as Fe9Mn) has the same Mn content of 9% like the steel 9% Mn from Table 1 (in the following designated as Fe9Mn + 2Ni1Ti1Mo0.15Al). But this reference steel lacks the additions of Ni, Ti, Al and Mo. Therefore, a precipitation hardening of reference steel Fe9Mn can be ruled out. Figure 7 compares the change of the ultimate tensile strength for both steels with holding time at 450°C. The strength is decreasing for steel Fe9Mn which can be attributed to a tempering effect. For steel Fe9Mn + 2Ni1Ti1Mo0.15Al the effect of precipitation hardening on strength prevails over the effect of tempering, resulting in a net increase of strength.

Up to 6 h the uniform elongation is increasing for both steels in a similar way (Fig. 7b). This coincides with only a small increase in strength for steel Fe9Mn + 2Ni1Ti1Mo 0.15Al. This indicates that in this time period the main effect increasing ductility is the tempering. After 6 h the more pronounced strength increase of steel Fe9Mn + 2-Ni1Ti1Mo0.15Al signifies an effective age hardening. This leads to a loss in ductility.

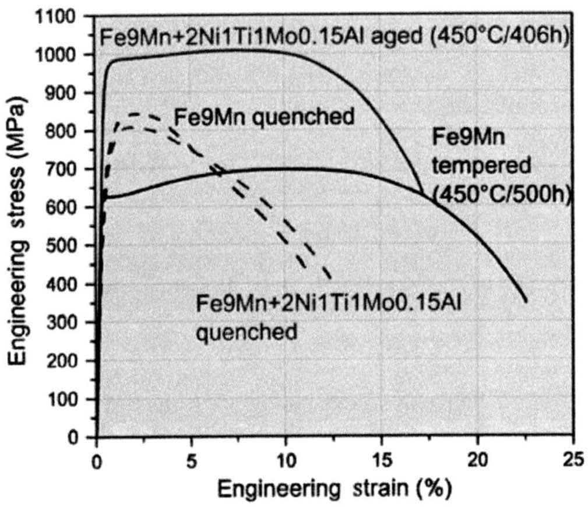

Fig. 8 Stress-strain curves for reference steel (9% Mn) and maraging steel with 9% Mn and additions of 2% Ni, 1% Ti, 1% Mo and 0.15% Al

With increasing time at 450°C the tensile strength of steel Fe9Mn + 2Ni1Ti1Mo0.15Al is increasing. For the reference steel Fe9Mn without precipitation hardening the ultimate tensile strength is decreasing with time at 450°C as a result of tempering. Up to 6 h at 450°C the uniform elongation of both steels are increasing in a similar way. After 6 h the uniform elongation of the steel Fe9Mn + 2-Ni1Ti1Mo0.15Al is decreasing which coincides with the increase of the ultimate tensile strength due to age hardening. The uniform elongation of the reference steel Fe9Mn continues to increase.

A comparison of stress–strain curves for both steels after quenching and after a long time holding time at 450°C is shown in Fig. 8. The curves for both steels in the as-quenched state (broken lines) are very similar. The tempering of steel Fe9Mn leads to an increase in ductility with a simultaneous decrease in strength. Steel Fe9Mn + 2Ni1-Ti1Mo0.15Al show a different behavior: After the holding at 450°C the strength is increased because the effect of precipitation hardening on strength surpasses the effect of tempering. The ductility is also increased but less than in the case of the reference steel Fe9Mn. This seems to be the addition of the beneficial effect of tempering and a reduction of ductility by the precipitation hardening. But because the precipitation hardening is only moderate in this steel, the holding at 450°C is resulting in a net increase of ductility. In the case of the standard 17% Ni-11% Co maraging steel the more pronounced precipitation hardening completely offsets a possible ductility increase by tempering. The strong effect of tempering on ductility observed here is surprising because the carbon content of the steels was in the range of 0.01% and the nitrogen content in the range of 0.001%.

In the as-quenched state the stress strain curves (broken lines) are similar for both steels. Tempering of Fe9Mn for 500 h at 450°C leads to a significant increase of elongation but loss of strength (solid black curve). For the age hardenable Fe9Mn + 2Ni1Ti1Mo0.15Al alloy the heat treatment 450°C/406 h is a combination of tempering and aging of the quenched martensite. This leads to a smaller increase of uniform and total elongation but a simultaneous increase of strength.

But from X-ray results it was found that the amount of the micro strain is reduced already in 6 h at 450°C to 50% of the initial value. Therefore it seems that micro creep and a possible recovery of the dislocation structure in the as-quenched martensite during holding at 450°C is responsible for the beneficial effect on ductility.

5 Conclusions

We presented a concept for the development of ultrahigh strength and at the same time ductile martensitic and austenitic-martensitic steels that is based on the precipitation of nano-sized particles via aging heat treatment and on the TRIP effect. The approach, hence, combines the TRIP mechanism with a maraging treatment. The alloy systems presented are Fe–Mn steels with a low-carbon martensitic matrix and elements for the formation of nano-precipitates (Ni, Ti, Al, Mo). Particularly the 12% Mn maraging TRIP steel and the 9% Mn maraging steel revealed a significant increase both in strength and elongation after the aging heat treatment. The unexpected increase in total elongation was mainly attributed to the effect of tempering of the as-quenched martensite. In combination with the moderate precipitation hardening of the very ductile low carbon martensite, which reduces ductility only slightly, a net increase of ductility can be achieved. By adding a TRIP effect the ductility can be increased even more.

References

1. J. Inoue, S. Nambu, Y. Ishimoto, T. Koseki, Scr. Mater. **59**, 1055 (2008)
2. S. Nambu, M. Michiuchi, Y. Ishimoto, K. Asakura, J. Inoue, T. Koseki, Scr. Mater. **60**, 221 (2009)
3. J.R. Patel, M. Cohen, Acta Metall. **1**, 531 (1953)
4. H.K.D.H. Bhadeshia, D.V. Edmonds, Metall. Trans. **10A**, 895 (1979)
5. G.B. Olson, M. Cohen, in *Mechanical properties and phase transformations in engineering materials*. ed. by S.C. Antolovich, R.O. Ritchie, W.W. Gerberich (Met. Soc. AIME, New Orleans, 1986) p. 367
6. M. Takahashi, H.K.D.H. Bhadeshia, Mater. Trans. JIM **32**, 689 (1991)

7. P.J. Jacques, E. Girault, T. Catlin, N. Geerlofs, T. Kop, S. van der Zwaag, F. Delannay, Mater. Sci. Eng. **A 273–275**, 475 (1999)
8. M. De Meyer, D. Vanderschueren, B.C. De Cooman, ISIJ Int. **39**, 813 (1999)
9. S. Traint, A. Pichler, K. Hauzenberger, P. Stiaszny, E. Werner, Steel Res. Int. **73**, 259 (2002)
10. P.J. Jacques, Curr. Opin. Sol. State Mater. Sc. **8**, 259 (2004)
11. S. Zaefferer, J. Ohlert, W. Bleck, Acta Mater. **52**, 2765 (2004)
12. R.F. Decker, J.T. Eash, A.J. Goldman, Trans. ASM **55**, 58 (2004)
13. R.F. Decker, *Source Book on Maraging Steels* (ASM International, Metals Park, 1979)
14. W. Sha, A. Cerezo, G.D.W. Smith, Metall. Trans. **24A**, 1251 (1993)
15. V.K. Vasudervan, S.J. Kim, C.M. Wayman, Metall. Trans **21A**, 2655 (1990)
16. C. Servant, N. Bouzid, Acta Metal. **36**, 2771 (1988)
17. R. Tewari, S. Majumder, I.S. Batra, G.K. Dey, S. Banerjee, Acta Mater. **48**, 1187 (2000)
18. A.J. Goldman, J. Manenc, Trans. ASM **58**, 645 (1965)
19. R.K. Ray, A.K. Seal, Scripta Metall. **10**, 971 (1976)
20. M. Alinger, G. Odette, D. Hoelzer, Acta Mater. **57**, 392 (2009)
21. G.R. Odette, M.J. Alinger, B.D. Wirth, Ann. Rev. Mater. Res. **38**, 471 (2008)
22. D.R. Lesuer, C.K. Syn, O.D. Sherby, Mater. Trans. **47**, 1508 (2006)
23. S. Yamashita, S. Ohtsuka, N. Akasaka, S. Ukai, S. Ohnuki, Phil. Mag. Let. **84**, 525 (2004)
24. M. Taneike, F. Abe, K. Sawada, Nature **424**, 294 (2003)
25. E. Courtois, T. Epicier, C. Scott, Micron **37**, 492 (2006)
26. A. Machova, Mater. Sc. Engin. **A 319**, 574 (2001)
27. L. Storojeva, D. Ponge, R. Kaspar, D. Raabe, Acta Mater. **52**, 2209 (2004)
28. R. Song, D. Ponge, R. Kaspar, D. Raabe, Z. Metallk. **95**, 513 (2004)
29. L. Storojeva, D. Ponge, D. Raabe, R. Kaspar, Z. Metallk. **95**, 1108 (2004)
30. R. Song, D. Ponge, D. Raabe, R. Kaspar, Acta Mater. **53**, 845 (2004)
31. R. Song, D. Ponge, D. Raabe, Scr. Mater. **52**, 1075 (2005)
32. R. Song, D. Ponge, D. Raabe, ISIJ Int. **45**, 1721 (2005)
33. R. Song, D. Ponge, D. Raabe, Acta Mater. **53**, 4881 (2005)
34. R. Song, D. Ponge, D. Raabe, J.G. Speer, D.K. Matlock, Mater. Sc. Engin. **A 441**, 1 (2006)
35. N. Tsuji, Y. Saito, H. Utsunomiya, S. Tanigawa, Scr. Mater. **40**, 795 (1999)
36. D. Raabe, D. Ponge, O. Dmitrieva, B. Sander, Scr. Mater. **60**, 1141 (2009)
37. K. Verbeken, N. Van Caenegem, D. Raabe, Micron **40**, 151 (2009)
38. G.P. Miller, W.I. Mitchell, J. Iron Steel Inst. **20**, 899 (1965)

The 3rd Generation Automobile Sheet Steels Presenting with Ultrahigh Strength and High Ductility

Wenquan Cao, Jie Shi, Chang Wang, Cunyu Wang, Le Xu, Maoqiu Wang, Yuqing Weng, and Han Dong

Abstract

In this study, research and development on the 3rd generation automobile steel, with targets of $R_m \times A$ no less than 30 GPa% at R_m level of 1–1.5 GPa, was carried out to fabricate high strength and high ductility steel by two methodologies, one is the medium manganese steels fabricated by intercritical annealing through austenite reverted transformation (ART-annealing) and another is the conventional carbon steels processed by quenching and partitioning(Q&P). The ultrafine grain sized austenite-ferrite duplex microstructure and the tempered martensite-fresh martensite-austenite multiphase microstructure were demonstrated based on the microstructure characterization in ART-annealed medium manganese steels and Q&P processed conventional carbon steels. In both heat treatment conditions, substantially enhanced ductility (30–40%) at ultrahigh tensile strength level (1–1.5 GPa) was obtained, which results in a significant improvement of the product of tensile strength to total elongation about 30–40 GPa%. Analysis on the work hardening behaviors and the relationship between microstructures and mechanical properties of the studied steels indicates that the greatly improved ductility results from the aid of the phase transformation induced plasticity (TRIP effects) and the ultrahigh strength stems from the hard matrix, such as the ultrafine grained duplex structure in ART-annealed steels and the martensite matrix in Q&P processed carbon steels. It is interesting to find that a strong dependence of the product of tensile strength to total elongation on the fraction of retained austenite phase of steels produced by both ART-annealing and Q&P processing techniques. It was proved that both ART-annealing and Q&P processes can be applied to fabricate the third generation automobile sheet steels offering ultrahigh strength and high ductility.

Keywords

Microstructure • Mechanical property • Automobile steel • Austenite reverted transformation • Quenching and partitioning • Product of tensile strength to ductility

1 Introduction

Due to the increasing demand for lightweight and safety steel products from automotive industry, material ability to absorb energy subject to loading are becoming the essential factor to meet the requirement of both energy saving and the impact safety improvement [1]. As an index of absorbed energy of materials, the product of ultimate tensile strength (R_m) and total elongation (A), *i.e.*, $R_m \times A$, has been

W. Cao (✉) · J. Shi · C. Wang · C. Wang · L. Xu · M. Wang · Y. Weng · H. Dong
Central Iron & Steel Research Institute, Beijing 100081, China
e-mail: caowenquan@nercast.com

Y. Weng et al. (eds.), *Advanced Steels*, DOI: 10.1007/978-3-642-17665-4_22,

applied to classify the automobile steel into the 1st generation automobile steel and the 2nd generation automobile steel [2]. The conventional automobile steels with $R_m \times A$ about 15 ± 10 GPa%, such as IF steel, dual phase steel, TRIP steel, was grouped into the 1st generation automobile steel, whereas the high alloyed steel with $R_m \times A$ about 60 ± 60 ± 10 GPa%, such as TWIP steel and other fully austenite steel, was grouped into the 2nd generation automobile steel. It is clear that the advantage of the 1st generation automobile steel is their low cost comparing with that of the 2nd generation automobile steel. The background mentioned above promoted worldwide study of development of the 3rd generation automobile steel with high $R_m \times A$ in between these two groups but with a relative low cost [3–6].

This situation of environmental consideration, energy crisis and safety requirements in automotive industry drives the tendency in automotive industries to develop high performance auto sheet steels. In 2000 in Europe, the Ultra Light Steel Auto Body (ULSAB) project organized, which has brought together steelmaking industry and automotive industry in world-wide scale, demonstrated the possibilities of lightweight automobile construction with high strength steel with comparable formability [7]. In 2007 in America, the third generation automobile steel conception aiming to develop high strength and high ductility steel at low cost was proposed by American researchers [8]. In the Center of Iron and Steel Research Institute of China (CISRI), which has half century research on the martensitic steel, launched out the fundamental research of third generation steel, including automotive sheet steel and the third generation low alloy steel being mentioned and discussed in this chapter, in 2008 after long time data analysis of the development of steel making industry and automotive industry [9].

For the conventional steel possessing primarily ferrite-based microstructures, such as interstitial-free steel (IF), dual phase (DP), transformation induced plasticity (TRIP), complex-phase (CP), and martensitic (MART), their $R_m \times A$ is only about 15 ± 10 GPa% named as the 1st generation steel in Fig. 1a [2, 10, 11]. Contrast with this low value, the second generation austenitic steel including twinning induced plasticity (TWIP) steels, Al-added lightweight steels with induced plasticity, and other fully austenite steel, their $R_m \times A$ is remarkably high up to 50 ± 10 GPa% named as the 2nd generation steel in Fig. 1a, which was thought to be the best material for cold forming and energy reservoir in crash but much more expensive and difficulty in fabrication [2, 12]. It is clear that the advantage of the 1st generation steel is their cost-effective but disadvantage is their low ductility, comparing with that of the 2nd generation steel. The background demonstrated above promoted worldwide study on development of 3rd generation steel with high $R_m \times A$ in between these two groups but relative low cost. In China in 2008, the government launched out the third phase 973 program on the high performance automobile steel, i.e., the 3rd generation automobile steel offering high strength and high ductility, with a target of $R_m \times A$ no less than 30 GPa% at strength level of 1–1.5 GPa [9]. In Fig. 1b, the product of $R_m \times A$ of different kinds of steel, such as conventional steel (IF steel, DP steel, Martensitic steel) [2], conventional TRIP steel [2], Nano-Bainitic steel [13], and TWIP and austenitic steel [2], is briefly summarized as a function of austenite volume fraction. It is interesting to be found from Fig. 1b that no matter TRIP effect or TWIP effect in the steel, generally the $R_m \times A$ linearly increases with increasing of austenite volume fraction from conventional steel to TWIP steel. The slope between $R_m \times A$ and the austenite volume fraction is ~0.65 GPa%/(1%γ), indicating the strong dependence of $R_m \times A$ on austenite volume fraction. It could be expected roughly that 20–40% metastable austenite is essential to obtain steel with $R_m \times A$ about 30–40 GPa%. Thus the BCC–FCC duplex structure with metastable retained austenite may be a promising way to design the new type automobile steel with

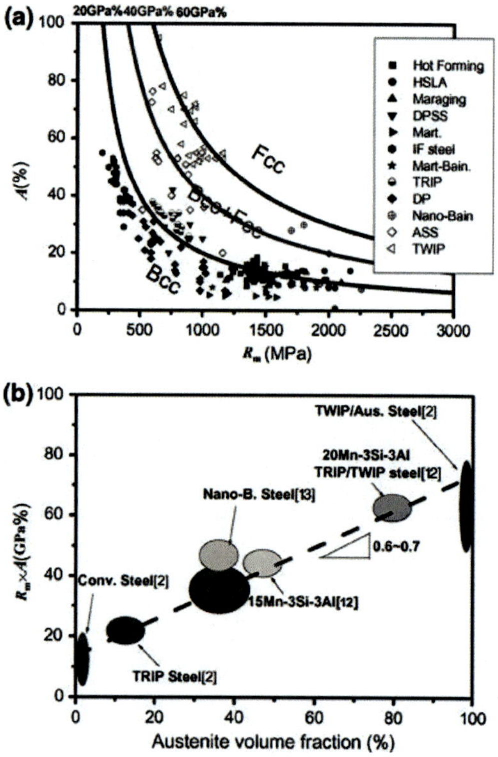

Fig. 1 Dependence of elongation to failure on tensile strength and residue austenite fraction of different steel **a** total elongation and tensile strength and **b** product of tensile strength to total elongation and austenite volume fraction

both ultrahigh strength and high ductility by means of TRIP/TWIP effects.

Recently, the conception of microstructure control of multi-phase, meta-stability and multi-scale, i.e., the M^3 structure control, was proposed by Han [9]. In this study, the M^3 microstructure is not only the characters of the desired microstructure but also the controlling approaches to get this structure. For example, to get the high strength, the microstructure should be controlled in different scales, such as prior austenite grain size scale, the packet size and the lath thickness scale control. These kinds of controlling of the microstructure were clearly demonstrated in ART-annealed medium manganese steel and the Q&P processed steels, which in turn gives microstructure of processed steels with M^3 features [14–17]. Both heat treatment methodologies can give not only the large fractioned austenite but also the hard matrix [6, 16, 18, 19].

For the ART-annealing, it is originally a technique to obtain a fully martensitic microstructure and then annealed it in the intercritical region (Ac_1–Ac_3) or subintercritical region (lower than Ac_1). ART-annealing of the Mn-bearing martensitic steel was demonstrated to be capable of fabricating the superfine grained austenite-ferrite duplex steel by austenite reverted transformation [6, 18, 19]. Excellent combination of strength and ductility was obtained by Miller [18], Morris [19], Han [6] and Jie [14, 15] in C/Mn steels with Mn of ~5% (wt%). The reported mechanical properties of Mn-steels processed by ART-annealing were significant improved up to ~800 MPa with total elongation ~30%, which is significantly higher than conventional TRIP steels [6, 18, 19]. However, so far few researches on the industry production have been reported. In addition, so far the few studies on the ART-annealing behavior/mechanism and the work hardening behavior has been reported, which may be strong affected by hot/cold deformation, work hardening behaviors of Mn-TRIP steel were reported in literatures [14]. For the quenching and partitioning (Q&P), it was first proposed by Speer in 2003 to produce steels by introducing carbon-enriched retained austenite into martensitic matrix [20, 21]. The Q&P process involves the partially quenching and follows tempering/annealing. The latter was called partitioning, in which carbon atoms were rejected from the developed martensitic phase and were absorbed by retained austenite phase. It was proved that Q&P process could be applied to increase the ductility of martensitic steel without losing its strength significantly. However, for the ultrahigh strength steel ($R_m > 1{,}500$ MPa), the ductility is still lower than 15%. Thus it still needs further research to improve the ductility of ultrahigh strength steel by tailoring Q&P heat treatment parameters, such as quenching temperature and time, partition temperature and time for different steel, to improve ductility larger than 20%.

In this study, both ART-annealing and Q&P processing were applied to develop the third generation automobile sheet steels with ultrahigh strength (1–1.5 GPa) and high ductility(>30% for 1 GPa and >20% for 1.5 GPa). The microstructure, mechanical properties and the relationship between them for the ART-annealed medium manganese steels and the Q&P processed conventional carbon steels will be demonstrated. The strengthening and ductility-enhancing mechanisms of the 3rd generation automobile steel will be proposed tentatively.

2 Experimental

Medium manganese steels (3–9%Mn and 0.01–0.4%C) were prepared by high frequency induction furnace in a vacuum atmosphere. The ingots were homogenized and forged into rods, and lastly cooled in furnace to room temperature. These forged rods were quenched into oil after austenization and then (sub)intercritical annealing by austenite reverted transformation (ART-annealing) in electrical box furnace with different time and finally air cooled to room temperature.

Compared with the medium carbon steels applied in ART-annealing process, the plain carbon steels with traced manganese content were considered in Q&P process. For the Q&P process, the carbon content should be a little bit high, usually higher than 0.2%. And some elements, such as Al and Si, were added to suppress the precipitation of carbides and stabilize the newly developed austenite. In order to meet this requirement, different kinds of steels with different carbon contents, 21C, 37C, and 41C were applied in this study, whose compositions were given in Table 1. These steels were melt in arc furnace and then hot rolled into plate with thickness about 16 mm. All specimens were austenized in a high temperature box furnace, then quenched into a salt bath for partially martensitic transformation, and finally austempered immediately in another salt bath (partitioning process), the samples were water quenched to

Table 1 Composition of the studied steels (wt%)

Steel	C	Si	Mn	Cr	Ni	Mo	V	Nb	Al	S	P
21C	0.21	1.75	0.290	1.03	2.86	0.31	0.08	0.049	0.020	0.0007	0.0060
37C	0.37	1.85	0.200	1.04	1.97	0.30	0.10	–	–	0.007	0.0072
41C	0.41	1.68	0.027	1.05	1.83	0.62	0.14	–	0.014	0.001	0.0050

room temperature. The Q&P heat treatment parameters for these steels were selected based on the Q&P principle. Particular interests will be focused on the relative high partitioning temperature with relative short time. Microstructure and mechanical properties of the designed steels were carefully examined by transmission electron microscopy (TEM), electron back scattered diffraction in scanning electron microscopy with field emission gun (EBSD-FEG/SEM), X-ray diffraction (XRD) and tensile test.

3 Results and Summary

3.1 Ultrafine Ferrite–Austenite Duplex Structure in Medium-Mn Steel Processed by ART-Annealing

As an example of the FeMnC-steel processed by ART-annealing, an ultrafine lamellar ferrite/martensite structure of Fe-0.2C-5Mn steel characterized by TEM was shown in Fig. 2. After 1 h annealing at 650°C, the dark/bright lamellar structure in the original Martensitic packets was clear with austenite lath parallel to the slightly coarsened Martensitic lath (Fig. 2a). The equiaxed austenite grains could be found in the packet boundary or original austenite grain boundary. After 1 h annealing, the thickness of austenite laths is about 0.2–0.3 μm. With increasing of annealing time, the BCC–FCC duplex microstructure with parallel ferrite laths and austenite laths still remains in Fe-0.2C-5Mn steel after 6, 12 and 48 h ART-annealing at 650°C as shown in Fig. 2b–d. In addition, it can be seen from Fig. 2 that increasing annealing time improves the uniformity of the microstructure and decreases the precipitation of carbides, suggesting the concurrence of austenite growth by thickening, precipitation dissolving in the interface region between precipitation and the coalescence martensitic lath. After 6 and 12 h annealing, the austenite thickness is about 0.3 μm, slightly thicker than that of 1 h annealed specimen. Even after 48 h annealing at 650°C, the thickness of both austenite lath and ferrite lath is only about 0.4 μm, indicating the high thermal stability at this high temperature. This may need further study on the mechanism of this kind of thermal stability.

It is very interesting that the thickness of both austenite lath and ferrite lath is still very fine after long time annealing. It was measured that the average thickness is about 0.4 μm for even after 144 h annealing, indicating the high thermal stability of the submicron grain sized Fe-0.2C-5Mn steel during ART-annealing process. Also with increasing of annealing time, no precipitation could be found in the studied regions, suggesting the transferring of carbon atoms from carbides to newly developed austenite.

The ultrafine lamellar microstructure was also observed through EBSD in FEG/SEM as presented in Fig. 3 as a function of annealing time. In this figure, the austenite phase was revealed by green color but the ferrite phase was

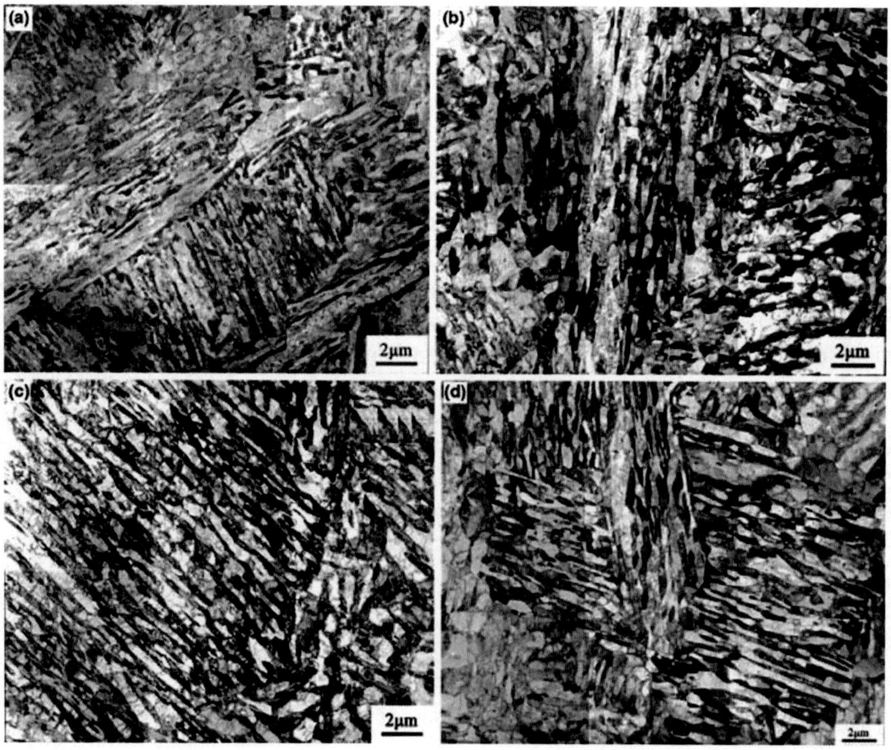

Fig. 2 Microstructure of Fe-0.2C-5Mn steel ART-annealed at 650°C with **a** 1 h, **b** 6 h, **c** 12 h and **d** 48 h

presented by the grey color. The blue colored lines depict the low angle boundaries with misorientation lying in between 3 and 15° and the black lines present the high angle boundaries with misorientation larger than 15°. It can be seen that the high angle boundary fraction increases with increasing annealing time when annealing carried out at 650°C.

It was proved that most of the high angle boundaries are prior austenite grain boundaries, packet boundaries and (sub)block boundaries when ART-annealing time shorter than 30 min. However, most of the high angle boundaries were replaced by phase boundaries when annealing time larger than 1 h ART-annealing. This transition means that most of the austenite nucleated in packet boundaries and (sub)block boundaries when ART-annealing time, which was revealed clearly in Fig. 3. It can be seen that the austenite grains gradually developed in the initial martensite microstructure. In the short ATR-annealing time, for example annealing time lower than 30 min at 650°C the nearly equiaxed austenite grains mainly could be found in the grain boundary or packet boundary as shown in Fig. 3a–c. Due to the low misorientation between martensitic laths, no clear martensitic lath structure could be identified in these shorted time annealed microstructure. When the annealing time was larger than 30 min at 650°C, the nucleation and growth of lath type austenite grains occur in between the martensitic lath as shown in Fig. 3c, d. Also it can be seen that the parallel lath typed structure becomes clear in these annealed microstructure, because the austenite nucleated in between the martensitic lath highlights the lath structure. After 6 h ART-annealing, the austenite grains nucleated and grew both in the boundary regions and in between the martensitic lath slightly coarsened with increasing annealing time as revealed in Fig. 3e, f.

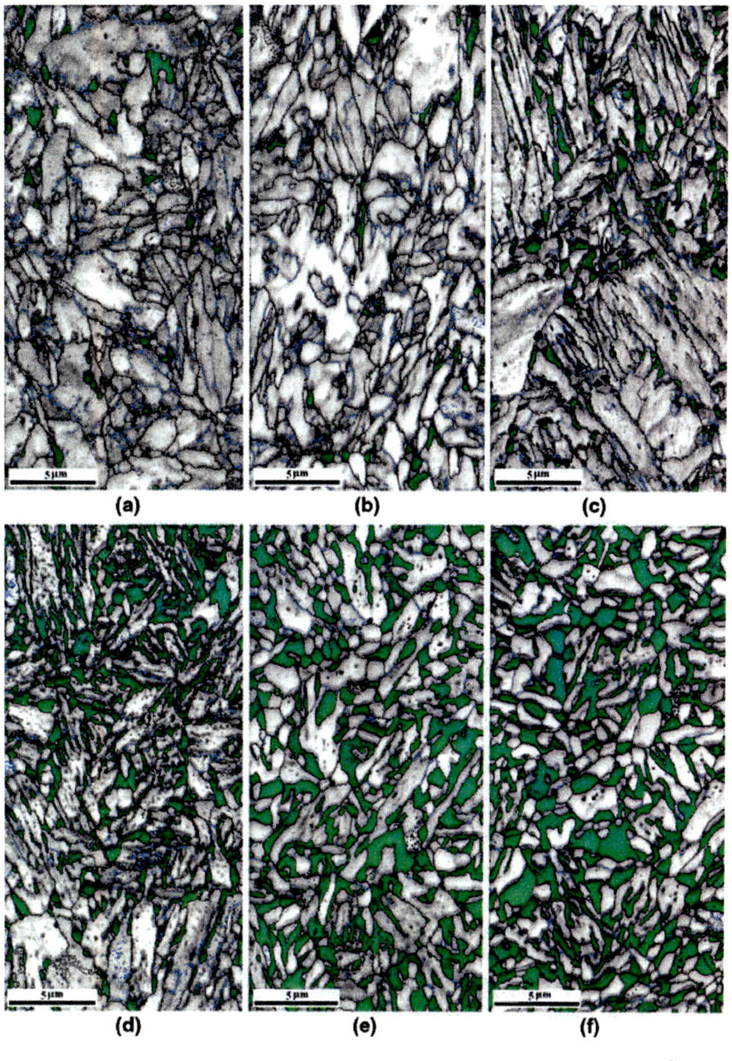

Fig. 3 Microstructural observation of Mn-TRIP steel ART-annealed at 650°C with different time **a** 1 min, **b** 5 min, **c** 30 min, **d** 1 h, **e** 6 h and **f** 12 h

3.2 Tensile Deformation Microstructure and Evolution of Austenite Faction of Fe-0.2C-5Mn Steel After ART-Annealing at 650°C with 6 h

Deformation microstructure of Fe-0.2C-5Mn steel after ART-annealing at 650°C with 6 h was characterized by TEM as shown in Fig. 4a, b. After tensile about 37% uniform tensile deformation, twined martensite could be distinguished from the annealed martensite. In Fig. 1, one type of martensite with twin plate direction perpendicular to the original austenite plates and another type martensite with twin plate's direction parallel to original austenite lath were presented. It can be seen that in one austenite only one packet could be found, which is different from the thermal stability controlled phase transformation. High dislocation density generated in the austenite neighbored martensite was evidenced by XRD measured results of 2.5×10^{15} m^{-2}. It may be expostulated that the high dislocation density in the martensite lath (ferrite) may be attributed to the volume expansion resulted from phase transformation from austenite to martensite.

In order to understand the deformation mechanism, the evolution of austenite volume fraction during tensile process at different deformation temperature was examined by XRD, which was given in Fig. 5. It can be seen that during deformation process the austenite gradually transforms into martensite, which evidenced by our TEM observation as given in Fig. 4.

The annealing time effects on the mechanical stability was measured by XRD as a function of annealing time for the ART-annealing time 5 min, 30 min, 6 and 144 h at 650°C. Similar result of the austenite transformation to martensite was found in 0.2C5Mn steels processed by ART-annealing with different annealing time at 650°C as given in Fig. 6. It can be seen that from Fig. 6a, b that after 6 h ART-annealing not only the starting austenite fraction does not change but their mechanical stabilities are almost same. The evolution austenite volume fractions as given in Figs. 5 and 6 follow an exponent law with true deformation strain as described in Eq. 1 [22, 23] were expected to be essential to interpret the mechanical properties.

$$f_{\gamma-\varepsilon} = f_{\gamma-0}\exp(-k\varepsilon) \quad (1)$$

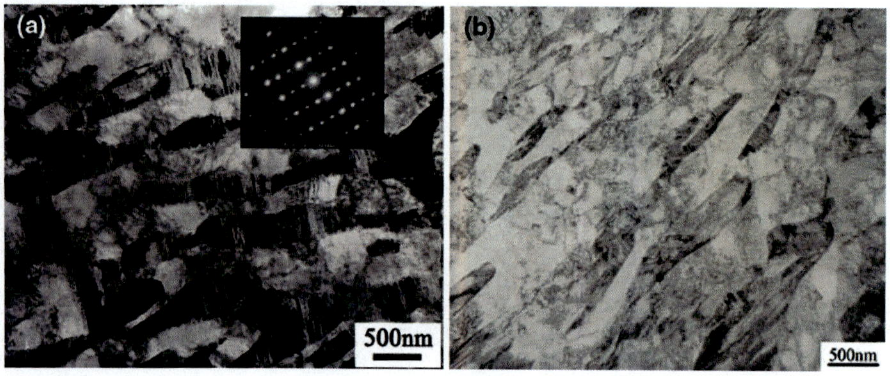

Fig. 4 Room temperature deformation structure at maximum uniform elongation (37%) of Fe-0.2C-5Mn steel after 6 h annealing at 650°C **a** one type of martensite with twin plates direction perpendicular to the original austenite plates with inserted diffraction pattern and **b** another type martensite with twin plates direction parallel to original austenite lath

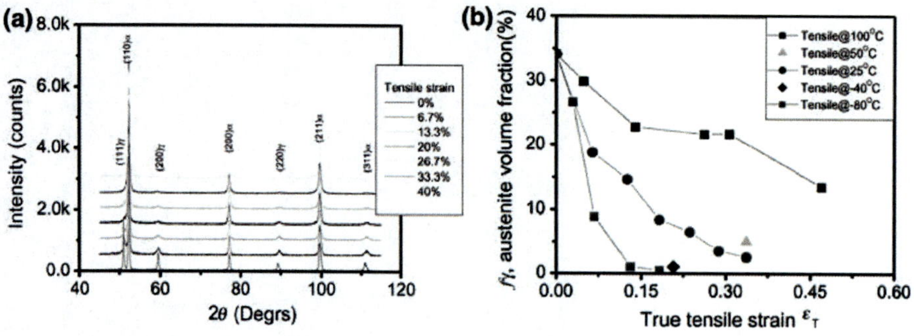

Fig. 5 Austenite evolution during tensile process **a** XRD intensity profiles during tensile process at room temperature and **b** The austenite volume fraction as a function of true tensile strain at different deformation temperature

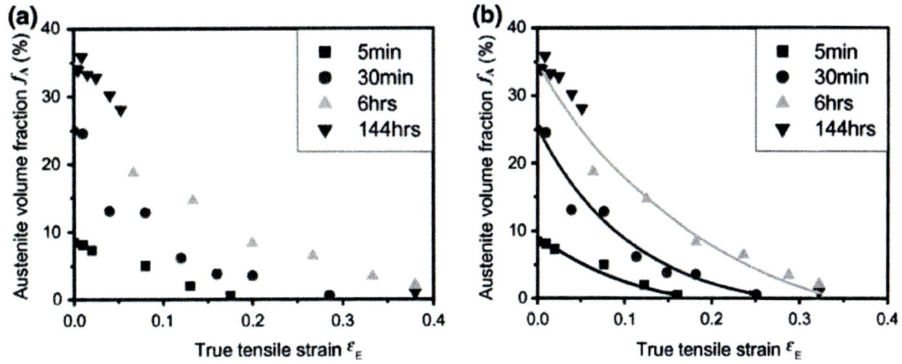

Fig. 6 The effects of ART-annealing time on the austenite mechanical stability **a** austenite fraction as a function of engineering strain and **b** austenite fraction as a function of true tensile strain

3.3 Mechanical Properties of Medium Manganese Steel Processed by ART-Annealing

The engineering tensile stress–strain curves of Fe-0.2C-5Mn after air cooling and annealing with different time were given in Fig. 7a. It can be seen from the engineering stress–strain curves that the that increasing annealing time from 1 min to 144 h increases the total elongation (A_T) significantly from 20 to 45%, decreases the yield strength from 830 to 500 MPa, whereas the ultimate tensile strength about 960 ± 30 MPa.

The true stress–strain curves of studied steel were given in Fig. 7b. It can be seen from the true stress–strain curves that the ultimate tensile stress increases from ~1,000 to 1,350 MPa with increasing annealing time from 1 min to 144 h, suggesting the increased work hardening with increasing of annealing. Interestingly, it can be seen that serrated or jerk flow feature can be observed from the stress strain curves, indicating dynamic strain aging and strong localized deformation during tension test in the steel after 1 h or long time annealing at ambient temperature (Fig. 7b). Furthermore, the true stress–strain curves of the long time annealed specimens assume "S" typed shape, which may indicate the phase transformation during tensile process. Deformation temperature effects on the tensile behaviors were carried out at different temperature ranged from −196 to 400°C, which were shown in Fig. 8.

From Fig. 8, it can be seen that the deformation behaviors of 0.2C5Mn steels were significantly affected by deformation temperature. It is quite interesting that with increasing deformation temperature, the tensile elongation first increases up to 60% elongation at 100°C and then decreases, which may be attributed to the deformation temperature effects on the mechanical stability of the austenite as shown in Fig. 6. Based on above results and our some other data (not given here), it can be seen that the mechanical properties of the medium manganese steels applied in this study were significantly improved by ART-annealing. Within the range of tensile strength from 800 to 1,500 MPa the total elongation can be high up to 30–45%. The product of tensile stress to total elongation in the tensile range 800–1,500 MPa is about 40 ± 10 GPa%, which means that the objective of the 3rd generation can be realized by ART-annealing in medium manganese steels and our pilot study indicates a promising results that ART-annealing of medium manganese steels could be produced in steel industry.

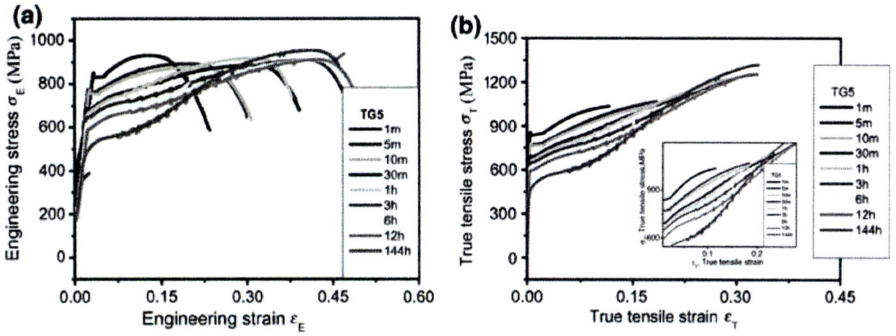

Fig. 7 The stress–strain curves of Fe-0.2C-5Mn processed by ART-annealing with different annealing time **a** engineering stress–strain curves and **b** true tensile stress–strain curves with inserted figure to reveal the dynamic strain aging phenomena

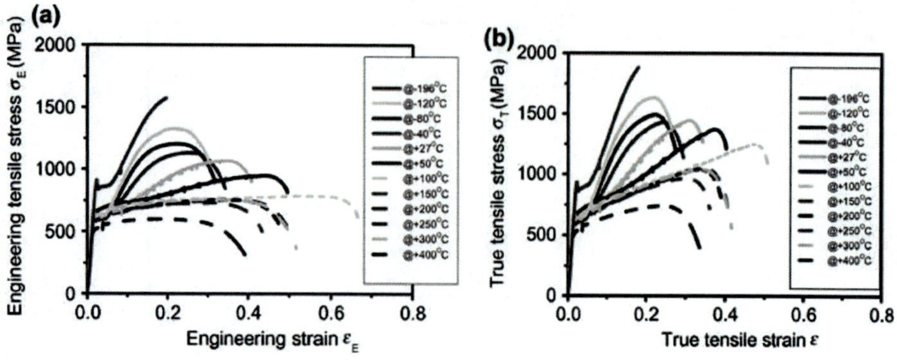

Fig. 8 Deformation temperature effects on the stress–strain curves of Fe-0.2C-5Mn processed by ART-annealing at 650°C with 6 h. **a** Engineering tensile stress and strain. **b** True tensile stress and strain

3.4 Microstructure of Plain Carbon Steels Processed by Q&P Process

3.4.1 Microstructure Observation by SEM, TEM and EBSD

As an example of the microstructure the steel of 21C was selected to study the microstructure evolution during Q&P process as shown in Figs. 9 and 10.

In Fig. 9 the microstructure of steel processed by first austenization at 900°C × 15 min, then quenched 250°C × 1 min and lastly partitioned at 300° with different time. In order to study the effect of partitioning temperature, the microstructure evolution of the steel partitioned 500°C was examined by SEM as well, which were given in Fig. 10. As shown in Fig. 9, the microstructure of steel partitioned at PT = 300°C assumes two phases, one is the easy-etched phase with precipitation of carbide and another is the hard-etched phase. It was demonstrated by Wang that the easy etched-phase is the first-martensite, which was formed during the quenching process. Whereas the hard-etched phase is mainly the fresh-martensite developed during the cooling process followed partitioning [24]. Furthermore, no big effect of partitioning time can be found in this studied range.

When the partitioning temperature increases up to 500°C, an equiaxed hard-etched phase and ultrathin hard-etched phase were easily observed (Fig. 10), which completely different from those observed in Fig. 9. It means that increasing the partitioning temperature increases the mobility of the interface between martensite and austenite during partitioning process. It could be expected that most of the ultrathin hard-etched phase is the austenite, which needs further examination by TEM and XRD. These temperature effects on the phase morphologies suggest that

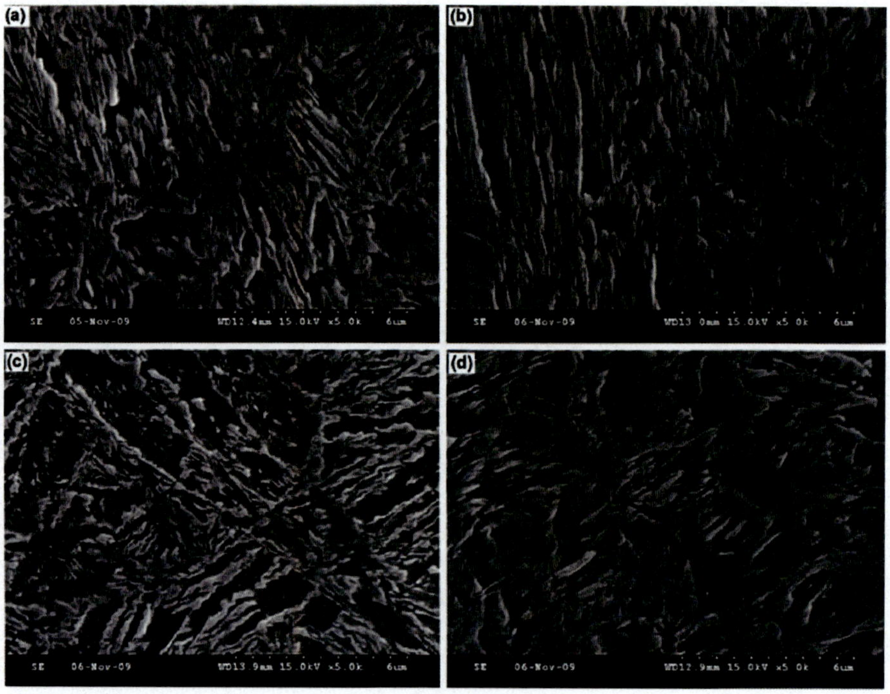

Fig. 9 The microstructure evolution of 21Csteel quenched at QT = 250°C and partitioned at PT = 300° with different time from 60 to 1800 s. **a** 60 s, **b** 300 s, **c** 600 s, **d** 1,800 s

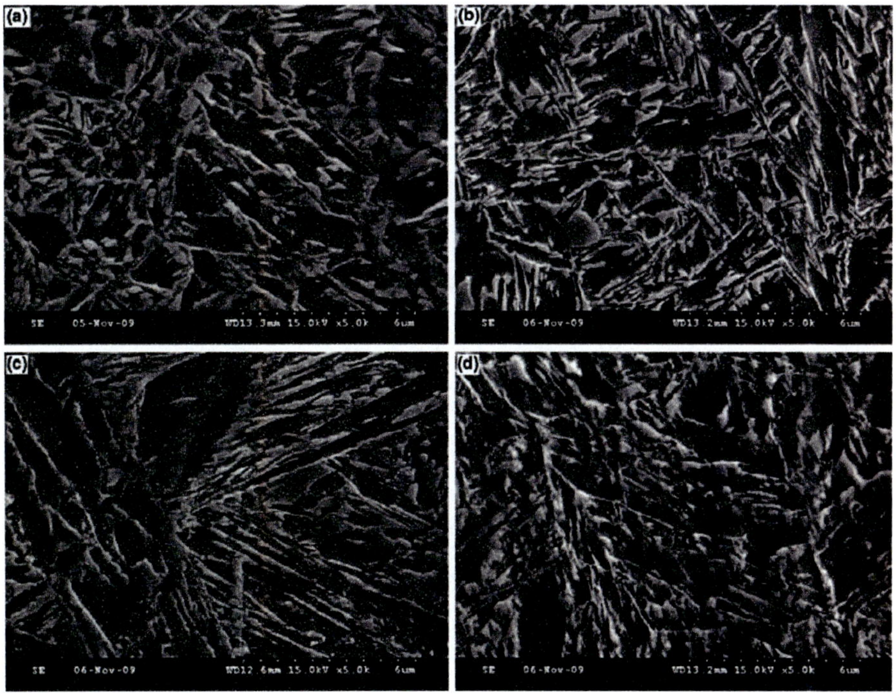

Fig. 10 The microstructure evolution of 21Csteel quenched at QT = 250°C and partitioned at PT = 500°with different time from 60 s to 1800 s. **a** 60 s, **b** 300 s, **c** 600 s, **d** 1,800 s

tailoring the Q&P parameters is very important to develop the required microstructures.

It can be seen from Figs. 9 and 10 that the microstructures were mainly composed of the first-martensite and some granular fresh martensite colonies, which was not significantly affected by partitioning time. However, with increasing of partitioning temperature, the fresh-martensite colony became much more equiaxed and much larger, implying the movement of interface between martensite and austenite. Furthermore, no partitioning time effects on the microstructure could be clearly found at a given partitioning temperature, which means that the microstructure change may take place in the very short time and thus further study should be carried out in the very short partitioning time in near future.

The hard-etched phase was further examined by TEM as shown in Fig. 11. It can be seen that not only from the morphology but also the size, the highlighted grains could be the fresh martensite as found in microstructure examined by SEM as shown in Figs. 9 and 10. The morphologies of austenite was examined by TEM as well as given in Fig. 12. The specimens were partitioned at 300 and 500°C with 5 min. It can be seen after Q&P process only small equiaxed austenite grains less than 0.2 μm and the thin austenite films can be retained. With increasing of partitioning temperature, the quantity of the thin austenite films increase. No large grain sized austenite can be found in the structure. With increasing partitioning temperature, more amount of the thin austenite lath could be observed. From Figs. 9, 10, 11, and 12 it can be seen that after Q&P process, the microstructure is composed of first martensite (easy-etched phase), fresh martensite(hard-etched phase) and austenite.

In order to examine the austenite distribution, EBSD mapping with relative small step size was carried out in the 21C steel after Q&P processing, which was shown in Fig. 13.

From Fig. 13a the fresh martensite could be identified as shown by the dashed line circles in the image quality map based on the size and the morphologies as observed in SEM. It can be seen from Fig. 13b that the austenite distribution is not uniform. It locates mainly in the easy-etched phase, i.e., the fist-martensite but not in the fresh martensite. This may tell us that the large austenite were not stable and will be transformed into fresh martensite completely. However the small sized austenite or austenite films, which developed in the first-martensite structure, will be stable and retained into room temperature. As predicted above that the stability is ascribed to either their small size or their high carbon concentration of austenite phase, which needs further discussion based on the size and carbon concentration measurement.

3.4.2 Austenite Volume Fraction and Carbon Concentration Examined by XRD

The austenite volume fraction and carbon concentration of 0.21C steel was measured by XRD to examine the effects of partitioning temperature and partitioning time. The calculated results were given in Fig. 14. It can be seen from

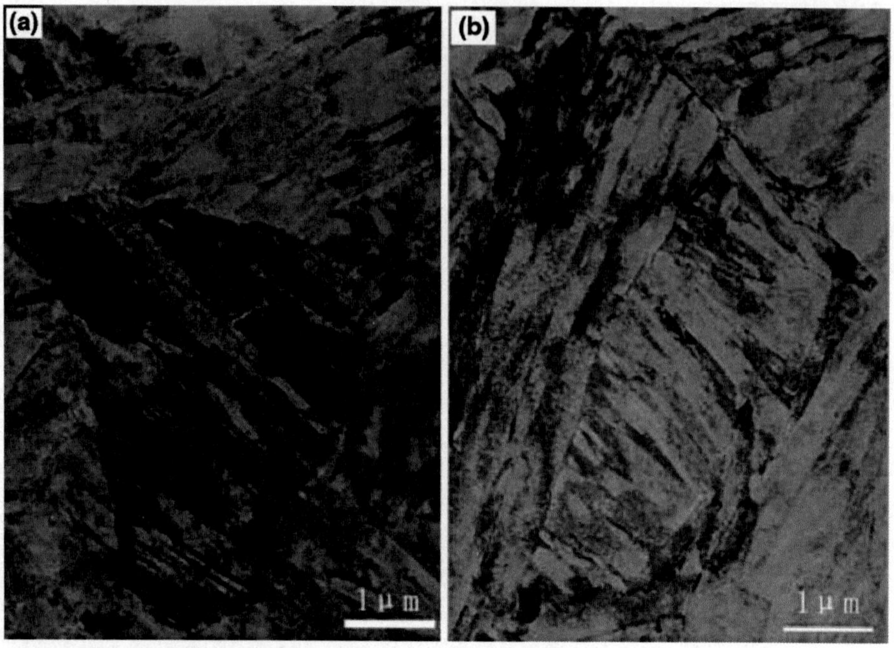

Fig. 11 The microstructure of 21C steel after Q&P process to show the morphologies of the fresh martensite **a** fresh martensite with triangle shape and **b** fresh martensite with quadrilateral shape

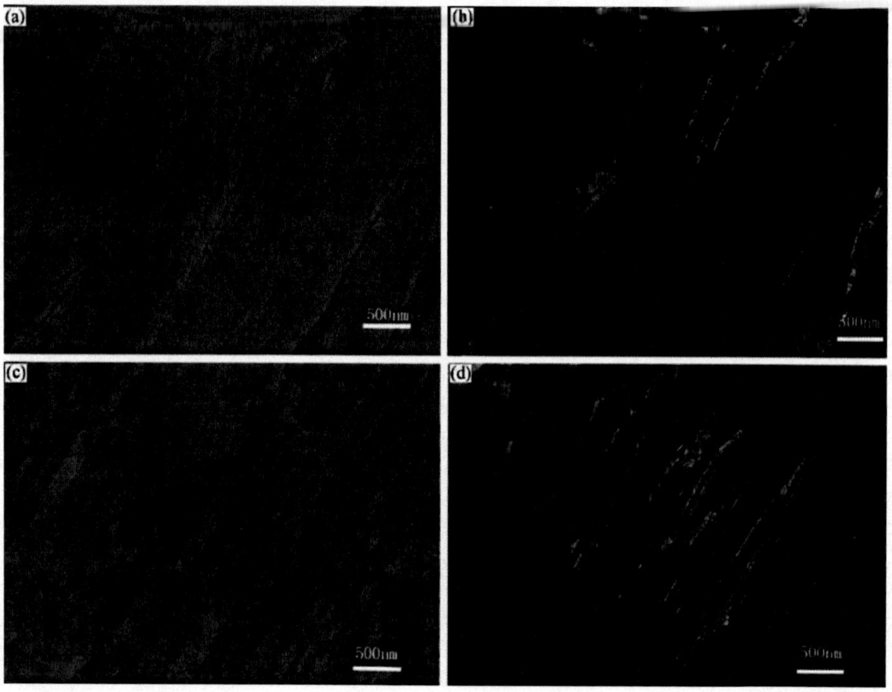

Fig. 12 Retained austenite morphologies in 21C steel processed with 5 min at different partitioning temperatures **a** bright field image of steel partitioned at 300°C, **b** dark field image of steel partitioned at 300°C, **c** bright field image of steel partitioned at 500°C and **d** dark field image of steel partitioned at 500°C

Fig. 14a increasing the partitioning time increases the austenite volume fraction slightly at 300, 400 and 500°C but increases it at 250°C. In addition the highest volume fraction as given in Fig. 14 occurs at 500°C at 1 min partitioning. In Fig. 14b the carbon concentration in austenite was given as a function of partitioning time at a given partitioning temperature. It can be seen that even though the volume fraction is high when partitioned at high temperature the carbon concentration is also very high. It was proposed that the ability of austenite phase to increase the ductility increases with the product of $f_A \times C\%$. Thus it is meaningful to plot the Q&P parameters effects on this product and in turn to predict the relative ductility as shown in Fig. 14c. From Fig. 14c it can be seen that the lowest value of $f_A \times C\%$ is at 2500°C and the highest value occurs at 500°C with annealing time about 30 s. From Fig. 14 it can be concluded

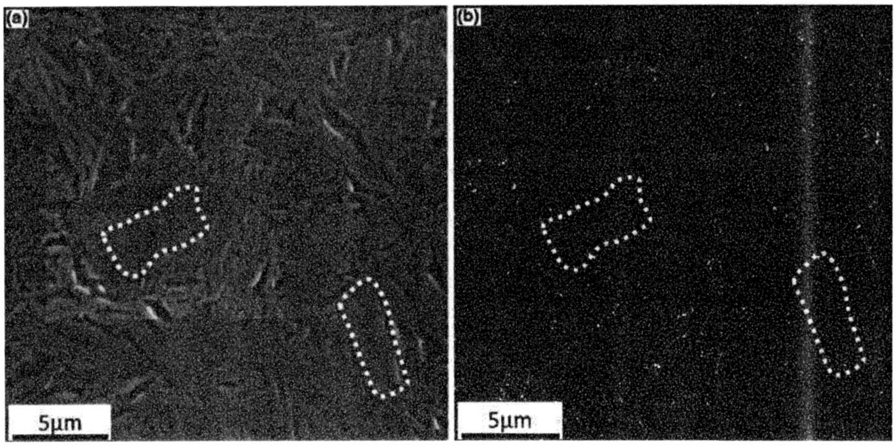

Fig. 13 EBSD mapping of 21C steel after Q&P processing **a** the image quality map, **b** austenite distribution map, in which the austenite grains were revealed in white

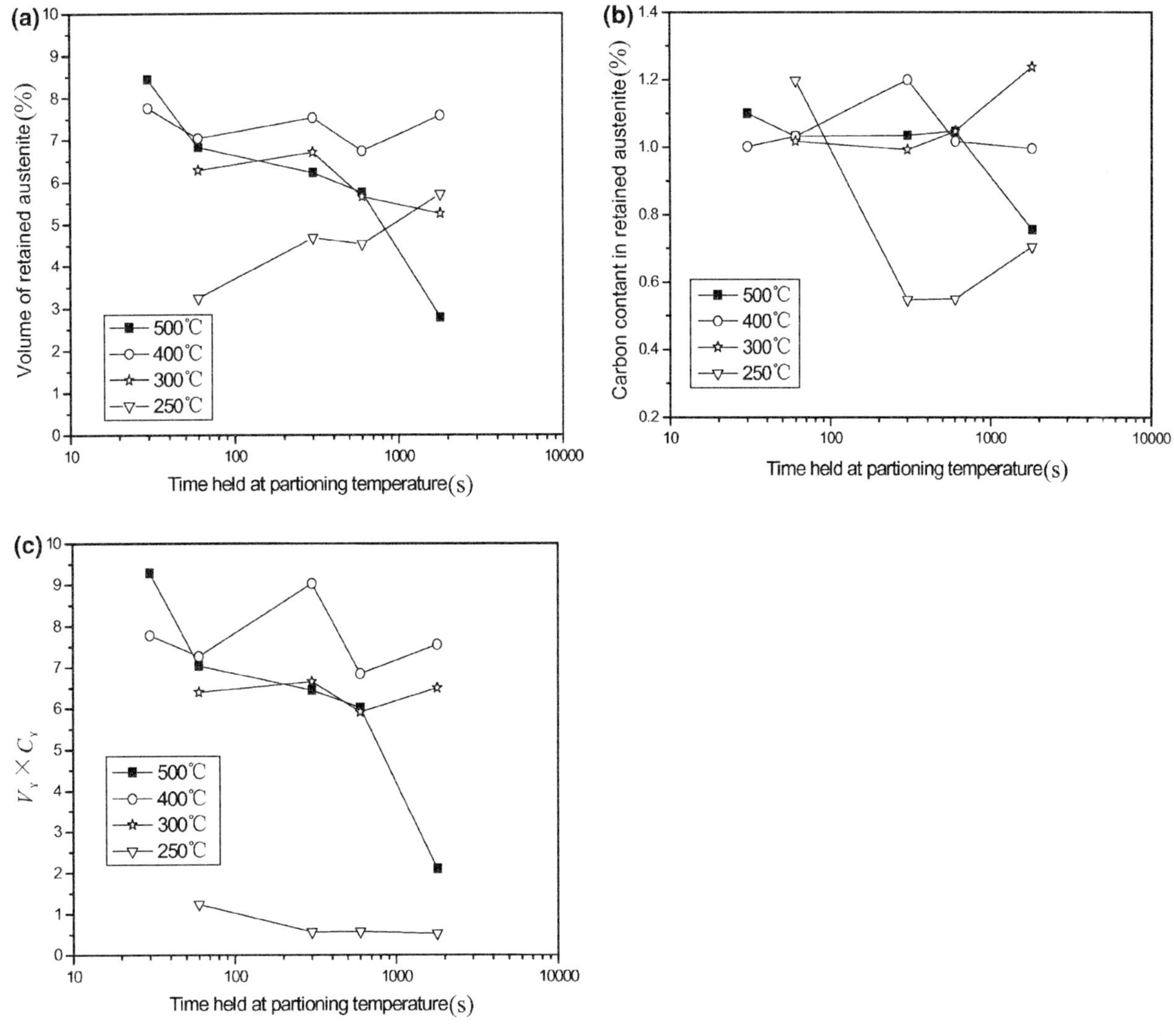

Fig. 14 Quenching and partitioning parameters effects on the austenite volume(f_A) in **a**, carbon concentration in austenite (C%) in **b** and their product (f_A × C%) in **c** for 21C steel

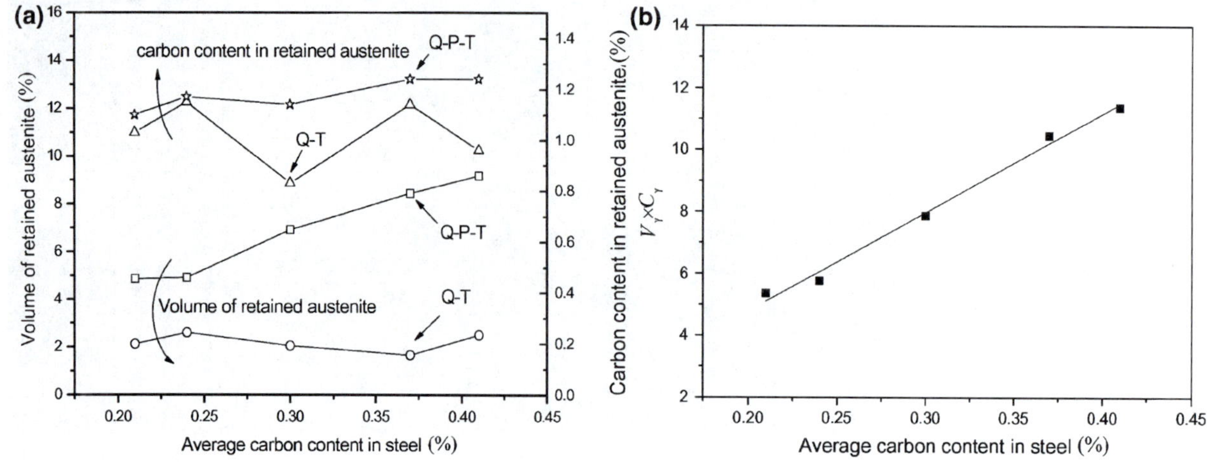

Fig. 15 Average carbon concentration effects on the austenite volume fraction (**a**) and its carbon concentration (**b**) in steels after Q&P process and QT process

that low temperature partitioning is not useful to improve the ductility and high temperature partitioning with longtime is also detrimental to the ductility. Thus further study should be focused on the high or relative high temperature partitioning. The underlined mechanism may be ascribed to the coupling behavior between carbon partitioning to austenite and the carbides precipitation. The partitioning temperature and time effects on the austenite volume fraction and its carbon concentration could be described as follows: (1) when partitioned at low temperature, the carbon partitioning rate is quite low, thus most of the austenite contain rather low carbon concentration thus it cannot be retained after partitioning. (2) However when partitioned at high temperature the carbon diffusion rate is high enough to partition into the austenite and stabilize it with very short time, during this short time the carbides precipitation cannot be carried. (3) But when partitioned at high temperature with relative long time, the carbides would precipitate from austenite and deteriorate its stability and result in relative low austenite fraction after partitioning.

The average carbon concentration effects on the austenite fraction, carbon concentration in retained austenite and $f_A \times C\%$ after Q&P and QT processing were shown in Fig. 15. It can be seen that the austenite volume fraction increases with increasing of the average carbon concentration but the carbon concentration almost does not increase with it. Compared with the QT process, the Q&P process is much more help to increase both retained austenite fraction but also its carbon concentration as shown in Fig. 15a. The $f_A \times C\%$ value increases with increasing average carbon concentration as shown in Fig. 15b, which implies that increasing average carbon concentration is useful to improve the mechanical properties in Q&P process.

As it was predicted above that high temperature partitioning with short time would result in a large fractioned austenite, which was tested in steels with different carbon content. The processing conditions and parameters were given in Table 2. The steels with different carbon contents were processed by Q&P at partitioning temperature measured by XRD were given in Table 2 as well. The retained austenite fraction after given Q&P process also increases from 12 to 27% with increasing carbon content, indicating the feasibility to obtain high fractioned austenite phase within martensitic matrix through heat treatment parameters control in Q&P process[17].

Based on the microstructure observation, phase identification and explanation on the thermal stability of austenite, it can be concluded here,

1. The microstructure of studied steels treated by Q&P process is mainly composed of three phases, *i.e.*, initial martensite, fresh martensite and retained austenite. The initial martensite assumes tempered martensite characteristics with

Table 2 Q&P heat treatment parameters

Steel	Austenization	Quenching	Partitioning	f_A (%)
21C-A	900°C × 15 min	330°C × 1 min	500°C × 1 minWQ	12.01
21C-B	900°C × 15 min	400°C × 20 min	500°C × 10 minWQ	11.98
37C-C	900°C × 15 min	250°C × 1 min	500°C × 2 minWQ	
37C-D	900°C × 15 min	350°C × 20 min	500°C × 1 minWQ	21.09
41C-E	900°C × 15 min	250°C × 1 min	500°C × 2 minWQ	27.39

larger sized martensite lath; The fresh martensite has quenched martensite characteristics with thinner thickness of martensite lath (0.1–0.2 μm) than that of initial martensite lath (0.2–0.3 μm),only one single set of packet existing in one fresh martensite grain, and higher carbon content and hardness than those of matrix.

2. The retained austenite exists in the thin film shape between martensite laths or in the blocky shape (about 0.1 μm) in the packet boundary and initial austenite grain boundary.
3. The higher retained austenite can be obtained using higher temperature quenching and high temperature and short time partitioning. In the studied range, increasing carbon content could increase the volume fraction and improve stability of retained austenite in steels processed by Q&P.

3.5 Mechanical Properties of Q&P Processed Steels

The stress/strain curves of the processed steels according to the heat treatment design as given in Table 3 were shown in Fig. 16. It can be seen from the tensile stress/strain curves that the yield behaviors of these steels could be classified into two groups, one is the high yield stress (>1,000 MPa) when long time austempering applied (21C-B and 37C-C) and another is the low yield stress (~700 MPa) in steels of 21C-A, 37C-D and 41C-E when they were austempered in a short time. The difference between these two yield behaviors may result from the different austempering behaviors at different time. As we know, during this partitioning process, stabilization of retained austenite, precipitation of second particles and the decreasing of dislocation density in the first developed martensite take place simultaneously. The different yield behaviors may be directly resulted from the different amount volume fraction. For steels of 21C-A, 37C-D and 41C-E, no or less second particles were developed in these steel due to the short time austempering or partitioning. However, for 21C-B and 37C-C, long time partitioning results in the precipitation of second particles.

In order to check the ductility contribution, i.e., the TRIP effects, the austenite volume fraction after different tensile were measured by XRD and the results was given in Fig. 16b. It can be seen that the austenite volume fraction gradually decreases with strain increasing, suggesting the TRIP effects during tensile process.

Comparing with the ductility (~10% total elongation and 2–5% uniform elongation) of the conventional martensitic steel, the steels studied in this study developed high ductility (~20% total elongation and 10–14% uniform elongation) by Q&P process as given in Table 3. Fig. 16b reveals the evolution of austenite volume fraction of 21C-A, 37C-D and 41C-E steels as a function of uniform tensile strain. It can be seen from this figure that the austenite volume fraction gradually decreases with strain increasing, implying the TRIP contribution to the improved ductility. It also can be found that increasing austenite increases the uniform elongation but decreases the non-uniform elongation (necking induced elongation). Further study on the

Table 3 Mechanical properties of studied steels shown in Fig. 16

Steel	R_m (MPa)	$R_{p0.2}$ (MPa)	A (%)	Agt (%)	f_A (%)	$R_m \times A$ (GPa%)
21C-A	1,310	768	20	10	12.01	26.2
21C-B	1,390	1,025	20.8	10.5	11.98	28.9
37C-C	1,570	1,235	20.0	10.0		31.4
37C-D	1,670	788	19.8	12.3	21.09	33.1
41C-E	1,835	740	19.0	14.0	27.39	34.9

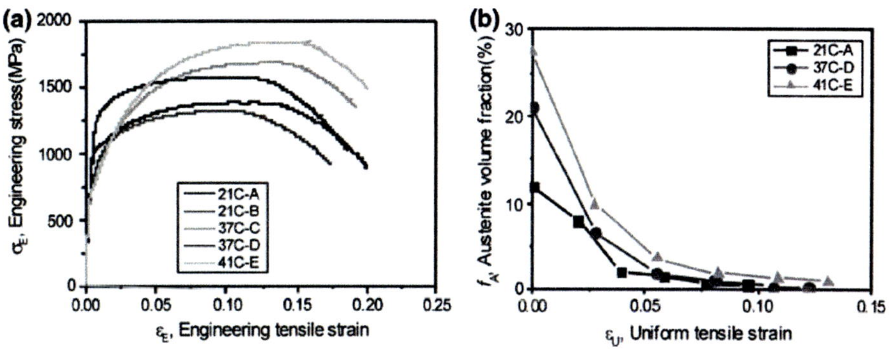

Fig. 16 Tensile curves and austenite fraction of Q&P processed steels. **a** tensile stress/strain curves and **b** evolution of austenite volume fraction as a function of tensile strain of 21C-A, 37C-D and 41C-E steels

mechanisms austenite volume contribution to uniform elongation and non-uniform elongation is necessary for the ductility-enhancing mechanism.

The stress–strain curves at different deformation temperature were given in Fig. 17. This steel was first processed at the condition of Q&P (900°C × 15 min-350°C × 10 min-500°C × 1–2 min, WQ) and then tensile deformed at different temperature. The engineering stress–strain curves were given in Fig. 11 and the mechanical properties were recorded in Table 4. It can be seen for 41C that at room temperature that the tensile stress is about 1,835 MPa and the total elongation is about 19%. However, when tensile deformation temperature increases up to 100°C, the tensile strength slightly decreases to 1,580 MPa but its total elongation significantly increases up to 39% as recorded in Table 4. This enhanced the mechanical properties with increasing deformation temperature may be directly ascribed to the deformation temperature effects on the mechanical stability of retained austenite, but needs to be studied further [17].

Based on the study on the Q&P processed carbon steels, it can be seen only the high carbon steels of 37C, 41C and 51C steel has the ability to arrive the objective to get total elongation high up to 20% elongation when these steels were processed by high temperature partitioning (500°C) with relative short partitioning time (1 min, which needs study in details further). This means high carbon steel is essential requirement to get ultrahigh tensile strength (R_m > 1,500 MPa) and high ductility (A > 20%).

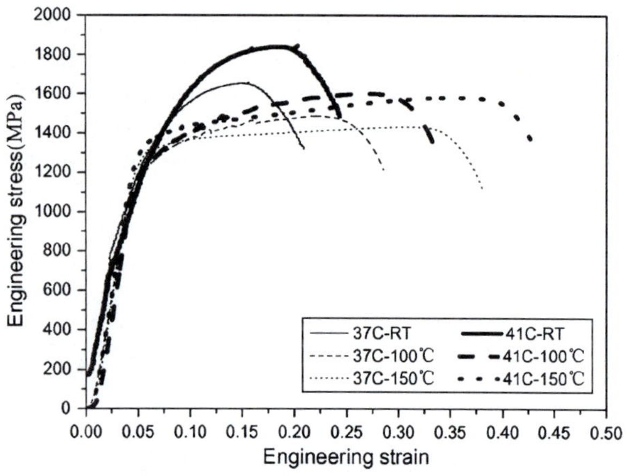

Fig. 17 Engineering tensile stress–strain curves at room temperature, 100 and 150°C of 37C and 41C steels processed by Q&P process

4 Discussion of the Work Hardening Behaviors, Strengthening and Ductility-Enhancing Mechanism of Multiphase Steels with Large Quantity Metastable Austenite Phase

4.1 Work Hardening Behaviors and TRIP Effects of Steels Processed by ART-Annealing and Q&P Treatment

In order to understand the enhanced ductility, the work hardening rates of typical medium manganese steels processed by ART-annealing and the carbon steels processed by Q&P heat treatment were calculated and shown in Fig. 18. The three stage work hardening behaviors of different ART-annealed steels were compared with the work hardening rate of IF steel and TWIP steel in Fig. 18a. Similar three stage hardening behaviors were also found in Q&P processed carbon steels as shown in Fig. 18b.

In addition, not only the three stage work hardening behaviors but also the ultrahigh work hardening rate of both ART-annealed steels and Q&P processed steels were found in Fig. 18. As it is well known, usually after a certain percent deformation, the work hardening rate is significantly lower and just above the flow stress. However, the work hardening rate of ART-annealed steel after 10% elongation is still higher than 4000 MPa, which close to the critical work hardening rate theoretically calculated from Kocks–Meching model [25]. This three stage work hardening behavior also indicates that the classical work hardening models, such as Hollomon equation [26], Swift equation [27], Voce equation [28], Ludwik equation [29] and Ludwigson equation [30] cannot be applied to simulate

Table 4 Mechanical properties of Q&P processed steels tensiled at different temperature

Steels	Temperature (°C)	Mechanical properties				
		R_m (Mpa)	$R_{p0.2}$ (Mpa)	A (%)	Z (%)	$R_m \times A$ (GPa%)
37C	RT(20)	1,670	788	19.8	45.0	33.1
37C	100	1,490	1,090	25.5	54.5	38.0
37C	150	1,600	1,110	35.0	60.5	56.0
41C	RT(20)	1,835	740	19.0	46.3	34.9
41C	100	1,430	1,160	29.0	52.0	41.5
41C	150	1,580	1,210	39.0	52.0	61.6

the flow behaviors of both ART-annealed medium manganese steels and Q&P processed carbon steels. Further study is needed to understand this abnormal work hardening behaviors and the substantially enhanced work hardening rate.

How to understand the three stage work hardening behavior and the observed ultrahigh work hardening rate as found in this study? It is well known, the Hollomon equation, $\sigma = A\varepsilon^n$, has long served as a model for characterization of work hardening behavior of many metals and alloys. However, latter people have showed that this relationship is inadequate to describe plastic-flow behavior of the other materials, such as the DP steel. Thus other different flow relationships such as Ludwik, Voce, Swift and Ludwigson were proposed [26–30]. Amongst these work hardening law, the Voce equation and Kocks–Mecking hardening law agrees very well and can be easily interpreted by dislocation hardening behavior and to simulate their stress–strain curves in single phased materials. The relation between the flow stress and the dislocation density is described by the so-called Taylor equation [25, 31],

$$\sigma = \sigma_0 + M\alpha G b \rho^{0.5} \tag{2}$$

where σ, σ_0, M, αG, b and ρ are the flow stress, frictional stress, a constant and dislocation density, respectively. Then the work hardening rate can be derived based on dislocation interaction as proposed by Kocks–Mecking,

$$\frac{d\rho}{d\varepsilon} = k_1\sqrt{\rho} - k_2\rho \tag{3}$$

$$\frac{d\rho}{d\varepsilon} = 0.5M\alpha G b \rho^{-0.5}$$

which gives the Voce hardening law

$$\frac{d\rho}{d\varepsilon} = k_1 - k_2(\sigma - \sigma_{0.2})$$

$$\frac{d\sigma}{d\varepsilon} = \Theta_0\left(1 - \frac{\sigma - \sigma_{0.2}}{\sigma_s - \sigma_{0.2}}\right), \Theta_0 \approx \frac{G}{20}$$

$$\sigma = \sigma_s - (\sigma_s - \sigma_{0.2})\exp\left(-\frac{\Theta_0}{\sigma_s - \sigma_{0.2}}\varepsilon\right) \tag{4}$$

For ART-annealed steel after long time annealing, three stages hardening phenomena were found during tensile deformation process. One interesting point is the extra hardening rate as shown in above Fig. 18, which was never found in the fully annealed steels and definitely ascribed to the different work hardening behavior of different phases. Usually if the work hardening behavior of polycrystalline was controlled by dislocation accumulation and recovery, the work hardening rate is normally lower than G/20 $\approx$ 4 GPa based on the Kocks–Mecking model [25]. However, as revealed in Fig. 18a the maximum work hardening rate for 0.2C7Mn steel is about 9 GPa, significantly higher than the theoretically derivation. This difference implies that the work hardening of ART-annealed 0.2C5Mn steel with duplex structure is not only controlled by dislocation but other mechanisms, such as the phase transformation and the phase interaction. In terms of the duplex structure of ferrite and metastable stable austenite, the work hardening is not only contributed from dislocation accumulation mechanism in austenite and ferrite as being described by Kocks–Mecking model, but the phase transformation from austenite to martensite mechanism and/or the coupling behavior among different phases.

Due to the gradually phase transformation, the volume fractions of both martensite phase and austenite phase are not constant but gradually change with deformation strain. Thus a dynamic composite model should be proposed to interpret the work hardening behaviors of the materials with multiphase structure and large fractioned austenite. As proposed by Embury and Bouaziz [32–34] (Eq. 5) and the work hardening rate could be derived to depend on both volume fraction and strength of different phases (Eq. 6).

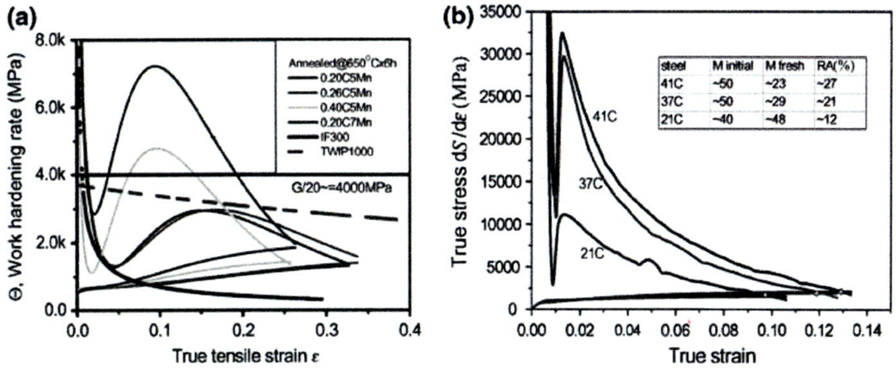

Fig. 18 Three stage work hardening behaviors of multiphase steels with large fractioned austenite phase **a** Medium manganese steels processed by ART-annealing and **b** Carbon steels processed by Q&P process at relative high partitioning temperature (500°C) with short time (1 min)

In Eqs. 5 and 6, σ_α, σ_γ and σ_M are the flow stress of ferrite, austenite and martensite, and f_α, f_γ and f_M are the volume fraction of ferrite, austenite and martensite.

$$\sigma_F = f_\alpha \sigma_\alpha + f_\gamma \sigma_\gamma + (1 - f_\alpha - f_\gamma)\sigma_M \tag{5}$$

$$\frac{d\sigma_F}{d\varepsilon} = f_\alpha \frac{d\sigma_\alpha}{d\varepsilon} + f_\gamma \frac{d\sigma_\gamma}{d\varepsilon} + f_M \frac{d\sigma_M}{d\varepsilon} + \frac{df_M}{d\varepsilon}(\sigma_M - \sigma_\gamma) \tag{6}$$

In stage-I ($0 < \varepsilon < 0.05$), f_M and σ_M is relative small, the work hardening rate is mainly contributed from the dislocation hardening in ferrite and austenite, $\frac{d\sigma_F}{d\varepsilon} \approx f_\alpha \frac{d\sigma_\alpha}{d\varepsilon} + f_\gamma \frac{d\sigma_\gamma}{d\varepsilon}$. With strain increasing, the work hardening contribution from ferrite and austenite decreases sharply after Stage-I and the hardening contribution from martensite increases with strain increasing. In stage-II ($\sim 0.05 < \varepsilon < \sim 0.15$), both f_M and $\sigma_M - \sigma_\gamma$ increase with strain increasing, thus $\frac{d\sigma_F}{d\varepsilon} \approx f_M \frac{d\sigma_M}{d\varepsilon} + \frac{df_M}{d\varepsilon}(\sigma_M - \sigma_\gamma)$. In stage-III, increasing strain would decrease $\frac{d\sigma_M}{d\varepsilon}$ and $\frac{df_M}{d\varepsilon}$, thus the work hardening rate would goes down. Above explanation reasonably explained the three stage work hardening behavior and the origin of the maximum work hardening rate as found in ART-annealed medium manganese steels and the Q&P processed steels as given in Fig. 18. Further study is needed to examine the flow behaviors of different phase aiming at reproducing the deformation stress–strain curves theoretically by fundamental study. Comparing with conventional TRIP steels, the studied steels presented a ultrahigh work hardening rate after a certain deformation as shown in Fig. 18, this significantly enhanced work hardening rate delays the necking during tensile process, which origins from the enhanced TRIP effects. Also due to the very high maximum work hardening rate (9 GPa) as shown in Fig. 18, the extra-ultra-high strength steel with relative uniform strain could be fabricated by multiphase control.

5 Strengthening and Ductility-Enhancing Mechanisms of Steels Processed by ART-Annealing and Q&P Treatment

In order to understand the mechanism of the enhanced mechanical properties, the uniform tensile elongation and the total elongation were shown as a function of austenite volume fraction in Fig. 19a for both ART-annealed medium-Mn steels and Q&P processed carbon steels. It can be seen that for ART-annealed medium-Mn steels both uniform and total tensile elongations increase with austenite volume fraction. However, for Q&P processed carbon steels, both uniform elongation and total elongation only slightly increases with increasing austenite fraction. Furthermore, at same austenite fraction level, both uniform elongation and total elongation of ART-annealed steels were significantly higher than that of Q&P process. This difference may be interpreted by the flow stress difference and the stability difference of austenite. It was observed that the tensile strength level Q&P processed steel is about 1.5–1.8 GPa, which is significantly higher than that of ART-annealed steel. Also the mechanical stability of Q&P processed steels is significantly lower than that of ART-annealed steel (see Figs. 5 and 6 and Fig. 16). However when the product of tensile strength to total elongation was shown as a function of the austenite fraction in Fig. 19b, it can be seen that the product almost linearly increases with austenite volume fraction no matter ART-annealed steels or Q&P processed steels. Both steels assume a similar increasing trend of the product increasing with austenite fraction. For comparison, the $R_m \times A$ values of typical 1st generation automobile steel [2], 2nd generation automobile steel [2], nano-bainite steel [13], 15Mn-3Si-3Al and 20Mn-3Si-3Al [12] were given in Fig. 19b. Thus it can be seen that the large austenite fraction is the main reason to the excellent mechanical properties as demonstrated in our study.

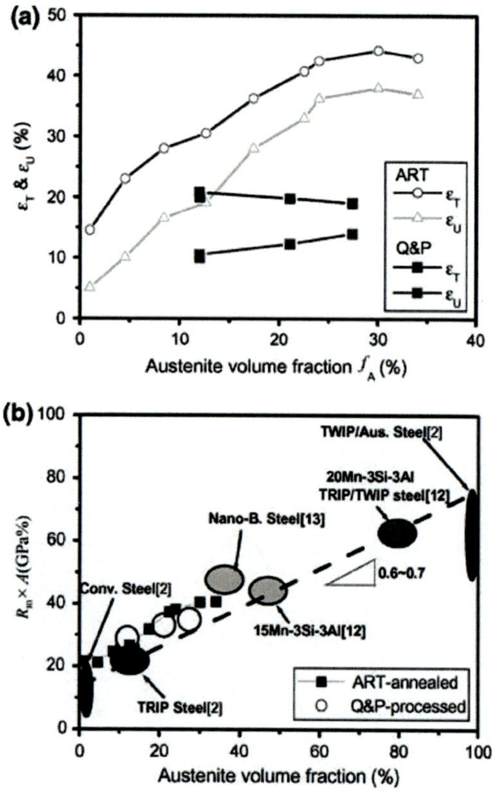

Fig. 19 Mechanical properties dependence on austenite volume fraction **a** relationship between elongation (ε_T and ε_U) and austenite fraction and **b** dependence of the product of $R_m \times A$ on austenite volume fraction

The high value of $R_m \times A$, 40 GPa% of the present studied steels may be explained by two main reasons. One is the enhanced TRIP effect as discussed above and the other may be the matrix characters, such as its ultrafine grained structure and its low stored energy as well. In order to understand the contribution of TRIP effect, the relationship between product of $R_m \times A$ of different kinds of steel, and the austenite fraction is re-analyzed in Fig. 19b. It can be seen, no matter TRIP effect or TWIP effect, generally the slope between $R_m \times A$ and the austenite volume fraction is ~0.6–0.7 GPa%/(1%γ), indicating the strong dependence of $R_m \times A$ on austenite volume fraction. Thus the contribution of austenite phase to $R_m \times A$ in Fe-0.2C-4.7Mn steel could be calculated to be ~18–21 GPa% but in 41C steel should be about 14.4–18.9 GPa%, almost half of the measured $R_m \times A$. In addition, the ultralow stored energy of the matrix due to the high temperature annealing may be another reason to the high value of $R_m \times A$. During annealing process at 650°C (higher than Ac1 of 630°C) most of the dislocations were depleted from the materials and thus providing a reservoir for high density dislocations produced by strain induced phase transformation and tensile deformation itself during further deformation process. As measured by XRD, the dislocation density before and after 37% uniform tension is about $\sim 3.6 \times 10^{13}\ \mathrm{m}^{-2}$ and $\sim 2.5 \times 10^{15}\ \mathrm{m}^{-2}$, respectively, which gives dislocation density increment about $2.46 \times 10^{15}\ \mathrm{m}^{-2}$, and a stored energy increment (ΔE_S) about 0.63 MJm^{-3} according to Eq. 8 [31]. The expended energy (E_{EXP}) in the matrix during tensile process could calculated to be ~22 GPa% according to Eq. 7 [31] with energy storage ratio $\chi = 0.02$. Thus $R_m \times A$ is roughly the sum of 18–21 GPa% from austenite phase and ~22 GPa% from the matrix, i.e., 39–43 GPa% for ART-annealed 0.2C-4.7Mn steel, which agrees very well with our experimental results. No measurement of dislocation density in 41C steel was carried out by XRD in this study, which needs further examination.

$$\Delta E_S = \alpha_1 \beta G b^2 \Delta\rho \quad (7)$$

$$E_{\mathrm{EXP}} = \Delta E_S / \chi \quad (8)$$

where α_1 is a constant about 0.5, β is the volume fraction of matrix, $\Delta\rho$ is the dislocation increment before and after tensile deformation, and χ is the energy storage ratio between the stored energy and the expended energy in the matrix.

Based on the above results and explanation, it can be seen that the ductility is not linearly proportional to the austenite because TRIP behavior was affected by many factors, such as the austenite fraction, austenite mechanical stability and the matrix strength. However, a strong dependence of $R_m \times A$ on austenite fraction was found, which could be related to the phase transformation from austenite to ferrite.

6 Simulation of the True Tensile Stress–Strain Behavior Under Some Assumption

The different parameters that influence the mechanical stability of the dispersed retained austenite and therefore the ductility of TRIP steels are numerous and often coupled, mainly: chemical composition, grain size and stress state of the surrounding matrix [35]. Considering the effects of chemical driving force, these parameters should include the deformation temperature. The optimization of such complex multiphase steels requires a detailed understanding and modeling on the mechanisms of phase transformation during mechanical testing. A common equation used to interpret the evolution of martensite as a function of plastic strain in TRIP assisted steels is described in Eq. 1, which is applied to analyze the relationship between austenite fraction and strain amplitude in Fig. 20. It can be seen that our experimental results can be fitted by Eq. 1 very well. The mechanical stability of austenite phase is not significantly affected by annealing time but strong affected by deformation temperature.

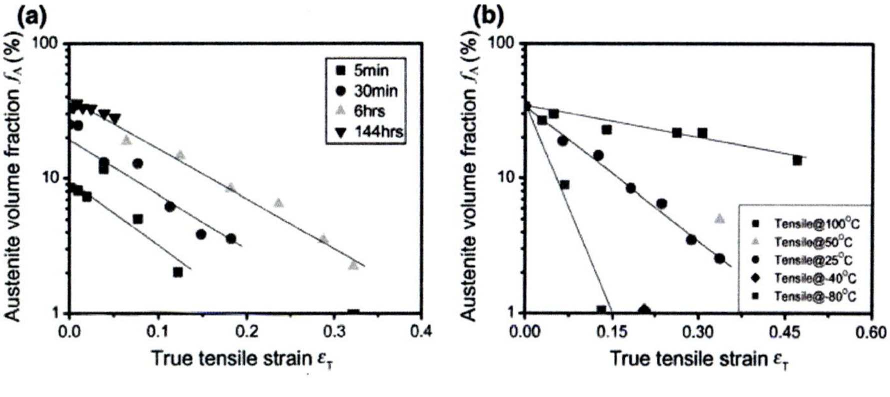

Fig. 20 The austenite fraction evolution as a function of true deformation strain **a** effects of annealing time on austenite fraction evolution and **b** deformation temperature effects on the austenite fraction evolution

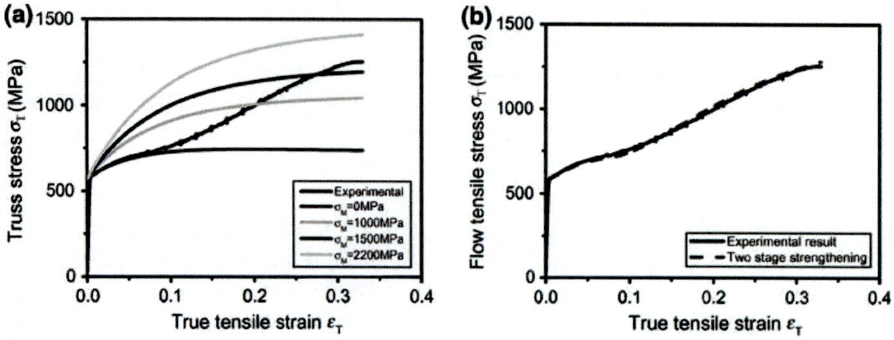

Fig. 21 The experimental and simulated results of the tensile flow curves of 0.2C5Mn steel after ART-annealing with 6 h. **a** Simulated stress–strain curves under the assumption of the constant martensite strength and **b** Simulated flow stress–strain curves under the assumption that the first stage deformation is controlled by ferrite and austenite deformation with no strength contribution from martensite before strain 0.08 and after this strain the stress stemming from ferrite, austenite and martensite but the stress increases with martensite fraction from Zero to saturated stress

Based on the three stage work hardening behavior as found in our study and the strain induced austenite phase transformation, a dynamic composite hardening model should be applied. According to the dynamic composite mixed hardening law as described in Eq. 5, the flow stress could be simulated under the reasonable assumption that the flow stress of both austenite and ferrite follow the Voce equation (Eq. 9), which could be calculated from flow stress–strain curves and the flow stress does not change during deformation process, i.e. [29],

$$\sigma_{\text{Ferrite}} = \sigma_s - (\sigma_s - \sigma_{0.2})\exp(-n_V\varepsilon) \quad (9)$$

Here the martensite flow stress is assumed to be constant, $\sigma_M = 0$, 1,000, 1,500, or 2,200 MPa.

Under this assumption and the austenite decay function of $f_{\gamma\text{-}\varepsilon} = 0.33\exp(-8\varepsilon)$ of 0.2C5Mn steel ART-annealed after 6 h, the flow stress could be simulated, which was plotted in Fig. 21a. It can be seen that based on Eq. 5 that the three stage hardening cannot be obtained. From this calculation, it also may be noted that the martensitic contribution to the flow stress is delayed to a certain strain, i.e., the transformation induced martensite at strain ε can play the stress contribution only after a critical strain $\varepsilon + \varepsilon_C$. Before this critical strain (ε_C) no flow stress contribution can be obtained, which may be the exact meaning of the delayed necking role of TRIP effects. This analysis of the critical strain agrees very well with the analysis of the three stage work hardening behavior. This retard hardening phenomena of martensite may be ascribed to its stress state transition from the compression state to tensile state due to the volume expansion from austenite to martensite during deformation process, $\frac{\Delta V_{\gamma-\alpha}}{V_\gamma} = 0.017 + 0.013C\%$

In this study, the retard hardening phenomena of martensite phase was applied to reproduce the stress–strain curves of 0.2C5Mn steel. First we assume reasonably that both ferrite and austenite follows the Voce equation during whole deformation process (Eq. 10),

$$\sigma_{F+A} = [\sigma_s - (\sigma_s - \sigma_{0.2})\exp(-n_V\varepsilon)](f_{\text{Ferrite}} + f_{\text{Austenite}}) \quad (10)$$

After a critical strain, $\varepsilon_C = 0.08$, the martensite starts its stress contribution (Eq. 11),

$$\sigma_{M,\varepsilon} = 0, \text{ when } \varepsilon < 0.08$$

$$\sigma_{M,\varepsilon} = kf^2_{\text{Maretnsite@}\varepsilon-0.08}, \text{ when } \varepsilon > 0.08 \quad (11)$$

Then the flow stress of the multiphase can be obtained by a linear addition as described in Eq. 5. The results were plotted in Fig. 21b, which agrees very well with experimental results. By means of calculated results in Fig. 21b, it is found that the flow stress of martensite is not constant, which varied with martensite fraction as a function of f_M^2. In Fig. 21b the flow stress were simulated in two range, first is the deformation of austenite and ferrite and then after 0.08 strain, the deformation is controlled by three phases, such as ferrite, austenite and martensite. It should be pointed out that that this kind of understanding needs to be tested both experimentally and theoretically further, which may help us to understand the exact role of phase transformation during deformation process.

7 Conclusion

In this section, the back grounds of 3rd generation automobile steel was briefly reviewed, the metallurgical and heat treat methodologies, such as ART-annealing and Q&P processing were applied to develop high strength and high ductility steels. The microstructure and mechanical properties of some studied steels were demonstrated. The main conclusion could be drawn as follows:

1. The limitation of energy resources, the stringent control on the environment pollution, the increase requirements safety standard, the easy formidability and low cost of car components had driven the world widely research and development on the high performance automobile steels offering both ultrahigh strength and high ductility.
2. The ultrafine grain sized austenite-ferrite duplex microstructure and the tempered martensite-fresh martensite-austenite multiphase microstructure were obtained in ART-annealed medium manganese steels and Q&P processed conventional carbon steels. It was proved that both ART-annealing and Q&P processes can be applied to fabricate the third generation automobile sheet steels offering ultrahigh strength and high ductility
3. In both heat treatment conditions, substantially enhanced ductility (30–40%) at ultrahigh tensile strength level (1–1.5 GPa) was obtained, which results in a significant improvement of the product of tensile strength to total elongation about 30–40 GPa%.
4. Analysis on the work hardening behaviors and the relationship between microstructures and mechanical properties of the studied steels indicates that the greatly improved ductility results from the aid of the phase transformation induced plasticity (TRIP effects) and the ultrahigh strength stems from the hard matrix, such as the ultrafine grained duplex structure in ART-annealed steels and the martensite matrix in Q&P processed carbon steels.
5. Both ART-annealing and Q&P processing were demonstrated to be capable of producing ultrafine austenite-ferrite duplex structure in medium manganese steel, which presenting ultrahigh strength and high ductility, which meet the objective of the governmental program of 3rd phase 973. But further study is essential to understand quantitatively the stress–strain curves and the corresponding work hardening behaviors of ART-annealed steels.

Acknowledgments This research is supported by National Basic Research Program of China (973 program) No. 2010CB630803 and National High-tech R&D Programs (863 programs) No. 2009AA03Z519 and No. 2009AA033401. Great thanks to Ms. Hui ZHAO and Nan LI for their efforts in the research and development of the 3rd generation automobile steel for their Bachelor course study.

References

1. K. Sugimoto, T. Iida, J. Sakaguchi, T. Kashima, ISIJ **9**, 902 (2000)
2. R. Heimbuch, Overview: Auto/Steel partnership. www.a-sp.org
3. R.H. Wagoner, Report: advanced high-strength steel workshop, Arlington, VA, USA, 22–23 October 2006
4. D.K. Matlock, J.G. Speer, in *Proceedings of the Third International Conference on Advanced Structural Steels*, 2006
5. O. Kwon, S.-K. Kim, G. Kim, in *Proceedings of the 21st Conference on Mechanical Behaviors of Materials*, 2007
6. H.N. Han, C.S. Oh, G. Kim, O. Kwon, Mater. Sci. Eng. A **499**, 462 (2009)
7. J. Pascal, Ph.D. Thesis, Louvain-la-Neuve University, Belgium, 1998
8. R. Heimbuch, Overview autosteel partnership. http://www1.eere.energy.gov/vehiclesandfuels/pdfs/merit_review_2008/lightweight_materials/merit08_heimbuch_3.pdf
9. Y.Q. Weng, H. Dong, Y. Gan, Conference of Dalian, 2009-9-10
10. K. Sugimoto, B. Yu, Y. Mukai, S. Ikeda, ISIJ **9**, 1194 (2005)
11. P. Jacques, Q. Furnemont, A. Mertens, F. Delannayz, Phil. Mag. **A81**, 1789 (2001)
12. G. Frommeyer, U. Brux, P. Neumann, ISIJ **43**, 438 (2003)
13. C. Garcia-Mateo, F.G. Caballero, ISIJ **45**, 1736 (2005)
14. J. Shi, X.J. Sun, M.Q. Wang, W.J. Hui, H. Dong, W.Q. Cao, Scr. Mater. **63**, 815 (2010)
15. J. Shi, W.Q. Cao, H. Dong, Mater. Sci. Forum **654–656**, 238 (2010)
16. W.C. Yu, Ph.D. Thesis, Central Iron & Steel Research Institute, 2010
17. W.Q. Cao, J. Shi, H. Dong, Mater. Sci. Forum **654–656**, 29 (2010)
18. R.L. Miller, Met. Trans. **A3**, 905 (1972)
19. M. Niikura, J.W. Morris Jr., Met. Trans. **A11**, 1531 (1980)
20. J.G. Speer, R.E. Hackenberg, B.C. DeCooman et al., Phil. Mag. Lett. **87**, 379 (2007)
21. J.G. Speer, D.K. Matlock, B.C. DeCooman et al., Acta. Mater. **51**, 2611 (2003)
22. K. Sugimoto, M. Kobayashi, S. Hashimoto, Metall. Mat. Trans. **23A**, 3085 (1992)
23. M.Y. Sherif, C. Garcia Mateo, T. Sourmail, H.K.D.H. Bhadeshia, Mater. Sci. Technol. **20**, 319 (2004)
24. C.Y. Wang, J. Shie, W.Q. Cao, H. Dong, Mater. Sci. Eng. A **527**, 3442 (2010)
25. U.F. Kocks, H. Mecking, Mater. Sci. **48**, 171 (2003)
26. J.H. Hollomon, Trans. AIME **162**, 268 (1945)
27. P. Ludwik, *Elemente der Technologischen Mechanik* (Springer, Berlin, 1909)
28. H.W. Swift, J. Mech. Phys. Solid **1**, 1 (1952)
29. E. Voce, J. Inst. Met. **74**, 537 (1948)
30. D.C. Ludwigson, Metall. Trans. **2**, 2825 (1971)
31. F.J. Humphreys, M. Hatherly, *Recrystallization and Related Annealing Phenomena*, 2nd edn. (Pergamon, 2004)
32. D. Embury, O. Bouaziz, Annu. Rev. Mater. Res. **40**, 243–270 (2010)
33. R.G. Davies, Metall. Trans. **9A**, 41 (1978)
34. R.G. Davies, Metall. Trans. **9A**, 671 (1978)
35. G.N. Haidemenopoulos, Ph.D. Thesis, MIT, 1988

Challenges Toward the Further Strengthening of Sheet Steel

K. Ushioda, J. Takahashi, S. Takebayashi, D. Maeda, K. Hayashi, and Y.R. Abe

Abstract

Sheet steel is a fundamental material used for various applications, such as automobiles, electrical appliances, building construction, and so forth. As such, the further strengthening of sheet steel is considered very important, as it has strong potential to help solve current global environmental issues through weight reduction and improvements in the efficiency of final products. However, at present, the capability of such steel has not been fully exploited. For further strengthening, the exploitation of material science together with application technologies is essential to overcome the abundance of inherent problems. In this paper, first of all, the recent progress concerning sheet steel used in automobiles is reviewed, taking into account history and technological problems, such as formability, fatigue properties, and hydrogen embrittlement. Since deformation and fracture behaviors are common subjects, focus is placed not only on strength, but also on the ductility (i.e., yielding, work hardening, and necking) of single- and complex-phase steels, on the fatigue properties of both base and welded steels, and on hydrogen embrittlement. Recent progress regarding the science involved and future scientific problems related to dislocation behavior in sheet steel are also discussed, together with prospects for the future.

Keywords

High strength steel • Sheet steel • Ductility • Fracture • Fatigue • Hydrogen embrittlement

1 Introduction

Steel products have greatly contributed to society as base materials for ships, architectural structures, line pipes, bridges, automobiles, and so forth. Technologies that further enhance the strength of steel will make extremely important contributions to the preservation of the global environment by enabling weight reduction of finished products that provide both improved efficiency and safety. While various types of steel sheet for automotive use, for instance DP (dual-phase) steel and TRIP (transformation-induced plasticity) steel, have already been developed in response to a social needs [1], the use of high-strength steel sheets has recently accelerated due to restrictions on CO_2 emissions. Various steel sheets with wide-ranging strength from 270 MPa to 1,470 MPa are now used in practical applications. While high-strength steel sheet accounted for only 30% of the steel sheet used in the 1990s, its share has now increased to 60%.

The strengthening of steels used in various applications has shown steady progress; however, the potential capabilities of steels have not been fully exploited [2]. As Maki pointed out [3], steels are very attractive materials having

K. Ushioda (✉) · J. Takahashi · S. Takebayashi · D. Maeda
Technical Development Bureau, Nippon Steel Corp., 20-1 Shintomi, Futtsu, Chiba, 293-8511, Japan
e-mail: ushioda.kohsaku@nsc.co.jp

K. Hayashi
Nagoya R&D, Nippon Steel Corp., 5-3 Tokai, Tokai, Aichi, 476-8686, Japan

Y.R. Abe
School of Mechanical Engineering, College of Science and Engineering, Kanazawa University, Kakuma, Kanazawa, 920-1192, Japan

Y. Weng et al. (eds.), *Advanced Steels*, DOI: 10.1007/978-3-642-17665-4_23,

many possibilities yet to be exploited. Meanwhile, further strengthening of steels is quite a challenge, with many technical problems to be overcome. Improvements in formability, fatigue properties, and hydrogen embrittlement are common challenges for sheet steels for automotive structure use [1].

This paper reviews recent progress in the steel science of sheet steels related to dislocation behavior, and discusses essential technical problems as well as future perspectives, focusing on ductility in terms of the stress–strain curve of single- and complex-phase steels, fatigue properties of both base and welded steel, and hydrogen embrittlement of high-strength steels.

2 Ductility (Stress–Strain Curve)

2.1 Yield Strength

The yield strength of steel is one of the most important properties that govern the formability and quality of parts.

Yield strength can be described by Eq. 1.

$$\sigma_y = \sigma_0 + k \cdot d^{-1/2} \qquad (1)$$

Here, σ_0 consists of the Peierls force, solid solution strengthening, precipitation/dispersion strengthening, and dislocation strengthening. The second term on the right-hand side expresses grain refinement strengthening. When different strengthening mechanisms work simultaneously, the additivity principle is not necessarily applicable, and considerations need to be made according to the strengthening mechanisms involved. For each strengthening mechanism, there still remain many unknown details. This section focuses on precipitation and grain refinement strengthening.

2.2 Precipitation Strengthening

Precipitation strengthening provides a greater hardening effect than other mechanisms, and so application of this mechanism, for instance exploitation of nano particles, is important [4]. As an example of precipitation strengthening, the study by Takahashi et al. [5] on Cu precipitates in steel is discussed. The three-dimensional atom probe (3DAP) technique was used to enable accurate measurement of hardening and rigorous discussion of strengthening mechanisms. The test sample, Fe-1.4%Cu alloy, was water quenched from 1,200°C and then re-heated to 500°C for isothermal aging treatment. Figure 1 shows 3DAP images with changing aging time. While nm-order Cu precipitates are identified, Cu in solid solution also co-exists. Therefore, for a rigorous calculation of precipitation strengthening, it is necessary to subtract the contribution of solid solution strengthening, which was evaluated to be 60·(Cu in solid solution (mass%)/1.4)$^{0.5}$ (MPa) determined by Takahashi. Figure 2 shows the aging-time dependency of this rigorously determined precipitation strengthening. Precipitation strengthening is maximized when several nm-size bcc-Cu precipitates are uniformly dispersed. The relationship between the shear stress (τ_C) and the break angle (ϕ_C) is expressed by Eq. 2 [6].

$$\tau_C = Gb/L[\cos(\phi_C/2)]^{3/2} \qquad (2)$$

Here, ϕ_C can be determined using an experimentally obtained inter-particle spacing (L) by 3DAP and τ_C by a tensile test. In addition, the maximum resistance force due to one particle, F_m, can be determined by the following Eq. 3.

$$F_m = 2T\cos(\phi_C/2) \qquad (3)$$

Here, line tension (T) is equal to $Gb^2/2$. The dependency of ϕ_C and F_m on particle size is shown in Fig. 3. Even though ϕ_C decreases with particle size, it has a large value greater than 100°, which indicates that the Cu particle has a weak pinning force and can be cut by dislocation, presumably because the precipitated Cu is small enough and softer than the matrix. As the size increases, the resistance force per particle also monotonically increases. This indicates that the precipitation strengthening due to Cu is mainly derived from the difference in elastic modulus [7], because if it was derived from the coherent strain around the

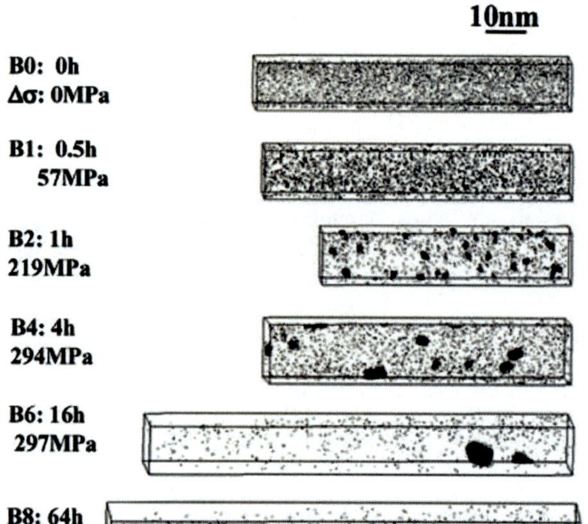

Fig. 1 3DAP images of precipitated Cu after aging for various times at 500°C [5]

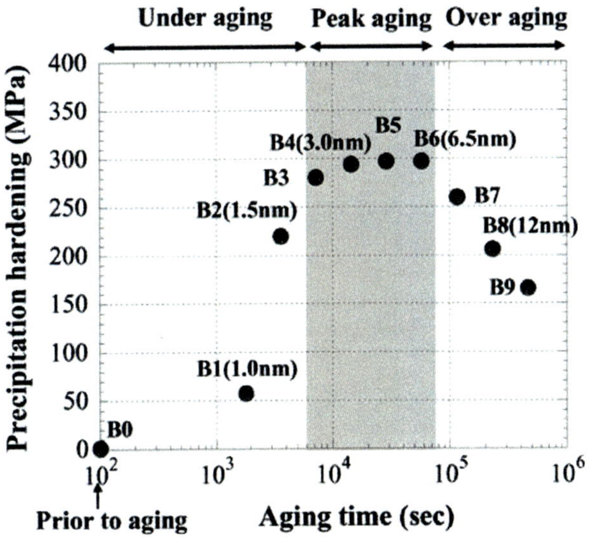

Fig. 2 Change in precipitation hardening with aging time at 500°C [5]

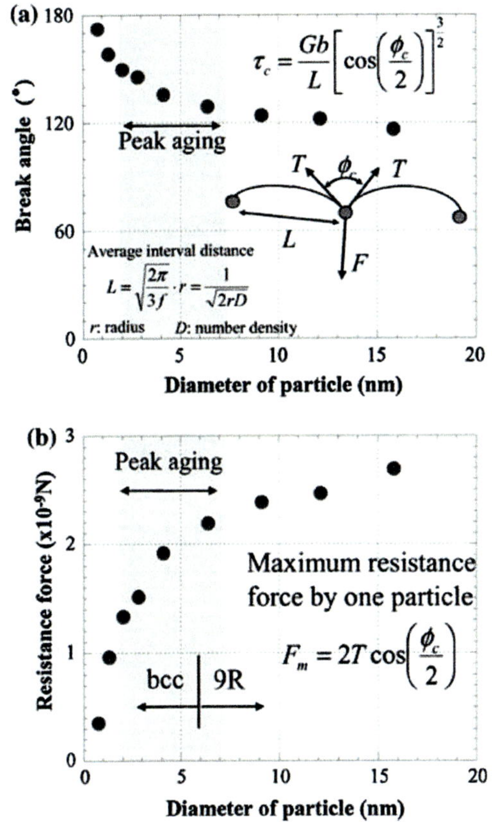

Fig. 3 Changes in **a** break angle (ϕ_C) and **b** resistance force due to one particle (F_m) with diameter of precipitates [5]

particles [8], an abrupt decrease in resistance force would be expected at several nm, where the coherent bcc-Cu changes to a semi-coherent 9R structure [9]. Moreover, if it were a result of a chemical effect, the dependency on particle size would be expected to be small [10].

For further exploitation of the precipitation strengthening mechanism, systematic studies are needed in the future on the relationship between the type of precipitates, ϕ_C, and resistance force per particle. Discussions have taken place recently on precipitation strengthening comparing different precipitates from the aspects of not only size but also rigidity and deformability of particles: Cu, VC [11], and TiC [12].

2.3 Grain Refinement Strengthening

According to recent studies, the Hall–Petch relationship holds true down to steel with a grain size of 0.1 μm as long as high-angle grain boundaries are concerned. However, coefficient k has yet to be investigated in detail.

Recently, Takeda et al. [13] reported that k increases remarkably as the total amount of C unfixed by metallic carbide increases. Here specimens were not subjected to temper-rolling. The value of k is about 100 MPa $\mu m^{1/2}$ in the case of interstitial-free steel, whereas it increases approximately sixfold when the total C content reaches 100 mass ppm. This was considered to stem from C segregated at the grain boundary during annealing and cooling, which contributes to strengthening the grain boundary.

Similarly, according to Spizig [14], when 0.05–0.1% of P is added to the base level of 0.01%P steel, the value of k increases about twofold, from 400 MPa $\mu m^{1/2}$ to 800 MPa $\mu m^{1/2}$ (the maximum value), whereas it decreases to 100 MPa $\mu m^{1/2}$ with further additions. It is worthwhile noting that the steels used contained approximately unfixed 50 mass ppm C. On the other hand, Sakuma et al. [15] and Fujiwara et al. [16] reported that using Ti-IF steel as a base material, the value of k is approximately 100 MPa $\mu m^{1/2}$ and P neither increases the value of k nor shows such a behavior after segregation treatment. Li [17] made a theoretical study of k considering grain boundary ledges as sources of dislocation. Taking into consideration these facts, it may be postulated that the function of grain boundary ledges as sources of dislocation may be suppressed by the segregation of interstitial atoms such as C, but not by substitutional atoms such as P [18]. However, the role of the ledge when impurities are present has not been sufficiently verified, and this is future work.

The yield phenomenon is commonly explained using a model of piled-up dislocations. However, only a few cases of verification have been reported. Ohmura et al. [19] studied the interactions between dislocation and grain boundary through in situ nano-indentation in a TEM using tempered martensitic steel of Fe-0.4%C. In the study performed, dislocations produced by the nano-indentation piled up against a low-angle (lath) boundary and were emitted from the boundary, whereas dislocations were absorbed into a high-angle (block) boundary. A grain boundary retards the movement of dislocations, as well as working as a place where dislocations are emitted or absorbed.

Nevertheless, it is necessary in the future first to accumulate systematic data on the role of grain boundaries with respect to dislocations and then to theorize about their role considering their characteristics, which is very important in exploring grain boundary engineering.

3 Work Hardening

Ductility is generally reduced when the strength of steel increases. However, steels having the same strength level exhibit widely varying elongation depending on the strengthening mechanism as shown in Fig. 4 [2]. Among various strengthening mechanisms, complex-phase steels such as DP steel and TRIP steel have a good strength-elongation balance. Furthermore, Fig. 4 indicates that ultrafine-grain steel (UFGS) with a grain size of 1.4 μm has a similar strength-ductility balance as DP steel after Tsuchida [20]. Since it is well known that UFGS shows only poor uniform elongation, this indicates that UFGS exhibits superior post-uniform elongation instead.

Factors affecting work hardening, in other words, evolution of the dislocation structure during deformation, are (1) solute atoms related to the cross-slip frequency of dislocation, (2) particle or second-phase dispersion work hardening proposed by Ashby [21] and Tanaka and Mori [22], (3) pre-existing dislocations that interact with the following dislocations resulting in characteristic mechanical behavior like the strain path effect [23], (4) grain boundaries, which have complex interaction with dislocations depending on grain boundary characteristics, (5) the TRIP effect, and (6) the TWIP effect.

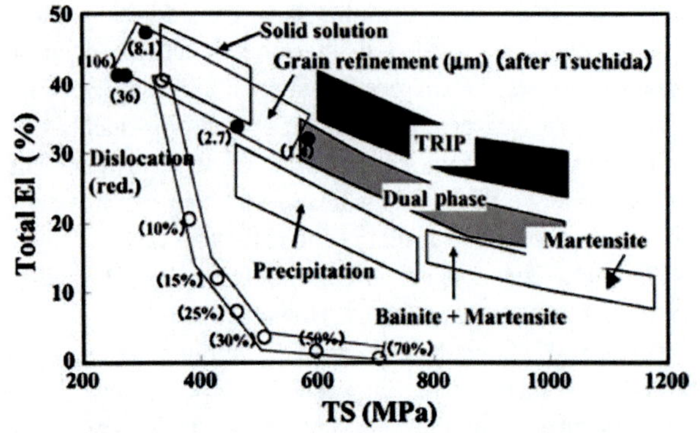

Fig. 4 Strength-ductility balance for various high strength steels with different strengthening mechanisms [2]

This section first introduces recent studies on the work-hardening behavior of single-phase ferritic steel and single-phase martensitic steel followed by complex-phase steel.

Figure 5 shows the influence of Si addition to Ti-IF steel on stress–strain curves. As already reported [24, 25], the work-hardening rate increases when Si is added. Moreover, the TEM micrographs in Fig. 6 indicate that the dislocation structure after stretching by 15%, which significantly depends on the crystal orientation of grains, becomes somewhat planar in grains with ND//<111> and fine, sharp dislocation walls are formed with Si addition, whereas in the specimen without Si addition, wide, diffused dislocation walls were observed with the indication of increased cross-slip frequency of dislocations. This is probably because Si in solid solution suppresses cross-slip due to lowering of the stacking-fault energy resulting in an increased work-hardening rate. However, no decisive, quantitative verification data on stacking-fault energy are available, and it is ultimately necessary to clarify the dislocation core structure with the presence of solid solute elements such as Si using cutting-edge observation techniques and first-principles calculation.

Medium- and high-carbon steels with a martensitic structure usually exhibit an elongation of only several percent, and hence their work-hardening behavior is not well known. Recently, Nambu et al. [26] prepared multilayered samples having a martensitic steel layer with 0.13%C sandwiched by the upper and lower SS304 layers, which exhibited an elongation of 60% in contrast to only 5% elongation for a single-layer monolithic martensitic structure. When a lath martensitic structure is stretched, the dislocation density determined by X-ray diffraction

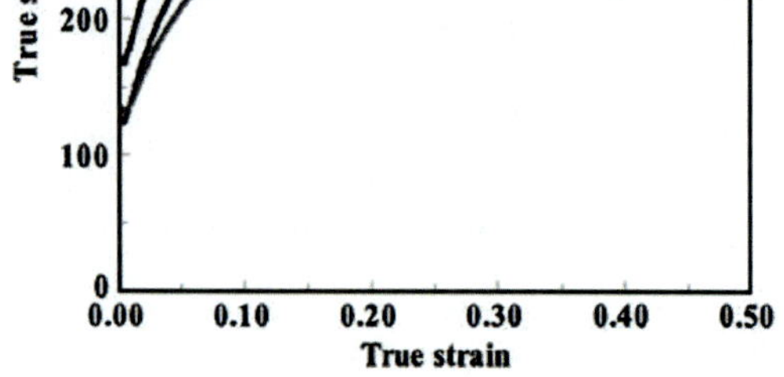

Fig. 5 True stress-true strain curves of Ti-IF steels with three different Si contents, 0%Si, 0.25%Si, and 0.75%Si, after cold rolling and annealing followed by rapid cooling. Specimens were not temper-rolled [2]

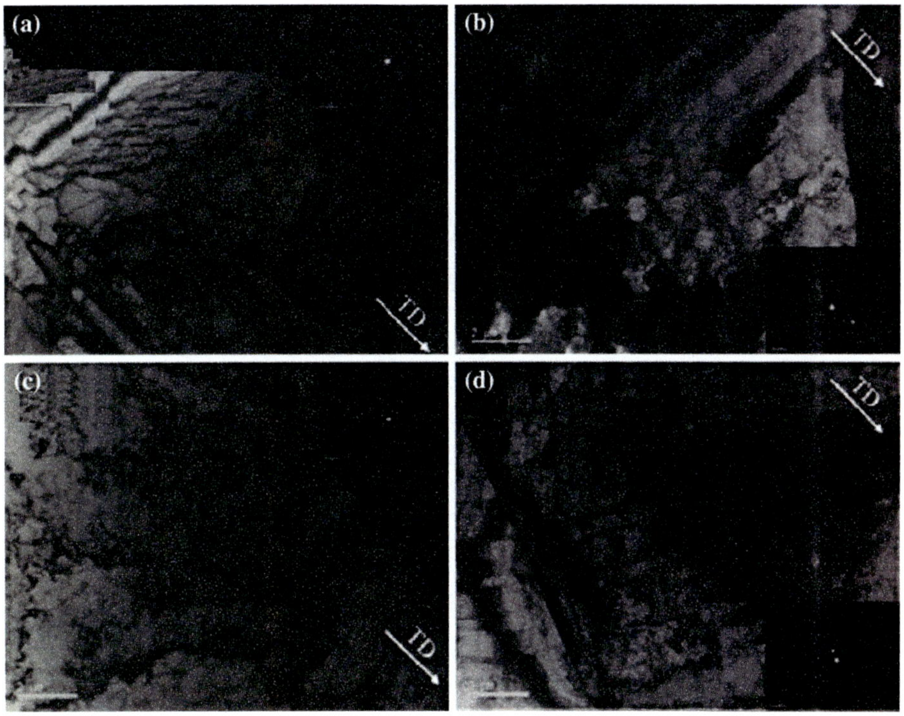

Fig. 6 TEM micrographs showing dislocation structures after stretching by 15%: **a** and **b**: E.B.//[100] and//[111] in 0%Si; **c** and **d**: E.B//[100] and//[111] in 0.75%Si. The tensile direction (TD) is parallel to the rolling direction (RD)

decreases during initial plastic deformation up to several percent due to the annihilation of statistically stored dislocation, it then remains unchanged up to a 20% strain, and it finally increases with further increases in strain as reported by Nakajima et al. [27] when martensite was cold-rolled. According to the in situ SEM observation of slip lines, primary slip occurs on {110} lath interfaces that have a large Schmid factor with the Burgers vector in the lath plane, and then secondary slip with the Burgers vector in the out-of-lath plane is observed across the lath interface. At this stage, the n value also increases from 0.07 to 0.12. TEM observation indicates that this kind of transition in the work-hardening behavior of martensitic steel leads to the destruction of lath boundaries and the commencement of restructuring of the tensile dislocation structure [26, 28].

Concerning the work-hardening behavior of the complex phase, there are theories for material containing hard second-phase particles by Ashby [21] and Tanaka and Mori [22]. They predicted an increased work-hardening rate due to the generation of dislocations necessary to be compatible with the rigid particles with different Young's moduli. Ballinger et al. [29] modified Ashby's theory for the work hardening of DP steel, and the work-hardening rate is expressed by Eq. 4,

$$d\sigma/d\varepsilon = 0.78kGb^{1/2}\varepsilon^{-1/2}(f/d)^{1/2} \quad (4)$$

Here, k is a constant of the order of 1, f is the volume fraction, and d is the average diameter of martensite. Equation 4 predicts that at a given strain, the work-hardening rate is proportional to $(f/d)^{1/2}$, which was verified by Ballinger et al. [29] and Kondo et al. [30]. However, more rigorous treatment of work hardening of DP steel is needed as future work, because in DP steel consisting of ferrite and martensite with mobile dislocations in the interface, martensite can be deformed plastically resulting in the stress and strain partitioning during deformation as experimentally verified by recent works using in-situ X-ray [31] or neutron diffraction [32]. Moreover, Ohtani et al. [33] demonstrated that a kind of grain subdivision takes place preferentially in ferrite grains when DP steel is subjected to tensile test. Such kind of grain subdivision was commonly observed in large ferrite grains, and it was enhanced in the area neighboring to the hard martensite. In this sense, there still remains a limitation in the practical application of the previous theories in terms of the strain range and plastic deformation of the hard second phase.

TRIP steels are well known for having excellent strength-ductility balance due to the deformation-induced martensitic transformation especially at the later stage of deformation. Therefore, the stability of retained austenite in terms of plastic deformation is essential. Figure 7a shows the change of tensile properties of the 0.14%C-1.9%Si-1.7%Mn steel specimen with austempering time at 400°C after intercritical annealing at 800°C [34]. Remarkable improvement of elongation was obtained after increasing the austempering time from 60 s to 500 s, which was confirmed by the change in behavior of the instantaneous

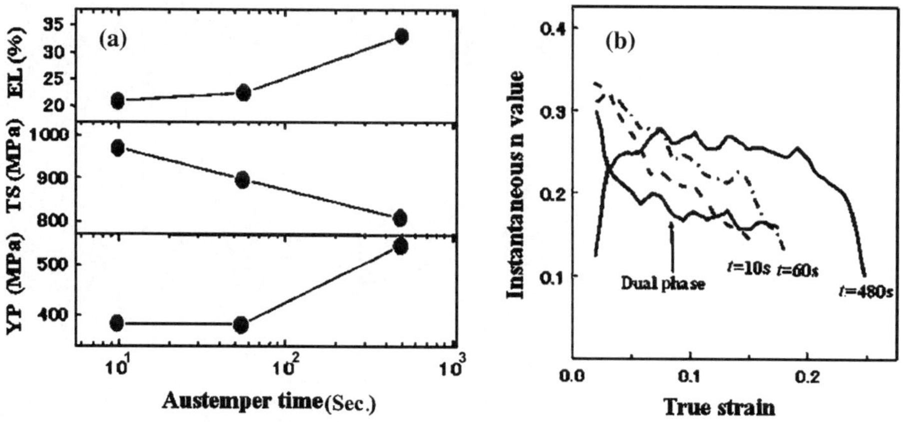

Fig. 7 **a** Tensile properties and **b** instantaneous *n* values for steels with 0.14%C-1.9%Si-1.7%Mn austempered for different times at 400°C after intercritical annealing at 800°C [34]

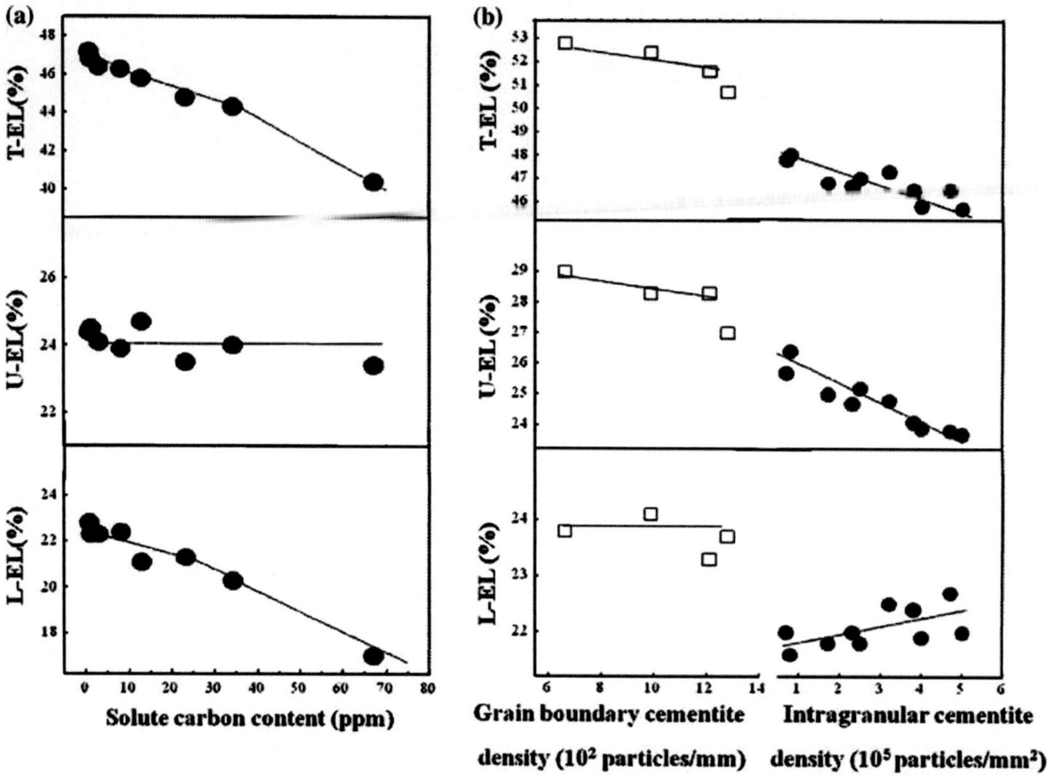

Fig. 8 Changes in L-El, U-El, and T-El with **a** the amount of C in solid solution and **b** the densities of grain boundary cementite and intragranular cementite keeping solute C zero in 0.02%C steel [36]

n value with strain (Fig. 7b). The high *n* value even at the later stage of deformation is a characteristic feature of TRIP steel caused by the TRIP effect of the plastically stable retained austenite due to enrichment of C to as high as 1.2%.

TRIP steel usually contains blocky, and lath-shaped morphologically different retained austenite. Matsuda et al. [35] investigated the influence of morphology of retained austenite on tensile properties. They concluded that lath-shaped retained austenite remained until a high-strain region, whereas the blocky austenite transformed into martensite in a low-strain region.

For the further improvement of ductility of complex-phase steels, the control of soft and hard phases is important in terms of the ductility of each phase, volume fraction, hardness ratio between the soft and hard phase, and 3D size and dispersion of the hard phase together with the optimum control of retained austenite.

4 Necking

Local (post-uniform) elongation after the maximum load is closely related to stretch flange forming and hole expansion forming, which has attracted a lot of attention in recent years. In general, complex-phase steels with good uniform elongation exhibit poor local elongation. In contrast, bainitic steels and ultrafine-grain steels that have poor uniform elongation exhibit good local elongation [1]. The determining factors of local elongation are the work-hardening rate, strain-rate sensitivity (*m* value), grain size, texture, and hard particles such as cementite for ferrite single-phase steel, and the characteristics of the hard phase such as the hardness difference between the soft and hard phases and the size and dispersion of the hard phase for complex-phase steels. This section discusses the influence of solute C and cementite [36] and the influence of grain size [20, 37] in ferrite single-phase steel. In addition, ductile failure in complex-phase steel is also discussed [38, 39].

Using Fe-0.02% C steel, two series of specimens were prepared for the tensile test: specimens with systematically changed solute C content and specimens with systematically changed grain boundry cementite desity as well as intragranular cementite density keeping solute C zero. Here, specimes were temper-rolled by 1.5% prior to tensile test. Figure 8 shows that solute C significantly deteriorates local elongation, but barely influences uniform elongation, and that intragranular cementite (size: 0.2–0.8 μm, spacing: 1–3 μm), relative to grain boundary cementite, impairs mostly uniform elongation but also local elongation somewhat. Moreover, as the cementite becomes finer, the amount of deterioration becomes smaller for local elongation and larger for uniform elongation. Figures 9 and 10 show optical and SEM micrographs of the voids and fracture surfaces observed in raptured specimens with C in solid solution (a), with C as precipitated cementites either intragranulary (b) or intergranulary (c) as well as specimens without C (d). Significantly large and many voids, which coagulate at the fractured tip, were observed in the specimen with C in solid solution (Fig. 9a). Moreover, the specimen has a feature of the fracture surface, where dimples were barely observed and a cleavage-like fracture surface was observed (Fig. 10a). On the other hand, it seems difficult for fine intragranular cementites to act as void nucleation sites as compared with relatively large grain boundary cementites (Figs. 9b, c and 10b, c); however, in the latter case, very fine dimples similar to those in the C-free specimen (Figs. 10c and d) were observed in the regions away from the grain boundary cementite. The deterioration of local elongation due to solute C is thought to be caused by the localization of strain due to negative strain-rate sensitivity by the occurrence of dynamic strain aging even at room temperature [36, 40]. The occurrence of dynamic strain aging was also confirmed by the fact that a small increase in YS was detected as compared with the relatively large increase in TS when the content of C in solid solution was increased (Fig. 11) [36]. In contrast, the

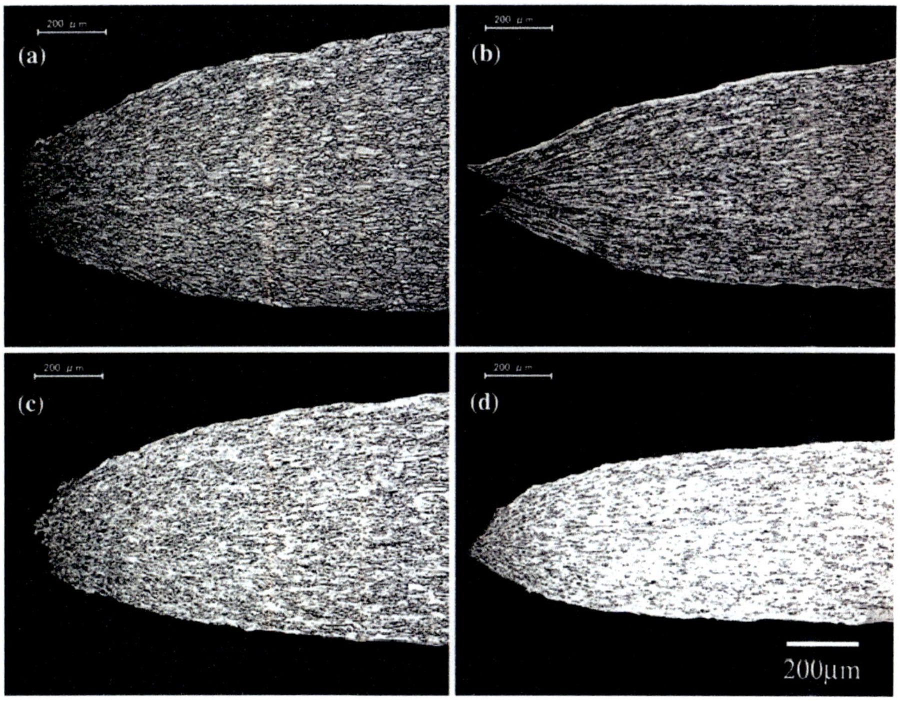

Fig. 9 Optical micrographs showing voids at the tip of ductile-fractured specimens of Fe-0.016%C: **a** C in solid solution quenched from 700°C, **b** intragranular cementite quenched from 700°C and aged at 300°C, **c** transgranular cementite air-cooled from 700°C to 550°C and subsequently furnace-cooled to 300°C. A micrograph of the specimen with 10 ppm C is shown in **d**, whose thickness is 0.8 mm, while the others are 1 mm thick

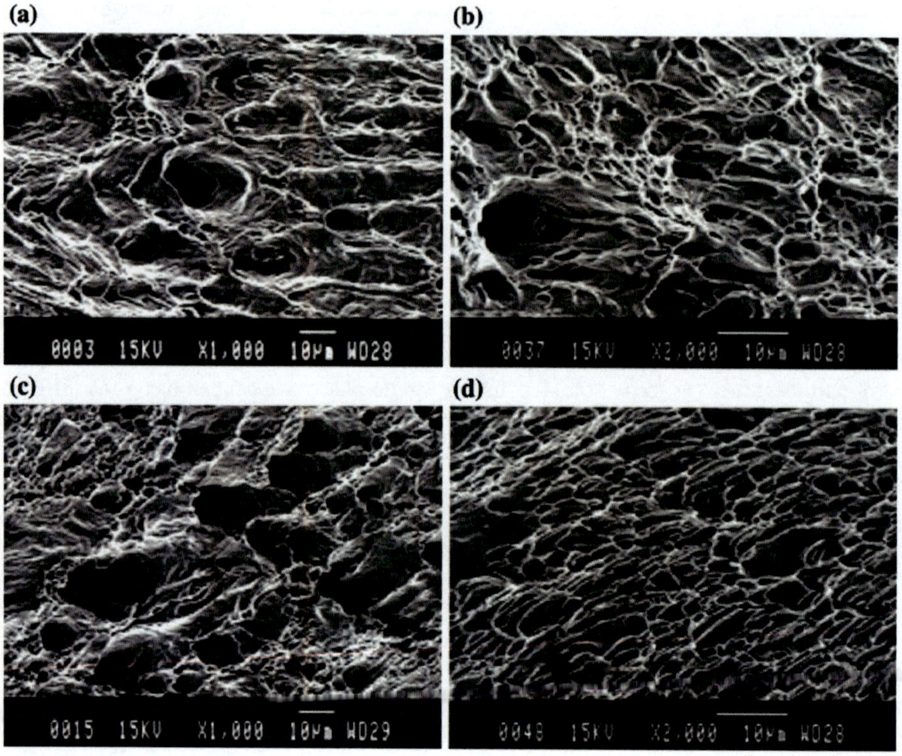

Fig. 10 SEM micrographs showing the fracture surface of specimens of Fe-0.016%C: **a** C in solid solution quenched from 700°C, **b** intragranular cementites quenched from 700°C and aged at 300°C, **c** transgranular cementite air-cooled from 700°C to 550°C and subsequently furnace-cooled to 300°C. Micrograph of the specimen with 10 ppm C is shown in (**d**)

deterioration of local elongation due to the cementite seems to be caused by ductile failure initiated from the cementite. However, since voids do not prefer to form around finer cementite, the small size of the intragranular cementite gives rise to a slight increase in local elongation. On the other hand, the prominent deterioration of uniform elongation with increased intragranular cementite density is associated with a large increase in YS by the Orowan mechanism as compared with the relatively small increase in TS (Fig. 11).

Tsuchida et al. [20] performed a tensile test on pure iron-based specimens having widely varying grain size from 1.4 to 106 μm, finding that local elongation increases as the grains become finer. From the fact that uniform elongation lessens when the grain size becomes smaller, still high work hardenability at the necking region is inferred to be one of the reasons together with homogenized strain due to the finer grains [41].

Many studies have been conducted on the local elongation of complex-phase steels [38, 39]. It has been reported that voids are mostly formed in the interface between the soft ferrite and the hard martensite phase in DP steel (Fig. 12). Consequently, DP steels usually have poor local elongation. However, this may be improved by the proper structure control of both ductile and hard phases. Matsuno et al. [38] studied the ductile failure of 590 MPa DP steel through experiments and two-dimensional pixel FEM analysis of uniaxial tension. They found that the ductile failure parameters $\left(D = \int_0^{\bar{\varepsilon}} (\sigma_{\max}/\bar{\sigma}) \mathrm{d}\bar{\varepsilon'}\right)$ proposed by Jeong [42] showed good agreement with the observed voids in the sense that voids are formed in the tensile direction near rather large martensite and coalesced usually in the direction of maximum shear stress (Fig. 12). Therefore, the ductile failure parameter might be exploited to optimize the structures of the soft and hard phases in terms of size, shape, hardness ratio as well as intervals of the hard phase.

A future challenge is to design the optimum 3*D* microstructure by exploiting advanced techniques for structure analysis together with computer simulation, and to develop a prediction model for the overall stress–strain curve.

5 Fatigue

5.1 Base Steel Fatigue

Fatigue strength improves with increasing base material strength. However, at high strengths beyond 780 MPa, the material becomes sensitive to the influence of inner inclusions and surface irregularities, and the fatigue strength does not always improve further. Mizui et al. [43] demonstrated that DP steel exhibits not only superior formability but also superior base material fatigue property, and promoted the application of DP steel for the wheel disks of automotives. They inferred that the ferrite phase

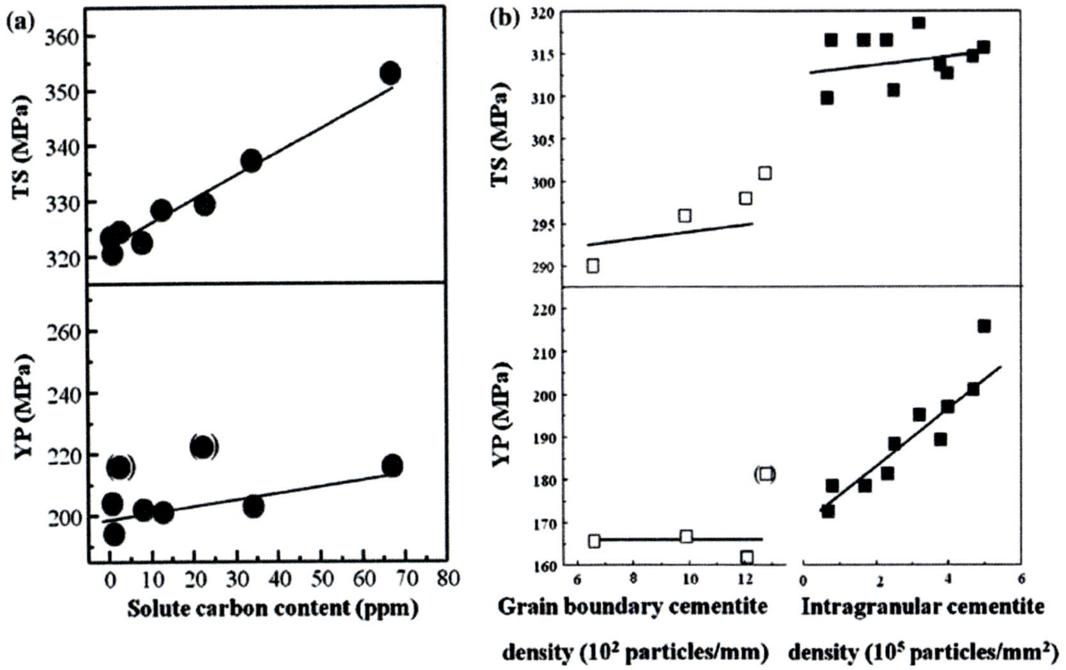

Fig. 11 Changes in YS and TS with **a** the amount of C in solid solution and **b** the densities of grain boundary cementite and intragranular cementite keeping solute C zero in 0.02%C steel [36]

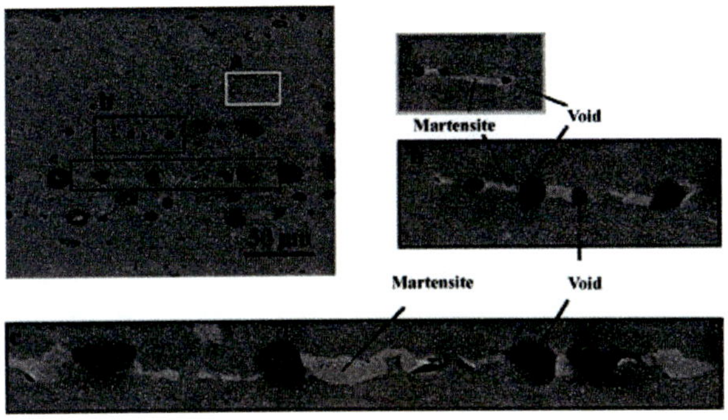

Fig. 12 SEM images showing voids formed in the interface between ferrite and martensite in 590 MPa DP steel

strengthened by the solute Si becomes hardened during cyclic deformation, and that the hard martensite diverts the cracks. Since the role of the ferrite phase is important, this section focuses on the influence of Si addition to ferrite in terms of the relationship between the evolution of dislocation substructure and fatigue crack behavior [24].

A cyclic bending test was conducted using Ti-IF steels with 0–1% Si (grain size: about 22 μm). The addition of Si improved the fatigue strength. The dislocation substructures were closely examined using a TEM. Figure 13 shows that in the 0% Si steel, a cell structure was formed, whereas in the 1% Si steel, a vein structure evolved after the high-stress amplitude cyclic test. The fine vein dislocation structure was considered to result in increased fatigue strength. Concerning the influence of Si addition on the evolution of dislocation substructure, it was inferred that the easy cross-slip brought about the cell structure in the 0% Si steel, whereas the vein structure developed in 1% Si steel because the Si addition made it difficult for cross-slip to take place, presumably due to the reduced stacking-fault energy. However, no clear stacking faults have been observed in Fe–Si alloy. Therefore, again, the interaction between dislocation and solute atoms like Si needs to be clarified.

Figure 14 shows a dislocation-free zone (DFZ) observed along the grain boundary in the 1%Si steel. The formation of a DFZ has also been reported in Cu [44]. In both cases,

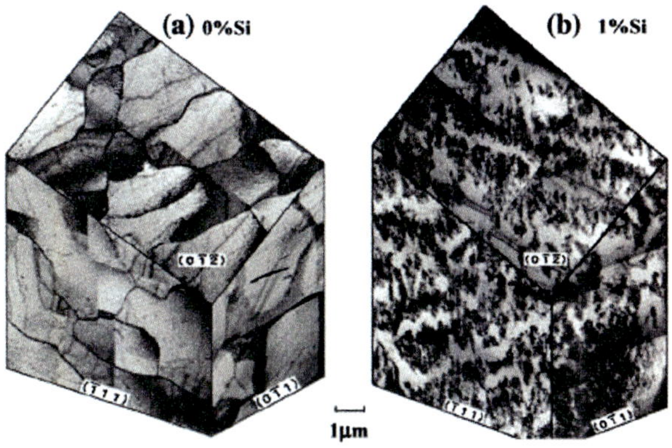

Fig. 13 TEM micrographs showing **a** the cell structure in 0% Si steel and **b** the vein structure in 1%Si steel which evolved during cyclic bending test [24]

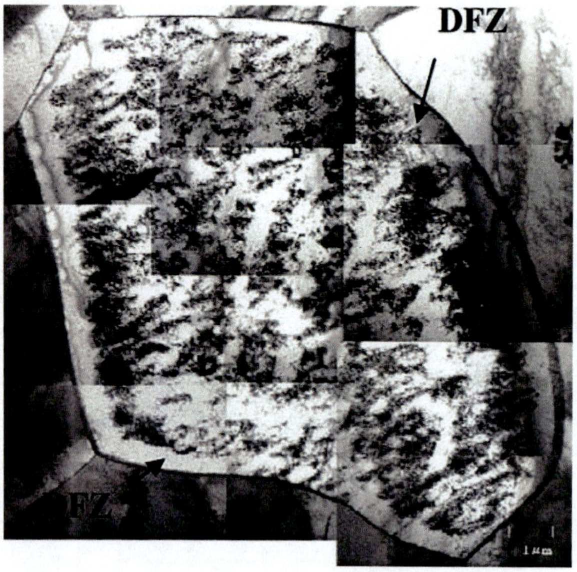

Fig. 14 TEM micrograph showing the dislocation free zone (DFZ) along grain boundaries in 1%Si steel which are formed in a grain with vein structure during cyclic bending test [24]

a DFZ was commonly formed when the vein structure developed. In the Si-added steel, grain boundary fracture was also observed immediately beneath the top surface, which was inferred to be closely related to the presence of a DFZ [24]. To explain the mechanism of DFZ formation, Luoh et al. [44] proposed that the operation of secondary slip occurs to satisfy the strain compatibility at the grain boundary, resulting in the disappearance of the dislocation dipole loops and dislocation walls. Since Si addition in steel suppresses cross-slip of screw dislocations, it is postulated that the vein structure consists of edge dislocations. Edge dislocations with opposite signs were inferred to be annihilated in the vicinity of grain boundaries leading to the formation of DFZ. Here, again the interaction between dislocation and grain boundary remains to be clarified.

6 HAZ Fatigue

Automotive underbody parts often use arc-welded joints. Cracks that usually occur at the toe of welded parts retard further strengthening of steels. Figure 15 indicates the importance of the radius of the curvature at the toe to improve the fatigue life of arc-welded parts, namely, the larger the better [45]. This is caused by the stress concentration or residual tensile stress at the toe of the welded parts. Therefore, besides the curvature control at the toe, the control of the chemical compositions of base steels [46] and wire [47] together with the introduction of compressive residual stress are considered to be preventive measures.

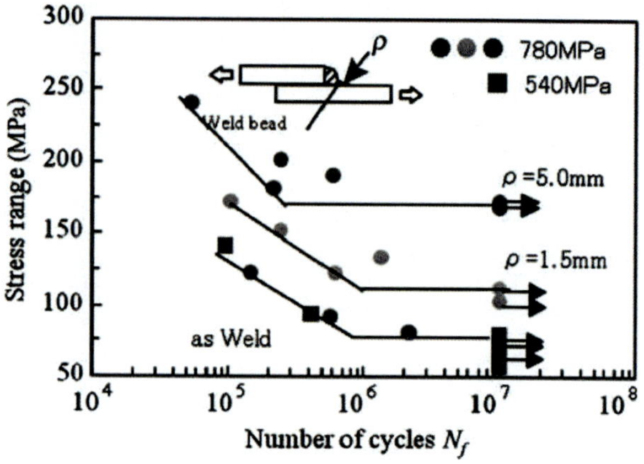

Fig. 15 Influence of radius of curvature at the toe on fatigue life of arc-welded high-strength steel [45]

Kasuya et al. [47] reported the development of new welding wire which provides the compressive stress due to low temperature phase transformation in welded part. For further strengthening, in addition to the material design itself, it is also important to work on the welding method and structure design to achieve overall optimization.

7 Hydrogen Embrittlement

When the strength level of steel exceeds 1,000 MPa, susceptibility to hydrogen embrittlement increases which poses a problem to be overcome for further strengthening. That is, under a given stress condition, if the amount of hydrogen entering into the steel from the environment (H_E) exceeds the critical diffusible hydrogen content (H_C), hydrogen cracking (grain boundary cracking) occurs. However, hydrogen embrittlement has been widely evaluated for wires and bolts, but has not been intensively evaluated for steel sheets. Hayashi et al. [48] have established a method for sheet products. As shown in Fig. 16, sheet samples are bent under controlled elastic stress calculated using elastic strain by the strain gauge, and hydrogen is electrically charged with different cathodic charge current densities

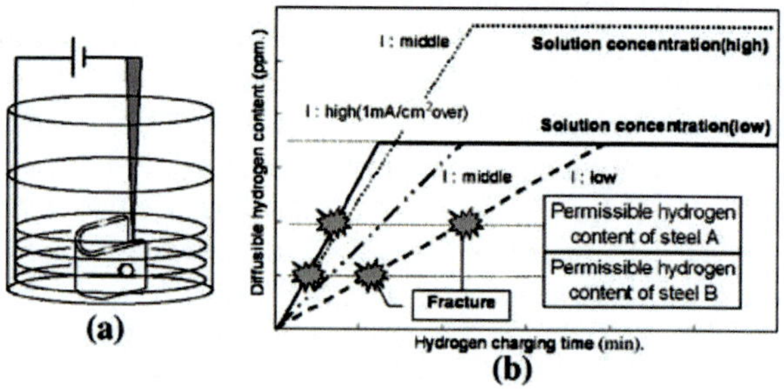

Fig. 16 Schematic presentation of **a** specimen set-up and **b** concept for the evaluation of hydrogen embrittlement of sheet steel [48]

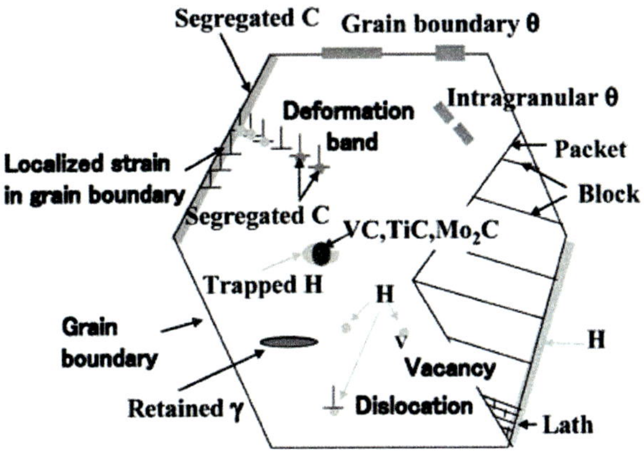

Fig. 17 Schematic presentation of the states of hydrogen and lattice defects in high strength steels [2], based on figure by Nagumo [49]

between 0.01 and 10 mA/cm^2 in an electrolyte of 3% NaCl solution with the addition of ammonium thiocyanate (NH_4SCN). Thermal desorption analysis (TDA) of hydrogen was used to measure the amount of hydrogen when rupture takes place. In a practical situation, close attention also needs to be paid to the welded area, which is normally the weakest point. This section reviews recent progress made on the mechanism of hydrogen embrittlement and discusses future issues.

Bearing the mechanism of hydrogen embrittlement in mind, Fig. 17 schematically depicts the relationship between the steel microstructure and the state of hydrogen, based on the figure by Nagumo [49]. In steels that are susceptible to hydrogen embrittlement, the base structures are martensite and tempered martensite, where fine precipitates such as VC and TiC and retained austenite are present. For instance, TiC acts as hydrogen trapping sites, which was first directly observed in TiC precipitation-hardened steel through atom probe tomography as shown in Fig. 18 [50]. Hydrogen embrittlement is thought to be mostly prior austenite grain boundary cracking. However, it is expected that C will usually segregate in the grain boundary, which suppresses the segregation of hydrogen and increases the strength of the grain boundary. If cementite precipitates at the grain boundary, it is expected that segregated C will decrease locally and stress concentration will occur at grain boundary cementites. While some models are proposed describing that hydrogen at the grain boundary lowers grain boundary strength and causes grain boundary embrittlement [51–53], it is necessary to clarify and develop a model of hydrogen embrittlement taking into account the state of C.

According to hydrogen-enhanced localized plasticity (HELP) theory, hydrogen accelerates dislocation mobility [54, 55], and it is thus inferred that strain is localized in the vicinity of the grain boundary, resulting in the localized slip lines observed at the grain boundary fracture surface. However, since the HELP theory was derived for the case of defect-free metals, whether or not this hypothesis is really applicable to the case of ultrahigh-strength steels where structures are very complicated with a high density of dislocations and point defects is arguable. Nagumo [56] proposed a model (hydrogen-enhanced strain-induced vacancy model, HESIV) in which point defects play an important role in hydrogen embrittlement in the sense that hydrogen stabilizes point defects and promotes vacancy coagulation leading to crack formation. Moreover, Fang et al. [57] proposed that stable retained austenite is expected to act as a hydrogen reservoir, but there remain many unknown positive and negative roles of retained austenite.

Essential progress in clarifying the mechanism exploiting advanced techniques is expected, which will bring about a new solution to hydrogen embrittlement.

8 Conclusions

Steels are attractive materials having many possibilities yet to be exploited. However, further strengthening of steels faces many technical problems. To achieve both high

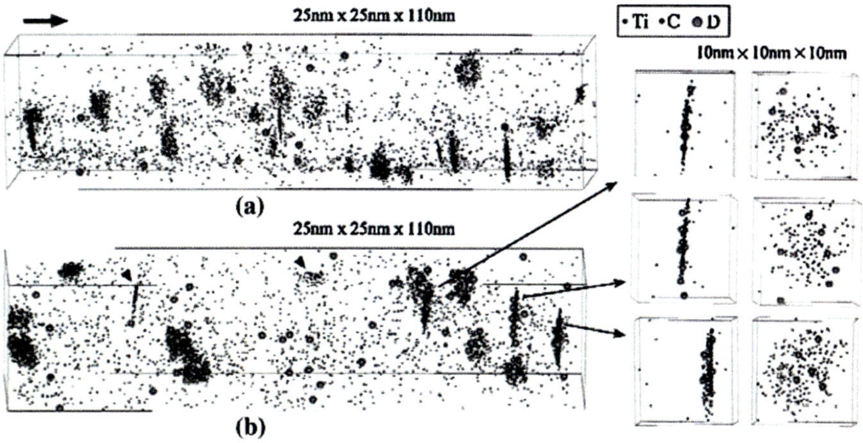

Fig. 18 3D elemental maps of deuterium-uncharged **a** and -charged steels **b** with nm-sized TiC precipitates. (The *bold arrow* indicates the analysis direction [50])

strength and formability, it is necessary to understand the essential nature of the interactions between dislocation and solute atoms, hard second phase, grain boundaries, and pre-existing dislocations, which govern yield strength, work hardening, and necking. Since a welded area always becomes the weakest part, the microstructure and properties of this part must be improved. It is also necessary to clarify the fracture mechanism in hydrogen embrittlement with the presence of C taken into consideration. It will be important to accumulate systematic experimental data and propose new mechanisms based on these data. On the other hand, recent technical progress in the 3D analysis of microstructures and computer science is remarkable, enabling multi-scale analysis ranging from nano-scale physical phenomena to product functions in the order of several 10 m. Utilization of these cutting-edge technologies is expected to result in discontinuous improvement of steel functionality, which can greatly contribute to solving global social problems and preserving the global environment.

Acknowledgments The authors are grateful for the helpful discussions with Prof. Emeritus T. Maki of Kyoto University, now an Executive Advisor to Nippon Steel Corp.

References

1. K. Ushioda, M. Takahashi, in *Asian Steel International Conference*, ISIJ, Fukuoka, vol. 114 (2006)
2. K. Ushioda, in *Proceedings of the 2nd International Conference on Steel Science* (*ISSS2009*), Kyoto, ISIJ, vol. 39 (2009)
3. T. Maki, Bull. Iron Steel Inst. Jpn. **13**, 544 (2008)
4. Y. Funakawa, T. Shiozaki, K. Tomita, T. Yamamoto, E. Maeda, ISIJ Int. **44**(11), 1945 (2004)
5. J. Takahashi, K. Kawakami, Y. Yamaguchi, CAMP-ISIJ **21**, 510 (2008)
6. L.M. Brown, R.K. Ham, *Strengthening Methods in Crystals* (Elsevier, Amsterdam, 1971), p. 9
7. K.C. Russel, L.M. Brown, Acta Metall. **20**(7), 969 (1972)
8. K. Osamura, H. Okuda, S. Ochiai, ISIJ Int. **34**(4), 359 (1994)
9. P.J. Othen, M.L. Jenkins, G.D. Smith, Phil. Mag. Lett. **64**(6), 383 (1991)
10. A. Kelly, R.B. Nicholson, Prog. Mater. Sci. **10**(3), 149 (1963)
11. Y. Imanami, M. Murakami, N. Nakada, CAMP-ISIJ **22**, 472 (2009)
12. Y. Kobayashi, J. Takahashi, K. Kawakami, CAMP-ISIJ **23**, 403 (2010)
13. K. Takeda, N. Nakata, T. Tsuchiyama, S. Takaki, ISIJ Int. **48**(8), 1122 (2008)
14. J.C. Spitzig, Metall. Mater. Trans. **3**, 1183 (1972)
15. Y. Sakuma, A. Itami, K. Ushioda, Memoirs of Research Committee on Improvement in the Strength and Reliability of Steels, ISIJ, 57 (1997)
16. M. Fujiwara, N. Nakata, T. Tsuchiyama, S. Takaki, CAMP-ISIJ **22**, 1279 (2009)
17. J.C.M. Li, Trans. Metall. Soc. AIME **227**(1), 239 (1963)
18. S.Takaki, Private communication
19. T. Ohmura, A.M. Minor, E.A. Stach, J. Mater. Res. **19**(12), 3626 (2004)
20. N. Tsuchida, Unpublished Work
21. M.F. Ashby, Philos. Mag. **21**(170), 399 (1970)
22. K. Tanaka, T. Mori, Acta Metall. **18**(8), 931 (1970)
23. S. Yonemura, S. Hiwatashi, A. Uenishi, Tetsu-to-Hagane **92**, 516 (2006)
24. K. Ushioda, S. Goto, S. Komatsu, ISIJ Int. **49**, 312 (2009)
25. Y. Kawasaki, T. Nakagaito, S. Kaneko, in *Proceedings of the 2nd International Conference on Steel Science* (*ISSS2009*), Kyoto, ISIJ, vol. 241 (2009)
26. S. Nambu, M. Michiuchi, Y. Ishimoto, Scripta Mater. **60**, 221 (2009)
27. K. Nakajima, F. Fujimura, H. Matsubayashi, Tetsu-to-Hagane **93**, 459 (2007)
28. S. Morito, S. Iwamoto, T. Maki, Int. Forum for the Prop. & Appl. of IF Steels (IF Steels 2003), ISIJ, Tokyo, 365 (2003)
29. N.K. Ballinger, T. Gladman, Met. Sci. **15**, 95 (1981)
30. D. Kondo, K. Kunishige, R. Ueji, Tetsu-to-Hagane **92**, 457 (2006)
31. Z.H. Cong, N. Jia, X. Sun, Metall. Mater. Trans. A **40A**, 383 (2009)
32. K. Asoo, Y. Tomota, S. Harjo, Memoirs of Research Committee on Cut Edge of Models for Predicting Microstructure and Mechanical Properties, ISIJ, 256 (2010)
33. S. Ohtani, T. Morikawa, K. Higashida, S. Hashimoto, Tetsu-to-Hagane **95**, 620 (2009)
34. A. Itami, M. Takahashi, K. Ushioda, ISIJ Int. **35**, 1121 (1995)
35. H. Matsuda, H. Noro, Y. Nagataki, Sci. Forum **638–642**, 3374 (2009)
36. S. Takebayashi, T. Katayama, K. Ushioda, J. JSTP **51**(7), 690 (2010)
37. N. Tsuji, Y. Ito, Y. Saito, Scripta Mater. **47**, 893 (2002)
38. T. Matsuno, D. Maeda, Y. Suwa, CAMP-ISIJ **22**, 1328 (2009)
39. K.S. Choi, W.N. Liu, X. Sun, Acta Mater. **57**, 2592 (2009)
40. M. Abe, Y. Kokabu, Y. Hayashi, J. Jpn. Inst. Met. **45**, 942 (1981)
41. K. Miyata, M. Wakita, S. Fukushima, in *Proceedings of the 2nd International Conference on steel Science* (*ISSS2009*), Kyoto, ISIJ, vol. 115 (2009)
42. S.H. Jeong, J.J. Kang, S.I. Oh, in *Proceedings of 5th ICTP*, vol. 631 (1996)
43. M. Mizui, M. Takahashi, Proc. of Mechanical Working and Steel Processing (MWSP) Conference, ISS, St. Lois, 57 (1991)
44. T. Luoh, C.P. Chang, Acta Mater. **44**, 2683 (1996)
45. H. Murasato, Unpublished Work
46. M. Uchihara, M. Kurita, T. Taka, Tetsu-to-Hagane **86**(5), 336 (2000)
47. T. Kasuya, K. Sasaki, Jpn. Weld. Soc. **27**, 158 (2009)
48. K. Hayashi, N. Fujita, S. Takebayashi, SAE paper 2005-01-0497
49. M. Nagumo, Hydrogen Embrittlement, Uchidarokakuroh, Tokyo, (2008)
50. J. Takahashi, K. Kawakami, Y. Kobayashi, T. Tarui, Scripta Mater. **63**, 261 (2010)
51. M.E. Eberhart, K.H. Johnson, R.M. Latanision, Acta Metall. **32**(6), 955 (1984)
52. M. Wang, E. Akiyama, K. Tsuzaki, Mat. Sci. Eng. **A398**, 37 (2005)
53. M. Yamaguchi, M. Shiga, H. Kaburagi, Science **307**(5708), 393 (2005)
54. H.K. Birnbaum, P. Sofronis, Mater. Sci. Eng. **A176**(1-2), 191 (1994)
55. I.M. Robertson, Eng. Fract. Mech. **68**, 671 (2001)
56. M. Nagumo, Mater. Sci. Tech. **20**(8), 940 (2004)
57. H.-S. Fang, Q. Li, B.-Z. Bai, Z.-G. Yang, Int. Forum for the Prop. & Appl. of IF Steels (IF steels 2003), ISIJ, Tokyo, 53 (2003)

Developments in High Strength Steels with Duplex Microstructures of Bainite or Martensite with Retained Austenite: Progress with Quenching and Partitioning Heat Treatment

David Edmonds, David Matlock, and John Speer

Abstract

Attention has been given to the development of new higher strength steel grades by employing duplex microstructures of bainitic or martensitic ferrite with retained austenite. Such microstructures rely first upon appropriate alloying to stabilise austenite against thermal decomposition, either isothermally to bainite or athermally to martensite. The alloying, typically by involving Si additions, promotes chemical stabilisation indirectly through suppression of carbide precipitation and hence increase of austenite carbon concentration. The most recent novel concept, aimed at higher strength martensitic grades, combines enhanced Si alloying with an interrupted quench, followed by an anneal to allow carbon migration from supersaturated martensite regions to untransformed austenite, a process which has become known as quenching and partitioning, or Q&P. Recent progress in understanding the detailed evolution of microstructure during the Q&P heat treatment, and the potential for exploration of the resulting microstructures in various steel grades, will be described.

Keywords

Steel heat treatment • Quenching and partitioning (Q&P) • Retained austenite • Bainite • Martensite • TRIP steels

1 Introduction

The recovery and transmission of oil and gas, and the need for safe light-weighting of vehicles for improved fuel efficiency and reduced carbon emissions, have probably been the main technology drivers for perhaps the most important steel developments in modern times, which have been in constructional steel plates and sections and automotive steels, respectively. Thus, microalloying combined with controlled processing have been applied to develop, on the one hand strength (with toughness and weldability) through grain refinement and carbide precipitation, and on the other hand, improved formability (with strength) by fixing carbon and nitrogen interstitials as carbonitrides. Many authors have recorded these successful developments, along with many of the milestone conferences and associated publications, e.g. [1–3].

A major reason for the dominance of iron and steel as engineering materials is the variety of microstructures and properties conferred by carbon. Thus it is interesting to note that the more modern steel developments referred to above have been brought about by the microalloying of reduced carbon steels, based solely upon ferritic microstructures, in order to achieve, in constructional steels improved weldability and toughness, or in automotive steels improved formability and reduced strain ageing. Furthermore, in more traditional quenched and tempered martensitic steels it is necessary to control the effects of carbon, which, although conferring strength can also lead to a variety of complications, e.g. quench cracking or temper embrittlement.

D. Edmonds (✉)
Institute for Materials Research, University of Leeds, Leeds, LS2 9JT, UK
e-mail: d.v.edmonds@leeds.ac.uk

D. Matlock · J. Speer
Advanced Steel Processing and Products Research Centre, Colorado School of Mines, Golden, CO 80401, USA

Y. Weng et al. (eds.), *Advanced Steels*, DOI: 10.1007/978-3-642-17665-4_24,

However, this paper introduces ways to use carbon, a relatively cheap alloying element, in a different novel sense which, it is suggested, opens up opportunities for developing new forms of steels and heat treatment. This is the recent emergence of steels with acicular microstructures based upon bainite or martensite containing retained austenite. These microstructures can contain carbides or can be carbide-free.

A primary objective in acicular steels with retained austenite, often overlooked, is to *prevent* carbide formation (hence, essentially they are "carbide-free"), thereby improving mechanical properties, not by *reducing* carbon, but by *removing* carbon from the acicular ferritic phase (where, in a general sense, it could be damaging to certain properties) and transporting it to the retained austenite phase. Initially the austenite phase acts as a potential carbon sink to protect the ferrite from any problems such as those mentioned above. Next, the carbon chemically stabilises the austenite against further decomposition. (It is interesting to note that this contrasts with conventional martensite tempering, where further decomposition of untransformed austenite is encouraged in order to stabilise the final microstructure.) In consequence, the refined acicular ferrite microstructure (which could be precipitation strengthened by allowing some controlled carbide formation) can promote strength, whilst the presence of the ductile fcc austenite second phase can potentially buffer crack propagation to enhance toughness, or alternatively, undergo mechanically-assisted transformation to increase work hardening and uniform elongation, and hence formability.

2 Carbide-Free Bainitic Steels

The bainite transformation concerns primarily the decomposition of austenite to bainitic ferrite and therefore it is not unreasonable to accept that bainite can form in carbon steels with or without carbide formation [4]. Thus, carbide-free bainite consists of bainitic ferrite interwoven with retained austenite (Fig. 1a, [5]). The austenite phase is retained due to its increased carbon concentration, resulting from the absence of carbide precipitation. This can result from suppression of cementite formation by alloying, usually with Si. It is argued that, although there are probably earlier origins, the carbide-free form of bainite in Si-containing steels, which likely led to subsequent industrial developments, can best be attributed to Hehemann and co-workers in the 1960s [6–9] where the lack of carbide precipitation but presence of retained austenite allowed a more thorough morphological and crystallographic study of the bainite transformation mechanism [5, 10]. This may have helped to revive commercial interest in the development of irons and steels based upon bainite, because although some questions of the transformation mechanism might remain unresolved [11] it is nevertheless possible to design appropriate microstructure/property combinations based upon bainitic microstructure. A notable example is so-called TRIP steel, discussed below, and also worth mentioning because of the similarity of the basic matrix microstructure, is austempered ductile iron (ADI) illustrated in Fig. 1b [12].

2.1 Early Experimental Carbide-Free Bainitic Steels

The carbide-free bainitic microstructures in some of the experimental steels developed during studies of the bainite transformation mechanism resulted in mechanical properties worthy of further exploration [13, 14]. Moreover, it was shown that because the extent of transformation to bainitic ferrite, and hence the fraction of retained austenite and its carbon content, could be predicted, it should be possible to design a duplex microstructure with specific properties [13]. In addition, the thermal and mechanical stability of the retained austenite phase were examined, which led to the suggested requirement for austenite with an interwoven thin film morphology (Fig. 1a) rather than a less stable blocky form for optimum toughness properties. Thus, good combinations of strength and toughness were observed, as shown by Fig. 2a, due to the refined duplex structure and also the suspected contribution from the separating second phase

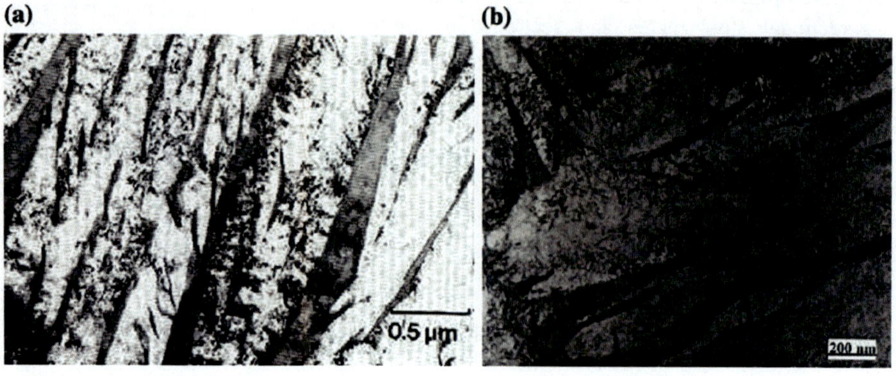

Fig. 1 Electron micrographs of the interwoven carbide-free bainitic ferrite and retained austenite in **a** steel [5] and **b** austempered ductile iron [12] (Retained austenite is the *darker grey phase*)

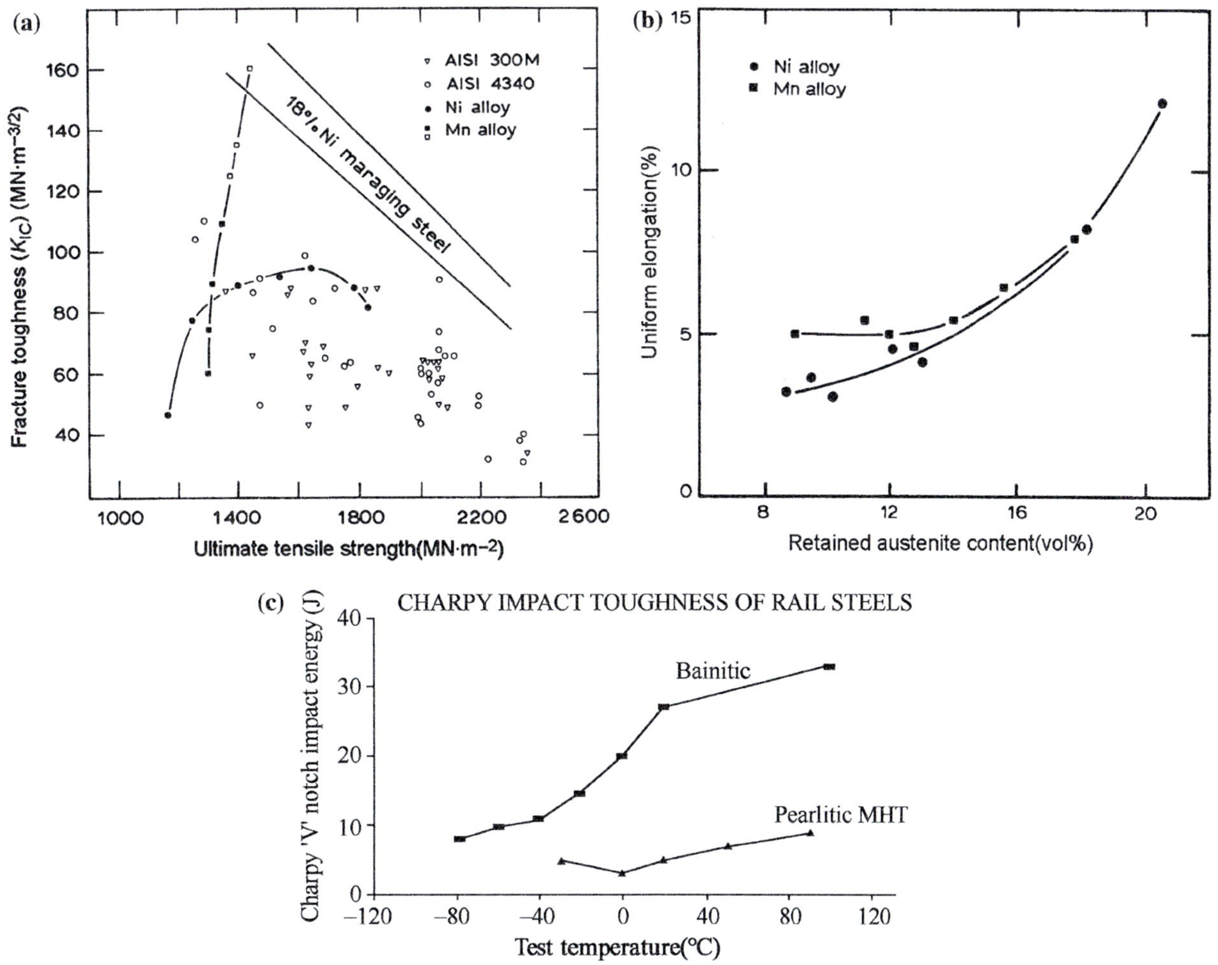

Fig. 2 Mechanical properties of experimental low-alloy steels with carbide-free bainitic microstructures: **a** ultimate tensile strength versus fracture toughness [14]; **b** uniform elongation versus retained austenite content [14]; **c** Charpy impact toughness compared with pearlitic rail steel [17]

austenite. Figure 2b also illustrates the increase in uniform elongation with retained austenite content, which would be important to formability. Another example is a rail track steel with greater than twice the fracture toughness, and improved wear rates, compared with conventional pearlitic rail steels [15–17]. Figure 2c compares the impact toughness of bainitic and pearlitic rails; it is claimed that the higher Charpy value should provide greater ability to withstand impact deformation and lower sensitivity to stress concentrations.

2.2 Transformation Induced Plasticity Steels

Perhaps the most important exploitation to date of duplex microstructures of carbide-free bainite containing retained austenite has been their application to improve formability at higher strength levels, giving rise to a new class of automotive sheet TRIP steels. The 'TRIP' acronym—from Transformation-Induced Plasticity—originated during the 1960s from the pioneering work of Zackay et al. [18, 19], which developed enhanced ductility and formability in high-alloy metastable austenitic steels (typically, 0.2–0.3C–8Cr–7–24Ni–4Mo–2Si) from the martensite transformation triggered by straining. This TRIP-assisted behaviour in automotive steels arises essentially from the presence of metastable retained austenite phase associated with carbide-free bainite in the microstructure. This form of bainite is induced in low C–Mn compositions (typically, 0.2–0.4C–1.0–1.5 Mn) by enhanced Si additions of around 1.5 wt% [20–25]. In these steels, usually intercritically annealed, the microstructure generally consists of intercritical ferrite plus bainite (and maybe martensite) (Fig. 3a). The bainite fraction is identical to that discussed above; carbide-free regions comprising bainitic ferrite interwoven with retained austenite. The steels develop enhanced formability from the mechanically-induced decomposition of the metastable austenite to martensite during processing in the way envisaged by Zackay and co-workers.

2.3 Nanobainite Steel

It has recently been found [26–29] that a very refined carbide-free bainite can be formed in steels with relatively high carbon concentrations (~1 wt%) after isothermal annealing for long times (up to several days) at relatively low temperatures (typically ~200°C). Extremely thin bainitic ferrite plates ~20–30 nm thick, separated by thin retained austenite films (Fig. 3b [30]), termed 'nanobainite', have resulted in hardness values ~690 HV, and yield and tensile strengths ~2.0 and 2.5 GPa, respectively. This has led to the development of hyper-strength or super bainitic Bainitic Alloys for Defence Applications (BADA) steels which have found potential applications in, for example, armour plate [31].

3 Carbide-Free Quenched and Partitioned Martensitic Steels

3.1 The Q&P Concept

A similar philosophy to exploit steel microstructures containing retained austenite, which builds upon the bainitic route described above, but is based upon a martensitic microstructure [11, 32–34] has emerged more recently. The different approach still employs Si alloying to stabilize austenite indirectly by suppressing cementite formation, but additionally, employs a modified heat treatment procedure. This is an interrupted quench to a temperature (QT) between the martensite-start (M_s) and martensite-finish (M_f) temperatures, which results in untransformed 'residual' austenite (Fig. 4, [33, 34]).

The interrupted quench is then followed by an annealing treatment (termed partitioning), either at, or above, the initial quench temperature (PT). As discussed above for carbide-free bainite formation it is anticipated that with enhanced Si alloying suppressing cementite precipitation, the residual austenite will be enriched with carbon expected to escape from the supersaturated martensite in which it has very low solid solubility. This novel heat treatment has been given the name 'Quenching and Partitioning' (Q&P), and is expected to produce a fine acicular aggregate of carbon-depleted and potentially carbide-free martensite laths interwoven with retained austenite stabilised by carbon-enrichment, as shown by Fig. 5, [35].

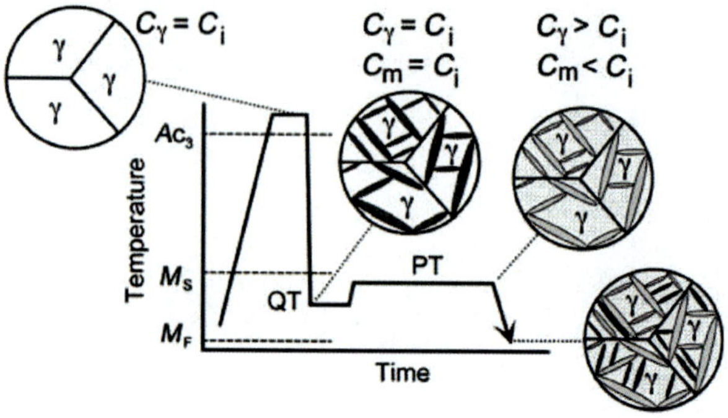

Fig. 4 Schematic of the 'Quenching and Partitioning' (Q&P) heat treatment procedure [33, 34]

3.2 Microstructural Evolution During Q&P Treatment

Detailed characteristics of the Q&P heat treatment process have been inaccessible for study by conventional metallographic methods because of the high temperatures involved. Consequently, the mechanisms of microstructural evolution have been interpreted based partly upon experimentation and partly upon empirical hypothesis, e.g. [11, 32–35]. However, the sequential steps of the heat treatment process have been studied more recently in alloys designed

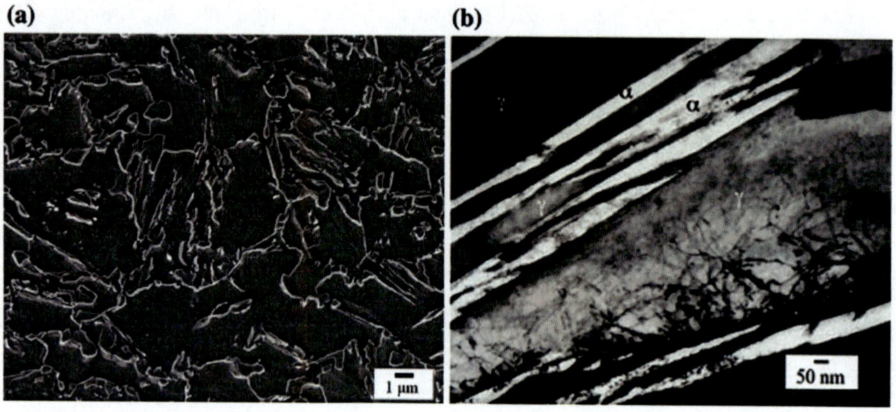

Fig. 3 **a** Scanning electron micrograph typical of a typical TRIP steel microstructure (Courtesy of Grant Thomas, Colorado School of Mines); **b** transmission electron micrograph of fine 'nanobainite' structure [30]

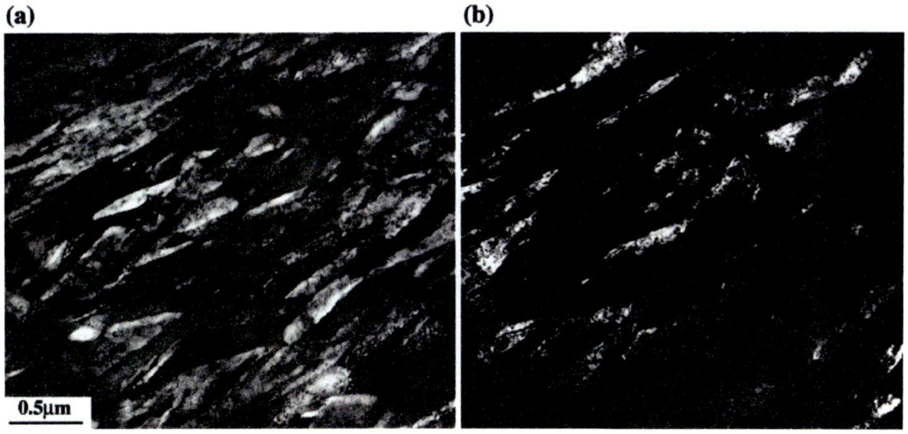

Fig. 5 Electron micrographs of Q&P microstructure in AISI 9260: QT 190°C, PT 400°C; **a** bright- and **b** dark-field pair using an austenite reflection [35]

to depress the M_s temperature to facilitate more direct experimental measurements [36, 37].

A more thorough examination of the possible microstructural changes after the first quench and during subsequent partitioning has been possible in a 0.64C–4.57Mn–1.30Si alloy, in which the M_s–M_f range spans room temperature. That the starting microstructure before the partitioning treatment is a mixture of as-quenched martensite and untransformed austenite is immediately guaranteed, which is not the case for a quench interrupted at a much higher temperature. Transformation of this quenched microstructure during and after heating to a partitioning temperature of 500°C has been followed in real time by neutron diffraction measurements using a furnace installed on the beam line. Thus, lattice and microstructural parameter changes are recorded in Figs. 6 and 7, respectively.

Release of carbon from supersaturated interstitial sites in martensite is indicated by the decay of the initial BCT martensite to a BCC or low c/a ratio BCT structure. Carbon 'trapped' in distorted interstitial sites around dislocations or

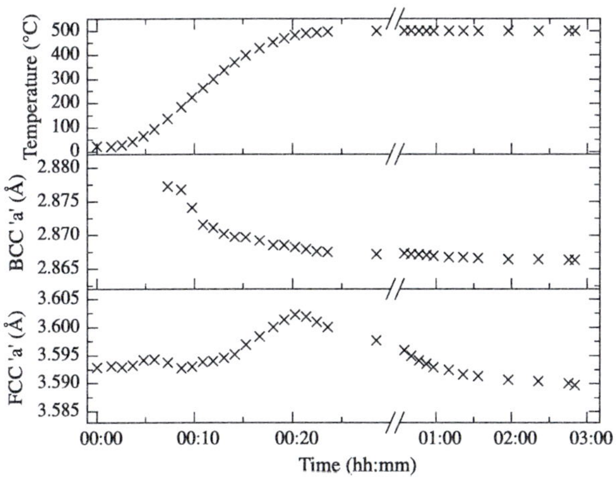

Fig. 6 Evolution of lattice parameters with respect to time, during heating up to, and partitioning at, 500°C (after [36, 37])

epsilon carbide precipitation could be expected [38] but the data collection time periods of ~1 min are too short to determine conclusively whether carbides are present. However, the simultaneous measurement of high lattice strains conforms to the high density of crystallographic defects expected in the body-centred lattice which should allow a 'trapping' mechanism to operate. Thus, in conformity with plain carbon steels [38], at temperatures not much above room temperature, equilibrium is believed to be in favour of 'trapping', whilst at higher temperatures epsilon carbide formation should become more favourable. However, the extent of this will depend upon the dislocation density, itself dependent upon the nature of the martensite, especially the fraction of lath or twinned martensite, also dependent upon the carbon concentration of the steel [38]. But at the higher temperatures of conventional Q&P treatment it is likely that this is a very short transient stage (the neutron diffraction data were collected over a relatively slow heating rate). At higher temperatures, where carbon mobility is increased, the alternative mechanisms of carbide precipitation and escape to the retained austenite phase, will compete with dislocation trapping, itself diminished by recovery of the dislocation substructure (as indicated by decreasing lattice strain with increasing temperature in Fig. 7).

Carbide precipitation at the partitioning temperature is suggested by the eventual decrease in carbon concentration of austenite with time (Fig. 7) and indeed, carbide peaks were subsequently detected by re-measuring at a much longer counting time after cooling the specimen back to room temperature. Fine-scale carbon atom clustering, or carbide formation, detected by very high resolution atom probe tomography, is illustrated in Fig. 8 for an experimental steel with composition representative of a TRIP-assisted sheet steel [0.19C–1.59Mn–1.63Si–0.036Al–0.013P (wt%)], austenitised at 900°C for 180 s, quenched and held at 240°C for 10 s, and partitioned at 400°C for 30 s [39–41]. A three-dimensional carbon atom map,

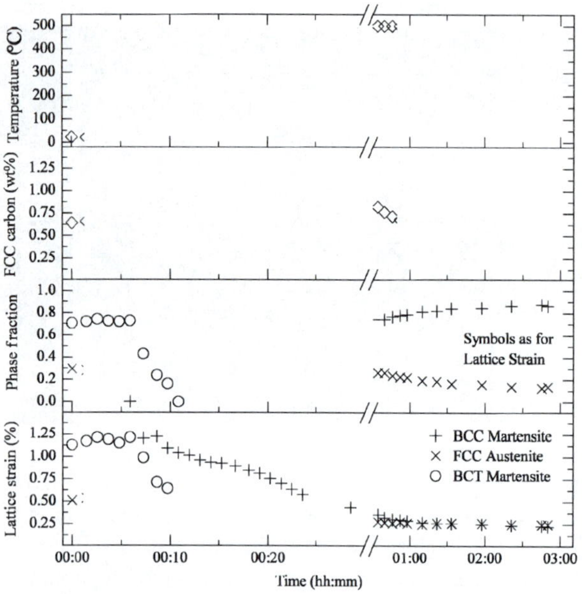

Fig. 7 Progression of microstructural parameters with respect to time, during heating up to, and partitioning at, 500°C (after [36, 37])

reconstructed from data collected from a region approximately 28 × 10^4 nm^3, is shown in Fig. 8a, and in this particular case, the size, shape, distribution and chemical composition of the carbon-rich regions, and the compositions determined, were consistent with that of epsilon carbide.

Excess carbon escaping from the martensite to retained austenite would be expected to contribute to an increase of the austenite lattice parameter and this is reflected in Fig. 6, aligned with its initial increasing carbon concentration in Fig. 7. Figure 7 also shows that after the specimen reaches 500°C the austenite phase fraction reduces with partitioning time, whilst the body-centred phase fraction increases. An explanation for this might be austenite decomposition to bainite, previously discussed as a potential alternative competing mechanism expected at longer partitioning times [35, 42]. In fact, as interest in the Q&P treatment as a potential industry procedure has increased, and various steel compositions have been subjected to various Q&P heat treatment schedules, reports of more complex microstructures containing bainitic constituents in the final microstructure are emerging [43–45]. It is important to recognize, however, that the fractions of austenite and (total) martensite appear essentially unchanged up to the peak of austenite enrichment, suggesting that carbon partitioning rather than bainite formation or interface migration is the operative mechanism at this point in the heating cycle.

The objective of the Q&P process is to stabilize austenite and so the basic theoretical model was based upon the simple assumption that 100% of the alloy carbon content is available for partitioning to untransformed austenite [33]. However, the experimentally measured austenite fraction was soon found to be lower than would be expected if this were the case [46]. The chief explanation for this has been attributed to competition with carbide formation, directly observable by transmission electron microscopy [35, 40, 47, 48], with the caveat that carbon 'trapping', and bainite formation at longer partitioning times, may also be involved. At intermediate partitioning temperatures it is most likely that carbide formation is the more dominant competing mechanism, rather than the preliminary carbon

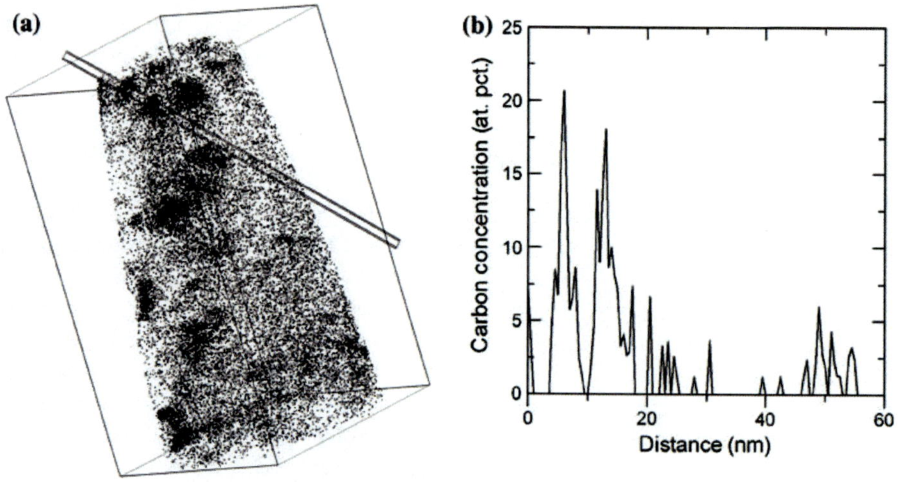

Fig. 8 Local electrode atom probe (LEAP®) results from Q&P microstructure after partitioning at 400°C for 30 s: **a** carbon atom map obtained from a volume 57 × 53 × 93 nm^3 showing carbon atom clusters or carbon-containing particles, and **b** the corresponding carbon concentration profile (along the 2 × 2 nm section indicated), with maximum peak carbon enrichment ~20 and ~18 at.%, or ~5.4 and ~4.6 wt%, respectively, for each carbon rich region traversed in this example [40]

trapping by the martensite substructure, which as observed by the experiments described above is likely to be more significant over much lower temperatures, an argument supported by a consideration of customary tempering behaviour [38].

However, further possibilities have been identified which may also influence the final austenite fraction. First, as mentioned above, the initial Q&P model assumed 100% carbon partitioning. In addition, it also incorporated idealized homogeneous partitioning across the whole austenite region, thus enabling an estimation of the optimum quench temperature required to give the appropriate fraction of martensite to provide just enough carbon to depress the M_s sufficiently to stabilize a maximum amount of retained austenite. However, subsequent calculations demonstrated that, owing to the vastly different diffusion rates between the rejecting martensite and accepting austenite, carbon partitioning across the untransformed austenite could be quite inhomogeneous, especially at short partitioning times [35, 47, 49]. Interestingly, when the local partitioning kinetics were taken into account, it was found that for particular conditions, the projected maximum retained austenite yield could be increased beyond that predicted initially for idealized partitioning [39, 40, 42, 47]. Figure 9a shows the predicted behaviour when carbon partitioning kinetics are incorporated and Fig. 9b verification from corresponding experimental measurements. Points falling above the solid idealized curve indicate a higher final retained austenite fraction.

Second, the initial model was also based upon the assumption that during partitioning the interface between the initial martensite region and untransformed austenite is stationary, precluding even short-range diffusion of iron and substitutionals; only interstitial carbon was allowed to diffuse. In this case of an immobile interface and conservation of iron and substitutionals between the two phases, the metastable ferrite/austenite equilibrium reached by the completion of carbon partitioning during the Q&P treatment was thus termed "constrained carbon equilibrium" or CCE [49–51]. It is perhaps implied that, should some of the retained austenite decompose by a bainite reaction during partitioning, probably likely to occur at longer partitioning times, this interface could perhaps migrate. However, it has also been suggested, and subsequently modelled, that unconstrained interface migration might occur, potentially at shorter times, under the thermodynamic driving force to equilibrate the chemical potential of iron between the two phases, as well as carbon, if the kinetic restriction on the diffusion of iron atoms is reduced, presumably at the higher temperature ranges of partitioning [52–54]. It has been indicated that, because the chemical potential of iron can be higher or lower in each phase, respectively, the interface could migrate either way, and interestingly, could theoretically move both ways during a single partitioning treatment [53].

Other possible complications which may influence the final volume of retained austenite have also been investigated. It has been suggested [55] that the presence of intercritical ferrite leading to the possibility of epitaxial ferrite forming during the interrupted quench could have more influence on the austenite retention than carbon diffusion from martensite to austenite during the partitioning anneal. In addition, dilatometric analysis of the Q&P process has identified the possibility of an isothermal martensite reaction at the quench temperature [56–58].

The Q&P treatment is clearly designed to produce a two-phase tempered martensite/retained austenite microstructure. In varying steel grades and compositions, more complex multiphase microstructures can arise, a straightforward example being the presence of intercritical ferrite in

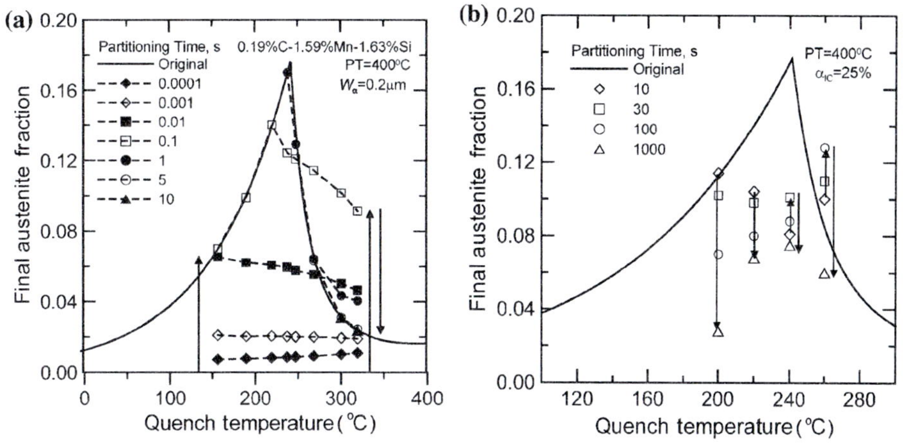

Fig. 9 Final austenite fraction as a function of quench temperature for a 0.19C–1.59Mn–1.63Si (wt%) alloy taking into account local carbon partitioning kinetics [42]; **a** predicted, and **b** measured. (The *solid curve* is the idealized calculated maximum austenite fraction as a function of initial quench temperature without incorporating carbon partitioning kinetics)

automotive steels upon which much of the early work has been focused, or the formation of bainite as a competing reaction as discussed above. Recent characterization studies have documented potential phase mixtures according to composition and Q&P heat treatment schedule [43–45, 59]. Alternatively, variation of the Q&P heat treatment concept has also been used to produce new forms of multiphase microstructure, one idea being to extend a low temperature partitioning or annealing treatment (~200°C) to deliberately austemper the higher carbon untransformed austenite fraction produced by the initial carbon partitioning to a refined low temperature bainite [60], the 'nanobainite' also discussed above. Work is likely to continue, however, on the underlying alloying fundamentals which ultimately govern the stability of the partitioned austenite fraction, for example, the substitution of Al for Si can result in retained austenite reduction whereas Mo added to a C–Mn–Si system can enhance austenite fraction [61].

In the lower partitioning temperature range, it has been fairly conclusively shown that alloying with Si has little or no effect upon formation of the metastable transition epsilon carbide. Copious formation of epsilon carbide is shown by Fig. 10 [35], which mitigates against full austenite stabilisation, but as suggested below, may alternatively be used to develop appropriate properties from the martensite fraction. Evidence for higher volume fractions of stabilised austenite [35] in the higher partitioning temperature range (when epsilon carbide is unstable), however, has been associated with suppression of cementite, and this behaviour is consistent with documented effects of Si on conventional martensite tempering reactions [62–65].

The main objective of partitioning described so far in the Q&P treatment is to supply carbon to the untransformed

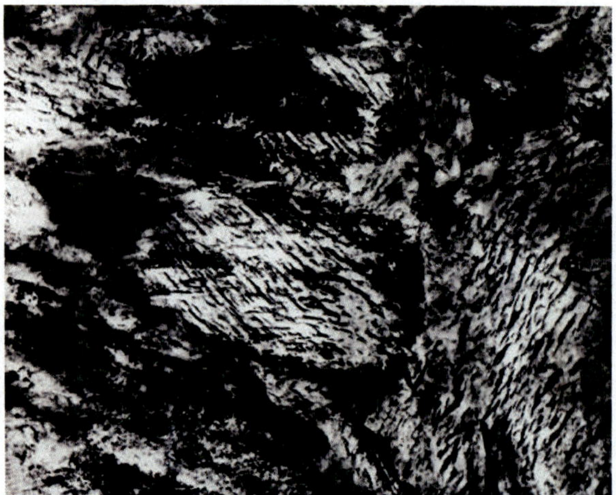

Fig. 10 Electron micrograph of Q&P microstructure illustrating epsilon carbide formation [35]

austenite phase from the supersaturated martensite. However, the "partitioning" heat treatment is also a tempering treatment of the martensite fraction. The martensite phase is thus expected also to temper in a conventional manner, moderated only by the well-known effect whereby Si can provide temper resistance. Consequently, an option embedded within the Q&P procedure to control the martensite composition to deliver temper hardness and strength, from deliberate temper carbide precipitation, has been pointed out previously [35, 66]. Figure 10 illustrates the fine-scale nature of the distribution of intermediate transitional carbides in the martensite fraction of Q&P treated steel, and although clearly a hindrance to the primary objective of achieving full austenite stabilization over a wide temperature range, this could nonetheless be traded to provide a useful strengthening increment to the martensite fraction. Thus, it has recently been reported [67] that a 0.41C–1.30Mn–1.27Si–1.01Ni–0.56Cr steel subjected to Q&P treatment has attained a tensile strength >2,400 MPa with elongation >10% from transitional carbide precipi-tation.

As mentioned above, the intermediate carbides forming over a lower partitioning temperature range are apparently not retarded by Si alloying, in fact, they are generally preserved, even to slightly higher tempering temperatures, aided by the suppression of cementite. Alternatively, strong carbide forming elements may be added to promote secondary hardening of the martensite fraction during partitioning. Thus, the Q&P terminology has been extended to emphasize this further option by developing 'tempered' Q&P steels (although this is expected to occur naturally during partitioning) [68–70]; for example, a base 0.2C–1.5Mn–1.5Si steel with 0.05Nb–0.13Mo additions subjected to the Q&P treatment has delivered a tensile strength of 1,500 MPa with 15% elongation and a 0.49C–1.2Mn–1.18Si–0.98Ni–0.21Nb steel a tensile strength >2,000 MPa with elongation >10%. Evidence of finescale NbC precipitation in the tempered martensite fraction has been detected [71, 72].

As mentioned previously, and ignoring competing reactions, the carbon available in the Q&P treatment for enriching untransformed austenite is decoupled from the initial "equilibrium" decomposition of the austenite by, say, austempering to bainite; instead, it is determined by the fraction of martensite produced at the interrupted quench temperature and the subsequent partitioning conditions. In consequence, there is control over the carbon concentration of the retained austenite phase, with potential to tune its thermal and mechanical stability. In addition, there is the as yet unexplored possibility of beneficial properties from exceptionally high carbon supersaturations, such as has been the subject of research on high alloy austenitic steels [73].

Furthermore, the nature of the martensite reaction, as compared with that of the bainite reaction, gives options for

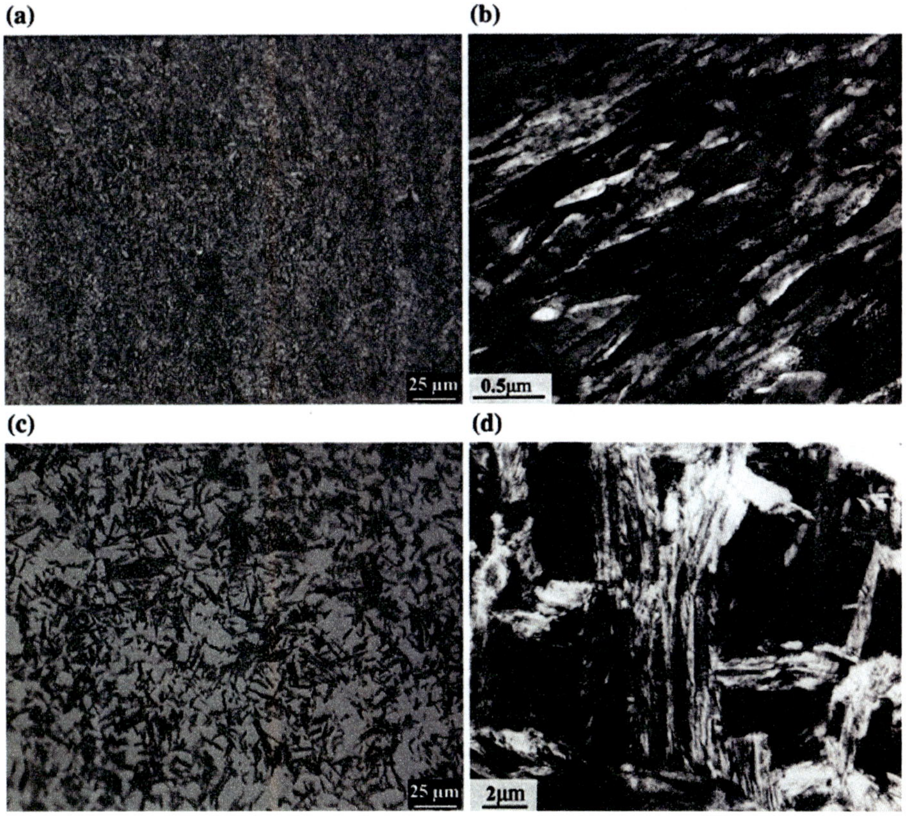

Fig. 11 Microstructures of experimental steel (0.60C–0.95Mn–1.96Si) given either, (**a**) and (**b**) a Q&P treatment with partitioning at 400°C for 120 s (for QT = 190°C), or, (**c**) and (**d**) an austempering treatment at 400°C for 120 s. (**a** and **c** are light micrographs; **b** and **d** are bright-field electron micrographs) [74, 75]

potentially achieving a more homogeneous and refined microstructure. Figure 11 demonstrates this in the same steel transformed via Q&P or austempering at the same temperature [74, 75]. The martensite/austenite structure is more homogenous and more finely divided in the Q&P structure as compared with the carbide-free bainite, which can also often contain more blocky regions of retained austenite which, if of lower carbon concentration, which is generally most likely, will be less beneficial. The Q&P microstructure can also be formed relatively quickly, limited only by the carbon diffusion rate during partitioning, whereas, as apparent from Fig. 11, bainitic microstructure will be determined by the overall bainite reaction kinetics involving and subject to, for example, the characteristic of incomplete reaction.

3.3 Commercial Steel Applications

3.3.1 Bar and Plate Steels

High hardness values >58 RHC in a 0.60 wt% high-Si AISI 9260 grade bar steel, which are intermediate between those achieved for fully-tempered martensite and bainite microstructures, can be achieved from microstructures produced by Q&P treatment [76]. This is an example where a beneficial contribution to hardness could be made from the numerous transition carbides obtainable at low partitioning temperatures in the martensite fraction in this steel [77], as discussed above. In addition, a Q&P treated steel with substantial retained austenite in the structure should possess potential for rolling contact fatigue applications [76, 78].

The mechanical properties of Si-added 0.10C–1.51Mn–1.48Si steel with microalloy additions of Nb, V, Ti and Mo have also been examined, and compared in the austempered, quenched and tempered and Q&P condition [79]. Superior impact toughness at higher yield strength levels was found after Q&P treatment compared with the other more traditional heat treatments when partitioning, austempering or tempering conditions were identical.

Good toughness and improved strength/elongation combinations have also been recorded for low alloy Ni-added steels subjected to Q&P treatment, extending strength levels towards 1,800–2,000 MPa with elongations of 5–10% [80, 81]. It has also been reported [82] that 0.20C–1.00Mn–1.50Si–3.00Ni steel subjected to Q&P treatment shows better resistance under impact abrasion after quenching and partitioning rather than quenching and tempering, thought to arise from the higher toughness and work hardening capacity generated by the deformation strengthening and toughening of the retained austenite phase. More recently, a Mn–Si–Cr alloying strategy for a 0.43C low alloy steel coupled with

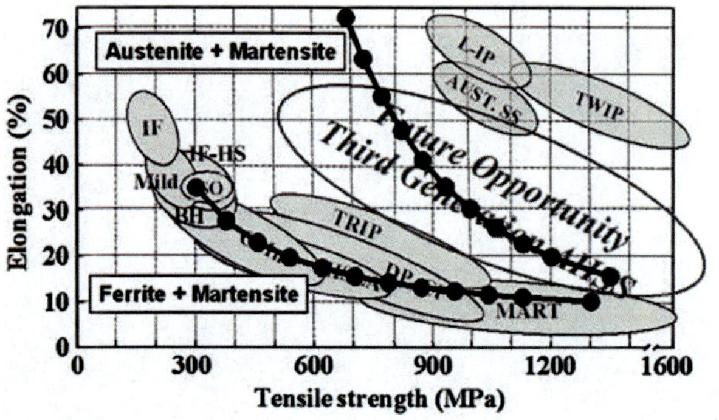

Fig. 12 Predicted potential for austenite/martensite to achieve property targets over ferrite/martensite mixtures for third generation advanced high strength sheet steels [84]

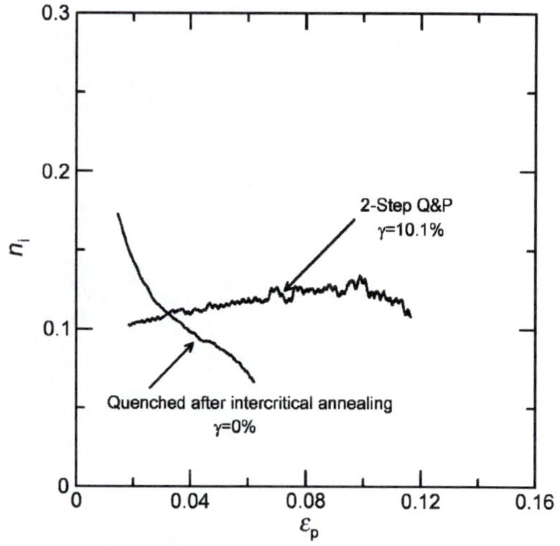

Fig. 13 Instantaneous *n* values versus plastic strain for a 0.19% C–1.59% Mn–1.63% Si steel composition, after intercritical annealing (dual-phase), followed by water-quenching, or after a Q&P heat-treatment of $QT = 240°C$; $PT = 400°C$ for 30 s (Intercritical ferrite, $\alpha_{IC} = 25\%$, for both conditions) [39, 47]

Q&P treatment has also demonstrated tensile strength levels >2,000 MPa and ductilities >10% [83].

3.3.2 Automotive Steels

Q&P treated steels with compositions similar to automotive steel grades but combining austenite with martensite, rather than with ferrite, have been proposed for third generation automotive steel (Fig. 12) [84]. The martensite fraction might be expected to enhance strength whilst, in addition, the Q&P microstructure should deliver ductility from TRIP-assisted behaviour, as demonstrated by Fig. 13 [39, 47]. In Fig. 13 instantaneous n values versus true plastic strain are compared for two treatments: one curve, exhibiting negative slope, is for a sample quenched to a dual-phase steel microstructure, whilst the second curve, showing positive slope, is for a Q&P heat-treated sample. Positive slope of instantaneous n versus plastic strain is suggestive of TRIP behavior [85]. More recently, it has been shown that austenite volume fraction after a Q&P treatment decreases with tensile strain, also consistent with TRIP behaviour, and that the steels have tensile strengths in the range 1,300–1,800 MPa with tensile elongation in the range 16–22% [86]. Thus, there have been a number of comparisons between different automotive steel heat treatments, generally dual phase, TRIP, quenched and tempered, with Q&P treated, with a consensus beginning to emerge of improved mechanical properties for the Q&P treatment [87–89], with the martensite fraction influencing strength and retained austenite controlling ductility. Thus, Q&P steel with, effectively, a tempered martensitic component, might at the very least fill a gap in the ductility/strength continuum: compared with DP and austempered TRIP steels Q&P may offer increased yield and tensile strength, whilst compared with martensitic steels, it may offer increased ductility.

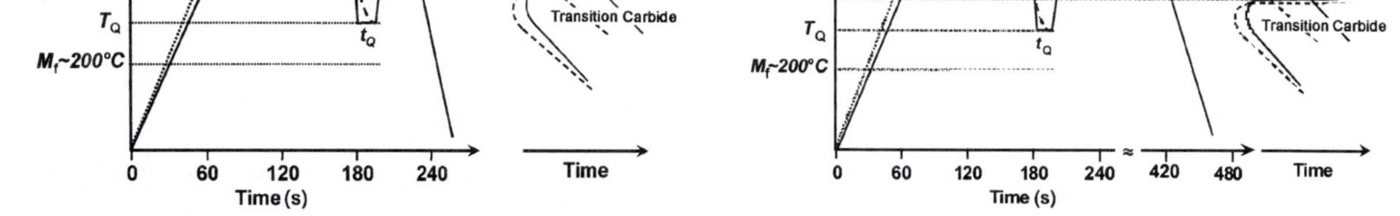

Fig. 14 Industrial potential of Q&P: **a** continuous galvanising and **b** continuous annealing lines [90]

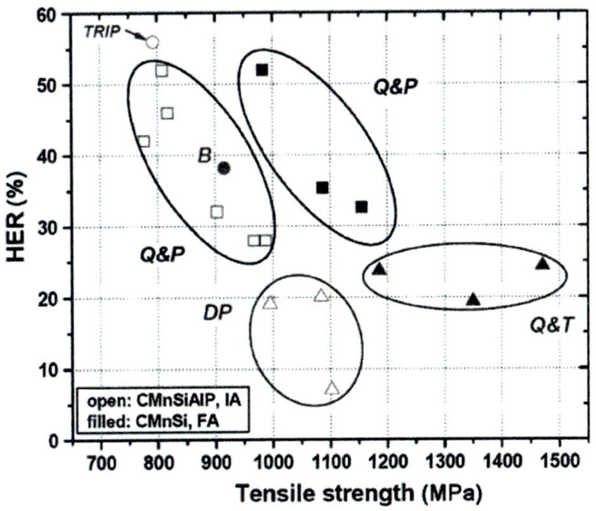

Fig. 15 Hole expansion ratio (HER) versus tensile strength for fully austenitised (*filled symbols*) and intercritically annealed (*open symbols*) Q&P microstructures [91]

Figure 14 illustrates schematically how the Q&P heat treatment, and required time/temperature/carbide-precipitation kinetics, could fit with continuous automotive sheet processing lines [90]. In addition, to assess automotive application better, comparative hole expansion tests have compared austempered TRIP, bainitic (B), dual phase (DP) and tempered martensite (Q&T) microstructures with Q&P microstructures [91]. As shown by Fig. 15 a better hole expansion ratio (HER) versus tensile strength performance of Q&P microstructure over these alternative steel conditions was found.

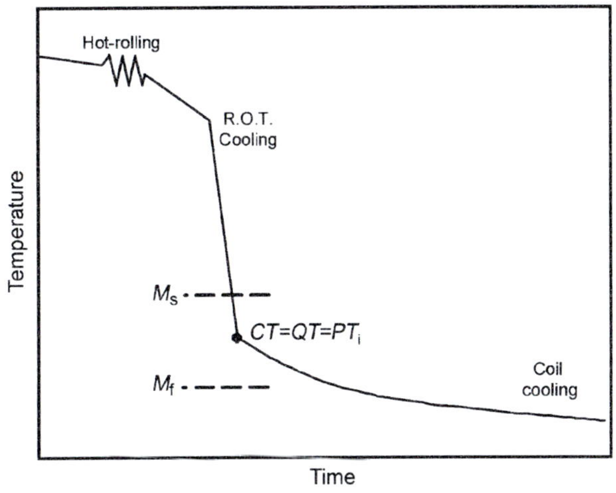

Fig. 16 Industrial potential of Q&P: hot-strip rolling [92]

3.3.3 Hot Strip Steels

The possibility of Q&P treatment on a hot-strip mill is suggested by the schematic of Fig. 16. After rolling reduction, accelerated cooling on the run-out table is the quenching stage, interrupted by coiling (*QT*) to give much reduced cooling in the coil, which can thus be approximated to a one-step partitioning treatment ($PT = QT$). The partitioning stage of this process is non-isothermal, albeit with a slow cooling rate, but initial simulations indicate significant levels of retained austenite and high strength levels [92].

4 Conclusions

Carbide-free bainitic microstructures containing carbon-stabilised retained austenite, first developed to analyse fundamental mechanisms of the bainite transformation, have also resulted in steels, and also cast irons, with useful combinations of mechanical properties. Similar carbide-free microstructures containing retained austenite, but based upon martensite, achieved by a combination of alloying and novel heat treatment procedures, have followed, with added potential for control over phase fraction, chemistry and properties of both martensite and austenite fractions. In particular, this novel concept of Quenching and Partitioning (Q&P) heat treatment provides a new means of expanding applications of martensitic steels, based upon potential improvements in combinations of strength, ductility, formability and toughness. In consequence, data are now emerging which show that promising combinations of properties are achievable by application of the Q&P concept. Furthermore, preliminary considerations also indicate that the novel but unconventional heat treatment cycle required could nonetheless fit with existing high-tonnage processing lines, especially for cold-rolled and hot-rolled sheet.

Acknowledgments A research programme linked to this article was conducted initially under an international collaboration supported in the authors' respective countries by NSF Grant # 0303510 (USA) and EPSRC Grant ref: GR/S86501 (UK). Support by sponsors of the Advanced Steel Processing and Products Research Center at the Colorado School of Mines is also acknowledged.

References

1. J.H. Woodhead, S.R. Keown, Am. Soc. Metals **15** (1986)
2. T. Gladman, *The Physical Metallurgy of Microalloyed Steels* (Institute of Materials, London, 1997)
3. G.E. Totten, L. Xie, K. Funatani (eds.), *Handbook of Mechanical Alloy Design* (Marcel Dekker, New York, 2004)
4. A.R. Marder, J.I. Goldstein (eds.), *Phase Transformations in Ferrous Alloys* (AIME, 1984)
5. H.K.D.H. Bhadeshia, D.V. Edmonds, Metall. Trans. **10A**, 895 (1979)

6. S. Matas, R.F. Hehemann, Trans. AIME **221**, 176 (1961)
7. J.P. Oblak, R.F. Hehemann, *Transformations and Hardenability in Steels* (Climax Molybdenum Co., 1967)
8. R.F. Hehemann, *Phase Transformations* (American Society for Metals, OH, USA, 1970)
9. R.F. Hehemann, K.R. Kinsman, H.I. Aaronson, Metall. Trans. **3**, 1077 (1972)
10. H.K.D.H. Bhadeshia, D.V. Edmonds, Acta Metall. **28**, 1265 (1980)
11. J.G. Speer, D.V. Edmonds, F.C. Rizzo, D.K. Matlock, Curr. Opin. Solid-State Mater. Sci. **8**, 219 (2004)
12. A.A. Honarbakhsh-Raouf, Dissertation, University of Leeds, UK (1997)
13. H.K.D.H. Bhadeshia, D.V. Edmonds, Metal Sci. J. **17**, 411 (1983)
14. V.T.T. Miihkinen, D.V. Edmonds, Mater. Sci. Technol. **3**, 422 (1987)
15. J.A. Charles, G.W. Greenwood, G.C. Smith (eds.), *Future Developments of Metals and Ceramics* (London, 1992)
16. J.K. Yates, Sci. Parliam. **53**, 2 (1996)
17. J.K. Yates, (1996) British Steel Track Products, Innovation in Rail Steel. http://www.msm.cam.ac.uk/Phase-Trans/Parliament.html
18. V.F. Zackay, E.R. Parker, Steel **161**, 73 (1967)
19. V.F. Zackay, E.R. Parker, D. Fahr, B. Bush, ASM Trans. Q **60**, 252 (1967)
20. I. Tamura, Metal Sci. **16**, 245 (1982)
21. O. Matsumura, Y. Sakuma, H. Takechi, Scripta Metall. **21**, 1301 (1987)
22. O. Matsumura, Y. Sakuma, H. Takechi, Trans. ISIJ **27**, 570 (1987)
23. P. Jacques, Y. Carnot, P. Harlet, J. Ladriere, F. Delannay, Metall. Trans. **29A**, 2383 (1998)
24. E. Girault, P. Jacques, P. Ratchev, J. Van Humbeeck, B. Verlinden, E. Aernoudt, Mater. Sci. Eng. **A273**, 471 (1999)
25. B.C. De Cooman, Curr. Opin. Solid State Mater. Sci. **8**, 285 (2004)
26. F.G. Caballero, H.K.D.H. Bhadeshia, J.A. Mawella, D.G. Jones, P. Brown, Mater. Sci. Technol. **18**, 279 (2002)
27. C. Garcia-Mateo, F.G. Caballero, H.K.D.H. Bhadeshia, J. De Phys. Colloque **112**, 285 (2003)
28. C. Garcia-Mateo, F.G. Caballero, H.K.D.H. Bhadeshia, ISIJ **43**, 1238 (2003)
29. F.G. Caballero, H.K.D.H. Bhadeshia, J.A. Mawella, D.G. Jones, P. Brown, Mater. Sci. Technol. **17**, 512 (2001)
30. H.K.D.H. Bhadeshia, in *Proceedings Conference New Developments on Metallurgy and Applications of High Strength Steels*, Buenos Aires, 3 (2008)
31. P.M. Brown, D.P. Baxter, in *Proceedings Conference Materials Science and Technology, Physical Metallurgy and Property Development* 433 (2004)
32. J.G. Speer, D.K. Matlock, B.C. De Cooman, J.G. Schroth, Acta Mater. **51**, 2611 (2003)
33. J.G. Speer, A.M. Streicher, D.K. Matlock, F.C. Rizzo, G. Krauss, Austenite Formation and Decomposition, 505 (2003)
34. D.K. Matlock, V.E. Brautigam, J.G. Speer, in *Proceedings Conference Thermec'03*. Trans Tech Publications, 1089 (2003)
35. D.V. Edmonds, K. He, F.C. Rizzo, B.C. De Cooman, D.K. Matlock, J.G. Speer, Mater. Sci. Eng. **A438–440**, 25 (2006)
36. T.D. Bigg, D.V. Edmonds, in *Proceedings 7th Pacific Rim conference Advanced Materials and Processing*. Mater Sci Forum, vol. 33 (2010) pp. 654–656
37. T.D. Bigg, D.K. Matlock, J.G. Speer, D.V. Edmonds, in *Proceedings conference Solid-Solid Phase Transformations in Inorganic Materials* (2010)
38. D. Kalish, E.M. Roberts, Metall. Trans. **2**, 2783 (1971)
39. A.J. Clarke, Dissertation, Colorado School of Mines, CO, USA (2006)
40. D.V. Edmonds, K. He, M.K. Miller, F.C. Rizzo, A. Clarke, D.K. Matlock, J.G. Speer, in *Proceedings conference Thermec'06*, Mater. Sci. Forum **4819**, 539–543 (2007)
41. A.J. Clarke, J.G. Speer, M.K. Miller, R.E. Hackenberg, D.V. Edmonds, D.K. Matlock, F.C. Rizzo, K.D. Clarke, E. De Moor, Acta Mater. **56**, 16 (2008)
42. A.J. Clarke, J.G. Speer, D.K. Matlock, F.C. Rizzo, D.V. Edmonds, M.J. Santofimia, Scripta Mater. **61**, 149 (2009)
43. M.J. Santofimia, T. Nguyen-Minh, L. Zhao, D.N. Hanlon, T.A. Kop, J. Sietsma, in *Proceedings conference New Developments in Advanced High-Strength Sheet Steels*, vol. 191 (2008)
44. M.J. Santofimia, L. Zhao, J. Sietsma, Metall. Trans. **40A**, 46 (2009)
45. H.Y. Li, X.W. Lu, X.C. Wu, Y.A. Min, X.J. Jin, Mater. Sci. Eng. **A527**, 6255 (2010)
46. F.L.H. Gerdemann, J.G. Speer, D.K. Matlock, in *Proceedings Conference Materials Science and Technology*, (2004)
47. D.V. Edmonds, K. He, F.C. Rizzo, A. Clarke, D.K. Matlock, J.G. Speer, in *Proceedings 1st conference Super-High Strength Steels* (2005)
48. K. He, D.V. Edmonds, J.G. Speer, D.K. Matlock, A. Clarke, F.C. Rizzo, in *Proceedings 16th. Microscopy Congress*, vol. 3, (2006) 1702
49. M. Hillert, J. Agren, Scripta Mater. **50**, 697 (2004)
50. J.G. Speer, D.K. Matlock, B.C. De Cooman, J.G. Schroth, Scripta Mater. **52**, 83 (2005)
51. M. Hillert, J. Agren, Scripta Mater. **52**, 87 (2005)
52. N. Zhong, X. Wang, Y. Rong, L. Wang, J. Mater. Sci. Technol. **22**, 751 (2006)
53. J.G. Speer, R.E. Hackenberg, B.C. De Cooman, D.K. Matlock, Phil. Mag. Lett. **87**, 379 (2007)
54. M.J. Santofimia, L. Zhao, J. Sietsma, Scripta Mater. **59**, 159 (2008)
55. M.J. Santofimia, L. Zhao, R. Petrov, J. Mater. Character **59**, 1758 (2008)
56. S.J Kim, H.S. Kim, B.C. De Cooman, in *Proceedings conference Materials Science and Technology*, MS&T'07, Steel Properties and Applications, 73 (2007)
57. S.J. Kim, J.G. Speer, H.S. Kim, B.C. De Cooman, in *Proceedings conference New Developments in Advanced High-Strength Sheet Steels*, 179 (2008)
58. D.H. Kim, J.G. Speer, H.S. Kim, B.C. De Cooman, Metall. Trans. **40A**, 2048 (2009)
59. H.Y. Li, X.J. Jin, in *Proceedings 7th Pacific Rim conference Advanced Materials and Processing*. Mater. Sci. Forum **86**, 654–656 (2010)
60. H. Li, X. Jin, Chin. J. Mech. Eng. **22**, 645 (2009)
61. E. De Moor, J. Penning, C. Fojer, A. Clarke, J.G. Speer, in *Proceedings Conference New Developments in Advanced High-Strength Sheet Steels*, 199 (2008)
62. W.S. Owen, Trans. ASM **46**, 812 (1954)
63. J. Gordine, I. Codd, J. Iron Steel Inst. **207**, 461 (1969)
64. W.C. Leslie, G.C. Rauch, Metall. Trans. **9A**, 343 (1978)
65. S.J. Barnard, G.D.W. Smith, A.J. Garratt-Reed, Sande J. Vander, in *Proceedings Conference Advances in the Physical Metallurgy and Applications of Steels*, 33 (1981)
66. D.V. Edmonds, in *Proceedings 6th conference Processing and Manufacturing of Advanced Materials, Thermec'09*, Mater. Sci. Forum **110**, 638–642 (2010)
67. H.Y. Li, X.W. Lu, W.J. Li, X.J. Jin, Metall. Trans. **41A**, 1284 (2010)
68. T.Y. Hsu, Z.Y. Xu, Mater. Sci. Forum **561–565**, 2283 (2007)
69. X.D. Wang, N. Zhong, Y.H. Rong, T.Y. Hsu, L. Wang, J. Mater. Res. **24**, 260 (2009)
70. N. Zhong, X.D. Wang, L. Wang, Y.H. Rong, Mater. Sci. Eng. **A506**, 111 (2009)
71. X.D. Wang, W.Z. Xu, Z.H. Guo, L. Wang, Y.H. Rong, Mater. Sci. Eng. **A527**, 3373 (2010)
72. Y. Wang, S. Zhou, Z. Guo, Y. Rong, in *Proceedings 7th Pacific Rim conference Advanced Materials and Processing*, Mater. Sci. Forum **37**, 654–656 (2010)

73. Y. Cao, F. Ernst, G.M. Michal, Acta Mater. **51**, 4171 (2003)
74. D.V. Edmonds, K. He, F.C. Rizzo, J.G. Speer, D.K. Matlock, in *Proceedings conference New Developments on Metallurgy and Applications of High Strength Steels*, 829 (2008)
75. K. He, D.V. Edmonds, J.G. Speer, D.K. Matlock, F.C. Rizzo, in *Proceedings 14th European Microscopy Congress, EMC'08*. vol. 2, Materials Science, 429 (2008)
76. F.L.H. Gerdemann, Dissertation, Aachen University of Technology, Germany (2004)
77. G. Krauss, Colorado School of Mines, CO, USA. Private communication (2009)
78. J.G Speer, A.J. Shutts, D.K. Matlock, D.V. Edmonds, F. Rizzo, E.B. Damm, in *Proceedings conference New Developments in Long and Forged Products: Metallurgy and Applications*, 191 (2006)
79. S.C. Hong, J.C. Ahn, S.Y. Nam, S.J. Kim, H.C. Yang, J.G. Speer, D.K. Matlock, Metals Mater. Int. **13**, 439 (2007)
80. F. Rizzo, A.R. Martins, J.G. Speer, D. Matlock, A. Streicher, B. De Cooman, in *Proceedings 5th conference Processing and Manufacturing of Advanced Materials*, Mater. Sci. Forum **4476**, 539–543 (2007)
81. A.R. Martins, F. Rizzo, D. Coelho, J.G. Speer, D.K. Matlock, K. He, D. Edmonds, in *Proceedings Conference Materials Science and Technology, MS&T'09*, 825 (2009)
82. C. Wang, J. Shi, S. Liu, S. Du, H. Dong, Chin. J. Mater. Res. **23**, 305 (2009)
83. B. Masek, H. Jirkova, D. Klauberova, L. Kucerova, D. Hauserova, in *Proceedings 7th Pacific Rim conference Advanced Materials and Processing*, Mater. Sci. Forum **94**, 654–656 (2010)
84. D.K. Matlock, J.G. Speer, *Structural Steels*, 774 (2006)
85. O. Yakubovsky, N. Fonstein, D. Bhattacharya, in *Proceedings conference Trip-Aided High Strength Ferrous Alloys*, 263 (2002)
86. W.Q. Cao, C.Y. Wang, J. Shi, H. Dong, in *Proceedings 7th Pacific Rim conference Advanced Materials and Processing*, Mater. Sci. Forum **29**, 654–656 (2010)
87. H.J. Jun, N. Fonstein, in *Proceedings Conference New Developments in Advanced High-Strength Sheet Steels*, 155 (2008)
88. J-W. Jin, S.H. Byun, S.B. Lee, S.I. Kim, C.S. Oh, N. Kang, K-M. Cho, in *Proceedings Conference New Developments in Advanced High-Strength Sheet Steels*, 169 (2008)
89. X. Wang, Z. Guo, Y. Rong, in *Proceedings 7th Pacific Rim conference Advanced Materials and Processing*, Mater. Sci. Forum **90**, 654–656 (2010)
90. B.C. De Cooman, J.G. Speer, Steel Res. Inst. **77**, 634 (2006)
91. E. De Moor, Dissertation, Universiteit Gent, Belgium (2009)
92. G.A. Thomas, J.G. Speer, DK Matlock, in *Proceedings conference New Developments in Advanced High-Strength Sheet Steels*, 227 (2008)

Development and Application of Q&P Sheet Steels

Li Wang and Weijun Feng

Abstract

Two important objectives of the automotive industry are the decrease in car weight and improvement in safety. High strength steels (HSS) especially advanced high strength steels (AHSS) and the third-generation HSS are the main measures to reduce automotive weight and improve safety in steel industry. Quenching and partitioning (Q&P) has been recently proposed by Speer as a fundamentally new way of producing martensite steels containing a considerable amount of retained austenite, it possesses good balance properties between tensile strength and elongation. In order to produce Q&P steel, a special thermal treatment is required, it consists of a two step thermal treatment. A special line designed by Baosteel especially for UHSS production was launched in March 2009, it is easy to implement the annealing cycle of Q&P. The industry trails with conventional C–Si–Mn TRIP780 composition by Q&P concept were carried out in this line. The results show that a good balance property is achieved, the microstructure and other properties are measured and discussed. A component of B pillar is made successfully by Q&P1000.

Keywords

Q&P • AHSS • C–Si–Mn • Baosteel

1 Introduction

Increasing requirements for automotive weight reduction and safety have led to continuous growth in application of high strength steels (HSS) especially advanced high strength steels (AHSS) and the third-generation HSS in steel industry. Quenching and partitioning (Q&P) steel has been recently proposed by Speer [1–3] as a fundamentally new way of producing martensite steels containing a considerable amount of retained austenite. In order to produce Q&P steel, a special thermal treatment is required, it consists of a two step thermal treatment. It is difficult to carry out with this special annealing cycle in the conventional continuous annealing line (CAL) or continuous galvanized line (CGL). The good balance properties between strength and ductility can be achieved of Q&P steel in the experiments of laboratory [4–6], however, the properties results of industrialized Q&P steel have never been reported.

In this paper, the steels with conventional C–Si–Mn TRIP780 composition were carried out by Q&P concept at special line designed by Baosteel especially for UHSS production. It shows that 1000MPa grade cold rolled (CR) can be achieved by C–Si–Mn TRIP780 composition. X-ray diffraction, optical and scanning electron microscopy have been applied to characterize the microstructures. It includes ferrite martensite and retained austenite. The formability of Q&P steel is also studied by hole-expanding and cold-bending test. A component of B pillar is made successfully by Q&P1000.

L. Wang (✉) · W. Feng
R&D Center, Baoshan Iron and Steel Co., LTD, 889 Fujin Road, Baoshan District, Shanghai, 201900, China
e-mail: lwang@baosteel.com

Y. Weng et al. (eds.), *Advanced Steels*, DOI: 10.1007/978-3-642-17665-4_25,

2 Experimental Procedure

The chemical composition of the C–Si–Mn steel used is shown in Table 1. Si addition is used to suppress the formation of carbide [7], and Mn is used to enhance the stability of austenite. The slabs were reheated to a temperature of 1,250°C and hot rolled to a thickness of 3.5 mm. The finishing temperature was about 860°C and coiling temperature was about 600°C. The thickness of cold rolled steel sheets is 2 mm.

The heat treatment was conducted in the special UHSS line on cold rolled sheet. Figure 1 shows the schematic of production line for UHSS in Baosteel, the main characteristics and advantages of this line are as follows:

(1) The tensile strength of 340–1,500 MPa for cold-rolled and 340–1,200 MPa tensile strength for hot-dip galvanizing sheet steel can be produced in this line. This unit particularly suitable for the production of ultra-HSS, it is the only one line for ultra-HSS in China.
(2) Production line with multi-functional, and produces cold-rolled sheets and hot-dip galvanizing sheets. Cold-rolled sheets can be produced by high concentrations of hydrogen cooling and water quenching. Hot-dip galvanized sheets can be produced by conventional cooling (5% H_2 + 95% N_2) and high concentration of hydrogen cooling, the highest concentration of hydrogen is about 75%.
(3) Laser welding machine to ensure that the ultra-HSS welding and the production quality and stable operation of the strip.
(4) The direct fire is adopted in the heating furnace; it is conducive to improve the zinc coating performance of HSS.
(5) The cooling rate of sheet is fast, when using more than 60% concentration of hydrogen for cooling, the cooling rate can be reach 140°C/s or more for 1 mm thick sheet; when the water quenching is used, the steel sheet cooling rate can exceed 500°C/s.
(6) Cold-rolled steel sheet surface coating has a good performance, especially for water quenching sheet steel. Because the sheet is pickled after water quenching, and the Mn-rich elements of surface will be reduced, thus further improving the corrosion resistance of cold-rolled steel sheet and painting performance.

The mechanical properties are measured using JIS5# in the transverse direction. The samples are two-step etched with 2% Nital and sodium metabisulfite ($Na_2S_2O_5 + H_2O$) [8] for optical microscopy analysis and etched with 2% Nital for observation using JEM-2100F Scanning Electron Microscopy (SEM). The amount of retained austenite is

Table 1 Chemical compositions of experimental steel (wt%)

C	Si	Mn	P	S	Al
0.2	1.5	1.8	0.010	0.004	0.046

Table 2 Mechanical properties of Q&P steels

Coil No.	YS (MPa)	TS (MPa)	ELu_{50} (%)	ELt_{50} (%)
1	675	981	13.5	21.2
2	647	995	13.2	20.9
3	625	1,032	11.6	18.0
4	857	1,061	9.1	15.1
5	851	1,065	10.2	17.8
6	773	1,097	7.7	13.5
7	1,029	1,098	7.9	16.1
8	908	1,117	6.9	11.9
9	1,107	1,223	4.4	9.3

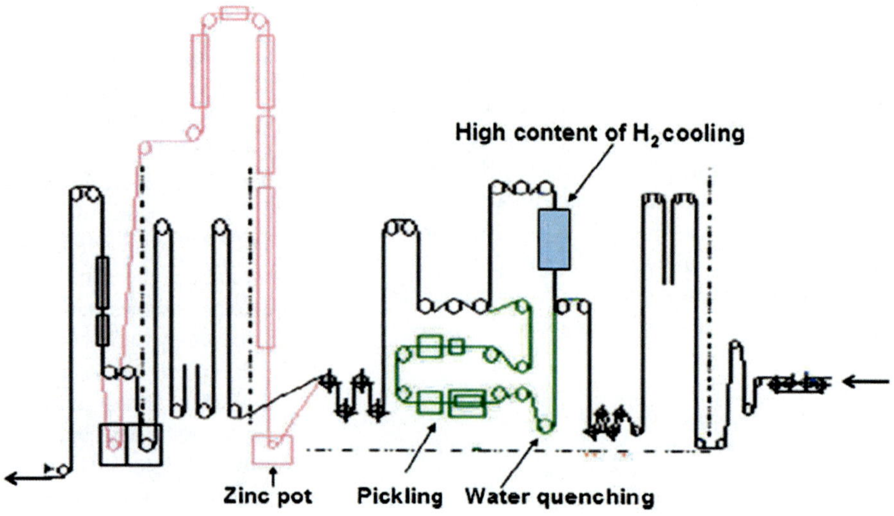

Fig. 1 The schematic of production line for HSS

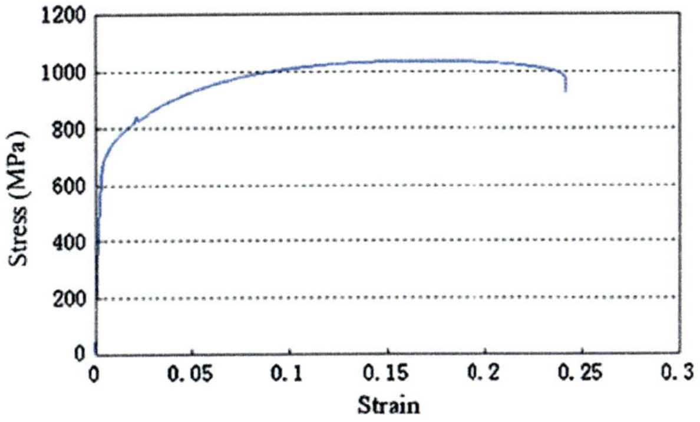

Fig. 2 The typical stress–strain curve of Q&P1000

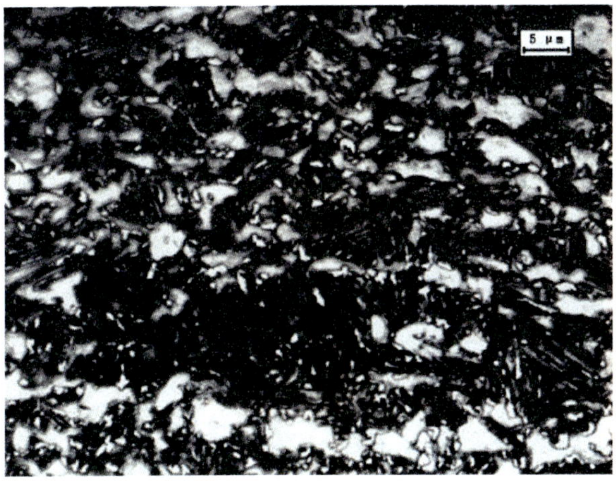

Fig. 3 Optical microstructure of typical Q&P1000

quantified by X-ray diffractometry using Mo–Ka radiation. Hole-punching and hole-expanding tests are conducted using disc specimens of 50 mm in diameter with oil lubricant. The cold-bending property is measured in the transverse and longitudinal direction.

3 Results and Discussion

3.1 Mechanical Properties

The mechanical properties of the TRIP780 steels treated by Q&P process in the special UHSS line of Baosteel are presented in Table 2, there are nine coils. With the typical TRIP780 steel composition, higher tensile strength (1000–1200 MPa grade) can be achieved by Q&P treatment, and the ductility is also good (9.3–21.2%). Comparison with the same grade of DP and Martensite sheet steels, Q&P steels possess excellent balance tensile strength and elongation (see Fig. 2). Although TRIP1000 possess good balance between tensile strength and elongation, the carbon content is very high, it is about 0.3 wt% [9]. So the weldability of Q&P steels are better than the same grade of TRIP steel.

3.2 Microstructure and Stability of Retained Austenite

Microstructures of Q&P1000 (Coil No. 5) is shown in Fig. 3. The microstructure includes ferrite (white), martensite (laths) and retained austenite (light near the grain boundary). With the increase of tensile strength, the amount of ferrite decrease.

SEM is carried out to observe the characteristics of retained austenite. Figure 4 shows the microstructure of sample from Coil No. 5. The large gray grains are ferrite while the brighter phases are martensite and retained austenite. The retained austenite usually appears as a small gray block with smooth and rounded edges at the laths martensite boundaries. It concludes that a considerable amount of retained austenite is obtained by Q&P treatment at industry continuously annealing line.

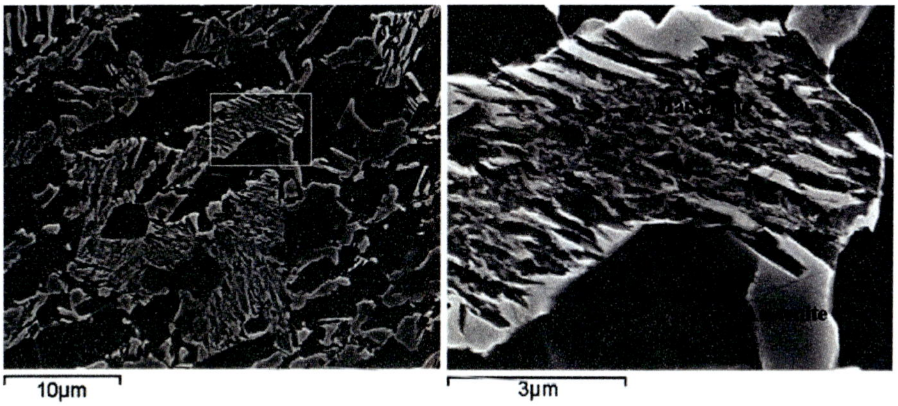

Fig. 4 Typical microstructure of Q&P1000

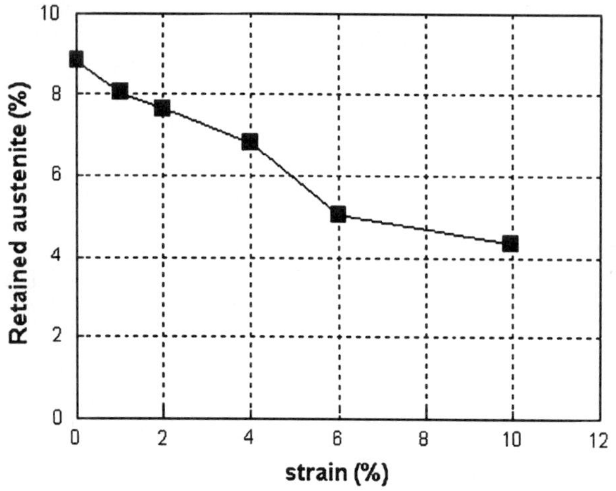

Fig. 5 The relationship between strain and retained austenite

Table 3 Hole expansion and bendability of Q&P1000

λ(%)	R min(⊥roll direction)	R min(‖roll direction)
30	1 t	1.5 t

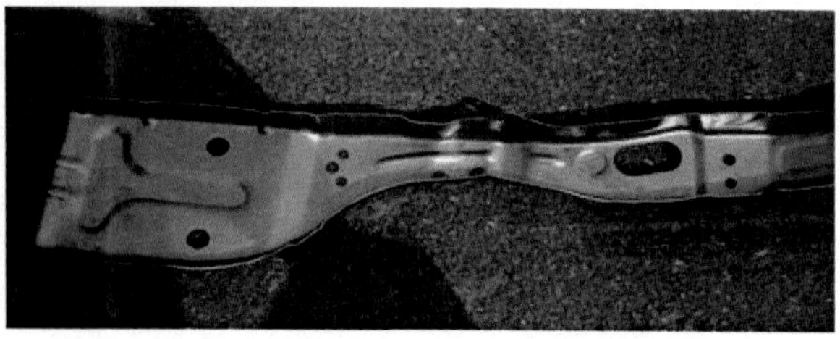

Fig. 6 The relationship between strain and retained austenite

Figure 5 shows the relationship between strain and retained austenite measured from Coil No. 5. The amount of retained austenite in annealed Q&P1000 steel is about 9%, with the increase of strain, the retained austenite of the sample decrease quickly. When the strain reach to 10%, the amount of retained austenite decrease to about 4%. This phenomenon is similar to TRIP steel.

3.3 Other Properties and Application

Table 3 shows the variations in hole expansion ratio and minimum relative bending radius of Q&P1000. The hole expansion ratio is about 30%, similar to the same grade DP steel. The minimum relative bending radius of perpendicular rolling direction is better than parallel rolling direction. Figure 6 shows the photograph of B pillar made by Q&P1000.

4 Conclusions

The steels with conventional C–Si–Mn TRIP780 composition are carried out by Q&P concept at a special commercial line designed by Baosteel, the following results are achieved:

(1) The good balance properties between strength and ductility can be achieved of C-Si-Mn TRIP780 steel by Q&P treatment. It possesses both higher tensile strength and excellent total elongation. The tensile strength is from 1,000–1,200 MPa and the elongation is from 9.3 to 21.2%.
(2) Q&P steels are composed of ferrite and carbon-depleted laths martensite and retained austenite.
(3) Comparison with the same grade of DP, Mart and TRIP steel, the Q&P steels possess good balance properties between high tensile strength and excellent stretch-flangeability.
(4) Q&P1000 are drawn successfully to a complex component (B pillar) of a domestic car.

References

1. J.G. Speer, A.M. Streicher, D.K. Matlock, F. Rizzo, G. Krauss, *Quenching and Partitioning: A Fundamentally New Process to Create High Strength TRIP Sheet Microstructures* Austenite Formation and Decomposition ed. by E.B. Damm and M.J. Merwin, (TMS, Warrendale, PA, 2003), pp. 505–522
2. J.G. Speer, D.K. Matlock, B.C. DeCooman, J.G. Schroth, Carbon partitioning into austenite after martensite transformation. Acta Mater. **51**, 2611–2622 (2003)
3. J.G. Speer, F.C. Rizzo, D.K. Matlock, D.V. Edmonds, The 'Quenching and Partitioning' Process: Background and Recent Progress, in *Proceedings of 59th Annual ABM Congress*, ABM, São Paulo, Brazil, 2004, pp. 4824–4836. Also published in Mater. Res. **8**(4), 417–423 (2005)
4. A.M. Streicher, J.G. Speer, D.K. Matlock, B.C. De Cooman, *International Conference on Advanced High Strength Sheet Steels for Automotive Applications Proceedings*, ed. by J.G. Speer, Quenching and Partitioning Response of a Si-Added TRIP Sheet Steel, AIST, (Warrendale, PA, 2004), pp. 51–62
5. J.W. Jin, S.H. Byun, S.B. Lee, S.I. Kim, C.S. Oh, N. Kang, K.M. Cho. *Mechanical Properties Comparison of DP, TRIP, Q&P Steels as a function of Heat Treatment Condition. In International Conference on New Developments in Advanced High-Strength Sheet Steel.* Florida, USA, 2008, p. 169
6. S.C. Hong, J.C. Ahn, S.Y. Nam, S.J. Kim, H.C. Yang, J.G. Speer, D.K. Matlock, Mechanical properties of high-Si plate steel produced by the quenching and partitioning process. Met. Mater. Int. **13**(6), 439–445 (2007)
7. M.F. Gallagher, J.G. Speer, D.K. Matlock, *Microstructure Development in TRIP-Sheet Steels Containing Si, Al, and P*, 44th MWSP Conference Proceedings, vol. XL, 2002
8. A.K. De, J.G. Speer D.K. Matlock, Color tint-etching for multiphase steels, Adv. Mater. Process, February 1, 2003
9. M. Takahashi, H. Yoshida, S. Hiwatashi, *Properties of TRIP Type High Strength Steels* International Conference on TRIP-Aided High Strength Ferrous Alloys, ed. by Prof. Dr. Ir. B.C. De Cooman, 2002, pp. 103–111

Microstructure and Mechanical Properties of Al-Added High Mn Austenitic Steel

Jae-Eun Jin and Young-Kook Lee

Abstract

Many studies about high Mn austenitic steels have been done due to their excellent mechanical properties such as tensile strength, uniform ductility, and wear resistance. In this paper, we tried to summarize the effects of Al addition on carbide precipitation, stacking fault energy (SFE), dynamic strain aging (DSA), tensile properties, and hydrogen delayed fracture in high Mn austenitic steels such as Hadfield and TWIP steels. The addition of Al suppressed the cementite precipitation due to the decreases in both activity and diffusivity of C in austenite. However, the κ'-carbide precipitation with a chemical formula $(Fe, Mn)_3AlC_x$ may occur in the high Mn and Al austenitic steel. The addition of Al linearly increased the stacking fault energy of austenite. The Al addition increased the strain hardening rate in the Hadfield steels owing to the increase in the DSA effect caused by the increased solute C content, while it decreased the strain hardening rate in the high Mn TWIP steel because of the decreases in mechanical twinning caused by high SFE and in DSA effect caused by slow diffusivity of C atoms. The addition of Al improved the resistant property to the hydrogen delayed fracture in the high Mn TWIP steels.

Keywords

Twinning-induced plasticity • Stacking fault energy • Strain hardening rate • Dynamic strain aging • Hydrogen delayed fracture

1 Introduction

Advanced high strength steels (AHSS) for automobiles have been developed due to the demands of excellent mechanical properties, high recycling efficiency, and lightweight for CO_2 saving [1]. Many studies have been done about the so-called the first-generation AHSS such as dual phase (DP), complex phase (CP), martensite, and transformation-induced plasticity (TRIP) steels, in which low-temperature transformation phases like bainite and martensite play the part of the strengthening. The first-generation AHSS possess the high strength of above 800 MPa at the expense of uniform ductility, resulting in the limited strength–ductility balance of about 20,000 MPa% [2].

High Mn austenitic steels, such as Hadfield [3–5], TRIP, twinning-induced plasticity (TWIP) [6–13], and shear/micro-band induced plasticity (SBIP/MBIP) steels [14], have an excellent combination of high strength of above 800 MPa and large uniform ductility of over 60% because of the high strain hardening rate by the formation of martensite, twins, or dislocation bands during the plastic deformation. They are generally called the second-generation AHSS. The strength–ductility balance of about 60,000 MPa% of the second-generation AHSS is three times higher than that of the first-generation AHSS. Especially, the high Mn TWIP steel has been spotlighted because of excellent mechanical properties.

J.-E. Jin · Y.-K. Lee (✉)
Department of Materials Science and Engineering,
Yonsei University, Seoul 120-749, Korea
e-mail: yklee@yonsei.ac.kr

Y. Weng et al. (eds.), *Advanced Steels*, DOI: 10.1007/978-3-642-17665-4_26,

Various researches on the Fe-Mn-C TWIP steels have been performed in terms of grain size, mechanical twinning, stacking fault energy (SFE), dynamic strain aging (DSA), and tensile properties [6–13]. Of them, in the present paper, especially the effects of Al addition on carbide precipitation, SFE, strain hardening, DSA, and hydrogen delayed fracture of the high Mn TWIP steel have been summarized.

2 Effect of Al on Carbide Precipitation

2.1 Cementite Precipitation

Figure 1 shows the carbide precipitation kinetics in Hadfield steels with (a) 1.25% C and 0% Al and (b) 1.75% C and 1.3% Al [3].

The start time of cementite precipitation in the Hadfield steel with 1.75% C and 1.3% Al was similar to that of the steel with 1.25% C and 0% Al, implying that the Al addition delayed the precipitation start of cementite in Hadfield steel. The Al addition also changed the morphology of carbides from thin plates to equiaxed blocks [3].

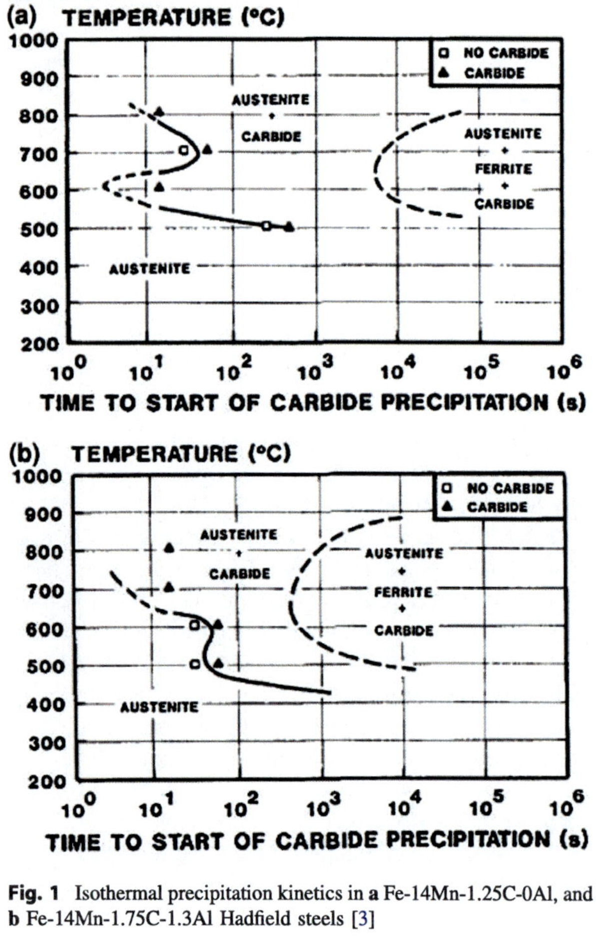

Fig. 1 Isothermal precipitation kinetics in **a** Fe-14Mn-1.25C-0Al, and **b** Fe-14Mn-1.75C-1.3Al Hadfield steels [3]

The effect of Al on the C activity in Hadfield steel was studied using a diffusion couple in the previous study [3]. Cylinders measuring 25 mm in diameter and 40 mm in length were machined from the Hadfield steel containing 1.5% C and different Al contents of 0 and 2.7%. One end of each cylinder was machined to be flat and perpendicular to the cylinder axis. The ends of two cylinders having different Al contents were joined by electron beam welding with a weld zone of 1.5 mm wide. Two couples were sealed in evacuated quartz capsules and annealed for 260 h at 1,100°C. Both as-welded and the 260 h annealed couples were sliced by lathe turning at intervals of 0.5 mm along the cylinder axis. The C concentration was analyzed by a combustion method from the lathe turnings.

The C concentration profile in an as-weld couple between Fe-14Mn-1.5C-0Al and Fe-14Mn-1.5C-2.7Al Hadfield steels was plotted against the distance from the weld as shown in Fig. 2a [3]. The C concentration on the high Al side was slightly lower than that on the low Al side. However, the C concentration on the high Al side became higher than that on the low Al side after annealing for 260 h at 1,100°C as shown in Fig. 2b. This result implies that the activity of C on the high Al side was lower than that on the low Al side. The ratio of the activity coefficient of C at the high Al side to that of the low Al side, which was determined by the ratio of the C concentration at the inner part of the high Al side, which is far from the weld, to that of the low Al side, was about 0.96. The C concentration at the inner part of each side was obtained by extrapolating the C profile to the distant position from the weld. The diffusivities of C atoms in both low Al and high Al sides, which were calculated by the Grube method [15], were 1.0 × 10^{-6} and 1.5 × 10^{-7} cm^2/s, respectively [3]. Therefore, it was realized that the Al addition decreases the activity and diffusivity of C atoms in high Mn austenitic steels.

2.2 κ′-Carbide Precipitation

Fe-30Mn-8Al-1.0C austenitic steel with supersaturated C atoms shows the age-hardening behavior in the temperature range of 700–950 K. The age-hardenability of the steel has been attributed to the precipitation of cubic κ′-carbides with a chemical formula (Fe, Mn)$_3$AlC$_x$. The κ′-carbide is a fcc-based L′1$_2$ ordered crystal structure, which resembles that of perovskite oxides, now with a carbon atom at the cube center, three iron/manganese atoms randomly at the face-centered positions, and an Al atom at the corner positions in the fcc-based unit cell. The ordered L′1$_2$ structure is similar to Ll$_2$, the γ′ ordered structure, as shown in Fig. 3 [16].

The overall sequence of the matrix precipitation process of κ′-carbides is [16]:

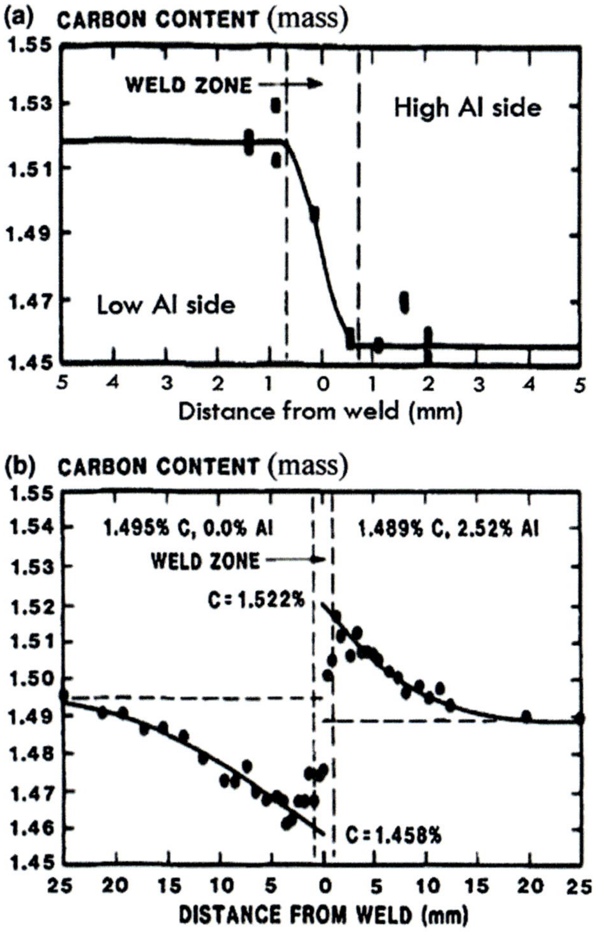

Fig. 2 The C concentration profiles from the weld part in **a** an as-welded couple and **b** the specimen annealed for 260 h at 1,100°C [3]

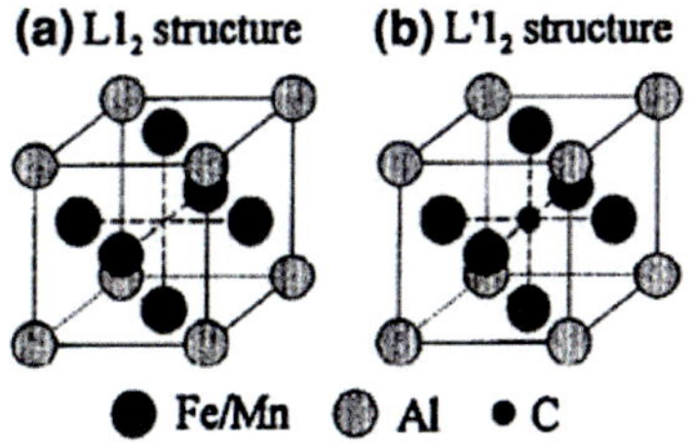

Fig. 3 Ordered a $L1_2$ and **b** $L'1_2$ structures in fcc alloys [16]

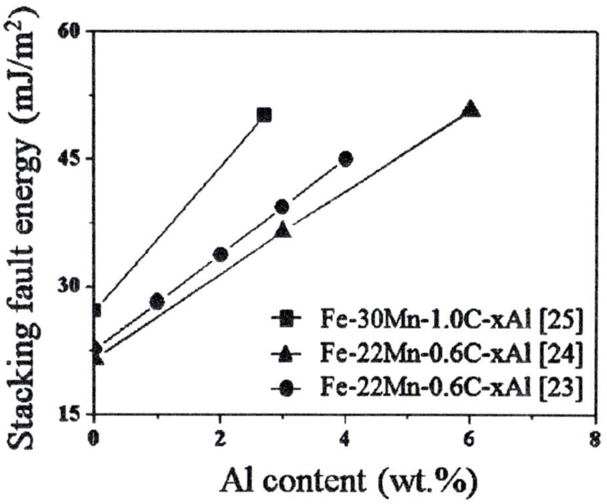

Fig. 4 Calculated stacking fault energies using a modified thermodynamic model based on the Olson–Cohen model [22–24] in the Fe-Mn-C-Al steels

C-supersaturated $\gamma \rightarrow \{100\}$ structure modulation consisting of C-rich and C-poor zone → coherent or partly coherent κ'-phase having $\{100\}$ habit planes → equilibrium κ'-carbide, (Fe, Mn)$_3$AlC.

3 Effect of Al on SFE

The deformation mechanism and mechanical properties of fcc steels are strongly related to the SFE. The mechanical twinning is reported to occur at the SFE range of about 18–45 mJ/m^2. When the SFE is less than 18 mJ/m^2 and the Gibbs free energy change of the γ-ε-martensite is negative, the γ-ε-martensitric transformation happens during the deformation, so-called the TRIP phenomenon. When the SFE exceeds 45 mJ/m^2, the plasticity and strain hardening are controlled solely by the glide of dislocations without any assistance of mechanical twins or martensite.

There are several calculation methods to evaluate the SFE. Early attempts for the SFE calculation were based on the electron theory in fcc metals [17]. Later, Cotterill and Doyama [18] calculated the SFE of copper by a variational method using a central-force approximation where a pairwise interaction between discrete atoms was represented by a Morse potential function. Olson and Cohen [19] proposed a thermodynamic model for the SFE calculation using a basic concept of classical nucleation theory with both volume and surface energies. There are many attempts for experimentally measuring the SFE value using extended nodes on the TEM image, electron diffraction of TEM, and the X-ray diffraction (XRD) [20, 21]. Figure 4 shows the calculated and measured SFE values in various Fe-Mn-C-Al steels. The Al addition increased the SFE value regardless of Mn and C contents.

4 Effect of Al on Tensile Properties

4.1 Effect of Al on DSA

The Al effect on the activation energy for the onset of the DSA phenomenon was quantitatively analyzed in the Fe-30Mn-1.0C-(0 and 2.7)Al TWIP steels [24]. The activation energy was calculated using the equation proposed by Nakada and Keh [25].

$$\dot{\varepsilon} = Kc^n \exp\left(-\frac{Q_M}{RT}\right)\varepsilon_c^m \qquad (1)$$

where $\dot{\varepsilon}$ is the strain rate, c is the atomic concentration of solute C, ε_c^m is a critical strain for the onset of serrations on the tensile curves, Q_M is the activation energy for onset of serrations, R is the gas constant, K, n, and m are empirical constants. Taking the logarithm of both sides of equation (2).

$$\ln\dot{\varepsilon} = \ln K + n\ln C - \frac{Q_M}{RT} + m\ln\varepsilon_c \qquad (2)$$

By plotting of $\ln\varepsilon_c$ vs $1/T$ at the same strain rate, the Q_M value can be determined from the slope of $Q_M/(R \times m)$. The activation energy for the onset of serrations of Fe-30Mn-1.0C-(0 and 2.7)Al TWIP steels were calculated to be 14.4 and 23.3 kcal/mol, respectively, implying that the Al addition suppressed the DSA behavior [24].

4.2 Al Effect in Hadfield Steel

It is known that the high strain hardening rate of the Hadfield steel without Al is attributed to the DSA and to the interruption of the dislocation glide path by stacking faults, and mechanical twins. However, the strain hardening rate of the Hadfield steel with Al was rather increased in spite of the decreases in the mechanical twins and stacking faults caused by the increased SFE with Al addition [3]. Accordingly, it is realized that the strain hardening of the Hadfield steel is caused mainly by the DSA. Nevertheless, there are no detailed explanations of the Al role on the stress–strain of the Hadfield steel. In order to improve the understanding of the role of dislocation sheet structures on the deformation response in the Hadfield steels with and without Al, Canadinc [26] performed tensile tests using a single crystal of the Hadfield steel made by the Bridgman method. Figure 5 shows the true stress–strain curves of single crystals of the Hadfield steels with and without Al, which were tensile strained in the <111>, <001>, and <123> directions at room temperature [26].

The experiments revealed that the stress vs. strain response depends strongly on the crystallographic orientation, and is changed significantly with Al addition for all crystallographic orientations. The addition of Al to the Hadfield steel increased the strain hardening coefficient for <111> and <001> tensile tests while the change in strain hardening along the <123> orientation was small. The deformation mechanisms depend on the crystallographic orientation and the alloy content. The mechanical twinning

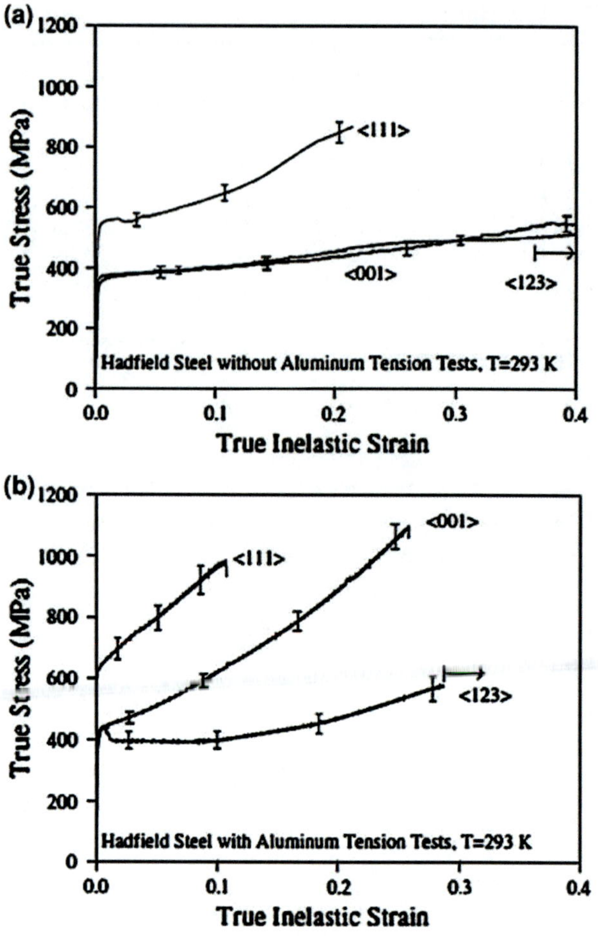

Fig. 5 True stress-true inelastic strain response of the Hadfield steel **a** without and **b** with Al, which were tensile strained in the <111>, <001>, and <123> directions at room temperature [26]

and dislocation slip coexisted for all crystallographic orientations in the Hadfield steel without Al, while the mechanical twinning was suppressed for both <111> and <001> orientations and some micro-twins were observed for the <123> orientation in the Al-added Hadfield steel. These results imply that the dislocation slip is a predominant deformation mechanism in the Al-added Hadfield steel. TEM observations exhibited many dislocation sheets with a high dislocation density, which formed on slip planes corresponding to the most active slip system. The dislocation sheets were arranged with two different directions in the <111> and <001> tensile specimens, while the dislocation sheets had a single direction in the <123> tensile specimen [26]. The high strain hardening coefficients in the single crystals of the Al-added Hadfield steel were attributed to the dislocation sheet arrangements, forming barriers to the dislocation glide.

Therefore, from the previous results, it is realized that the strain hardening of the Al-added Hadfield steel is

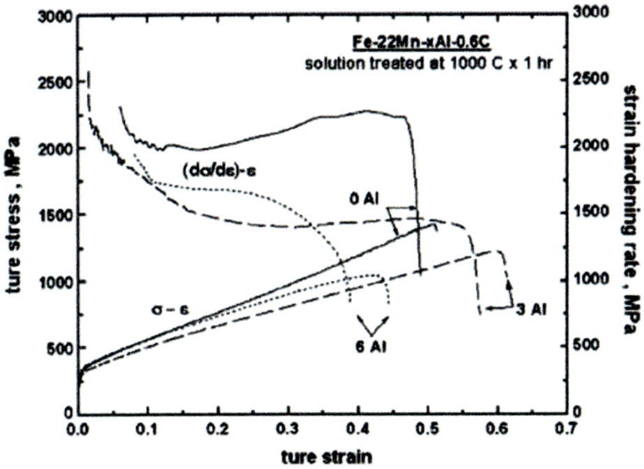

Fig. 6 True stress (σ)–strain (ε) and strain hardening rate ($d\sigma/d\varepsilon$) curves of the Fe-22Mn-0.6C-(0, 3, 6) Al TWIP steels [23]

attributed to mainly the obstacles of DSA and dislocation sheets to the glide of dislocations.

4.3 Al Effect in TWIP Steel

Figure 6 shows the true stress (σ)–strain (ε) and strain hardening rate curves of the Fe-22Mn-0.6C-(0, 3, 6)Al TWIP steels [23]. Yield strength (YS) was high in a following order of the 6 Al, 0 Al, and 3 Al steels. However, true strain showed the opposite tendency. The highest YS of the 6 Al steel was probably due to the finest austenitic grain size, while tensile strength (TS) of 6 Al TWIP steel was the lowest, indicating different strain hardening behaviors in the three steels.

The 0 Al TWIP steel possessed the highest strain hardening rate beyond 2000 MPa at strain of 0.1. It increased to 2300 MPa until a strain close to necking. In the case of the 3 Al TWIP steel, the strain hardening rate gradually decreased up to a true strain of 0.2 and then remained almost constant about 1500 MPa up to the necking. The strain hardening rate of the 6Al TWIP steel continuously decreased from the onset of yielding and remained nearly unchanged shortly and then rapidly decreased at high strains. The highest strain hardening gave the 0 Al TWIP steel to the highest TS in spite of low YS. The different strain hardening behaviors with Al addition were attributed to deformation modes depending on both Al concentration and stacking fault energy.

4.4 Comparison of Al Effect on Strain Hardening Between Hadfield and TWIP Steels

The strain hardening rate was raised with Al addition to the Hadfield steel [3, 26], whereas it was decreased in the high Mn TWIP steel [27]. This difference in strain hardening behavior between Hadfield and TWIP steels is probably due to the different strain hardening mechanisms. The major strain hardening mechanism of the Hadfield steel is known to be the DSA, although there are some mechanical twins in some Hadfield steels. The Al addition increased the C concentration in austenite of the Hadfield steel by suppressing the carbide precipitation, resulting in more interstitial–substitutional (IS) pairs. The increase of IS pairs enhanced the pinning effect on the movement of dislocations, increasing the DSA effect and strain hardening rate in the Hadfield steel [3]. Meanwhile, both DSA and TWIP effects on strain hardening rate were operated in the high Mn TWIP steel. The Al addition increased the SFE of austenite in the high Mn TWIP steel, resulting in the deceleration of mechanical twinning. The DSA effect was also reduced due to the decrease in C diffusivity. These are why the strain hardening rate was decreased with Al addition in high Mn TWIP steel.

4.5 Hydrogen Delayed Fracture

Figure 7 shows the occurrence of the hydrogen delayed fracture phenomenon after cupping deformation [28]. The hydrogen delayed fracture occurred in the high Mn TWIP steel without Al addition. However, the hydrogen delayed fracture did not occur even after more than 1,000 days after

Fig. 7 Comparison of the resistance to hydrogen delayed fracture in the high Mn TWIP steels with and without Al [28]

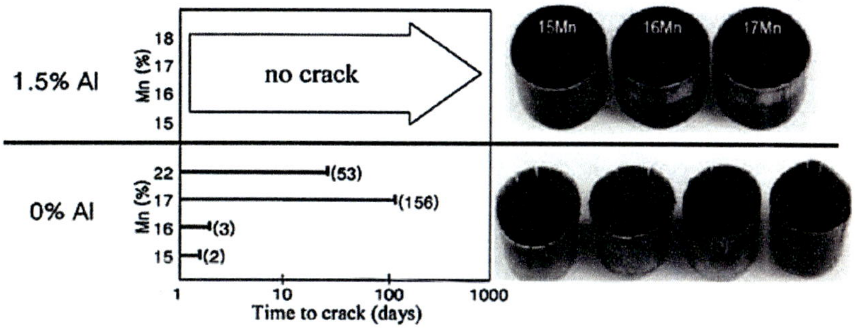

cupping deformation in the Al-added TWIP steel [28]. It is clear that the Al addition improved the resistance to the hydrogen delayed fracture in the high Mn TWIP steels. However, the reason is not clearly proved yet.

5 Summary

The Al addition delayed the precipitation kinetics of cementite in the Fe-high Mn-C austenitic steel because of the increase in the solubility of C by reducing the C activity and the reduction in the diffusivity of C. High Mn austenitic steels with super-saturated C and high concentration of Al exhibited κ'-carbides with a chemical formula (Fe, Mn)$_3$AlC$_x$, causing the age-hardening in the temperature range of 700–950 K. The Al addition increased the stacking fault energy of austenite in the high Mn austenitic steels, decreasing the probability of mechanical twinning. The strain hardening rate in the Hadfield steels was increased with Al addition owing to the increase in solute C atoms, while it was decreased in the high Mn TWIP steel because of the decreases in mechanical twinning and dynamic strain aging effect. The resistant property to hydrogen delayed fracture in the high Mn TWIP steels was greatly improved, however the reason is not clear yet.

References

1. M. Militzer, Science **298**, 975 (2002)
2. L. Bracke, K. Verbeken, L. Destens, J. Penning, Acta Mater. **57**, 1512 (2009)
3. B.K. Zuidema, D.K. Subramanyam, W.C. Leslie, Met. Trans. **A18**, 1629 (1987)
4. D. Canadinc, H. Sehitoglu, H.J. Maier, Mater. Sci. Eng. **A454–455**, 662–666 (2007)
5. B. Hutchinson, N. Ridley, Scripta Mater. **55**, 299 (2006)
6. O. Grässel, L. Krüger, G. Frommeyer, L.W. Meyer, Int. J. Plast. **16**, 1391 (2000)
7. L. Lu, Y.F. Shen, X.H. Chen, L.H. Qian, K. Lu, Science **304**, 422 (2004)
8. D. Barbier, N. Gey, S. Allain, N. Bozzolo, M. Humbert, Mater. Sci. Eng. **A500**, 196 (2009)
9. J.G. Sevillano, Scripta Mater. **60**, 336 (2009)
10. L. Chen, H.S. Kim, S.K. Kim, B.C. Decooman, ISIJ Inter. **47**, 1804 (2007)
11. S. Allain, J.P. Chateau, O. Bouaziz, S. Miget, N. Guelton, Mater. Sci. Eng. **A387–389**, 158 (2004)
12. S. Kang, Y.S. Jung, J.H. Jun, Y.K. Lee, Mater. Sci. Eng. **A527**, 745 (2010)
13. J.E. Jin, Y.K. Lee, Mater. Sci. Eng. **A527**, 157 (2009)
14. J.D. Yoo, K.T. Park, Mater. Sci. Eng. **A496**, 417 (2008)
15. W.L. Grube, J.G. Gay, Met. Trans. **A9**, 1421 (1978)
16. W.K. Choo, J.H. Kim, J.C. Yoon, Acta Mater. **45**, 4877 (1997)
17. 17. A. Seeger, *Defects in crystalline solids* (The Physical Society, London, 1955) p. 328
18. R.M. Cotterill, M. Doyama, Phy. Rev. **145**, 465 (1966)
19. G.B. Olson, M. Chohen, Met. Trans. **A7**, 1897 (1976)
20. S. Curtze, V.T. Kuokkala, Acta Mater. **58**, 5129 (2010)
21. K. Kapoor, D. Lahiri, I. Batra, S. Rao, T. Sanyal, X-ray diffraction line profile analysis for defect study in Cu-1 wt% Cr 0.1 wt% Zr alloy. **54**, 131 (2005)
22. A. Dumay, J.P. Chateau, S. Allain, S. Miget, O. Bouaziz, Mater. Sci. Eng. **A483–484**, 184 (2008)
23. K.T. Park, K.G. Jin, S.H. Han, S.W. Hwang, K. Choi, J.S. Lee, Mater. Sci. Eng. **A527**, 3651 (2010)
24. T. Shun, C.M. Wan, J.G. Byrne, Acta Metall. **40**, 3407 (1992)
25. Y. Nakada, A.S. Keh, Acta Metall. **18**, 437 (1970)
26. D. Canadinc, H. Sehitoglu, H.J. Maier, Y.I. Chumlyadov, Acta Mater. **53**, 1831 (2005)
27. K. Chung, K. Ahn, D.H. Yoo, K.H. Chung, M.H. Seo, S.H. Park, Formability of TWIP (twinning induced plasticity) automotive sheets. Int. J. Plast. **27**, 52–81 (2011)
28. Y. Kim, N. Kang, Y. Park, I. Choi, G. Kim, S. Kim, K. Cho, J. Kor. Inst. Met and Mater. **46**, 780 (2008)

Microstructure and Property Control of Advanced High Strength Automotive Steels

Lin Li

Abstract

Martensite phase transformation and microstructure features in DP, TRIP and TWIP steels are discussed. For the obtaining of better properties, details in controlling the production of DP and TRIP steels are described based on the understanding of the stability of microstructures. Substitution of P for Al and/or Si in TRIP steel is evaluated from the aspects of thermodynamic and kinetic estimation. The minimum fast cooling rate and optimal over-aging temperature and time has been calculated for TRIP steel. Stack fault and its effect on phase transformation in high Mn steel are evaluated in the relationship with composition, strain, and heat treatment.

Keywords

DP • TRIP • TWIP • Microstructure • Phase transformation

1 Introduction

HSS (High Strength Steel) has received great interest in automobile industry due to the fact that the substitution of HSS for mild steel makes significant progress in energy saving, environment protection and passenger safety. ULSAB (Ultra Light Steel Auto Body) [1] partnership project, sponsored by more than 30 steel enterprises, aimed at building up modern cars adopting HSS for the structural parts as much as possible. In the following project, the ULSAB Advanced Vehicle Concept for mid-size passenger cars (ULSAB-AVC), only HSS is specified as shown in Fig. 1 [2].

Recently TRIP (Transformation Induced Plasticity) steel has been considered as the most promising structural material in the rank of HSS since it exhibits the best combination of high strength and high ductility [2].

Another tremendous progress was the manufacturing of DP (Dual Phase) steel on the modern Continuous Annealing Line (CAL), which occupies quite large portion in the selection of ULSAB–AVC project as shown in Fig. 1. Because of the high tensile strength that they both have, these two steels are mainly used in the bump shield parts such as A, B pillar and door reinforced rod, etc. High strength of these steels comes from the same source, i.e., martensite transformation. However, the mechanisms in the two steels are totally different as well as the controlling points in the process. The details will be described in this paper.

2 Processing of DP and TRIP Steels [3]

There existed some similarities and differences in the processing of low-alloy DP and TRIP steels whatever in the hot rolling or cold rolling process. In the hot rolling process as shown in Fig. 2, the cooling rate for DP steels must be low enough, or a holding period has to be inserted to ensure the transformation of about 80% austenite to ferrite and sufficient carbon enrichment of austenite to take place. Then the

L. Li (✉)
School of Materials Science & Engineering, Shanghai University, Shanghai 200072, China
e-mail: liling@shu.edu.cn

Y. Weng et al. (eds.), *Advanced Steels*, DOI: 10.1007/978-3-642-17665-4_27,

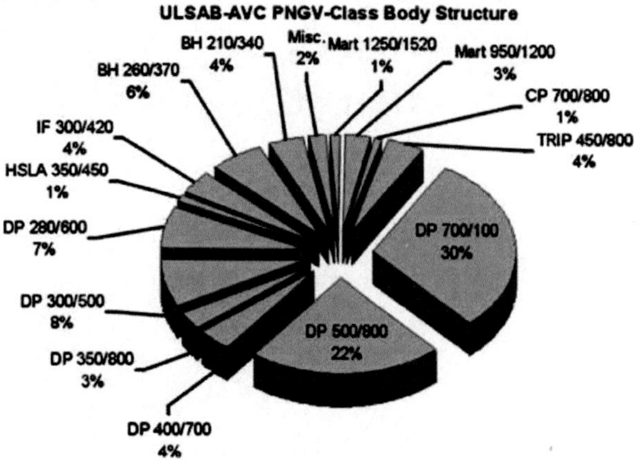

Fig. 1 Material selection in ULSAB-AVC project [2]

cooling rate should be high enough to avoid the formation of pearlite and bainite and to ensure the formation of martensite at low temperatures of about 250°C.

For TRIP steels a lower cooling rate can be applied since austenite is more stable and pearlite formation is delayed during the whole process due to the special alloying concept. Coiling is carried out in the temperature range of bainite formation at about 500°C or lower. The final microstructure comprises about 60% ferrite, 30% bainite, and 10% retained austenite. The meta-stable austenite there does not transform to martensite, since the carbon enrichment during ferrite and bainite transformation shifts the M_s to a much lower temperature.

In the cold rolling process, the sheet steel has to undergo a special heat treatment named as continuous annealing (CA) as shown in Fig. 3. Low-alloy TRIP steels are subjected to a two steps heat treatment, i.e., intercritical annealing in the temperature range between 760 and 820°C, then cooled with a rather fast cooling rate to bainite transformation zone between 350 and 500°C for isothermal annealing, after soaking for several minutes the steels are cooled to room temperature with final microstructure of about 60% ferrite, 30% bainite, and 10% retained austenite. Carbon enrichment

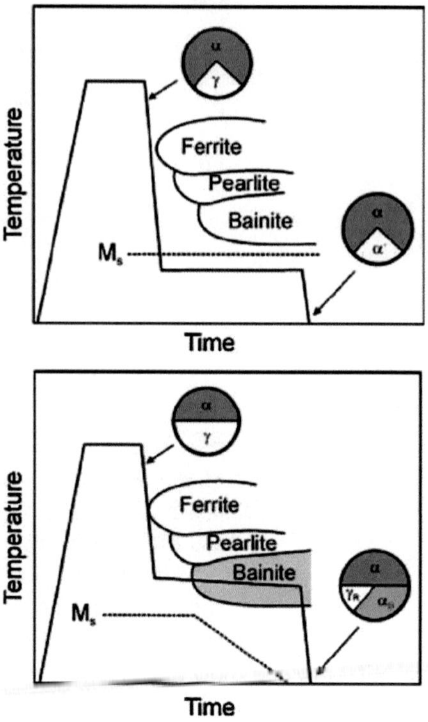

Fig. 3 Temperature–time-schedule for cold-rolled DP (*top*) steels and TRIP (*bottom*) [2]

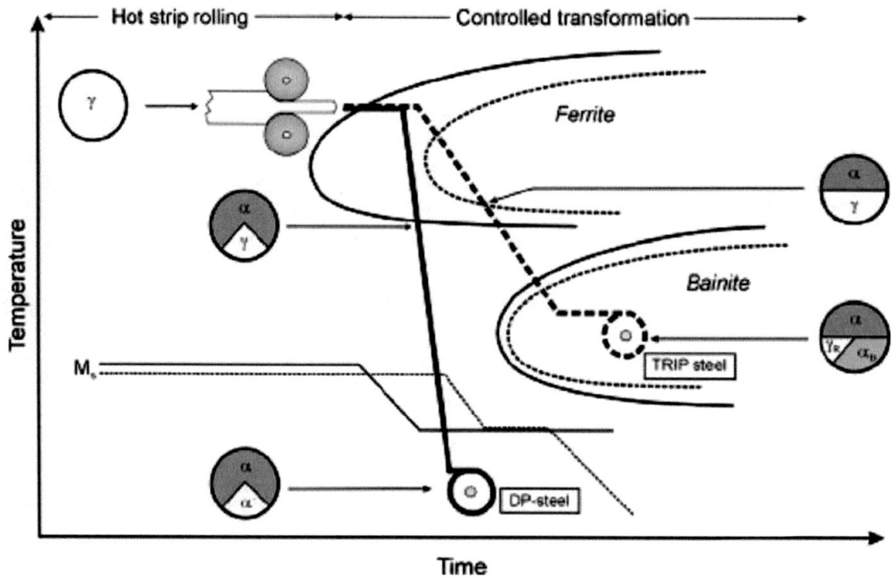

Fig. 2 Temperature–time-schedule for hot-rolled TRIP and DP steels [2]

of austenite takes place during the intercritical annealing phase, then in bainite transformation period, which enables a further carbon enrichment to occur. The latter step is quite necessary to maintain a certain amount of meta-stable retained austenite, which is stable enough and can exist at very low temperature. Otherwise, the carbon enrichment of austenite is not sufficient leading to the transformation from unstable austenite to martensite during cooling down to room temperature and deteriorates the mechanical property. For the DP steels, a two steps heat treatment including intercritical annealing and overageing treatment is applied. Carbon enrichment of austenite also takes place during the intercritical annealing. Afterwards, the steels are cooled to the overaging temperature with a sufficient fast rate in the temperature range between 220 and 350°C. During the phase, no upturn of temperature is permitted which leads at that moment to martensite tempering and strength of the steel lowers.

3 New Design of Composition and Constituent of TRIP Steels

As it is known that a certain amount of quite stable austenite is required for TRIP steel through the CAL process. For low-alloy TRIP steels, the most efficient element to increase austenite stability is carbon. Si is another important element, which leads to the increase of carbon content in austenite and prevent the precipitation of cementite. However, high Si content causes problems in steel production such as strong oxide layer, poor surface characteristics and poor coating feasibility [4]. In order to replace Si partially and/or totally, efforts were made with Al instead [5]. The substitution was successful but only limited at low Al content. High Al content in the steel may cause clogging during continuous casting due to precipitation of aluminum oxide.

In the periodic table of elements, Al, Si, and P are located nearby in the same period, these elements have similar electronic structures and may have similar properties. This comes true especially to Si and P since they both are semi-metals. It is then natural to consider P as another substitution for Si. P can, on one hand, depress the carbon activity in cementite, prevent cementite precipitation and increase the stability of austenite. On the other hand, this element has a strong tendency to segregate at grain boundary and to induce cold brittleness. Segregation formulation was developed by Guttmann and Mclean in a ternary system (Fe, M, I) [6] based on the sublattice model in melt salts and stoichiometric phases contributed by Hillert and Staffansson [7]. Li et al. extended the segregation equation into a five-element system [8] and generalized it into a multicomponent system [9] as follows:

$$y_I^g/y_V^g = \left(y_I^B/y_V^B\right)\exp\left\{\left[\Delta^0 G_I + 1/c^g \sum_{i\neq A} y_i^g \beta_{iI}^g - 1/c^B \times \sum_{i\neq A} y_i^g \beta_{iI}^g - 1/c^g \sum_{J\neq I} y_J^g \beta_{A:JI}^g + 1/c^g \sum_{J\neq I} y_J^g L_{A:JV}^g + 1/c^B \sum_{J\neq I} y_J^B \beta_{A:JI}^B - 1/c^B \sum_{J} y_J^B \beta_{A:JV}^B\right]/RT\right\} \quad (1)$$

$$y_M^g/y_A^g = \left(y_M^B/y_A^B\right)\exp\left\{\left[\Delta^0 G_M + 1/a^g \sum_{i} y_I^g \beta_{MI}^g - 1/a^B \sum y_I^B \beta_{MI}^B - 1/a^g \sum_{i\neq M} y_i^g L_{i:MV}^g + 1/a^B \sum_{i\neq M} y_i^B L_{i:MV}^B + 1/a^B \sum_{i\neq A} y_i^g L_{Ai:V}^g - 1/a^B \sum_{i\neq A} y_i^B L_{Ai:V}^B\right]/RT\right\} \quad (2)$$

where variable y indicates site fraction in sublattice, g grain boundary, B matrix, V vacancy, A base metal, J and I represent impurity, i and M metal element, c and a fraction of sites available in grain boundary or matrix, $\Delta^0 G_I$ and $\Delta^0 G_M$ are the intrinsic segregation Gibbs free energy of impurity I and metal element M, $L_{A:JI}^{B(g)}$ stands for the interaction energy between J and I in one sublattice when another sublattice is fully occupied by element A, $\beta_{MI}^{B(g)}$ means the interaction energy between elements in different sublattices.

Following references [9, 10], assuming that interaction energy keeps the same on grain boundaries and in matrix and omitting the interaction between impurities and vacancy and between metal element and iron [10], a series of parameters can be listed [9, 10] as follows: $a^g = c^g = 0.5$, $a^B = 0.75$, $c^B = 0.25$, $\Delta^0 G_P = 47$ kJ/mol, $\Delta^0 G_{Mn} = 8$ kJ/mol, $\beta_{MnP} = 12.5$ kJ/mol, $L_{Fe:PC} = -9$ kJ/mol. With all the data stated above and the composition of an assumed steel with enough high phosphorus (C = 0.15%, Mn = 1.6%, Si = 0.3%, P = 0.07%), the equilibrium segregation amount of phosphorus on grain boundaries at 400°C is calculated as 32%, which is a value high enough to cause temper brittleness.

The soaking temperature at the over-aging process in processing cold-rolled TRIP steel is set in the temperature range from 350 to 480°C. However, the soaking time is quite short, normally for 3–5 min, being far from equilibrium condition. Thus, the estimated equilibrium segregation amount of phosphorus is not suitable in describing the over-aging process of TRIP steels. Still following Mclean with his kinetic approach for a binary system [11], the amount of segregative element on grain boundaries can be expressed as follows:

$$\left(y_i^{\phi}(t) - y_i^{\phi}(0)\right) / \left(y_i^{\phi} - y_i^{\phi}(0)\right) \approx 2\left(X_i^{\alpha}/y_i^{\phi}d\right)(D_i t/\pi)^{1/2} \tag{3}$$

where $y_i^{\phi}(t)$ is the grain boundary coverage of element i at time t, $y_i^{\phi}(0)$ is its initial value and y_i^{ϕ} the equilibrium value. D_i is the bulk diffusion coefficient of i, d the grain boundary thickness and X_i^{α} the mole fraction of i in matrix. Taking $d = 10^{-7}$ cm as usual, $t = 5$ min $= 300$ s and $T = 673$ K according to the actual process condition, $X_P^{\alpha} = 1.387 \times 10^{-3}$ (i.e., wt $= 0.07\%$), $y_i^{\phi}(0) = 0$; according to Thermo-calc [12], $D_P = 30 \times D_{Fe}$, $D_{Fe} = D_0\exp(-Q/RT)$, $D_0 = 0.5$ cm^2/s, $Q = 240$ kJ/mol, thus $D_P = 3.53 \times 10^{-18}$ [12].

The segregated P after 300 s soaking at grain boundary is:

$$y_P^{\phi}(300\,\text{sec}) \cong 2(X_P^{\alpha}/d)(D_P t/\pi)^{1/2} = 0.051\% \tag{4}$$

which is a small amount and would not be of any damage in grain boundary for the TRIP steel.

Mechanical properties of the designed steel were measured at low temperatures (–20°C, –60°C) and no substantial difference was found between the results obtained at low temperatures and room temperature. Cold fracture test was taken with the structural part assembles of automobiles and no sign of cold brittleness was detected.

4 Determination of the Minimum Fast Cooling Rate and Bainite Transformation Temperature and Time

For the cold rolling TRIP steels, there are several steps for carbon enrichment in the production as shown in Fig. 3. The first step is carried out during the intercritical annealing, at which carbide is dissolved, austenite formed at the carbon rich zone and all the elements redistribute in different phases. In the production line, i.e., Continuous Annealing Line (CAL), it is normal to complete the carbon diffusion. The next step is to cool the sheet steel to the bainite transformation zone.

Alloy concept and cooling rate are the factors affecting the precipitation of cementite and process controlling. Lower cooling rate causes precipitation of cementite and deteriorates the mechanical properties of TRIP steels but higher cooling rate causes deformation of sheet steel. Therefore, a minimum fast cooling rate must be determined with which the cementite precipitation is efficiently inhibited and at the same time, there should not exist any drastic cooling and no deformation or distortion of sheet occurs.

The precipitation of cementite could be referred as "diffusion controlled phase transformation" as it goes depending on element concentration variation. Diffusion controlled phase transformations in a more generalized way have been extensively studied by Ågren [13]. The author suggested local equilibrium at the moving interface between interacting phases was always to be established and it must obey the flux balance equation [13]. In addition, the "number-fixed frame of reference" was taken with respect to the substitutional elements while the vacancy exchange mechanism of diffusion was predominated by the lattice-fixed frame of reference [13].

Substantial support for the diffusion calculation comes from a dynamic database MOB and the program Dictra in Thermo-calc software package, though a thermodynamic database SSOL contributed to the calculation too [12].

However, that understanding pearlite growth kinetics in a lamellar aggregate model needs still sets of parameters such as surface tension, optimum growth rate factor and diffusion coefficient on boundaries for elements. The first two were offered in Ref. [14] and adopted in Ref. [15]. For the last one, only listed was boundary diffusion coefficient of Mn [14]. Reference [15] has deduced such data for P and Si. The calculated result for pearlite growth is shown in Fig. 4 where the vertical axis represented the volume fraction of pearlite, and horizontal axis the cooling time (in second).

In Fig. 4 one can find there is no substantial growth of pearlite in 3 s. In the calculation, the starting temperature of fast cooling is set as 700°C, the assigned cooling rate is 30°C/s, after 3 s the steel is cooled to 610°C, which is out of the zone of pearlite transformation. Then this assigned cooling rate was referred as the minimum fast cooling rate for the steel with the given composition (C = 0.15%, Mn = 1.6%, Si = 0.3%, P = 0.07%) in line production.

There is another important isothermal process in cold-rolled TRIP steel production, i.e., over-aging treatment. During this process, part of the austenite transforms to

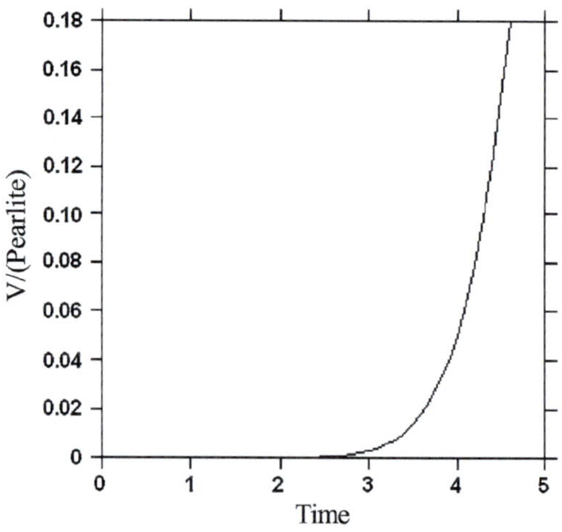

Fig. 4 Pearlite growth kinetics of a designed TRIP steel [15]

bainite and the retained austenite gets more enriched in carbon content, which makes austenite more stable and the steel is of better TRIP effect. Over-aging temperature can be determined through extensive experimental tests or the estimation of T_0 temperature at which the Gibbs free energy of γ-phase (austenite) equals to that of α phase (bainite). Problem arises from the fact that after a short time of intercritical annealing, neither γ nor α-phase attains full equilibrium. If one calculates the phase diagram with an equilibrium condition, the obtained composition (point B in a thick line in Fig. 5) does not represent the real case. If calculating the diagram with para-equilibrium assumption (the *thin line* passing point A in Fig. 5), which presumes only carbon gets equilibrium during intercritical annealing, but other elements remain unchanged in γ- and α-phases, the result also differs from the actual one.

Therefore, the present author [15] suggested to choose a middle value between T_B (equilibrium condition) and T_A (para-equilibrium condition) as over-aging temperature and took 420°C for bainite transformation in the line production of the designed TRIP steel.

The second step for carbon enrichment is taken in the bainite transformation range, i.e., the over-age process. For a proposed mechanism for formation of Widmannstatten ferrite or the ferrite phase of the bainite structure, in the light of Hillert and his cooperators' approximation [15], the ferrite plate formed is embedded in an austenite matrix and the phase boundary is immobile. Through putting ferrite and austenite in two different cells using Thermo-calc, the estimated carbon contents in austenite after aging at 440 and 500°C for various time (quite short) are shown in Figs. 6 and 7. Calculation was performed with the TRIP steel at the composition: C = 0.13%, Mn = 1.54%, Si = 0.47% and P = 0.066%. It was obvious that there is no substantial difference in the enrichment of austenite at 440 and 500°C

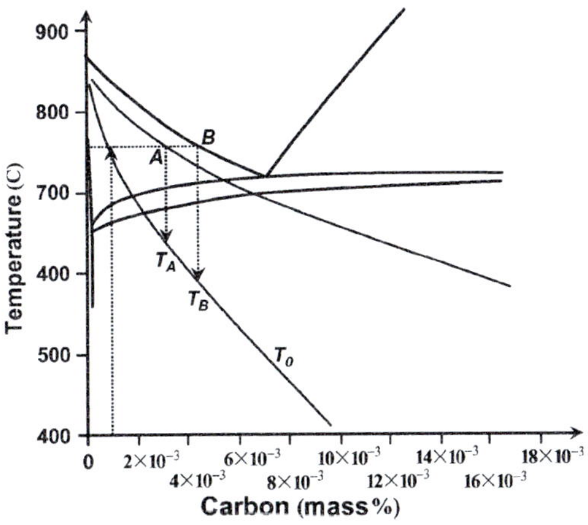

Fig. 5 Schematic diagram of over-aging temperature determination [15]

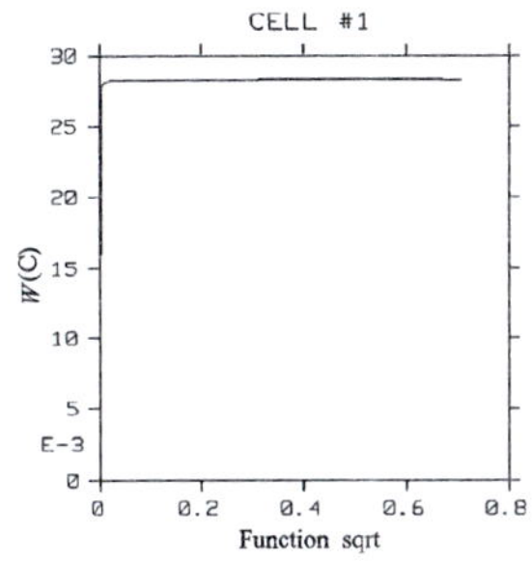

Fig. 6 Carbon enrichment in austenite after various time over-aging treatment at 440°C, longitudinal axis representing carbon content and transverse axis the square root of time (s)

and one could conclude that the second step of carbon enrichment was complete.

5 Connection Between TRIP and TWIP Steels

Different compositions of High Mn steels were designed and the effect of composition on microstructures and mechanical properties are discussed. The sample compositions are listed in Table 1.

Sample steels were melted in vacuum induction furnace. After mold casting, forging, hot rolling and cold rolling, the steels were cut into 200 mm × 100 mm × 1.2 mm sheet blanks, then heated to 1,050°C and soaked for 30 min, then quenched in oil or air-cooled to room temperature.

Austenite volume fraction values of steels after oil quenching were measured by using X-ray diffractometer (D/MAX-RC). It is shown in Table 2, when Mn content exceeds 25.41% (steel 3), nearly full austenite structure is obtained.

Microstructures of steels 1, 3, and 5 were examined and shown in Figs. 8, 9 and 10. In Fig. 8, it is seen that after annealing and quenching, steel 1 exhibits similar microstructures (martensite, ferrite, and austenite) but different morphology of twins and different grain size. In Fig. 9, steel 3 shows always austenite and annealing twins no matter what process was applied. Austenite grains in steel 3 after annealing are larger, twins developed much stronger and twin boundaries were much straighter than those after quenching. In Fig. 10, there exist some twins in the shape of lens or semi-lens, which might be caused by the limited

Table 1 Sample compositions of high Mn steels (mass%)

No.	Mn	Si	Al	C	Fe
1	14.69	2.60	2.29	0.0067	Balance
2	19.85	2.65	2.31	0.0066	Balance
3	25.41	2.55	2.28	0.0053	Balance
4	28.06	2.55	2.25	0.0076	Balance
5	32.32	2.30	3.79	0.024	Balance

Table 2 Volume percent of constituents of steels after oil quenching

No.	γ (%)	α', ε (%)
1	52.8	47.2
2	88.0	12
3	98.4	1.6
4	98.5	1.5
5	96.2	3.8

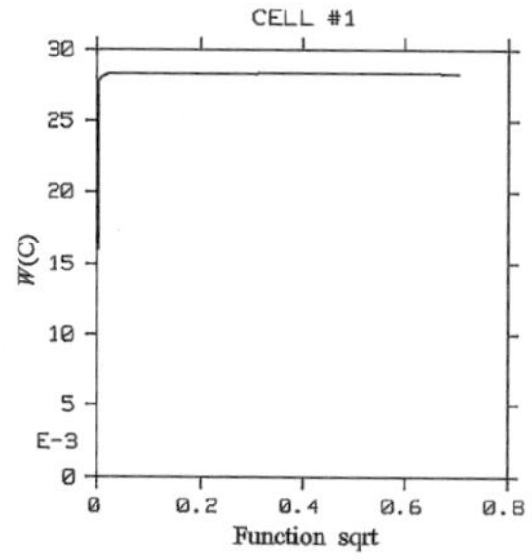

Fig. 7 Carbon enrichment in austenite after various over-aging treatment at 500°C, longitudinal axis representing carbon content and transverse axis the square root of time (s)

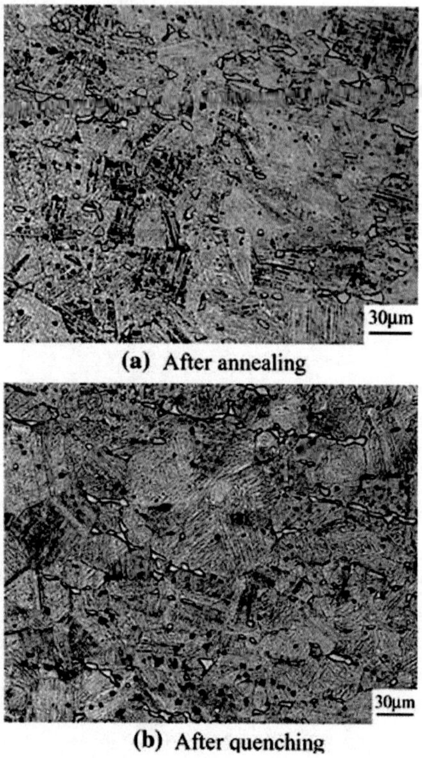

(a) After annealing

(b) After quenching

Fig. 8 Optical micrographs of steel 1 after annealing and quenching

(a) After annealing

(b) After quenching

Fig. 9 Optical micrographs of steel 3 after annealing and quenching

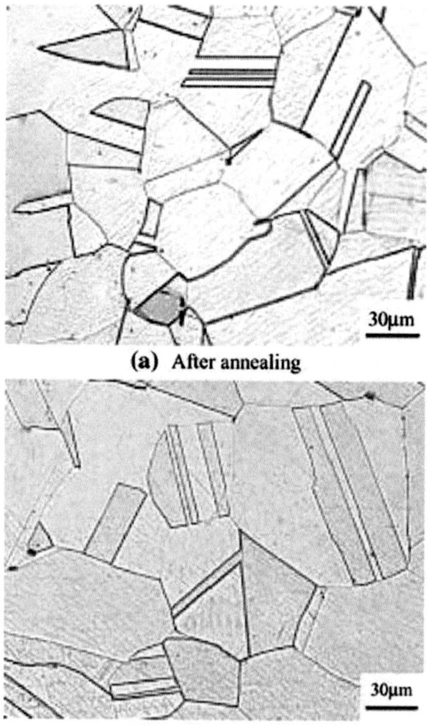

(a) After annealing

(b) After annealing

Fig. 10 Optical micrographs of steel 5 after annealing and quenching

time growth of the twin core (the stack faults formed during recrystallization).

There exists austenite in all the steels after quenching. Austenite in steel 1 was about 50% in volume percent and exhibiting big and thin film with very less dislocations in it (see in Fig. 11a). Austenite in steel 2 had two types of morphology, one being the same as in steel 1 and another is

Fig. 11 TEM micrographs of austenite and its indexing

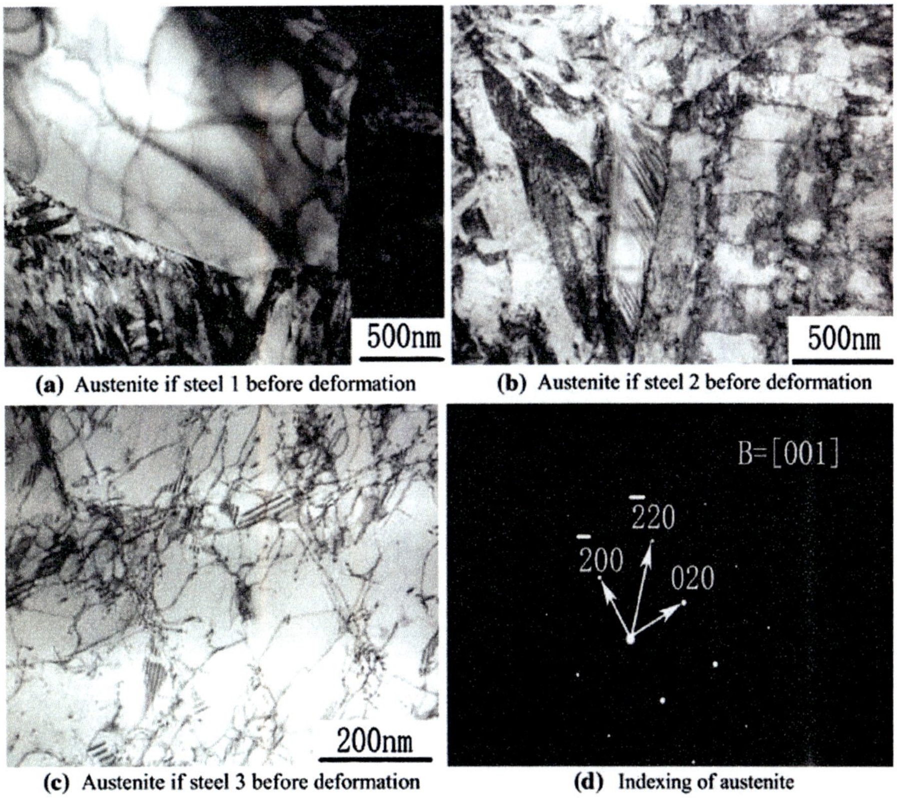

smaller and surrounded by lath martensite. There is a large amount of stack faults in the austenite as shown in Fig. 11b. The substructure of austenite in steel 3 after quenching is of dislocations and stack faults too (see Fig. 11c).

Stack fault or stack fault energy is an important factor in high Mn steels that governs whether martensite transformation or twin growth can take place. According to Schramm and Reed [17], the stack fault energy (SFE) γ was listed as:

$$\gamma = \frac{Gb^2}{8\pi d_e}\frac{2-v}{1-v}\left(1+\frac{2v}{2-v}\right) \quad (5)$$

where G is shear modulus, b Burgers vector, d_e the width of stack fault, v Poisson's ratio. It was obvious from Eq. 5 that γ is in the inverse proportion to d_e, i.e., the wider the stack fault, the lower the stack fault energy. In Fig. 12, the stack fault width in steel 1 is about three times wider than that in steel 3, SFE in the former must be much lower than the latter.

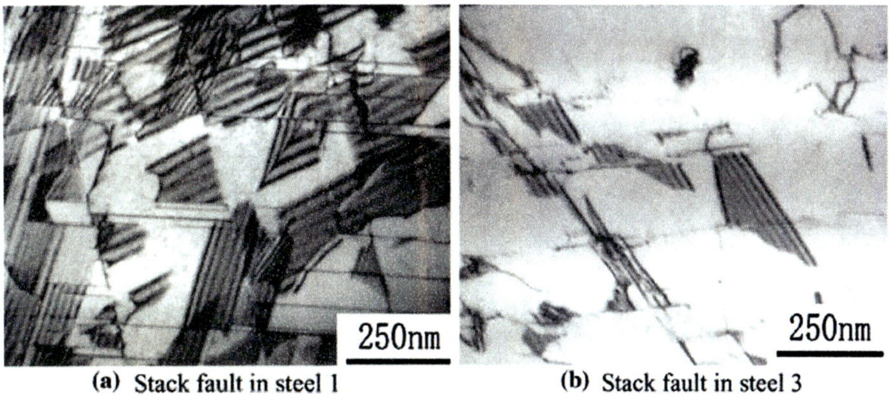

Fig. 12 TEM micrographs of stack faults in steels 1 and steel 3 before deformation

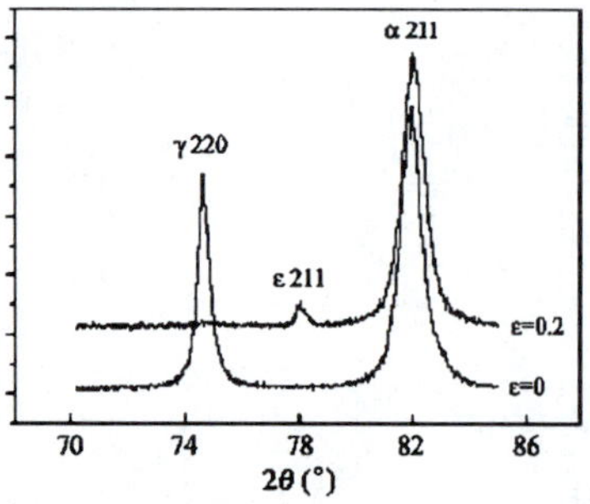

Fig. 13 XRD spectra of steel 1 before and after deformation [18]

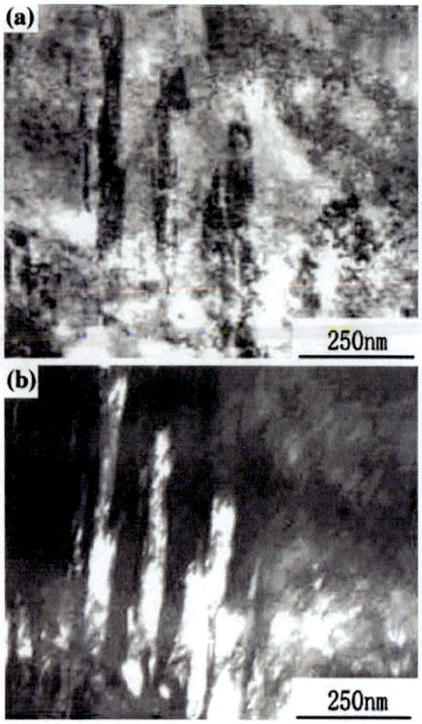

Fig. 14 TEM images showing the morphology of ε martensite in steel 1 after deformation: **a** Bright field **b** Dark field [18]

The microstructures in steel 1 after quenching are austenite and martensite. For further understanding the change in microstructure after deformation, samples with and without tensile deformation were tested by using X-ray diffractometer. The result is shown in Fig. 13 which indicates that before deformation there was a big amount of austenite but after that austenite peak disappeared but ε and α' martensite peaks existed. TEM observation also revealed that after deformation there appeared ε martensite and a lot of dislocations in α' martensite (see Fig. 14). However, there is no sign of appearance of deformed twins after the deformation. It is then clear martensite transformation takes place during deformation both in the way of $\gamma \rightarrow \varepsilon$ and $\gamma \rightarrow \varepsilon \rightarrow \alpha'$. The strengthening mechanism should be transformation induced plasticity (TRIP) for steel 1.

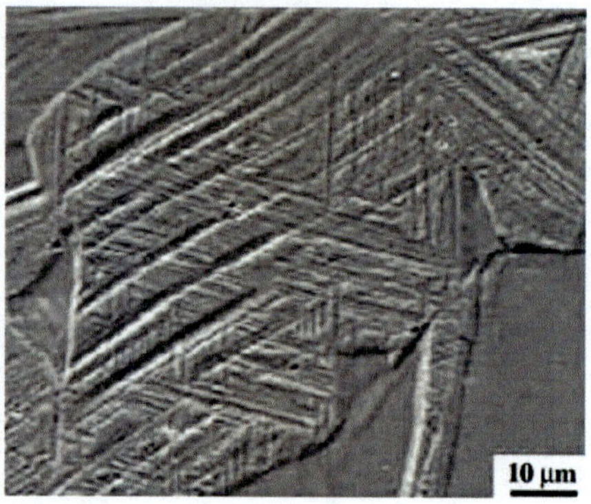

Fig. 15 SEM micrograph showing deformation twins of steel 3 after deformation [18]

In steel 3 there are two constituents, i.e., austenite and annealing twins after quenching, while after further deformation there appear deformed twins as shown in Fig. 15. However, there might exist more than one twin system in one grain as shown in Fig. 16. When one twin is interrupted by the grain boundary, another twin induces due to the increasing stain and starts to grow in between the former and to form stage-like morphology. But if two twin systems start simultaneously, the grid-like morphology would appear. It follows from the above results that steel 3 is strengthened with the twin induced plasticity (TWIP) effect.

Phase transformation amount under strain for steels 1, 2, and 3 was detected and the results for steels 1 and 2 are shown in Fig. 17. The true strain was obtained through

Fig. 16 TEM image showing deformation twins in steel 3 after deformation

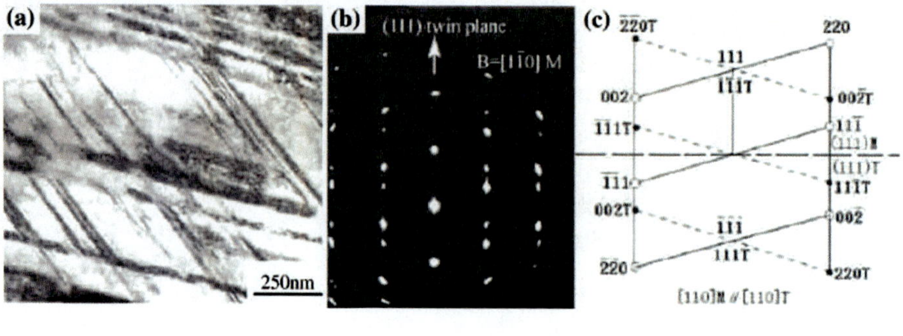

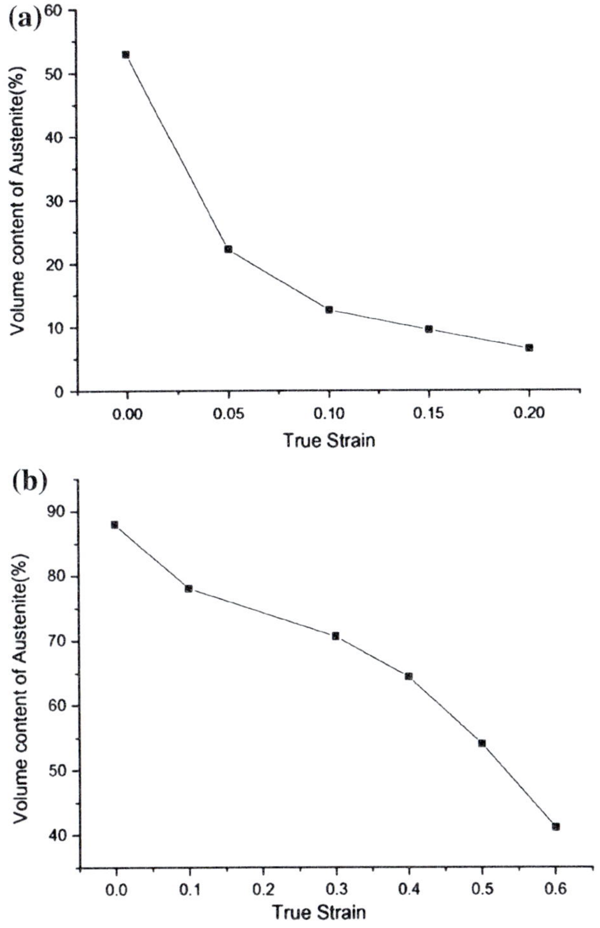

Fig. 17 Phase transformation amount under various strain for steels 1 and steel 2

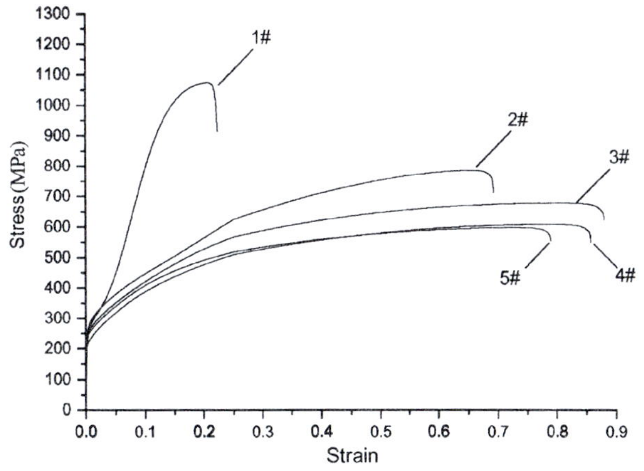

Fig. 18 Strain–stress curves of steels 1–5

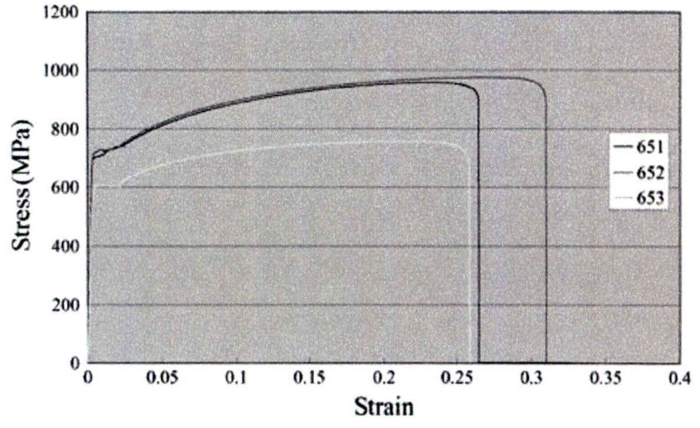

Fig. 19 Strain–stress curves for a low Mn steel after TRIP treatment

comparison of length change in tensile sample before and after deformation. After tensile test, the transformed austenite was measured by using XRD.

Figure 17a indicates that steel 1 deformed just in the way of TRIP: a small amount of strain could cause a big portion of martensite transformation from austenite. In Fig. 17b, the same amount of strain did not bring as much transformation as in Fig. 17a since part of the strain energy was used to deform the twins. For steel 3, no matter how much strain was applied, there was no reduction of austenite, i.e., no transformation taking place.

Interesting phenomena could be encountered in the tensile test. Various results are shown in Fig. 18, where steel 1 acted just as the typical TRIP steel with high strength but a bit low elongation since its ductility came mainly from transformation. Steel 2 in Fig. 18 exhibits better elongation but lower strength because its ductility comes from both transformation and twins. Steel 3 has much better ductility but the lowest strength since there is no contribution of martensite transformation that increases the strength efficiently. However, high Mn content there brings high stack fault energy and strengthens the twin inducing plasticity. Steels 4 and 5 act in the same way as steel 3.

It is also interesting if one carefully design the composition and the corresponding process, hence the distribution of stack faults and twins, the surface energy, etc., it is possible to find a steel with excellent mechanical properties (see in Fig. 19) of low Mn content steel.

6 Conclusion

Understanding the martensite phase transformation features and stability of microstructures are the key points in controlling the properties of DP, TRIP, and TWIP steels. For the obtaining of better properties, details in controlling the production of DP and TRIP steels are described. Substitution of P for Al and/or Si in TRIP steel is evaluated from the aspects of thermodynamic and kinetic estimation. The minimum fast cooling rate and optimal over-aging temperature and time has been calculated for TRIP steel. Stack fault and its effect on phase transformation in high Mn steel are evaluated in the relationship with composition, strain, and heat treatment. By applying thermodynamic estimation and accurate experiments, new steels with excellent properties can be produced.

Acknowledgments The author is thankful to the grants from 115 Project, 973 Project (No. 2010CB630802) and NSFC (No. 50934011). The valuable supports and discussions from Professor Z. Y. Xu, Associate Professors R. Y. Fu, Y. L. He, W. Shi, and M. Zhang are acknowledged.

References

1. ULSAB Advance Vehicle Concepts, *Technical transfer dispatch #6* (ISIJ, Brussels, 2001)
2. W. Bleck, in Proceedings of the International Conference on TRIP-added high strength ferrous alloys (Ghent, Belgium, 2002), p. 13
3. L. Li, Conclusion report of "Production technology of DP, TRIP steels", 115 Scientific Project Supported by China Science & Technology Ministry, 2009
4. L. Li, B.C. De Cooman, P. Wollants, Y.L. He, X.D. Zhou, J. Mater. Sci. Technol. **2**, 135 (2004)
5. M.D. Meyer, D. Vanderschueren, B.C. De Cooman, ISIJ Inter. **8**, 813 (1999)
6. W.C. Johnson, J.M. Blakely (eds.), *Interfacial Segregation* (Metals Park, Ohio, 1979)
7. M. Hillert, L.I. Staffansson, Acta Chem. Scand. **24**, 3618 (1970)
8. L. Li, L. Delaey, P. Wollants, O. Van Der Biest, J. Chim. Phys. **90**, 305 (1993)
9. L. Li, L. Delaey, P. Wollants, O. Van Der Biest, J. Mater. Sci. Technol. **(3)**, 238 (1996)
10. M. Guttmann, Ph. Dumoulin, M. Wayman, Metall. Trans. **13A**, 1693 (1982)
11. D. Mclean (ed.), *Grain Boundaries in Metals* (Clarendon Press, Oxford, 1957)
12. B. Sundman, B. Jansson, J. Andersson, Calphad **2**, 153 (1985)
13. J. Ågren, ISIJ Inter. **3**, 291 (1992)
14. B. Jonsson, TRITA-MAC-0478, Internal Report, Div. Phys. Metall., Royal Inst. Technol., S-10044, Stockholm, Sweden (1992)
15. L. Li, B.C. De Cooman, R.D. Liu, J. Vleugels, M. Zhang, W. Shi, J. Iron Steel Res. Int. **6**, 37 (2007)
16. M. Hillert, L. Höglund, J. Ågren, Acta Metall. Mater. **41**, 1951 (1993)
17. R.E. Schramm, R.P. Reed, Metall. Trans. **6**, 1345 (1975)
18. Q. Li, R.G. Xiong, J.R. Chen, R.Y. Fu, L. Li, Trans. Mater. Heat Treat., **2**, 52 (2008)

Microstructure and Mechanical Properties of a TRIP Steel Containing 7 Mass% Mn

Seong-Jun Park, Chang-Seok Oh, and Sung-Joon Kim

Abstract

A TRIP steel containing 7 mass% Mn was annealed at various temperature for 10 min and changes in microstructure and mechanical behavior due to different annealing temperature were investigated. Excellent mechanical properties of high tensile strength over 1,200 MPa with high elongation over 20% could be obtained for proper annealing temperature range. Volume fraction and mechanical stability of austenite were highly dependent on annealing temperature, which shows a transition from stable to metastable behavior of austenite. Effects of austenite grain size and solute element contents in austenite grains on stability were discussed.

Keywords

TRIP steel • Mn • Austenite stability

1 Introduction

Deformation-induced martensitic transformation of metastable austenite plays an important role in improving strength-ductility balance of high strength steels. A recent study on mechanical balance of high strength multiphase steel emphasized that high stability as well as sufficient fraction of austenite would be essential for desirable elongation with high tensile strength [1]. Austenite fraction and its stability can be increased by austenite stabilizing elements (C, Mn, Ni etc.). It has been reported that steels with 5 ~ 7 mass% Mn content (Mn TRIP steel) can have considerably high austenite fraction of 20 ~ 40% after intercritical annealing which enables them to exhibit excellent mechanical balance [2–5]. However, batch-type annealing conditions that consist of prolonged annealing time and relatively low annealing temperature have been commonly adapted to Mn TRIP steels because austenite stability decreases considerably at high annealing temperature. High Mn content of Mn TRIP steel enhances hardenability of intercritical austenite in Mn TRIP steel, therefore it is difficult to control the fraction and stability of the austenite by a subsequent heat treatment following the intercritical annealing as in case of conventional TRIP steels. Continuous annealing was successfully applied to a Mn TRIP steel containing 3 mass% Al [6, 7], which yielded tensile strength of 850 ~ 1,050 MPa with elongations higher than 20%. In this study, continuous annealing was tested with a TRIP steel containing 7 mass% of Mn and 1 mass% of Al. Effects of annealing temperature on mechanical properties were investigated in terms of fraction and stability of austenite phase.

2 Experimental

An alloy having composition of 0.12C-7.2Mn-0.5Si-1.0Al in mass percent was prepared by vacuum induction melting, followed by hot and cold rolling to fabricate 1 mm thick cold rolled sheets. The cold-rolled sheets were annealed using an infrared heating furnacc at various temperatures (550 ~ 800°C) for 10 min with heating and cooling rates

S.-J. Park (✉) · C.-S. Oh · S.-J. Kim
Korea Institute of Materials Science, Gyeongnam, Changwon 641-831, Korea
e-mail: hyega@kims.re.kr

Y. Weng et al. (eds.), *Advanced Steels*, DOI: 10.1007/978-3-642-17665-4_28, of 10°C/s. Microstructures were observed by scanning electron microscope (SEM) equipped with electron back-scatter diffraction (EBSD) attachment. Samples for SEM/EBSD were mechanically polished with colloidal silica suspension for final polishing stage. The fraction of retained austenite was determined by XRD using Cu K_α radiation with graphite monochromator. Specimens were prepared by mechanical polishing followed by chemical polishing in a 10%HF + H_2O_2 solution. Integrated intensities of $(200)_\alpha$ $(211)_\alpha$, and $(220)_\gamma$ $(311)_\gamma$ reflections were used for the determination of the phase fractions of ferrite and austenite [8]. Mn and Al contents in austenite were measured by TEM/EDS from 20 ~ 30 austenite grains and average values were calculated. C content in austenite was calculated using lattice parameter of austenite(a_γ) measured by XRD and following equation [9].

$$a_\gamma\left(\mathring{A}\right) = 3.578 + 0.033[\text{mass\%C}] + 0.00095[\text{mass\%Mn}] + 0.0056[\text{mass\%Al}] \quad (1)$$

3 Results and Discussion

The change of tensile strength and total elongation as a function of annealing temperature is shown in Fig. 1. In low temperature region of 550 ~ 640°C, tensile strength decreases and total elongation increases slightly as annealing temperature increases. Meanwhile, in the temperature region of 650 ~ 750°C, significant changes can be observed both in tensile strength and total elongation. Tensile strength starts increasing and total elongation starts decreasing after a

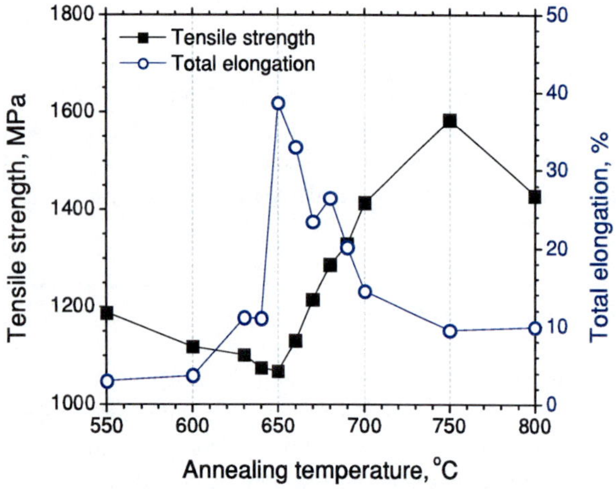

Fig. 1 Effect of annealing temperature on tensile strength and total elongation

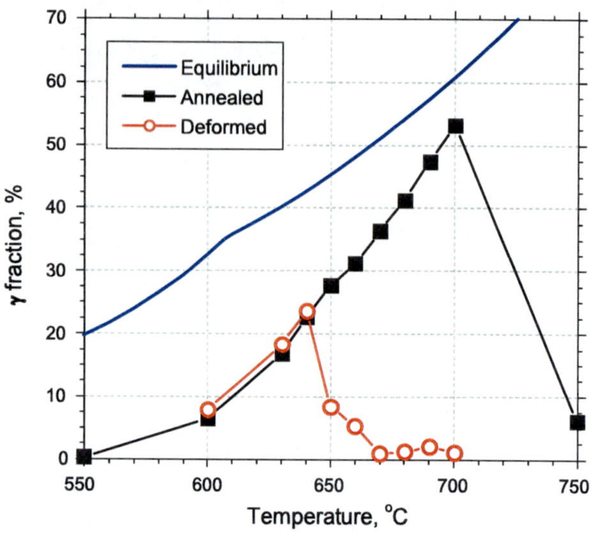

Fig. 2 Fraction of retained austenite before and after tensile test

sharp increase at 650°C. These changes are caused by deformation induced martensitic transformation of austenite, which can be noticed in Fig. 2.

Figure 2 shows austenite fraction measured by XRD from annealed samples. As annealing temperature increases, fraction of retained austenite increases getting closer to equilibrium fraction. Fraction of retained austenite after tensile test was also measured from uniformly elongated part and compared in Fig. 2. Values of austenite fraction before and after tensile deformation are almost the same below annealing temperature of 640°C, while they show considerably different values in annealing temperature range of 650 ~ 700°C. Martensitic transformation of austenite during tensile deformation makes the austenite fraction lower than that before tensile test, and the gap represents total amount of martensite formed. Retained austenite fraction in a sample annealed at 750°C is less than 10% due to martensitic transformation of austenite during cooling after annealing. The increase of tensile strength in annealing temperature region of 650 ~ 750°C is caused by the increase of martensite fraction. The peak point of total elongation at 650°C indicates that retained austenite have optimum stability for persistent martensitic transformation during tensile test. The enhancement of mechanical balance due to TRIP effect can be also seen in Fig. 3 which shows relationship between tensile strength and total elongation. The data points grouped by the bigger ellipse in Fig. 3 are from samples annealed at 650 ~ 750°C that have TRIP effect due to metastable austenite. Meanwhile, the smaller ellipse groups samples annealed at 550 ~ 640°C that have no TRIP effect due to stable austenite.

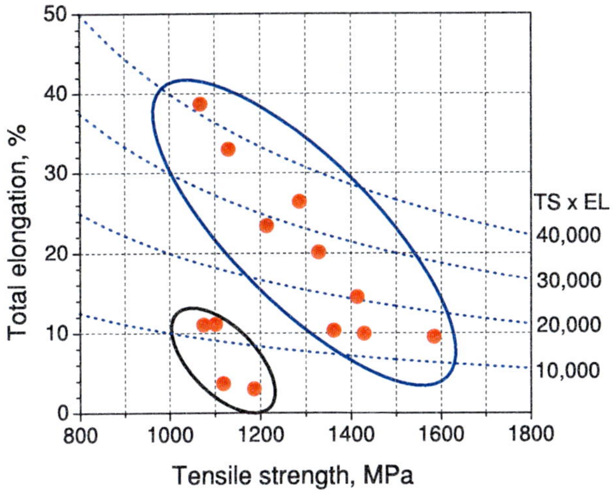

Fig. 3 Relationship between tensile strength and total elongation

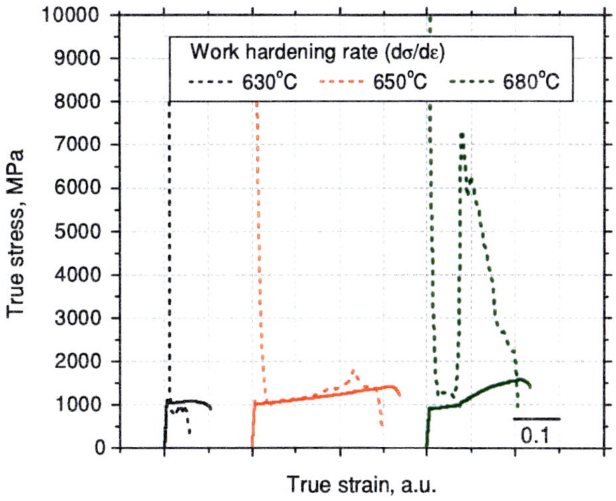

Fig. 5 Work hardening rates

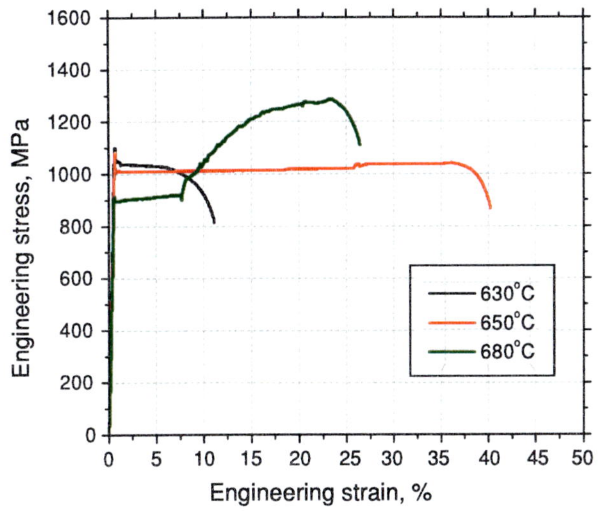

Fig. 4 Stress-strain curves

Figure 4 shows stress–strain curves of samples annealed at 630, 650 and 680°C. The sample annealed at 630°C breaks with negligible work hardening. In case of the sample annealed at 650°C, yield stress is almost same with the sample annealed at 630°C, but total elongation is drastically enhanced. Work hardening rate of each sample is represented in Fig. 5. Work hardening rate of the sample annealed at 650°C shows a slow but persistent increase after strain of 0.05, while that of sample annealed at 680°C shows a steep increasing and decreasing pattern which is caused by lower stability of austenite. Austenite stability is dependent on grain size and chemical composition of austenite, especially C content.

The fine microstructure composed of annealed martensite matrix and austenite grains formed by reverse transformation can be observed in phase map shown in Fig. 6. Black region represents pixels undefined in EBSD analysis due to low pattern quality of annealed martensite matrix. The average grain size of austenite was 0.16 ~ 0.24 μm increasing with annealing temperature as shown in Table 1. C content in austenite was also calculated using lattice parameter shown in Fig. 7. The Al and Mn contents in austenite measured by EDS are shown in Table 1 as well as the calculated C content. C content in austenite is 0.432 mass% in the sample annealed at 630°C. It decreases with annealing temperature and becomes 0.169 mass% after annealing at 680°C. It is reasonable to attribute the dilution of C concentration to the increase of austenite fraction at elevated annealing temperature. As C dilution is more significant than austenite grain growth, it seems to be the main reason for the sensitive change of austenite stability and tensile properties shown in Figs.1 and 2. The measured lattice parameter of austenite in Fig. 7 decreases with increasing annealing temperature, which also can be explained by C dilution in austenite.

4 Summary

In this study, a TRIP steel containing 7 mass% Mn was annealed under continuous annealing condition and changes in mechanical properties and austenite stability according to annealing temperature were investigated. Austenite stability varies with annealing temperature so that transition from stable to metastable austenite behavior was observed. TRIP effect due to metastable austenite resulted in enhanced mechanical balance. In optimized range of annealing temperature, high total elongation over 20% with high tensile strength over 1.2 GPa could be obtained. The sensitive change of austenite stability was attributed to C dilution in austenite which follows increase of austenite fraction at elevated annealing temperature.

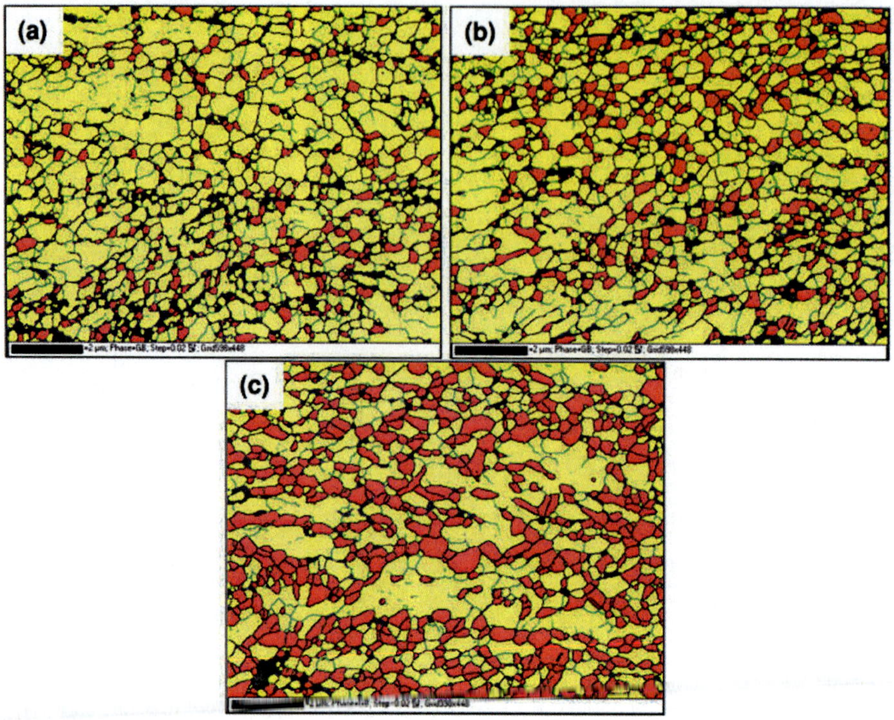

Fig. 6 Phase map of samples annealed at **a** 630°C, **b** 650°C and **c** 680°C for 10 min (BCC:, FCC: ●, Black area is undefined region.)

Table 1 Chemical composition and volume fraction of austenite in annealed samples

Annealing temperature (°C)	Al (mass%)	Mn (mass%)	C (mass%)	f_γ (%)	D (μm)
630	0.58	8.64	0.432	16.8	0.16
650	0.61	9.29	0.315	27.6	0.18
680	0.62	10.58	0.169	41.3	0.24

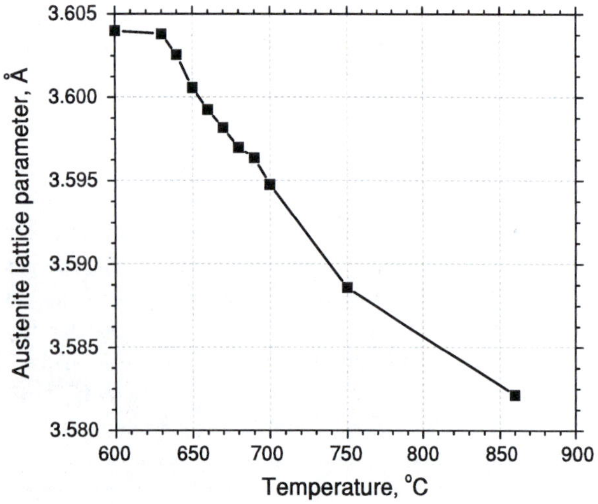

Fig. 7 Lattice parameter of austenite after annealing

Acknowledgments This study was supported by a grant from "Fundamental R&D Program for Core Technology of Materials" funded by the Ministry of Knowledge Economy, Korea.

References

1. D.K. Matlock, J.G. Speer, *Proceedings of 3rd International Conference on Advanced Structural Steels*, Gyeongjoo, p. 744 (2006)
2. R.L. Miller, Metall. Trans. **3**, 905 (1972)
3. H. Huang, O. Matsumura, T. Furukawa, Mater. Sci. Tech. **10**, 621 (1994)
4. T. Furukawa, H. Huang, O. Matsumura, Mater. Sci. Tech. **10**, 964 (1994)
5. H. Takechi, JOM **60**, 22 (2008)
6. J.-M. Jang, S.-J. Kim, N.H. Kang, K.-M. Cho, D.-W. Suh, Met. Mater. Int. **15**, 909 (2009)
7. D.-W. Suh, S.-J. Park, T.-H. Lee, C.-S. Oh, S.-J. Kim, Metall. Mater. Trans. A **41**, 397 (2010)
8. C.F. Jatczak, SAE Technical Paper Series 800426 (1980)
9. D.J. Dyson, B. Holmes, J. Iron Steel Inst. **208**, 469 (1970)

Part IV

Advanced High Strength Low Alloy Steels

Development of High Strength and High Performance Linepipe and Shipbuilding Steels

Ki Bong Kang, Ju Seok Kang, Jang Yong Yoo, Dong Han Seo, In Shik Suh, and Gyu Baek An

Abstract

Advanced TMCP products, such as X80/X100 linepipe steels and YS 460MPa grade shipbuilding steel, have been developed by POSCO. In order to apply X80/X100 linepipe to strain-based design (SBD) concept, the main issue is to prevent strain aging after thermal coating. In case of YS 460MPa shipbuilding steel, challenging study is necessary to increase plate thickness up to 80 mm without deterioration of strength and toughness. It was found that solute carbon started to segregate at dislocation tangles in coarse ferrite during and after UOE pipe forming, resulting in strain aging. And it was shown that the optimum microstructure of X80 and X100 steels was the mixture of acicular ferrite, bainitic ferrite and fine ferrite. Those X80 and X100 steels revealed good strain hardening resistance as well as reasonable compressive strain capacity. The developed YS 460MPa grade steel plate for shipbuilding showed a good mechanical property at quarter and mid-thickness and also an excellent weldability. It was confirmed that the developed steel had a high CTOD value both HAZ and base metal to prevent an initiation of brittle crack and a good crack arrestablity to suppress long distance propagation of brittle crack.

Keywords

Strain-based design • Strain aging • CTOD • Brittle crack • Linepipe • Shipbuilding

1 Introduction

The needs for advanced steel plates having high strength and high performance have been gradually increased. In oil and gas industry, materials for pipeline under harsh environment are increasingly required. Buried pipelines are subject to a number of loading conditions. These include internal pressure caused by the action of the fluids they convey, axial forces induced by thermal effects, and bending caused by differential soil movements. Recently, differential soil movements are taken an important consideration in the design and assessment of buried pipelines. When a buried steel pipeline is subject to the increasing curvatures arising from differential geotechnical movements, eventually it will buckle locally. Therefore, recent pipeline design requires that pipeline has the high deformation resistance to local buckling. Under this background, POSCO have developed X80/X100 grade linepipe steels with higher deformability.

In the field of shipbuilding business, the size of container ship has been gradually increased for mass transportation and cost reduction in the shipping industry. The high strength steel plate above YS 400 MPa was demanded as the size of container ship was increased above 10,000 TEU (twenty equivalent unit). The upper deck structure of the ship called "hatch coaming" was expected to use YS 460MPa grade with 80 mm thickness to build the large container ship above 10,000 TEU. On the other hand, the higher strength and thicker plate can easily cause the brittle

K. B. Kang (✉) · J. S. Kang · J. Y. Yoo · D. H. Seo · I. S. Suh · G. B. An
Technical Research Laboratory, POSCO, Pohang, Korea
e-mail: poscokgb@posco.com

Y. Weng et al. (eds.), *Advanced Steels*, DOI: 10.1007/978-3-642-17665-4_29,

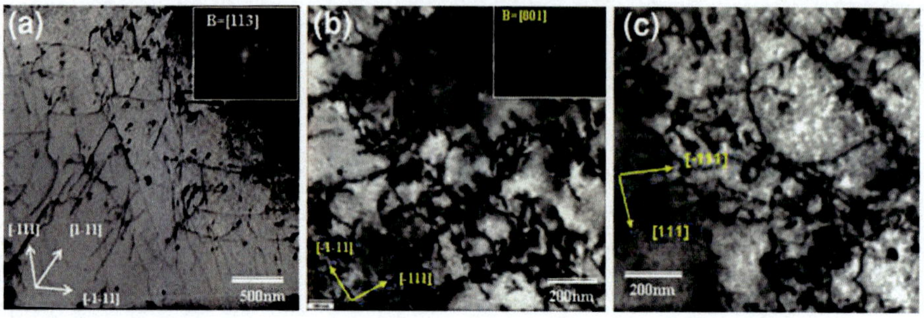

Fig. 1 TEM micrographs of coarse ferrite in steel plate (**a**), UOE pipe (**b**) and coated pipe (**c**)

fracture, compared with lower strength and thinner plate. Possibility for brittle fracture can cause the big disasters with certain critical condition such as low temperature and high applied stress in the shipping industry. Consequently, steel plate should possess good toughness not only in base metal but also in HAZ. Recently, POSCO developed YS 460MPa grade thick plate having high fracture toughness and good weldability.

In the present paper, microstructure and mechanical properties of X80/100 grade plate for strain-based design (SBD) and YS 460MPa grade thick plate for shipbuilding are introduced.

2 Linepipe Steels for Strain-Based Design

Strain-based design concept can be applied to buried linepipes which transport natural gas from production to consumption via permafrost and/or seismic region. In addition to the requirements for stress-based design linepipes, the SBD linepipes demand large strain capacity on longitudinal direction to prevent pipe fracture by unexpected ground movement [1–3]. Thus, one of most important mechanical properties for SBD linepipes is deformability. Unfortunately, the steel pipes reveal discontinuous yielding behavior after anti-corrosion coating process, so-called occurrence of strain aging. Then deformability of the linepipes significantly degraded [4–6]. Thus, we investigated the exact mechanism of strain aging in linepipe steels and based on it we tried to fabricate SBD X80 and X100 linepipe steels.

2.1 Strain Aging Process in Linepipe Steels: The Atomic Scale Observations

Strain aging in ferritic steels is known to cause by the segregation of interstitial atoms such as carbon and nitrogen to dislocations [7]. It is reported that Cottrell–Lomer dislocation reaction occurred during straining and heating of ferritic steels [8]. Then, interstitial atoms might segregate severely at edge dislocation since it has larger strain energy than that of screw dislocation. Thus, we tried to observe any changes in dislocation structure and solute carbon distribution during pipe making process. That is (1) steel plate, (2) UOE pipe and (3) coated pipe.

The microstructures of investigated steel plate were consisted of coarse ferrite and bainite. Among them we focused on the change of dislocation structure and solute carbon distribution in coarse ferrite rather than bainite. Since highly dislocated bainite is known to be less sensitive to strain aging. Figure 1 shows the dislocation structure in coarse ferrite of each pipe making process. Dislocation configuration in coarse ferrite of steel plate is a set of straight lines (Fig. 1a). In contrast, curved dislocations are tangled in coarse ferrite of UOE pipe (Fig. 1b, c). And the dislocation structure in the coarse ferrite of coated pipe is very similar to that of UOE pipe. It means that dislocations were actively glided and interwoven each other during UOE piping process but they were not moved during coating process. However, the most dislocations in coarse ferrite of every pipe making steps were $\vec{b} = \frac{a}{2}\langle 111\rangle_{\text{screw}}$ type. Although some dislocation tangles contained $\vec{b} = a\langle 001\rangle_{\text{edge}}$ components but pure edge type dislocation was not observed. That means Cottrell–Lomer dislocation reaction of $\frac{a}{2}\langle 111\rangle_{\text{screw}} + \frac{a}{2}\langle 111\rangle_{\text{screw}} = a\langle 001\rangle_{\text{edge}}$ was not actively occurred during pipe making process.

The actual carbon distribution in each pipe making process was analyzed by using three-dimensional atom probe and the results are illustrated in Fig. 2. In case of steel plate specimen (Fig. 2a), carbon atoms were homogeneously dispersed in coarse ferrite matrix. However, some carbon segregated areas were observed in the specimens of UOE pipe and coated pipe (Fig. 2b, c). Carbon concentration profiles along line A–B and line C–D in Fig. 2b, c were strengthened the carbon segregation phenomenon. The highest carbon level in the coated pipe was 0.35 at.% but that of UOE pipe was 0.14 at.%. It can be interpreted like below. During UOE process dislocation density increases and dislocation tangles formed. It means the site for solute carbon infiltration in coarse ferrite drastically increased. Then, carbon segregation to dislocation tangle could occur

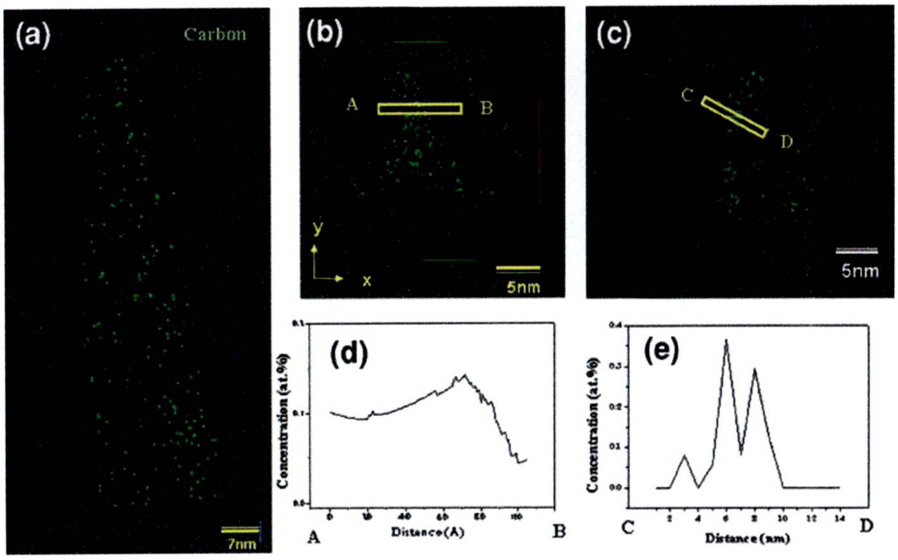

Fig. 2 Carbon atom maps of polygonal ferrite in **a** as-plate, **b** UOE pipe and **c** coated pipe. And carbon content profiles along **d** line AB and **e** line CD in (**b**) and (**c**), respectively

Table 1 Chemical compositions of X80/X100 steels (mass%)

Steel	C	Si	Mn	P	S	Others	Ceq
X80	0.05–0.07	≤0.2	≤1.8	≤0.01	≤0.001	Mo, Ni, Cu, Ti, Nb, V	0.42–0.44
X100	0.05–0.07	≤0.2	≤2.0	≤0.01	≤0.001	Mo, Ni, Cu, Cr, Ti, Nb, V	0.46–0.48

even in the coarse ferrite of UOE pipe. The segregated carbon atoms may inhibit the movement of dislocations during coating process, resulting in similar dislocation configuration in coarse ferrite of both UOE pipe and coated pipe. During coating process further carbon segregation to dislocations and/or dislocation tangles can occur since carbon atoms easily diffuse to dislocation cores during the process—the steel surface heated up to 250°C. It is almost impossible to prevent strain aging if the line pipe steel contained coarse ferrite. Thus, we designed the microstructure of SBD X80 and X100 steels as a mixture of acicular ferrite, bainitic ferrite and fine ferrite except for coarse ferrite.

2.2 Strain-Based Design X80/X100 Steels

The chemical composition of SBD X80 and X100 steels are summarized in Table 1. The steel plates of 19.8–25 mm thickness were manufactured by TMCP process with 250–300 mm thick slabs. In view of austenite conditioning for smaller grain size in steel plate, controlled rolling was applied in the non-recrystallization region. After rolling, for the purpose of obtaining appropriate microstructures, the plate was cooled down with the high cooling rate.

The optical micrographs of X80 and X100 steel were shown in Fig. 3. In case of X80 the microstructure consists of acicular ferrite and fine ferrite under the size of 7–8 μm. Bainitic ferrite having lath-like and/or granular morphology as well as fine ferrite were observed in X100 steel as shown in Fig. 3b.

To evaluate the mechanical properties and strain aging resistance, the pipes of 1219.2 mm diameter and 1 m length were made with X80 and X100 steel plates by UOE simulator. And then, the UOE pipe was heat treated at 250°C for 5 min to simulate anti-corrosion coating process.

The mechanical properties of UOE pipe before and after coating simulation (denoted as aging) are summarized in Table 2.

After coating simulation, both yield and tensile strength of L-direction were increased, however the increment of yield strength was larger than that of tensile strength, resulting in high yield to tensile ratio of 0.92 and 0.90 in X80 and X100 aged pipe, respectively. The increase of yield strength and decrease of uniform elongation is well known strain aging phenomenon [9]. In the investigated steel pipe, however, the uniform elongation was slightly increased rather than to be decreased. The deviation of mechanical properties of original plate can be a reason of this phenomenon [4, 10], but the exact reason is still unclear. Although the yield strength increased by coating process, both of aged steel pipes showed round house type yielding curves as shown in Fig. 4. If the pipe reveals discontinuous yield behavior, strain can be localized to reach bucking or wrinkling strain [11]. Thus, the investigated steels were expected to have good bucking resistance.

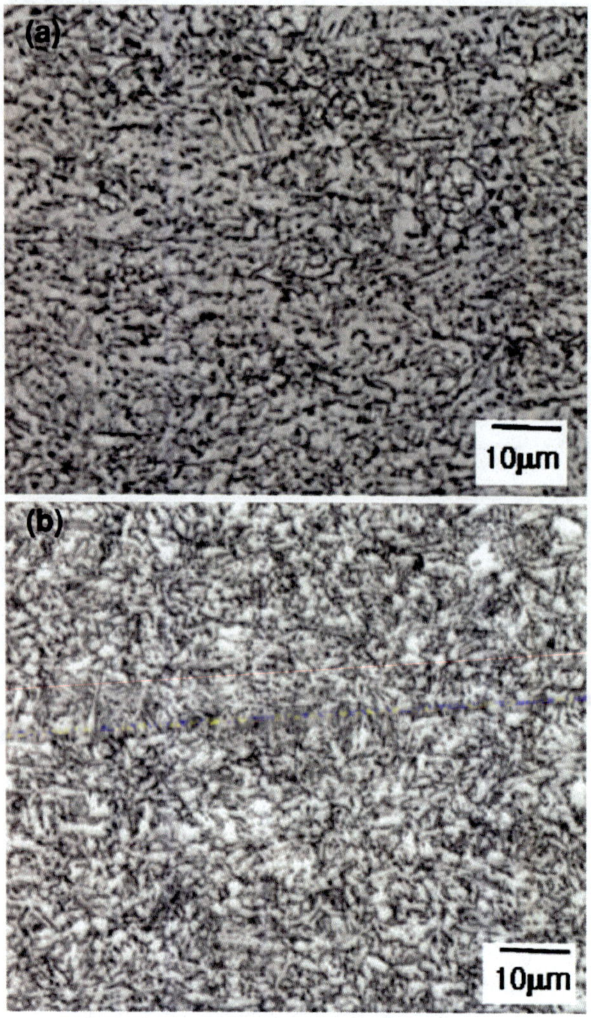

Fig. 3 Optical microstructures of tested steel plate **a** X80 steel, **b** X100 steel

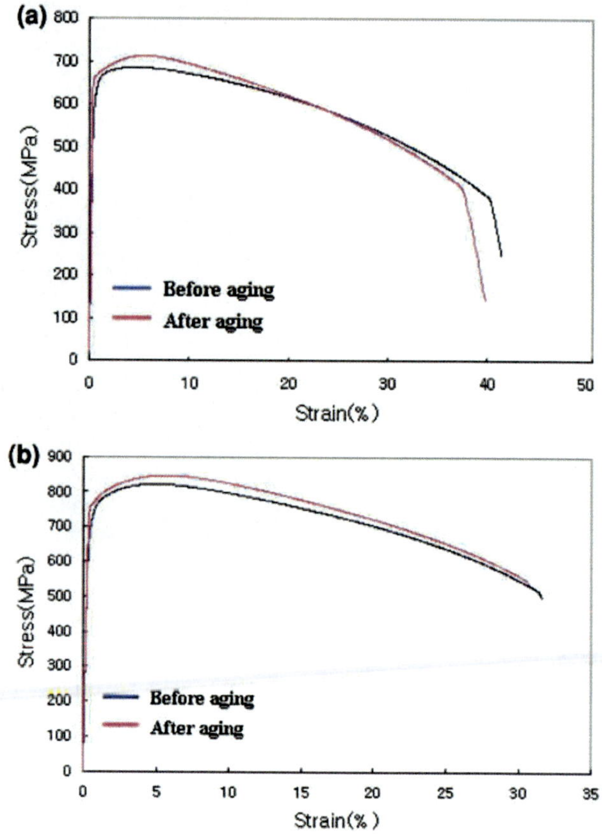

Fig. 4 Stress–strain curves of **a** X80 and **b** X100 UOE pipes before and after thermal aging at 250°C for 5 min

The compressive buckling tests on pipe with 4,750 mm length were carried out by using Universal Testing Machine (UTM). The test results are shown in Table 3. In both cases (X80 and X100), the critical compressive strain slightly decreased after aging process. However, these compressive strain values with D/t is similar or a little bit higher than that of prediction value by DNV OS F101 (2000) without internal pressure [12, 13]. Thus, our X80 and X100 steels can be acceptable for SBD application.

3 YS 460MPa Grade Steel Plate for Shipbuilding

3.1 Development Target and Concept

Target properties, which are referred to International Association of Classification Societies (IACS) specification, are shown in Table 4. Good fracture toughness, such as crack tip opening displacement (CTOD) and crack

Table 2 Mechanical properties of UOE pipes

Grade	Direction	Aging	YS0.5% (MPa)	TS (MPa)	YR	Uniform elongation (%)	CVN[a] (−20°C) (J/cm^2)	DWTT[a] (−20°C) (%)
X80	L	Before	602	697	0.86	5.6	332	85
		After	662	716	0.92	6.3		
X100	L	Before	700	822	0.85	5.2	295	97
		After	758	846	0.90	6.2		

[a] CVN and DWTT tests were performed at T-direction

Table 3 Results of full scale buckling test

Grade	Aging	OD	t	D/t	Critical compressive strain (%)
X80	Before	700	25.4	27.6	3.56
	After	711	25.4	28	2.71
X100	Before	711	19.8	35.6	2.08
	After	711	19.8	35.6	1.73

Table 4 Target properties of developed steel

Steel grade	Thickness (mm)	Base metal properties				Welded joint properties	
		YS (MPa)	TS (MPa)	El. (%)	vE−40 (J)	TS (MPa)	vE−40 (J)
EH47	80	≥460	570–720	≥17	≥64 (Avg.)	≥570	≥64 (Avg.)

Table 5 Chemical compositions and mechanical properties of developed steel

Thick (mm)	Chemical compositions (mass%)							Mechanical properties				
	C	Si	Mn	P	S	Ceq	Others	Location	YP (MPa)	TS (MPa)	El.(%)	vE−40°C (J)
80	≤0.09	≤0.3	≤2.0	≤0.01	≤0.003	0.44	Ni, Cu, Cr, Nb, Ti, Al	1/4 t	500	604	24.7	380
								1/2 t	470	607	24.4	407

arrestability, was required in both base metal and HAZ for the developed steel plate. A steel plate with heavy gauge has been developed by well controlled micro-alloying and TMCP technology to get high strength, good toughness at low temperature and good weldability.

TMCP is a well-known process to attain high strength and good toughness for steel plate with relatively low alloy composition [14, 15]. The excellent properties of TMCP steel resulted from a fine microstructure obtained by controlled rolling and accelerated cooling. However, the beneficial effect of TMCP is relatively small in the thick plates because they generally experience lower rolling reduction and slower cooling rate, compared to thin plates. Also, thick plates produced by TMCP sometimes show large variation in the mechanical properties in the thickness direction. It is caused by an inhomogeneity of plastic deformation and cooling rate in the thickness direction.

New accelerated cooling equipment, allowed a fast cooling rate and low final cooling temperature (FCT) for a thick plate, was installed. This equipment makes it possible to produce high strength thick plates without adding a large amount of alloying elements. On the other hand, the developed steel plate produced under the unique controlled rolling technology to get a good mid-thickness property and homogeneity in the thickness direction [16].

3.2 Properties of Base Metal

The mechanical properties of developed steel were summarized in Table 5 with its chemical composition. Alloy composition was designed an optimum value for good properties, based on a low C–Si–Mn steel with micro-alloying. Mechanical properties were satisfied the requirement of IACS specification with some margin. The strength difference between quarter- and mid-thickness was insignificant. It was believed that this results from the microstructure control technology uniformly with thickness direction. Figure 5 shows Charpy impact transition curve with test direction and specimen location. Ductile brittle transition temperature (DBTT) was around −90 and −70°C at the quarter- and mid-thickness, respectively. This good Charpy impact toughness was due to fine acicular ferrite

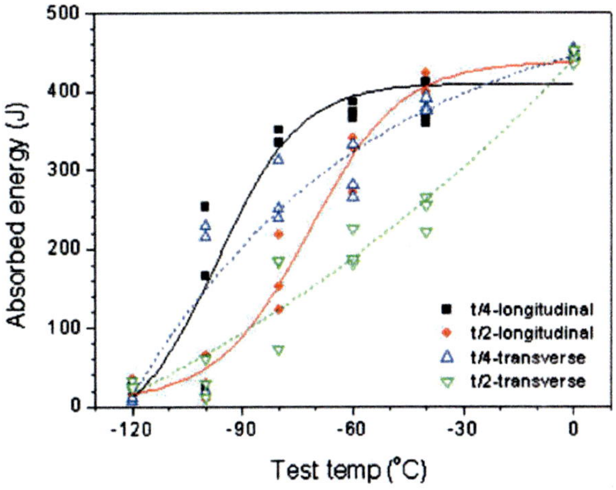

Fig. 5 Charpy transition curve of developed steel

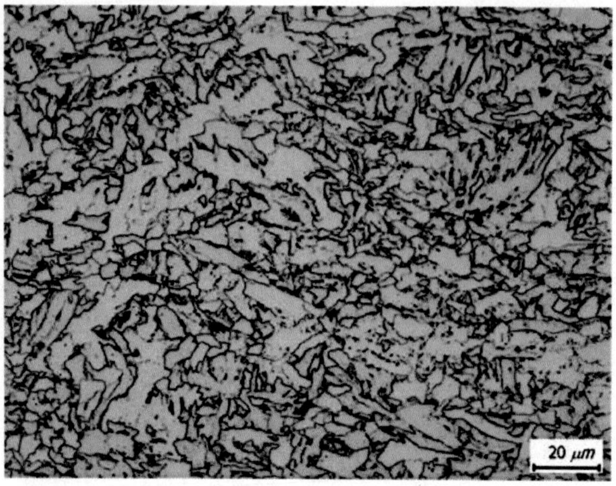

Fig. 6 Typical microstructure of developed steel

microstructure (see Fig. 6) obtained from the precise control of rolling sequence and accelerated cooling.

3.3 Weldability and Properties of Weld Joint

The sensitivity to weld cracking of developed steels was evaluated by the oblique Y-groove as specified in JIS Z 3158. Test plates were welded at the heat input of 17.2 kJ/cm with a welding material of AWS A5.29 E81T1-Ni2. There was no observation of any cold cracking at room temperature. In addition, maximum hardness test was carried out at the heat input of 17.2 kJ/cm in accordance with JIS Z 3101. Maximum hardness was measured at 248 HV. It was pointed out from these results that the developed steel has good weldability.

Flux arc welding (FCAW) and submerged arc welding (SAW) are extensively used in the shipyard. Welded joints of the developed steel plates were, therefore, evaluated by two welding methods whose detailed conditions were shown in Table 6. Figure 7 shows microstructure of samples welded by FCAW and SAW. Table 7 shows the tensile properties and Charpy impact toughness at −40°C. The tensile strength surpassed minimum requirement of 570 MPa for FCAW and SAW. On the other hand, Charpy impact absorbed energy was satisfied for all test positions and notch locations.

3.4 CTOD and Crack Arrestability

The high strength and thicker plate can easily cause the brittle fracture, compared with low strength and thinner plate [17]. Brittle fracture of ship made of high strength and thick plate ship can cause the big disasters with certain critical condition such as low temperature and high applied stress. Steel plate should firstly have a high CTOD value at HAZ and base metal to prevent an initiation of brittle crack. Also, base metal should have capability to arrest brittle crack to prevent long propagation of brittle crack.

CTOD test was carried out at −10 and −30°C according to BS7448. The CTOD specimen was prepared with full

Table 6 Welding condition for each welding method

Thickness	Welding method	Angle of groove (°)	Root gap (mm)	Consumable	Current (A)	Voltage (V)	Speed (cm/min)	Heat input (kJ/cm)
80	FCAW	30	6	SF-36E (φ 1.2 mm)/ CO_2 100%	300	32	30	20
	SAW	30	6	S-460Y (φ 4.8 mm)/ H14	800	36	35	50

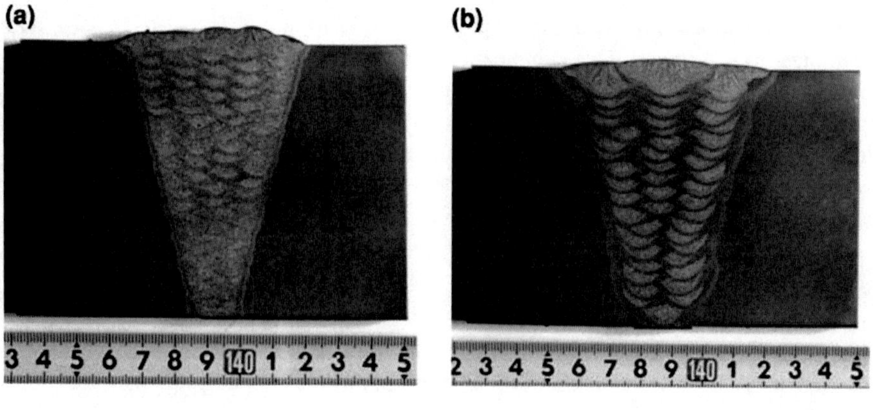

Fig. 7 Microstructure of samples welded by FCAW and SAW

Table 7 Mechanical properties of welded joint

Thickness	Welding method	YS[a] (MPa)	TS[a] (MPa)	Fractured position	vE–40°C (J)[a] Position	Weld metal	Fusion line	FL + 2	FL + 5	FL + 20
80	FCAW	497	610	Base metal	Face	57	107	280	287	287
					Center	98	150	164	173	139
					Root	77	125	282	288	294
	SAW	493	606	Base metal	Face	129	224	246	275	291
					Center	81	180	280	219	239
					Root	129	227	294	301	287

[a] An average of three tests

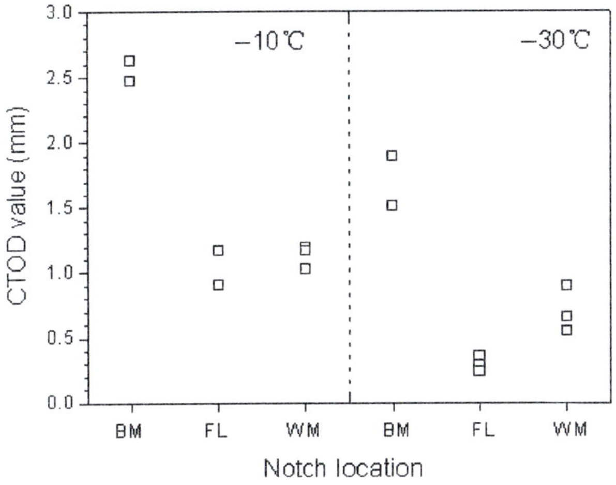

Fig. 8 CTOD values of base metal and weld joint

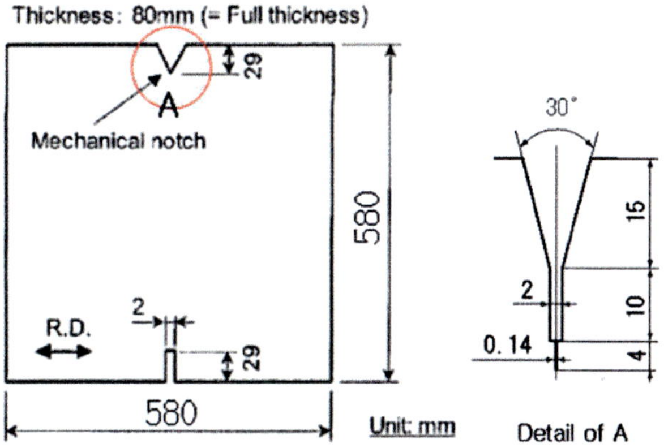

Fig. 9 Geometry of ESSO test specimen

thickness of B × B type and taken in a base plate and a welded joint of FCAW. Figure 8 shows CTOD values at various notch locations (BM: base metal, FL: fusion line, WM: weld metal). CTOD values of all notch locations are high as above 0.9 mm at −10°C, the design temperature of a container ship. It was believed from these results that the developed steel has good resistance to brittle crack initiation.

The arrestabililty on brittle crack propagation was evaluated by the ESSO test of temperature gradient type using a 3,000 ton large scale tensile test machine. A specimen geometry and notch detail was illustrated in Fig. 9. Figure 10 shows crack pass and fracture appearance after testing under the condition of an applied stress of 235 N/mm^2. For this experiment, the brittle crack was arrested after propagation up to 287 mm position where temperature was −10.3°C. The stress intensity factor (K_{ca}) of 7,960 N/mm$^{1.5}$ was obtained at −10.3°C from this result. Test results of three different applied stresses were plotted in Fig. 11. Result shows K_{ca} value of 7,643 N/mm$^{1.5}$ at −10°C, the design temperature of a container ship. It was confirmed that the

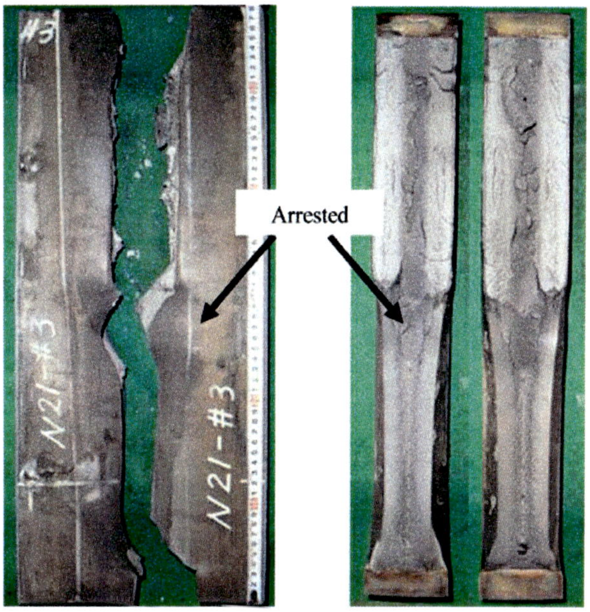

Fig. 10 Crack pass and fracture appearance

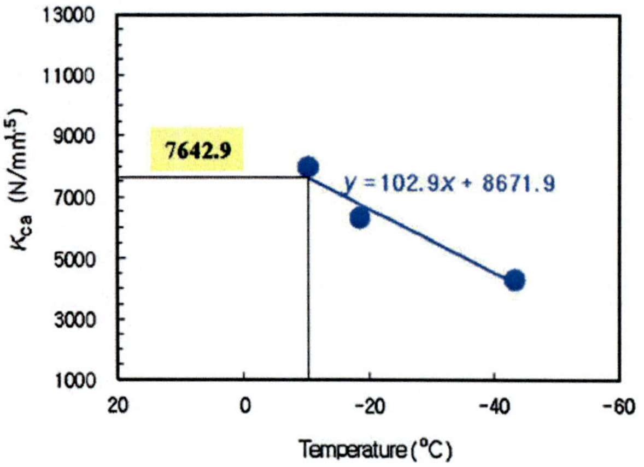

Fig. 11 ESSO test result of base plate

developed steel had a good arrestability on brittle crack propagation above 6,000 N/mm$^{1.5}$, the class NK guideline on brittle crack arresting design.

4 Conclusions

(1) During UOE piping process, dislocation density increased and tangled dislocations were formed, and its configuration was not changed during aging process. Solute carbon atoms were homogeneously distributed in the coarse ferrite of steel plate. However, carbon segregated region exhibited coarse ferrite in UOE pipe and coated pipe. The carbon atoms might be segregated in the vicinity of tangled dislocations in UOE pipe, and carbon segregation to dislocations and/or dislocation tangles were accelerated by coating process.

(2) The mechanical properties of investigated steel pipes after aging successively satisfy API standard in X80 and X100 UOE pipe. Though the yield to tensile strength ratio of aged pipe increases to 0.92 and 0.90 for X80 and X100, respectively, stress–strain curves of both aged pipes showed round house type yielding behavior. And compressive strain capacity of aged X80 and X100 pipes was 2.71 and 1.73%, respectively. These results satisfy some guideline for compressive strain in strain-based design linepipe.

(3) The YS 460MPa grade steel plate for shipbuilding has been developed by micro-alloying and TMCP technologies. It was confirmed through experiments that the developed steel had a high CTOD value at both HAZ and base metal to prevent the initiation of brittle crack and good crack arrestablity to suppress long distance propagation of brittle crack.

References

1. A.B. Dorey, D.W. Murray, J.J.R. Cheng, G.Y. Grondin, Z.J. Zhou, *Proceedings of the 19th International Conference on Offshore Mechanics and Arctic Engineering*, ASME, OMAE99, PIPE-5022, 1999
2. W. Mohr, *Pipe Dreamer's Conference* (Yokohama, 2002), pp. 629–643
3. A. Glover, B. Rothwell, *Proceedings of the International Pipeline Technology Conference* (Ostend, Belgium, 2004), pp. 65–79
4. Y. Shinohara, T. Hara, E. Tsuru, H. Asahi, N. Doi, *Proceedings of 24th International Conference OMAE*, Halkidiki, Greece, ASME, OMAE2005-67055, 2005
5. Y. Shinohara, T. Hara, E. Tsuru, H. Asahi, N. Doi, N. Ayukawa, M. Murata, *Proceedings of 7th International Offshore and Polar Engineering Conference* (Lisbon, Portugal, ISOPE, 2007), pp. 2949–2954
6. C. Timms, D. DeGeer, M. McLamb, *Proceedings of 24th International Conference OMAE*, Halkidiki, Greece, ASME, OMAE2005-67401, 2005
7. A.H. Cottrell, B.A. Bilby, Proc. Phys. Soc. **A 62**, 49–62 (1949)
8. S.T. Mandziej, Mater. Sci. Eng. A **164**, 275 (1993)
9. A.K. De, S. Vandaputte, B.C. De Cooman, Scripta Materilia **41**(8), 831 (1999)
10. D.H. Seo, J.Y. Yoo, W.H. Song, K.B. Kang, *Proceedings of 7th International Pipeline Conference*, Calgary, Canada, IPC2008-64220, 2008
11. N. Suzuki et al., NKK Giho 44–49 (1999)
12. B. Liu, X.J. Liu, H. Zhang, *Proceedings of 7th International Pipeline Conference*, Calgary, Alberta, Canada, IPC2008-64030, 2008
13. J. Wolodko, D. Degeer, *Proceedings of 25th International Conference on Offshore Mechanics and Artic Engineering*, Hamburg, Germany, OMAE2006-92173, 4–9 June 2006
14. S. Imai, *General Properties of TMCP Steels, Proceedings of 12th International Offshore and Polar Engineering Conference* (ISOPE, Kitakyushu, 2002) pp. 392–396
15. M. Toyosada, Characteristics of TMCP Steels and Their Welded Joints Used for Hull Structures, *Proceedings of 12th International Offshore and Polar Engineering Conference* (ISOPE, Kitakyushu, 2002), pp. 385–391
16. Y. Okayama, H. Yasui, K. Hara, Y. Ueshima, F. Kawazoe, S. Umeki, H. Kato, M. Hoshino, *Production of High Quality Extra Heavy Plates with New Casting Equipment*, Nippon Steel Technical Report, Nippon Steel Corporation, No. 90, pp. 59–66
17. T. Inoue, T. Ishikawa, S. Imai, T. Koseki, K. Hirota, M. Tada, H. Kitada, Y. Yamaguchi, H. Yajima, Crack Arrestability of Heavy-Thick Shipbuilding Steel, *Proceedings of the 16th ISOPE Conference*, 2006, p. 132

MoNb-Based Alloying Concepts for Low-Carbon Bainitic Steels

Hardy Mohrbacher, Xinjun Sun, Qilong Yong, and Han Dong

Abstract

Low-carbon bainitic steels offer a solution to produce strip and plate products with unique properties in terms of strength and toughness. MoNb-based alloying concepts have a great potential to further develop and optimize such steels. Mo can promote the transformation of acicular ferrite and bainite through its retardation effect on the proeutectoid ferrite transformation, and therefore improve the strength and toughness of low alloy steels. An optional alloying element is boron. Special attention is focused on the cross effects of Mo with the microalloying elements with regard to hardenability, grain size control and precipitation strengthening. Important metallurgical effects like recrystallization, microstructure and precipitation behavior are analyzed and related to mechanical properties. Different processing options for strip and plate mills are being discussed in that respect. Finally, some examples of applications of Mo in high grade pipeline steel, high strength engineering machinery steel, fire-resistant steel and hot-rolled strip for automobile are briefly introduced.

Keywords

High strength steel • Recrystallization • Grain refinement • Precipitation • Hardenability • Accelerated cooling • Microalloying • Bainite

1 Introduction

Increasing attention is being paid to the economic advantages that high-strength low-alloy steels have to offer. These advantages include lower cost structural components, increased resistance to brittle failure, economies during construction and transportation as a result of lower cost in handling lighter sections, fewer man-hours of welding and lower electrode consumption as a result of lighter sections (Fig. 1). The advantages listed are of primary interest to the transportation and materials handling industry where the ratio of payload to dead weight load is of paramount importance. Fringe benefits resulting from this are greater speeds, less fuel consumption per load and smaller sized propulsion units. In addition to strength properties, structural steel selection is concerned with ease of forming, welding and other fabrication procedures. Service conditions require that the steel exhibits good toughness at the temperature of service, thus the candidate steel should possess adequate impact resistance at the lowest temperatures anticipated in service.

Traditional high strength structural steel is produced based on a carbon-manganese alloy concept having a ferritic–pearlitic microstructure and obtained by either normalizing or thermomechanical rolling. Such steel covers a yield strength range of up to around 460 MPa. To make the desired strength, different strengthening mechanisms are employed. The base strength originates from solid solution strengthening mainly obtained by manganese and silicon

H. Mohrbacher (✉)
NiobelCon bvba, Schilde, Belgium
e-mail: hm@niobelcon.net

X. Sun · Q. Yong · H. Dong
Central Iron and Steel Research Institute, Beijing 100081, China

Y. Weng et al. (eds.), *Advanced Steels*, DOI: 10.1007/978-3-642-17665-4_30,

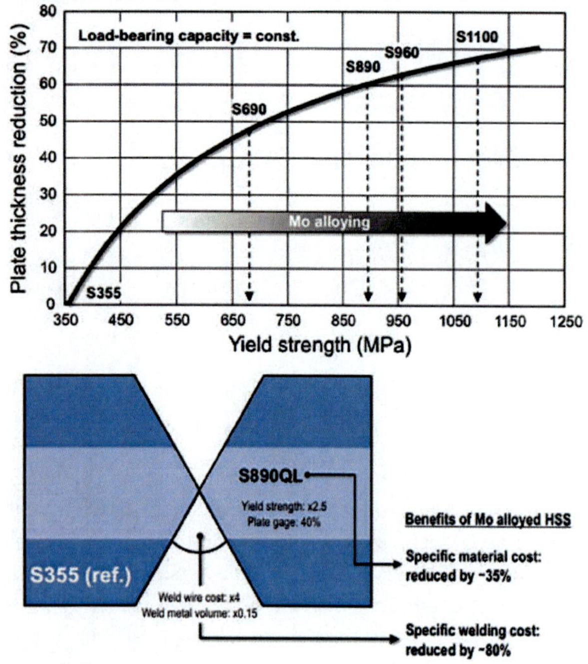

Fig. 1 Benefits of using high strength steel with regard to steel consumption and processing efforts

bulk alloying. The most important contribution to increasing the strength is grain refinement and the most effective way to achieve this is by microalloying in combination with thermomechanical rolling. Grain refinement is the only strengthening mechanism that also improves toughness. Niobium is in that respect by far the most effective element followed by titanium. The dispersion of fine precipitates, typically carbides or nitrides of the microalloying elements further increases the strength.

In modern high strength structural steels with strength levels above 460 MPa, it is necessary to modify the nature of the ferrite matrix and to avoid pearlite formation. One method is to force the austenite-to-ferrite transformation to occur at temperatures below 700°C and thereby increase the dislocation density and refine the subgrain size. The resulting microstructure is bainite or degenerated ferrite. The two alloying elements that prominently assist this transformation strengthening are molybdenum and boron. To a lesser extent chromium and niobium are also effective in that respect. Mo addition has been playing an important role in the development of low alloy steels. In 1957, Irvine and Pickering [1] found that a small amount of Mo and B additions can inhibit the formation of proeutectoid ferrite, but has little effect on the bainite transformation kinetics. Thus, bainite could be acquired by a wide range of cooling rate, consequently increasing the strength and toughness of steels. Since then, Climax Molybdenum Company has developed lower C steel with the addition of Mo–B, the so-called Climax steel. Its typical composition is 0.04% C-2.0% Mn-0.4% Mo-0.05% Nb [2]. Due to the lower carbon equivalent, its weldability and toughness was greatly improved. The effect of Mo addition on transformation strengthening in low alloy steel was then continuously developed and applied. In addition to transformation strengthening, there are several other effects of Mo addition in low alloy steels as follows:

(1) to improve the solid solubility of microalloying elements Nb, V, and Ti in austenite and therefore to promote precipitation of microalloying carbonitride in ferrite.
(2) to increase the thermal stability of microalloying carbonitride and thus to improve the high temperature performance.
(3) to delay the recrystallization of austenite and hence to enlarge the region of non-recrystallization of austenite in the rolling process. These effects above were to some extent applied in low alloy steel production.

2 Effects of Molybdenum and Niobium During Hot Rolling

Adding molybdenum to (microalloyed) steel has important effects during all stages of the hot rolling process as is schematically shown in Fig. 2. Thereby Mo acts directly as a solute atom and indirectly by influencing the behavior of microalloying elements such as Nb, Ti and V. The addition of Nb to low carbon steel significantly retards the rate of static recrystallization (SRX). For instance, by microalloying of 0.04% Nb the time for 95% recrystallization (t_{95}) at 1,060°C is around 20 s, whereas it takes over 50 s to complete the SRX if the Nb content of the steel is increased to 0.095%. Also an increase in the Mo content from 0.1 to 0.6% leads to a significant retardation of the SRX kinetics (Fig. 3). Since it is the aim of roughing rolling to obtain a homogeneous, fully recrystallized austenite microstructure this effect of Nb and Mo has to be taken into account when designing the rolling schedule. Taking the maximum interpass time in the roughing mill being 20 s, full recrystallization must occur within that period, i.e., t_{95} must be less than the interpass time. This demand determines the temperature level of roughing rolling. On the other hand, the slab discharge temperature has to be considered that has typically maximum values of 1,150 and 1,250°C or plate mills and strip mills, respectively. Thus the processing window, where fully recrystallizing roughing rolling passes can take place, is 80–180°C for the 0.1% Mo-0.04% Nb alloy, whereas it is reduced to 30–130°C for the 0.6% Mo-0.04% Nb alloy.

From stress–strain curves obtained in multi-pass tests, the interpass fractional softening (FS) was determined following a procedure described by Liu and Akben [3]. From the plot of fractional softening vs. the inverse temperature (Fig. 4) two

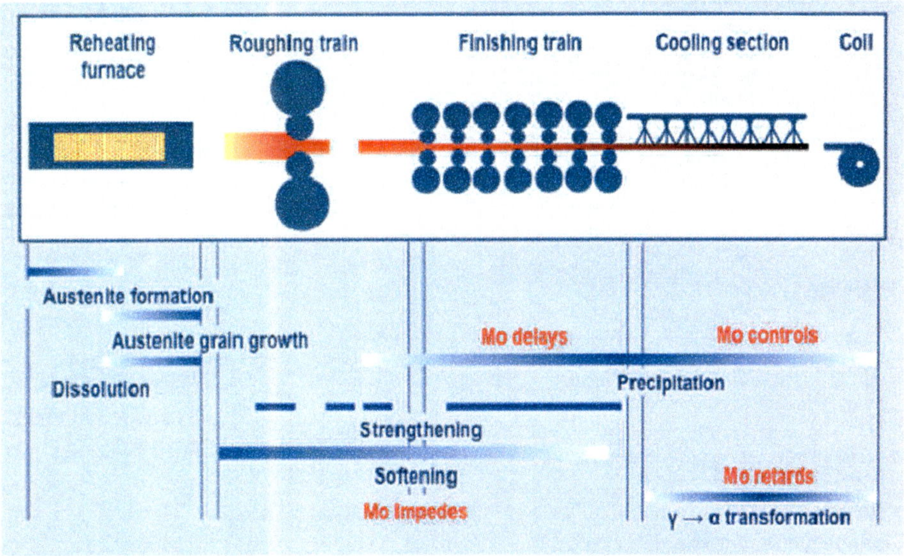

Fig. 2 Principal effects of Mo alloying during the hot-rolling stage

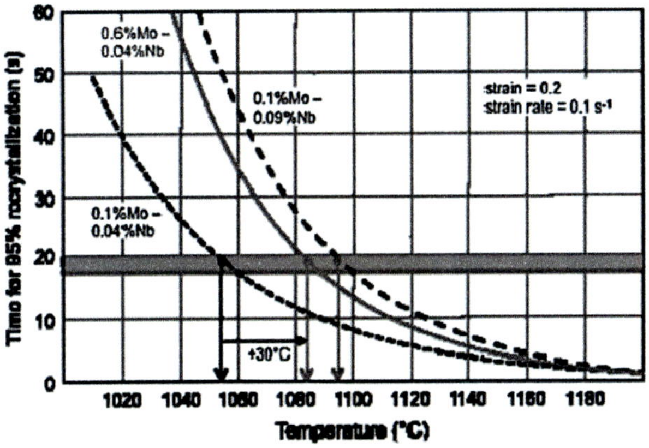

Fig. 3 Influence of Mo–Nb alloy combinations on static recrystallization at high temperature

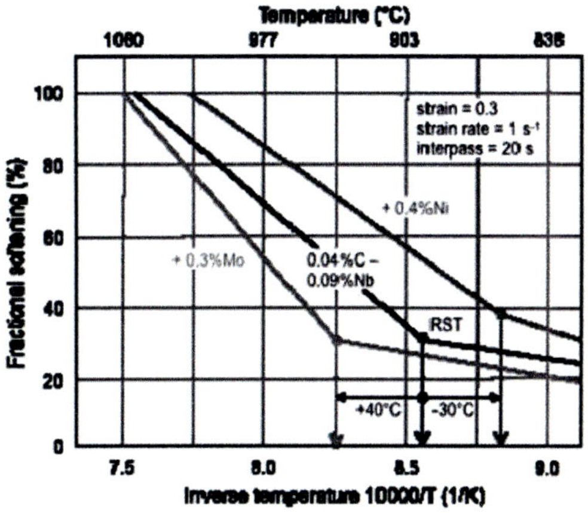

Fig. 4 Effects of Mo and Ni addition to a low-C high-Nb base alloy on fractional softening behavior

characteristic temperatures with respect to recrystallization can be derived. The recrystallization limit temperature (RLT) indicating the temperature below which softening is less than 100%, and the recrystallization stop temperature (RST) indicating the temperature below which no softening is observed between deformation passes. Figure 6 shows the behavior for 0.04% C-0.09% Nb steel as a reference having a RST of around 900°C. Adding 0.3% Mo to such a steel has little influence on the RLT but further raises the RST by about 40 to around 940°C. The influence of Mo can be understood as retarding recrystallization by solute drag. On the contrary adding 0.4% Ni reduces the RLT as well as the RST significantly. Since the recrystallization retarding effect of Nb saturates above 0.06%Nb [4] addition of Mo can effectively help to further raise the RST with respect to non-recrystallizing finish rolling without negative impact on the RLT with respect to recrystallizing roughing rolling.

Earlier experiments performed by Akben et al. [5] revealed the effect of Mo additions on the dynamic recrystallization (DRX) of microalloyed steels based on hot compression tests. When microalloyed steels are deformed above the solution temperature of their respective carbonitrides, the addition of Mo leads to a distinct retardation in the initiation of dynamic recrystallization. The solute retarding effect of Mo alone is intermediate between that of Nb, which has the greatest and that of V, which has the least effect on an equal atom fraction basis. The relative influence of these elements in solution is consistent with the relative magnitudes of their atomic size and electronic differences with respect to γ-iron. When such steels are deformed below the solution temperatures of their respective carbonitrides, in situ precipitation of small particles results in a further component of retarding recrystallization. In this case the addition of Mo involves two opposing effects. One is an

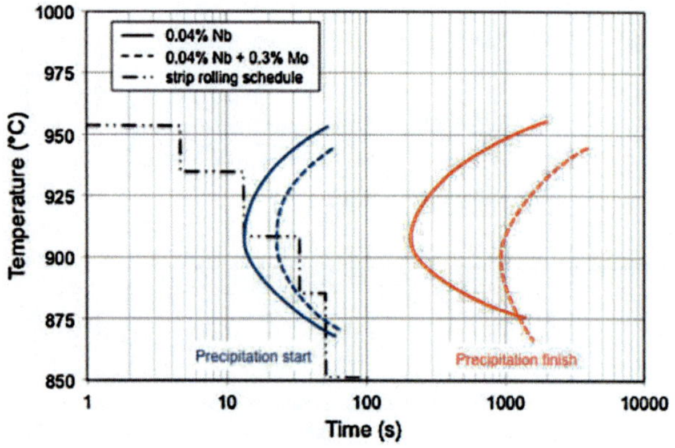

Fig. 5 Influence of Mo addition on the dynamic precipitation behavior of Nb (schematic hot strip rolling schedule)

increased retardation of recrystallization due to its effect as a solute. The other is a decrease in the amount of precipitation due to a reduced activity of C and N by Mo. It was observed that the onset of precipitation of Nb in a 0.05% C-0.04% Nb steel takes twice as long once 0.3% Mo is added as shown in Fig. 5. It is evident that especially during hot strip rolling with short interpass time a large portion of Nb can be retained in solid solution. Following a calculation method described in the literature [6], the effect of Mo on precipitation behaviour of Nb (C, N) in austenite was calculated. The calculation results are shown in Fig. 6 [7]. It is seen that the precipitation start time is delayed by nearly one order of magnitude after adding 0.14% Mo. Thus, more microalloying elements can be preserved in austenite.

Microalloying elements prevailing in solid solution after finish rolling have the capability of reducing the austenite-to-ferrite transformation temperature (Fig. 7). This effect is most pronounced for solute Nb followed by that of solute Ti while solute V has only a weak influence. The effect becomes more significant when accelerated cooling is applied. Such a delay of transformation effectively results in a further grain refinement due to enhanced ferrite nucleation and reduced grain growth. This delay increases with the cooling speed (Fig. 7). Secondly, solute microalloying elements have the potential to precipitate during or after the phase transformation to a much finer particle size than that of precipitates formed in austenite. The finer the particle size the higher is the strengthening effect as will be discussed later in detail.

3 Effects of Mo Addition on $\gamma \rightarrow \alpha$ Phase Transformation in Low Carbon Low Alloy Steels

The main effect of Mo addition on $\gamma \rightarrow \alpha$ phase transformation is to significantly delay proeutectoid ferrite transformation, resulting in obtaining bainite. The reasons are the coupled effects of solute drag and decreasing the carbon diffusivity at the interface, which result from the strong tendency to segregation of Mo at the interface in steels.

Figure 8 exemplifies the hardenability effect of molybdenum by adding different levels of Mo to a constant low carbon Mn-Cr base alloy [8]. Each alloy was heated to a temperature of 50°C above the Ac_3 temperature and held for 10 min. Afterwards the alloy was cooled at various rates and the microstructure was evaluated. Adding 0.25% Mo to the base composition significantly delays pearlite formation and lowers the transformation temperature. Simultaneously, the bainite field is largely extended. As such, practically at all technically relevant cooling rates a ferriti–bainitic microstructure is obtained. Increasing the Mo content further to 0.5% leads to a delay of ferrite formation and a complete suppression of pearlite formation. At cooling rates of above 30 K/s a fully bainitic microstructure exists.

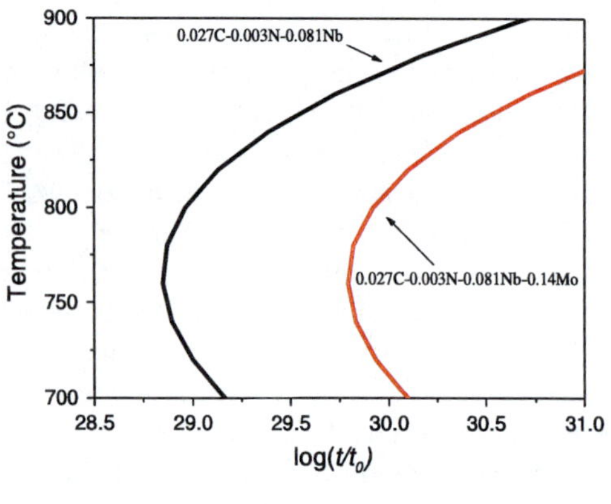

Fig. 6 Calculated effects of Mo on Nb (C, N) precipitation PTT curve from austenite

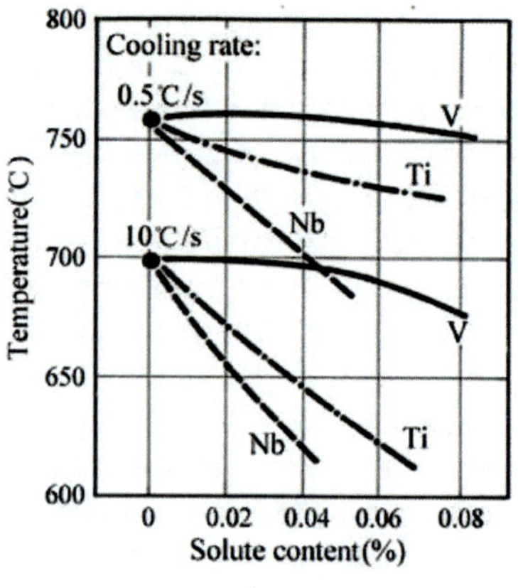

Fig. 7 Effect of solute microalloying elements on $\gamma \rightarrow \alpha$ transformation temperature

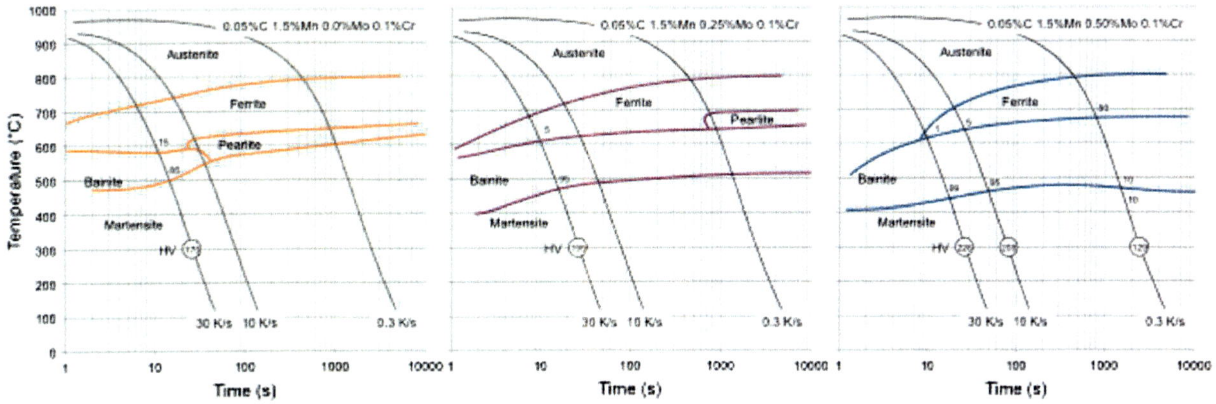

Fig. 8 Effect of Mo alloying on the transformation behavior of a low-carbon base alloy (heating to $Ac_3 + 50°C$—holding for 10 min without deformation—cooling at various rates)

More detailed analysis of the bainitic phase revealed that the block size decreases with increasing Mo content, whereas the misorientation angle between bainite laths as well as the dislocation density increases. These effects result in an increased hardness (strength) of bainite as the Mo content is raised. Increasing the cooling rate at a constant Mo content leads to the same effects. Simplified this means that raising the Mo content in such low carbon steel can substitute for lack of cooling rate. This is relevant to mills equipped with less powerful accelerated cooling devices. On the other hand, Mo helps to achieve bainitic transformation and sufficient strengthening for heavier gauged strip or plate material.

Figure 9 shows the continuous cooling transformation (CCT) curves of two different test steels (0.056Nb steel, 0.057 Nb + 0.23 Mo steel) with deformation. Austenitization, austenite deformation and CCT measurement were carried out by Gleeble. It can be seen in Fig. 9 that a fully bainitic microstructure was obtained in 0.057 Nb + 0.23 Mo steel at the cooling rate exceeding 1°C/s. However, in 0.057 Nb steel without Mo addition fully bainitic microstructure was only obtained when the cooling rate exceeding 50°C/s. Furthermore, the transformation start temperature of 0.056 Nb steel (Ar_3) is more sensitive to the cooling rate than the Mo-added steel (B_s). Figures 10 (0.056 Nb) and 3 (0.057 Nb + 0.23 Mo) show the microstructure forming at different cooling rates [9]. As shown in Fig. 10, when the cooling rate was lower than 10°C/s, the microstructure was polygonal ferrite and a small amount of quasi-polygonal ferrite; when the cooling rate was higher than 10°C/s, granular bainite is formed. However, as shown in Fig. 11, granular bainite formed at the cooling rate higher than 1°C/s in 0.057 Nb + 0.23 Mo steel; when the cooling rate higher than 10°C/s, lath bainite was acquired. Figure 12a summarized the results of Vickers hardness of test steels at different cooling rates. It is seen that the hardness of the Mo-added steel was higher than the Mo-free steel. Furthermore, the higher the Mo addition, the greater is the hardness increase under accelerated cooling. As shown in Fig. 12b, the hardness (or strength) of test steels increased linearly as the transformation start temperature decreased. Combined with the CCT curve in Fig. 9, it is

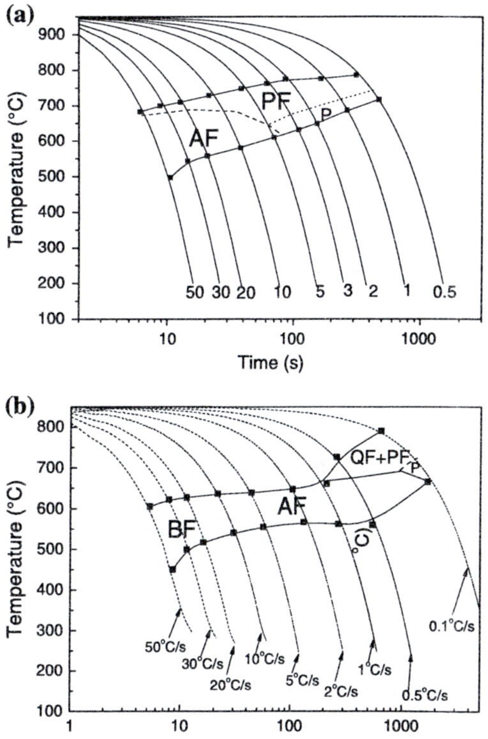

Fig. 9 Dynamic CCT curve of experimental steels (austenitized at 1,200°C for 5 min, 40% compression at 850°C): **a** 0.056Nb steel: 0.04 C-0.26 Si-1.65 Mn-0.056 Nb-0.014 Ti, **b** 0.057 Nb + 0.23 Mo steel: 0.062 C-0.25 Si-1.53 Mn-0.057 Nb-0.23 Mo-0.019 Ti [9]

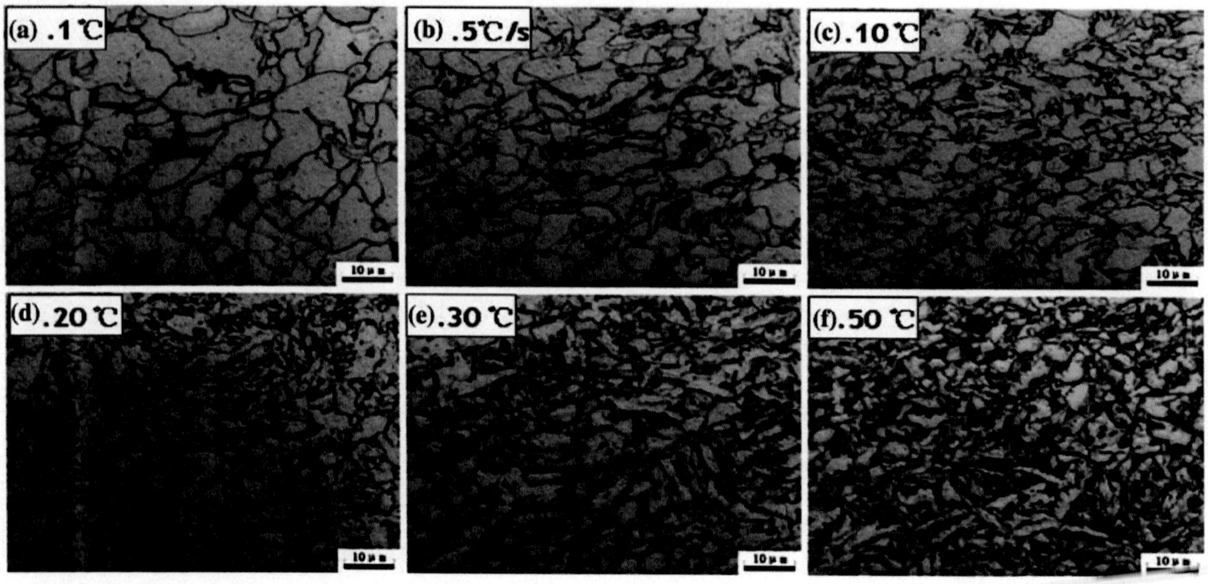

Fig. 10 Optical micrographs showing the microstructure obtained under different cooling rates in 0.056 Nb steel [9]

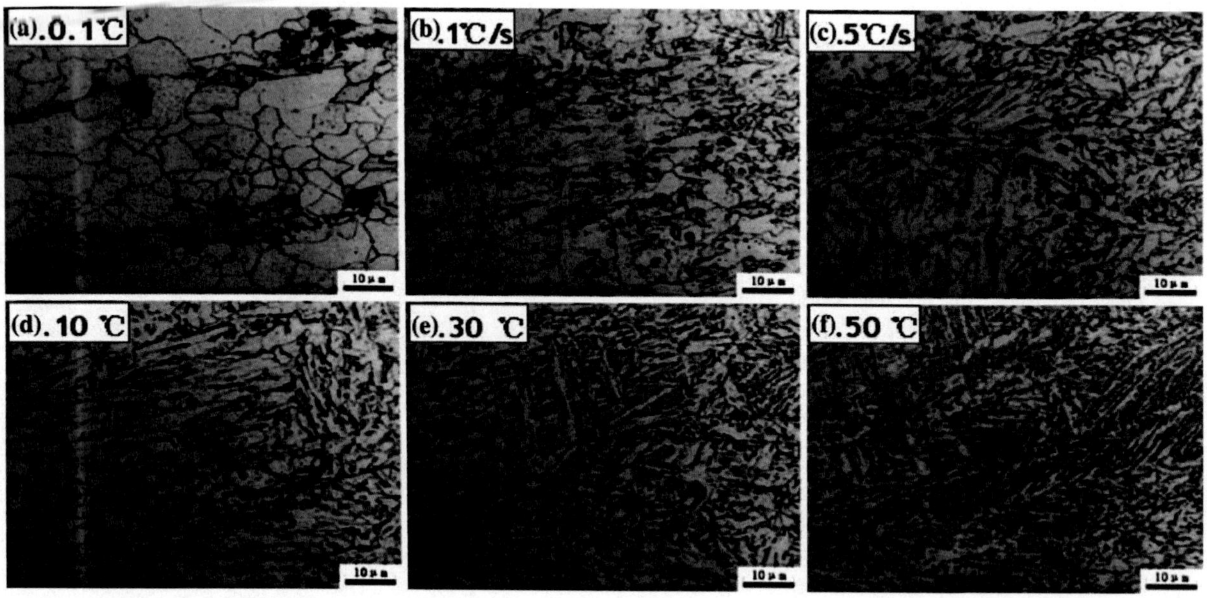

Fig. 11 Optical micrographs showing the microstructure obtained under different cooling rates in 0.057 Nb + 0.23 Mo steel [9]

indicated that Mo not only increases the strength of steel, but also decreases the heterogeneity of microstructure and performance along the thickness of the thick plate, where the cooling rate varies along the thickness.

More recently, low-carbon steel alloys with Nb additions of up to 0.1% came into use for industrial applications. Combining a low-carbon high-niobium alloy with Mo leads to a synergetic effect on the transformation behavior. For this alloy pearlite formation is suppressed and the onset of ferrite formation is significantly delayed (Fig. 13a). The critical cooling rate to obtain full bainite formation in the 0.04% C-1.4% Mn-0.1% Nb-0.3% Mo alloy is around 10 K/s. Modifying this alloying concept by increasing carbon and manganese levels, reducing the molybdenum content and adding chrome results in a further improved hardenability (Fig. 13b). The critical cooling rate for full bainite transformation in this alloy variant is in the order of only 4 K/s. In yet another alloy variant molybdenum was substituted by a

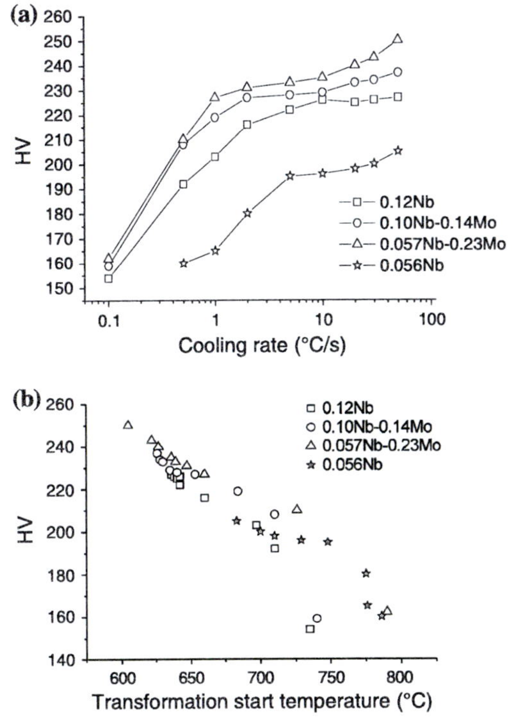

Fig. 12 **a** Variations of Vickers hardness (HV) with cooling rate, **b** relationship of HV and transformation start temperature [9]

combination of nickel and copper. Although this alloy also results in fully bainitic microstructure at cooling rates of above 4 K/s, several differences in the transformation behavior can be observed (Fig. 13c). The start temperature of bainite transformation is increased by around 50°C. At slower cooling rate (1 K/s) such as occurring by air-cooling, no bainite is formed anymore but a small fraction of pearlite appears, which is detrimental to toughness. At high cooling rates (95 K/s), as they can occur after welding, martensite formation is possible. These three examples particularly underline the effect of molybdenum in promoting bainite and avoiding pearlite formation, respectively. With regard to mechanical properties, all three alloys reach a strength level of 650 MPa tensile and 550 MPa yield under suitable rolling conditions and fully bainitic microstructure.

For continuously cooled molybdenum-boron steels, the main effect of bulk alloying elements such as manganese or chrome is to lower the temperature at which the transformation to bainite begins. In carbon-manganese steels, the addition of molybdenum in combination with solute titanium or niobium reduces the critical cooling rate for producing bainite, i.e., suppresses transformation to polygonal ferrite. Boron is a very powerful hardenability element that is added to steel in minute amounts, usually not more than 50 ppm. As such it is used in (ultra-) low carbon bainitic steels as well as in quench-hardening steels. The effectiveness of boron to provide hardenability lies in its segregation to the austenite grain boundary where it obstructs the formation of grain boundary ferrite at transformation temperature. However, it was shown that only the coupled addition of B with Mo or Nb improves the hardenability as shown in Fig. 14 [10].

Boron increases the hardenability only when it is in solid solution. Since B is a strong nitride former, Ti is usually added at around stoichiometric ratio (Ti = 3.4 × wt%N) to protect B. However, B can still be lost by forming a complex $Fe_{23}(C,B)_6$ precipitate. This happens in the austenite grain boundary particularly when increased amounts of B and C are present due to segregation. Asahi [11] and Hara et al. [12] have identified this phenomenon for steels with ultra-low as well as over-peritectic carbon contents. In both cases the addition of Mo to the alloy improved the effectiveness of B since intragranular Mo-C cluster forming reduces the carbon diffusion into the austenite grain boundary (Fig. 15). Niobium microalloying was found to

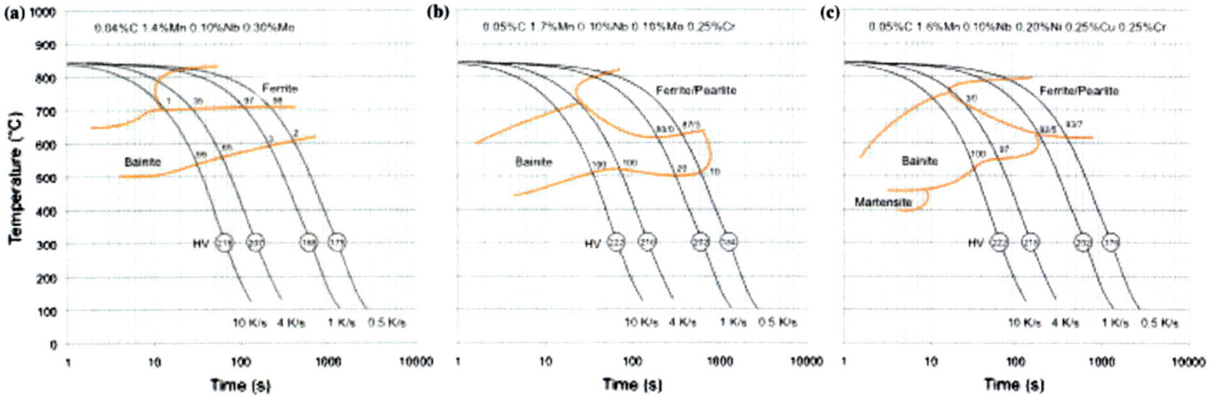

Fig. 13 Transformation behavior of various low-carbon alloys of similar P_{cm} and strength level (holding at 1,200°C for 10 min—deformation j = 0.6 at 1,050°C—deformation j = 0.6 at 850°C—cooling at various rates)

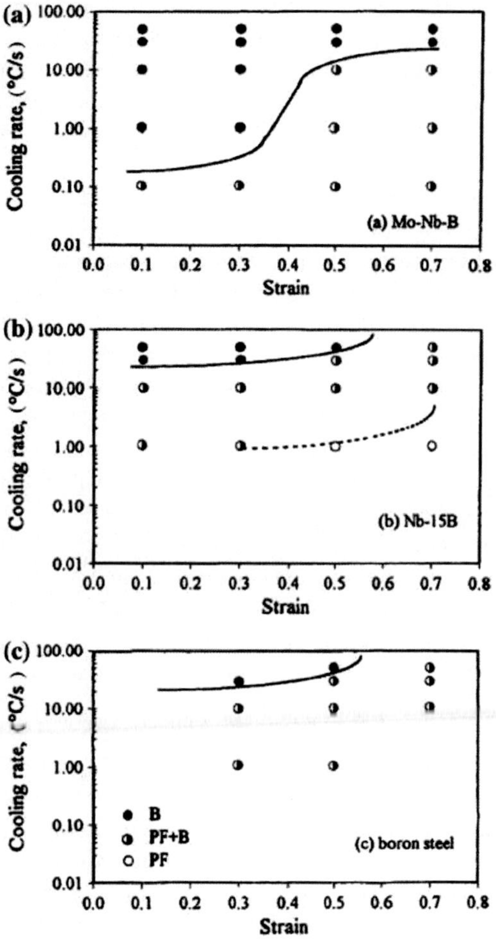

Fig. 14 The critical strain and cooling rate to avoid the formation of polygonal ferrite in different steels: **a** Mo-Nb-B, **b** Nb-15B, **c** B-only [10]

have a similar effect of avoiding $Fe_{23}(C,B)_6$ precipitation, which however is rather related to NbC precipitation removing solute carbon.

Besides of rendering a better effectiveness of boron in preventing grain boundary ferrite formation, molybdenum acts as a hardenability element by itself. This Mo effect additionally enhances the B effect as becomes evident from Fig. 16 [12]. Compared to the Nb-B steel where only B acts as hardenability agent the Mo–B steel shows lower transformation temperature as well as higher hardness at any cooling rate.

4 Precipitation in Ferrite

It has been postulated that microalloying elements would not precipitate solely but co-precipitate with Mo in Mo-alloyed steels, i.e., (M, Mo) (C, N) with a NaCl-type fcc structure [13]. Figure 17 shows a high resolution TEM image and EDS spectra after completion of ferrite transformation in a steel with 0.027 C-0.20 Si-1.11 Mn-0.081 Nb-0.14 Mo-0.003 N (wt%). The steel was held at a ferrite transformation temperature of 750°C for 316 s after austenite deformation. It is seen that the precipitates are very fine with a disc-like morphology of about 2 nm in diameter. It is thus indicated that they were precipitated from ferrite. When microalloying elements precipitate from ferrite, Nb(C, N) always holds Baker–Nutting relationship with ferrite. The mismatch between the precipitate and the matrix varies with direction. In order to minimize the total interfacial energy, the shape of precipitate should be disc-like, with the flat side parallel to the (100) lattice plane [6]. The EDS analysis confirmed that the precipitated phase contains Mo, illustrating Mo and Nb precipitates together from ferrite.

In order to predict the chemical composition of particles at different precipitation temperature, it is necessary to build the corresponding thermodynamic model. Microalloy precipitation containing Mo can be assumed to be regular solution consisting of microalloying carbide MC, nitrides

Fig. 15 Schematic of boron grain boundary (GB) segregation and precipitation in B and Mo-B steel

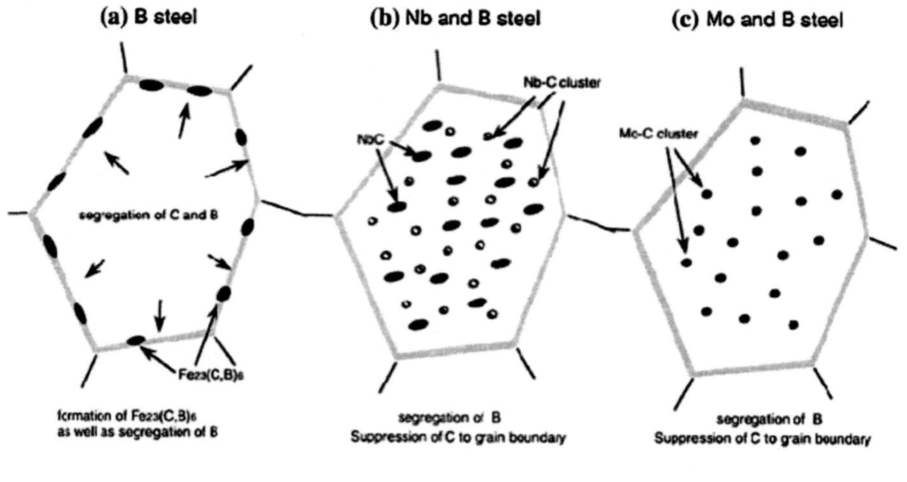

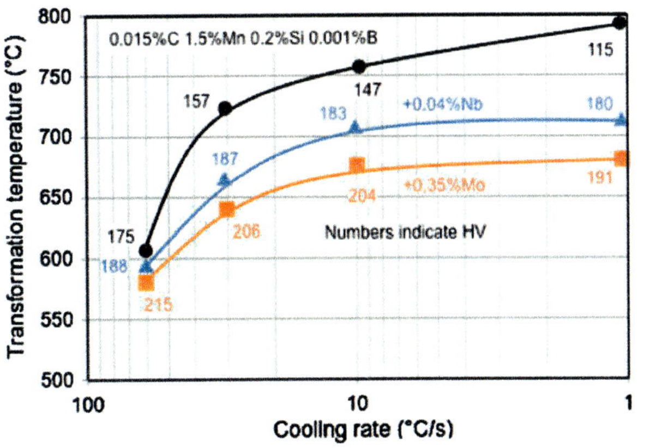

Fig. 16 Effect of cooling rate on transformation temperature and hardness in ULCB steel with B, Nb-B and M-B alloying

MN and molybdenum carbide MoC, which have the same crystal structure—NaCl type face-centered cubic. According to the double sublattice model of regular solution, M and Mo occupy a sublattice but C and N occupy the other sublattice. The complex precipitate can be expressed as (M_{1-x}, Mo_x) (C_{1-y}, N_y), where x and y, respectively, denote Mo and N mole fractions in their sublattice ($0 \leq x \leq 1$, $0 \leq y \leq 1$). The related thermodynamic calculation methods can be found in the literature [14]. Figure 18 shows the calculated results of the test steel above. It is seen that the calculated results agree very well with the EDS analysis results at different temperatures.

Figures 19 and 20, respectively, show the calculation results of Mo site fraction in its sublattice in the Nb and Ti microalloyed steels with different Mo contents. According to those two figures, the site fraction of Mo increases with the increasing of Mo content but with the decreasing of precipitation temperature. More Mo precipitates at the lower temperatures. Such complex precipitation inevitably increases the volume fraction of precipitates. On the other hand, the diffusion coefficient of Mo is smaller than that of the other microalloying elements. Therefore, both the growth of precipitates and coarsening after the completion of precipitation should be inhibited. Consequently, the size of the complex precipitates is relatively smaller. Hence, it is expected that the complex precipitates containing Mo have a greater precipitation hardening effect than the simple microalloy precipitates.

5 Thermal Stability of Precipitates

Coarsening of Mo-containing co-precipitates requires the diffusion of both Mo and other microalloying elements from the ferrite matrix to the ferrite/precipitate interface. Since the diffusion rate of Mo is slower than that of other microalloying elements, the coarsening rate of precipitates is controlled by the diffusion of Mo and is thus slow. Figure 21 presents a group of TEM images and EDX data showing the particle morphology and compositions of interface precipitation for single Ti-, Ti-Nb- and Ti-Mo-bearing steel after isothermal holding at 750°C for 60 min [15]. Figure 22 shows the average particle sizes measured for the three steels with different holding times. For the three steels, the contents of Ti, Nb and C are 0.20, 0.04 and 0.10%, respectively. It is clear that the particles containing Mo and Ti are much finer and more thermally stable than those containing Ti and Nb and only containing Ti. The above-mentioned characteristics of Mo-containing steel can weaken the coarsening tendency of precipitate during the coiling process of hot strip production, and can also help to improve the properties of such steels being exposed to higher temperature for some time.

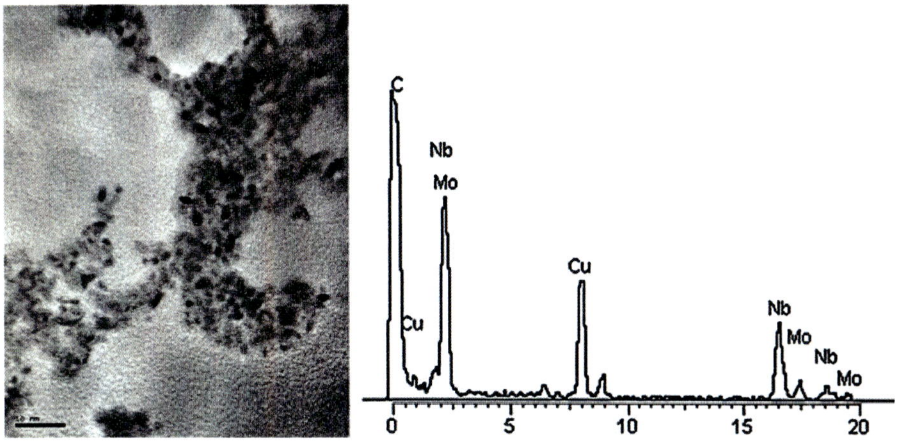

Fig. 17 High resolution TEM image and EDS spectra after completion of ferrite transformation at 750°C for 316 s after austenite deformation in 0.027 C-0.20 Si-1.11Mn-0.081 Nb-0.14 Mo-0.003 N (mass%) steel [7]

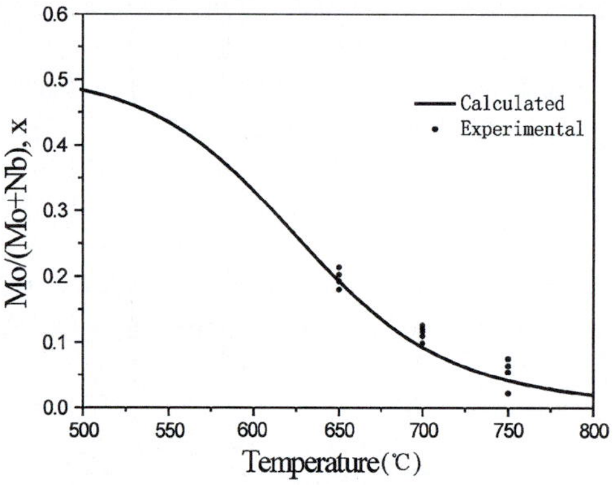

Fig. 18 Relationship of Mo site fraction in complex precipitates and precipitation temperature in an experimental steel alloy [7]

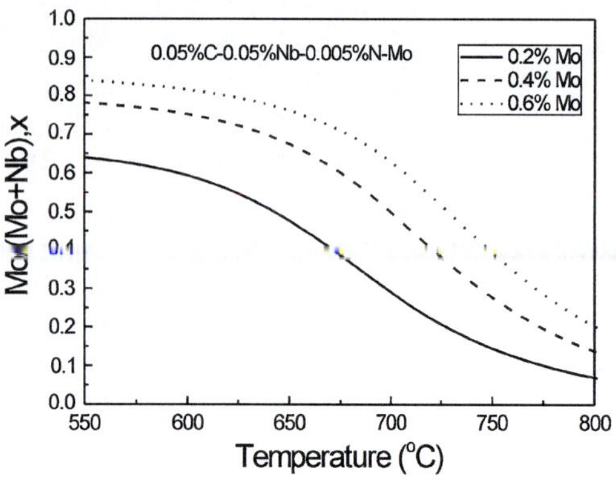

Fig. 19 Relationship of Mo site fraction in complex precipitates and precipitation temperature in Nb microalloyed steels with different Mo contents

For particular applications it is not favorable to have pearlite in the final microstructure. Therefore, it is important to delay the pearlite formation to times longer than the isothermal holding period. Molybdenum alloying is efficient in achieving this delay as discussed above. During austenite decomposition so-called interphase precipitation occurs by a mechanism of periodical particle nucleation at the moving austenite/ferrite interface during transformation [16]. The austenite side of the moving phase boundary enriches with carbon and microalloying elements due to their lower solubility in ferrite. At certain intervals supersaturation leads to precipitation manifesting itself by row arrangement of precipitates (Fig. 23). Since the entire process is diffusion controlled, interphase precipitation is most effective at a comparably slow cooling rate. In practice, coiling in a hot strip mill naturally provides isothermal holding since the cooling rate in the coil is slow (around 30°C/h). In fact the coil is still at high temperature after precipitation is complete and this promotes particle coarsening by Ostwald

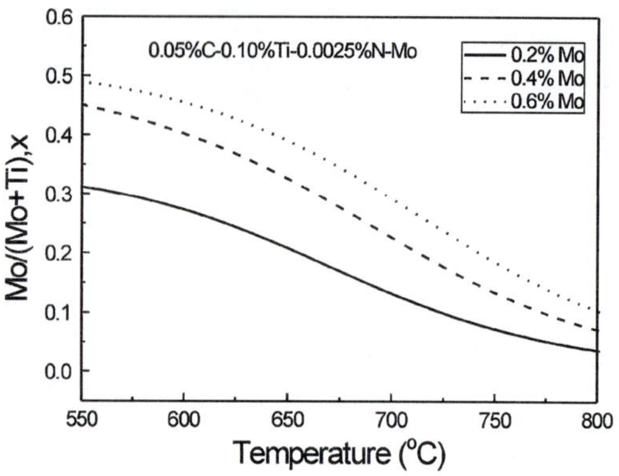

Fig. 20 Relationship of Mo site fraction in complex precipitates and precipitation temperature in Ti microalloyed steels with different Mo contents

ripening. In this respect molybdenum is found to inhibit particle coarsening by segregating to the particle matrix interface and blocking the diffusion of carbide forming elements between particle and matrix. This mechanism results in a much-reduced loss of particle strengthening even for extended isothermal holding at 650°C (Fig. 24) [17]. In reality, the temperature continuously drops by natural coil cooling. At temperatures below 550°C diffusion is so limited that particle coarsening does not take place effectively. This situation would be reached in the example of Fig. 14 after around 12,000 s. The molybdenum alloyed steel shows after that time almost no loss in hardness (strength). In absence of molybdenum, particle coarsening of TiC would lead to a hardness drop of approximately 30 HV (corresponding to 100 MPa strength).

If the coiling temperature is set below the range allowing efficient interphase precipitation and/or rapid cooling is applied, microalloying elements will remain in solution and can only precipitate during a secondary heat treatment.

Figure 25 summarizes the strengthening effects by transformation and/or precipitation hardening for the Mo-Nb and Mo-V alloy system ,respectively [18]. The precipitation strengthening, having a maximum in the range of 630–650°C, adds to the strengthening effect in the matrix. The latter one increases with decreasing coiling temperature. For the Mo-Nb system this strength increase continues below 600°C, whereas it levels of at that temperature for the Mo-V system. This is because the synergetic effect of Mo and Nb effectively promote the forming of a bainitic microstructure providing transformation hardening. It appears that the gain in transformation hardening nearly equals the loss of precipitation hardening at lower coiling temperature. Hence, the total strength being the sum of the two strengthening components remains nearly constant. This constitutes a considerable advantage as in a hot strip mill one can encounter considerable drifts of the actual

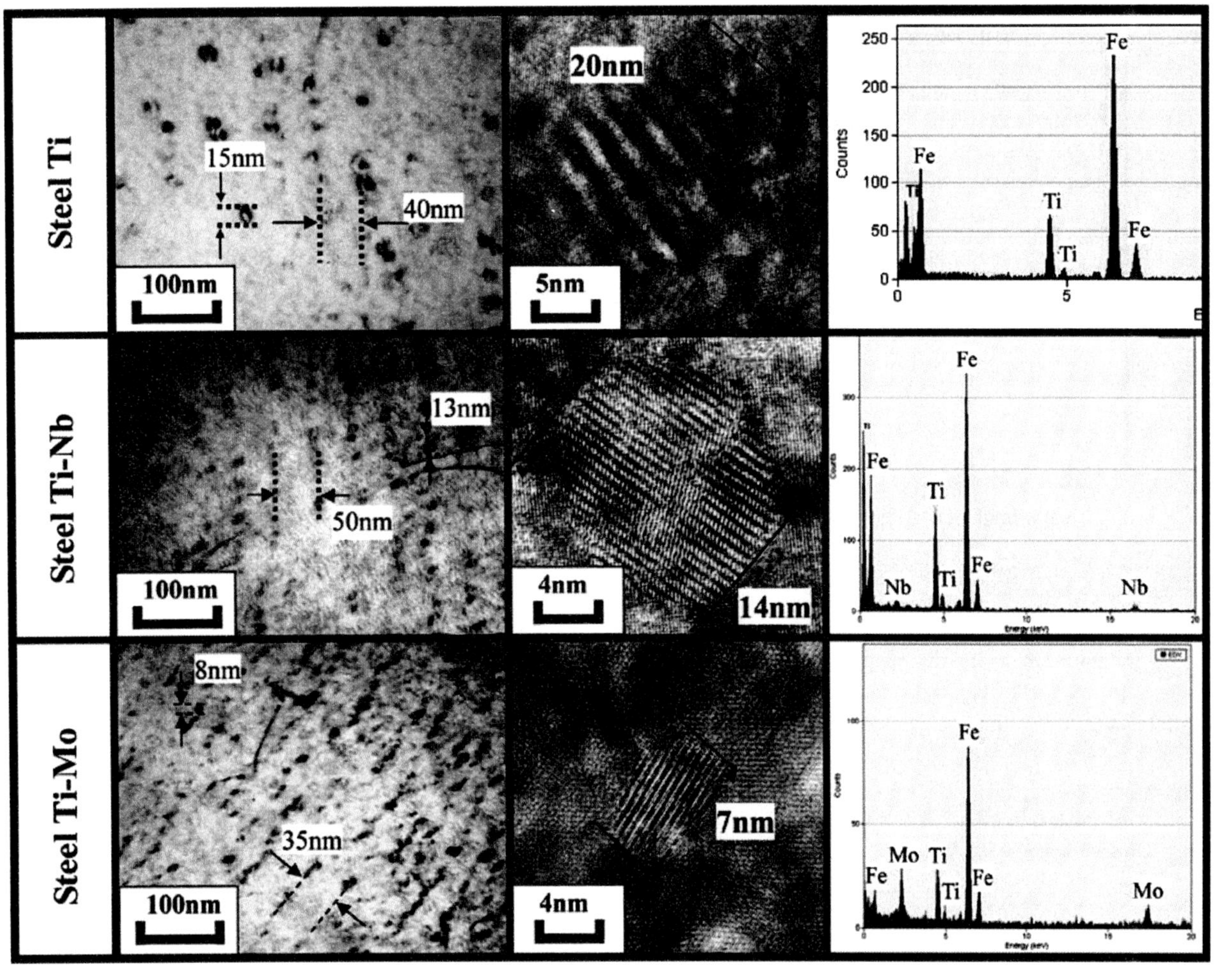

Fig. 21 TEM images and EDX data showing the particle morphology and compositions of interface precipitation for single Ti-, Ti–Nb- and Ti-Mo-bearing steel after isothermal holding at 75 C for 60 min

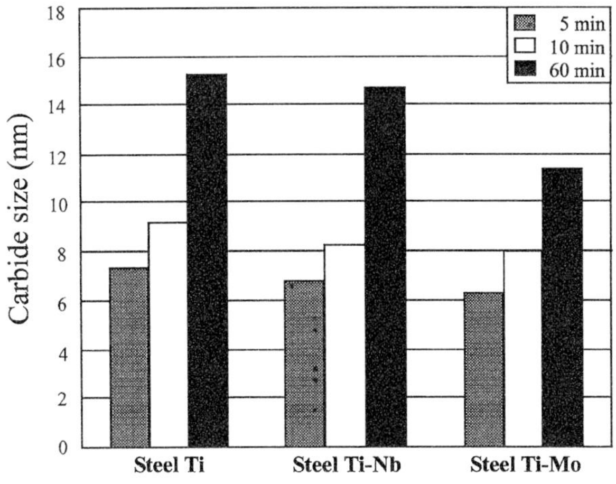

Fig. 22 Average particle sizes measured for the three steels with different holding times [15]

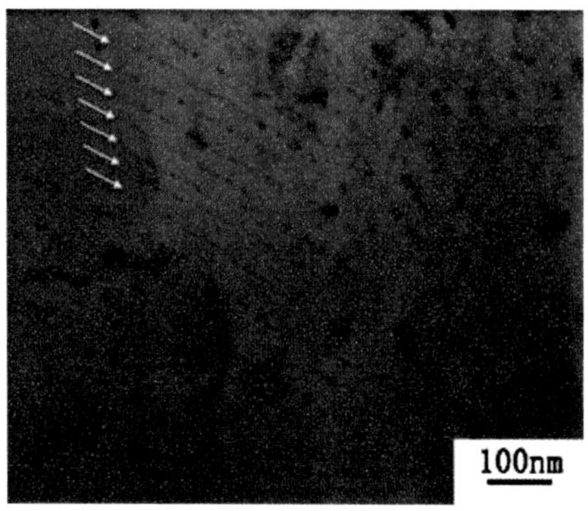

Fig. 23 Example of row-like precipitate arrangement (*arrows*) in a 0.05% C Mo-Nb-Ti steel coiled at 630°C

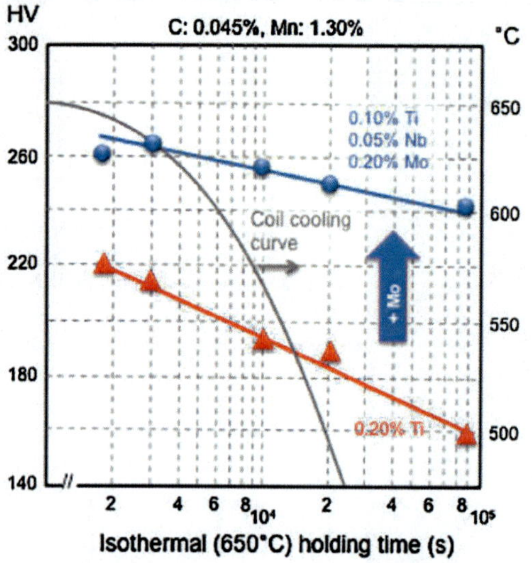

Fig. 24 Influence of isothermal holding at 650°C on precipitation hardening (typical coil cooling curve is indicated)

coiling temperature along the strip. In the Mo-V system such a temperature drift would necessarily reflect in substantial scattering of the final yield strength.

6 Applications of MoNb-Based Low-Carbon Steels

Thus far Mo has been widely applied in the production of high-grade pipeline and structural steel, high strength engineering machinery steel and fire-resistant steel. Most of the pipeline steels with grade X70/X80/X100 are alloyed with 0.10–0.30% Mo, especially for heavier gage material. The microstructure of pipeline steel is composed of a mixture of acicular ferrite and granular bainite. With the increase in strength grade, the microstructure is refined and the fraction of GB is increased. Recently, low-carbon NbMo steel has been extensively used for spiral pipe production in China's West-East-Gas-Pipeline projects [19].

Combined Mo-B addition is used for X120 steel to further improve the hardenability, and the microstructure obtained consists of lath bainite (or lower bainite) [20]. Combined Mo-B addition is also used for non-quench typed engineering machinery steel and the microstructure is composed of granular bainite or lath bainite. All these steels largely make use of the phase transformation strengthening resulted from the enhancement of B on the bainite transformation.

The application of Grade S500M3z [YS 500–580 MPa, TS 600–750 MPa, CVN (transv., mid-thickness) at 40°C $\geq$ 60 J (each individual test $\geq$42 J)] in parts of the topside of an offshore platform gave sufficient weight reduction in comparison with Grade S420. The main cost benefit of that weight reduction was achieved as it enabled installation of the topside with a single operation [21]. A CuNiMoNb-steel with a CEIIW of about 0.42 in combination with a DQST process is applied to produce plates with granular bainite constituents in thicknesses up to 80 mm. With that concept adequate toughness properties can be achieved even at mid thickness position.

Combined addition of Mo-Nb has been adopted for the fire-resistant steel, which can improve the high temperature properties of the steel, making use of fine and highly thermal stable particles containing Mo. As a consequence softening and resulting collapse of the steel structure is delayed [22].

In automotive chassis and frame applications good cold formability (bending, hole expansion and stretch flanging) at simultaneously high yield strength are desired properties. Hot rolled strip steel with fine-grained microstructure and

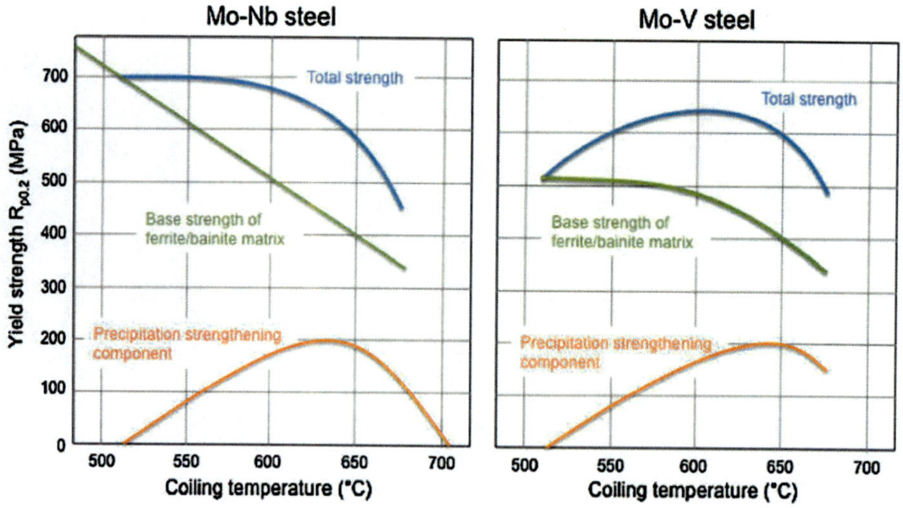

Fig. 25 Effect of coiling temperature on yield strength in low-carbon Mo-Nb and Mo-V steel

substantial precipitation hardening was found to fulfill such requirements. Under the optimum processing conditions (i.e. coiling at 630–650°C), the matrix remains fully ferritic. The strength level of 780 MPa (yield) or higher can be achieved by low-carbon steel (0.03–0.05%) and alloying a combination of Nb, Ti and Mo. JFEs proprietary grade "nano HiTEN" uses Ti for precipitation hardening and Nb for grain refinement [23]. Other approaches work with high Nb addition and Mo [24]. In either case Mo alloying controls the size and thus the strengthening effect of the precipitates. Furthermore, it avoids the formation of pearlite which would be particularly detrimental to the hole expansion performance.

7 Conclusions

Molybdenum as an alloying element in low-carbon steels has powerful effects with regard to the recrystallization behavior and hardenability. Its full potential, however, is developed when molybdenum interacts with other alloying elements leading to synergetic effects. In combination with Cr, Ni or B the hardenability is much increased. Interaction with the microalloying elements results in clearly improved precipitation strengthening and increased thermal stability of precipitates. Particularly in combination with Nb, molybdenum shows a very interesting potential in producing some of the highest strength steels for structural, pipe and also automotive applications. A prerequisite of utilizing these advantages is the design of suitable processing route, which necessitate a thorough understanding of molybdenum's metallurgical effects.

More recently, a trend in China's steel industry tends to reduce the addition of expensive alloying elements by means of low temperature rolling and super-rapid cooling using modern and powerful mill equipment. However, regardless of its alloying cost Mo still plays an irreplaceable and unique role in the production of thick plate. Mo can overcome the inhomogeneity of microstructure and properties resulted from the inhomogeneous deformation and cooling to a great extent in thick plate and thus improve the strength and toughness of steel. This is the reason why Mo is nevertheless widely applied in the production of high-grade heavy gage pipeline steel. In this respect it is necessary to enhance the research works on physical metallurgy of Mo with emphasis on thick plate steel. Nb improves the steel's properties when added in comparatively small amounts and is thus a quite cost attractive alloying element. Cross effects between Mo and Nb lead to unique properties that cannot be achieved by either element alone. Therefore, it is of particular interest to profoundly study such cross effects of Mo with microalloying elements as their enhanced application can result in better as well as more cost efficient solutions.

References

1. K.J. Irvine, F.B. Pickering, J Iron Steel Inst. **187**, 292 (1957)
2. X.L. He, C.J. Shang (eds.), *High Performance Low Carbon Bainite Steel—Chemistry, Processing, Microstructure, Property and Application* (Metallurgical Industry Press, Beijing, 2008)
3. W.J. Liu, M.G. Akben, Can. Metall. Q. **26**, 145 (1987)
4. L.J. Cuddy, Thermomechanical Processing of Microalloyed Austenite TMS-AIME, Warrendale, PA,(1984), p.129
5. M.G. Akben, B. Bacroix, J. Jonas, Acta Met. **29**, 111 (1981)
6. Q.L. Yong, *Secondary Phases in Structural Steels* (Metallurgical Industry Press, Beijing, 2006)
7. J.C. Cao, Dissertation, Kunming University of Science and Technology, 2006
8. M. Masimov, N. Kwiaton, Euromat 2009, Glasgow
9. J.H. Yang, Dissertation, University of Science and Technology Beijing, 2009
10. D.Q. Bai, S. Yue, T.M. Maccagno, J.J. Jonas, ISIJ Int. **38**, 371 (1998)
11. H. Asahi, ISIJ Int. **10**, 1150 (2002)
12. T. Hara, H. Asahi, R. Uemori, H. Tamehiro, ISIJ Int. **8**, 1431 (2004)
13. K. Miyata, T. Omura, T. Kushida et al., Metall. Mater. Trans. A. **34A**, 1565 (2003)
14. H. Luo, L.P. Karjalainen, D.A. Porter et al., ISIJ Int. **3**, 273 (2002)
15. C.Y. Chen, Dissertation, National Taiwan University, 2007
16. R.W. Honeycombe, Trans. AIME. **7A**, 915 (1976)
17. Y. Funakawa, K. Seto, Tetsu-to-Haganes **93**, 49 (2007)
18. A.P. Coldren, R.L. Cryderman, N.L. Semchysheh, *Strength and Impact Properties of Low-Carbon Structural Steel Containing Molybdenum*. Climax Molybdenum Company
19. Z. Tian, Q. Liu, J. Zeng, *Proceedings of the International Steel Technologies Symposium*, Kaohsiung, Taiwan 2008
20. H. Asahi et al., *Proceedings 13th International Offshore & Polar Engineering Conference*, Honolulu, 19, 2003
21. M. Piette, E. Dubrulle-Prat, Ch. Perdrix, V. Schwinn, A. Streisselberger, K. Hulka, *Proceeding of the 4th International Conference on HSLA Steels*, 2000
22. Y. Mizutani et al., Nippon Steel Technical Report 2004, **90**, 45 (2004)
23. Y. Funakawa, T. Shiozaki, K. Tomita et al., ISIJ Int. **44**, 1945 (2004)
24. W. Haensch, C. Klinkenberg, *Proceedings 2nd International Conference. On Thermomechanical Rolling*, Liège, 155 (2004)

Vanadium in Bainitic Steels: A Review of Recent Developments

Yu Li and David Milbourn

Abstract

Recent works at KIMAB and Arcelor Research indicate that the mechanical properties of bainitic steels can be improved with V-microalloying. This is potentially important since there is expected to be a shift from existing precipitation hardened HSLA steels towards bainitic types as demand arises for higher strength. Although the bainitic transformation temperature is too low to promote precipitation strengthening, it seems that fine V(C, N) particles can effectively prevent the recovery of the dislocation structure of the bainite during coiling of hot-rolled strip product, giving an increase in strength. There are also indications that vanadium combined with nitrogen in bainitic steel promotes the formation of lower-bainite, referring to a bainitic microstructure with smaller crystallographic units delineated by high-angled boundaries, with associated increase in strength and good toughness. Recent investigations on bainitic rail steels in Japan and Europe demonstrate that vanadium is also useful to give fine laths microstructures and improve wear resistance.

Keywords

Vanadium microalloying • Bainitic steel • Bainitic microstructure • Mechanical properties

1 Introduction

Bainitic steels are gaining in industrial importance as more arduous service conditions drive up steel property requirements, and the avoidance of heat treatment is essential for economic competitiveness. Examples of medium carbon bainitic steels have been seen in forging and rail steels, and developments in strip and plate steels are moving rapidly towards increasingly high strength levels with low carbon contents. Up to yield stress levels of about 600 MPa, strengthening can be realised by ferrite grain refinement and microalloy precipitation as in traditional HSLA steels. Higher strength steel necessitates substructure strengthening as in bainite or tempered martensite. This development poses a challenge to vanadium (as well as to niobium and titanium) since the transformation temperatures are too low to form the particles that can provide direct strengthening. There is, however, some evidence that vanadium alloying allows more of the structural strengthening to be retained in bainitic strip steels. Vanadium is well known to contribute both hardness and structure stability in quenched and tempered engineering steels. Lessons may be learned from these technologies that can be applied to new grades of structural steels.

2 Use of Vanadium in Bainitic Steels

2.1 Vanadium in Bainitic Automotive Sheet

In order to develop vanadium microalloyed bainitic steel combining high strength and good resistance to damage for the automotive products, two series of vanadium microalloyed steels have been investigated by ArcelorMittal (France) [1] to study the effect of vanadium and nitrogen on bainitic

Y. Li (✉) · D. Milbourn
Vanitec Limited, Westerham, Kent, UK
e-mail: yu.li@vanitec.org

Y. Weng et al. (eds.), *Advanced Steels*, DOI: 10.1007/978-3-642-17665-4_31,

Table 1 Chemical composition of the steels investigated in ArcelorMittal France (wt.%)

Steel	C	Si	Mn	P	Ti	V	N
0V	0.087	0.3	1.89	0.002	/	0.002	/
3V	0.087	0.3	1.89	0.002	/	0.032	/
9V	0.09	0.305	1.91	0.002	/	0.09	/
18V	0.086	0.3	1.89	0.002	/	0.18	/
0VN	0.100	0.315	1.92	0.002	0.001	0.002	0.0110
3VN	0.100	0.315	1.92	0.002	0.001	0.029	0.0115
8VN	0.100	0.315	1.92	0.002	0.001	0.075	0.0128
15VN	0.098	0.310	1.92	0.002	0.001	0.153	0.0153

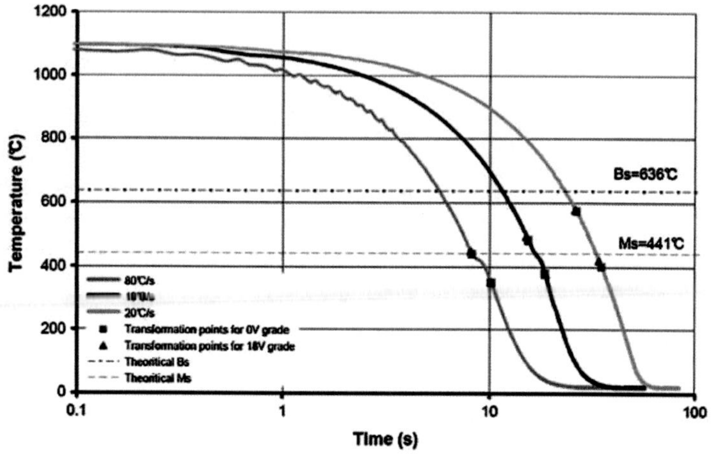

Fig. 1 CCT diagram for Steels 0V and 18V

microstructure obtained after isothermal treatments. The first one (0–18V) contained no nitrogen and the second one (0–15VN) had around 100 ppm nitrogen. The chemical composition is given in Table 1.

In order to define the complete isothermal treatments to obtain ideal bainitic microstructures, CCT data have been generated for the steels. For producing CCT curves, a series of dilatometric experiments have been performed in a deformation dilatometer where the samples ($\Phi = 6$ mm, $L = 10$ mm) first were homogenised for 5 min at 1,100°C and cooled at different cooling rates (20, 40 and 80°C/s). The results for Steels 0V and 18V are presented in Fig. 1. The CCT experiments showed that there was no significant effect of vanadium on the starting and ending bainitic transformation temperature. The three cooling rates resulted in different transformation products. At 80°C/s, the transformation occurred in martensite domain, at 40°C/s, the transformation took place in both the bainitic and martensitic domains, and at 20°C/s, almost all the transformation occurred in the bainitic region.

The isothermal treatments were carried out to produce specimens with fully bainitic microstructures for subsequent detailed metallographic characterisation and the determination of mechanical properties. The samples for isothermal treatments were heated at 1,100°C for 5 min to ensure that all vanadium was in solid solution before transformation. The samples were then cooled to 500 or 550°C at a cooling rate of 60°C/s and held for 1 h before being quenched with helium.

Hardness measurements showed a clear strengthening effect of vanadium (Fig. 2). The increase in hardness was up to 35HV in the steels without nitrogen and 40HV in the steels with nitrogen added.

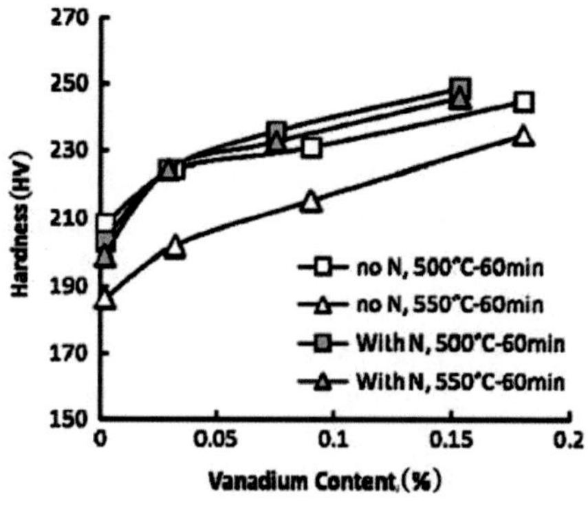

Fig. 2 Effect of vanadium and nitrogen on the hardness of steels after isothermal treatments at 500 and 550°C

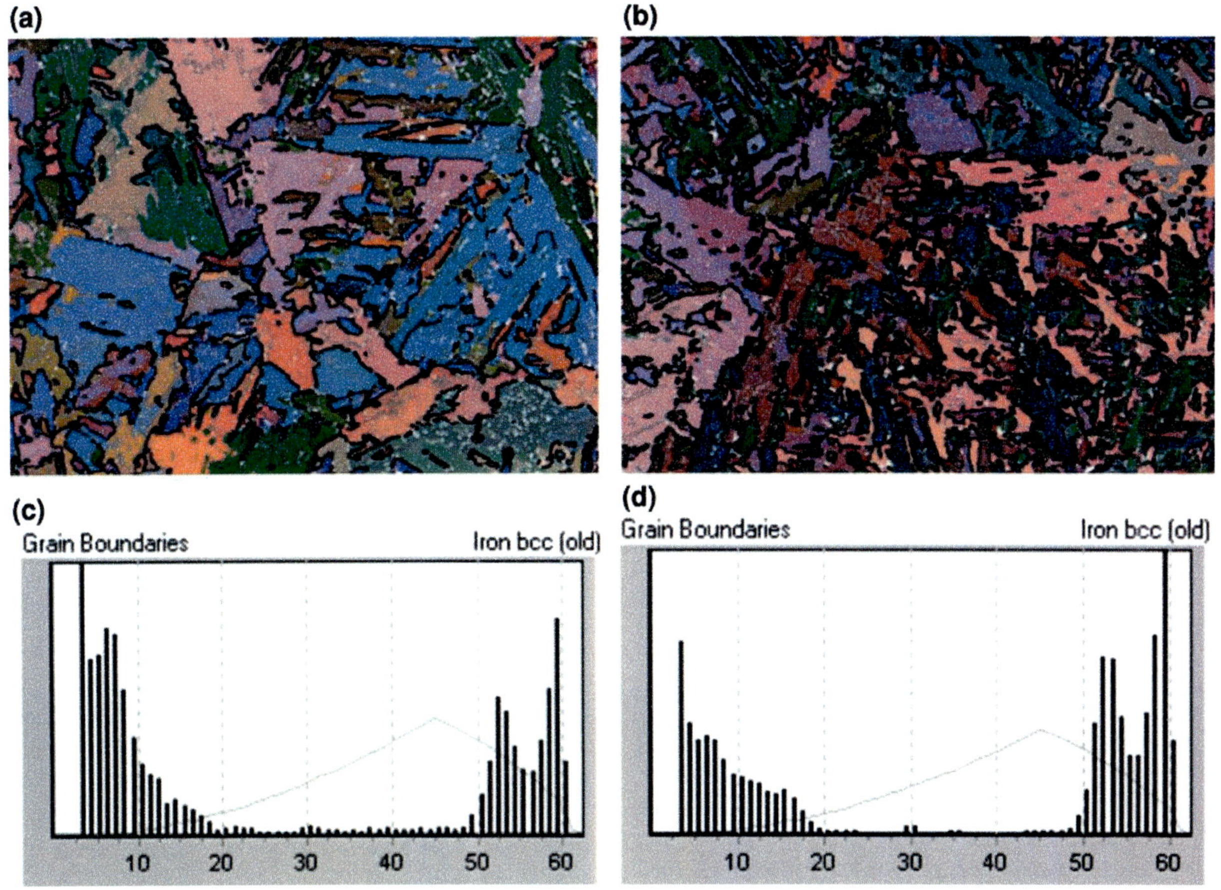

Fig. 3 EBSD maps of **a** Steel 0V and **b** Steel 15V after isothermally transformed at 500°C. Colours represent crystallographic orientation. Misorientation distributions of **c** Steel 0V and **d** Steel 15VN

Unlike the nitrogen-free steels, the hardness of the steels containing nitrogen did not appear to depend on the holding temperature. Within the group of steels with nitrogen added, similar hardness levels were obtained after isothermally transforming at 500 and 550°C, except in the case of the steel with no vanadium for which a slight softening was observed.

Scanning electron microscopy (SEM) and Electron back scattering diffraction (EBSD) analysis showed that vanadium alone did not have a significant effect on the refinement of the bainitic microstructure. However, combined with nitrogen, it was clearly shown that increasing the vanadium content resulted in a significant refinement of the microstructure (Fig. 3a, b). Comparison of the misorientation distributions of Steels 0VN and 15VN showed an increase in high-angled and decrease in low-angled misorientations in the higher vanadium steel. This change could reflect a shift from upper-type bainite to lower-type bainite in the vanadium and nitrogen containing steel according to the classification proposed by Zajac et al. [2], which is favourable for good toughness. It was suggested that increasing the vanadium content, combined with enhanced nitrogen, activates the nucleation of the smallest crystallographic entitles in the austenitic grains, leading to microstructural refinement.

2.2 Vanadium in Hot Rolled Strip

A recent study in KIMAB on a series low carbon (~0.04%), vanadium (~0.08%) microalloyed high strength (YS ≥ 600 MPa) bainitic hot strip-simulated steels (Table 2) has also demonstrated that vanadium microalloying effectively prevents the recovery of the bainitic ferrite and leads to retention and even to enhancement of the strength of the bainitie during coiling [3]. The steels with 1.4%Mn, 1.0%Cr and 0.25%Mo were designed to provide adequate base harden ability to produce a fully bainitic microstructure in 8 mm hot rolled strip steel cooled at a rate of 30°C/s after finish rolling and subsequently coiled at 400°C. Some variations were also made in order to find out what margins exist before ferrite is formed and to

Table 2 Chemical composition of the steels investigated in KIMAB (wt.%)

Steel	C	Si	Mn	Cr	Mo	V	Al	N	Bs (°C) [4]
1.0Cr0.25Mo (Ref.)	0.043	0.12	1.40	1.03	0.25	0.001	0.038	0.010	600
1.0Cr0.25MoV	0.037	0.11	1.44	0.96	0.26	0.081	0.03	0.012	602
1.0Cr0.25MoVN (low-Al)	0.045	0.11	1.38	1.03	0.26	0.078	0.003	0.025	600
1.4Cr0.12MoVN (low-Al)	0.045	0.10	1.35	1.38	0.12	0.08	0.003	0.021	590
1.4Cr0.12MoVN	0.043	0.10	1.50	1.36	0.12	0.078	0.023	0.024	579
2.0Cr0.12MoV0.038N	0.041	0.11	1.31	2.00	0.12	0.073	0.033	0.038	551
2.0Cr0.12MoVN	0.048	0.10	1.37	2.00	0.12	0.079	0.026	0.020	544

what extend the expensive molybdenum can be replaced by less expensive chromium. Nitrogen (0.010–0.038%) is added to promote fine V(C, N) precipitation.

The steels were melted and cast as 40 mm × 40 mm × 200 mm ingots and after reheating at 1,150°C for 1 h, were hot rolled to 20 mm plate. Specimens (Φ5.0 mm × 10 mm) for the dilatometric experiments were prepared from the plates to determine the CCT curves and to simulate the last stage of hot strip rolling and coiling. All dilatometer samples were first reheated at 1,150°C for 30 min and quenched to room temperature.

For CCT determination they were then reheated at 1,150°C for 3 min, cooled to 1,000°C and deformed 25% to refine the austenite grain structure, followed by cooling at 2°C/s to 900°C and thereafter cooled at 5, 10, 30, 60, and 100°C/s to room temperature. The CCT curves are displayed in Fig. 4. By comparing the transformation curves of the reference steel 1.0Cr0.25Mo (C2), with the three CrMoV steels, 1.0Cr0.25MoV (A5), 1.0Cr0.25MoVN (low-Al) (E1), 1.4Cr0.12MoVN (low-Al) (F1), it can be seen that they are all similar and that the observed variations fall within the experimental scatter.

The thermomechanical controlled processing of 8 mm hot strips was simulated by hot deformation dilatometry by first reheating the speciments in the dilatometer in vacuum at a rate 5°C/s to 1,150°C and holding for 3 min and subjecting to the following hot deformation schedules: cooling 2°C/s → 1,000°C/def.25%—cooling 2°C/s → 930°C/def.25%—cooling 2°C/s → 900°C—30°C/s →coiling at 400–600°C—10°C/h to RT.

The effects of composition and coiling temperature on the yield strength of the hot strip steels are displayed in Fig. 5. In addition, the yield strength of Steels 1.0Cr0.25Mo, 1.0Cr1.25MoV and 1.0Cr0.25MoVN (low-Al) quenched from 900°C at a rate of 30°C/s to room temperature are also given in this figure. The results in the figure show that vanadium microalloying produced higher yield strength than the reference steel for all the coiling temperatures in the range of 400–550°C. This effect is more significant at 400°C and it diminishes with increasing coiling temperature, but for Steel 1.0Cr0.25MoV the effect is reduced only marginally. At 400°C coiling temperature, the strength is ~680 MPa for the reference steel 1.0Cr0.25Mo, and it is increased to ~750 MPa for Steels 1.0Cr0.25MoV and 1.4Cr0.12MoVN, and to ~780 MPa for Steels 1.0Cr0.25MoVN (low-Al) and 1.4Cr0.12MoVN (low-Al). The Al content of the steels could affect of the amount of free nitrogen, which is available for vanadium, since more or less of the nitrogen is tied up as AlN at high temperatures in austenite, in turn, this would affect the

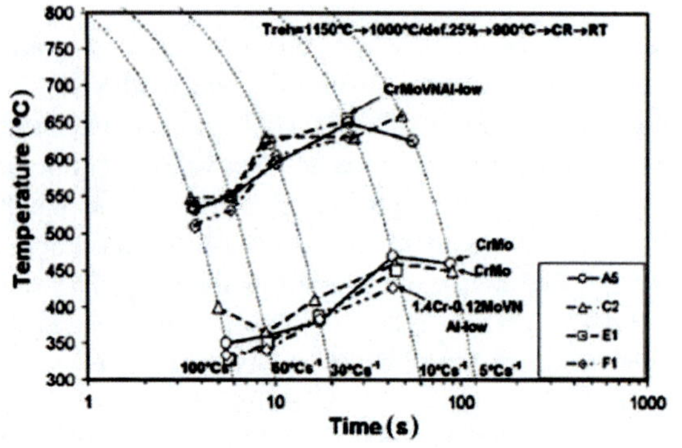

Fig. 4 Effect of V, Mo, Cr and N in the bainitic Cr-Mo-V steels on the transformation characteristics

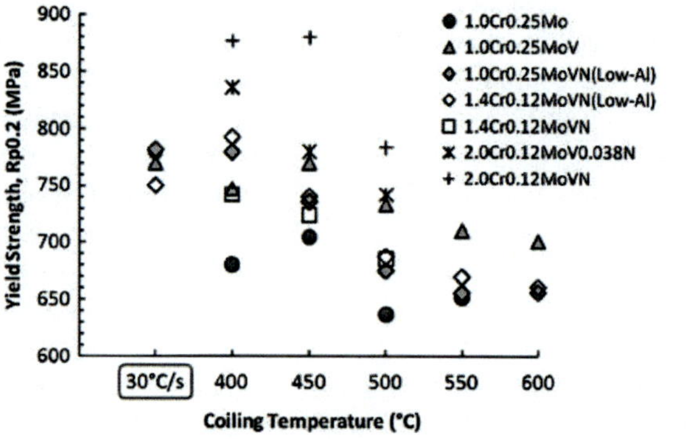

Fig. 5 Effect of composition and coiling temperature on the yield strength

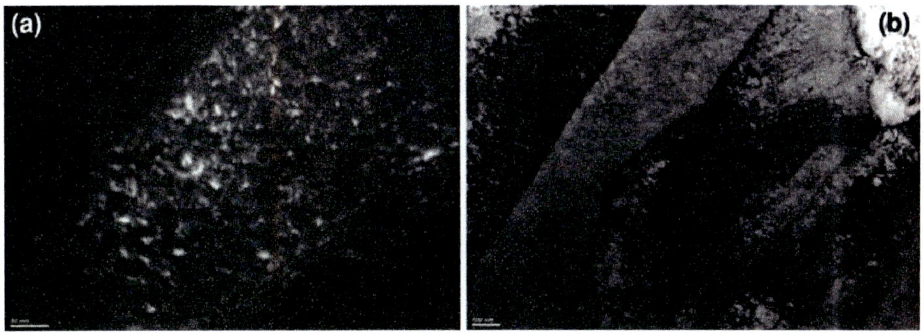

Fig. 6 TEM micrographs showing **a** high density of fine V (C, N) particles on the dislocations and **b** high dense dislocations of the bainitic microstructure after coiling at 450°C for steel CrMoV

Table 3 Composition of the vanadium containing bainitic rail steels

	C	Mn	Si	Cr	Mo	V	B	Reference
Nippon Steel	0.28	1.21	0.30	1.65		0.08		[5]
Thyssen Stahl AG	0.40	0.7	1.5	1.1	0.8	0.1		[7]
Corus (B320)	0.15–0.25	1.40–1.70	1.00–1.50	0.30–0.70	0.10–0.20	0.10–0.20	0.002–0.004	[6]

ability of V–N to influence the recovery of bainitic structure during coiling. For the 2%Cr steels, which had enhanced bainitic hardenability, the strength values were much higher, especially for the high nitrogen containing steel (2.0Cr0.12MoV0.38N). The extra strength in the 2%Cr steels was considered to be attributed to their lower bainitic transformation temperatures.

When bainite is formed during cooling to the coiling temperature or in the coil, it would normally be expected to recover and soften significantly during the coiling and prolonged holding at elevated temperatures after coiling. Figure 5 confirms that the reference steel CrMo, without vanadium, underwent a substantial softening during coiling, leading to a reduction of strength from 780 to 680 MPa. However, for the vanadium containing steels, the softening was retarded. For example, after coiling at 400°C, the strength was retained in Steel 1.0Cr0.25MoVN (low-Al) and even hardening (750–790 MPa) was obtained in Steel 1.4Cr1.2MoVN (low-Al). This indicates that vanadium, especially in combination with nitrogen is capable of eliminating the softening during coiling.

Transmission electron microcopy (TEM) showed that a fine, dense V(C, N) precipitation on dislocations and a preserved high density dislocation bainitic structure in the vanadium containing steels. An example is given in Fig. 6. It was concluded that the vanadium microalloyed bainitic steels retained their strength primarily by an effective retardation of recovery of dislocation structure by fine V(C, N) precipitates. Precipitation strengthening contributed only to a lesser degree.

2.3 Vanadium in Bainitic Rail Steel

Speeds and axle loads of trains have increased on railways in recent years, resulting in greater wear and rolling contact fatigue, and bainitc steels are emerging as a potential new material to replace conventional pearlitic steels for rails used in these more severe service conditions. Recent investigations in Japan (NKK) [5] and Europe (Corus, Thyssen Stahl) [6, 7] have confirmed that rail tracks made of bainitic steels, containing vanadium, have very high strength, good wear resistance and excellent rolling contact fatigue resistance, and are easy to weld. Some of the commonly used vanadium microalloyed bainitic rail steels are given in Table 3.

Bainitic rail steels with carbon content of 0.15–0.45% and different alloying elements contents are described in the literature. The alloying elements molybdenum and boron delay the formation of ferrite and pearlite and accordingly promote the formation of bainite. Furthermore, bainite transformation can be controlled by nickel, chromium and manganese, which decrease the bainite start-temperature and thus improve the strength and to some extent the toughness as well [7, 8]. Vanadium is commonly used in bainitic rail steel to increase strength and wear resistance, and improve fracture toughness [9]. Bainitic rails produced from the Corus developed grade, B320, were installed in the Euro-tunnel in January and March 2006, and have shown no evidence of cracks at all [10] after 3 years of service and 286 million tonnes of traffic, where standard rails have already been ground due to fatigue crack.

3 Conclusions

(1) Vanadium has no significant effect upon bainitic transformation temperature.
(2) Vanadium, especially with nitrogen addition in the bainitic steels results in higher strength.
(3) V–N microalloying is capable of preventing softening during coiling that occurs in steels without V–N addition. This eliminated softening is mainly resulted from an effective prevention of the recovery of dislocations of the bainitic structure and to a less extent due to V(C, N) precipitation strengthening.
(4) Vanadium combined with nitrogen in bainitic steel promotes the formation of lower-bainite.
(5) Vanadium is useful to gain fine laths microstructure in bainitic rail steels and improves wear resistance.

References

1. D. Barbier, L. Chapuis, T. Iung, *Project Report 7 to Vanitec* (ArcelorMittal Maizières Research SA, France, 2009)
2. S. Zajac, J. Komenda, P. Morris, P. Diericksz, S. Materia, F. Penalba Diaz, Final Report for the ECSC, no. 4267 (2003)
3. T. Siwecki, J. Eliasson, R. Lagneborg, B. Hutchinson, ISIJ Int. **50**, 760 (2010)
4. W. Steven, A.G. Haynes, J. Iron Steel Inst. **183**, 349 (1956)
5. H. Kageyama, M. Ueda, K. Sugino, U.S. Patent 5,382,307, 1995
6. F. Monfort, Lincolnshire Iron and Steel Institute, Young Members paper (2007)
7. A. Kern, H. Schmedders, A. Zimmermann, 39th MWSP Conf. Proc., ISS, XXXV, 1998
8. J. Pacyna, J. Achiev. mater.manuf. Eng. **1**, 19 (2008)
9. N. Tsunekage, K. Kobayashi, H. Tsubakino, Mater. Sci. Tech. **7**, 847 (2001)
10. Fatigue resistance extends life expectancy, Rail-News.com, 18 Aug 2010

Nanostructural Engineering of TMCP Steels

Peter D. Hodgson, Ilana B. Timokhina, Hossein Beladi, and Subrata Mukherjee

Abstract

This paper reviews current research that aims to understand and manipulate the properties of advanced steels through control of the nanostructure, with an emphasis on thermomechanically produced steels. While the concepts of such strengthening mechanisms have been understood and used for many years, it is now possible through advanced characterisation methods to gain detailed insights. Similarly, our knowledge of phase transformations supported by improved modelling also allows us to design microstructures that have length scales in the range of 10–100 s of nanometres. Examples are provided that relate to the application of this concept in Advanced High Strength Steels of precipitation hardening, bake hardening, the development of ultrafine ferrite microstructures, nanoscale dispersed multiphase steels and nanoscale bainite.

Keywords

Steel • Microstructural engineering • Thermo mechanical processing • Bainite • Atom probe tomography • Precipitation

1 Introduction

Steel is still the structural material of choice in many industries. This is largely due to its ready availability, relatively low cost, low cost volatility and the unique property packages that can be obtained through changes in composition and processing route.

From an energy perspective there are major benefits if the properties can be obtained directly after hot rolling without any further heat treatment. This is the basis for thermomechanical controlled processing (TMCP) where the aim has been to control the evolution and state of the austenite grain structure and to then control the subsequent phase transformation during hot rolling. Steel has a major advantage over most competitive metals as it has a very high Hall–Petch slope. Hence any reduction in the ferrite grain size leads to significant enhancements in strength and, as it turns out, toughness. Therefore, thermomechanical processing has been used in many different types of rolling mills to produce ferrite grain sizes down to a few microns, leading to major increases in product performance.

More recently in advanced high strength steels (AHSS) for the automotive industry the objective is to obtain steels with a fine ferrite microstructure but containing other micron scale particles of secondary phases such as martensite, retained austenite and bainite. Increasingly, these microstructures are becoming more complex and their evolution is determined by a relatively complex series of phase transformations. It has also been shown [1] that some of the phases will have nanoscale dimensions and their composition is also determined by the sequence of phase transformations.

There have always been nanoscale structures that have impacted on the product performance of steel. Pearlite, precipitation hardening and bainite are examples of this. What has changed is our ability to be able to quantify

P. D. Hodgson (✉) · I. B. Timokhina · H. Beladi · S. Mukherjee
Centre for Material and Fibre Innovation, Deakin University, Waurn Ponds 3217, Australia
e-mail: peter.hodgson@deakin.edu.au

Y. Weng et al. (eds.), *Advanced Steels*, DOI: 10.1007/978-3-642-17665-4_32,

them in a reasonable amount of time and with reasonable statistics through techniques such as FEGSEM-EBSD and Atom Probe Tomography (APT).

The objective of this paper is not to ignore that steel already relies on nanoscale features but to focus on recent research more deliberately aimed at engineering the micro-structure at this scale length. This is, in many ways, still a work in progress as much of the work is still observing what is happening as the design and control is yet to be developed. Figure 1 shows how there are various fundamental factors that can be controlled to influence the final microstructure. In many cases we are really talking about atomistic effects rather than nanoscale effects, but the same tools (e.g. APT) are still used. Grain boundary engineering can take two different forms. One is to change the surface energy which can suppress phase transformations (e.g. the role of boron and Nb), while the second is to change the Hall–Petch slope (e.g. carbon). The arrangement of primary phases mostly refers to the austenite to ferrite transformation. As this transformation progresses, carbon is rejected into the remaining austenite and there are major differences between transformation from an equiaxed austenite and a pancaked austenite in terms of the arrangement of the ferrite and the size and nature of the austenite. After the formation of ferrite it is possible to either form pearlite, which has a controlled nanostructure, or bainite which is of increasing interest these days. As will be shown in the following sections, these phases can also have a wide range of carbon contents and then also clusters and/or precipitates.

2 Ultra Grain Refinement of Ferrite

The past decade has seen considerable interest in the development of processes to produce sub-micron grain sizes. These range from very simple thermomechanical processes [2], through to complex multistage processes [3]. The activation of dynamic strain induced transformation (DSIT) has been widely studied in Japan, Korea, China and Australia [2].

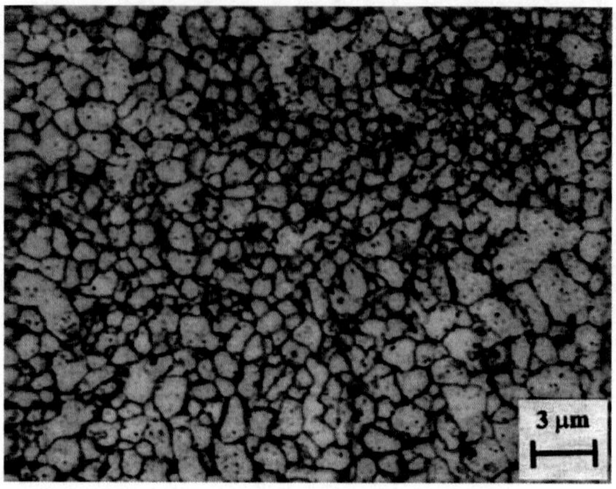

Fig. 2 Grain structure of DSIT 0.06 C steel

Here the aim is to deform the steel below the equilibrium transformation temperature but above the Ar3, such that deformation is taking place in undercooled austenite. The resulting structures can be very fine—typically around 1 μm (Fig. 2). This can easily lead to a doubling of yield strength and has broken the typical minimum grain size level seen in most steels of around 5 μm with a conventional ferrite microstructure [4]. However, as is now well documented these simple microstructures have very limited work hardening capacity and hence limited ductility.

They also require very large reductions and hence, high rolling loads, which has made it difficult to implement in practice. However, in a systematic study of the effect of holding time between deformations it was found (Fig. 3) that as long as the grain size after the first strain induced transformation grew by only a factor of two or less then in the second pass increasing strain will introduce more strain induced ferrite and the original ferrite, remained fine [5].

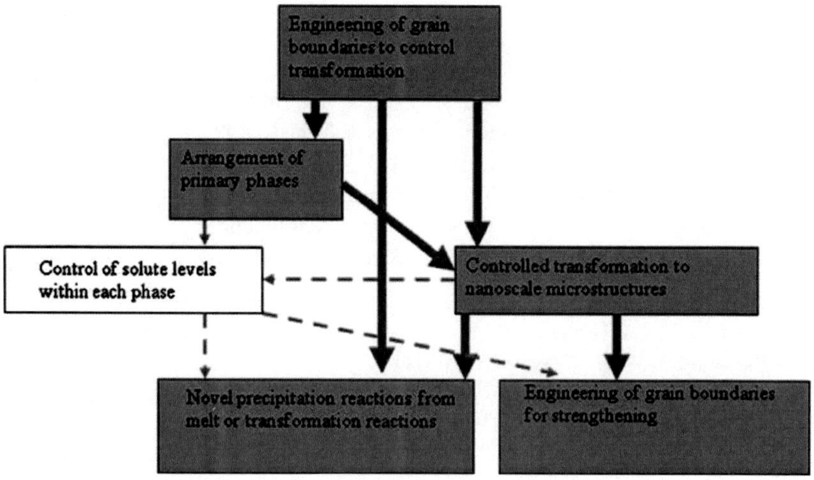

Fig. 1 Integrated approach to the engineering of the steel microstructure and nanostructure

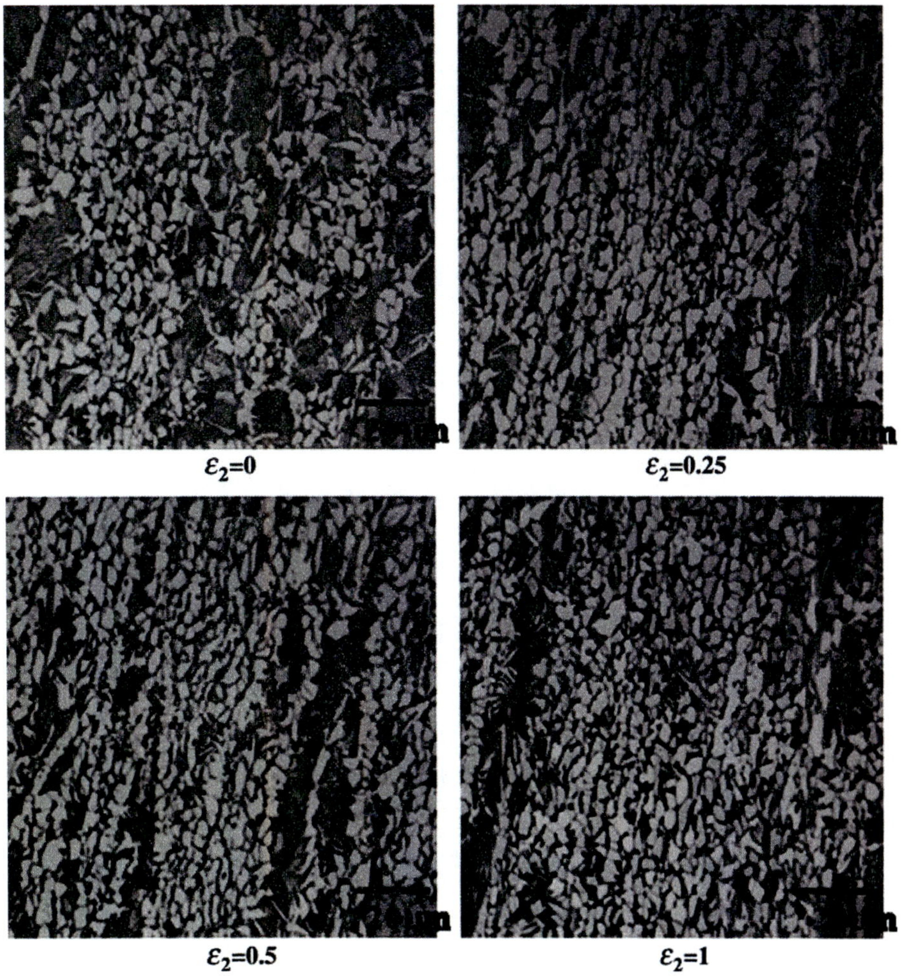

Fig. 3 The effect of the second pass deformation on microstructural evolution of the steel deformed at a strain of 1 at 760°C held for 5 s

However, if it grew more (Fig. 4) then the ferrite needed to fragment and large strains were required to re-establish the ultrafine structure. These results suggest that by controlling the interpass time and/or cooling between deformations to minimise ferrite growth it is possible to build-up an ultrafine ferrite structure over a number of deformations. This is shown in Fig. 5 for two cases of four strains of 0.25 and 0.5.

Others have come to similar conclusions and now there have been recent attempts to develop the DSIT for commercial applications, although this has required a major redesign of the rolling process to minimise the time between deformations and to have almost instantaneous cooling after the final stand [6].

3 Transformation to Nonferritic Microstructures

The above followed the traditional concepts of increasing strength by refining the ferrite grain size and in reality the microstructure scale is not really at the nano level except for the more extreme cases. However, it is possible to control the transformation to other microstructures. This has been shown most elegantly in the recent papers by Bhadeshia and co-workers [7] where a thermodynamic approach was taken to design a composition where high temperature transformations could be avoided but where it was then possible to form bainite without martensite. This led to a nano bainitic composite microstructure, where the layers of bainite were separated by layers of retained austenite. The steel produced an amazing combination of strength (e.g 2 GPa) and ductility.

Our group has also been exploring this alloy and other similar concepts. Some examples will be described here based on three approaches. The first repeated the original work of Bhadeshia with a high C content alloy (Steel 1, Table 1). In the second approach a leaner composition (Steel 2) was used and the first stage of the TMCP schedule involved partial transformation prior to holding at the bainite transformation temperature. For the examples shown here, both of these approaches used a bainite transformation temperature of 350°C; the transformation time was reduced

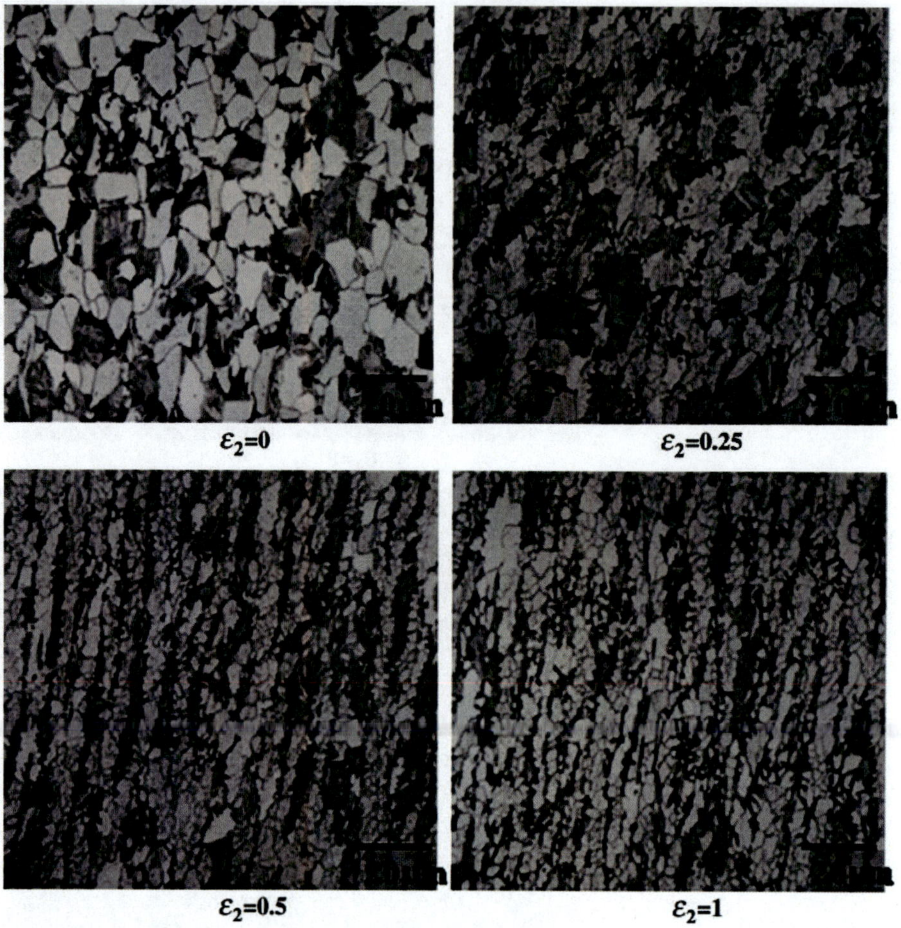

Fig. 4 The effect of the second pass deformation on microstructural evolution of the steel deformed at a strain of 1 at 760°C held for 40 s

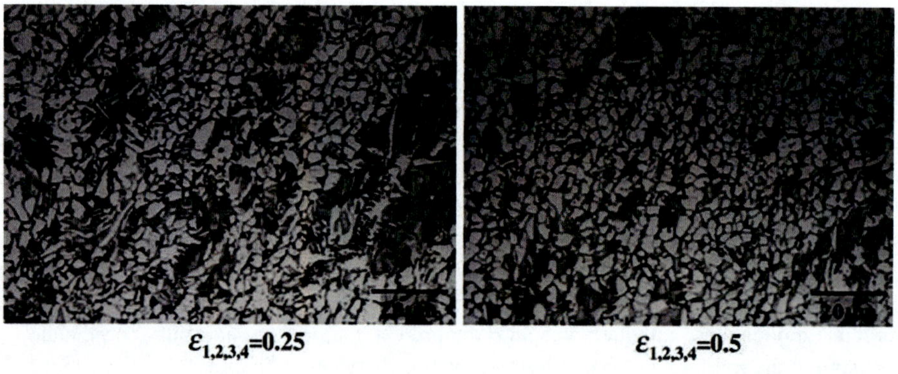

Fig. 5 The effect of pass strain on the grain refinement through successive deformations

from 1 day to 2 h in moving from Steel 1–2. The third approach (Steel 3) was based on a simple TMCP approach utilising controlled rolling with partial transformation to ferrite and then a coiling simulation at 470°C.

The tensile properties show that each approach leads to similar initial strength levels but that the ductility is markedly changed (Fig. 6) with the best ductility arising from the leanest steel with the highest bainite transformation temperature.

TEM of the bainite phase for each approach (Fig. 7) shows that for the 350°C isothermally transformed steel there is a relatively aligned structure of alternating bainitic ferrite and retained austenite with bainite lath widths from 200 to 400 nm. The retained austenite (RA) was significantly coarser in Steel 2; it is important to remember that RA is the residual from the transformation, and hence this implies a lower fraction of bainitic ferrite. Also this steel had $\sim$20% ferrite from the high temperature transformation.

Table 1 Composition of nanobainite steels

		C	Si	Mn	Al	Co	Cr	Mo	Ni	Cu	Nb
Steel 1	wt%	0.8	1.51	1.98	1.06	1.6	0.98	0.24	0.1	0.1	–
	at%	3.5	2.9	1.9	2.08	1.4	1.0	0.13	0.09	0.08	–
Steel 2	wt%	0.3	1.96	2	–	–	0.1	0.31	0.01	0.01	–
	at%	1.2	3.8	1.9	–	–	0.1	0.17	0.01	0.01	–
Steel 3	wt%	0.2	1.2	1.5	0.57	–	–	0.3	–	0.02	0.04
	at%	0.95	2.3	1.5	1.15	–	–	0.165	–	0.02	0.02

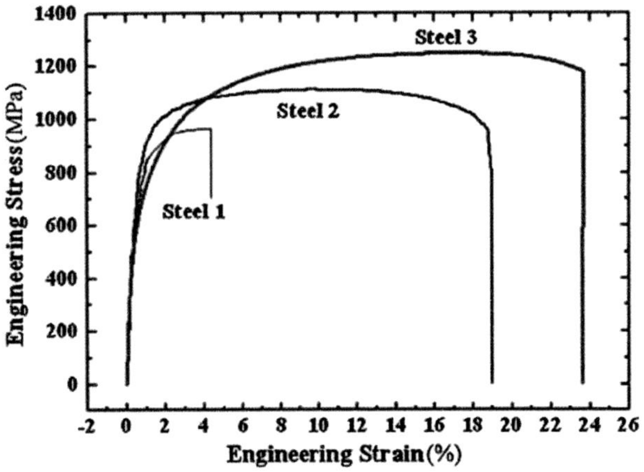

Fig. 6 Mechanical properties of the nano-bainitic steels

The poor ductility of Steel 1 is believed to be due to other undesirable phases that were present in the structure and the austenite was over stabilised and did not transform during deformation. For Steel 2 there was clear evidence of the austenite transforming to martensite during deformation.

Steel 3 had coarser bainite (200–500 nm) and austenite films/islands (200–400 nm) along with ~15% polygonal ferrite. The high strength is somewhat surprising but emphasises the importance of having the correct level of stability of RA so that it transformed gradually, leading to an optimum work hardening compared with the other two steels.

The atom probe maps (Fig. 8) provide a great deal of insight into the levels and distribution of C in the various phases, as well as clear evidence of the formation of various carbide phases. The main point of current interest is to understand the various carbide phases being seen in the bainitic ferrite to gain insights into the transformation mechanisms and understand their effects on the mechanical behaviour.

Throughout these studies the importance of all aspects of steel processing and the transformation sequence have become increasingly clear. The deformation of the austenite will distribute the sites and rate of proeutectoid ferrite. In effect, this then sets up the subsequent islands of austenite and their carbon contents prior to subsequent transformation. Also for the high temperature deformation under conditions where pancaking of the austenite occurs dislocations are introduced into the austenite and will remain in this phase through the subsequent transformation sequences.

It will also potentially introduce fine strain induced precipitates that will remain in the microstructure and while these are unlikely to affect the properties or subsequent

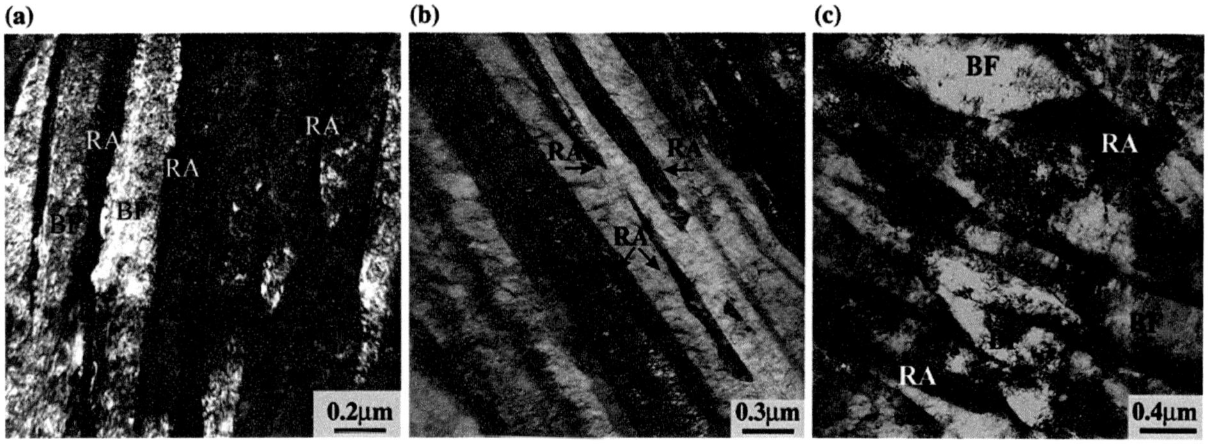

Fig. 7 TEM of microstructure of **a** Steel 1, **b** Steel 2 and **c** Steel 3

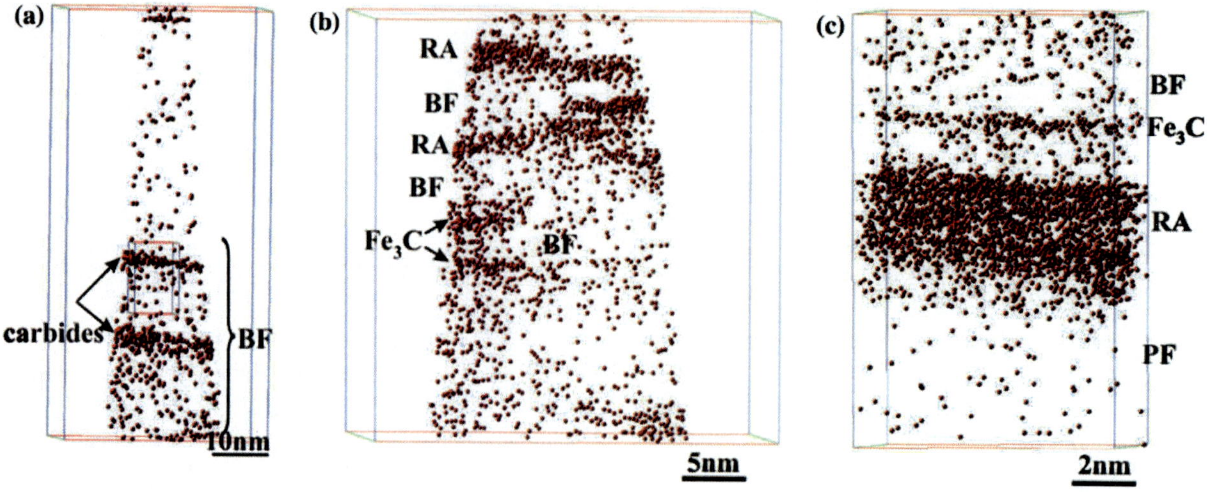

Fig. 8 APT maps of **a** Steel 1, **b** Steel 2 and **c** Steel 3—RA = retained austenite, PF = polygonal ferrite and BF = Bainitic ferrite

transformation events they will be present in further analysis of the microstructure and may confuse the nature and formation of precipitates in the subsequent phases.

The formation of proeutectoid ferrite leads to carbon rejection into the austenite, and hence the amount will depend on the amount of ferrite formed. As shown in the examples above managing this aspect of the transformation can have a very profound effect on the final microstructure and properties. Hence the role of austenite conditioning in TMCP is still very important in these complex microstructure steels.

The above APT maps and other work soon to be published question whether the bainite can really be said to be carbide free. There is some debate about whether the high carbon areas are precipitates or just clusters but most of our work does suggest a wide range of carbides being present in the bainitic ferrite.

For Steel 1 under the current conditions the average number density of C-rich clusters was $\sim 7 \times 10^{11}$ m^{-3}. Most of the fine carbides had a plate-like shape with average thickness of 3.3 ± 0.9 nm. The carbon level in the fine particles varied from ~20–80 at%C, with the balance Fe. Most of these carbides had a composition close to that of low-temperature $Fe_{32}C_4$ carbides and Fe_3C. The bainitic ferrite formed in Steel 1 had a higher carbon content (0.35 ± 0.05 at%) compared to the para-equilibrium level, which suggests diffusionless type of transformation of austenite to bainite. However, this is still an area of active investigation and debate.

For Steels 2 and 3 there was also a high level of C in the bainitic ferrite. While these steels are seen by many as futuristic even the new TMCP TRIP steels have complex nanoscale microstructures (Fig. 9) that can be highly manipulated to achieve a range of properties.

In this case there is a retained austenite island between proeutectoid ferrite grains, which has partially transformed to martensite. In other areas of the microstructure the retained austenite is between bainitic ferrite laths, similar to the microstructures above. Where the retained austenite is located will have an effect on the carbon level, and hence the chemical stability. At the same time the nature of the phase(s) around the retained austenite will affect strain partitioning and the mechanical stability. Therefore, with these complex microstructures it is possible to "engineer" the retained austenite to maximise the TRIP effect so that it occurs progressively over a wide strain range. This will provide the optimum work hardening behaviour and strength.

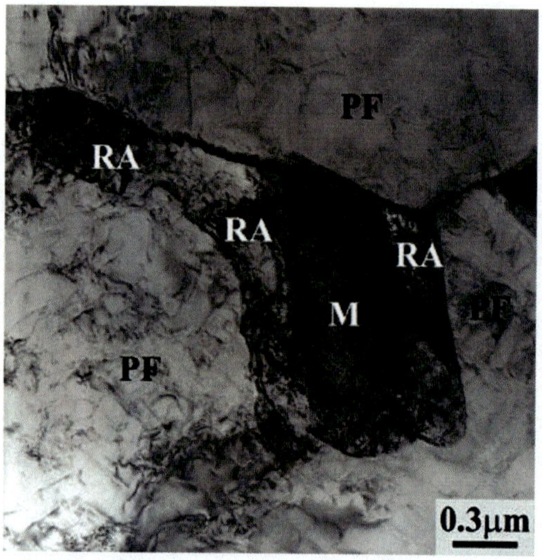

Fig. 9 TEM of a TMCP TRIP steel showing the nanoscale distribution of the various phases

4 Precipitation Hardening

High strength low alloy (HSLA) steels have relied on the formation of nanoscale precipitates since they were first developed. However, recent work in Japan [8] has shown that the levels of precipitation hardening can be radically improved through careful design of the alloy and the production process. From what they have presented, they are able to radically refine the size and spacing of the precipitates.

This suggests also a strong interaction between the transformation and the precipitation. If we consider interphase precipitation as the optimum form for hardening then the rate of movement of the interphase has to be at a speed that is not too slow or the precipitates will be coarse and not too fast or it will not occur and precipitation will occur later in the matrix. Also the “step size” of the moving interface will determine the interparticle spacing. We have also used APT to study precipitation [9] in these steels.

More recently we have been examining the effect of TMCP conditions (i.e. retained strain and cooling rate) and composition on the formation of interphases precipitates. For example, increasing the retained strain affects the rate of transformation and the spacing of the interphase rows (Fig. 10). One of the major limitations with TEM is the magnetic effects of steel which can make it difficult to detect these precipitates, particularly if there is a high background dislocation density. The range of tilts angles that can be used is restricted. In Fig. 10 it is a simple microstructure and while the row spacing is clear the spacing within rows and the detailed shape of the precipitates is difficult to determine.

In Fig. 11 the detailed APT analysis of the condition in Fig. 10b is shown. Here it is possible to obtain much greater insight into the nature of the precipitates in terms of size, spacing, shape and composition.

As APT allows the precipitate/cluster to be viewed much more clearly in 3D (e.g. the box can be rotated in all directions to view the precipitate or cluster) we are now seeing a much more complex range of precipitate morphologies that may appear as simple round objects under TEM. In this work we believe it is also possible to distinguish pre-precipitation clusters that have been shown in non-ferrous alloys [10] to have the potential for significant strengthening.

5 Solute Effects

Our recent work on AHSS has focussed on the arrangement and nature of phases and their detailed composition using APT and other techniques. At the same time we have been interested in how TMCP and the transformation sequence affect not only the elemental distribution but also the bake hardening behaviour. This is due to the fact that the distribution of C and the internal dislocation structures vary and we, like others, have observed very different levels of bake hardening strength increase (Fig. 12a).

With APT we have been able to examine the local carbon levels in the ferrite and then also the clustering along linear defects that we assume to be dislocations (Fig. 12b). With this it is clear that the carbon supersaturation level within the ferrite can be altered to maximise ageing. It is also possible to follow the formation of carbides during ageing.

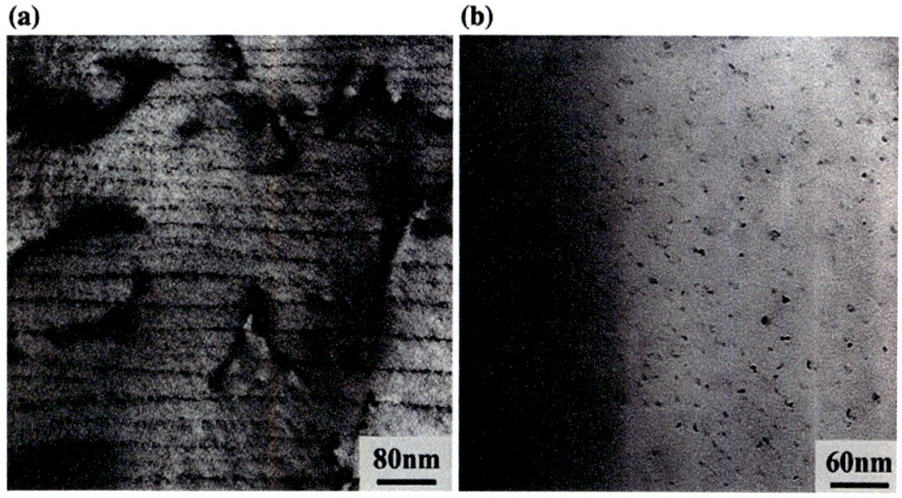

Fig. 10 TEM images of interphase precipitation in a Ti-Mo microalloyed TMCP steel after holding at 650°C for 1 h with **a** 0 and **b** 1.0 retained strain prior to transformation

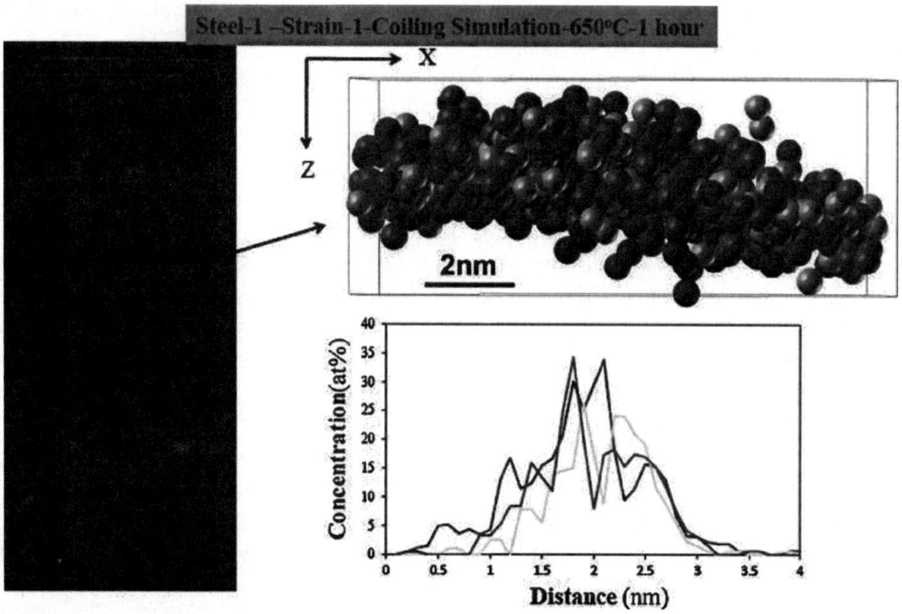

Fig. 11 APT map showing an individual precipitate and its composition and arrangements of precipitates

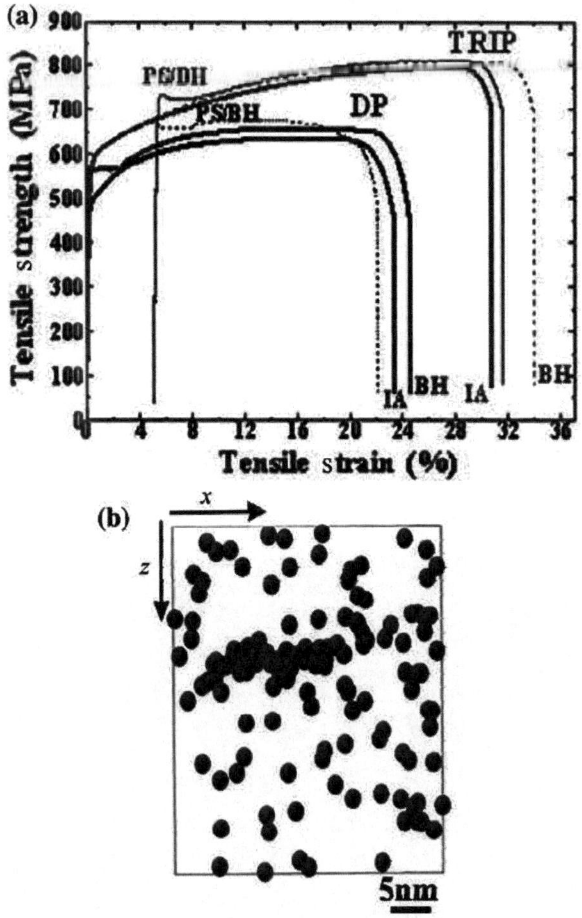

Fig. 12 Bake hardening behaviour of a dual phase (DP) and transformation induced plasticity (TRIP) steels after pre-strain (PS) and bake hardening, **a** stress–strain behaviour and **b** C distribution along a dislocation

6 Conclusions

The work presented here highlights the potential for the engineering of steel structure at all scale lengths to improve the properties; particularly strength and ductility. Atom Probe Tomography can now be used to sample relatively large volumes in reasonable time and therefore is more amenable to obtaining new insights. This supported by conventional TEM and FEG-SEM with advanced EBSD can allow us to more fully understand the relationships between processing and structure.

Acknowledgments Much of the work reported here has been supported by the Australian Research Council, including the Federation and Laureatte Fellowships for PDH. Some of the unpublished APT work has been in collaboration with Professor Simon Ringer and his group at Sydney University.

References

1. I.B. Timokhina, P.D. Hodgson, E.V. Pereloma, Metallurgical Mater. Trans. A **35A**, 2331 (2004)
2. H. Beladi, G.L. Kelly, P.D. Hodgson, Int. Mater. Rev. **1**, 14 (2007)
3. N. Tsuji, R. Ueji, Y. Minamino, Scr. Mater. **2**, 69 (2002)
4. R. Priestner, P.D. Hodgson, Mater. Sci. Technol. **8–10**, 849 (1992)
5. A. Shokouhi, P.D. Hodgson, Mater. Sci. Technol. **23**, 1233 (2007)
6. K. Miyata, M. Wakita, S. Fukushima, M. Eto, T. Sasaki, T. Tomida, Mater. Sci. Forum **539–543**, 4698 (2007)
7. H.K.D.H. Bhadeshia, Mater. Sci. Eng. A **481–482**, 36 (2008)
8. Y. Funakawa, T. Shiozaki, K. Tomita, T. Yamamoto, E. Maeda, ISIJ Int. **44**, 1945 (2004)
9. I.B. Timokhina, P.D. Hodgson, S.P. Ringer, R.K. Zheng, E. Pereloma, Scr. Mater. **56**, 601 (2007)
10. S.P. Ringer, K. Hono, Mater. Charact. **44**, 101 (2000)

Research of Low Carbon Nb-Ti-B Microalloyed High Strength Hot Strip Steels with Yield Strength ≥700 MPa

Hongtao Zhang, Chengbin Liu, and Ganyun Pang

Abstract

The effects of different compositional alternatives of Nb-Ti-Mo, Nb-Ti-Mo-B, and Nb-Ti-B with the basic composition of 0.08C-1.8Mn were studied aiming at optimization of steel composition and hot rolling and cooling technology of a microalloyed high strength hot strip steels with YS ≥700 MPa. In the work, four experimental steels with different combinations of micro-amounts of Nb, Ti, and B are compared. Moreover, two experimental steels, i.e. the steel of Nb-Ti and the steel of Nb-Ti-B were subjected to CCT diagram experiment in order to find the effects of sub micro-amount B in this type of steels. It has been found that sub micro-amount B addition brings about significant influence on the tensile properties, microstructures, and phase transformation characters of this kind of steels. It has been also found that as coiling temperature after hot rolling, 600°C is better than 570°C when strip steel is applied under as hot rolled condition. If strip steel is applied under as annealed condition, 570°C is the preferable coiling temperature leading to larger strength increment and higher elongation after annealing. Boron microalloying is more suitable for as annealed strip steels. The CCT diagram study shows that sub micro-amount B addition strongly lowers the transformation start temperature and transformation stop temperature and obviously increases hardness at fast cooling rates (44–1°C/s). B suppresses ferrite transformation during continuous cooling at comparatively fast cooling rates promoting the granular bainite formation and increasing steel hardness. It is common that for this type of steels that in both B free and B containing steels, at certain cooling rates martensite islands is prone to occur. B shifts the martensite islands phase presence from faster cooling rates to slower cooling rates. In the case of as hot rolled high strength strip steel, martensite islands are to be avoid; while for the as annealed high strength hot strip steel, martensite islands are beneficial leading to further increase of strength level and to the improvement of elongation value after annealing.

Keywords

High strength hot strip steels • Low carbon steels • Microalloying • CCT diagram

1 Introduction

As the globe is requiring low carbon economy, for automotive industry and many other moving vehicle machinery industry, the weight reduction, fuel reduction, and reduction of toxic emission are badly demanded. As the most popular steel materials concerned, higher strength steel products are more demanded. Among the flat steel products concerned,

H. Zhang (✉) · C. Liu · G. Pang
Institute of Structural Materials Research, CISRI, Beijing 100081, People's Republic of China
e-mail: zhanghongtao321107@yahoo.com

Y. Weng et al. (eds.), *Advanced Steels*, DOI: 10.1007/978-3-642-17665-4_33,

the category of high strength hot rolled strip steels are more cost-effective compared to cold rolled high strength strip steels. Therefore, hot rolled high strength strip steels are popular for application in industry whenever the surface condition is satisfactory.

The extra high strength hot rolled strip steels with yield strength ≥700 MPa are considered to be a kind of hot rolled strip steels of the highest strength applied under as hot rolled condition without further processing. Therefore, the hot rolled high strength strip steels with YS ≥700 MPa belong to the category of modern advanced steel materials.

The industrial trial production of this kind of steels shows that there exist still many problems, e.g. insufficient strength, low ductility and weak toughness when the strip thickness is larger than 10 mm. It is therefore considered that the steel composition and processing technology need further investigation and adjustment.

In consequence, a laboratory research work has been carried out aiming at optimization of steel composition and of processing parameters of controlled hot rolling and controlled cooling. The paper introduces the research results and wages a discussion on the possible ways of compositional and processing improvement.

2 Experimental Materials and Laboratory Research Procedure

2.1 Experimental Heats and Chemical Compositions of Experimental Steels

In industrial practice, a micro-alloy system of Nb-Ti is applied for the basic steel composition 0.08C-1.8Mn-0.1Mo. For increasing hardenability, element of boron is a very effective while it is a very cheap element. Boron is very effective in controlling microstructure especially when very low carbon steels are concerned. In addition, there had been a lot of research work on the synergic effect of Nb-B combination and of Mo-B combination [1–6]. Therefore, in our research boron is paid special attention. Three alloy combinations, i.e. Mo-Nb-Ti, Nb-Ti–B, and Mo-Nb-Ti–B, are tested. Simultaneously, 0.05C and 0.08C are tried in order to see the possibility of ductility and toughness improvement by lowering C content.

Four experimental heats were designed. Laboratory heats of 25 kg each were melted in vacuum induction furnace. The chemical compositions of these four experimental heats are listed in Table 1.

As shown in Table 1, chemical composition of heat 1# is basically the chemical composition of 0.08C-1.76Mn-Nb-Ti-Mo, that is basically used in the industrial production. Compared to heat 1#, in Heat 2# Mo is omitted while boron is added. Heat 3# with 0.057-1.74Mn-Nb-Ti-Mo-B and heat 4# with 0.052C-1.80Mn-Nb-Ti–B, both are of lower C content. As it is suggested, C content in the steel higher than 0.04% is unnecessary, since the extra C forms only pearlite nodes which do not contribute to yield strength increase [6]. In addition, lower C content is also beneficial to hot ductility improvement during continuous casting.

2.2 Laboratory Experimental Procedure

The four heats were cast into four ingots of 25 kg weight each then subjected to hot forging. Four hot forged slabs were of 63 × 105 × 130 mm in sizes. Then the experimental slabs were subjected to laboratory hot rolling and cooling simulating the industrial controlled hot strip rolling and controlled laminar cooling and coiling. The whole experimental procedure is schematically shown in Fig. 1.

Hot rolling was conducted on laboratory flat rolling mill with working rolls of 300 mm in diameter. Slab reheating was done in an electric furnace. As for the controlled cooling simulation, a special water-cooling devise was used for this work. For coiling simulation, another electric furnace heated to temperature of 570 or 600°C beforehand was applied. After hot rolling, the rolled strip was subjected to water-cooling at cooling speed ~40°C/s and stop cooling temperature was controlled at around 570 or 600°C by using an photo-electric thermometer. The strip after cooling down to around 570 or 600°C was then put into the furnace ready-heated to 570 or 600°C. All the strips were kept at 570 or 600°C for 1 h and then cooled down inside and with the shut-off furnace. The whole process of strip controlled rolling and controlled cooling and coiling simulation then was finished.

Compared to the practical hot rolling in hot strip rolling mill, the total thickness ratio of slab to strip, when 6 mm thick strip was rolled, t_{slab}/t_{strip} in the laboratory rolling is

Table 1 Chemical compositions of experimental steels

Heat no.	Chemical composition/(mass%)										
	C	Si	Mn	P	S	Nb	Ti	Mo	B	Al	N
1#	0.080	0.18	1.76	0.0064	0.007	0.058	0.17	0.14	~	0.037	0.0027
2#	0.077	0.19	1.75	0.0053	0.007	0.060	0.14	~	0.0013	0.030	0.0026
3#	0.057	0.20	1.74	0.0066	0.010	0.060	0.14	0.13	0.0018	0.020	0.0026
4#	0.052	0.20	1.80	0.0062	0.004	0.053	0.12	~	0.0014	0.024	0.0027

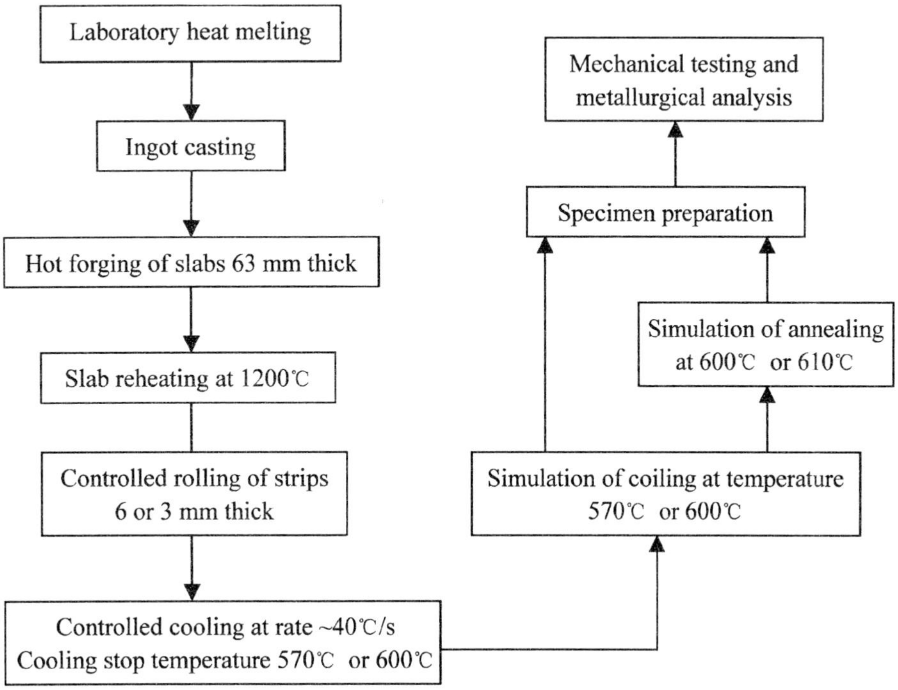

Fig. 1 Schema of laboratory experiment

still small, i.e. $t_{slab}/t_{strip} = 63/6 = 10.5$; while that in actual production the thickness ratio is $t_{slab}/t_{strip} = 230/10 = 23$ for 10 mm thick strip, and 28.75 for 8 mm thick strip. When 3 mm thick strips were rolled in laboratory, $t_{slab}/t_{strip} = 21$ is close to the practical rolling.

In practical production, the as hot-rolled strip steel can be further annealed in order to explore potential of further precipitation strength increase. Therefore, the 6 mm thick strip steels were subjected annealing at 600°C that is higher than coiling temperature 570°C. 6 mm thick strips were heated to 600°C and kept for 2 h and cooled inside and together with the furnace. The 3 mm thick strips were then subjected to annealing at 610°C, since their coiling temperature was already 600°C. The as annealed 6 and 3 mm thick strips were then subjected tensile testing.

3 Experimental Results and Discussion

3.1 Tensile Strength Properties of as Hot Rolled Experimental Steels

Tensile property results in transverse direction of 6 mm strips of four experimental steels are listed in Table 2.

It is seen from Table 2 that steel 1# and steel 4# show similar strength level with two specimens while steel 2# and 3# show some variation of strength level between the two specimens.

Secondly, we see that steel 2# has reached the required strength level and only one specimen of steel 3# reaches the required strength level while the other specimen does not. That means there is lack of strength uniformity of steel 3#.

Table 2 Tensile properties of 6 mm strips of experimental steels

Heat no.	Specimen no.[a]	R_{eL}	R_m	R_{eL}/R_m	A (%)	A_g (%)[b]
1# 08-Mo	1-1	625	685	0.91	15.0	5.0
	1-2	635	685	0.93	12.0	2.0
2# 08-B	2-1	730	765	0.95	11.5	2.0
	2-2	815	850	0.96	12.5	3.0
3# 05-Mo-B	3-1	635	705	0.90	8.5	2.5
	3-2	815	865	0.94	9.5	2.0
4# 05-B	4-1	575	680	0.84	16.0	7.0
	4-2	585	680	0.86	15.0	6.5

[a] Transverse tensile specimens
[b] Homogeneous elongation

Table 3 Tensile properties of 3 mm thick hot rolled strips of experimental steels

Heat no.	Specimen no.[a]	R_{eH} ($R_{0.2}$)	R_m	$R_{0.2}/R_m$	A_5 (%)	A_g (%)[b]
1# 08-Mo	1-1	720	745	0.96	17.5	5.0
	1-2	720	735	0.98	18.5	6.0
2# 08-B	2-1	610	675	0.90	14.5	3.5
	2-2	635	690	0.92	13.0	3.5
3# 05-Mo-B	3-1	615	705	0.87	20.5	8.0
	3-2	625	710	0.88	22.5	8.5
4# 05-B	4-1	690	710	0.97	17.0	6.0
	4-2	675	710	0.95	17.5	5.5

[a] Transverse tensile specimens
[b] Homogeneous elongation

When we look at the yield ratio of the steels, it is seen that steel 1#, 2#, and 3#, all three steels show higher yield ratio values in the range of 0.91–0.96 while steel 4# shows the value in the range of 0.84–0.86.

The tensile property results of the 3 mm thick strips of experimental steels are listed in Table 3.

From Table 3 it is seen that only the strength level of Steel 1# almost reaches the steel requirements and those of Steel 2#, 3#, and 4# are lower than the required level.

3.2 Analysis of Microstructures of as Hot Rolled Experimental Steels

3.2.1 Microstructures of 6 mm as Hot Rolled Strip Steels

Optical micrographs of 6 mm strip samples of four experimental steels are shown in Fig. 2. From the figure it can been seen that Steel 1# and Steel 4# show basically acicular ferrite and granular bainite, while Steel 2# and Steel 3# show basically lath bainite plus some granular bainite. Since lath bainite is formed at lower temperatures leading to higher strength properties, while acicular ferrite and granular bainite are transformed at higher temperatures exhibiting lower strength level. Strength properties and microstructures are in good relationship. In the case of 6 mm strip rolling, the 08C-Nb-Ti-B and 05C-Nb-Ti-Mo-B steels show higher hardenability than the 08-Nb-Ti-Mo and 05C-Nb-Ti-B steels. Extra micro-amount of B showed effective influence in raising hardenability of this kind of steels.

3.2.2 Microstructures of 3 mm as Hot Rolled Strip Steels

From Fig. 3, it can be seen that 3 mm thick strip Steel 1# shows a microstructure of acicular ferrite plus granular bainite with ~40% of acicular ferrite. The rest Steel 2#, 3#, and 4# all show microstructure of granular bainite and lath bainite. That means the B addition has obvious influence eliminating the formation of acicular ferrite.

However, if the strength properties are considered, addition of B did not bring any effectiveness in strengthening, although B eliminates ferrite formation.

3.3 Summary of Mechanical Properties and Microstructures of as Hot Rolled Experimental Steels

From the experimental results of as hot rolled strip steels, some general summary points can be drawn as follows:

1. In the case of 6 mm strips, when coiling temperature was 570°C, Steel 1# (08C-Mo-Nb-Ti) and Steel 4# (05C-Nb-Ti–B) show insufficient strength values, although they possess acicular ferrite and granular bainite microstructures. On the other hand, Steel 2# (08C-Nb-Ti–B) and Steel 3# (05C-Mo-Nb-Ti–B) show high strength values meeting strength requirement, but their elongation values are low to meet its requirement. These two steels possess basically lath bainite microstructures. This unfavorable situation might have come from the small hot rolling reduction, i.e. $t_{slab}/t_{strip} = 63/6 = 10.5$ that is far insufficient compared to practical production.
2. In the case of 3 mm strips, when coiling temperature was 600°C, only the strength level of Steel 1# reaches basically the requirements and strength values of Steel 2#, 3#, and 4# are not up to the required level. It can be considered that the coiling temperature 600°C is a bit higher than required. The lack of Mo content for Steel 2# and Steel 4# and lower C content of 0.05% for Steel 3# and Steel 4# are in some respects responsible for the insufficiency of strength. In this case, Steel 1# shows ~40% acicular ferrite in the microstructure, while Steels 2#, 3#, and 4# show microstructures composed of granular bainite and lath bainite without acicular ferrite. B addition

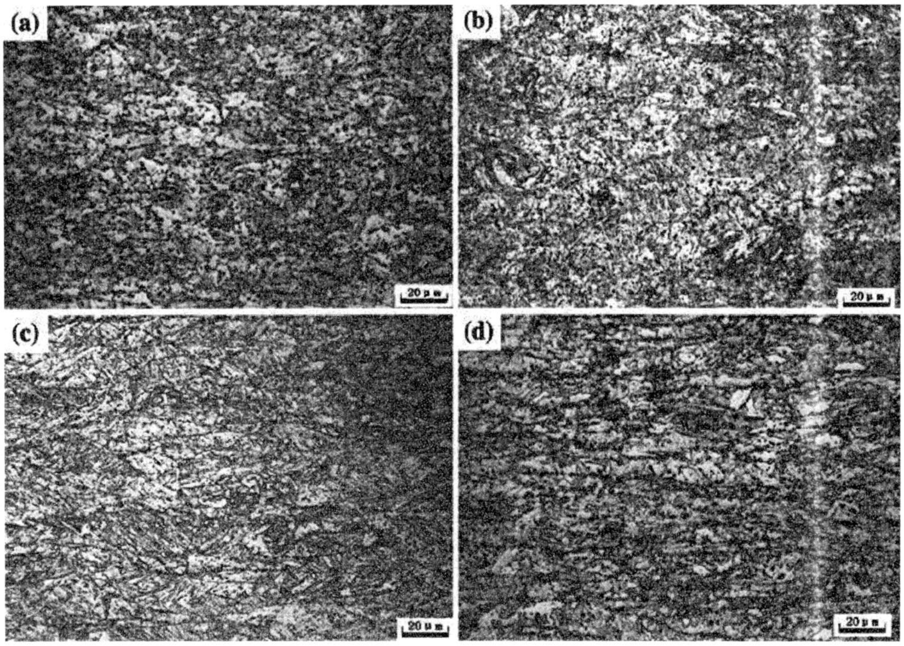

Fig. 2 Optical micrographs of 6 mm strips of four experimental steels: **a** Steel1#; **b** Steel 2#; **c** Steel 3#; **d** Steel 4

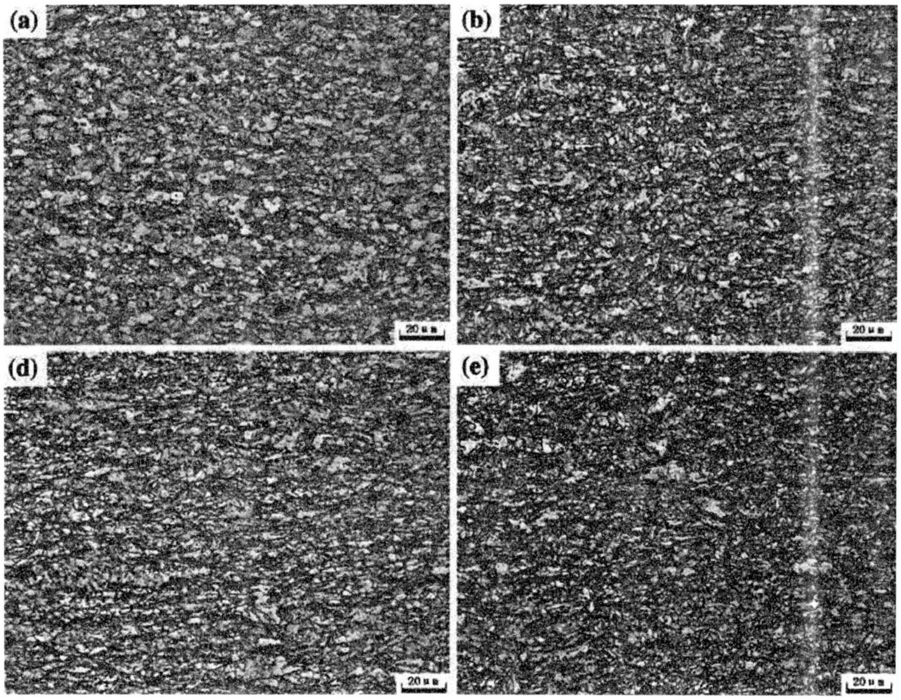

Fig. 3 Optical micrographs of 3 mm strips of four experimental steels: **a** Steel1#; **b** Steel 2#; **c** Steel 3#; **d** Steel 4#

exerts obvious influence eliminating the formation of acicular ferrite.

3. Acicular ferrite is favorable to strength increase compared to granular bainite plus lath bainite in this kind of high strength hot strip steels microalloyed with Nb + Ti, possibly it is due to more efficient precipitation of Ti and Nb carbonitrides in acicular ferrite.

3.4 Annealing Treatment Experiment of as Hot Rolled Strips

3.4.1 Annealing Treatment of 6 mm Strip Steels

Since the experimental as hot rolled strip steels have been subjected to fast cooling after finish rolling and before coiling, i.e. in the temperature range of 870–570°C (6 mm)

Table 4 Tensile properties of 6 mm experimental steels as hot rolled and then annealed at 600°C

Heat no.	Spec. no.[a]	$R_{0.2}$	R_m	$R_{0.2}/R_m$	A_{50} (%)
1# 08-Mo	1-1	780	820	0.95	22.0
	1-2	790	825	0.95	18.0
2# 08-B	2-1	830	845	0.98	17.5
	2-2	835	850	0.98	15.0
3# 05-Mo-B	3-1	880	900	0.97	18.0
	3-2	815	845	0.96	18.5
4# 05-B	4-1	820	840	0.97	16.5
	4-2	850	860	0.98	16.5

[a] Transverse tensile specimens

or to 600°C (3 mm), the precipitation of carbonitrides of microalloying elements is not sufficient under as rolled condition. Therefore, annealing treatment of hot rolled high strength strip steels should help to explore the precipitation potential. Annealing treatment was carried out. Annealing temperature was 600°C for 6 mm strips and 610°C for 3 mm strips and annealing time was 2 h for both strips.

The results of annealing of 6 mm strip steels have been obtained. Here for comparison, tensile property values and schematic figures are shown in Table 4 and Figs. 4, 5, and 6 respectively.

From Table 4 and Figs. 4, 5, and 6 some clear results can be noted as follows:

1. The strength level including YS and UTS has been obviously increased by 600°C annealing. The YS values of all the four experimental steels are far above the level 700 MPa and the UTS values of all four steels are all above 800 MPa. The YS/UTS ratio values have also increased by annealing due to precipitation hardening.
2. The elongation values of the four steels have been also improved due to annealing. That means annealing process not only leads to strength increase but also to the elongation improvement.
3. From the above-mentioned effectiveness of annealing, it could be anticipated that for this kind of steels the defined chemical compositions and applied rolling technology is good for high strength hot rolled strip steels under rolling + annealing condition.

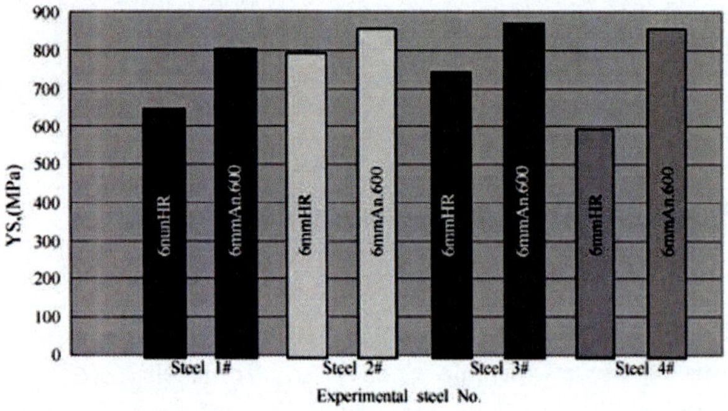

Fig. 4 YS values of 6 mm strips as hot rolled and annealed at 600°C

3.4.2 Annealing Treatment of 3 mm Strip Steels

Since almost all four experimental 3 mm strip steels have not satisfactorily reached the specified strength level, esp. the three experimental steels containing B failed in strength properties, annealing treatment of hot rolled 3 mm strip

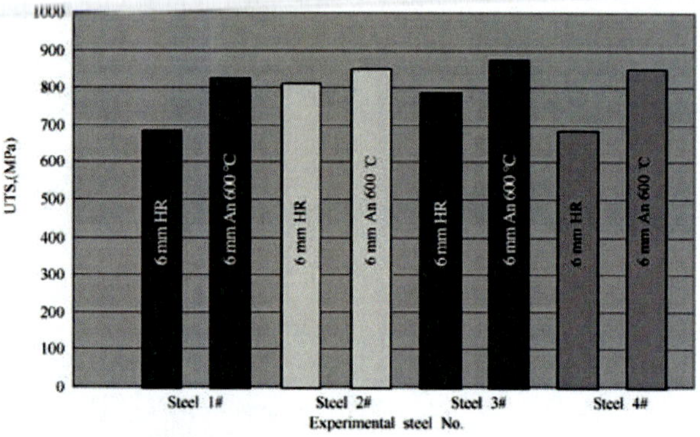

Fig. 5 UTS values of 6 mm strips as hot rolled and annealed at 600°C

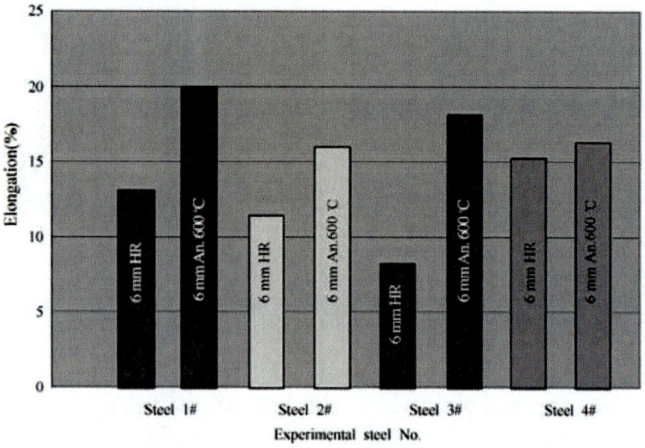

Fig. 6 Elongation values of 6 mm strips as hot rolled and annealed at 600°C

Table 5 Tensile properties of 3 mm strip steels as annealed at 610°C

Heat no.	Spec. no.[a]	$R_{0.2}$	R_m	$R_{0.2}/R_m$	A_{50} (%)	A_g (%)[b]
1# 08-Mo	1-1	715	750	0.95	14.5	7.0
	1-2	730	750	0.97	16.0	8.0
2# 08-B	2-1	720	745	0.96	11.5	1.5
	2-2	705	750	0.94	10.5	3.5
3# 05-Mo-B	3-1	750	785	0.95	16.0	7.5
	3-2	760	790	0.96	16.5	8.0
4# 05-B	4-1	745	765	0.97	13.5	6.0
	4-2	750	765	0.98	—	—

[a] Transverse tensile specimens
[b] Homogeneous elongation

steels was suggested for the purpose of increasing strength level by enhancing the precipitation hardening.

As the annealing temperature concerned, it was proposed to be 610°C. Although 610°C is only 10°C higher than the coiling temperature, it was anticipated that certain amount of precipitation hardening will take place during the annealing.

As annealing parameters, annealing temperature was 610°C annealing time was 2 h. 3 mm strips after annealing were subjected to tensile testing. Tensile testing results are listed in Table 5 as follows:

The data of tensile properties of 3 mm as hot rolled and annealed strip steels in Table 5 can be schematically shown in Figs. 7, 8, and 9.

The influence of annealing on the properties of 3 mm hot rolled strip steels can be summarized as follows:

1. From Table 5, Figs. 7 and 8, it is seen that annealing at 610°C which is only 10°C higher than coiling temperature, exerts strong effect in increasing strength level. The YS and UTS of all four experimental steels are up to the specified levels.
2. From Fig. 9 it can be seen that the elongation values after annealing of all four experimental steels are inferior to the respective values of as hot rolled 3 mm strips. However, when comparing Steel 1# and Steel 3#, it is seen that sub-micro addition of B improves elongation values of strips under as hot rolled condition and of those after annealing.
3. The elongation values of 3 mm strips decreased due to annealing, while in the case of 6 mm strips, annealing brought about increase of elongation values for all four steels. This may be because in the case of 6 mm when

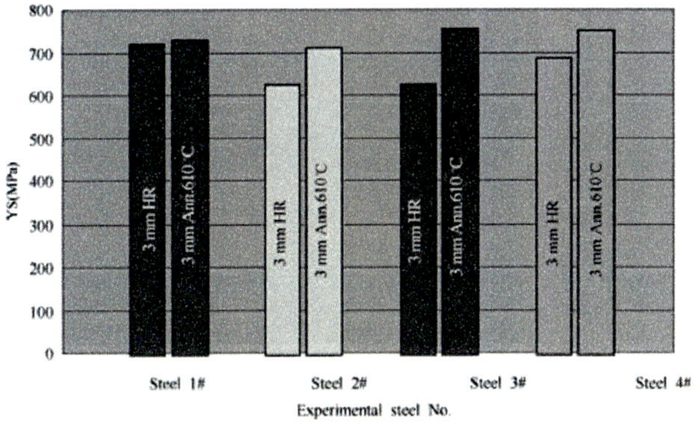

Fig. 7 YS values of 3 mm strips as hot rolled and annealed at 610°C

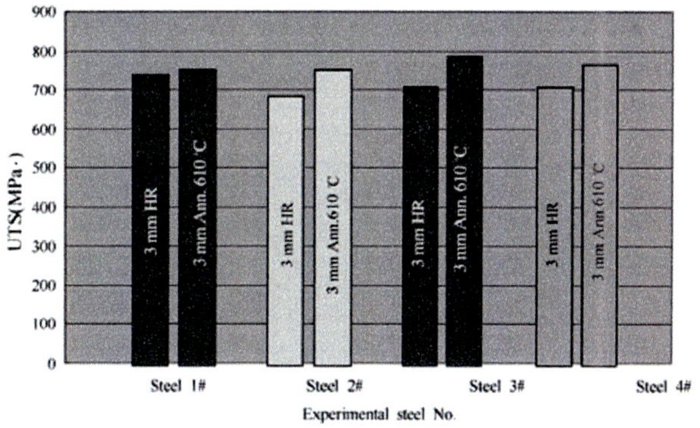

Fig. 8 UTS values of 3 mm strips as hot rolled and annealed at 610°C

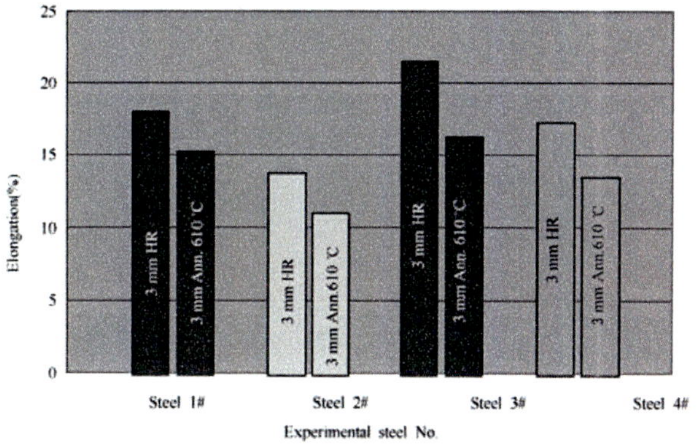

Fig. 9 Elongation values of 3 mm strips as hot rolled and anncaled at 610°C

coiling temperature was as low as 570°C, during annealing at 600°C effect of dislocation restructuring prevails effect of precipitation leading to the improvement of elongation value.
4. In general, annealing at proper temperature for this kind of high strength hot rolled strip steels is a worthwhile technological measure in exploring the steel strength potential.

3.4.3 Comparison of Tensile Properties of 6 and 3 mm as Annealed Strip Steels

It should be informative that the tensile properties of both 6 and 3 mm as annealed strip steels are compared. The respective results are graphically shown in Figs. 10, 11, and 12.

When the application condition of high strength strip steels is as hot rolled plus annealed, the tensile property data could be of reference concerning the compositional optimization and processing technology improvement. From Fig. 10 it is seen that YS of 6 and 3 mm strips and of all four experimental steels are above the specified 700 MPa; while YS values of 6 mm strips of four experimental steels are all higher than those of 3 mm strips of respective experimental steels. The similar situation is seen in Fig. 11. The UTS values of 6 mm strips of all four experimental steels are higher than those of 3 mm strips of respective steels. That means in the case of 6 mm strips lower coiling temperature, 570°C, and lower annealing temperature 600°C bring about larger strength increments due to annealing than in the case of 3 mm strips where coiling temperature is 600°C and annealing temperature 610°C.

From Fig. 12 it is seen that elongation values of 6 mm strips of all four experimental steels are higher than those of 3 mm strips of respective experimental steels. The reason for this needs further investigation. Nevertheless, all elongation values of both 6 and 3 mm strips of all four experimental steels meet the specification requirement except for 3 mm strip of Steel 2#.

General speaking, it can be summarized that if annealing treatment is applied in the production, the processing parameters of 6 mm strips are superior to those of 3 mm strips, i.e. coiling temperature 570°C and annealing temperature 600°C are better than respective 600 and 610°C for 3 mm strips.

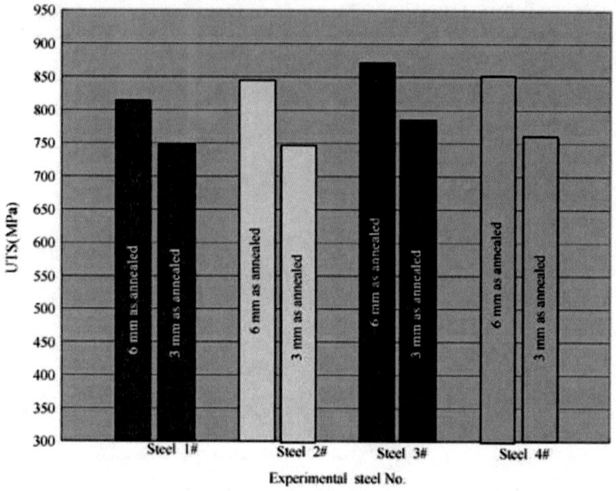

Fig. 11 Comparison of UTS values of 6 and 3 mm as annealed strip steels

3.5 Summary of Current Laboratory Research of High Strength Hot Rolled Strip Steels

From the results of as hot rolled 6 and 3 mm experimental strip steels and the results of as hot rolled plus annealed 6 and 3 mm respective strip steels, some summary remarks which are useful for practical production and also are of

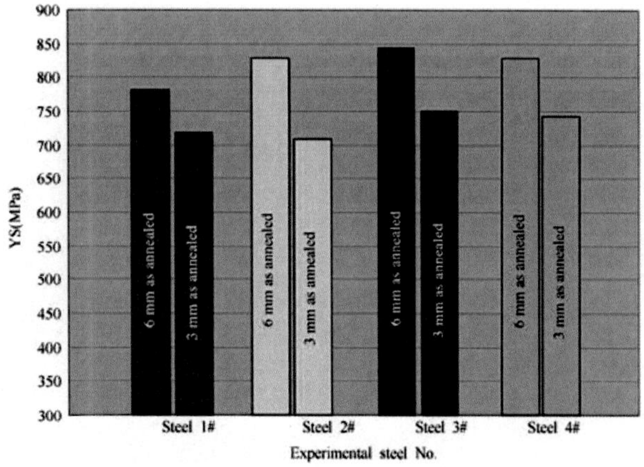

Fig. 10 Comparison of YS values of 6 and 3 mm as annealed strip steels

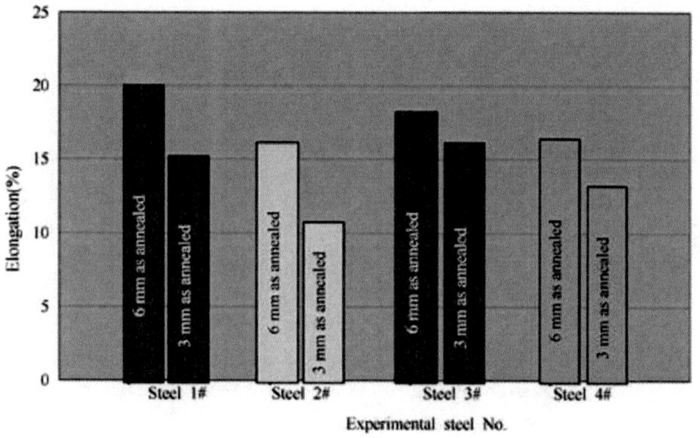

Fig. 12 Comparison of elongation values of 6 and 3 mm as annealed strip steels

important reference for future study of this kind of steels, can be drawn as follows:

1. For the grade of high strength hot strip steel with YS $\geq$700 MPa, the basic composition, 0.08C-0.2Si-1.8Mn-0.15Mo-0.06Nb-0.14Ti is basically feasible. However, if higher ductility is demanded, its carbon content can be reduced from 0.08 to 0.05% without strength decrease. Sub micro-amount of B can be added improving both strength and ductility maybe due to synergetic effects of combined Nb-B and/or Nb-Mo alloying, especially when C content is as low as 0.05%.
2. As the technological parameters of hot rolling concerned, when the strip steel is applied under as hot rolled condition, the coiling temperature 600°C is better than 570°C mainly because CT 600°C ensures necessary ductility level. When giving consideration to the strength requirement, in order to guarantee strength level to lower a bit lower coiling temperature could be a good adjustment.
3. When annealing treatment is realized, coiling temperature should be as low as 570°C. From the property results of as annealed 6 mm strip steels, it can be concluded that coiling temperature 570°C and annealing temperature 600°C give rise to the obvious increments of both strength and ductility values. In the case of 3 mm strip steels, coiling temperature 600°C and annealing temperature 610°C led to enough strength increments but to lower ductility level to all experimental steels.
4. Presence of martensite islands in the microstructure of as hot rolled strip steels should be avoid, since it leads to lower yield strength possibly under required level of 700 MPa. This problem needs careful further work. The γ-α phase transformation behaviors at continuous cooling of the steels have to be studied and the ways of controlling microstructures have to be found, especially the ways of avoiding martensite islands in the microstructures of as hot rolled strip steels. If the strip steel is applied under as annealed condition, there is no worry about the presence of martensite islands in microstructure under as hot rolled condition.

4 Industrial Trial Production of Mo-Nb-Ti Hot Strip Steel

High strength hot strip steel with Mo-Nb-Ti alloy system has been put into industrial production. Steel heat was melted in 180 t BOF and refined in LF and then cast into 230 mm thick slabs. Slabs were reheated to 1,200°C and hot rolled in a hot strip mill of HSM 2250 into 8 and 10 mm thick strips. After finish rolling, strips were water-cooled on the laminar cooling table and then coiled at coiling temperature 570°C. The chemical composition of industrial heat is listed in Table 6. The tensile properties of hot rolled 10 mm thick strip and 8 mm thick strip are shown in Table 7.

From Table 6 it is seen that the industrial melted heat is of 0.08C-1.8Mn-0.1Mo-0.06Nb-0.13Ti composition. From Table 7, some special characters concerning the tensile properties of as hot rolled 10 and 8 mm thick strips can be indicated as follows: (1) The longitudinal yield ratio of 10 mm thick strip is 0.84 while that of 8 mm thick strip is 0.93. If we compare the relevant optical micrographs of 10 and 8 mm strips respectively in Figs. 13 and 14, it can be understood that the low yield ratio of 10 mm strip is corresponding to the existence of martensite islands in its microstructure (Fig. 13). While in the micrograph of 8 mm strip (Fig. 14), acicular ferrite and small fraction of pearlite are seen; (2) The YS value of transverse specimen is much lower than that of longitudinal specimen in both cases of 10 and 8 mm thickness. In practice, longitudinal YS value 700 MPa is required. Therefore, the YS values of 10

Table 6 Chemical composition of industrial strip steel

Element/(mass%)									
C	Si	Mn	P	S	Al	Mo	Nb	Ti	N
0.084	0.14	1.82	0.011	0.001	0.069	0.095	0.062	0.13	0.0061

Table 7 Tensile properties of industrial produced strips

Thickness (mm)	Specimen direction	Tensile properties					
		R_{eH} (Mpa)	R_m (Mpa)	R_{eH}/R_m	A_5 (%)	A_g^a (%)	A_n^b (%)
10	Long.	695	825	0.84	21.5	12.0	9.5
	Trans.	745	805	0.92	16.5	–	–
8	Long.	690	740	0.93	21.5	9.0	12.5
	Trans.	730	790	0.92	18.0	–	–

[a] Homogeneous elongation
[b] Local elongation

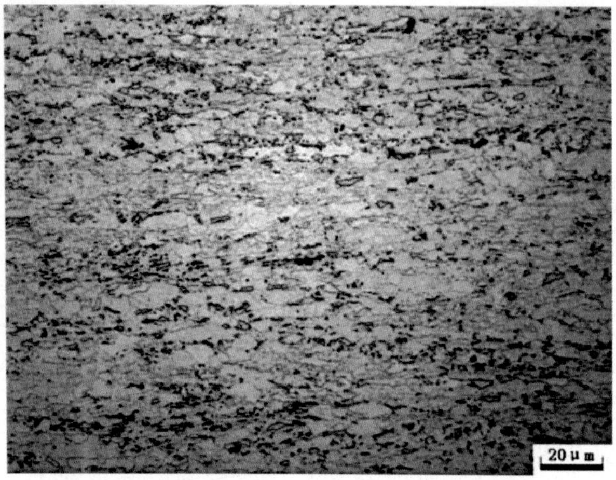

Fig. 13 Optical micrographs of industrial 10 mm thick strip at positions of ¼ thickness

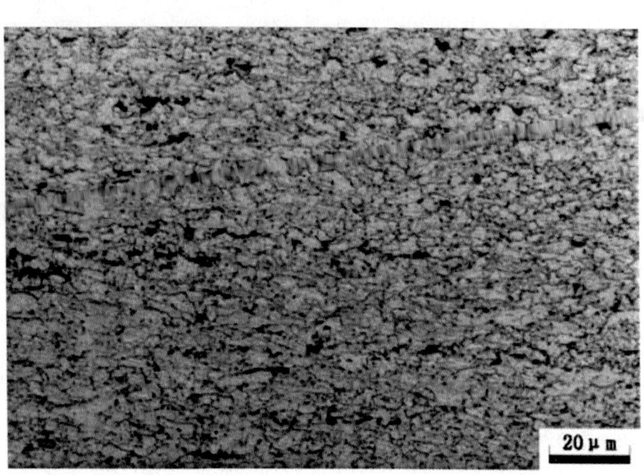

Fig. 14 Optical micrographs of industrial 8 mm thick strip at positions of ¼ thickness

and 8 mm strips are a bite insufficient, i.e. 10 or 5 MPa lower than required. Some compositional or hot rolling schedule adjustments are needed for future production in order to guarantee the strength level.

5 CCT Diagram Determination and Analysis for Steels of Mo-Nb-Ti and Mo-Nb-Ti–B Alloy Systems

5.1 Purpose of Determination of CCT Diagram of Steel 08Mn2MoNbTi and Steel 05Mn2MoNbTiB

The effect of B element on the γ–α phase transformation behaviors at continuous cooling of the steels has been studied. CCT diagrams of an industrial produced steel 08Mn2MoNbTi and of boron containing steel, experimental Steel 3#, 05Mn2MoNbTiB were determined and corresponding microstructures at various continuous cooling rate were analyzed.

It is considered that comparison of these two CCT diagrams is very important. Revealing and understanding of the physical metallurgical nature of the sub micro-amount B in the steel is of great interest to us, since microalloying with B is esteemed as a promising measure in providing synergetic function of Nb-B and/or Mo-B in low carbon steels [1–6]. However, in our laboratory work on the hot strip steels it has been shown that B addition did not work in raising the strength level in case of 3 mm strip while in case of 6 mm B addition did increase the strength level but deteriorate the elongation value because of formation of lath bainitic microstructure [7]. The systematic knowledge of phase transformation behaviors and transformation products i.e. resulted microstructures during various continuous cooling process of austenite are very important. These data are useful not only for defining heat treatment regimes and for further fabricating but also for the controlling of hot rolling, cooling and coiling in hot strip rolling mills.

5.2 Experimental Materials and CCT Diagram Determination Procedure

Thermal dilatation measuring method was applied for determination of CCT diagram. Dilatation testing specimen is of cylinder shape of $\varphi 3 \times 10$ mm. Experimental materials are taken from 8 mm strip industrially produced and from 3 mm laboratory rolled hot strip of experimental Steel 3#. The chemical composition of industrial strip steel is as follows (in mass%): C 0.084, Si 0.14, Mn 1.82, Nb 0.062, Ti 0.13, Mo 0.096, Al 0.069, N 0.006, p 0.011, S 0.001 and that of laboratory Steel 3# is as follows (in mass%): C 0.057, Si 0.20, Mn 1.74, Nb 0.06, Ti 0.14, Mo 0.13, B 0.0018, Al 0.020, N 0.0026, p 0.0066, S 0.010.

The experiment was conducted on the thermal dilatation testing equipment, FORMASTER-D. Autenitizing temperature was 920°C, holding time 5 min. For each specimen, respective cooling rate was applied. Due to the different cooling rates, different thermal dilatation curves were obtained. From each thermal dilatation curve, the critical temperatures, i.e. the phase transformation starting temperature and finishing temperature were determined.

In the case of B-containing steel, at some cooling rates transformation starting temperature and finishing temperature for ferrite and for bainite are distinct. For example, as it is shown in Fig. 15a, when cooling rate is 1.8°C/s, the dilatation curve shows that the ferrite transformation starts at 700°C and ends at 650°C and bainite transformation starts at 560°C and ends at 510°C. There is a unique area between 650 and 560°C where there is no phase

Fig. 15 Dilatation curves during continuous cooling at rate 1.8°C/s for **a** steel 05Mn2MoNbTiB and **b** Steel 08Mn2MoNbTi

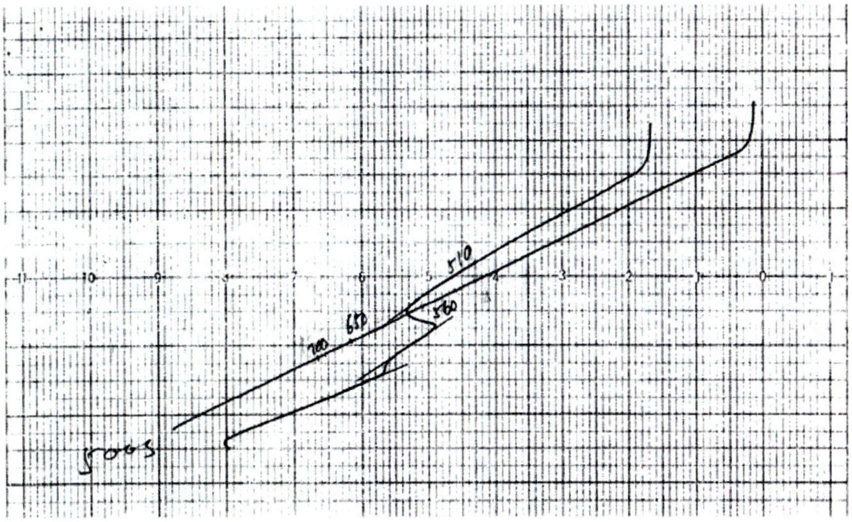

(a) Steel 05Mn2MoNbTiB

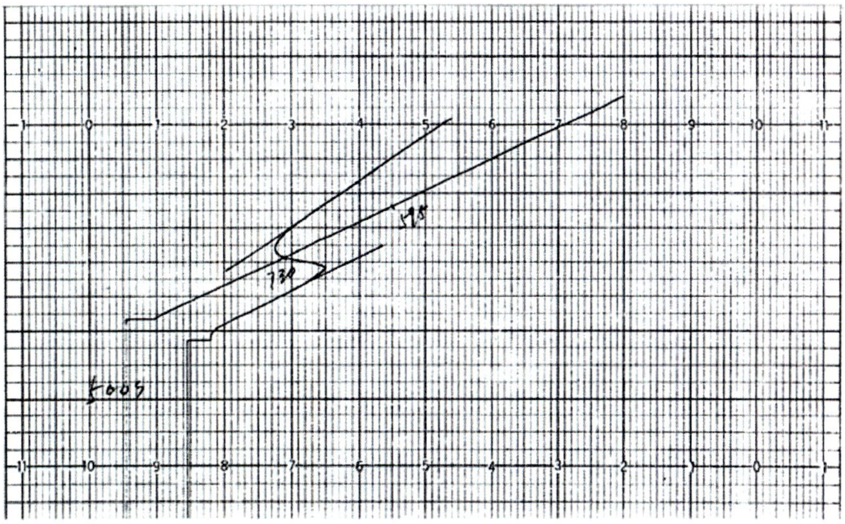

(b) Steel 08Mn2MoNbTi

transformation in between. For comparison, the dilatation curve at cooling rate of 1.8°C/s of steel 08Mn2MoNbTi without B (See Fig. 15b), there is only one transformation area starting at temperature 730°C and ending at 595°C.

By connecting the respective transformation critical points, the CCT diagram is basically obtained. This diagram has to be verified and elaborated by microstructural observation and analysis of the transformation products at each cooling rate. The tested specimens were sectioned transversely and mounted. The specimens were etched with Nital. Optical micrographs of 500 × magnification were taken. By comparing the dilatation curves with respective micrographs, some dotted curves were added in order to divide the area of ferrite and bainite transformation from the area of ferrite and pearlite transformation in the CCT diagram.

5.3 CCT Diagrams of Steel 08Mn2MoNbTi and Steel 05Mn2MoNbTiB

The determined CCT diagram of steel 08Mn2MoNbTi is shown in Fig. 16 and the optical micrographs at various cooling rate are shown in Fig. 17. The CCT diagram of steel 05Mn2MoNbTiB in Fig. 18 and its optical micrographs at various cooling rate are shown in Fig. 19.

As it is seen from Fig. 16 that two solid curves represent the loca of transformation starting temperature

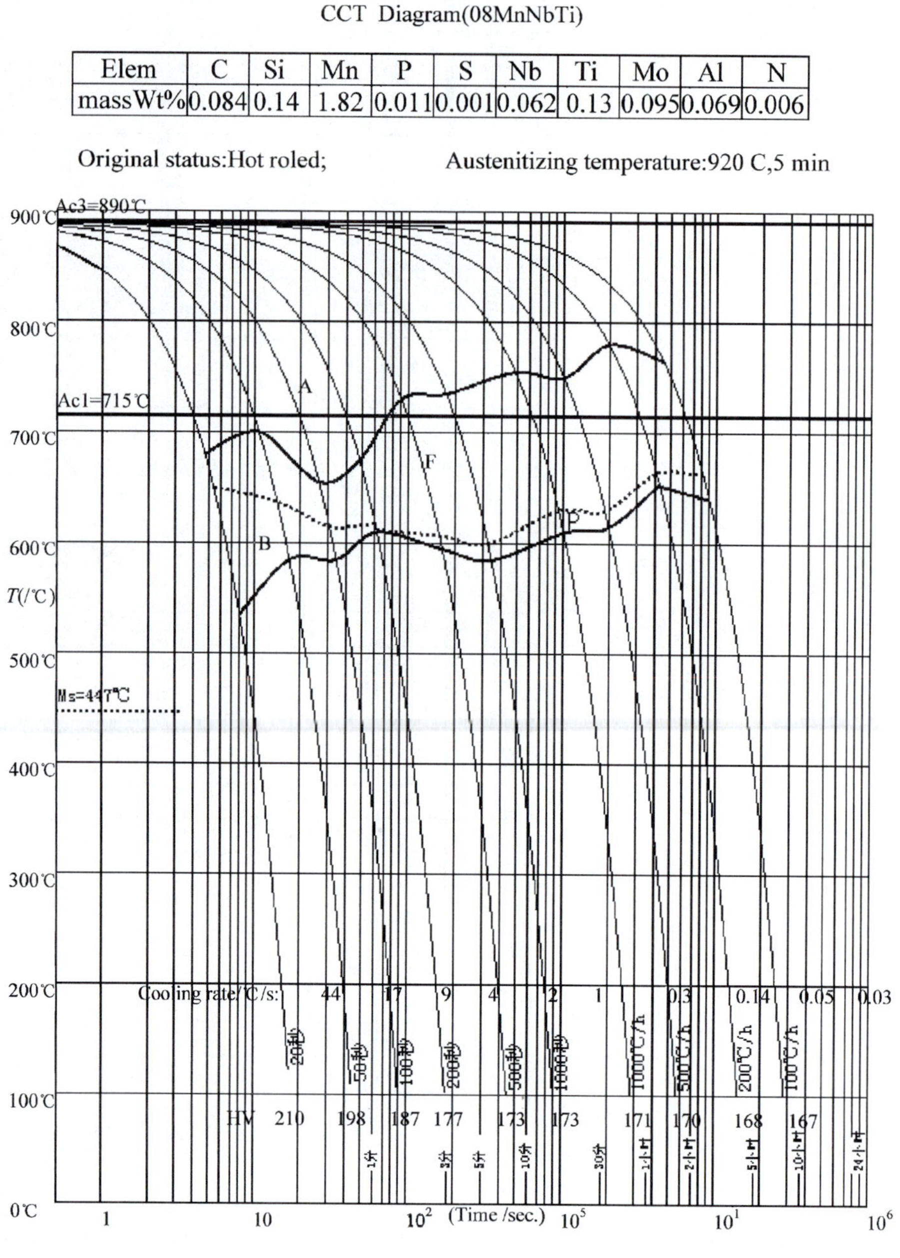

Fig. 16 CCT diagram of steel 08Mn2MoNbTi

and the transformation finishing temperature respectively. A dotted curve located in between represents the starting temperature of bainite and/or pearlite. The dotted curve is estimated according to the respective microstructure components. From Fig. 18 it is seen that also two solid curves represent the loca of transformation start temperature and finish temperature. In Fig. 18, there is a distinct area between cooling rates of 4 and 1°C/s where ferrite transformation area and bainite transformation area are distinctly separate. The start temperature and finish temperature of bainite and those of pearlite were determined according to respective thermal dilatation curves. The other two areas, one of faster cooling rates and the other with slower cooling rates show continuity of ferrite transformation with bainite transformation and of ferrite transformation with pearlite transformation respectively.

In each CCT diagram, there are shown also ten cooling curves of different cooling rates. In addition, respective hardness HV value is indicated at each cooling curve.

For detail comparison, the transformation critical temperatures at different cooling rates of the two experimental steels are listed in Table 8. Moreover, the description of microstructural components and hardness HV values at

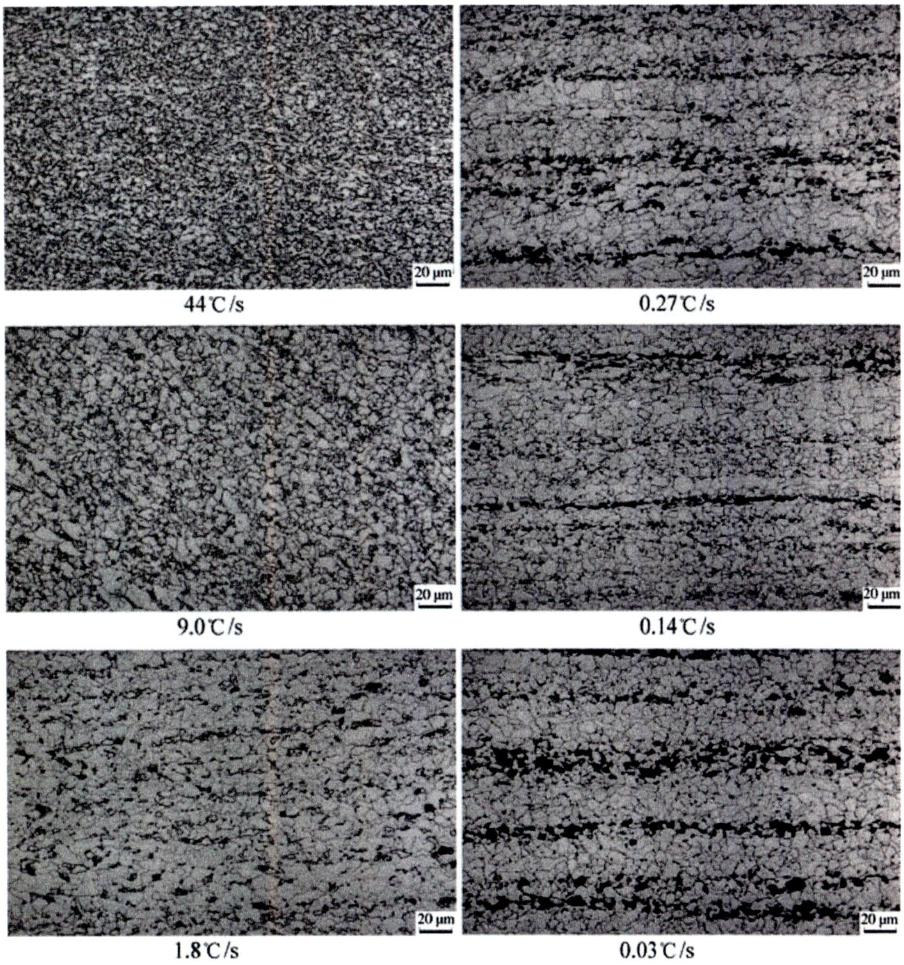

Fig. 17 Micrographs of resulted microstructures of phase transformation at various cooling rate of steel 08Mn2MoNbTi

different cooling rates for the two steels are also listed in Table 8.

From Table 8, the effect of sub micro-amount B on CCT behavior of this kind of steel can be seen in the following respects:

(1) At fast cooling rates, B-containing steel significantly lowers the ferrite transformation start temperature and the transformation finish temperature too. Let us bear in mind that steel 08Mn2MoNbTi contains C 0.084%, while the steel 05Mn2MoNbTiB contains lower C, i.e. 0.057% and the lower C content should increase the ferrite transformation start temperature. The experimental data confirm the strong effect of B in retarding ferrite transformation.

(2) At the median cooling rates (4–1°C/s), in the case of the B containing steel, the ferrite transformation area and the bainite transformation area are separate, while the B-free steel shows only one continuous transformation area which can not be definitely divided into ferrite area and bainite area. Experimental dilatation curves clearly show this difference (See Fig. 15).

5.4 Summary of CCT Diagram Study

Effects of sub micro-amount B on the phase transformation of Nb-Ti microalloyed high strength hot strip steel can be summarized as follows:

1. B suppresses ferrite transformation during continuous cooling. This effect of suppressing ferrite formation is stronger as cooling rate gets higher. B obviously shifts the bainite transformation to lower cooling rate until 1°C/s. At the same time, B lowers bainite transformation stop temperature ~100°C. Therefore, as it is found that B increases hardness in the cooling rate range of 44–1°C/s. It can be esteemed that in the hot strip rolling production when laminar water-cooling is applied, the steel hardenability is strongly increased by sub micro-amount B.

Fig. 18 CCT diagram of steel 05Mn2MoNbTiB

CCT Diagram(05Mn2MoNbTiB)

C	Si	Mn	Mo	P	S	Nb	Ti	B	Al	N
0.057	0.20	1.74	0.13	0.0066	0.01	0.06	0.14	0.0018	0.02	0.0026

Original status: Hot Rolled Austenitizing Temperature:920℃,5 min

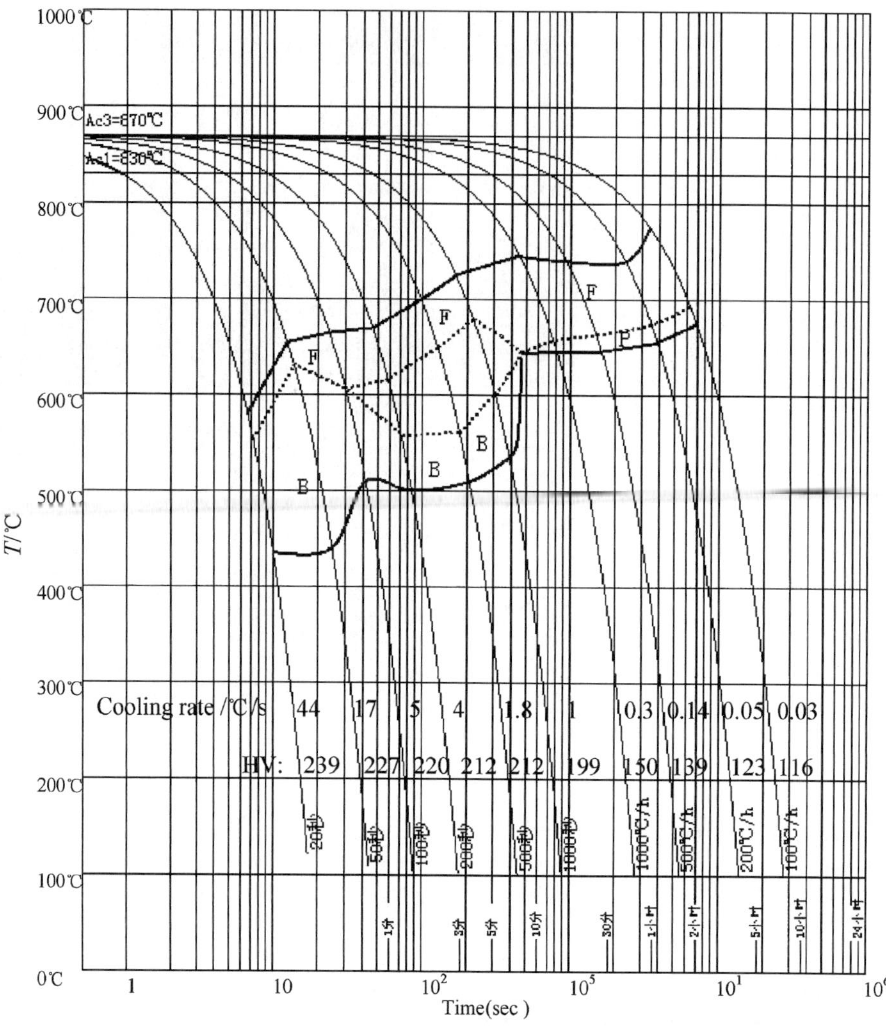

2. B separates the transformation range into two separate areas, one is the ferrite range and the other bainite range, when cooling rate is intermediate, e.g. 4–1°C/s. While in the case of B-free steel, there is no such evidence of two ranges of phase transformation.
3. When cooling rate is very slow, e.g. 0.3–0.03°C/s, B has no significant influence on the transformation range to ferrite and pearlite. On the contrary, B leads to coarser grain size of ferrite and leading to much lower hardness at very slow cooling rate.
4. If annealing process is applied, B has significant effect in increasing strength level of both yield strength and UTS.

Through the CCT diagram determination and at the same time the microstructural observation and hardness measurement of steels 08Mn2MoNbTi and 05Mn2MoNbTiB, the phase transformation behaviors of two experimental steels at different cooling rates have been in quite detail understood.

6 Conclusions

Through laboratory research on the low carbon Nb, Ti, B microalloyed high strength hot rolled strip steels and industrial trial production of 08Mn2MoNbTi hot strip steel

Table 8 Microstructures and respective hardness HV values of 08Mn2MoNbTi (steel 08) and 05Mn2MoNbTiB (steel 05B)

Steel	Cooling rate/(°C/s)	Microstructure	Hardness HV
08	44	F + B + MI	210
05B		GB	239
08	17	F + B + MI	198
05B		F + GB + MI	227
08	9	F + B + MI	187
05B		F + GB + MI	220
08	4	F + MI + P	177
05B		F + GB + P + MI	212
08	1.8	F + MI + P	173
05B		F + GB + P + MI	212
08	1.0	F + banding P + MI	173
05B		F + GB + P + MI	199
08	0.3	F + banding P	171
05B		Coarse F + banding P	150
08	0.14	F + banding P	170
05B		Coarse F + banding P	139
08	0.05	F + banding P	168
05B		Coarse F + banding P	123
08	0.03	F + banding P	167
05B		Coarse F + banding P	116

F ferrite, *GB* granular bainite, *P* Pearlite, *MI* martensite islands

and CCT diagram study of both steels 08Mn2MoNbTi and 05Mn2MoNbTiB, conclusions can be drawn as follows:

1. For as hot rolled condition, when coiling temperature was 570°C, sub micro-amount B addition brings about strength increase but lower elongation values in the case of 08Mn2NbTiB and 05Mn2MoNbTiB both showing lath martensite. While in the case of 08Mn2MoNbTi and 05Mn2NbTiB both possess low strength level. The microstructure of 08Mn2MoNbTi is composed of acicular ferrite and distributed individual carbides while the microstructure of 05Mn2NbTiB is composed of acicular ferrite, distributed martensite islands, and individual carbides as well. The Latter steel shows low yield ratio, $R_{eL}/R_m = 0.85$ which is due to the martensite islands in its microstructure.
2. When coiling temperature was 600°C, only the strength level of 08Mn2MoNbTi reaches basically the requirements and strength values of B containing 08Mn2NbTiB, 05Mn2MoNbTiB, and 05Mn2NbTiB are not up to the required level, i.e. 700 MPa. As respective microstructures concerned, steel 08Mn2MoNbTi shows ~40% acicular ferrite in its microstructure, while the rest three experimental steels show microstructures composed of granular bainite and lath bainite. However, the microstructure of 05Mn2MoNbTiB contains distributed martensite islands besides granular bainite and lath bainite leading to low yield ratio, $R_{eL}/R_m = 0.87$. In the case of 08Mn2MoNbTi, acicular ferrite is favorable to strength increase compared to granular bainite in this kind of high strength hot strip steels microalloyed with Nb + Ti. It might be due to more efficient precipitation of Ti and Nb carbonitrides in acicular ferrite at coiling temperature 600°C.
3. Annealing treatment is an important process for improvement of this kind of steels. The results of annealing treatment at 600°C of 6 mm experimental strip steels and those of annealing treatment at 610°C of 3 mm experimental strip steels show that coiling temperature 570°C is more proper for the as annealed steels since there are all-round improvement of strength and elongation values. B containing steels show higher strength level under as annealed condition. Boron microalloying is more suitable for as annealed strip steels.
4. CCT diagram study shows that sub micro-amount B addition strongly lowers the transformation start temperature and transformation stop temperature at fast cooling rates (44–1°C/s) promoting the granular bainite formation and increasing steel hardness. While at slow cooling rates (0.3–0.03°C/s), B increases pearlite transformation temperature leading to coarse ferrite grains and pearlite nodes and thus reduces hardness. Sub micro-amount B separates the phase transformation range into two parts, one is the ferrite range and the other bainite range, when cooling rate is intermediate, e.g. 4–1°C/s.

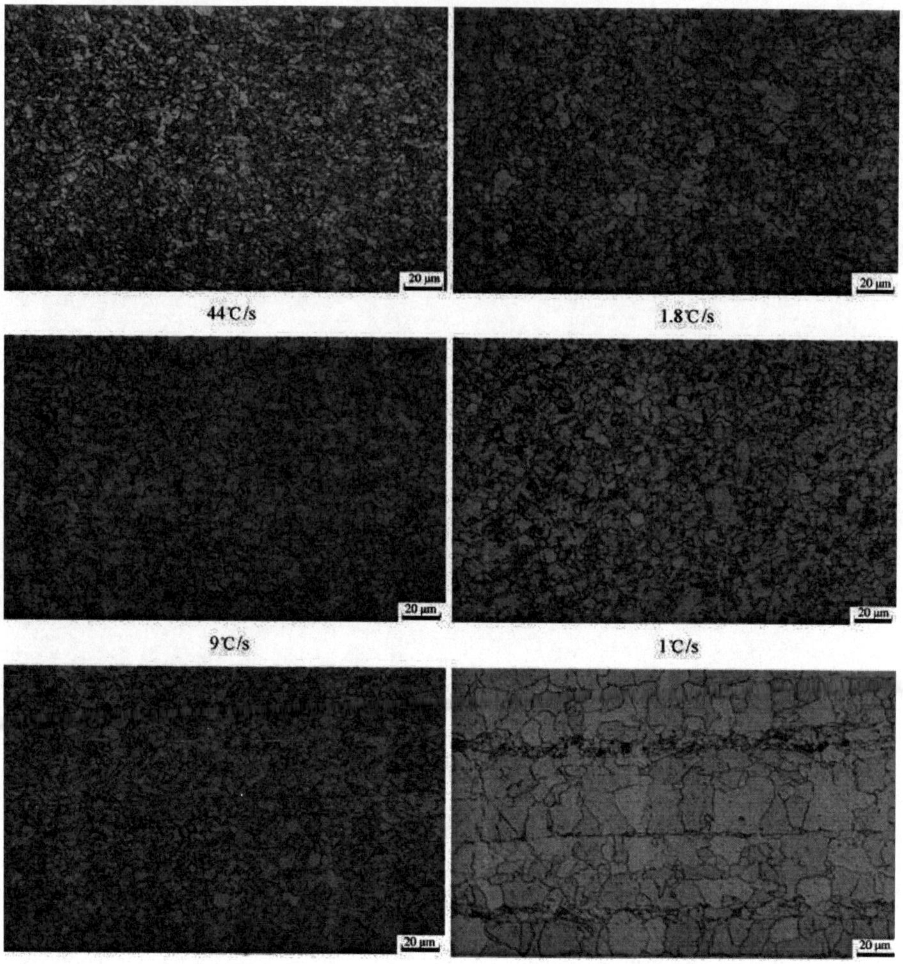

Fig. 19 Micrographs of resulted microstructures of phase transformation at various cooling rate of steel 05Mn2MoNbTiB

While in the case of B-free steel, there is no such phenomenon.

5. It is common for this type of steels that in both B free and B containing steels at certain cooling rates martensite islands is prone to occur. In the case of B free steel, martensite islands exist at faster cooling rate than in the case of steel with B. Sub micro-amount B shifts the martensite islands phase presence from higher cooling rates to lower cooling rates (median cooling rates). In the case of as hot rolled high strength strip steel, martensite islands are to be avoid; while for the as annealed high strength hot strip steel, martensite islands are beneficial leading to further increase of strength level and to the improvement of elongation value after annealing.

Acknowledgments The authors of the paper are obliged to CBMM, Brazil for its financial and technical support to our research project.

References

1. H. Tamehiro, Trans. ISIJ **27**(2), 120 (1987)
2. H. Tamehiro, Trans. ISIJ **27**(2), 130 (1987)
3. H. Tamehiro, Trans. ISIJ **25**(1), 54 (1985)
4. T. Hara, H. Asahi, R. Uemori, H. Tamehiro, ISIJ Int. **44**(8), 1431 (2004)
5. H. Mohrbacher, in *Proceedings of International Seminar on Applications of Mo in Steels*, Beijing, 2010, pp. 74–95
6. CISRI Internal Research Progress Report No. 1 (2008)
7. M. Klein et al., in *Microalloying for New Steel Processes and Applications, Conference Proceedings*, ed. J.M. Rodriguez-Ibabe et al., Trans Tech Publications Ltd., 2005, pp. 543–550

Mechanical Properties and Microstructure of X80 Hot-Rolled Steel Strip for the Second West-East Gas Pipeline

Junhua Kong, Lin Zheng, Lixin Wu, Xiaoguo Liu, and Liwei Li

Abstract

Increasing the steel grade and gas pressure in pipeline project means the crack arrest toughness for pipe needs to enhance greatly. In this paper, effect factors of X80 toughness were studied. Through commercial production of X80 steel, it was found that the content of carbon or sulphur influenced impact energy obviously. Large numbers of statistical data showed that impact energy would well satisfy the requirement of X80 hot-rolled steel strip for the Second West -East Gas Pipeline if [C] was between 0.035 and 0.065 wt.% and [S] below 30 ppm. Through EBSD analysis, it was concluded that the effective grain size was smaller, toughness would be better. Fine and equably dispersed M-A structure would improve the strength, while structure with strip or sharp angel would deteriorate the toughness of X80 steel. Development of X80 strips for Chinese first full-scale burst test was also introduced in the paper.

Keywords

Pipeline • X80 • Chemical composition • Microstructure • Toughness

1 Introduction

Transporting oil and gas with long-distance pipeline is more economic, safe and efficient than any other means. Increasing the steel grade and gas pressure will enhance transportation efficiency and reduce construction cost of pipeline project.

Since the first application of X80 steel in Germany in 1985 [1], more and more efforts were concentrated on the research of its weldability, strength and toughness, crack control behavior over the past twenty years.

In general, increasing the steel grade and gas pressure in pipeline project means the crack arrest toughness for pipe needs to enhance greatly. In all strengthening methods, only fine grained one does good to strength and toughness compared with the others. On the other side, as the wall thickness of steel increases, it is more difficult to get fine grain and good toughness.

X80 steel has the properties of high strength, good toughness, weldability and corrosion resistance, its study and production represented the general and technology level of an iron & steel enterprise.

WISCO has supplied more than 4,50,000 tons of X80 hot-rolled steel strips for the Second West-East Gas Pipeline during the past two years. Through constant producing and improving, especially in the production of X80 steel for the Full-scale Burst Test, what to influence and how to control toughness were summarized and well understood. This experience would give some reference for the research and production of high toughness pipeline steel.

2 Specification of X80 Hot-Rolled Strip for the Second West-East Gas Pipeline

Pipe diameter for the Second West-East Gas Pipeline was 1219 mm, and gas pressure 12 MPa. SSAW pipe took up 56% and LSAW 44% in the demand of the project.

J. Kong (✉) · L. Zheng · L. Wu · X. Liu · L. Li
Research and Development Centre of WISCO, Wuhan, 430080, China
e-mail: junhuakong@yahoo.com.cn

Y. Weng et al. (eds.), *Advanced Steels*, DOI: 10.1007/978-3-642-17665-4_34,

Table 1 Requirements of chemical composition of X80 steel for the Second West-East Gas Pipeline(wt.%)

Elements	C	Si	Mn	P	S	Nb	V	Ti	Cu	Mo	Ni	Cr	Pcm
Requirement for heat	≤ 0.08	≤ 0.40	≤ 1.85	≤ 0.020	≤ 0.004	≤ 0.11	≤ 0.06	≤ 0.025	≤ 0.30	≤ 0.35	≤ 0.45	≤ 0.40	≤ 0.23
Requirement for product	≤ 0.09	≤ 0.42	≤ 1.85	≤ 0.022	≤ 0.005	≤ 0.11	≤ 0.06	≤ 0.025	≤ 0.30	≤ 0.35	≤ 0.50	≤ 0.45	≤ 0.23

Table 2 Requirements of tensile properties of X80 SSAW pipe body for the Second West-East Gas Pipeline

Yield strength	Tensile strength	Yield ratio	Elongation
Rt0.5/MPa	Rm/MPa	Rt0.5/Rm	A/%
555 ~ 690	625 ~ 827	≤0.92 Permission 5% ≤ 0.93	As API SPEC 5L

Note Properties of the strips must assure SSAW pipe conforming to the specification

The requirement of chemical composition and mechanical properties for X80 strips were stated as below.

(1) Chemical composition (Table 1)

(2) Tensile properties (Table 2)

(3) Fracture toughness (Tables 3, 4)

Toughness was the key factor to influence crack and fracture of pipeline. Enough toughness could prevent crack initiate and propagation. In order to guarantee the safety and reliability, toughness must be enhanced corresponding with the increase of the strength. That was to mean minimum toughness of the steel must be higher than crack arrest toughness requirement.

3 Actual Properties of X80 Hot-Rolled Strips for the Second West-East Gas Pipeline

3.1 Statistical Result of [S] Content of X80 Steel

Statistical data of 352 heats of X80 steel showed the average sulphur content was 8 ppm.

Table 3 Requirements of Charpy Impact Test of X80 strips for the Second West-East Gas Pipeline (sample size: 10 × 10 × 55 mm)

Test temp.	SA/%		KV_2/J	
	Single	Average	Single	Average
−20°C	≥80	≥90	≥180	≥240

Table 4 Requirements of DWTT of X80 strips for the Second West-East Gas Pipeline

Test temp.	Single	Average
−15°C	≥70%	≥85%

3.2 Mechanical Properties of Commercial Produced Strips

During commercial production, in order to enhance yield strength, the contents of main alloy elements and rolling process modified and optimized, especially finish rolling temperature and pass reduction, and the proportion of different phase in the microstructure changed.

Figures 1, 2, 3 were the statistical results of tensile properties of the strips sampling from 30 degree to the

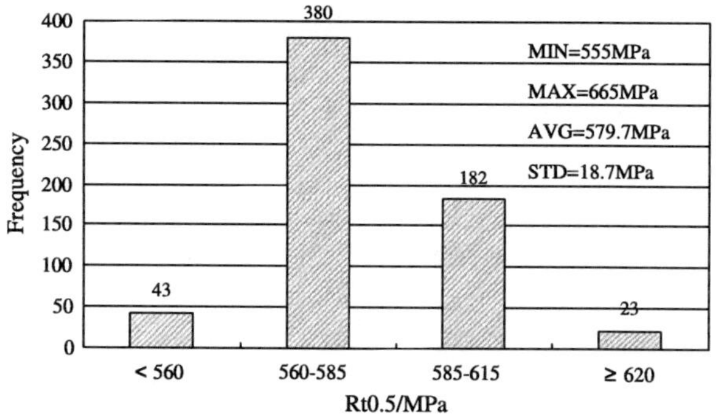

Fig. 1 Histogram of Rt0.5 of X80 strips with thickness 18.4 mm in WISCO (n = 628)

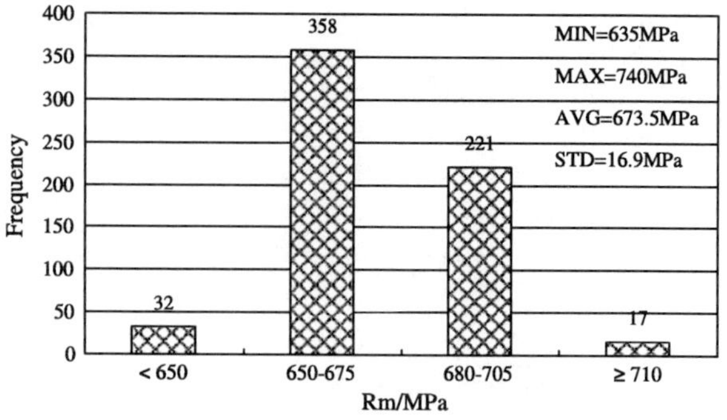

Fig. 2 Histogram of Rm of X80 strips with thickness 18.4 mm in WISCO (n = 628)

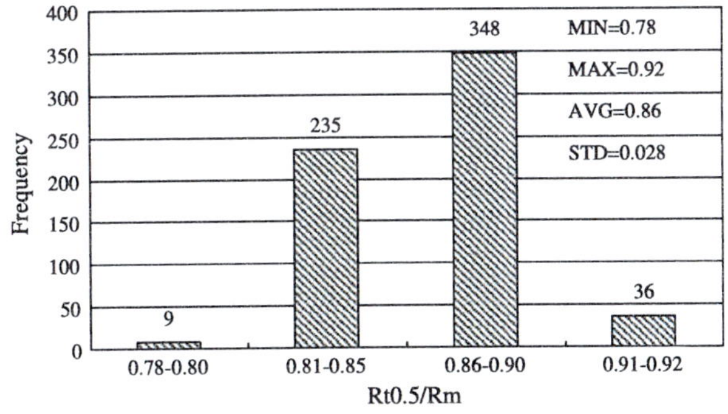

Fig. 3 Histogram of Rt0.5/Rm of X80 strips with thickness 18.4 mm in WISCO (n = 628)

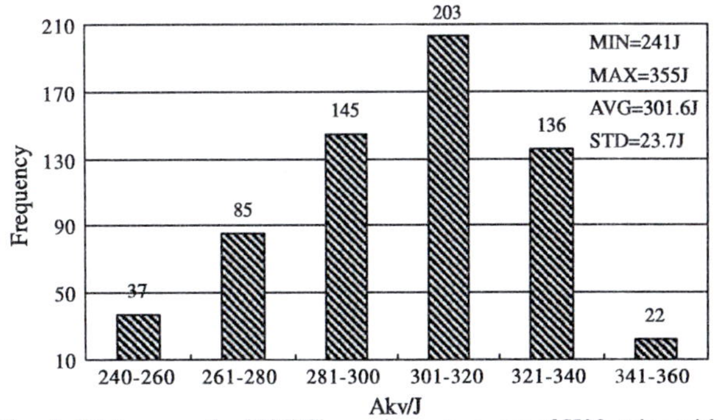

Fig. 4 Histogram of −20°C Charpy impact energy of X80 strips with thickness 18.4 mm (n = 628)

rolling direction with thickness 18.4 mm. From the results it could be seen, the average yield strength was 580 MPa, there was 25 MPa margin compared with the specification. The average tensile strength was 674 MPa, and the yield ratio was 0.86. All the properties could well satisfy the requirements.

Figure 4 was the result of Charpy impact energy tested at −20°C. The average energy was 301.6 J.

4 Analysis of Key Factors Influencing Toughness of X80 Strips for the Second West-East Gas Pipeline

Chemical composition and microstructure were the key factors influencing toughness. Low carbon, high Mn microalloyed steel with proper content of Cu, Mo, Ni was adopted. The composition could ensure obtaining acicular ferrite in a wide range of cooling rate.

4.1 Composition

Through 3102 samples analysis of the relationship between composition and impact energy, carbon and sulphur contents were found effecting toughness obviously.

4.1.1 Carbon

As the carbon content increased in pipeline steel, toughness reduced and weldability deteriorated [2, 3]. On the other hand, the ability to resist HIC and SSC dropped. So carbon content usually went down as the steel grade increased in pipeline steel.

Figure 5 showed in different surphur content steels, the impact energy tended to decrease as carbon content increased.

Figure 6 showed linear correlation between carbon content and average impact energy of X80 strips. As carbon content increased, average energy decreased directly.

From the curve we could conclude, if carbon content was more than 0.075 wt.%, impact energy would be difficult to achieve the minimum requirement of 240 J.

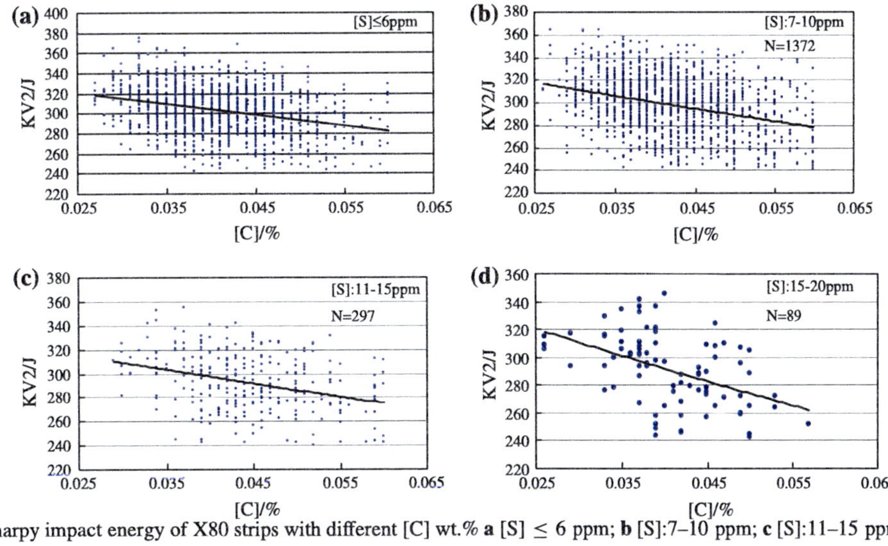

Fig. 5 −20°C Charpy impact energy of X80 strips with different [C] wt.% **a** [S] ≤ 6 ppm; **b** [S]:7–10 ppm; **c** [S]:11–15 ppm; **d** [S]:16–20 ppm

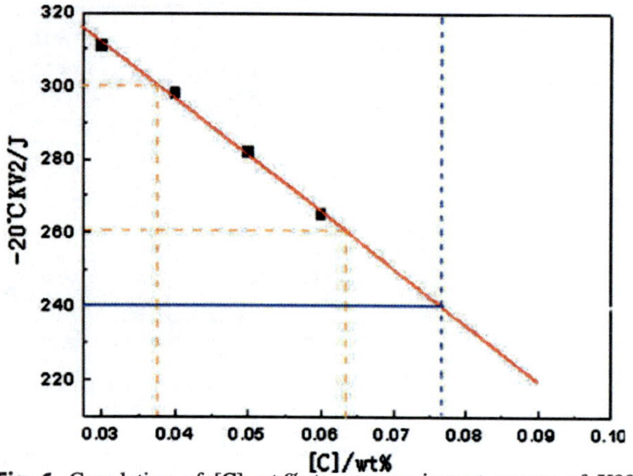

Fig. 6 Correlation of [C] wt.% to average impact energy of X80 strips

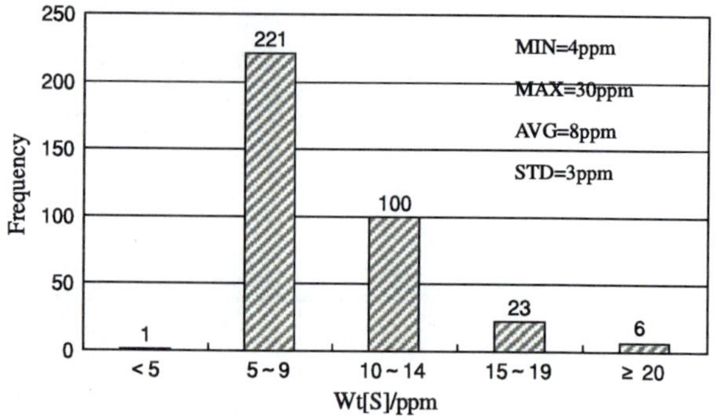

Fig. 7 Histogram of [S] content of X80 steel in WISCO(n = 352)

It was better and economic to control carbon content between 0.035 and 0.065 wt.%. Continuous reduction of carbon content would lead to the increase of steel-making cost.

4.1.2 Sulphur

Sulphur could combine with Mn and become long-bar MnS inclusion. The presence of the inclusion would deteriorate transverse impact energy due to its ductile property along rolling direction. Reduction of sulphur content would improve toughness greatly [4]. Proper calcium treatment process was adopted to control sulfide inclusion shape and enhance transverse fracture toughness.

Because the shape and distribution of nonmetallic inclusion affected toughness directly, sulphur content was restricted severely, and inclusion level was also restricted. WISCO constantly developed clean steel making technology during X80 production for the Second West-East Gas Pipeline. Through controlling sulphur content from molten iron and scrap steel to refining, the average [S] could be reduced to 8 ppm for commercial production (Fig. 7).

Figure 8 showed in different carbon contents, the impact energy tended to decrease as surphur content increased, but the increment was not so big.

From Fig. 9 we could conclude that in order to ensure the impact energy satisfy the specification of X80 steel sulphur content would be better below 30 ppm.

4.2 Microstructure

Acicular ferrite combined with polygonal ferrite was the typical microstructure of X80 steel. As increase of cooling rate M-A structure formed. The size and volume percentage of M-A would affect the strength and toughness of the steel [5].

Fig. 8 −20°C Charpy impact energy of X80 strips with different [S] wt.%
a [C]:0.025–0.034%;
b [C]:0.035–0.044%;
c [C]:0.045–0.054

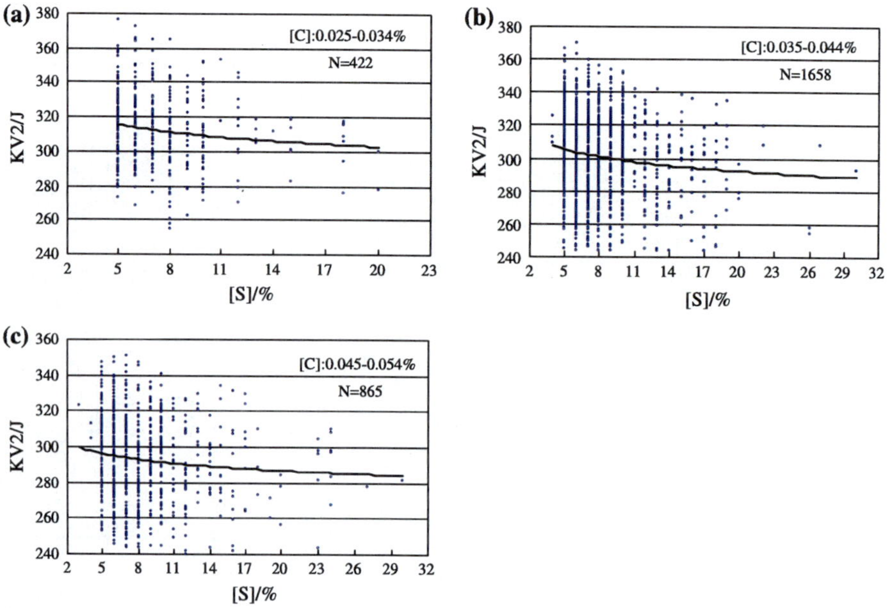

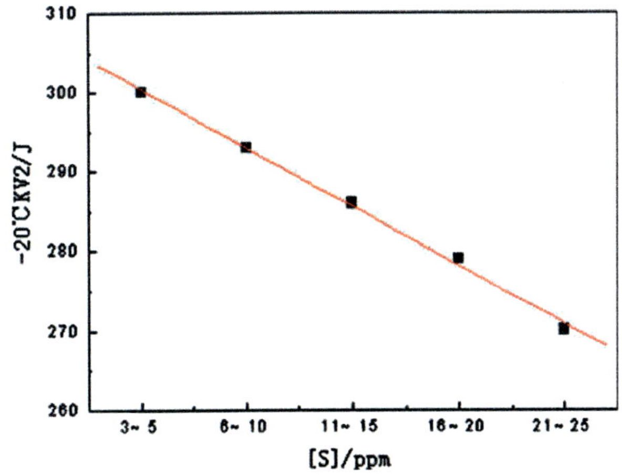

Fig. 9 Correlation of [S] wt.% to average impact energy of X80 strips

Acicular ferrite transformed at moderate temperature. It was obtained through special chemical composition, large transformed cumulation and controlled cooling. The steel with AF microstructure had better weldability, HIC resistance and perfect toughness than ferrite and pearlite steel. It could fully satisfy the crack arrest requirement for high pressure transporting gas pipeline.

4.2.1 Influence of Effective Grain Size

The effective grain of acicular ferrite was ferrite laths. It was known that finer lath, higher strength and toughness. Because in acicular ferrite, crack propagation was prevented strongly by the interwoven nonparallel ferrite laths, strength and toughness of the steel were enhanced effectively [6, 7].

Figure 10 and Table 5 showed that yield strength and impact energy both enhanced as the effective grain size of X80 strips decreased.

4.2.2 Influence of M-A Structure

Research indicated that crack usually turning and changing direction while encountering with M-A structure. M-A held up crack propagation intensively with its residual austenite reduced the tip stress of crack and consumed part of spread power.

The shape and distribution of M-A structure affected toughness. Fine and equably dispersed M-A would improve the strength, while structure with strip or sharp angel would deteriorate the toughness of X80 steel obviously.

Figure 11 and Table 6 showed that abundant presence of strip and sharp angle M-A would deteriorate toughness remarkably.

Fig. 10 Analysis of X80 strip in EBSD

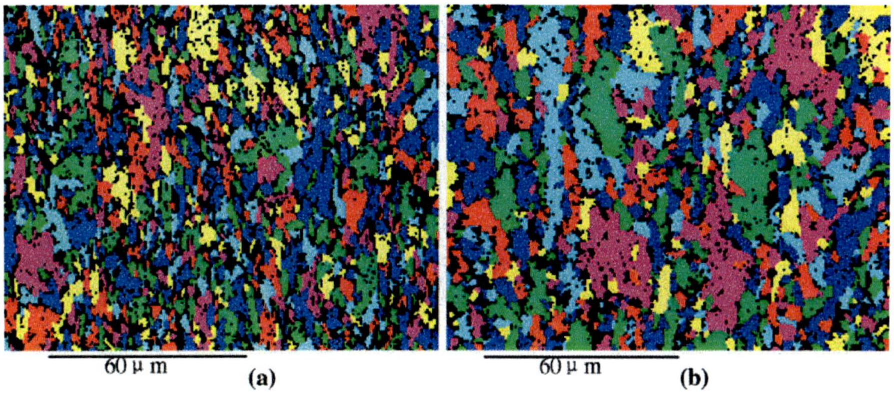

Table 5 Effective grain size of X80 strips and corresponding mechanical properties

Fig. no.	Effective grain size/μm	−20°C Charpy impact test		Tensile test			
		Single KV_2/J	Average KV_2/J	Rt0.5/MPa	Rm/MPa	Rt0.5/Rm	A_{50mm}/%
Figure 10a	3.27	304,297,300	300	575	695	0.83	41
Figure 10b	4.35	242,250,246	246	550	705	0.78	40

Table 6 M-A structures and the corresponding mechanical properties of X80 strips

Fig. no.	−20°C Charpy impact test		Tensile test			
	Single KV_2/J	Average KV_2/J	Rt0.5/MPa	Rm/MPa	Rt0.5/Rm	A_{50mm}/%
Figure 11a	295,285,278	286	595	685	0.87	41
Figure 11b	253,181,146	193	580	675	0.86	39

Table 7 Target toughness of X80 SSAW pipes (Test Temp.: + 10°C)

Target KV_2/J					
190	220	230	240	245	260

X80 specification of the Second West -East Gas Pipeline except impact toughness

(2) *Target toughness* The special target toughness of X80 SSAW pipes was asked as Table 7.

5 Development of X80 Strips for Chinese First Full-Scale Burst Test

WISCO began to develop X80 strips needed for Chinese first full-scale burst test in May, 2009. Good results were got after several trial production and modification.

5.1 Requirements for the Material

(1) *General requirement* Properties of the coils and SSAW pipes for full-scale burst test should satisfy with the

5.2 Results

Through modifying chemical composition and production process, WISCO supplied seven coils for the test. Seven SSAW pipes were selected after pipe making and testing.

Table 8 indicated that difference between actual impact energies of selected pipes and target toughness was no more than 10 J. WISCO supplied all the strips needed for Chinese first X80 full-scale burst test SSAW pipes.

Figure 12 was the photos taken from Chinese first X80 full-scale burst test in Italy in March this year.

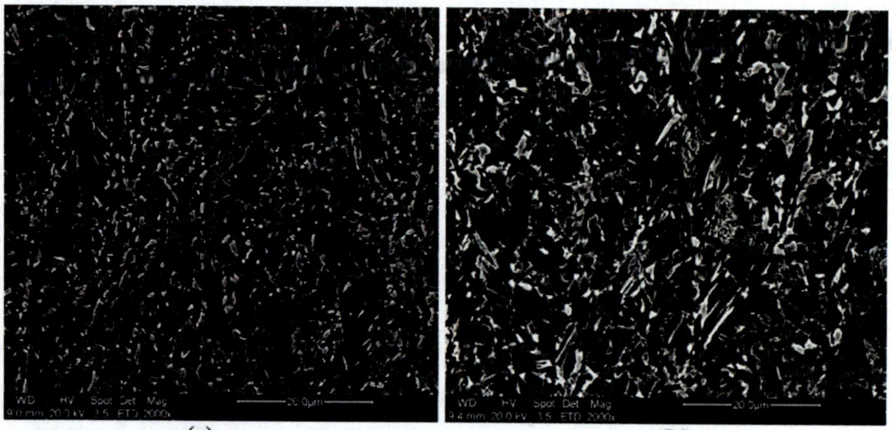

(a) (b)

Fig. 11 SEM micrograph of X80 strips for the Second West -East Gas Pipeline

(a) (b)

Fig. 12 X80 full-scale burst test (**a**) Explosion scene (**b**) Fracture stopped place of SSAW pipes

Table 8 Selected SSAW pipes for Chinese first X80 full-scale burst test

Coil no.	Selected pipe no.	10°C target KV_2/J	10°C actual KV_2/J
92068046	331414	245	254
92048579	329116	280	280
92065370	381561	260	266
92065371	381564	245	253
92061332	381568	230	229
	381570	220	211
92061331	381572	190	198

6 Conclusion

(1) Large numbers of statistical data showed that impact energy would well satisfy the requirement of X80 hot-rolled steel strip for the Second West -East Gas Pipeline if [C] was between 0.035 and 0.065 wt.% and [S] below 30 ppm.

(2) Finer effective grain size, better toughness and higher yield strength. Fine and equably dispersed M-A structure would improve the strength, while structure with strip or sharp angel would deteriorate the toughness of X80 steel obviously.

(3) Through proper chemical composition and production process, target toughness of strips for X80 full-scale burst test could be got.

Acknowledgements Authors are grateful for Dr. Liu Jibin, Zhou Guifeng, Mr. Guo Bin, Mr. Liu Changming, Mr. Zheng Hua's corporation in the experiments. Authors also appreciate the support of Ma Jingbin from Baoji Pipe Corp. and Huo Chunyong, Ji Lingkang, Liyang etc. from TGRC.

References

1. M.K. Graf, H.G. Hillenbrand, K.A. Niederhoff, Production of large-diameter line pipe and bends for the world's first long-range pipeline in grade X80(GRS 550). In: Proceedings of 9th Biennial Joint Technical Meeting on Line Pipe, Houston (1993)
2. L. Meyer, H. de Boer, *Welding of HSLA Structural Steel* (ASM, Metals Park, Ohio, 1978), pp. 42–62
3. A.K. Lis, J. Lis, L. Jeziorski, Advanced ultra-low carbon bainitic steels with high toughness. J. Mater. Proc. Technol. **64**, 255–266 (1997)
4. Yu. Dingfu, Development of low sulphur steel. Iron Steel **22**(11), 52 (1987)
5. I.A. Yakubtsov, J.D. Boyd, Bainite transformation during continuous cooling of low carbon microalloyed steel. Mater. Sci. Technol. **17**, 296–301 (2001)
6. D.V. Edmonds, R.C. Cochrane, Structure property relationships in banitic steels [J]. Metall. Trans. A **21 A**, 1527–1540 (1990)
7. ZHAO Ming-chun, SHAN Yi-yin, XIAO Fu-ren et al., Study on formation and strength and toughness behavior of acicular ferrite in a pipeline steel. Mater. Sci. Technol. **9**(4):356–358 (2001)

Refinement of Prior Austenite Grain in Advanced Pipeline Steel

Chengjia Shang and Chengliang Miao

Abstract

A series of fundamental and applied investigations were carried out to develop high grade pipeline steel with high Mn high Nb design, and it mostly focused on that the static and dynamic recrystallization behaviors of high Mn high Nb pipeline steel. Various experimental methods were adopted, which include stress relaxation tests, physical metallurgical modeling analysis, the etching of prior austenite grains and TEM observation of precipitates. According to the results, new control rolling technology is bring forward under high Mn high Nb approach, as a consequence, fine and homogeneous prior austenite grains can be generated by complete static recrystallization prior to finish rolling, and the coarsening can be suppressed largely by drag effect, through proper finish rolling process, prior austenite grain will be flatten fully and uniformly to higher flow stress or S_v, and the mixed grains structure caused by partial dynamic recrystallization may be avoided. The physical metallurgy principle for refinement of prior austenite grain through rough rolling and pancake through finish rolling can be adopted for refinement the austenite grain in high Mn high Nb structural steel.

Keywords

Structural steel • High Mn high Nb • Static recrystallization • Dynamic recrystallization • Grain refinement • Strain accumulation

1 Introduction

High Mn-high Nb approach for X80 pipeline is being applied in Chinese pipeline projects. The diameter of both spiral and UOE/JCOE pipe is 1219 mm with thickness of 18.4 and 22 mm, respectively [1]. In order to meet the requirement of high performance, various metallurgical phenomena of high Mn-high Nb pipeline steel need to be concerned during hot rolling and accelerated cooling. No matter what rolling process is employed, grain refinement and strain accumulation are the keys for good properties of steel. In general, it is influenced by recovery of dislocation, static and dynamic recrystallization of austenite grain and precipitates. However, how to refine austenite grain and accumulate strain effectively are still not so clear under high Mn-high Nb approach. Many literatures [2] had proved that the increasing of Nb content in steel is the most effective method to retard SRX compared with using other micro-alloyed elements. Based on this characteristic, high temperature processing (HTP) was applied in the production pipeline steel, and it is suggested to non-recrystallization deformation that 50–100°C or more need increased for start temperature of finishing rolling [3, 4]. High temperature rolling is more efficient in industry, on the contrast, faster recovery of dislocation and grain growth at higher temperature in fact has not obvious benefit for higher S_v level. Moreover, besides the effect of Nb, SRX behavior always

National Basic Research Program of China (973 Program) (No. 2010 CB19630801).

C. Shang (✉) · C. Miao
School of Materials Science and Engineering, USTB-CBMM-CITIC Nb-bearing Steel Research Lab, University of Science and Technology Beijing, Beijing 100083, China
e-mail: cjshang@ustb.edu.cn

Y. Weng et al. (eds.), *Advanced Steels*, DOI: 10.1007/978-3-642-17665-4_35,

Table 1 Chemical compositions of experimental steels

	C	Mn	Nb	Si	N (ppm)	Ni + Ti + Cu + Cr
s1	0.04	1.71	0.012	–	≤ 40	–
s2	0.04	1.76	0.063	–	≤ 40	–
s3	0.04	1.72	0.10	–	≤ 40	–
Industrial high Mn	0.04	1.75	0.095	0.22	≤ 40	<1.0 wt%

depends on many factors [5, 6] in the complex rolling conditions, e.g. precipitation, initial grain size, reduction, interval time, etc. Finer initial grain or larger reduction can accelerate recrystallization, and the addition of high Mn [7] can retard precipitation, as well as, it can influence recrystallization behavior of high Nb steel, moreover, dragging force (compared with precipitation) would not be strong enough to retard recovery and recrystallization [7]. Therefore, non-recrystallization temperature range of high Mn-high Nb system should be reconsidered. The other important point during hot rolling is strain accumulation or high S_v. For high Nb steel, the recovery between passes would be retarded due to drag effect of soluble Nb, as a consequence, accumulated strain will be higher, and it would increase the possibility of DRX and meta-dynamic recrystallization (MDRX), especially in the finish rolling of hot strip mill (HSM). Therefore, in industrial production, actual static and dynamic recrystallization behaviors in different conditions should be considered as HSM and plate mill (PM), which have different metallurgical characteristics, respectively.

Therefore, what this paper mostly focused are how to refine prior austenite grains further homogenously, and how to obtain uniform pancake grains and larger strain accumulation.

2 Static Recrystallization and Grain Refinement

2.1 Mechanism of SRX and Grain Refinement in High Mn High Nb Steel

Four kinds of low carbon steels were employed for the stress relaxation tests, as shown in Table 1, s1, s2 and s3 are laboratory steels with different Nb content from 0.012 to 0.1 wt%, which were used to investigate the effect of niobium on recrystallization and precipitation behaviors, and the possibility of grain refinement. All thermo-simulation tests were accomplished by Gleeble-simulator, and cylinder specimens of ∅8 mm × 12 mm were prepared for all experiments, and all were treated by solution treatment (1,200°C for 1 h and then quenched) before simulation test. Figure 1 presents the processes of stress relaxation test, after 25% reduction with 1 s^{-1} strain rate at different temperature, the specimens were held for 400 s in the case of strain as a constant, and the stress-time curves were recorded. Actually, stress relaxation method [8] can supply enough data to analyze the whole softening behavior in discretional given condition, and it can be used to characterize recovery or recrystallization behaviors, solute dragging and precipitation pinning. It is considered that more exact results for isothermal recrystallized kinetics can be obtained by stress relaxation method [9]. Moreover, the specimens were quenched at different time during holding period to examine the evolution of prior austenite grains and the precipitates. All morphology of austenite grains in different cases was etched by special etchant at 55–60°C, as shown in Table 2, and film specimens obtained from carbon replica were observed by TEM to examine the conditions of precipitates.

Table 2 Method of etching prior-austenite grain boundaries

Etchant	Picric acid + distilled water + detergent + CCl_4 + NaCl
Etching temperature	50–60°C, using constant temperature oven
Time	180 s or so

2.1.1 Morphology of Austenite Grains and Precipitation Behaviors Above 1,000°C

Figure 2 shows the stress relaxation curves of different Nb steels at 1,050 and 1,000°C. According to Fig. 2a, it is not found any precipitation hardening at 1,050°C, and recrystallization softening occurs for all three kinds of steels after deformation, but the onset times of recrystallization are different, 0.10Nb steel has the longest incubation period,

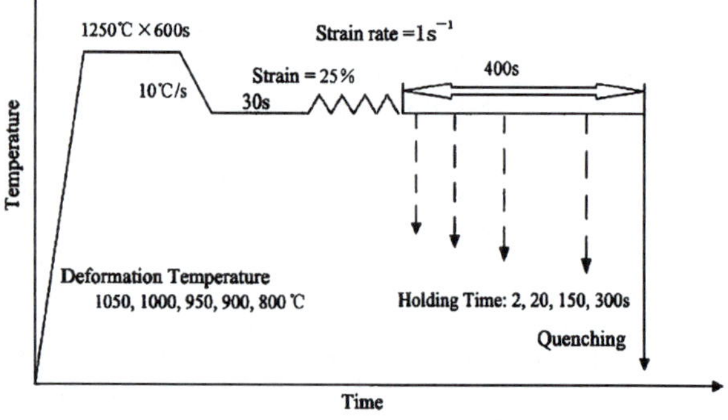

Fig. 1 Schematic diagram of stress relaxation tests

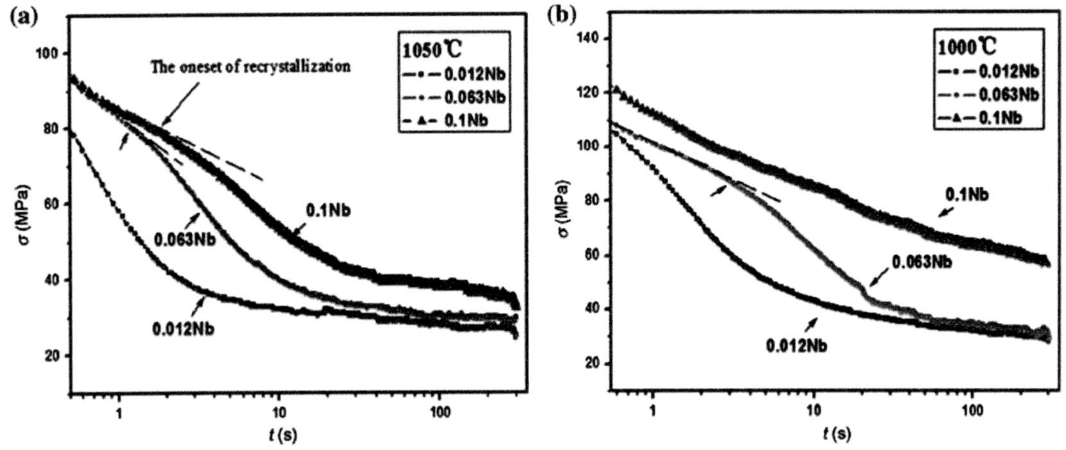

Fig. 2 Stress relaxation curves of various Nb content samples at 1,050°C (**a**) and 1,000°C (**b**)

less than 2 s, but the time for 0.012Nb steel is shorter than 0.5 s, moreover, 0.10Nb steel need longer time to finish whole recrystallization process compared with other two steels. From the morphology of austenite grains in Fig. 3, partial recrystallization occurs in 0.10Nb steel at 2 s, besides, it is evident that most grains in 0.10Nb steel have finish complete recrystallization at 30 s, and it also is coincide with the results of stress relaxation curve at 1,050°C. Compared the morphology of austenite grains of 0.10Nb and 0.063Nb steel at 300 s, it is very obvious that

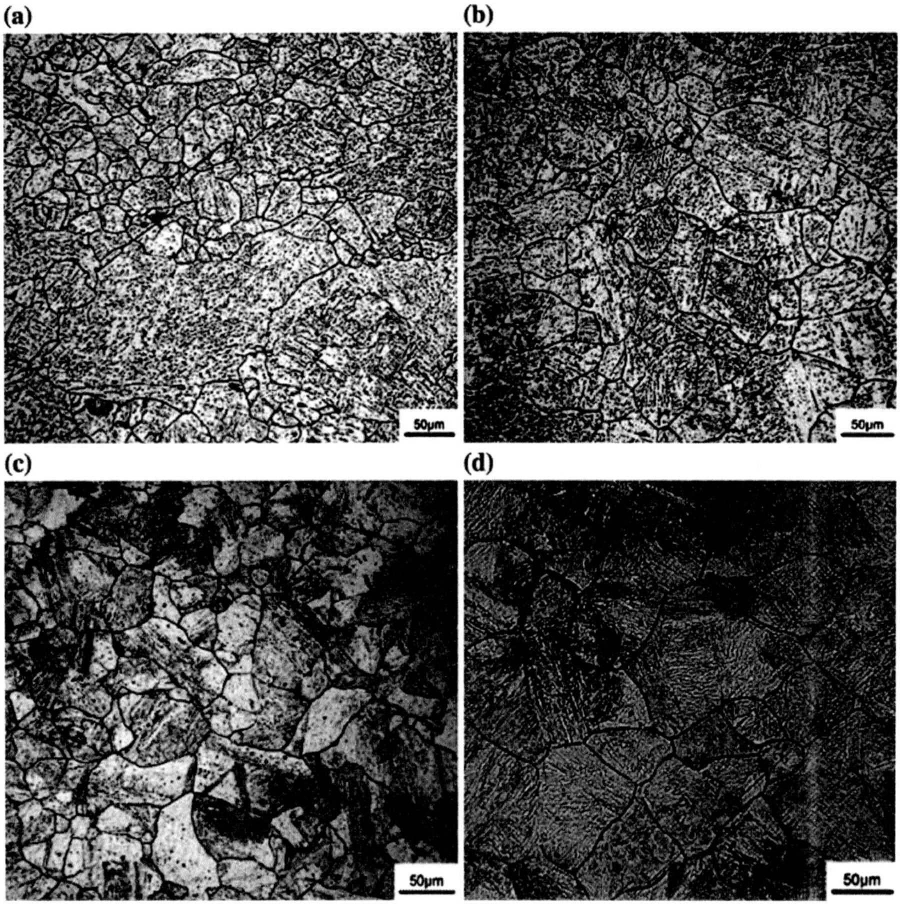

Fig. 3 Morphology of austenite grains compressed at 1,050°C and holding for **a** 2 s, **b** 30 s, **c** 300 s of 0.100Nb steel, and **d** 300 s of 0.063 Nb steel

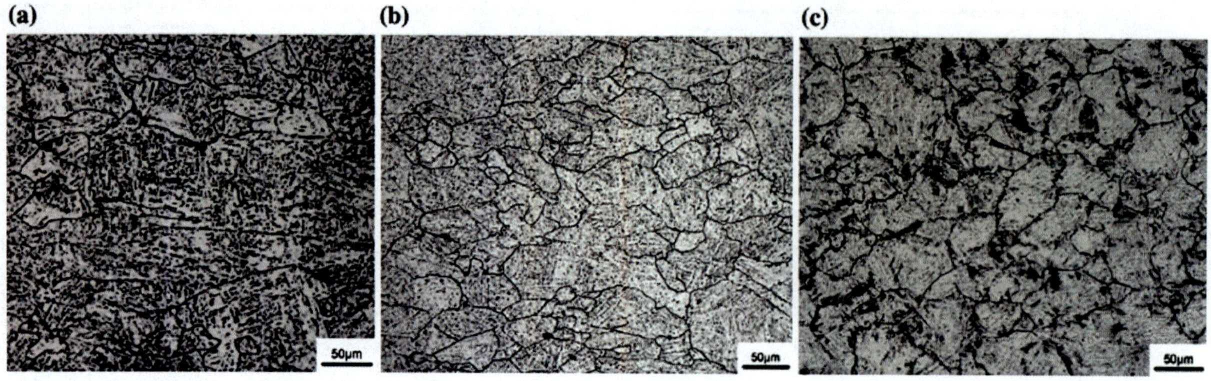

Fig. 4 Morphology of austenite grains after compression at 1,000°C and holding for **a** 30 s, **b** 300 s of 0.10Nb steel and for **c** 300 s of 0.063Nb steel

the coarsening rate of austenite grain is low in higher Nb steel, due to more intense solute drag effect.

Likewise, Fig. 2b presents obvious recrystallization softening for 0.063Nb steel and 0.012Nb steel at 1,000°C, and incubation time of recrystallization increased as the temperature decreasing, but 0.012Nb steel still starts recrystallization before 0.5 s. It is interesting for the curve of 0.10Nb steel at 1,000°C that no obvious accelerated softening by recrystallization, as well as no suppressed softening by precipitation. Actually, from TEM observation on the specimen of 0.10Nb steel after 300 s holding, available strain-induced precipitated particles were not found, combined the etching results of austenite grains shown in Fig. 4, there is a very slow recrystallization process, which also is time dependent, as the case at 1,050°C.

Stress relaxation curves can reflect recrystallization softening and precipitation hardening behavior effectively [10]. Once the recrystallization starts, the softening can be accelerated, presented by the increasing of curve rate, moreover, if most grains finish recrystallization process, the softening rate will be reduced to the minimum. However, slow static recrystallization cannot generate obvious softening acceleration, as the case of 0.10Nb steel at 1,000°C, although the recrystallization happens, the curve rate is not changed obviously compared with the softening behavior by recovery.

2.1.2 Morphology of Austenite Grains and Precipitation Behaviors Below 1,000°C

Know from the conditions above 1,000°C, in high Mn high Nb steel, the strain-induced precipitation plays useful effect on the recovery and recrystallization by degrees with decreasing of temperature. From Fig. 5, the curves of 0.10Nb steel at 950 and 900°C both show the softening rates are reduced by strain-induced precipitation at 5 and 3 s, respectively. And it is also proved by TEM observation that the precipitation is not found in the specimens of 0.10Nb steel at 2 s from 950 to 850°C, and after holding 20 s,

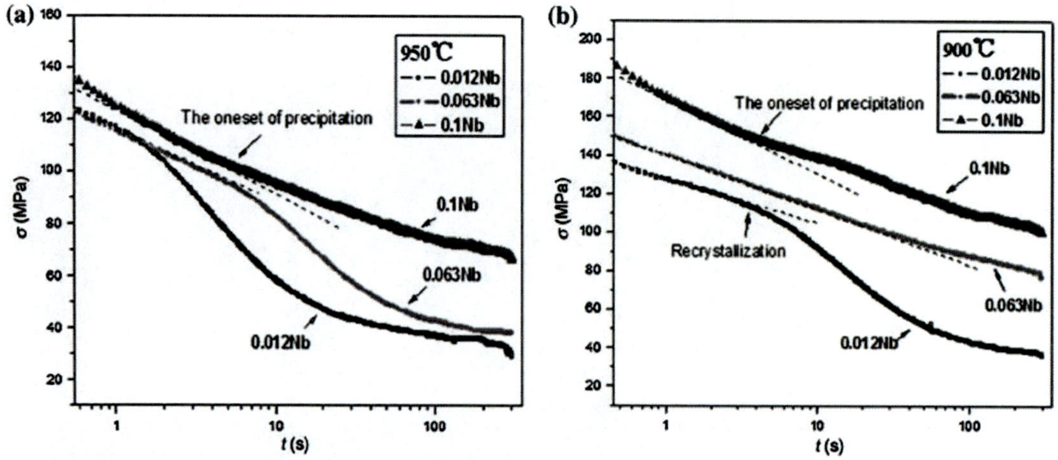

Fig. 5 Stress relaxation curves of various steels at **a** 950°C and **b** 900°C

particles with size ≤50 nm was observed, as shown in Fig. 6a. However, complete static recrystallization still finish within 100 s for 0.012Nb steel above 900°C and 0.063Nb steel at 950°C, and 950°C also is special temperature for 0.063Nb steel, namely, it is not found that obvious precipitation hardening and softening acceleration by recrystallization, like the case of 0.10Nb steel at 1,000°C, moreover, Fig. 6c shows no available precipitation is generated before 150 s for 0.063Nb steel, nevertheless, some precipitates can be found at 300 s for 0.063Nb steel, but actually, these particles have not any effect on the recrystallization behavior that has completed previous to the occurrence of precipitation, and just can influence the growth and coarsening of recrystallized grains.

The etching results of austenite grains, as shown in Fig. 7, indicate partial recrystallization occurs during holding period of 0.1Nb steel at 950°C, but it seems that this recrystallization is not time dependent. Combined with precipitation observation, it can be deduced that precipitation halt recrystallization at 950°C, and the evolution of grains after the occurrence of precipitation is a coarsening process of austenite grains which contain fine recrystallized grains and pancake unrecrystallized grains. In addition, as shown in Fig. 7c, d, it not only causes mixed grain structure, but also more serious grain coarsening compared with the case of 0.063Nb steel at 950°C. Figure 8 illustrates no recrystallization behavior take place within 150 s for 0.1Nb steel at 900°C. So traditionally, 900°C belongs to non-recrystallization temperature of 0.10Nb steel in industrial rolling, which inter-pass time is about 5–100 s, but for 0.063Nb is partial recrystallization temperature, and complete recrystallization temperature for 0.012Nb.

In low Mn Nb-bearing steel, the onset of precipitation is very fast, less than 1 s [7, 11]. Comparing the characteristics of precipitation in high Mn Nb-bearing steel, the addition of high Mn cause the delay of precipitation, and suppressing recrystallization is the main effect of precipitation. Moreover, Fig. 9 shows more and finer precipitates can be found in higher Nb steel, there should be more nucleation sites in high Nb steel. Based on stress relaxation curves, the nose temperature of precipitation for 0.10Nb steel is 900°C, and the time is 3 s or so.

2.2 Recrystallization of Industrial High Mn High Nb Steel

The recrystallization behaviors of a commercial high Nb pipeline steel also were examined. Its chemical compositions are shown in Table 1. The softening behavior was studied by traditional double compression method [12], interval time between passes and deformation temperature were shown in Table 3, 25% reduction and 1 s^{-1} strain rate in each pass were adopted in this experiment. Stress relaxation method also was employed to further research on recovery, recrystallization and precipitation, and its schematic of processes was shown in Fig. 10.

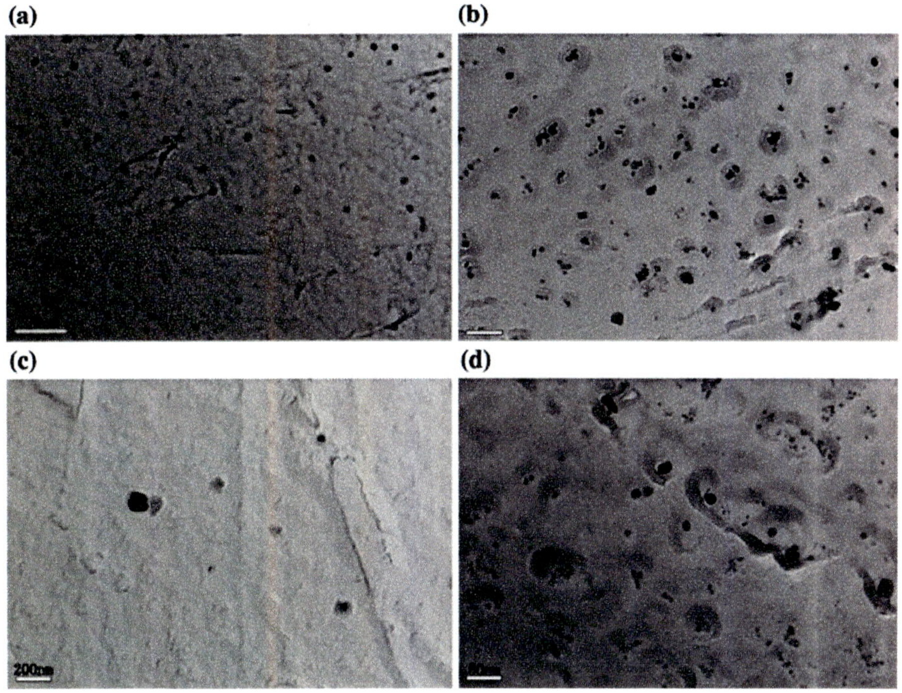

Fig. 6 Morphology of precipitates after compressed at 950°C, holding for 20 s (**a**), 300 s (**b**) of 0.10Nb steel and for 150 s (**c**), 300 s (**d**) of 0.063Nb steel

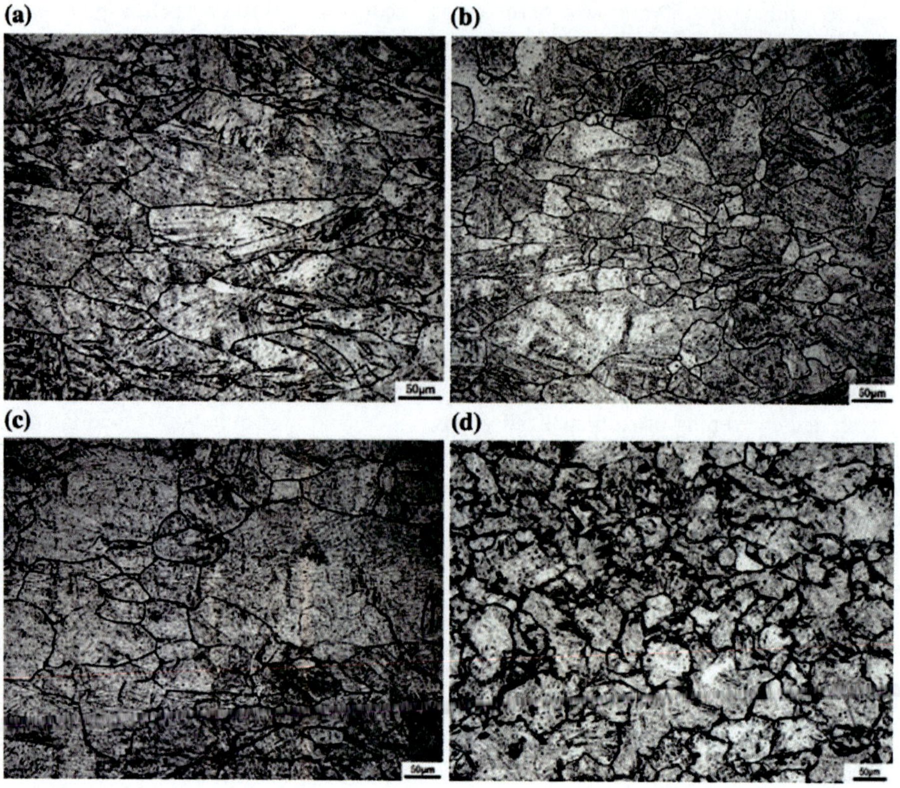

Fig. 7 Morphology of austenite grains after compression and holding at 950°C for **a** 2 s, **b** 20 s, **c** 300 s of 0.10Nb steel and 300 s of 0.063 Nb steel

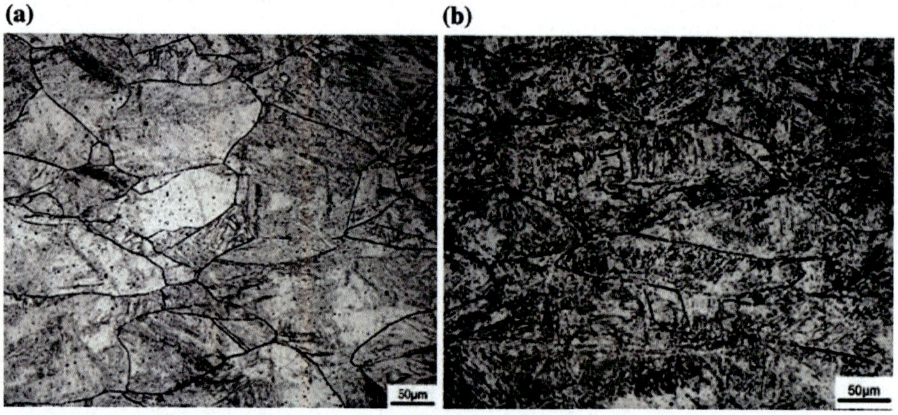

Fig. 8 Morphology of austenite grains after compression and holding at 900°C for **a** 2 s, **b** 150 s of 0.10Nb steel

2.2.1 Softening and Static Recrystallization

Softening fraction-time curves at different temperatures based on double-passes compression are shown in Fig. 11. The softening percentage does not exceed 25% after holding 500 s at 980°C, and after holding 500 s at 1,000°C, recrystallization percentage is only 55%, but holding 10 s at 1,050°C, softening percentage close to 65%. That is to say, austenite grains with 25% prior-deformation can softening completely above 1,050°C within a few seconds, that means complete recrystallization should be take place at 1,050°C. But if temperature decreases below 1,000°C, the softening will be delayed markedly, and the occurrence of softening need long time to incubation. So based on these results, it is no doubt that 1,000°C is contained in the temperature region of non-recrystallization or partial recrystallization from traditional viewpoints.

Stress-relaxation curves of high Mn high Nb steel were measured at different temperatures, and the thermo-simulation process as shown in Fig. 10. As shown in Fig. 12a, b, stress level in the beginning period fall quickly at 1,100, 1,050°C, due to occurrence of recovery and recrystalliztion. When static recrystallization occurs, the relaxation time hardly reached 1 s, and recrystallization stop times also are similar with the results of double compression tests. Once

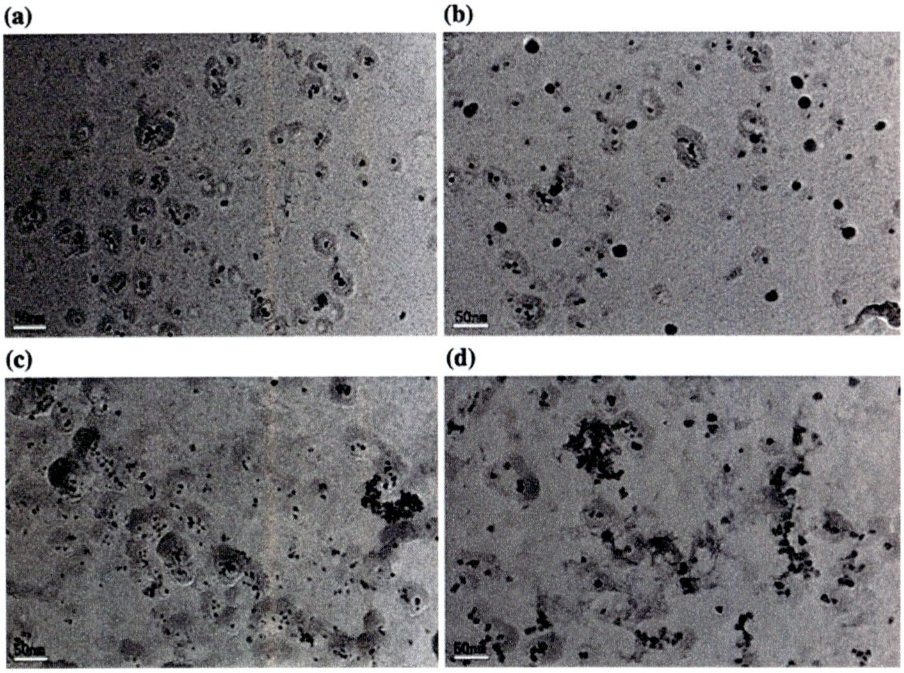

Fig. 9 Morphology of precipitates after compression and holding for 300 s of **a** 0.063% Nb at 850°C, **b** 0.063% Nb at 900°C, **c** 0.1% Nb at 850°C, **d** 0.1% Nb at 900°C

Table 3 Different interval time in the double compression test

Temperature (°C)	Delay time between each pass (s)			
1,050	1	10	40	
1,000	10	40	100	500
980	10	40	100	500

relaxation temperature was decreased to 1,000°C, as shown in Fig. 12c, visible softening behavior caused by recrystallization does not appear within 100 s, and hardening platform resulted from strain-induced precipitation also is not observed clearly in this curve, that is to say, there are not available precipitated particles that can play their effect on dislocation. Figure 12d indicated strain-induced precipitation occurs after 25% prior deformation at 900°C, and hardening behavior happened after relaxing 4 s. Moreover, the sample deformed at 950°C is good example to reveal partial static recrystallization, as shown in Fig. 13, Firstly, the softening is accelerated by partial static recrystallization (sector I in Fig. 13) after deforming at 950°C, but after delaying for several seconds, softening behavior was suppressed (sector II in Fig. 13) and hardening taken place subsequently by occurrence of precipitation. The softening at 950°C is influenced by two kinds of mechanism, softening by partial recrystallization and precipitation hardening. The occurrence of precipitation may be the main reason which may result in partial recrystallization.

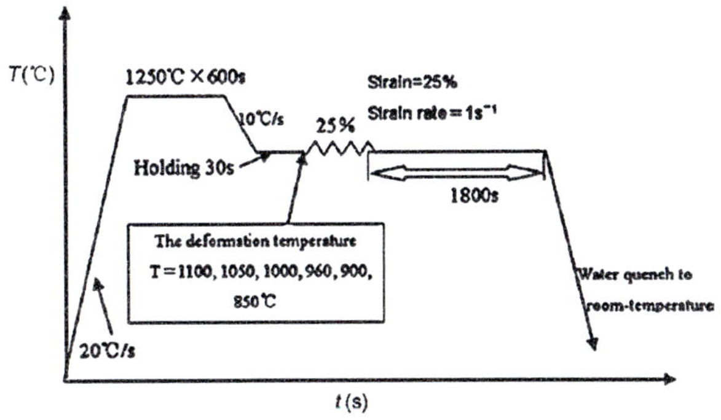

Fig. 10 Thermal simulation technology of stress relaxation tests

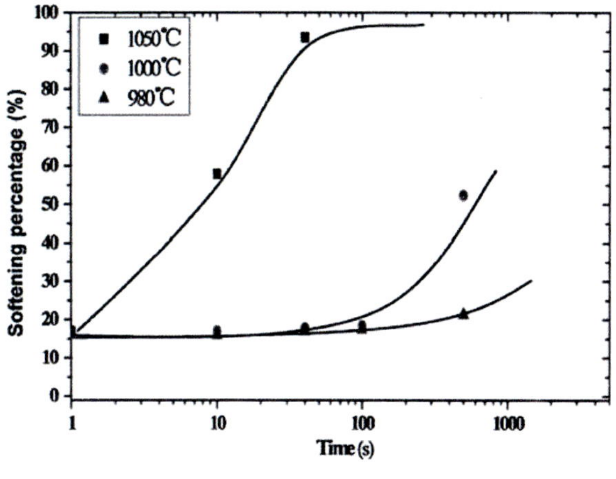

Fig. 11 Softening-temperature–time curves of high Mn-high Nb pipeline steel at various temperatures

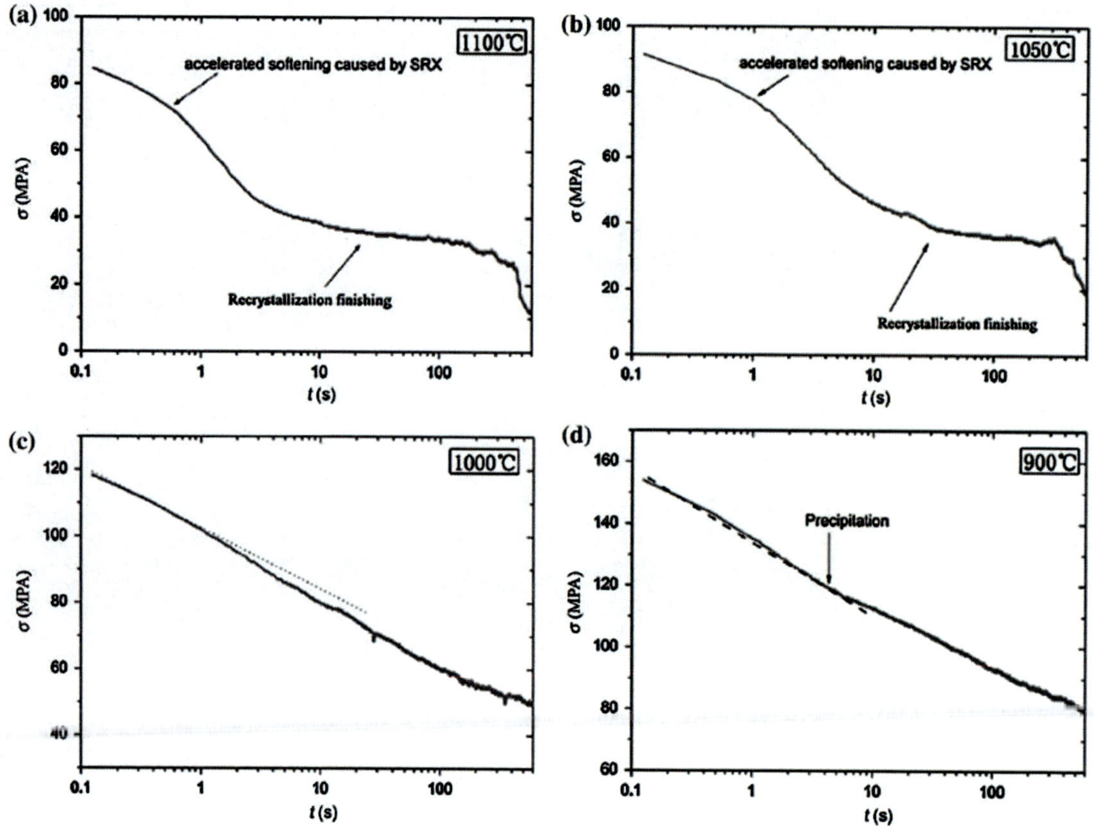

Fig. 12 Stress-relaxation curves of high Nb-bearing steel at **a** 1,100°C, **b** 1,050°C, **c** 1,000°C, **d** 900°C

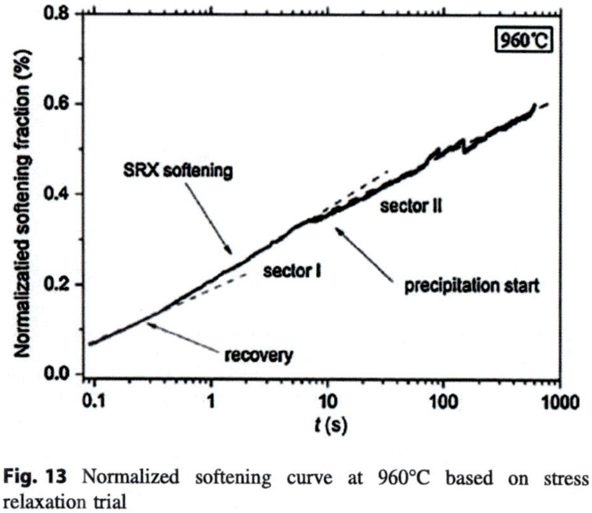

Fig. 13 Normalized softening curve at 960°C based on stress relaxation trial

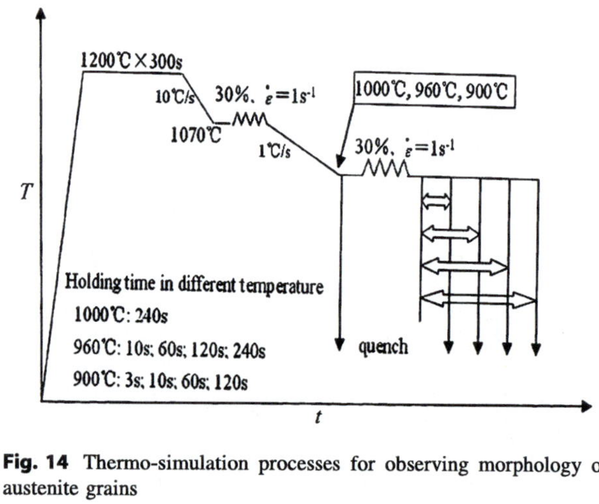

Fig. 14 Thermo-simulation processes for observing morphology of austenite grains

2.2.2 Refining and Flattening Prior-Austenite Grain

Thermo-simulation tests were adopted for industrial high Mn high Nb steel, as shown in Fig. 14, and statistic distributions of grain size of a part of samples were analyzed by software *Image Pro* based on grains database in five different regions of each tested sample.

Figure 15 shows the morphology of static recrystallizated grains after 30% deforming at 1,070°C, it can be seen that average size of completely recrystallized austenite grains is more than 80 μm. The morphology of prior

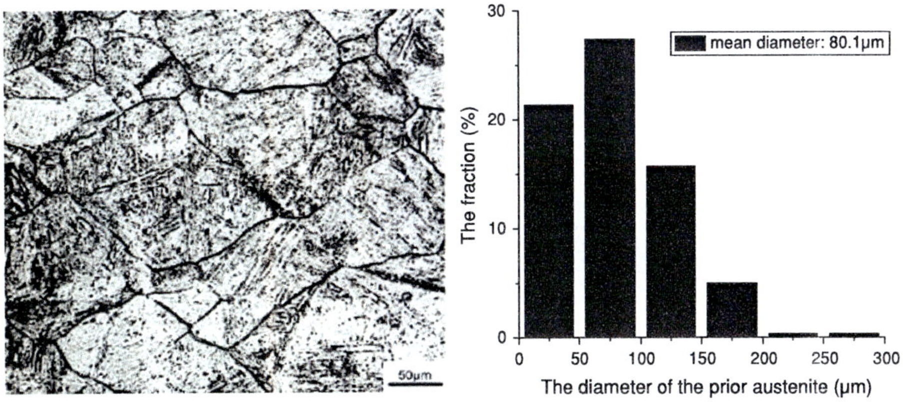

Fig. 15 Morphology of the prior-austenite grain and statistic distribution of grain size after 30% deforming at 1,070°C

austenite grains after deformed and holding for 10, 60 and 240 s are illustrated in Fig. 16, and statistic results of grain size were shown in Fig. 17. The sample holding 10 s (Fig. 16a), austenite grains are homogenous and very fine, the average grain size is only about 19 μm, namely, most austenite grains have finished SRX. Even holding 240 s at 1,000°C, the average grain size is still less than 28 μm. Growth rate of recrystallized grains is relative much slow comparing with of 1,070°C. It can be explained that the slow coarsening rate of recrystallizated grains is resulted from intense drag effect.

Mixed grain structure can be observed in the specimen deformed at 960°C, as shown in Fig. 18a, pancake and recrystallized austenite grains are co-existed. However, at 900°C, there is only flatted austenite grains after deforming in all holding time (Fig. 19), and it means there is not static recrystallization at 900°C within 60 s after deforming.

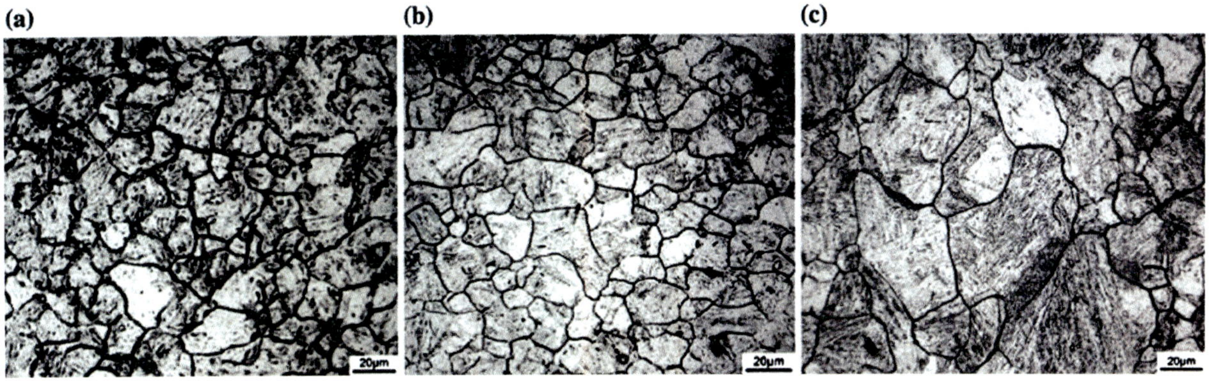

Fig. 16 Stress relaxation curves of high Nb industrial steel after deformed at 1,000°C for 30% and holding for **a** 10 s, **b** 60 s, **c** 240 s

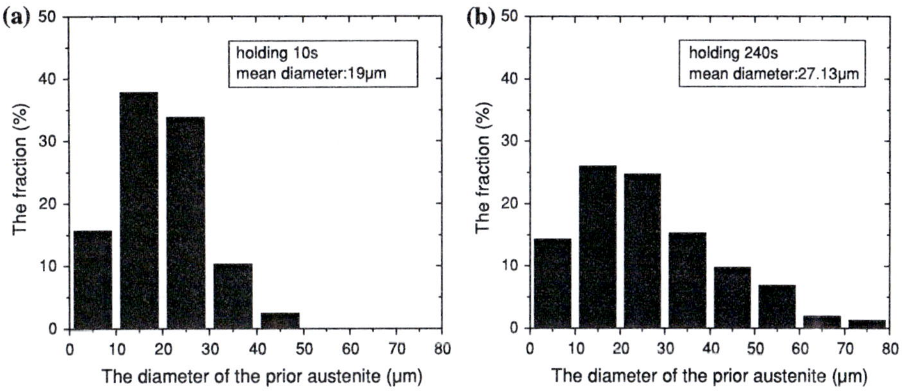

Fig. 17 Distribution of the prior austenite grain size in different conditions, **a** 30% deforming at 1,000°C and holding 10 s, **b** 30% deforming at 1,000°C and holding 240 s

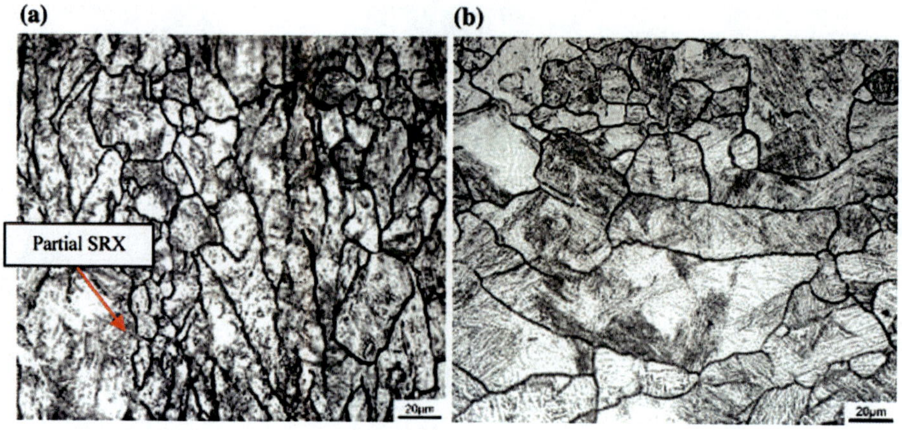

Fig. 18 Morphology of austenite grains after 30% deforming at 960°C and holding for **a** 10 s and **b** 120 s

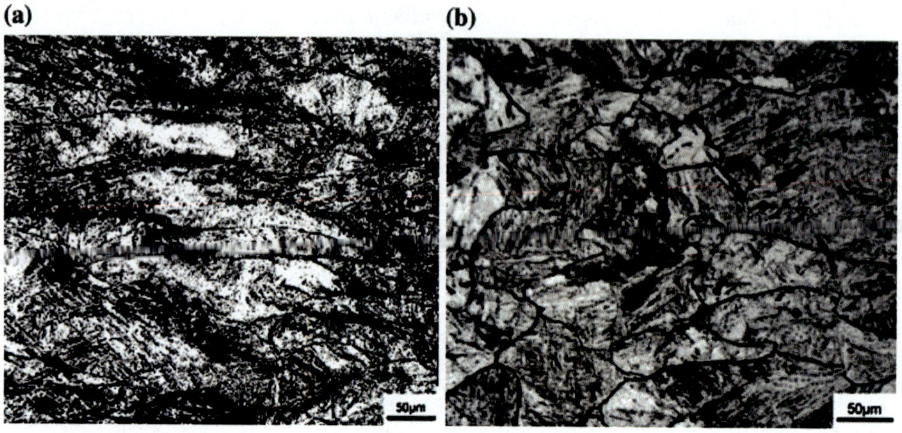

Fig. 19 Morphology of austenite grain after 30% deforming at 900°C and holding for **a** 3 s and **b** 60 s

3 Dynamic Recrystallization and Strain Accumulation

3.1 The Effect Nb on Kinetics of DRX

In the present study, steels of two types were employed too. Steels of the first type were S2 and S3 from Table 1 to singly investigate the effect of Nb on flow stress and DRX, the influence of other micro-alloyed elements can be eliminated. The steels of the second type were obtained from industrial X80 pipeline steels (as shown in Table 4), which are to understand strain accumulation behavior during multi-pass rolling at relative high temperature. Cylinder specimens of ∅8 mm × 10 mm for the research on DRX of single pass were prepared prior to testing in a Gleeble-3500 thermo-mechanical simulator, and this simulated process is shown in Fig. 20. Specimens were reheated to 1,100°C for 120 s and then cooled to the test temperature (900–1,100°C) at 10°C/s. After 30 s of soaking at this temperature, the samples were compressed with 60% deformation at constant strain rates of 0.05–2 s^{-1}, and then the samples were quenched immediately.

Table 4 Chemical compositions of investigated steels

ID	C	Mn	Nb	Si	Mo + Cu + Al + Ti + Ni + Cr
I2	0.044	1.83	0.092	0.18	0.79
I4	0.048	1.81	0.067	0.20	0.77

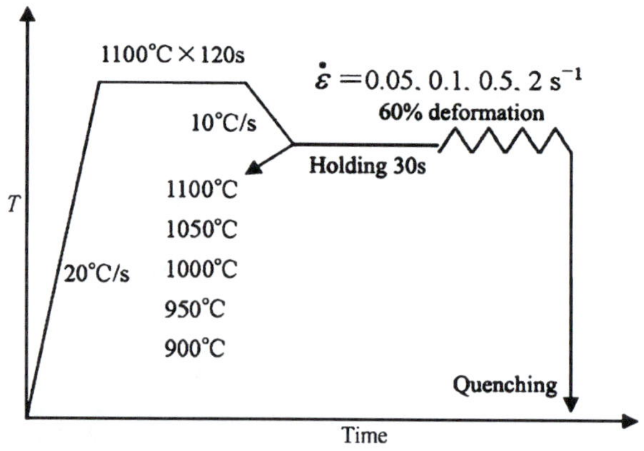

Fig. 20 Schematics of simulated processes for a single pass with 60% deformation

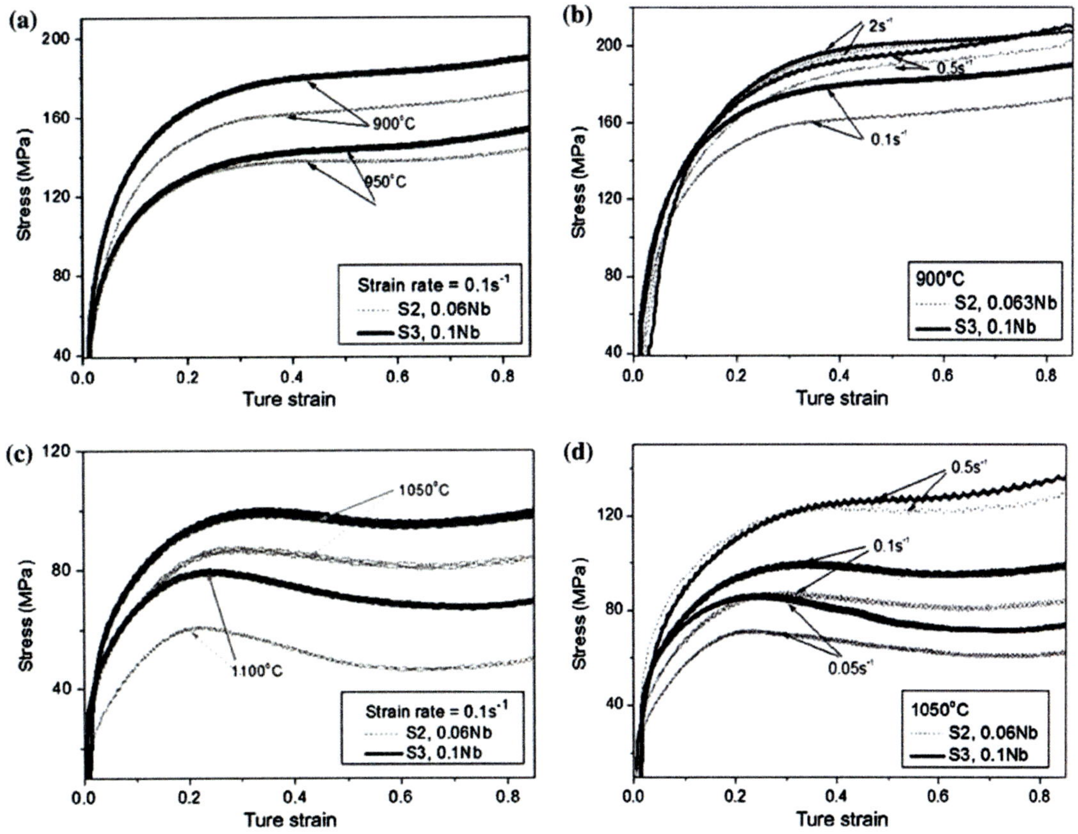

Fig. 21 Flow stress curves of sample S2 (0.06 wt%) and S3 (0.1 wt%) **a** low temperatures (900, 950°C) and strain rate of 0.1 s^{-1}, **b** different strain rates and low temperature (900°C), **c** high temperatures (1,050, 1,100°C) and strain rate of 0.1 s^{-1}, **d** different strain rates and high temperature (1,050°C)

To correlate hot flow stress behaviors with the actual rolling, the stress–strain curves with different strain rates at low and high temperature are shown in Fig. 21. In the case of the deformation at low temperature, the discrepancy of stress level between high and low steels is enlarged as decreasing of temperature, and large strain rate is benefit for higher stress value, in evidence, high Nb has larger advantage in the deformation at low temperature, especially in rolling with low strain rate. At high temperature, slow strain rate can accelerate the onset of DRX, but the addition of Nb content can raise peak stress (σ_p) of DRX, and the discrepancy will be widened with increasing of deformation temperature. Generally, plastic deformation is a thermally activated process [13], the influence of temperature and strain rate on deformation behavior can be described by Zener–Hollomn parameter (*Z*) shown in Eq. 1 [14, 15]. Based on the thermo-simulation database, deformation activation energy (Q_{def}) and other parameters (A, α, n) all can be determined trough the relationship of $\ln(\sinh(\alpha\sigma_p))$, $1/T$ and ln(strain rate) in Eq. 2 [15, 16], the calculated results of these parameters are shown in Table 5. Therefore, final DRX kinetics [17, 18] of different Nb steels can be characterized by the equations in Table 6, therein, the ratios of $\varepsilon_p/\varepsilon_c$ and σ_p/σ_c were determined by the evolution of working hardening rate θ ($d\sigma/d\varepsilon$) [19, 20], illustrated in Fig. 22. The parameters of DRX kinetics coincide with the other results of related investigations very well [21].

Table 5 Calculated results of parameters of DRX kinetics

Steel	Q_{def} (kJ/mol)	A	α	n
S2	327.812	0.87×10^{12}	0.0104	3.27
S3	339.582	2.9×10^{12}	0.0096	4.86

Table 6 DRX kinetics of different Nb steels

Steel	S2	S3
Z	$Z = \dot{\varepsilon} \times \exp\left(\frac{327812}{RT}\right)$	$Z = \dot{\varepsilon} \times \exp\left(\frac{339582}{RT}\right)$
$\varepsilon_p = A^*\ln Z - B$	$\varepsilon_p = 0.0664\ln(Z) - 1.54$	$\varepsilon_p = 0.0895\ln(Z) - 2.23$
$\varepsilon_c = M^*\varepsilon_p$	$\varepsilon_c = 0.52\varepsilon_p$	$\varepsilon_c = 0.54\varepsilon_p$
$\sigma_p = N^*\sigma_c$	$\sigma_p = 0.837\sigma_c$	$\sigma_p = 0.895\sigma_c$

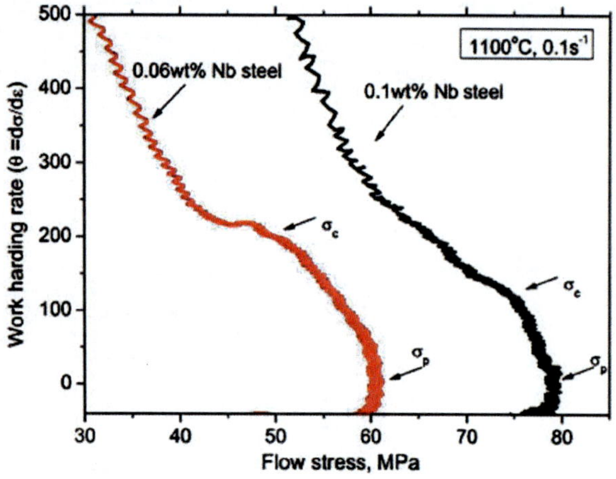

Fig. 22 Working hardening-stress curves of different Nb steels at 1,000°C and strain rate of 0.1 s^{-1}

$$Z = \dot{\varepsilon} \times \exp\left(\frac{Q_{def}}{RT}\right) = A(\sinh(\alpha\sigma_p))^n \quad (1)$$

$$\ln\dot{\varepsilon} + \frac{Q_{def}}{RT} = \ln A + n\ln(\sinh(\alpha\sigma_p)) \quad (2)$$

where A, α and n are constant.

3.2 Strain Accumulation and DRX

3.2.1 Effect of Nb Content on Strain Accumulation

Figure 23 illustrates the simulated process of multi-pass rolling (plane strain) by Gleeble-3800. Therein, the reductions and pass intervals are based on actual hot strip rolling of X80 pipeline steel. Strain–stress data were recorded for all simulated tests, and mean flow stress (MFS) values were calculated [22] in the analysis for this process.

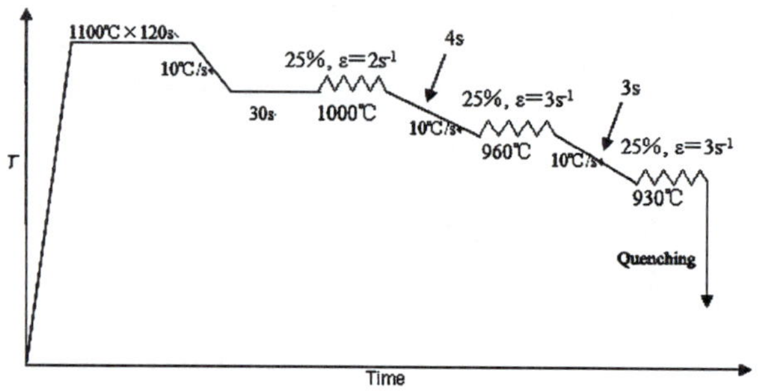

Fig. 23 Schematics of simulated process for multi-passes plane strain deformation

Figure 24 illustrates the stress–true strain curves and the evolution of mean flow stress in multi-pass plane strain deformation. There is a stress peak in the second pass for industrial high Nb steel, and also, its MFS value is decreased in the second pass. Decreasing of MFS value implies occurrence of DRX/MDRX during rolling. However, MFS value of medium Nb steel increases steadily during rolling. The softening caused by DRX is just found in high Nb steel, therewith, final stress values of these two steels are close. According to DRX kinetic equations deduced in the last section, the critical accumulation strains for the onset of DRX were calculated under the conditions with different Nb content, as shown in Table 7. Actually, high Nb steel has larger critical strain value, so the occurrence of DRX can be explained by higher strain accumulation under high Nb content. Moreover, this DRX will generate serious mixed grains structure, and it will worsen toughness of steel [23].

3.2.2 Effect of Deformation Temperature and Strain Accumulation on DRX

Plane strain thermo-simulation machine was used to investigate strain accumulation of industrial high Mn high

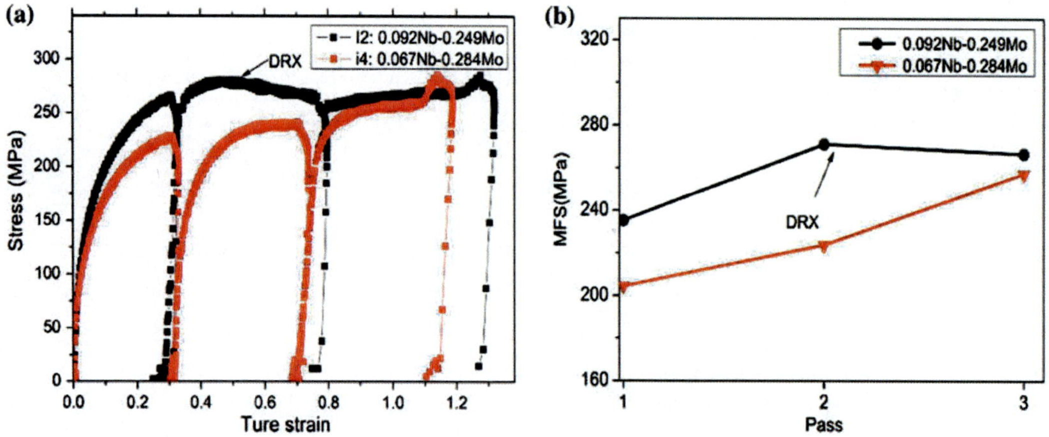

Fig. 24 a Ture stress–strain curves and b mean flow stress of multi-pass rolling

Table 7 Critical accumulation strain values for DRX happening

Steel	The first pass (25% reduction at 1,000°C, 2 s^{-1})	The second pass (25% reduction at 960°C, 3 s^{-1})
S2 (0.063 Nb)	0.292	0.341
S3 (0.10 Nb)	0.383	0.453

Nb steel (Table 1), the thermo-simulation process based on industrial actual rolling of X80 strip was shown in Fig. 25, it focus on the effect of rolling temperature. Two finish rolling start temperatures were chosen, 940 and 910°C, respectively, and true stress–strain curves were obtained.

In high Mn high Nb steel, the kinetics of stain-induced precipitation Nb(C,N) is putted off by higher Mn content [7]. The speed of finish rolling is very quick in industrial tandem rolling (the interval time is very short), so there is hardly any available precipitated Nb(C,N) in passes interval, but drag effect is still very strong. In this case, the strain accumulation in passes will easily attain to a higher lever. Once accumulated-strain exceed to the critical strain, dynamic recrystallization will take place [24]. Cho's research results [25] indicated dynamic recrystallization is much more possibility with increasing Mn content in Nb-bearing steel.

Figure 26 gives stress–strain curves measured by plane strain machine. Table 8 gives actual strain rate of each pass in this experiment, which is similar in different processes. In the first case that finishing rolling starts at 910°C, stress values increases steadily. But when start temperature is raised to 940°C, transform trend of the curve is changed, and stress values of latter three passes are on the same level. The characteristic of stress–strain curves in the third pass were in accord with the type of dynamic recrystallization, the same results can be obtained while the format of diagrams is changed to mean flow stress (MFS), decreasing of MFS means occurrence of DRX [26], as shown in Fig. 27. In dynamic recrystallization, the relationship between critical strain value and deforming temperature is inverse relation [27]. That is to say, critical strain value is decreased with temperature increasing. If the rolling process (include rolling reduction, strain rate, etc.) is improper, dynamic recrystallization is easier to take place in high temperature deforming process.

Table 8 Actual strain rate of each pass in the plane strain test

SRT/pass	1	2	3	4	5
910°C	0.49	0.51	0.41	0.39	0.55
940°C	0.59	0.41	0.49	0.55	0.67

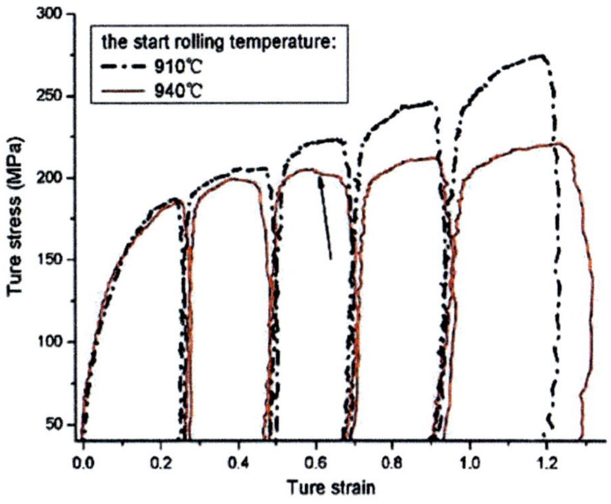

Fig. 26 True stress–strain curves in different processes of plane strain simulating

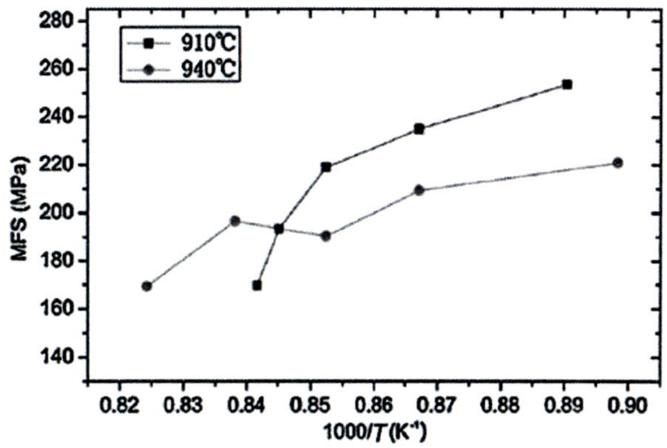

Fig. 27 Mean flow stress (MFS) in different processes of plane strain simulating

Fig. 25 Process of plane strain test

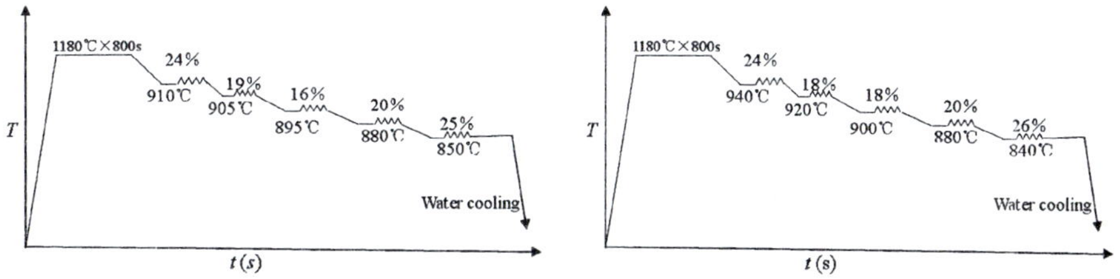

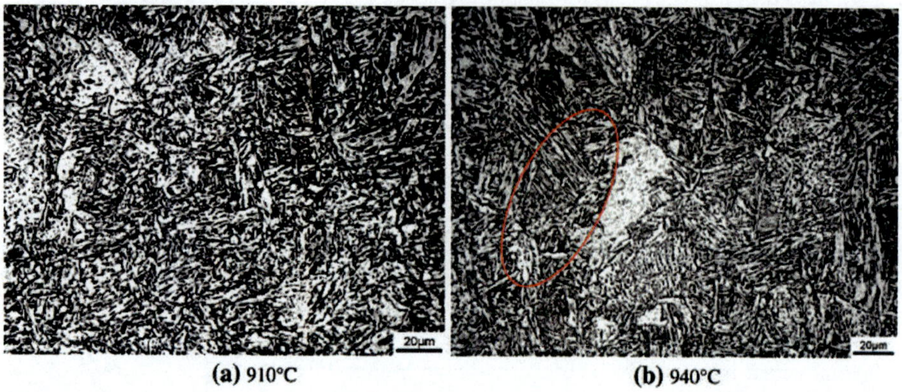

Fig. 28 Morphology of microstructure etched by 4% Nital: **a** 910°C and **b** 940°C

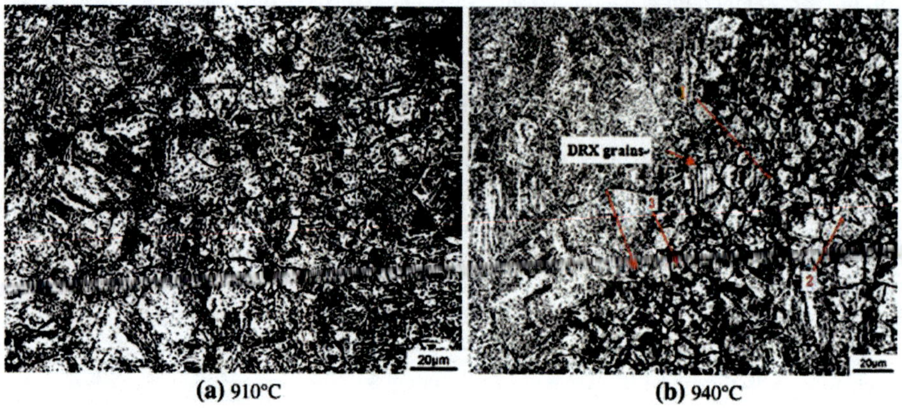

Fig. 29 Morphology of prior-austenite grain (vertical direction of rolling): **a** 910°C and **b** 940°C

Optical microstructures of these two kinds of simulated processes are given in Fig. 28. Both of similar microstructure include lath bainite and a small quantity of granular bainite. Although these optical microstructures etched by 4% Nital are similar, the difference between prior-austenite grains is obvious. Prior austenite grains are shown in Fig. 29a are homogeneous, which start rolling temperature is 910°C. But austenite grains are non-uniform in Fig. 29b, and there is a grain layer which is composed by small size grains among some rough grains, and the form is like a necklace, which should be caused by occurrence of partial dynamic recrystallization. Moreover, nanohardness values of these austenite grains with various sizes are different, as shown in Table 9, which were measured by Nano Indenter II. The hardness shown that nanohardness of the smaller grains is less than 1.2 GPa, but the large grain is more than 3.8 GPa, so the smaller grains must be recrystallized grains with low dislocation density, but the large one is the ausformed grains. These results also can prove the happening of partial DRX. Generally, complete dynamic recrystallization refines grain size in evidence, but partial dynamic recrystallization leads to serious mixed grain in size and loss of dislocation density. That would be not benefit to accumulate strain in the austenite resulting in yield strength decreasing.

Table 9 Average nanohardness of austenite grains with various sizes at 940°C

Grain type	1 (diameter $\leq$ 5 μm)	2 (diameter $\geq$ 30 μm)
Average nanohardness (GPa)	1.12	3.88

Combining these above characteristics of high Mn high Nb steel, a skeleton map can be outlined. Deforming above 1,050°C is to break coarse austenite grains, but due to absence of precipitates and solute drag effect is not very strong in this temperature region, growth and coarsening behavior is fast. Controlling rough rolling to refine austenite grains further, complete recrystallization will happen after deforming within the temperature range nearby 1,000°C, dragging effect from solute can retard the growth and coarsening of grain effectively, homogeneous and fine grain structure can be obtained after this critical rough rolling. Partial recrystallization must be avoided during finish rolling, and decreasing rolling temperature and control proper reduction in each pass, pancake austenite grain with small width will be obtained. Whether in plate or in strip rolling, high S_v always is one of aims to acquire good properties for the steel, and the density of dislocation can be reflected by flow stress level [28]. Actually, the flow stress behaviors during and after deformation both are key factors to determine final density of dislocation. Whereas, the influence of Nb content on flow stress behaviors during deformation have various characteristics depended on temperatures and strain rates. In actual rolling, plate rolling has the metallurgical characteristics of low strain rate and large deformation, thus, high Nb steel can exhibit larger advantage, suppress DRX behavior at high temperature effectively, and higher stress level or S_v is easier to obtain at low temperature. However, in hot strip rolling, larger strain rate will be carried out, so the influence of Nb content on flow stress behavior will be weaken during finish rolling process. The main influenced factors for final stress level are the softening and hardening behaviors after deformation, which has close correlation with solute Nb content and precipitates [29]. Plentiful solute Nb can prohibit large recovery between passes, so high Nb content can suppress softening between pass intervals, and it may cause higher strain accumulation. Although DRX of high Nb steel need higher critical stress value, strain accumulation still can exceed it, the onset of DRX for high Nb steel is easier during hot strip rolling (finish rolling process) with relative high start rolling temperature. However, by decrease the finish rolling temperature, the DRX of high Mn high Nb steel can be restrained, and higher and steady MFS can be achieved, as a consequence, homogeneous grain structure and high S_v will be obtained.

4 Conclusion

(1) In general, due to the addition of high Mn, the precipitation of high Mn high Nb steel is delayed, the available precipitation occurs after 3–4 s around 950°C. After deforming at this temperature region, the recrystallization is no longer time dependent, because the available precipitated particles can quite halt recrystallization, after that, the growth and coarsening of grains actually include recrystallizated fine grain and unrecrystallizated deformed grain, it can generate mixed grain structure. But in the absence of precipitates, the recrystallization behavior of high Mn high Nb steel is time dependent, and as the decreasing of temperature, the recrystallization kinetics need longer incubation time and longer finish time.

(2) The characteristic of high Mn high Nb creates a new temperature window around 1,000°C, it is not appeared in stress relaxation curve of this temperature that obvious accelerated softening by slow recrystallization behavior of 0.1Nb steel, moreover, it is not found that available precipitates occur at this temperature. Refining grain size previous to the deformation around 1,000°C, complete recrystallization of 0.1Nb steel will finish within a short time, and higher solute Nb can slow down the mobility of recrystallized grain boundary, and suppress grain coarsening effectively. Homogenous and fine grains will be generated prior to the finish rolling in which grains will be flattened fully. This improvement of prior austenite grain condition has large positive effect on final properties of the steel.

(3) The results of multi-pass rolling indicate, high Nb can suppress the softening between passes, and it may increase the possibility of partial DRX caused by over high strain accumulation during hot strip finish rolling. Therefore, start temperature of finish rolling should lower than 940°C to flatting austenite grain fully and to accumulate strain or S_v.

Acknowledgments The authors wish to thank the funding support from CBMM, CITIC, and the collaboration from Chinese steel companies (TISCO, Sha-steel, SHOU-steel, etc.)

References

1. The technical conditions of the hot rolling steel plate for the Second West to East Gas Pipeline project. China Petroleum Enterprise Standardization (2007)
2. L.J. Cuddy, *Thermomechanical Processing of Microalloyed Austenite* (TMS, Warrendale, 1982), p. 129
3. K. Hulka, P. Brodignon, J.M. Cray, Niobium Technical Report-No1/04 (2004)
4. D.G. Stalheim, K.R. Barnes, D.B. McCutcheon, in *International Symposium of Microalloyed Steels for the Oil and Gas Industry*, CBMM/TMS (2006)
5. H.S. Zurob, C.R. Hutchinson, Y. Brechet, G. Purdy, Acta Mater. **50**, 3075 (2002)
6. H.S. Zurob, C.R. Hutchinson, Y. Brechet, G. Purdy, Mater. Sci. Eng. A **382**, 64 (2004)
7. H.S. Zurob, G. Zhu, S.V. Subramanian, G. Purdy, ISIJ Int. **45**(5), 713 (2005)
8. L.P. Karjalainen, Mater. Sci. Technol. **6**, 557 (1995)
9. J. Kliber, I. Schindler, J. Mater. Process. Technol. **60**, 597 (1996)
10. W.J. Liu, J.J. Jonas, Metall. Trans. A **19A**, 1430 (1988)
11. C.L. Miao, C.J. Shang, G.D. Zhang, G.H. Zhu, H. Zurob, S.V. Subramanian, Front. Mater. Sci. China **4**, 197 (2010)
12. A. Laasraoui, J.J. Jonas, Metall. Trans. A **22**, 151 (1991)
13. A.I. Fernandez, P. Uranga, B. Lopez, Mater. Sci. Eng. **361**, 367 (2003)
14. C. Zener, J.H. Hollomon, Am. Soc. Metals Trans. **33**, 163 (1944)
15. S.F. Medina, C.A. Hernandez, Acta Mater. **44**, 137 (1996)
16. J.R. Cao, Z.D. Liu, S.C. Cheng, G. Yang, J.X. Xie, Acta Metall. Sin. **43**, 35 (2007)
17. S.V. Subramanian, G. Zhu, C. Klinkenberg, Mater. Sci. Forum **475–479**, 141 (2005)
18. G. Amn, S. Nakajima, M. Miyahara, ISIJ Int. **32**, 261 (1992)
19. E.I. Poliak, J.J. Jonas, ISIJ Int. **43**, 692 (2003)
20. E.I. Poliak, J.J. Jonas, Acta Mater. **44**, 127 (1996)
21. L. Zhang, W. Yang, Z. Sun, J Univ. Sci. Technol. Beijing **14**, 1 (2007)
22. T.M. Maccagno, J.J. Jonas, S. Yue, B.J. Mcrady, R. Slobodian, D. Deeks, ISIJ Int. **34**, 917 (1994)

23. T.X. Cui, C.J. Shang, C.L. Miao, W.G. Xue, Y.T. Hu, Iron and Steel **44**, 55 (2009)
24. A.I. Fernández, P. Uranga, B. López, Mater. Sci. Eng. A **361**, 367 (2003)
25. S.H. Cho, K.B. Kang, J.J. Jonas, ISIJ Int. **41**, 63 (2001)
26. J.X. Dong, F. Siciliano, J.J. Jonas, ISIJ Int. **40**, 613 (2000)
27. K. Minami, F. Siciliano, T.M. Maccagno et al., ISIJ Int. **36**, 1507 (1996)
28. Y. Atsuhiko, F. Takashi, F. Masaaki, O. Kentaro, M. Hirofumi, ISIJ Int. **36**, 474 (1996)
29. S.V. Subramanian, G. Zhu, H.S. Zurob, G.R. Purdy, G.C. Weatherly, J. Patel, C. Klinkenberg, R. Kaspar, *Thermo-mechanical Processing: Mechanics, Microstructure and Control* (University of Sheffield, England, 2004)

Part V
Specialty Steels

Grain Boundary Hardening and Single Crystal Plasticity in High Nitrogen Austenitic Stainless Steels

Markus O. Speidel

Abstract

It is well known that nitrogen in solid solution can strongly enhance both, solid solution hardening and grain boundary hardening (fine grain hardening) of steels with a face centered cubic crystal lattice. The present work shows a quantitative relation between the critical resolved shear stress of single crystals and the yield strength of polycrystals of such solid solutions. This represents one more step towards understanding the strength of this excellent class of materials.

Keywords

Single crystals • Stainless steel • Polycrystals • High-nitrogen steels

1 Yield Strength of Polycrystals and Single Crystals

The yield strength of polycrystalline face centered cubic solid solutions depends clearly on the average diameter of the crystallites, the "grain size". This has been known for half a century [1]. A recent example is shown in Fig. 1.

Note in Fig. 1 that the yield strength, $R_{p0.2}$, depends on the grain size according to the well known Hall–Petch relation: the yield strength increases linearly with the inverse square root of the grain size. In such a diagram, it is easy to extrapolate to infinite grain size, since this corresponds to the intersection with the strength axis. We have labeled this intersection R_o. In the case of the particular austenitic stainless steel shown in Fig. 1, $R_o = 180$ MPa. In the present paper we will address two questions:

(1) How does the yield strength of a polycrystal, extrapolated to infinite grain size, relate to the critical resolved shear stress of a single crystal of similar composition?

(2) How do these strength values depend on the nitrogen content and the temperature in austenitic stainless steels?

In order to address these questions we have done a large number of experimental studies of the kind shown in Fig. 1. The schematic results of such studies are shown in Fig. 2.

For the present work the R_o data are the only ones of interest, and many such R_o data, that is yield strength extrapolated to infinite grain size, will be analyzed. R_o is often called the friction stress, since it is assumed that this is the stress for dislocations to overcome the resistance to slip (friction) without contribution of the grain boundaries.

In a first step we recapitulate what has been found in 1965 [1], for solid solution alloys crystallized in the face centered crystal lattice, using the schematic Fig. 3.

Note in Fig. 3 that R_o, a tensile yield stress without any contribution of grain boundaries is not simply just twice the critical resolved shear stress of a single crystal. This is because in true polycrystals (with, say, more than 20 crystallites in the specimen cross section) there is an orientation distribution of the crystallites and thus it is not just the slip system under 45° to the tensile axis which becomes operative. Thus, slip begins not yet macroscopically when the most favorably oriented slip system in one crystallites starts to operate with a Smith factor of 2, but only when the Taylor factor 3.06 is satisfied [1]. And there will always be

M. O. Speidel (✉)
Swiss Academy of Materials Science, Birmenstorf, Switzerland
e-mail: speidel@materialsacademy.com

Y.Weng et al. (eds.), *Advanced Steels*, DOI: 10.1007/978-3-642-17665-4_36,

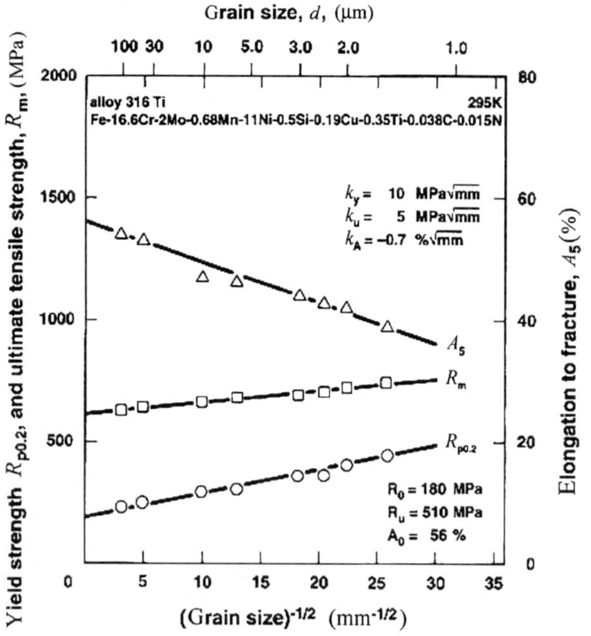

Fig. 1 Effect of grain size on the mechanical properties of an austenitic stainless steel [2]

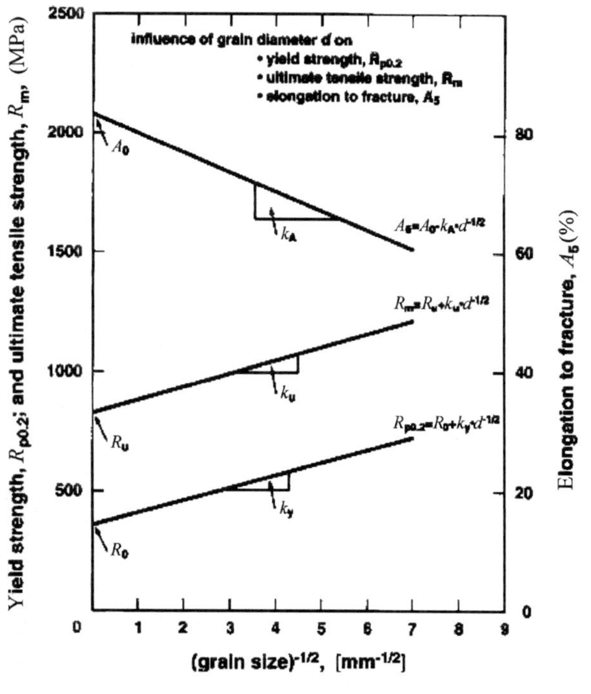

Fig. 2 Schematic evaluation of the kind of data shown in Fig. 1

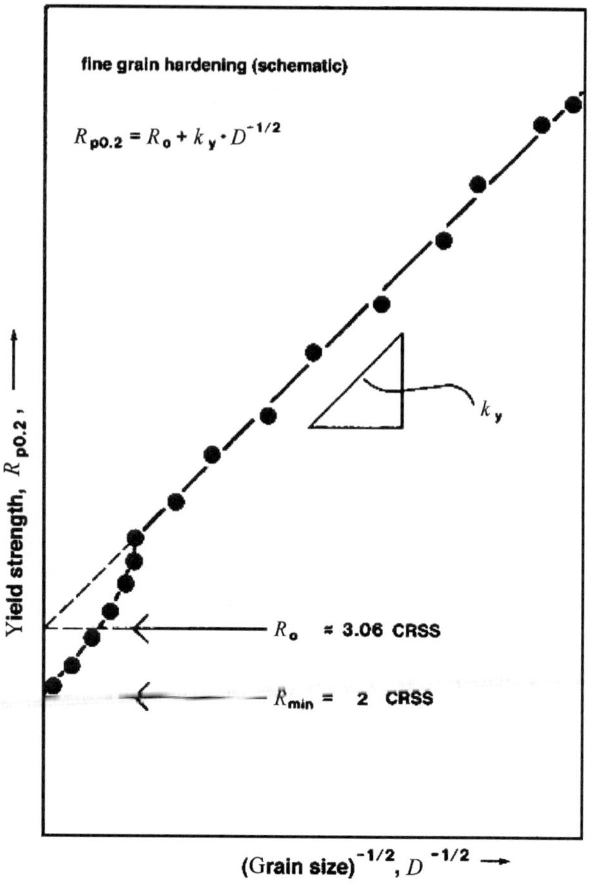

Fig. 3 Relation between the effect of grain size on yield strength and the critical resolved shear stress of single crystals in fcc solid solutions

polyslip in a true polycrystal from the yield point on because of the compatibility requirements at the grain boundaries. For this reason, R_0 corresponds to 3.06 CRSS of a single crystal oriented for polyslip.

If there are only a few crystallites in the cross section of a polycrystal, yield can occur at lower tensile stresses than 3.06 CRSS, because then we have no longer a material with all the restrictions of a true polycrystal. This is seen schematically in Fig. 3. The realization goes like this: assume we have just only one crystallite in the polycrystal cross section. This is labeled a "bamboo" crystal. The crystallites may then still have a statistical orientation distribution. But macroscopic yield will begin when the stress starts to activate the most favorably oriented slip system, that is at 2 CRSS [1].

2 Austenitic Stainless Steels with Nitrogen and Carbon

In Fig. 4 we show the results of an analysis of the grain size effect on the yield strength of a significant number of austenitic stainless steels with nitrogen in solid solution. Only the R_0 data points are shown. These indicate the solid solution

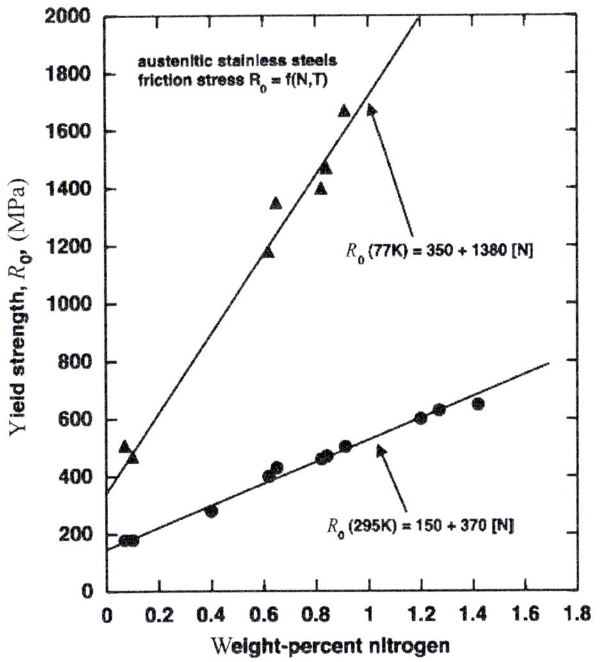

Fig. 4 Solid solution hardening of austenitic stainless steels by nitrogen polycrystal data extrapolated to infinite grain size

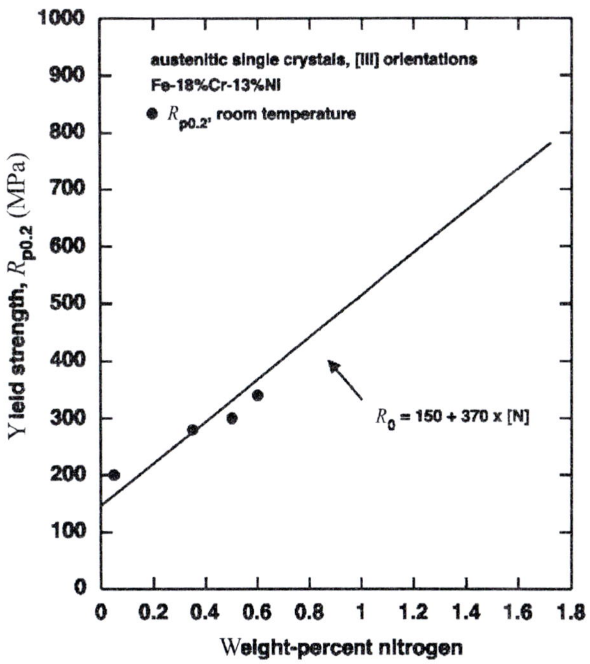

Fig. 5 Single crystal yield strength data from [3] compared to the line and equation from Fig. 4 (Note the close similarity between our polycrystal work, Fig. 4 and the single crystal work of Barannikova and Zuev [3])

hardening of the stainless steels by nitrogen extrapolated to infinite grain size.

Note the quantitative expressions for solid solution hardening in Fig. 4.

In a next step, this is to be compared with single crystal data of similar alloys.

Fortunately, in recent years, two teams of the Russian Academy of Science have investigated and published single crystal plastic behavior of austenitic stainless steels with high nitrogen contents in solid solution [3, 4]. From their work we have first extracted the yield strength data of single crystals oriented for polyslip. The crystal geometry of plastic shear for the single crystals mentioned in Fig. 5 is such that it facilitates multiple slip as soon as the yield point is attained.

Next we compare the critical resolved shear stresses CRSS of a slightly different austenitic stainless steel with our polycrystal data, as shown in Fig. 6.

The conclusions from the comparison shown in Fig. 6 are far-reaching and satisfactory.

First, the analysis presented in Figs. 2 and 3 really works. This appears to be the appropriate way to compare single crystal strength and polycrystal strength in a quantitative manner, and it has been established already in 1965 [1].

Second, there is good agreement between various research groups concerning solid solution hardening by nitrogen in fcc iron-based solid solutions. Even on the very strong effect of temperature there is quantitative agreement.

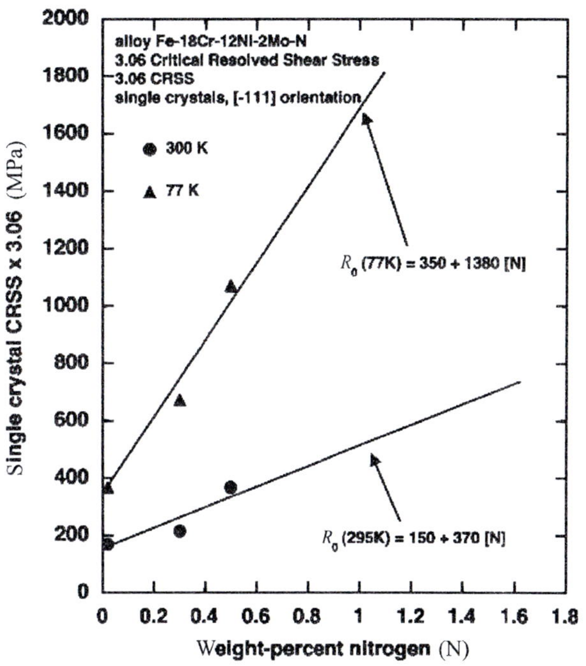

Fig. 6 Single crystal critical resolved shear stress data from the work of Kireeva et al. [4] (the CRSS data have been multiplied by 3.06 and compared with the equations and the lines from Fig. 4)

Once this work is crowned by measurements at 4.2 K, there will be finally a sound basis for comparisons with solid solution hardening theories.

3 Summary and Conclusions

Yield strengths of polycrystals and critical resolved shear strengths of single crystals are quantitatively related and one can be converted into the other by the analysis presented here which was established already in 1964.

Using this analysis a very high degree of agreement concerning solid solution hardening of high-nitrogen austenitic stainless steels is observed between the work at the Russian Academy of Science and the Swiss Academy of Materials Science.

Further work at 4.2 K is urgently needed to prepare a basis for the judgement of solid solution hardening theories. This is because in the alloy systems investigated both, the effect of nitrogen and the effect of temperature are extremely strong and thus a perfect experimental check for theoretical explanations.

References

1. W. Köster, M.O. Speidel, Z. Metallkunde **56**, 585 (1965)
2. M.O. Speidel, M. Zheng-Cui, in *High-Nitrogen Austenitic Stainless Steels. HNS 2003, High Nitrogen Steels* (vdf Hochschulverlag AG, ETH Zürich, 2003), ISBN: 3-7281-2891-0
3. S.A. Barannikova, L.B. Zuev, in *The effect of Interstitial Impurity Content in Austenitic Steel Monocrystals. HNS 2009, High Nitrogen Steels*, (MISIS, Moscow, 2009), p. 189
4. I.V. Kireeva, Y.I. Chumlyakov, A.V. Tverskov, N.V. Luzginova, in *The Effect of Nitrogen on Twinning in Single Crystals of Austenitic Stainless Steels. HNS 2009, High Nitrogen Steels*, (MISIS, Moscow, 2009), p. 213

Unexplored Possibilities of Nitrogen Alloying of Steel

Jacques Foct

Abstract

In contrast with highly nitrogen alloyed stainless steels such as classical HNS, an optimized, partial or total replacement of carbon by nitrogen does not seem to have been frequently studied. Many arguments based on: (1) usual concepts of physical metallurgy, (2) the observed impact of interstitial alloying on steel properties, (3) cost efficiency and the demand of sustainable development, (4) an overview of steel sorts likely to become nitrogen upgraded are examined. Probable scenario of the evolution of steel-making including nitrogen alloying concern is discussed.

Keywords

Advanced steel • Nitrogen alloying • Steel processing • Simulation • Physical metallurgy • Interstitial

1 Introduction

Rather than unexplored the promises of nitrogen alloying have probably been unexploited, it is endeavored below to identify probable cases for which nitrogen addition could be beneficial and for which reason. In comparison with usual articles devoted to well-established experimental results, unambiguous characterizing, accurate calculations, the present article suffers similar limits to conjecture compared to demonstration. Although speculative, the question of an extension of nitrogen alloying to cases which seem to have been partially or totally neglected is suggested by many concepts, data, analogies, simulations, etc. It is aimed here to examine and to discuss the following points likely to initiate and stimulate the design of innovative nitrogen alloyed steels:

- Specific role of nitrogen on physical metallurgy of transition metals alloys.
- Brief overview of the properties of high nitrogen steels (HNS) [1–39].
- Cost effectiveness of alloying and limits of ore resources.
- Nitrogen alloying and the large steel grades diversity.

In conclusion a scenario of steel-making evolution and its impact on nitrogen as chosen alloying element is discussed.

2 Basic Characteristics of Physical Metallurgy of N Interstitial Alloying

By far carbon is the most important alloying element of steel [39], not only because of its role played in the reduction of iron ores but also because of its interstitial behavior in the iron-based solid solution and compounds. C and N, close relatives in the periodic chart, may often be substituted to each other according to a result which is never neutral concerning the properties of the steel. Therefore optimal amount and optimal ratio of C vs. N are to be identified especially if the property under consideration is strongly interstitial dependant. Devoted experimental, computational, and conceptual research studies shed light on the interactions of C and N [32–39] with each other, with other

J. Foct (✉)
Laboratoire de Metallurgie UMET, Universite Lille 1, Villeneuve-d'Ascq, France
e-mail: jacques.foct@univ-lille1.fr

Y. Weng et al. (eds.), *Advanced Steels*, DOI: 10.1007/978-3-642-17665-4_37,

light elements such as H, He, B, with lattice defects (vacancy, dislocation, twin, stacking fault, grain and phase boundaries), with substitution elements. Despite well identified limits of computational techniques (such as ab initio, molecular dynamics, Monte-Carlo, phase field, finite elements, thermo-calc, dictra, etc. very interesting results were established such as: trends to ordering of N atoms in BCC-like structure in contrast with clustering of C, different level of repulsion interaction for C–C and N–N in FCC, interaction interstitial—vacancy, activation energy for diffusion, energy of interstitial—dislocation configuration. It was clearly established that the stability of interstitial compounds of transition metals is decreasing with the "d" electron density, close to the middle of the 3d series, typically for Fe, the increase of 2p electron of the interstitial atom resulting from the replacement of C by N is characterized by an increasing electron density at the Fermi level as proven by electronic structure calculation as well as electron spin resonance. These results correspond to a metal-interstitial Fe-I bonding which is less ionic-covalent with N than with C and therefore more metallic. Although the present comment is very schematic and intuitive it shows that in comparison with C, N alloying may dramatically modify some atomic scale phenomena such as diffusion, structure of interstitial precipitates, ordering, interface and interphase energies, stacking fault energy, spinodal decomposition, etc. No one of the above mechanisms is neutral concerning thermo-mechanical treatment of the steel and of the resulting properties as shown in the model case of HNS: mechanical resistance, corrosion, recovery and recrystallisation, precipitation, hardening. Initially the design of HNS did not resulted from atomic scale calculation, but on transposition of steel making expertise with the support of basic thermodynamics of alloys. Calculation of phase diagram performed thanks to dedicated codes such as thermo-calc is now likely to guide more accurately engineers to the target. Despite the success of calculation and of simulation it never substitutes to pilot experiments, because any code may suffer some limit such as neglected magnetism, large number of parameters, validity of approximations, uncertainties in experimental input data, the preliminary computational approach, likely to spare time as well as money, is necessary.

3 Impact of High Nitrogen Alloying Exemplified by HNS

Segregation of nitrogen at the grain boundaries of low alloyed steel after air blowing which induced brittleness, was avoided, about 60 years ago, with the oxygen blowing process. Meanwhile it still induced some perverse reserves against nitrogen alloying which became surmounted for chromium alloyed stainless steel. Iron based alloys containing around 20 at.pc. (Cr + Mn) made possible to increase the nitrogen solubility limit to about 4 at.pc. In order to increase further the amount of dissolved nitrogen the chemical potential of N in the liquid as well as in the gas phase had to be raised either under high pressure (20–40 bar) in the "Pressure electro-slag re-melting" process (PESR), or thanks to atomic N originating from plasma or decomposition nitrides [2, 18]. The principles of these processes are known but the industrial accomplishment demands additional abilities cares and expenses. The actual demonstration of the pertinence of high nitrogen alloying was given by the steel P900 made by Krupp VSG thanks to the efficient action of late G. Stein [1, 20] and of his co-workers. High mechanical resistance of non-magnetic retaining rings for electric power plants, pieces for aircrafts and space ships, stress corrosion-resistant parts for naval applications as well as for chemical industry exemplify the beneficial role of nitrogen alloying as quoted in a well developed technical and scientific literature. The main grounds for the N induced upgrading of HNS may be roughly summarized: nitride precipitates appear less prone to ripening than carbides, the reasons for this being related with a smaller inter-phase energy and a closer compatibility of crystallographic structure of particles and matrix; a more efficient influence of interstitial alloying on grain size strengthening through the nitrogen dependence of the K coefficient of the Hall–Petch law; electron density at the Fermi level likely to diminish stacking fault energy but also to reduce the height of the Peierls barrier. Sometimes observations appeared difficult to decipher and even contradictory, in low cycle fatigue the observed planar structure of dislocations suggesting localization of deformation and a reduction of life time was in fact shown to be beneficial [28, 29].

Another phenomenon, interesting to consider, results from a synergistic influence of interstitial C and N [16, 35, 36] atoms which suggests that up to certain level of concentration the possible detrimental effect of an excessive amount of a single interstitial type can in fact be opposed by the effect of the other: for example clustering of C inhibited by ordering of N.

Considering the influence of N on corrosion, general as well as localized such as stress corrosion, pitting corrosion, corrosion erosion, fatigue corrosion, etc. two reasons for improvement are identified: less sensitization around nitrides precipitates than around carbides, second a modification of the chemistry of the passive film.

This résumé of the action of N in HNS has been deliberately chosen to be qualitative and intuitive because the transfer to other groups of steel grades is likely to be initiated by rather general considerations which will precede simulation and calculation necessary to focus the experimental conditions and pilot development. The validation of generic improvement resulting from N alloying does not mean that the nature of the involved mechanisms and properties coincide whatever are structure, composition and thermal treatment. It was just verified that under suitable modus operandi austenitic, ferritic, martensitic and duplex, nitrogen alloying has been proven to induce dramatic improvement.

4 Exportation of Nitrogen Alloying to Other Steel Grades

The main feature of interstitial alloying of metals is well known never to be neutral, therefore the idea to fabricate these materials containing H, C or N may not be considered as original. The actual obstacle to study more systematically the effect of introduction of nitrogen in steel results from the up-hill chemical potential barrier towards the liquid and solid metallic phases. The first step is therefore to increase the N solubility thanks to Cr + Mn alloying of the steel. Other nitride former elements are either too strong (Ti, Zr-V, Nb) or expensive (Mo, W). Interstitial alloying does not act independently of substitutional addition and of the making processing which is continuously evolving.

It is foreseeable that the future grades would contain Cr + Mn [19, 20] as main substitutionals, very small amount of strong nitride formers, and possible association N + C interstitials of which the concentration ratio would be optimized, the total N weight concentration being usually below 0.5% which corresponds to about 2.5 at.pc., an interstitial concentration which roughly corresponds to 15% of metallic atoms touched at the first interstitial neighborhood as a consequence of the octahedral interstitial site occupation. Other elements such as B, Al, Si could also be added for grounds of similar influence with the nowadays made grades. Meanwhile the interaction of these elements with N leads to covalent BN, AlN and Si_3N_4 nitrides and do not increase the N solubility in the steel matrix.

In the single phase steels the role of N is expected mainly to contribute to solution hardening through N as well as N–Me pairs solid solution, an effect which is known as among the few largest in comparison with other alloying elements, except B and C. In addition the presence of N in the single phase matrix is likely to constitute a tool to optimize grain boundary engineering (recrystallization and creep) thanks to the interstitial–grain boundary interaction [23–34].

The main interest of N as well as of C results from their influence on all types of phase transformations which concern steel: ordering, clustering, shear and diffusion transformations (martensitic, bainitic, spinodal, precipitation, ripening, ageing, strain ageing, tempering, etc.). In consequence, once mastered the future steel making processes and identified the pertinent thermo-mechanical treatments, nitrogen alloying could generate original high performance steels in the following groups: Bainitic, Dual, martensitic, marageing, TRIP, TWIP [39]. The mastering of combined influence of manganese and nitrogen constitutes an interesting challenge. Manganese can be considered as an awkward partner of material scientists: polymorphism of complex structure, multiplicity and variability of atomic bonding and many other features of this difficult character. Meanwhile Mn was proven to offer unique opportunities for steel alloying as shown by Hadfield steel and more recently by Nickel-free anti-allergenic N, Cr and Mn alloyed HNS.

High rate cold-working of austenitic HNS has been shown to lead to extremely high resistance stress, near over 3000 MPa [7], value comparable to piano wire steel obtained by drawing eutectoid carbon steel. Taking into account that this sort of steel is not only utilized for music instruments but also in much more widely spread tire industry, some attempt to fabricate similar wires from the Fe–Fe_4N eutectoid have been achieved. Although this operation did not seem to meet yet a complete success, it probably constitutes a promising step towards perlitic-like nitrogen steel. Nitrogen enrichment of hard steel and tool steel, is suggested by the properties of transition metal nitrides and carbonitrides, it is also approached by gradient structure resulting from surface treatment or mechanical alloying. The success of these achievements is a consequence of the swift diffusion of interstitial atoms.

Among the numerous problems met by steel industry, the increase of resistance to hydrogen as well as to irradiation-induced brittleness determines many strategic options. In both cases the evolution of H interstitial atoms distribution or of point defects resulting from neutron irradiation (self-interstitials, vacancies, cascade) originates detrimental influence on mechanical resistance. Because N interstitial as well as C lead to strong interaction with H as well as with point defects the presence and the structure of the interstitial alloying elements play in both cases a key role (beneficial or detrimental) on the kinetics evolution of the resistance. New grades devoted to nuclear power plant (vessel steel) containing optimized interstitial concentration should emerge.

5 Conclusion Suggested by the Possible Impact of the Evolution of Steel Making on Nitrogen Alloying

Steel industry is characterized by a succession of technical breakthroughs which may be partially masked by the multi-centenarian period during which it took place. Among these steps forwards, blast furnace, Bessemer process, stainless steel, continuous casting, oxygen blowing, (HNS?), etc. Most of the time progress was pulled by strategic, economical, political constraints as well as scientific and technical innovation. Similar challenge of the present time resulting from: globalization, shortage of natural resources, fuel as well as metallic and non-metallic ores (Li, F, P, Ni,

Zn, Sn, U, etc.) increase the driving force for R&D on steel making. This industry which produces about 1.3 billion tons per year is energy-vorous. The iron and steel activity corresponds to about 3.2% of global greenhouse gas emission which is becoming higher than 30 billion metric tons equivalent CO_2. These data are not ignored by steel producers. Since many years R&D studies which are dealing with processes aimed to spare energy, such as hydrogen reduction of ore, smelting reduction, near-net-shape-casting, limitation of coke consumption, reduction in the liquid phase in order to increase kinetics, etc. are strategic subjects.

These projects should result to dramatic breakthroughs such as steel factory downstream to nuclear power plant producing the energy necessary to melt as well as hydrogen for reduction of metallic ore. The confluence of the demands of economics and of environment protection, are likely to suggest new steels from the less rare alloying elements C, N, Cr, Mn, which constitute the tool box of updated nitrogen alloyed grades.

Because the mid- as well as long-term demands and needs are unavoidable all industrial sectors will have to reinforce R&D and it would be surprising that in the future dramatic changes of iron and steel industry the usage of nitrogen alloying could be omitted.

References

1. J. Foct, A.Hendry (eds.), *High Nitrogen Steels*, HNS88(The Institute of Metals London Publishers, ISBN 2901462_45_4) (1989)
2. G. Stein, H. Witulski (eds.), *High Nitrogen Steels*, HNS 90Verlag Stahleisen mbH, Duesseldorf De., ISBN3-514-00452-8 (1990)
3. V.G. Gavriljuk, V.M. Nadutov (eds.), *High Nitrogen Steels* (Kiev, Ukraine, 1993)
4. M. Kikuchi, Y. Mishima (eds.), *Special Issue on High Nitrogen Steels* (ISIJ International, The Iron and Steel Institute of Japan, 1996)
5. H. Hänninen, S. Hertzman, J. Romu (eds.), *High Nitrogen Steel* (Trans Tech Publications Ltd, Finland, 1998)
6. R. Baldev, M. Speidel, U. Kumachi Mudali, V.S. Srinivasan (eds.), *High Nitrogen Steels* (Trans. Indian Inst. Met., Chennai, 2002)
7. M.O. Speidel, C. Kowanda, M. Diener (eds.), *High Nitrogen Steels* (Institute of Metallurgy, ETH Zürich, 2003)
8. N. Akdut, B.C. De Coomand, J. Foct (eds.), *High Nitrogen Steels* (International Conference, Ostend, 2004)
9. H. Dong, J. Su, M. Speidel (eds.), *High Nitrogen Steels* (Metallurgical Industry Press, Beijing, 2006)
10. M.O. Speidel, P.J. Uggowitzer, Ergebnisse der Werkstoff-Forschung, Band 4 (Verlag der Schweizerischen Akademie der Werkstoff-Wissenschaften, 1999)
11. H. Dong, J. Su, M.O. Speidel, in proceedings of International Conference on High Nitrogen Steels, 2006
12. J. Kunze, *Nitrogen and Carbon in Iron and Steel* (Akademie Verlag, Berlin, 1990)
13. N. Grigorova, *Carbonitrides and High Speed Steels—Chemical Phase analysis* (Interlsoft, Sofia, 1995)
14. T. Rashev, *High Nitrogen Steels, Metallurgy under Pressure* (Publishing House of the Bulgarian Academy of Sciences, Sofia, 1995)
15. T. Murata, M. Sakamoto, *Nitrogen-Alloyed Steels—Fundamentals and Applications* (AGNE Publishing Inc., 1997)
16. V.G. Gavriljuk, H. Berns, *High Nitrogen Steels* (Springer-Verlag, Berlin, 1999)
17. U. Kamachi Mudali, B. Raj (eds.), *High Nitrogen Steels and Stainless Steels* (Narosa Publishing House, Chennai, 2004)
18. P. Perrot, J. Foct, Techniques de l'Ingénieur **275**, 1 (2003)
19. H.J.C. Speidel, M.O. Speidel, (Institute of Metallurgy, ETH Zürich) **7**, 101 (2003)
20. G. Stein, J. Menzel, H. Dörr, (The Institute of Metals London Publishers, ISBN 2901462_45_4) **1**, 32 (1988)
21. J. Romu, H. Hänninen, (Trans Tech Publications Ltd, Finland) **5**, 673 (1998)
22. F. Vanderschaeve, R. Taillard, J. Foct, Steel Res. **64**, 221 (1993)
23. J. Keichel, G. Gottstein, J. Foct, Mat. Sci. Forum **318–320**, 785 (1999)
24. G. Stein, V. Diehl, (International Conference, Ostend) **8**, 421 (2004)
25. J. Keichel, J. Foct, G. Gottstein, ISIJ Int. **11**, 1788 (2003)
26. A. Nyilas, B. Obst, in 1, 194 (1988)
27. P. Müllner, C. Sollenthaler, P.J. Uggowitzer, M. Speidel, Acta. Metall. Mater. **42**, 2211 (1994)
28. J.B. Vogt, J. Foct, G. Robert, J. Dhers, Metall. Trans. **22A**, 2385 (1991)
29. J.B. Vogt, B.A. Saadi, J. Foct, Z. Metallkd. **90**, 323 (1999)
30. J. Foct, N. Akdut, G. Gottstein, Scripta Metall. Mater. **27**, 1033 (1992)
31. J. Rawers, D. Govier, D. Cook, **4**, 958 (1996)
32. V.G. Gavriljuk, V.N. Shivanyuk, J. Foct, Acta Mater. **51**, 1293 (2003)
33. H.-Y. Ha, T.-H. Lee, C.-S. Oh, S.-J. Kim, Scripta Mater. **61**, 121 (2009)
34. M. Ojima, M. Ohnuma, J. Suzuki, S. Ueta, S. Narita, T. Shimizu, Y. Tomota, Scripta Mater. **59**, 313 (2008)
35. V. Gaviljuk, B. Shanina, HTM J. Heat Treat. Mat. **65**, 189 (2010)
36. C. Domain, C.H. Becquart, J. Foct, Phys. Rev. B **144112-1**, 12 (2004)
37. C.S. Becquart, C. Domain, J. Foct, Phil. Mag. **85**, 533 (2006)
38. J. Foct, C. Cordier, Hyperfine Interact. **190**, 15 (2009)
39. N. Akdut, B.C. De Cooman, H.S. Kim, Proceedings of First International Conference on Interstitially Alloyed Steels, 459 (2008)

High-Nitrogen Steels: the Current State and Development Trends

Anatoly G. Svyazhin, Jerzy Siwka, and Ludmila M. Kaputkina

Abstract

Production expansion of high nitrogen steels, both with low nickel content and without nickel, in various structural classes will be the main trend.

Keywords

High nitrogen steels • Nitrogen optimal concentration • Current state • Future

The significant application of nitrogen as an alloying element commenced in the 1980s of the past century. The steels produced then, which contained 0.5–1.0% of nitrogen, were called high-nitrogen steels (HNS) or nitrogen "hyperequilibrium" steels. At such its contents, nitrogen imparts unique properties to the steel; for example, stainless high-nitrogen steels are characterized by high strength and high corrosion resistance at the same time, and therefore the high-nitrogen steels have initiated a new branch in physical metallurgy. Since 1988, international conferences devoted to these steels have been regularly held. More than 20 years have already passed since the first HNS conference (1988, Lille, France). During this time, a new branch in metallurgical science and international cooperation between researchers and engineers involving in the problems of HNS have been established.

A considerable progress has been made in recent years in the understanding of alloying steels with nitrogen under normal and high pressures and in studying the nature of the processes of forming the structure and properties of HNS, and new areas of application of these steels have been set out. The commercial-scale production of high-nitrogen steel products for the power industry, transport, the chemical and food industries, and medicine has grown. It should be emphasized, however, that the use of HNS still does not meet the potentials for improving the properties of the steels, which result from their nature.

Nitrogen, as an austenite-forming element, is a substitute for nickel. The nickel prices are growing at an accelerated rate. At least two causes of this phenomenon can be mentioned, namely: the increased demand for nickel for production of nickel-based special alloys, and the systematic growth in the production of the classical 18-8 stainless steel, though in many instances this steel could be substituted with a high-nitrogen steel with a low nickel content, or even with this type of steel without nickel. In addition, nickel has proved to be an allergen, which also favours the spreading of use of nickel-less steels. In this situation, the basic trend will be an unavoidable spread of the production and use of economically advantageous nitrogen-alloyed steels with low nickel content or steels not containing this expensive metal, of various structural types (austenitic, martensitic, and multiphase).

With any method of introducing nitrogen to the liquid steel the nitrogen content will be determined by the pressure of the nitrogen gaseous phase. Hence, the nitrogen pressure may be the basis for the most generalized classification of nitrogen-alloyed steels. According to the classification proposed earlier [1], the alloying of steels with nitrogen can be divided into three groups, depending on the nitrogen pressure during their smelting: micro-alloy steels with nitrogen,

A. G. Svyazhin (✉) · L. M. Kaputkina
National University of Science and Technology "MISiS", Moscow, Russia
e-mail: svyazhin@misis.ru

J. Siwka
Czestochowa University of Technology, Cestochowa, Poland

Y. Weng et al. (eds.), *Advanced Steels*, DOI: 10.1007/978-3-642-17665-4_38,

nitrogen steels and high-nitrogen steels. At a pressure equalized with ambient pressure, nitrogen steels and micro-nitrogen steels can be smelted. The difference between them is determined by the chemical composition of the steels. Micro-nitrogen steels have a ferritic matrix. The solubility of nitrogen in ferrite is much lower than in liquid metal. Therefore, in order to avoid blistering, the nitrogen content in micro-nitrogen steels prior to their casting must be lower than the equilibrium content for $p_{N_2} = 0.1$ MPa. For example, in low-alloy steels this is normally not more than 0.02–0.03% N.

Nitrogen steels contain Cr, Ni, Mn and crystallize following the austenite mechanism. The solubility of nitrogen in alloyed austenite is higher than in liquid metal. For this reason, in austenitic steels smelted under normal pressure the nitrogen content can be as high as from 0.4 to 0.5%.

High-nitrogen steels, in international terminology called HNS, have nitrogen contents higher than the equilibrium content for $p_{N_2} = 0.1$ MPa, and therefore, specialized hyperbaric reactors are required for the production of these steels. The nitrogen content in austenitic high-nitrogen steels may be higher than 1.0%.

Under its normal condition, nitrogen is a gas. This makes it different from other alloying elements. The solubility of nitrogen in liquid metal, the phases α and γ, is variable. Therefore, two technological tasks are performed in the production of nitrogen-alloyed steels: to achieve the required nitrogen content of the finished metal without nitrogen blistering during metal solidification, and to employ the appropriate heat treatment method ensuring the required structure of the steel and the correct nitrogen distribution among phases. The nitrogen content should not be too high, but optimal in order to prevent nitrogen not only from blistering, but also from excessive precipitating in the form of nitrides during crystallization and in the solid state in the high temperature range (Fig. 1). The more nitrides and the higher their precipitation temperature, the larger and the less soluble they are during successive treatments—heating, which means the redundancy of the excess nitrogen. Moreover, rational selection of composition with respect to the remaining alloying elements is necessary. At every stage of treatment, the formation of large, sparingly soluble nitrides must be avoided, because their contribution to hardening is low, and the adverse effect on plasticity and impact strength is relatively high.

Upon cooling at a rate of 2,000–3,800 K/s, these steels in their as-cast state have an austenitic structure. It can be seen in Fig. 1 that the higher nitrogen pressure during steel smelting, and thus the higher nitrogen content in the steels concerned, the relatively greater part of nitrogen occurs in nitrides, and the smaller in the solid solution after solidification. It has been established previously that in order to dissolve high-temperature and crystallizing nitrides, higher temperatures and long times of holding at those temperatures are needed, which induces negative phenomena, such as grain growth and denitriding of the surface layer. For steel Cr18Ni10, 30 min of holding at 1,200°C will be needed, with the depth of the "denitrided" surface layer reaching 1 mm [2]. For these reasons, steels with alloying nitrogen require the precise determination of the concentration ratio of nitrogen and other alloying elements, as well as increased crystallization rates. During solidification of industrial ingots, the cooling rates amount to several Kelvin per second. Hence, the fraction of nitrogen forming nitrides during solidification is considerably higher than the one indicated in Figs. 1 and 2. Respective ingot zones are characterized by different cooling rates during solidification and, consequently, the magnitudes of nitrogen distribution between the solid solution and nitrides are different, with a particularly great difference appearing for two-phase steels (Table 1).

At a cooling rate of up to 50 K/s, these steels in an as-cast state have an austenitic-martensitic structure. The fraction of phases and the amount of nitrogen contained in them depend on the local cooling rates. The higher cooling rate, the less martensite and more nitrogen in the solid

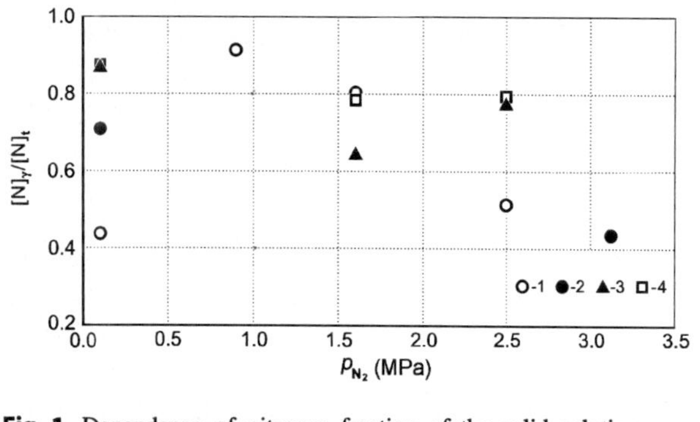

Fig. 1 Dependence of nitrogen fraction of the solid solution γ on nitrogen pressure. Steels: *1*, *2* Cr18Ni10; *3* Cr15Ni7Cu; *4* Cr15Ni7Cu-MoV. Cooling rate (K/s)*1*, *3*, *4* 3,800; *2* 2,000

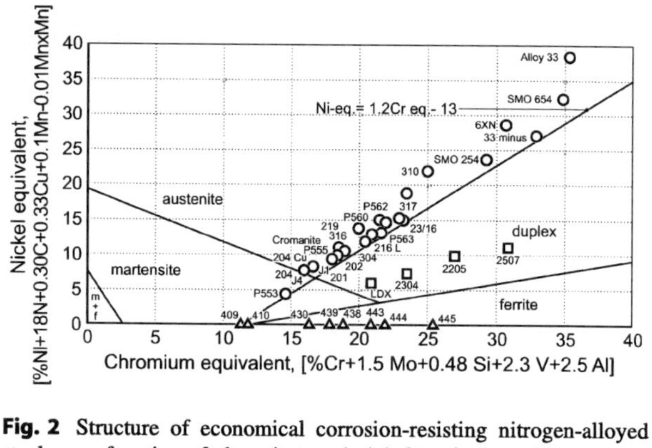

Fig. 2 Structure of economical corrosion-resisting nitrogen-alloyed steels as a function of chromium and nickel equivalents [5]

Table 1 Chemical and phase compositions of two-phase steels—the as-cast structure

Parameter	Cooling rate (K/s)		
	50	8	2
05Cr15Ni5CuMo			
$[N]_t$ (%)	0.21	0.20	0.20
$[N]_\gamma$ (%)	0.21	0.11	0.07
$[N]_\alpha$ (%)	20	30	40
02Cr15Ni5CuV			
$[N]_t$ (%)	0.14	0.14	0.13
$[N]_\gamma$ (%)	0.16	0.11	0.08
$[N]_\alpha$ (%)	0.11	0.09	0.08
η_α (%)	40	50	60

solution, and thereby less nitrides. Therefore, the structure is very inhomogeneous within the volume of metal. In the zones with low cooling rates, high-temperature nitrides occur in large quantities. In this situation, in respect of the as-cast structure of a slow-solidifying ingot, it is not possible to regulate the phase composition and to avoid the precipitation of large quantities of high-temperature nitrides. This should be taken into account during subsequent thermoplastic working. The correct selection of deformation cycles and heat treatment, and in particular the application of thermomechanical working, will enhance the effectiveness of alloying steels with nitrogen [3].

The International High-Nitrogen Steels Conference 2009 was held in Moscow on 6–8 July 2009. The previous conferences were held in France, Germany, Ukraine, Japan, Finland, Sweden, India, Switzerland, Belgium and China.

The conference was attended by delegates from 17 countries, representing virtually all leading scientific schools dealing with HNS problems. The problems of the thermodynamics and kinetics of nitrogen alloying, phase transitions, the evolution of the structure and properties of steel, austenitic, ferritic and two-phase steels, production and application of HNS, and the corrosion and surface treatment of these steels were discussed. It has been established that the production of steels with "superequilibrium" nitrogen is concentrated in companies, such as Boehler, Sandvik and Energietechnik Essen for the purposes of special products intended for the power industry, transport and the chemical industry, where the production costs of these steels are offset by unique properties that can be achieved by carrying out alloying with nitrogen in a liquid metal state under high pressure. A new area of application of these steels in a wide scale might be the manufacture of pipes for submarine petroleum and gas pipelines from austenitic stainless steels with the "super-equilibrium" nitrogen content [4]. The basis for this argument has been provided by the statistical data on the operation of the petroleum and gas pipeline mains in the USA in the years 2002–2003. According to these data, losses due to failures are in 40% of instances caused by metal corrosion [4].

The most advantageous economic effect resulting from using nitrogen as an alloying element, from the present day's point of view, is gained for smelting steel at atmospheric pressure, as there is no need for using any special equipment. A large number of economical corrosion-resisting nitrogen-alloyed steels with a nitrogen content from 2 to 4% have been developed and applied to date in many areas [5].

As an alternative to high-quality materials, such as super-austenitic nickel-based stainless steels, the Sandvik company [6] has developed two-phase steels smelted at atmospheric pressure, which distinguish themselves by high corrosion resistance, excellent mechanical properties, good weldability and a relatively low price (Table 2).

A new factor stimulating the production of nitrogen-alloyed steels at atmospheric pressure has appeared to be the C + N alloying system. Joint C + N alloying makes it possible to enhance strength, impact resistance and plasticity, while retaining sufficiently high corrosion resistance.

Table 2 Chemical composition of two-phase steels (wt%) [6]

Grade	UNS	C_{max}	Cr	Ni	Mo	N	PRE[a]
Sandvik SAF 2507	S32750	0.03	25	7	4	0.3	42.5
Sandvik SAF 2707 HD	S32707	0.03	27	7	5	0.4	48
Sandvik SAF 3207 HD	S33207	0.03	32	7	3.5	0.5	50

For two-phase stainless steels, resistance to pitting corrosion is proportional to the pitting resistance equivalent, PRE (PRE = % Cr + 3.3% Mo + 16% N)

[a]Materials for pipes

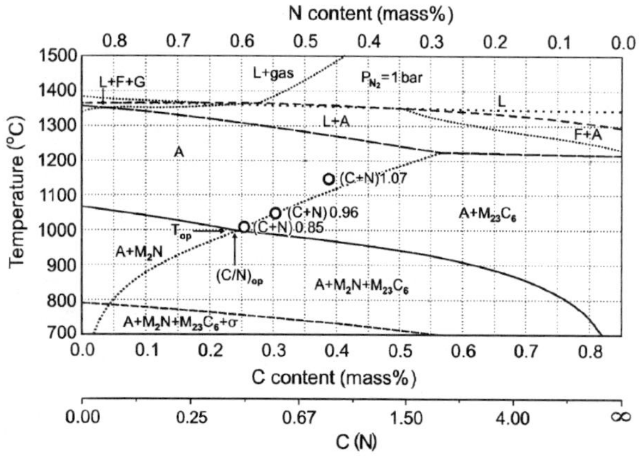

Fig. 3 Phase diagram for steel Cr18Mn18 with C + N=0.85 and C/N = 0.43. T_{op} and $(C/N)_{op}$—optimal values. The temperatures of the start of precipitation of $M_{26}C_6$ carbides are indicated by cross marks [7]

works [7, 8] provide the results of investigation and development of steel Cr18Mn18 with a C + N content in the range from 0.85 to 1.07%. With the optimal ratio of C to N, a stable homogeneous austenitic structure can be produced in these steels, with a yield stress of approximately 600 MPa, i.e. three times greater than for the standard CrNi stainless steel, and with an elongation of approximately 70% (Fig. 3).

For nitrogen-alloyed steels it is possible to extend the region of effective application by means of enhancing the special (functional) properties. The effect associated with introducing nitrogen to steel increases as a result of applying high-temperature thermomechanical treatment. For example, a number of high-strength and corrosion-resisting steel based on the following alloying system: 15–17% Cr; 5–10% Ni; 2% Cu; 1–2% Mo; 0.1–0.2% V; 0.12–0.15% N have been developed at the Moscow Institute for Steels and Alloys. These are economical steels in terms of their contents of alloying elements and in terms of their production technology, because they can be smelted at normal pressure. Depending on the nitrogen content and the thermomechanical treatment regime, these steels may exhibit a structure varying from austenitic to martensitic, while having the complex of mechanical properties corresponding to those structures, as shown in Fig. 4.

The strength level achieved after the high-temperature thermomechanical treatment of austenitic, martensitic and two-phase aged nitrogen steels is higher by 1.5–2 times compared to similar steels not containing alloying nitrogen, with the same toughness and plasticity. By changing the HTMT cycle, the region of correlation between σ and ψ, as

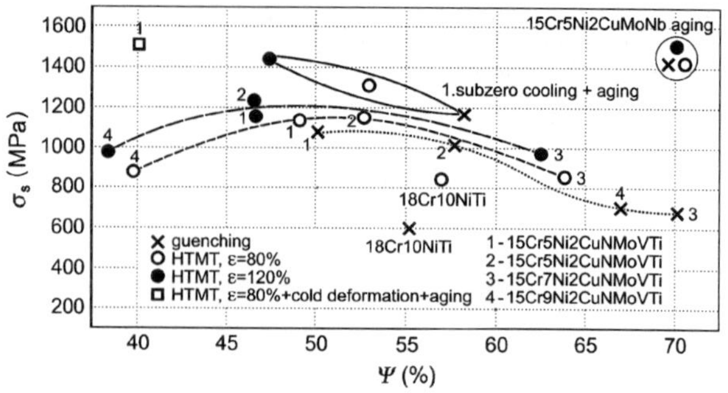

Fig. 4 Diagram of the constructional strength of nitrogen steels ([N] = 0.12–0.15%) after different treatments. *HTMT* High Temperature Mechanical Treatment. Structure: *1–3* austenite + martensite; *4* austenite

necessary for different applications, can be established (Fig. 4). The highest strength is achieved by steels with a martensitic matrix, which are hardened by nitrides and carbidonitrides. At the same time, they retain their corrosion resistance in moderately aggressive environments, as in the case of steel Cr18N10.

In nitrogen-alloyed steels it is technically possible to introduce precipitation hardening with nano-size nitrides and carbidonitrides to the technological process of their production.

Thus it will be possible to produce a high-strength thermally stable state in an industrial scale, while maintaining or enhancing the effects of structural and substructural, including nano-structural, hardening. In this case, the properties (capabilities) of a material of complex chemical composition are largely used.

References

1. A.G. Svyazhin, Hutnik-Wiadomosci Hutnicze **8**, 264 (1996)
2. L.M. Kaputkina, A.G. Svyazhin, J. Siwka, V.G. Prokoszkina, Hutnik-Wiadomosci Hutnicze **7**, 357 (2007)
3. A.G. Svyazhin, L.M. Kaputkina, Izv. VUZ. **10**, 36 (2005)
4. Ts. Rashev, Ch. Andreev, L. Jekova et al., High Nitrogen Steels, HNS 2009. Proceedings of the 10th International Conference, 2009, p. 287
5. M. Speidel, High Nitrogen Steels, HNS 2009. Proceedings of the 10th International Conference, 2009, p. 121
6. G. Chai, U. Kivisakk, S. Ronneteg, High Nitrogen Steels, HNS 2009. Proceedings of the 10th International Conference, 2009, p. 67
7. H. Berns, S. Riedner, V. Gabriljuk, High Nitrogen Steels, HNS 2009. Proceedings of the 10th International Conference, 2009, p. 129
8. V. Gavriljuk, B. Shanina, A. Tyschenko et.al., High nitrogen steels, HNS 2009. Proceedings of the 10th international Conferences, 2009, p. 140

Development of Stainless Steels with Superior Mechanical Properties: A Correlation Between Structure and Properties in Nanoscale/ Sub-micron Grained Austenitic Stainless Steel

S. Rajasekhara, L. P. Karjalainen, A. Kyröläinen, and P. J. Ferreira

Abstract

A review of the structure–property–performance relationship in a technologically important cold-rolled and annealed metastable austenitic stainless steel (AISI 301LN SS) is presented. AISI 301LN SS is cold-rolled to 63% reduction and subsequently annealed at 600–1,000°C from 1 to 100 s. Cold-rolled and annealed samples are studied through X-Ray Diffraction (XRD), Superconducting Quantum Interference Device (SQUID), transmission electron microscopy (TEM) and tensile testing to understand the morphology of the cold-rolled AISI 301LN and the annealed martensite to austenite reversion, the formation of nano/submicron grain sizes and the mechanical properties achieved. Tests show that cold-rolled samples annealed at 600 and 700°C exhibit partial $\alpha' \rightarrow \gamma$ reversion, while for the case of 800–1,000°C annealing treatments, the reversion from α'-martensite to γ-austenite is almost complete, along with rapid austenite grain growth. Tensile tests performed on AISI 301LN nano/submicron grained SS reveal a high yield strength of ~700 MPa, which is twice the typical yield strength of conventional fully annealed AISI 301LN SS. An analysis of the relationship between yield strength and grain size in these nano/submicron grained SS indicates a classical Hall–Petch behavior, despite the temperature dependence observed due to an interplay between fine grained austenite, solid solution strengthening, precipitate hardening and strain hardening.

Keywords

Austenitic stainless steels • Ultra-fine grains • Phase transformations • Mechanical properties

1 Introduction

Almost 1.1B tons of steel was consumed worldwide in 2006 [1] representing a global economy of ~\$0.8 trillion [1]. Following the current trend in globalization, the demand for steel continues to be strong owing to increased consumption from emerging economies (Brazil, Russia, India and China) and a continued demand from developed countries to maintain/upgrade their infrastructural needs [1]. Considering the importance of this industry, one of the goals of scientists and engineers working in this field has been to develop alloys that exhibit both high strength and high ductility. This combination of properties would allow scientists and engineers working in industries, such as Oil & Gas, Construction, Food, Machinery and Automotive to design safer, durable, more fuel-efficient (in case of the automotive industry) structures. Unfortunately, until the 1910–1930s all of the abovementioned industries were using only a few steel grades, namely high carbon steel and medium carbon steel, with a tensile strength of ~450 MPa

S. Rajasekhara · Prof. P. J. Ferreira (✉)
Materials Science and Engineering Program,
The University of Texas, Austin, TX 78712, USA
e-mail: ferreira@mail.utexas.edu

L. P. Karjalainen
Department of Mechanical Engineering,
The University of Oulu, 90014, Oulu, Finland

A. Kyröläinen
Outokumpu Stainless Oy, 95400, Tornio, Finland

Y. Weng et al. (eds.), *Advanced Steels*, DOI: 10.1007/978-3-642-17665-4_39,

and a total elongation of $\sim$30% [2]. These steels were adequate for simple engineering structures but became inadequate as design complexities grew and safety requirements became more stringent.

The 1940–1970s saw rapid progress in steel manufacturing, as well as research, resulting in the development of advanced ferrous alloys with enhanced properties. Of particular importance was the development of various families of stainless steels (SS) (austenitic, ferritic, martensitic, duplex stainless steels), dual-phase steels (steels with a combination of ferrite and martensite phases), and TRIP steels (steels with a good combination of strength and ductility, achieved through phase transformations). More recently, further developments have produced steels that have either high strength and poor ductility (Complex phase and Martensite steels) or low strength and high ductility (IF Steels) (Fig. 1).

Clearly, there is a need to develop Fe-based alloys that have both exceptional strength and ductility. With these novel materials, engineers will be able to design complex structures that can sustain natural environments (such as humid air and corrosive media), be durable and lightweight. In this regard, SS are good candidates due to their excellent corrosion resistance properties, but they lack the combination of high strength and ductility. As we uncover exciting potential applications for SS, the motivation for this work is centered around the following question: Is it possible to develop commercial SS with high strength and high ductility?

One approach to achieve this objective is to produce ultra-fine grained SS. The basis for this approach is based on the Hall–Petch equation [9–11] given by:

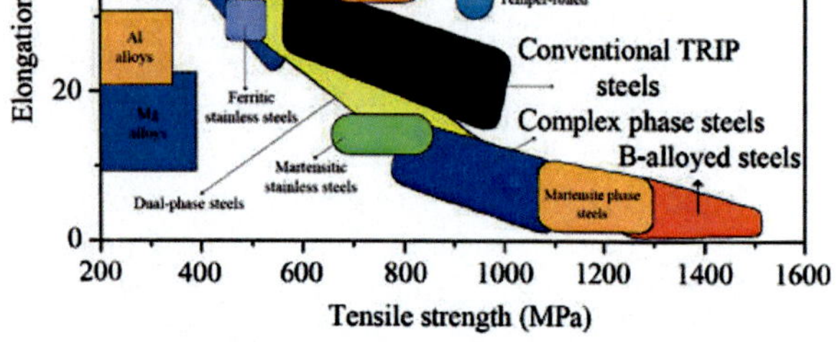

Fig. 1 Elongation until fracture (%) versus tensile strength for a variety of alloys (adapted from [3–8])

$$\sigma_y = \sigma_0 + kd^{-1/2} \tag{1}$$

where σ_0 is the friction stress in the absence of grain boundaries, k is a constant and d is the grain size. Thus, the yield stress of a material increases with smaller grain size because pile-ups in fine grained materials contain fewer dislocations, the stress at the tip of the pile-up decreases and thus, a larger applied stress is required to generate dislocations in the adjacent grain.

Studies in the past have demonstrated this concept in a low nickel (LNi) non-commercial SS, where a $\sim$25% increase in tensile strength was observed when the austenite grain size was decreased from 50 to $\sim$3 μm [12]. Recent studies have demonstrated commercial SS with submicron grains sizes of $\sim$0.8 μm [13, 14], and nanoscale grain sizes of $\sim$100 nm in non-commercial SS with significant improvements in yield strength [15].

The abovementioned studies employ a thermo-mechanical treatment wherein a metastable SS is initially heavily cold-rolled to produce deformation induced martensite. This martensite phase consists of regions of lath martensite and ultra-fine dislocation-cell type martensite [16–22]. Subsequently, the heavily deformed SS is annealed to produce ultra-fine austenite grains. These studies have emphasized that in order to obtain nano/sub-micron SS it is crucial to be able to attain deformation-induced martensite upon cold rolling. While the cold rolling reduction per rolling pass is an important consideration to obtain deformation-induced martensite in SS, the stability of austenite determines to a large extent the amount of deformation-induced martensite that can be obtained after cold rolling reduction. In particular, metastable austenitic SS are ideal materials for this process.

In this work, several commercial metastable austenitic SS grades—AISI 301LN, 301, 304 and 316—whose composition is shown in Table 1, were initially considered as potential candidates. To identify the SS with most promise, the amount of deformation induced martensite as a function of cold reduction was obtained for AISI 301 and 301LN SS, while previous research conducted by Shrinivas and Murr [20] demonstrated that for 60% cold reduction, negligible fraction of martensite was produced in AISI 316 SS, whereas AISI 304 SS exhibited approximately 40% martensite (Fig. 2b). Ferritoscope measurements indicated that AISI 301LN SS exhibited almost 100% martensite after 60% cold-rolling reduction and therefore, it was considered the best choice for the thermo-mechanical processing described above.

In this paper, the phase fraction, morphology and mechanical properties of the cold-rolled and annealed AISI 301LN SS were analyzed with a combination of techniques, to develop a comprehensive understanding of the interplay between structure, properties and performance for this alloy.

Table 1 Composition of various metastable austenitic SS (wt%)

	C	N	Cr	Ni	Mn	Si	Cu	Mo
AISI 301	0.1	0.07	16.7	6.3	1.18	1.06	0.24	0.65
AISI 301LN	0.02	0.15	17.3	6.5	1.29	0.52	0.2	0.15
AISI 304	0.07	–	19.12	10.43	1.4	0.59	–	0.14
AISI 316	0.04	–	17.85	13.13	1.77	0.21	–	2.99

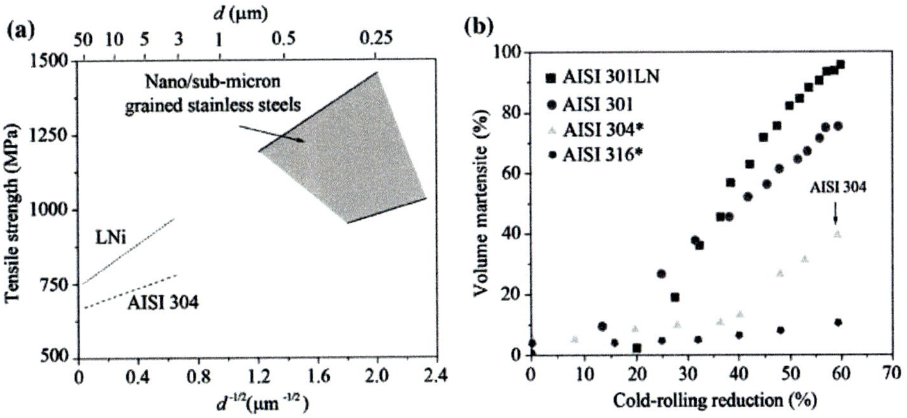

Fig. 2 **a** The relationship between grain size and tensile strength in an low nickel (LNi) alloy and AISI 304 SS [12]. The gray band shows the envisaged strength enhancement due to grain boundary refining for nano/sub-micron grained SS. **b** Volume percentage of martensite as a function of cold reduction for various commercial SS grades. *Data for AISI 304 and 316 SS was adapted from Shrinivas and Murr [20]

2 Experimental

2.1 Materials and Methods

A 165 mm thick AISI 301LN type stainless steel slab, with the composition shown in Table 1, was produced by continuous casting at Outokumpu Stainless Oy, Tornio, Finland. The slab was solution annealed at 1,100°C for 1 h to dissolve any secondary phase precipitates and subsequently hot rolled to ~1.5 mm thickness. Samples from this 1.5 mm thick sheet were then cold-rolled through 19 consecutive passes in a four-calendar rolling mill at the Research Center of Outokumpu Stainless Oy to achieve a cold-reduction of 63%. Cold-rolled samples of dimensions ~100 × 5 × 0.8 mm^3 were heat-treated in a Gleeble-1500 thermo-mechanical simulator at the University of Oulu, Oulu. The samples were heated at a heating rate of 200°C/s and annealed at 600–1,000°C for 1–100 s and forced-air-cooled at a cooling-rate of about 200°C/s.

2.2 Characterization of the Cold-Rolled and Annealed AISI 301LN SS

The phase fraction of the cold-rolled and annealed AISI 301LN SS was determined by X-ray diffraction (XRD) analysis and by measuring the saturation magnetization per unit volume of these samples when subjected to an external magnetic field in a SQUID instrument. The fractional amount of austenite (paramagnetic) versus martensite (ferromagnetic) that are present in a sample is obtained from the magnetization data, which is related with the X-Ray data to obtain phase fraction information, as discussed elsewhere [23, 24].

The microstructure of these samples was analyzed by transmission electron microscopy (TEM). The annealed samples were first thinned to ~100 μm by mechanical polishing. Subsequently, these samples were thinned by jet-polishing [23]. Two different TEMs, a JEOL 200CX and a JEOL 2010F, were used in this work to identify (i) the martensite morphology in the cold rolled sample, (ii) the morphology of the austenite formed after reversion, (iii) the austenite grain size and (iv) secondary phase precipitates. The JEOL 200CX is ideal for viewing large areas at low magnification, whereas the JEOL 2010F is ideal for studying the microstructure at higher resolutions.

The Image J software (National Institute of Health, Bethesda, MD) was used on the scanned TEM negatives to estimate the austenite grain size of annealed samples. A statistically significant sample of 100 grains were located in the scanned images and measurements were taken along the longest direction within the grain and its perpendicular, from which the mean was calculated and used as the average grain size $\bar{d}$ for a particular sample. The nano/sub-micron grains (30–200 nm) measured through this technique exhibit an error of approximately 10%, while the larger austenite grains (>1 μm) have an error of just 4%.

Finally, the mechanical properties of these samples were tested on a Zwick tensile testing apparatus at Outokumpu Stainless Oy, Finland [24].

3 Results and Discussion

3.1 Martensite/Austenite Phase Fractions of Cold-Rolled and Annealed AISI 301LN SS

The X-ray Diffraction spectrum of the 63% cold-rolled (CR) AISI 301LN sample is shown in Fig. 3a. The sample consists of primarily martensitic (α') with strong (211), (110) and (200) α' peaks and a weak (220) austenite (γ) peak (Fig. 3a), which may be due to the retained austenite that failed to convert to martensite upon cold-rolling deformation. Furthermore, CR samples exhibited a saturation magnetization approximately 944 emu/cm^3 [23]. Knowing the peak intensities obtained from this X-ray diffraction and employing the procedure described elsewhere [23], the volume percentage of martensite in the CR sample was determined to be approximately 95.7%. A similar procedure was applied to the annealed samples and the results are shown as a function of temperature in Fig. 3b. At 600°C, hardly any reversion to austenite occurs for samples annealed for 1 and 10 s. However, partial reversion to austenite is shown in samples annealed at 600°C for 100 s and 700°C for 1 s. At higher annealing temperatures and longer annealing durations, the annealed samples exhibit almost complete reversion to austenite (Fig. 3b).

3.2 Morphology and Grain Size of Cold-Rolled and Annealed AISI 301LN SS

The 63% CR AISI 301LN SS sample prior to annealing shows regions of dislocation cell-type martensite and lath-type martensite, which are characterized by the presence of dislocation forests and ultra-fine martensite [23], and confirmed by the ring-like selected area diffraction pattern (SADP) (Fig. 4a), while lath-type martensite regions are confirmed by the spot-like SADP (Fig. 4b). Samples annealed at 600°C for 1 and 10 s exhibit dislocation-cell type and lath-type martensite, which is confirmed by the ring-like and spot-like patterns (Fig. 5a, b). Martensitic regions are bounded by dislocation walls and forests (Fig. 5b). Diffraction streaks and spots corresponding to austenite (for instance (113)γ in Fig. 3a, and 220γ in Fig. 5b) are also observed, which may indicate the formation of new austenite grains within the martensite matrix. Nevertheless, the overall microstructure of the samples annealed at 600°C for 1 and 10 s still resemble the overall microstructure of the CR sample. In fact, these annealing conditions are still insufficient to significantly drive the $\alpha' \rightarrow \gamma$ reversion. When the annealing duration is increased to 100 s, evenly distributed equiaxed nano/sub-micron grains of austenite within the martensitic matrix are visible, which is confirmed by the austenitic ring-like reflections (Fig. 5c). Combining this result with the phase fraction analysis (Fig. 3b), it is clear that cold-rolled AISI 301LN SS partially revert to austenite when annealed at 600°C for 100 s [23].

Samples annealed at a higher temperature of 700°C for 1 s exhibit regions with equiaxed nano/sub-micron grains as well as fine martensite, which is confirmed by the presence of ring-like diffraction patterns from both phases (Fig. 6a). This morphology along with the phase fraction results reveals partial reversion from martensite to austenite (approximately 35% austenite) (Fig. 3b). TEM images of samples annealed for longer annealing times at 700°C show

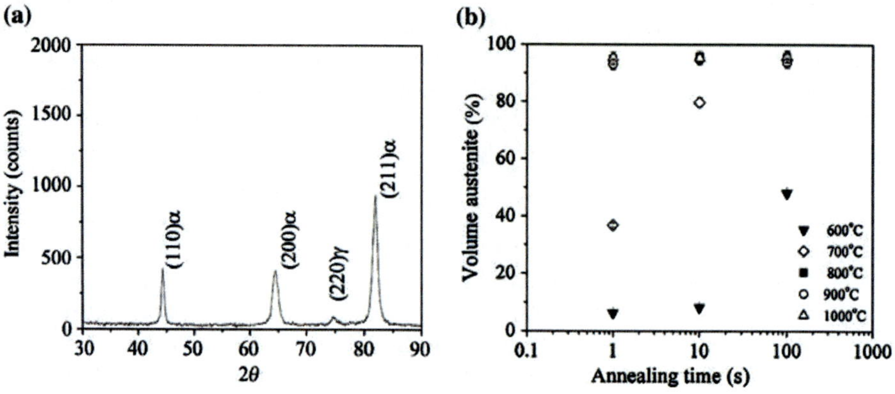

Fig. 3 a X-ray diffraction analysis of 63% cold rolled AISI 301LN SS, **b** volume % of austenite obtained after annealing a 63% cold-rolled AISI 301LN SS at different temperatures and times [23]

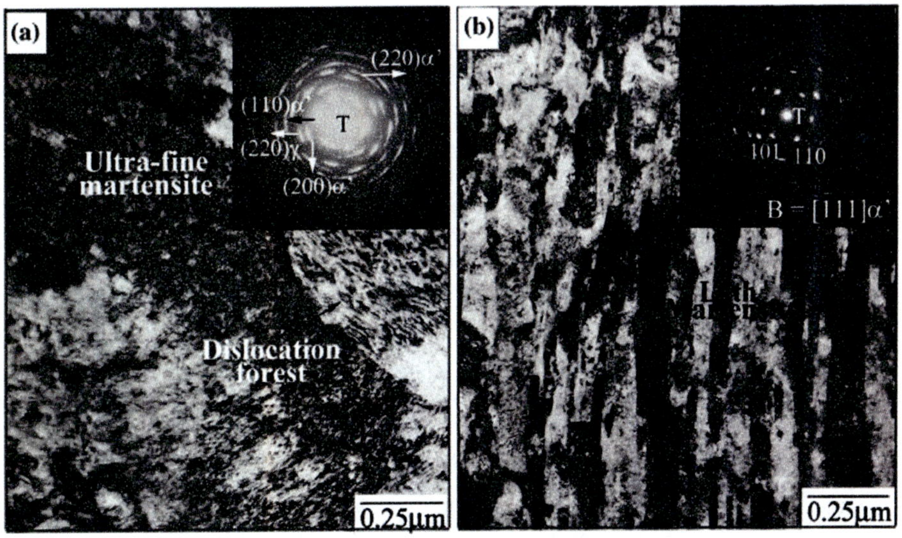

Fig. 4 TEM images of 63% cold rolled AISI 301LN SS; **a** region with dislocation-cell martensite and corresponding SADP, **b** region with heavily deformed lath-martensite and corresponding SADP [23]

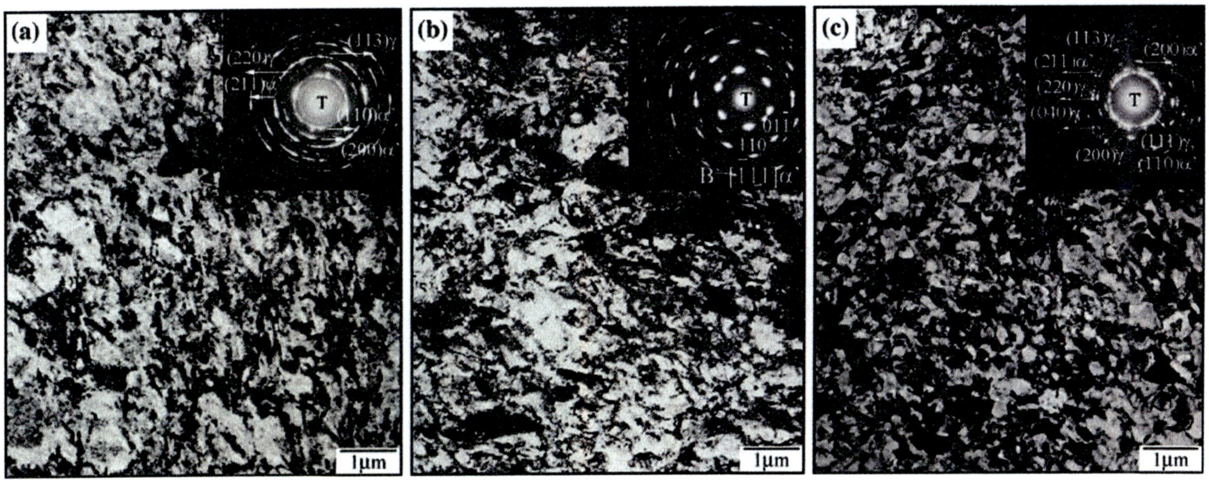

Fig. 5 TEM images of 63% cold rolled AISI 301LN SS annealed at 600°C for **a** 1 s, **b** 10 s, and **c** 100 s. Diffraction patterns corresponding to these images are shown in the image insets [23]

the formation of larger grains (Fig. 6b, c). The corresponding SADP changes from being almost ring-like (in the case of 1 s anneal: Fig. 6a) to a more spot-like pattern for the 10 and 100 s annealing (Fig. 6b, c) confirming this tendency. The sample annealed at 10 s show the presence of a shear band, most likely to be an austenitic shear band, inherited from the cold rolled sample. For both 10 and 100 s annealing, there is a significant distribution of grain sizes [23].

Relatively defect-free and equiaxed austenitic grains are observed in samples annealed at 800°C for all the three annealing durations (Fig. 6d–f). Samples annealed at 800°C exhibit a large grain size distribution with a mixture of nano/sub-micron grains and large grains, which is likely due to two reasons: (i) presence of different types of martensite (dislocation cell-type and lath) in the cold rolled sample leading to different rates for austenite nucleation and (ii) nucleation of austenitic grains at lower annealing temperatures, which grow in size as the annealing temperature is rapidly brought to 800°C. Samples annealed at 900 and 1,000°C for all the three annealing durations exhibit austenitic grain structures with apparently narrower grain size distribution, which rapidly grow in size as the annealing duration is increased from 1 to 100 s (Fig. 7) [23].

The average grain sizes for all annealing treatments are shown in Fig. 8a. Samples annealed at 600 and 700°C hardly exhibit any grain growth (Fig. 8b), while samples annealed at 800, 900 and 1,000°C show a dramatic increase in grain size as the annealing duration is increased to 100 s (Fig. 8a). The normalized range of grain sizes present in a given annealed sample (ratio of the difference between the maximum and the minimum grain size in a sample size of 100 grains to the mean grain size) as a function of annealing

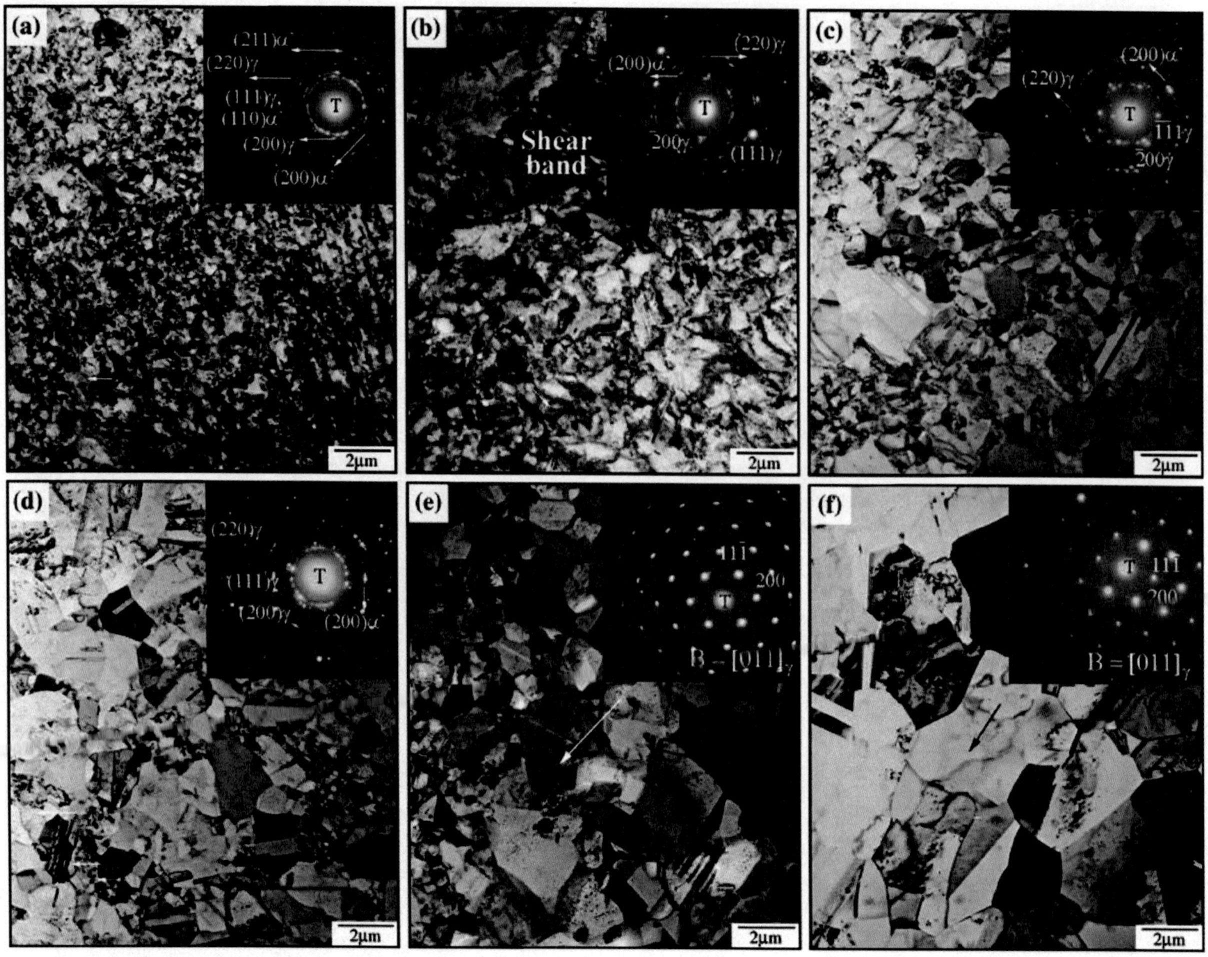

Fig. 6 TEM images and diffraction patterns corresponding to cold rolled AISI 301LN SS annealed at 700°C for **a** 1 s, **b** 10 s, **c** 100 s, and 800°C for **d** 1 s, **e** 10 s, and **f** 100 s [23]

conditions was also analyzed. We note that for 1 s annealing duration, the samples at 600 and 700°C show a grain size variation that is considerably less than for samples annealed at 800, 900 and 1,000°C (Fig. 8c). Interestingly, the grain size variation peaks at 800°C and decreases with increasing annealing temperature. At 10 s of annealing, the grain size variation of samples annealed at 600, 700, 900 and 1,000°C are comparable, but samples annealed at 800°C exhibits the most grain size variation (Fig. 8d). However, when samples are annealed for longer duration of 100 s, the normalized grain size range values for all annealing temperatures are similar (Fig. 8e) [23].

4 Mechanical Properties of Annealed AISI 301LN SS

Figure 9 shows the yield strength of samples annealed at various temperatures. As expected, the yield strength decreases as the annealing temperature increases. The high strength of samples annealed at 600°C for 1–10 s is due to the presence of martensite. However, samples annealed at 600°C for 100 s and 700°C for 1 s still exhibit high strengths, due to a mixture of martensite and austenite. Thus, although there is a dramatic increase in austenite fraction, the strength is only slightly decreased. The samples annealed at 800°C for 1 s are austenitic in nature and show a yield strength of approximately 700 MPa that is approximately twice the strength of conventional annealed AISI 301LN SS (approximately 350 MPa), while exhibiting a relatively high ductility (approximately 35%)

5 Discussion

5.1 Influence of Annealing Conditions on the Austenite Phase Fraction

Samples annealed at 600°C for 1 and 10 s show a relatively small austenite phase fraction (approximately 6%), which is

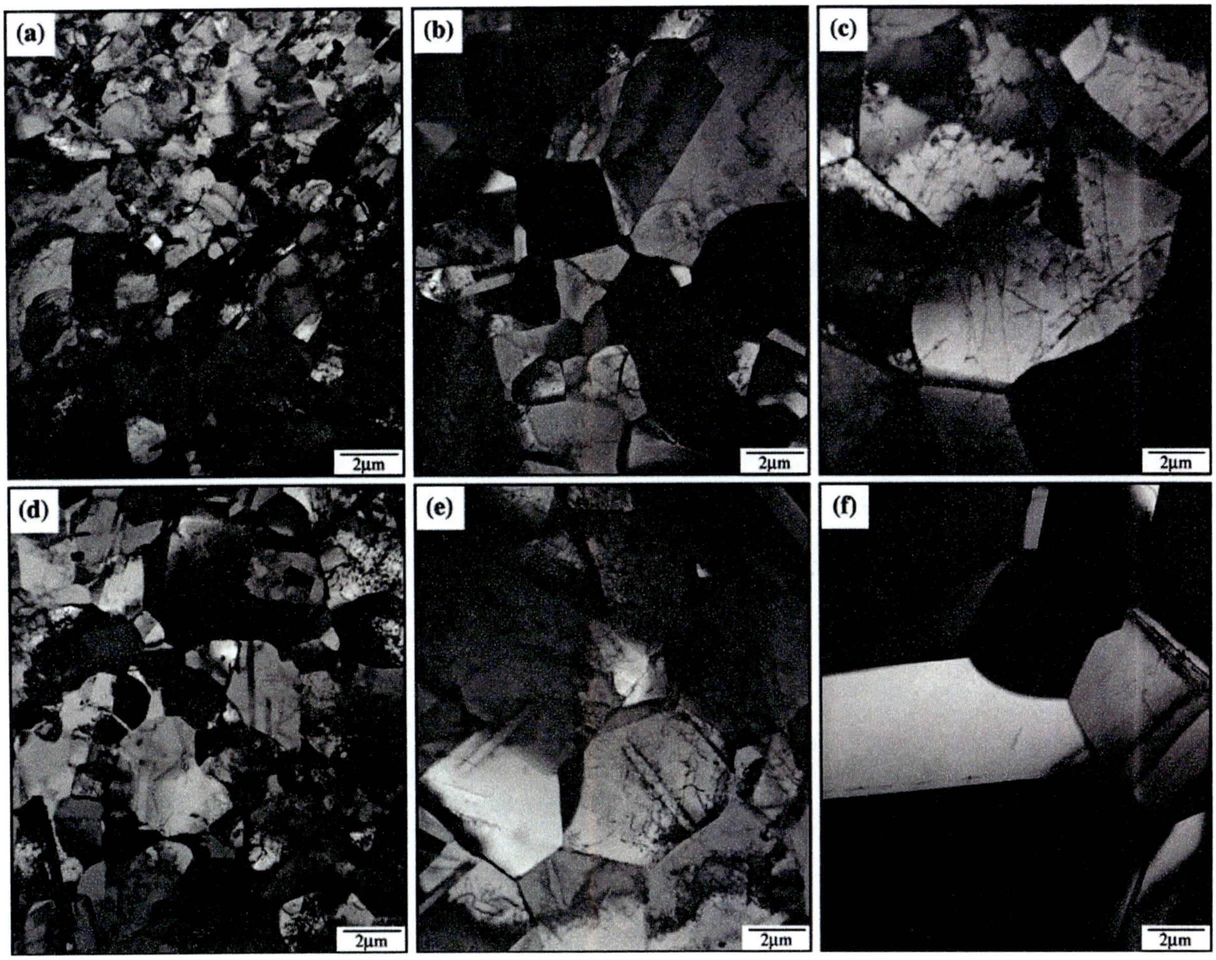

Fig. 7 TEM images and diffraction patterns corresponding to cold rolled AISI 301LN SS annealed at 700°C for **a** 1 s, **b** 10 s, **c** 100 s, and 800°C for **d** 1 s, **e** 10 s, and **f** 100 s [23]

comparable to the austenite phase fraction present in the CR sample (approximately 4%). The microstructure present in these samples is also similar to that present in the CR sample. Therefore, these annealing conditions are not sufficient to drive the $\alpha' \rightarrow \gamma$ reversion. However, samples annealed for 100 s at 600°C, or 1 s at 700°C exhibit partial $\alpha' \rightarrow \gamma$ reversion leading to a nano/submicron austenite and martensite microcstructure (Figs. 3b, 5c and 6a). This data indicates that the $\alpha' \rightarrow \gamma$ reversion should be driven by a diffusion mechanism where an activation energy barrier prevents austenite nucleation at low annealing temperature and short annealing durations, but higher annealing temperatures or longer annealing durations result in austenite nucleation and growth. A higher annealing temperature provides more thermal energy to overcome the activation energy for nucleation, while longer annealing durations promote diffusion driven kinetics allowing the reversion process to proceed towards completion [34]. Evidence of this concept is clearly visible in samples annealed at 700°C, where 1 s of annealing may be just sufficient to overcome the activation energy barrier for austenite nucleation, while diffusion driven kinetics at longer annealing durations of 10 and 100 s take $\alpha' \rightarrow \gamma$ reversion to completion (Fig. 6b, c). As expected, samples annealed at higher annealing temperatures of 800, 900 and 1,000°C all show almost complete reversion to austenite phase (approximately 93% austenite).

5.2 Influence of Annealing Conditions on Grain Size Variation

As reported earlier, samples annealed at 600 and 700°C for all annealing durations show a narrow grain size variation, while annealing at higher temperatures result in a broad variation in grain sizes. The presence of different kinds of nucleation sites with different morphologies, such dislocation-cell martensite, dislocation walls, and lath-type martensite seems to be responsible for such variation.

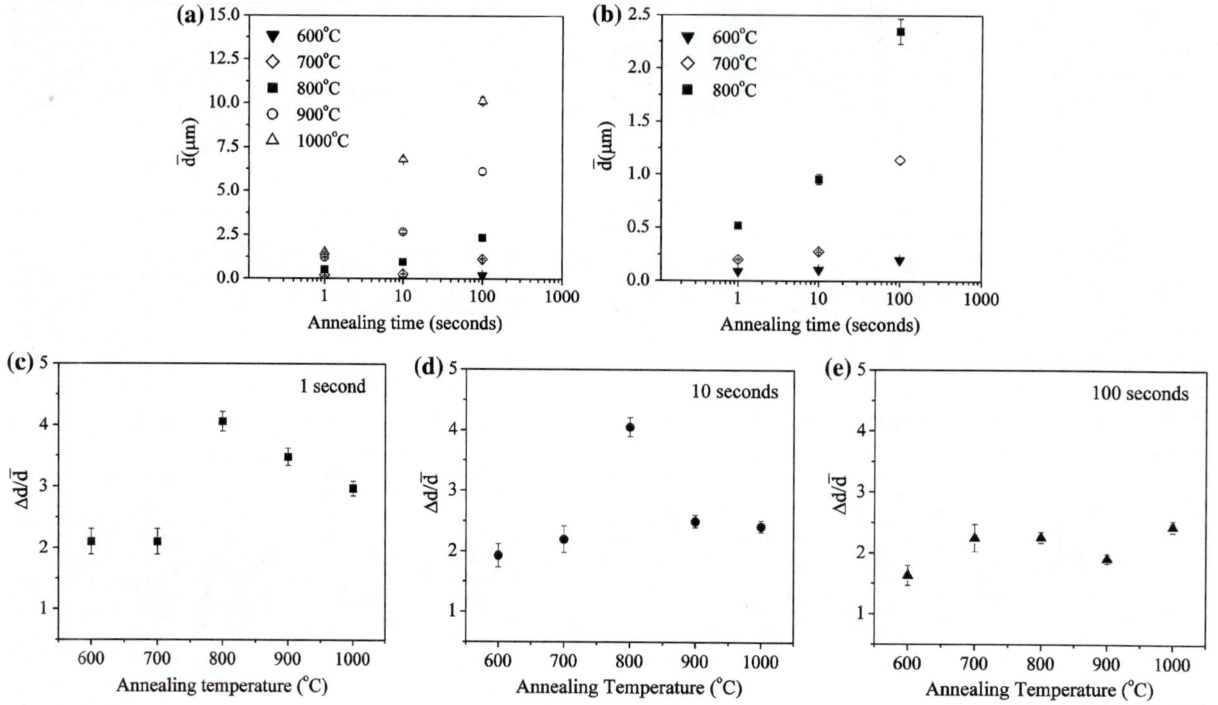

Fig. 8 Grain size measurements for AISI 301LN SS samples as a function of time and temperature: **a** annealed at 600–1,000°C, **b** enlarged view of the samples annealed at 600–800°C. **c–e** Normalized grain size in AISI 301LN SS samples annealed at 600, 700, 800, 900 and 1,000°C for 1, 10 and 100 s $\Delta d = d_{max} - d_{min}$ [23]

Fig. 9 **a** Yield strength, and **b** tensile ductility of AISI 301LN SS subjected to various annealing conditions [24]

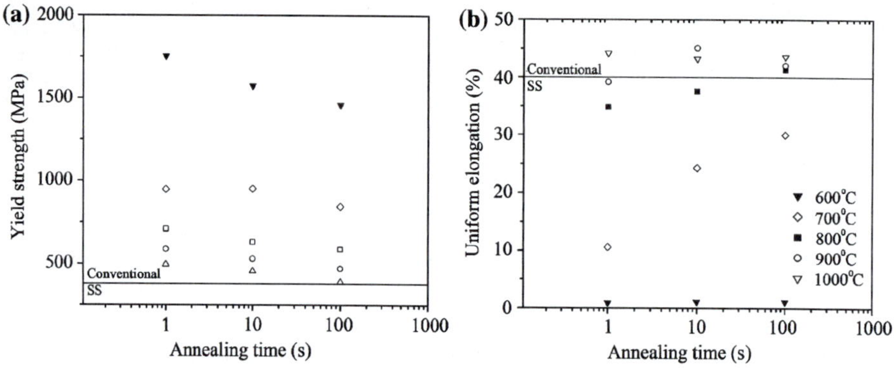

Johannsen et al. [25] have shown that the nucleation of austenite on dislocation-cell type martensite is easier to occur than on lath-type martensite. Thus, for a fixed annealing time during the $\alpha' \rightarrow \gamma$ reversion, the austenite nucleates first on dislocation-cell martensite, followed by grain growth, whereas austenite nucleation on deformed lath-type martensite takes place later leading to smaller grains.

Moreover, according to the research done by Tsuji et al. [16], Ueji et al. [17] and Hansen [26], the deformed lath-type martensite boundaries comprise of incidental dislocation boundaries (IDBs) and geometrically necessary boundaries (GNBs) and ultra-fine grains preferentially nucleate at the GNBs. In the context of the results shown herein, it can be proposed that upon annealing the heavily cold-rolled AISI 301LN SS, austenite nucleates on the dislocation-cell type martensite. Once these nucleation centers are consumed, the austenite grains preferentially nucleate at GNBs present within the deformed lath-type martensite. The sequential nature of austenite nucleation and growth is reflected in the wide grain size variation in samples annealed at 800, 900, and 1,000°C, since the smallest austenite grains measured correspond to the austenite that nucleated last (Fig. 7a). However, annealing at longer durations allows the smaller grains to grow thus reducing the grain size variation (Fig. 7c).

Since extremely short annealing durations are studied in this work, we believe that 100 s of annealing may not be sufficient to completely consume the IDBs present in the deformed lath-martensite, which then leads to tempered martensite in the annealed samples. Detailed observations of the microsctructures of the samples annealed at 700 and 800°C confirm the presence of small pockets of martensite, which is presumably tempered martensite (Fig. 11). In fact, Tomimura et al. [27] and Takaki et al. [28] have found pockets of tempered martensite present at austenite grain triple junctions in thermo-mechanically treated steels.

5.3 Mechanism for the Martensite → Austenite Reversion

From the analysis of the microstructure, the morphology of the phases present and the phase fraction of all annealed AISI 301LN SS samples (Fig. 3), we can infer that the $\alpha' \rightarrow \gamma$ reversion is characterized by: (1) a wide annealing temperature range (600–1,000°C) where the reversion occurs, (2) the formation of defect-free equiaxed austenitic grains which grow in size with time, (3) a sequential use of various types of martensite nucleation sites, leading to a wide grain size distribution, and (4) formation of secondary phase precipitates. These features are typical of a diffusion-type reversion mechanism, in contrast to a shear-type reversion mechanism where: (1) reversion occurs over a small temperature range of ~50 K [25, 27, 28], (2) nucleation is time independent, (3) the formation of defect-laden austenite grains retaining the parent martensite morphology occurs [24, 25, 27, 28], and (4) absence of austenite grain growth is observed [27, 28]. Clearly, there is experimental evidence that seem to confirm the diffusion-type phase transformation mechanism in cold-rolled and annealed austenitic SS. However, currently there is no quantitative analysis of this phase transformation. In addition, several fundamental concepts involved in this phase transformation, such as: (i) the type of the diffusing species driving the phase transformation and (ii) the nature of austenite/martensite grain boundary junctions where the transformation occurs are not yet known. An understanding of these factors is critical for tailoring the composition of metastable austenitic SS grades, as well as for optimizing the cold-rolling procedure and subsequent annealing to promote an efficient $\alpha' \rightarrow \gamma$ transformation, which can lead to the production of ultra-fine grained SS.

Therefore, we have recently applied the general phase kinetics relationship proposed by Erukhimovitch and Baram [29, 30] to quantitatively address the $\alpha' \rightarrow \gamma$ phase transformation in this SS [31]. This approach assumes that the embryos of a new phase nucleate randomly from the time-dependent parent phase. In the case of the $\alpha' \rightarrow \gamma$ phase transformation, this relationship may be expressed as:

$$\ln(1-\xi_\gamma(T,t)) = -\frac{4\pi}{3}\int_0^t v^3(T,t)J(T,t)(t-\tau)^3 d\tau, \quad (2)$$

where $\xi_\gamma(T, t)$ is the time (t) and temperature (T) dependent austenite phase volume fraction, $v(T, t)$ the austenite grain growth velocity, $J(T, t)$ the austenite phase nucleation rate and τ th incubation time for the formation of the first austenite nuclei. For simplicity, we assume the nucleation rate to be independent of annealing duration. Eq. 2 thus reduces to:

$$\ln(1-\xi_\gamma(T,t)) = -\frac{4\pi J(T)}{3}\int_0^t v^3(T,t)(t-\tau)^3 d\tau \quad (3)$$

While the term $J(T)$ will be discussed later, the integral in Eq. 3 may be solved by using an explicit time and temperature dependence relationship for $v(T, t)$, which is given by [32]:

$$v(T,t) = \frac{(kt)^{\frac{1}{n}}}{nt} \quad (4)$$

where k is the austenite grain growth parameter, n the austenite grain growth exponent, and t the annealing time. The grain growth parameter k is a measure of the material's grain boundary mobility, which depends on annealing temperature and grain growth activation energy, while the grain growth exponent n depends on annealing temperature, grain orientation, texture, residual strain and soluble impurities.

According to classical nucleation theory, the time-dependent nucleation rate $J(T, t)$ in solid-state transformations is given by [32, 33]:

$$J(T,t) = Z\beta C_1 \exp\left(-\frac{\Delta G^*}{k_B T}\right)\exp\left(-\frac{\tau}{t}\right) \quad (5)$$

where Z is the Zeldovich non-equilibrium factor that accounts for the nuclei that exceed the critical nucleus size, β the rate at which atoms are added to the critical nucleus, C_1 the concentration of available nucleation sites, ΔG^* the activation energy for heterogeneous austenite nucleation, and τ the incubation time. We assume the nucleation rate to be constant within 1–100 s of annealing and thus, the time-independent nucleation rate may be written as:

$$J(T,t) = Z\beta C_1 \exp\left(-\frac{\Delta G^*}{k_B T}\right) \quad (6)$$

The relationships for Z and β have been developed by Johnson et al. [33], which have been modified for the specific case of austenite nucleation as shown below:

$$Z(T) = \frac{V_{\alpha' Fe}[\Delta G_v(T)]^2}{8\pi\sqrt{k_B T K_j \sigma_{\alpha'/\gamma}}} \tag{7a}$$

$$\beta(T) = \frac{16\pi\sigma^2_{\alpha'/\gamma} D(T) L_j}{a^4_{\alpha'}(\Delta G_v(T))^2} \tag{7b}$$

where $V_{\alpha' \mathrm{Fe}}$ is the average volume of an iron atom in the martensite phase, $\Delta G_v(T)$ the Gibbs free energy change per unit volume of martensite to austenite phase transformation, K_j and L_j are geometric parameters that depend on the α'-phase and γ-phase interfacial energies, $D(T)$ the temperature-dependent diffusion coefficient of the species that govern the martensite to austenite transformation, $\sigma_{\alpha'/\gamma}$ the martensite-austenite interfacial energy, and $a_{\alpha'}$ the martensite lattice parameter.

To determine $J(T)$ in Eq. 7a, b, we further need to estimate C_1, the concentration of nucleation sites available for austenite nucleation, which depends on the type of nucleation sites considered. On the basis of several studies on cold-rolled SS grades and Fe–X alloys [23–25, 35, 36], we have assumed that the austenite nucleates on martensite grain boundaries. The concentration of martensite grain boundaries C_1^{gb}, available for austenite nucleation is estimated according to [34]:

$$C_1^{gb} = C_0 \frac{d^{gb}_{\alpha'}}{\bar{d}_{\alpha'}}, \tag{8}$$

where C_0 is the total equilibrium concentration of nucleation sites available, $d^{gb}_{\alpha'}$ the martensite grain boundary width, and $\bar{d}_{\alpha'}$ the average martensite grain size.

Finally, to determine $J(T)$ in Eq. 7, the activation energy for heterogeneous nucleation $\Delta G^*(T)$ must be calculated. $\Delta G^*(T)$ depends on the volume of the austenite nucleus, the grain boundary area of the α'/γ interface, the grain boundary area of the eliminated α'/α' interface and the volume free energy change $\Delta G_v(T)$. Thus, $\Delta G^*(T)$ may be expressed as [37]:

$$\Delta G^*(T) = \frac{4(b\sigma_{\alpha'/\gamma} - a\sigma_{\alpha'/\alpha'})^3}{27c^2(\Delta G_v(T))^2} \tag{9}$$

where a is the eliminated martensite grain boundary area during the $\alpha' \rightarrow \gamma$ phase transformation, b the new austenite/martensite grain boundary area, c the new austenite nucleus volume [37], and $\Delta G_v(T)$ the volume free energy change due to the martensite to austenite transformation. The parameters a, b, and c depend on the geometry of the nucleating grain (2-grain, 3-grain or 4-grain austenite/martensite junctions). On the basis of Eqs. 7–10, Eq. 5 may be rewritten as:

$$\xi_\gamma(t,T) = 1 - \exp\left(-\frac{4\pi Z(T)\beta(T) C_1^{gb} \exp\left(-\frac{\Delta G^*(T)}{k_B T}\right) k_n^{\frac{3}{n}} t^{\frac{3}{n}+4}}{3n^3}\right) \tag{10}$$

where all the terms retain their previous definitions.

The values of C_1^{gb}, $\Delta G^*(T)$, $Z(T)$, $\beta_{\mathrm{Cr}}(T)$, and $\beta_{\mathrm{N}}(T)$ have been used in Eq. 10 to determine the percentage of austenite phase at the annealing temperatures of 700–1,000°C for the rate limiting cases of nitrogen and chromium diffusion; when austenite nucleates at 2, 3, and 4-grain austenite/martensite junctions (Fig. 10). The experimental data is also plotted for comparison.

Based on the calculations performed, if the $\alpha' \rightarrow \gamma$ phase transformation is driven via nitrogen diffusion, two distinct observations may be made. For the annealing temperatures of 900 and 1,000°C, calculations show that complete $\alpha' \rightarrow \gamma$ transformation occurs irrespective of the type of the α'/γ junction geometry, which agrees with the available experimental data (Fig. 10a–c). However, for the lower annealing temperatures of 700 and 800°C, the results show that the nature of the α'/γ geometry plays an important role in the $\alpha' \rightarrow \gamma$ transformation kinetics. The following observations can be made: (i) the phase transformation is sluggish relative to the experimental data when austenite nucleates at 2-grain junctions (Fig. 10a), (ii) the $\alpha' \rightarrow \gamma$ phase transformation is faster and agrees reasonably well with the experimental data when austenite nucleates at 3-grain austenite/martensite junctions (Fig. 10b), and finally (iii) for the case of austenite nucleation at 4-grain martensite junctions, our calculations predict the $\alpha' \rightarrow \gamma$ transformation to be considerably faster than the experimental observations (Fig. 10c).

For comparison, calculations were also performed for the case of chromium driven diffusion. In this case, the phase transformation is considerably sluggish relative to experimental data, when austenite nucleates at 2 and 3-grain α'/γ junction geometries (Fig. 10d, e). On the other hand, the phase transformation kinetics is considerably faster relative to the experimental observations when austenite nucleates at 4-grain junctions (Fig. 10f). Clearly, in all the three cases just discussed, the calculated phase transformation data with chromium diffusion as the rate limiting factor does not agree with the experimental observations.

The results also show that $\alpha' \rightarrow \gamma$ transformation, regardless of the nature of the diffusing species, strongly depends on the geometry of available nucleation sites. In particular, it is sluggish when austenite nucleates at 2-grain martensite junctions and faster when it nucleates at 3-grain

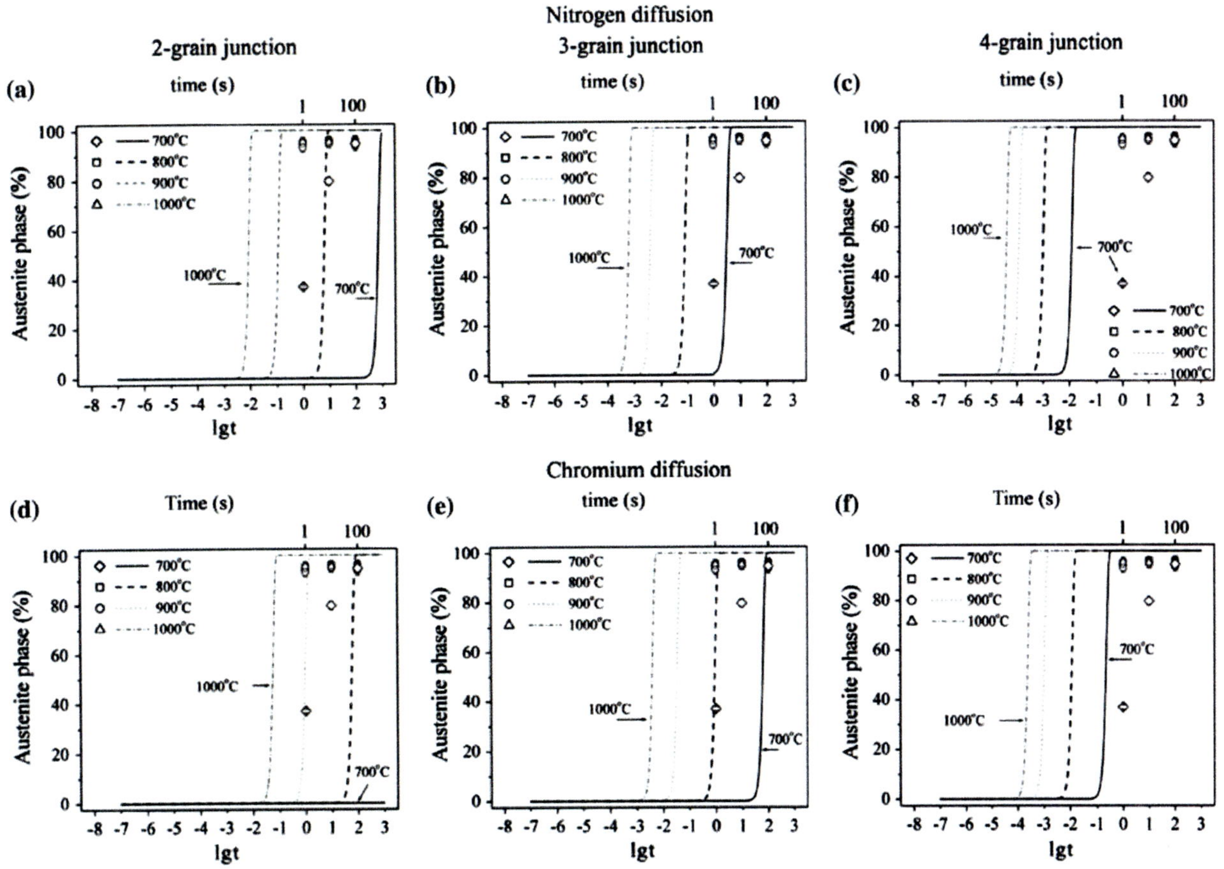

Fig. 10 Calculated (curves) and experimental (scatter points) austenite phase percentage as a function of annealing time and temperature. **a–c** show the phase percentage when nitrogen diffusion is assumed to be the rate limiting step, whereas **d–f** show the phase percentage when chromium diffusion is assumed to be the rate limiting step

or 4-grain martensite junctions (Fig. 10). This is expected because the austenite phase fraction, given by equation (10), depends strongly on the activation energy for heterogeneous nucleation $\Delta G^{*}(T)$, which is large at 2-grain martensite junctions relative to that at 3-grain or 4-grain martensite junctions (Fig. 2). In summary, our calculations show that nitrogen diffusion controlled austenite nucleation at 3-grain martensite junctions appears to reasonably agree with the experimental data (Fig. 10b).

5.4 Influence of Annealing Conditions on the Mechanical Properties

From Fig. 9, it is clear that the yield strength for samples annealed at 800°C is significantly higher than those annealed at 900 and 1,000°C, even when all these samples exhibit the same austenite phase fraction. The yield strength can be correlated with the grain size obtained, according to the Hall–Petch relationship In this fashion, a k-value of $\sim$274 MPa $\mu m^{1/2}$ and an offset stress of $\sim$250 MPa were calculated (Fig. 11a). These values differ considerably from those obtained by Schino et al. [12–14] for AISI 301 SS and AISI 304 SS. In particular, these investigations reported higher k values ($\sim$400 MPa $\mu m^{1/2}$) for both SS, and a deviation from the Hall–Petch relationship for AISI 301 SS below a grain size of 3 μm. Thus, we believe that there may be fundamental reasons, other than grain size, that may be responsible for a low k value. Possible reasons may include (i) the presence of retained tempered martensite after annealing, which was not considered by Schino et al., (ii) the presence of nitrogen in AISI 301LN SS, which has a different strengthening effect from that of carbon, which is present in AISI 301 SS and AISI 304 SS and (iii) the formation of CrN nitrides in AISI 301LN SS, that may be present in AISI 301 SS and AISI 304 SS. Another aspect in establishing the correlation between the yield strength and grain size is the fact that the microstructure of AISI 301 LN SS is dependent on the annealing temperatures used in this work. Thus, it is relevant to consider the Hall–Petch behaviour separately for each temperature. This analysis is shown in Fig. 11b, where a marked influence of the

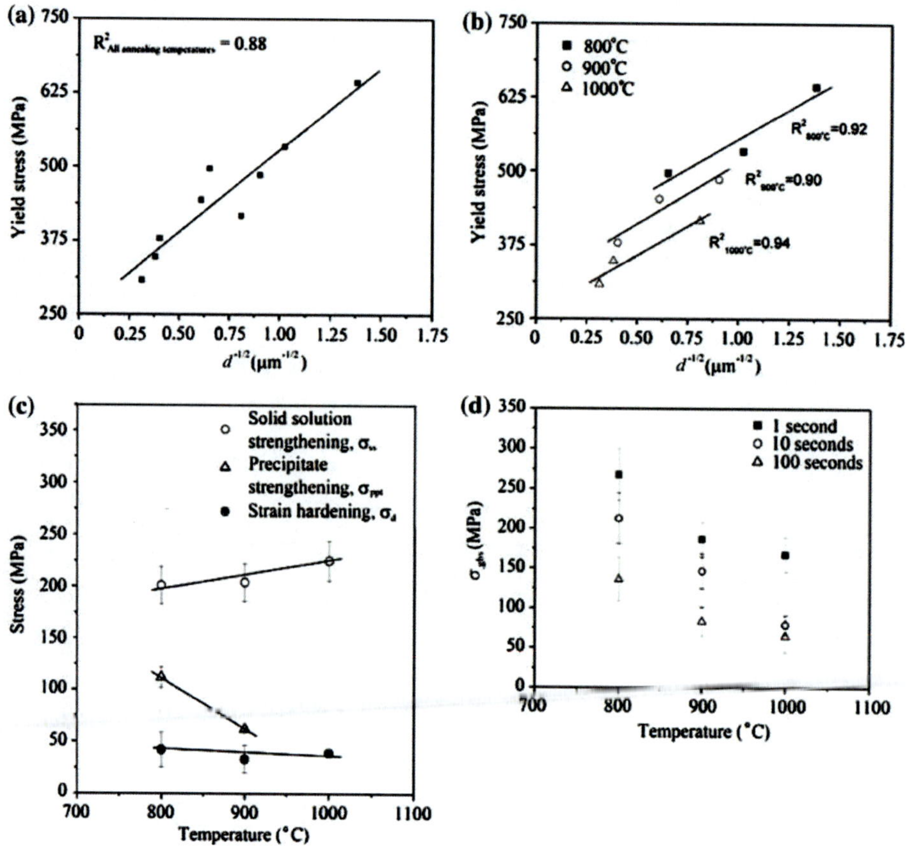

Fig. 11 Hall–Petch relationship between austenite grain size and yield strength, assuming **a** temperature independent and **b** temperature dependent conditions. The R^2 values are given for comparison. **c** Strengthening mechanisms operating in AISI 301LN SS samples annealed at various temperatures, **d** grain boundary strengthening as a function of annealing conditions in AISI 301LN SS samples [24]

annealing temperature on the offset stress σ_o (intercept with the y-axis) is observed. On the other hand, it is seen that there is no significant effect of annealing temperature on the Hall–Petch coefficient k.

To further understand the temperature effects on the Hall–Petch equation, we considered that the offset stress can be written, in general, as the sum of several strengthening mechanisms, namely (i) precipitate strengthening, (ii) strain hardening and (iii) solid solution strengthening. This can be expressed as:

$$\sigma_o = \sigma_d + \sigma_{ppt} + \sigma_{ss} \quad (11)$$

where $\sigma_d = 2\mu b\sqrt{\rho}$ is the strengthening contribution due to dislocation–dislocation interaction (strain hardening), μ is the shear modulus, b is the Burgers vector and ρ is the dislocation density; σ_{ppt}= $(0.84\tau_{ppt}b)/LC_\varphi$ [38] s the strengthening contribution due to the presence of precipitates, τ_{ppt} is the critical resolved shear stress required to move dislocations past a random array of precipitates, L is the distance between the precipitates and C_φ is the Schmid factor. Finally, σ_{ss} is the strengthening contribution due to solid solution. If we assume that the samples annealed at 1,000°C are fully annealed and have no precipitates the only strengthening contribution is due to solid solution. With these assumptions, σ_{ss} in 1,000°C annealed samples could be calculated. For the 800 and 900°C samples, the calculation of σ_{ss} is not as trivial because CrN nitrides are present at these temperatures, leading to a depletion of chromium and nitrogen from the austenitic matrix. A loss of nitrogen from the austenite matrix, due to nitride formation, will thus affect the strengthening due to solid solution. Hence, it is necessary to quantify the amount of nitrogen present in the precipitates to be able to assess σ_{ss}. This can be done by estimating the volume fraction of nitrides in a given sample, according to the existing number and size of nitrides per unit volume, determined through TEM analysis [24]. Subtracting the reduction in $\sigma_{ss,}$ due to nitrogen depletion from the calculated $\sigma_{ss,}$ for the fully annealed alloy, yields the σ_{ss} values for the samples annealed at 800 and 900°C. Now consider the strengthening contribution σ_{ppt} due to the presence of CrN nitrides. The results indicate a dramatic decrease in precipitate strengthening as the annealing temperature is increased from 800 to 900°C due to the absence of nitrides at higher temperatures.

Finally, the strain hardening σ_d can be calculated by subtracting the values of σ_{ss} and σ_{ppt} from the offset stresses σ_o obtained for each annealing temperature (Fig. 11b).

The results show that the strain hardening contribution obtained for the 800 and 900°C samples is approximately 40 MPa and decreases slightly for higher annealing temperature (Fig. 11c). Based on this value, an approximate dislocation density was estimated to be approximately 1×10^{12} m^{-2} for both samples which compares well with that of a fully annealed polycrystalline alloy. This indicates that the retained austenite that may have failed to recrystallize does not seem to play an important role in strain hardening. In general, it is clear that for samples annealed at 800°C a large offset stress primarily due to solid solution and precipitate strengthening can be observed. However, as the annealing temperature is increased, the contribution of precipitate strengthening to the offset stress is reduced and solid solution strengthening—mainly due to the presence of nitrogen—becomes the predominant effect. As shown in Fig. 11c, strain hardening has a minor influence on yield strength at all annealing temperatures considered. Attention is thus focused on the remaining strengthening mechanisms. At the lowest annealing temperature of 800°C and shortest annealing time of 1 s, the presence of ultra fine austenitic grains contribute the most to the yield strength via grain boundary strengthening (Fig. 11d). However, as temperature increases and/or time progresses, grain growth occurs. Thus, grain boundary strengthening is diminished (Fig. 11d), while solid solution strengthening becomes the predominant factor (Fig. 11c). Furthermore, at the annealing temperature of 800°C, the presence of nitrides contributes substantially to yield strength through precipitate strengthening. This contribution decreases at the annealing temperature of 900°C, and eventually plays no role at 1,000°C, the highest annealing temperature tested (Fig. 11c). From this discussion, one can see that different strengthening mechanisms gain prominence at different annealing times and temperatures.

6 Conclusions

To summarize, a commercially important metastable austenitic AISI 301LN SS has been thoroughly analyzed and the interplay between structure, property and performance correlations have been addressed. In particular, we found that:

1. Cold rolled samples consist of regions of dislocation cell-type martensite, which are characterized by the presence of dislocation walls and forests, and regions of heavily deformed lath-type martensite.
2. Diffusion-driven $\alpha' \rightarrow \gamma$ reversion kinetics is responsible for partial $\alpha' \rightarrow \gamma$ reversion to nano/submicron austenite (~200 nm) for samples annealed at 600°C for 100 s and 700°C for 1 s, while complete $\alpha' \rightarrow \gamma$ reversion to submicron to ultra-fine grains in samples annealed at higher temperatures.
3. Austenite nucleation from martensite is sequential where dislocation cell-type martensite is consumed first, followed by austenite nucleation on geometrically necessary grain boundaries of heavily deformed lath-type martensite. This sequence of events is responsible for a narrow austenite grain size variation obtained at lower annealing temperatures and a broader grain size variation at higher temperatures.
4. Short annealing duration of 1–100 s at 600–800°C produces nanoscale (~20 nm) *fcc* chromium nitride precipitates within the austenite grains.
5. Interstitial nitrogen atoms enable fast diffusion kinetics due to their high diffusivities, thereby driving the $\alpha' \rightarrow \gamma$ phase transformation.
6. The geometry of austenite/martensite grain junctions is significant. The results indicate that the $\alpha' \rightarrow \gamma$ transformation appears to occur at the austenite/martensite 3-grain junctions. Heavy cold-rolling of a metastable austenitic SS makes available these heterogeneous nucleation sites and thus an important step for an efficient $\alpha' \rightarrow \gamma$ transformation.
7. Strengthening in this alloy is an interplay between fine-grained austenite, solid solution strengthening and strain hardening.

References

1. *World Steel in Figures*. The International Iron and Steel Institute (2006)
2. The Materials Information Society, *ASM Handbook: Ferrous Alloys* (ASM International, Metals Park, 1994)
3. *European steel technology platform, Strategic research agenda; A vision for the future of the steel sector*, ESTEP—Belgium, December (2005)
4. *Steel grades, properties and global standards*, Outokumpu Stainless Oy
5. M.C. Somani, L.P. Karjalainen, M. Koljonen, P. Aspegren, T. Taulavuori, A. Kyröläinen, in *Proceedings of the 5th European Congress, Stainless Steel Science*, (2005) p. 37
6. M.C. Somani, L.P. Karjalainen, A. Kyröläinen, T. Taulavuori, Mater. Sci. Forum **539–543**, 4875 (2007)
7. M.C. Somani, L.P. Karjalainen, P. Juntunen, S. Rajasekhara, P.J. Ferreira, A. Kyröläinen, in *International Symposium on Ultra-fine Grained Steels* (2005)
8. P. Juntunen, M.C. Somani, L.P. Karjalainen, A. Kyröläinen, in *International Symposium on Advanced Steels* (2007)
9. A.H. Cottrell, Trans. Am. Inst. Metall. Eng. **216**, 192 (1958)
10. E.O. Hall, Proc. Phys. Soc. **64B**, 747 (1951)
11. N.J. Petch, Philos. Mag. **1**, 186 (1956)
12. A. di Schino, J.M. Kenny, M.G. Mecozzi, M. Barteri, J. Mater. Sci. **35**, 4803 (2000)
13. A. di Schino, I. Salvatori, J.M. Kenny, J. Mater. Sci. **37**, 4561 (2002)
14. A. di Schino, M. Barteri, J.M. Kenny, J. Mater. Sci. Lett. **21**, 751 (2002)
15. Y. Ma, J.-E. Jin, Y.-K. Lee, Scripta Mater. **52**, 1311 (2005)
16. N. Tsuji, R. Ueji, Y. Minamino, Y. Saito, Scripta Mater. **46**, 305 (2002)

17. R. Ueji, N. Tsuji, Y. Minamino, Y. Koizumi, Acta Mater. **50**, 4177 (2002)
18. M.W. Bowkett, S.R. Keow, D.R. Harries, Met. Sci. **16**, 499 (1982)
19. J. Singh, J. Mater. Sci. **20**, 3157 (1986)
20. V. Shrinivas, L.E. Murr, Metall. Mater. Trans. **26A**, 661 (1995)
21. A. Kumar, L.K. Singhal, Metall. Trans. **20A**, 2857 (1989)
22. P.L. Mangonon Jr., G. Thomas, Metall. Mater. Trans. **1A**, 1577 (1970)
23. S. Rajasekhara, L.P. Karjalainen, A. Kyröläinen, P.J. Ferreira, Mater. Sci. Eng. A **527**, 1986 (2010)
24. S. Rajasekhara, P.J. Ferreira, L.P. Karjalainen, A. Kyröläinen, Metall. Mater. Trans. **38A**, 1202 (2007)
25. D.L. Johannsen, A. Kyröläinen, P.J. Ferreira, Metall. Trans. **37A**, 2325 (2006)
26. N. Hansen, Metall. Mater. Trans. **32A**, 2917 (2001)
27. K. Tomimura, S. Takaki, Y. Tokunaga, ISIJ Int. **31**, 1431 (1991)
28. S. Takaki, K. Tomimura, S. Ueda, ISIJ Int. **34**, 522 (1994)
29. V. Erukhimovitch, J. Baram, Phys. Rev. B **50**, 5854 (1994)
30. V. Erukhimovitch, J. Baram, Phys. Rev. B **51**, 6221 (1995)
31. S. Rajasekhara, P. J. Ferreira, Acta Mat. **59**, 738 (2011)
32. K.S. Chan, J.K. Lee, G.J. Shiflet, K.C. Russell, H.I. Aaronson, Metall. Mater. Trans. **7A**, 1016 (1978)
33. W.C. Johnson, C.L. White, P.E. Marth, P.K. Ruf, S.M. Tuominen, K.D. Wade, K.C. Russell, H.I. Aaronson, Metall. Mater. Trans. **6A**, 911 (1975)
34. D.A. Porter, K.E. Easterling, *Phase Transformations in Metals and Alloys*, 1st edn. (Chapman & Hall, London, 1990)
35. W. Huang, M. Hillert, Metall. Mater. Trans. **27A**, 480 (1996)
36. K.C. Russell, Acta Metall. **17**, 1123 (1969)
37. P.C. Clemm, J.C. Fisher, Acta Metall. **3**, 70 (1955)
38. F.R.N. Nabarro, *Dislocations in solids* (Dover Publications, New York, 1987)

Advanced Heat Resistant Austenitic Stainless Steels

Guocai Chai, Jan-Olof Nilsson, Magnus Boström, Jan Högberg, and Urban Forsberg

Abstract

UNS S31035 is a newly developed austenitic stainless steel with the highest creep strength among the commercial available heat resistant grades for the next generation of coal fired power plants. Alloy 800HT is a well developed material that has been recommended as a candidate material for generation IV nuclear power plants. In this chapter, several advanced heat resistant austenitic stainless steels are reviewed, but the focus will be on these two materials. The influences of composition on the structural stability and on the creep behavior are discussed. The creep mechanisms at different temperatures and loading conditions have been identified. The interaction between dislocations and precipitates and their contribution to the creep rupture strength are discussed. Different models have been used to evaluate the long-term creep behavior of the grades. Finally, highly alloyed composite tube products for different corrosive steam boiler applications are introduced.

Keywords

Heat resistant steels • Austenitic stainless steels • Creep strength • High temperature corrosion resistance

1 Introduction

Global increase in energy consumption requires more energy production. Meanwhile the concern on the environmental impact from energy production is continuously increasing. Although combustion processes generate carbon dioxide, coal-fired thermal power generation is still one of the most important methods in the medium to long-term future to satisfy this demand, as coal is available at a competitive price and often is the single domestic energy. However, the biggest challenge facing coal-fired power plants is to improve their energy efficiency. This can be accomplished by increasing the maximum steam temperature and the steam pressure. Conventionally, the heat efficiency of coal-fired power plants has stayed at around 41% in the super critical (SC) condition with a temperature of 550°C and pressure of 24.1 MPa. In order to attain a power generating efficiency of about 43%, ultra super critical (USC) conditions with a steam temperature at about 600°C should be reached. By increasing the temperature from 550 to 600°C (at most present power plants) to 650–700°C (at next generation power plants), the power plant efficiency can be increased from 36% to more than 50% and the CO_2 emission can be reduced about 30% [1].

In a power plant with an operating temperature below 500°C, ferritic or martensitic heat resistant steels are the dominant materials for steam generation and austenitic steels as boiler. In USC conditions, however, these materials, including the higher chrome steels, do not have sufficient creep rupture strength and resistance to high temperature corrosion. Austenitic stainless steels are therefore used for different applications. Although nickel

G. Chai (✉) · J.-O. Nilsson · M. Boström · J. Högberg · U. Forsberg
R&D Center, Sandvik Materials Technology, 811 81 Sandviken, Sweden
e-mail: guocai.chai@sandvik.com

Y. Weng et al. (eds.), *Advanced Steels*, DOI: 10.1007/978-3-642-17665-4_40,

base alloys can be used, they are comparatively expensive. Austenitic stainless steels are Fe–Cr–Ni alloys with chromium content higher than 13 mass% and an austenitic structure at room temperature. In order to obtain the required properties, other alloying elements are often added. They can be either interstitial elements such as carbon and nitrogen or substitutional elements such as Mn, Mo, W, Cu, Al, Ti, Nb, V, etc. These alloying elements are classified as either ferrite stabilizers or austenite stabilizers depending on whether they promote a ferritic structure or an austenitic structure. Their contributions can be evaluated using the notation of chromium and nickel equivalents as presented below [2].

$$Ni_{eq} = [Ni] + [Co] + 0.5[Mn] + 30[C] + 0.3[Cu] + 25[N](mass\%) \quad (1)$$

$$Cr_{eq} = [Cr] + 2.0[Si] + 1.5[Mo] + 5.5[Al] + 1.75[Nb] + 1.5[Ti] + 0.75[W](mass\%) \quad (2)$$

Nb, Ti and V are also called stabilizing elements, which can greatly improve the creep strength of austenitic stainless steels, mainly by precipitating fine carbides or carbonitrides. Besides, addition of these elements can stabilize the alloy against intergranular corrosion. Table 1 shows the compositions of some heat resistant austenitic stainless steels, mainly AISI 300 series alloys. Alloy 800HT is an austenitic nickel–iron–chromium alloy. This alloy is characterized by high creep strength and very good resistance to oxidation. Super austenitic stainless steels 253MA or UNS S30815 and 353MA or UNS S35315 are austenitic chromium–nickel steels alloyed with nitrogen and rare earth metals. They have high creep strength and very good resistance to isothermal and, above all, cyclic oxidation.

Figure 1 shows the creep strength of some austenitic stainless steels. The austenitic stainless steels Alloy 800HT, S20815 and S35315 show high creep strengths at temperature higher than 800°C.

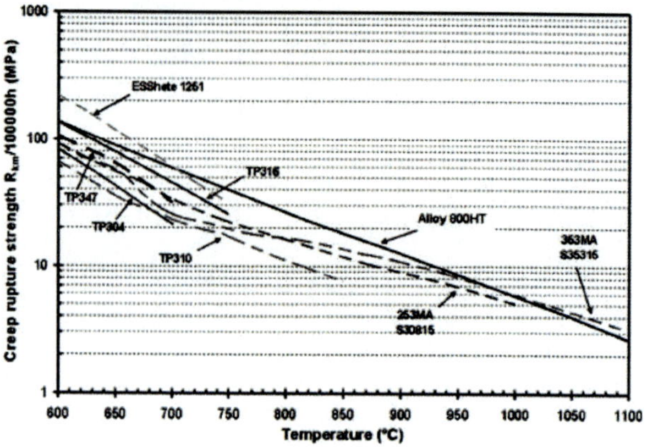

Fig. 1 Creep rupture strengths of some austenitic stainless steels at 100,000 h [3]

The materials used for the next generation power plants are required to have even higher yield strength at elevated temperature, creep strength (typically 100,000 h rupture strength of around 100 MPa at the metal temperature) and high temperature corrosion resistance [4]. As shown in Fig. 1, no current heat resistant alloy can meet these requirements. Therefore, in the EU-project "Advanced (700°C) PF Power Plant—AD700" under the Joule-Thermie programme, a subproject to develop new advanced heat resistant materials for next generation power plants has been organised [5]. One of the aims was to develop materials for superheaters and reheaters in USC boilers for use at temperatures up to 700°C. One material successfully developed is the austenitic stainless steel grade UNS S31035 (Sanicro 25) [6]. This material provides very high creep strength and good corrosion resistance at high temperatures. The purposes of this chapter is to give an overview on this newly developed and other heat resistant materials, especially Alloy 800HT that is a candidate material for generation IV nuclear power plants.

Table 1 Nominal composition of austenitic heat resistant stainless steels (mass%)

Grades	C_{max}	Si	Mn	Cr	Ni	Mo	Others
304H	0.04	0.4	1.3	18.5	9.5		
309H	0.07	0.5	1.7	22.5	14		
310H	0.06	0.75	1.5	24.5	21		
316H	0.05	0.4	1.7	17	12	2.6	
321H	0.05	0.4	1.3	17.5	10.5		
347H	0.06	0.4	1.8	17.5	11		Nb, Ti
Alloy 800	0.030	0.5	0.6	20	32		Al, Ti
Esshete 1250	0.1	0.5	6.3	15	9.5	1.0	V, Nb, B
253MA or UNS S30815	0.08	1.6	0.8	21	11		N, REM
353MA or UNS S35315	0.07	1.6	1.5	25	35		N, REM

REM rare earth metals

Table 2 Nominal composition of UNS S31035 (mass%)

C_{max}	Si	Mn	Cr	Ni	W	Co	Cu	Nb	N	Fe
0.1	0.2	0.5	22.5	25	3.6	1.5	3.0	0.5	0.23	Bal.

To protect environment is an international key issue. Municipal waste incineration is a process to utilize the waste for energy production or "waste to energy". This technology becomes more interesting now than ever. However, the corrosive environment in incinerators is a major problem for energy production efficiency and lifetime of incinerators. This can be improved by using composite tube materials. In this chapter, the heat resistant composite tube materials will also be reviewed.

2 Material Development and Microstructure of Grade UNS S31035

Since the UNS S31035 austenitic stainless steel was aimed for use in super-heaters and reheaters with metal temperatures up to 700°C for the next generation of pulverized coal-fired power plants, the alloy was designed to have high creep strength and corrosion resistance, good micro structural stability and fabric ability. The chemical composition is shown in Table 2.

The alloy was designed as a stabilized austenitic stainless steel by addition of niobium. The high temperature strength is mainly attained by precipitation strengthening with stable nano precipitates. Addition of niobium was proposed to form precipitates of MX (M = Nb, Cr or W, X = N or C) carbides or carbo-nitrides that can lead to an increase of creep strength and consequently suppress the precipitation of other phases [7]. Fine uniformly distributed intergranular MX can greatly increase the creep strength [8, 9]. Copper is added to form copper rich nanoparticles that increase the creep strength [10, 11]. The alloy is also solution-strengthened by small additions of W and Co. To achieve a good hot corrosion resistance the Cr content is high. Elements such as Si and Mo promoting sigma phase are kept low, while Ni and N that can suppress the formation of sigma phase [12] are added to reach sufficient structural stability and good fabricability.

Figure 2 shows some fine precipitates observed in this newly developed S31035 austenitic stainless steel that can contribute to the improvement of the creep strength. Both intra- and inter-granular $M_{23}C_6$ precipitates can be observed (Fig. 2a). In the grain boundaries, they have a $\langle 100\rangle_\gamma // \langle 100\rangle_{M_{23}C_6}$ coherent relationship to the austenite matrix. Laves phase was observed both randomly within the grains but also ordered on what appears to be former twin boundaries (Fig. 2b). Both coherent Laves phase precipitates with a $[100]_{Laves}//[100]_\gamma$ orientation relationship, and incoherent Laves phase precipitates were observed. In the aged material [13] the Laves phase are needle shaped but rather small. In the creep tested materials, these particles are fine and isometric. Copper rich nanoparticles can also be observed. They are round with a size up to 50 nm (Fig. 2c). In this material, a dense distribution of about 10 nm large precipitates was observed (Fig. 2d). Due to the limitation of the TEM, these particles could not be identified, but they are probably MX carbides or carbonitrides. Similar particles have been identified in other analyses [7].

The presence of the MX nanoparticles in the material was observed by TEM EDS mapping. Fig. 3 shows the possible elements in the particles within the grains. These particles are probably niobium carbonitride, or MX precipitates.

3 Creep Properties and Strengthening Mechanisms

3.1 Creep Properties

Up to now (August 2010), the longest time to rupture for this alloy is approximately 45,000 h and some samples are still running after 65,000 h. Figure 4 shows the results of

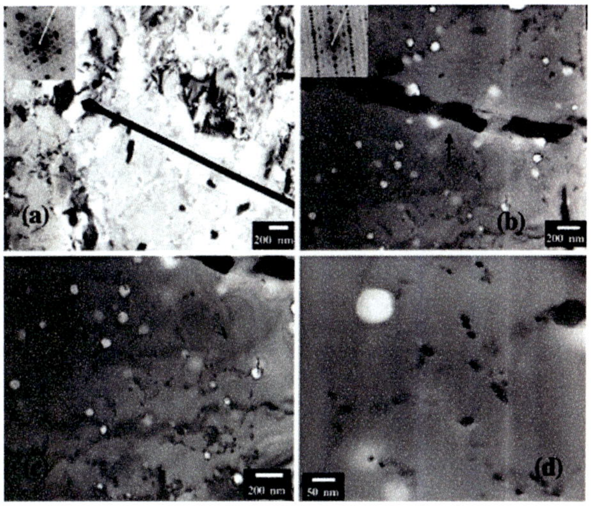

Fig. 2 Fine precipitates that contribute to the creep strength of S31515 austenitic stainless steel. **a** $M_{23}C_6$, 700°C for 1,000 h. **b** Laves phase, 700°C for 30,000 h. **c** Copper-rich phase, 700°C for 30,000 h. **d** Nanoparticles, 700°C for 30,000 h

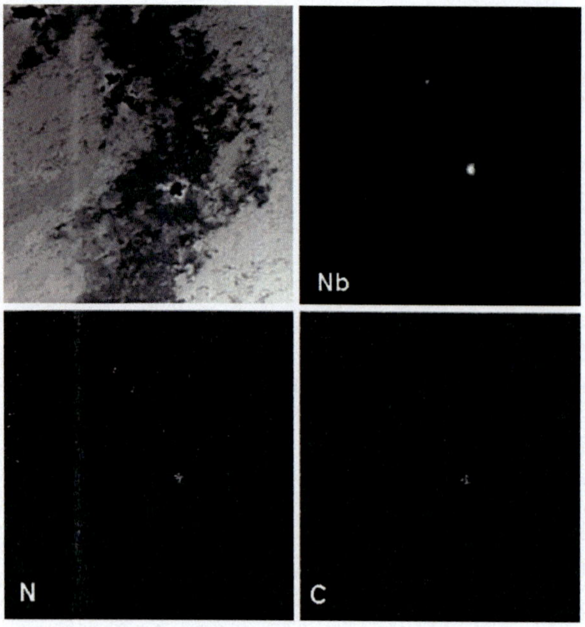

Fig. 3 Analysis of TEM EDS mapping for the particles within the grains in the material tested at 650°C

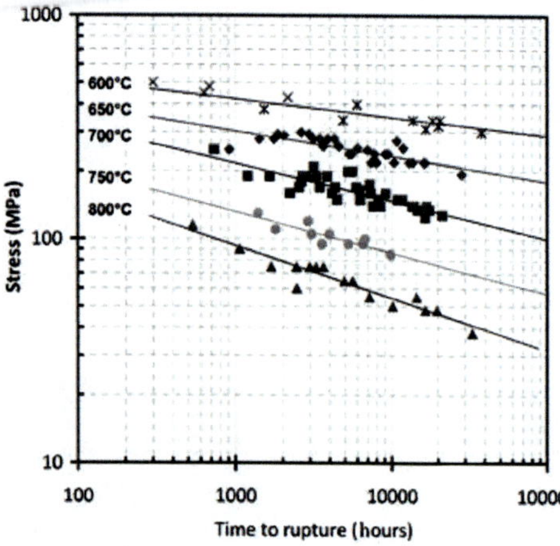

Fig. 4 Creep strength versus rupture time of UNS S31035 austenitic stainless steel and the linear regressions in the creep data

the creep in rupture plots for the UNS S31035 material. With linear least squares regression to extrapolate the creep rupture data, the predicted 10^5 h creep rupture strength at 700°C is 104 MPa.

As we know, pressure vessel components are normally designed for a long service time such as 200,000 h. Since creep is a very slow process, creep data for such times are not available for design. The design data are therefore obtained by extrapolation from the creep data of shorter tests. Different models and approaches have been developed [14–18]. One recently proposed procedure for extended extrapolation of creep rupture data is the free temperature model that allows for extrapolation by more than a factor of three in time [16]. The procedure is based on a time–temperature parameter (TTP), which has the general mathematical form:

$$P_{\mathrm{TTP}} = v(\mathrm{T})\log(t_r) + w(\mathrm{T}) \tag{3}$$

where P_{TTP} is the time–temperature parameter, t_r is the time to rupture, and v (T) and w (T) are functions of temperature. In the proposed procedure, these functions are assumed to be polynomials in T. The TTP in Eq. 3 is in this model referred to as the free temperature model (FTM). The free temperature model is well suited for austenitic stainless steels where the temperature dependence of the creep rupture curve is non-monotonous [16].

The master curve is expressed by the creep stress as a function of polynomial in the TTP:

$$\log(\sigma) = \sum_{j=0}^{n_P} a_j \mathrm{TTP}^j \tag{4}$$

The coefficients a_j are fitted to the creep rupture data and the stress–rupture time relations are derived. The reason for using a polynomial in log (t_r) rather than log (σ) which is the more common approach, is that it has been shown that this improves the accuracy in extended extrapolated values [19]. The extrapolation was performed in a Matlab program. The coefficients to the polynomials were derived by a non-linear least squares fit to the creep data.

To extrapolate the creep rupture data, the polynomials in Eq. 3 were both set to order 3 and the master curve, Eq. 4, was set to a second order polynomial. Figure 5a shows the master curve for the free temperature model. The curve is based on the creep data of this alloy at 550, 600, 650, 700 and 800°C.

The extrapolation was performed three times with rupture data up to 40,000 h where the evaluation satisfies the post-evaluation tests (PATs) and other criteria proposed by the European Collaborative Creep Committee (ECCC) [20, 21]. Figure 5b shows a comparison of the extrapolation with rupture data up to 40,000 h and the experimental data. They are comparable. The error in the extrapolated value can be estimated as the difference between the curves in Fig. 5a for the full dataset and the one including data up to 15,000 h rupture time. At 700°C and 100,000 h it is 3 MPa. Thus, the result of the extrapolation for this condition is (99 ± 3) MPa.

As discussed previously, the predicted 10^5 h creep rupture strength at 700°C is 104 MPa using linear least squares regression to extrapolate the creep rupture data. This value is higher than that of the free temperature model. Linear regression is, however, not a good method to use for extrapolation since almost every material shows a drop in creep strength at longer times.

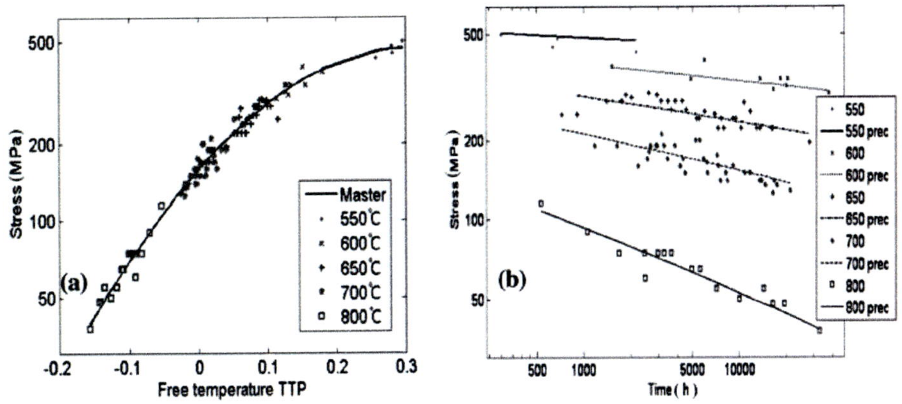

Fig. 5 Extrapolation of the creep data. **a** Master curve for UNS S31035 material. **b** Correlations between the extrapolation with rupture data up to 40,000 h and the experimental data

Figure 6 shows a comparison of the creep rupture strength of S31035 alloy with some commercially available austenitic stainless steels [21, 22]. The nominal data of S31035 indicates creep rupture strength of 95 MPa at 700°C/100,000 h, which is higher than that of the others. For example TP310NCbN and UNS 30432, which are commonly used today, show a creep rupture strength of 66 MPa, respectively, 68 MPa at 700°C/100,000 h.

4 Strengthening Mechanism

In the temperature range up to 700°C, one of the main creep strengthening mechanisms is the interaction between dislocations and precipitates [21]. Figure 7 shows two examples how interaction between the dislocations and precipitates in a UNS S31035 creep specimen tested with 210 MPa at 700°C and with a rupture time of 3,153 h. Moving dislocations at the nano-sized particles can be seen. Around the intergranular precipitates, the dislocation density is high which indicates that they function as obstacles for the dislocation movements. This increases the creep strength. These nano-sized particles were identified as $M_{23}C_6$ and Laves phases using electron diffraction. In the dislocation dense area, dislocation walls have formed [21].

In Fig. 7, it can also be observed that smaller nanoprecipitates such as copper rich particles and MX particles are more effective as obstacles for the dislocation movements. However, they have different mechanisms for dislocation crossing. For the copper rich nanoparticles, dislocations cross the particles mainly by climb/bypass of unit dislocations (Fig. 8a). For the MX nanoparticles, deformation might occur by shearing of partial dislocations. The shadow around the particle in Fig. 8b is believed to be dense dislocations.

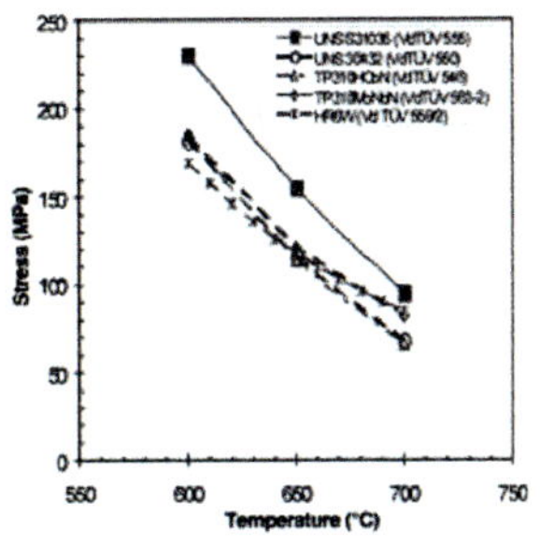

Fig. 6 A comparison of creep rupture strength of different austenitic stainless steels at 100,000 h

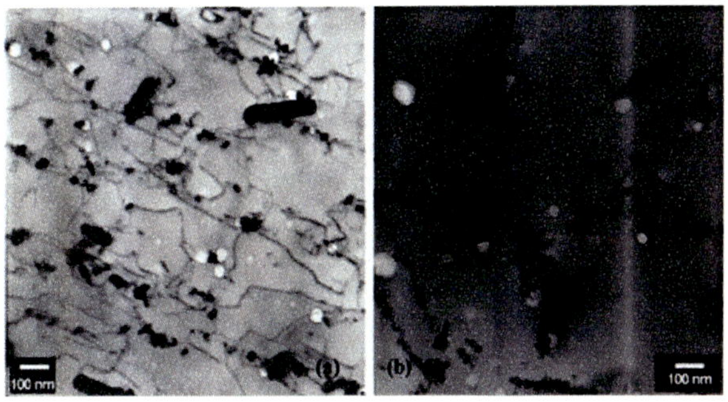

Fig. 7 TEM image showing the interactions between the dislocations and nano-sized precipitates in a UNS S31035 creep specimen tested at 210 MPa and 700°C and with a rupture time of 3,153 h

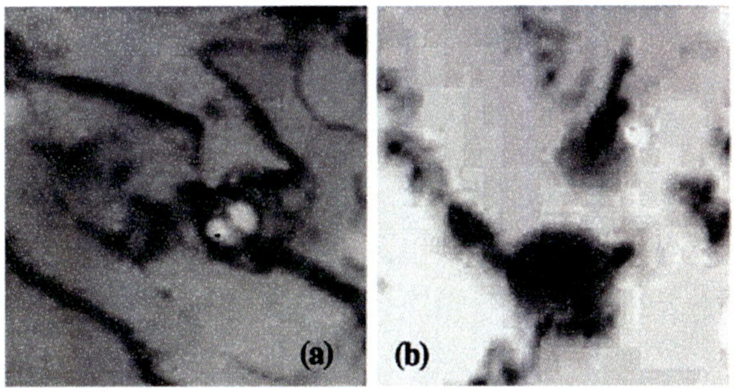

Fig. 8 Mechanisms for dislocation crossing the particles. **a** Copper rich nanoparticle. **b** MX nanoparticle

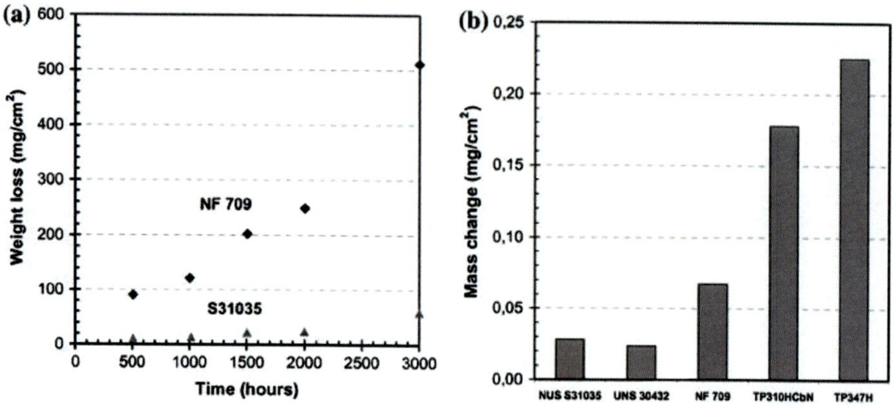

Fig. 9 High temperature corrosion properties of heat resistant stainless steels **a** Hot corrosion tests at 700°C up to 3,000 h. **b** Oxidation in steam at 700°C for 1,000 h

4.1 High Temperature Corrosion Resistance

The results from the hot corrosion test [6, 23] show that UNS S31035 can develop a more protective oxide layer than the commercially available grade NF 709 and thus has much lower corrosion rate (Fig. 9a). The characteristics of the corrosion on UNS S31035 is localized corrosion with some pits especially after long exposure times whereas the reference grade shows a more uniform corrosion front with a great loss of thickness. The oxidation test in steam also shows a low oxidation rate for UNS S31035 (Fig. 9b). The average weight change for the UNS S31035 samples exposed for 1,000 h at 700°C is only 0.028 mg/cm^2. The visual examination show a very thin oxide layer formed on the surfaces. The oxide thickness measured is approximately 0.8 μm. The discontinuous oxidation test in air gives similar results [23].

Test installations with UNS S31035 tubes have been made in five different boilers in Europe. The biggest trial, in Scholven has been in operation since mid-2005. The steam data for this test rig are maximum 670°C/256 bar. Ring samples were taken after 16,000 and 35,000 h service. They were still in very good condition. After 35,000 h, the material evaluation indicated a thin oxide, less than 15 μm, at the steamside, and the oxide was about 100 μm thick at the fireside. No significant internal oxidation and no

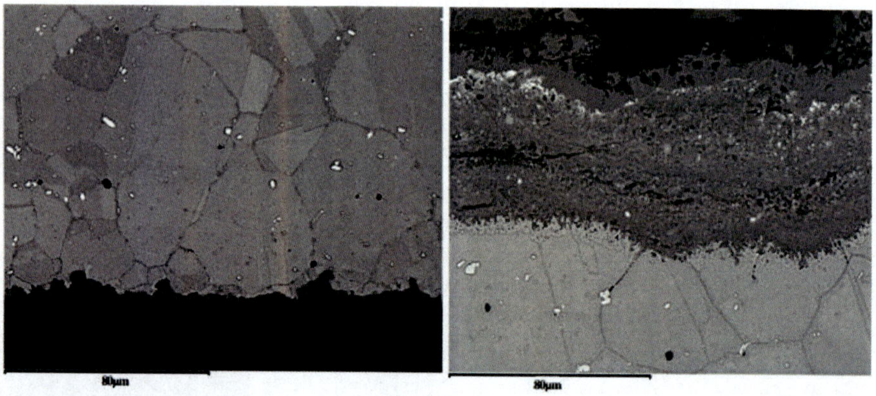

Fig. 10 Oxide formed on the inner (*left*) and outer (*right*) surface after exposure in a boiler for 16,000 h (scale 80 μm)

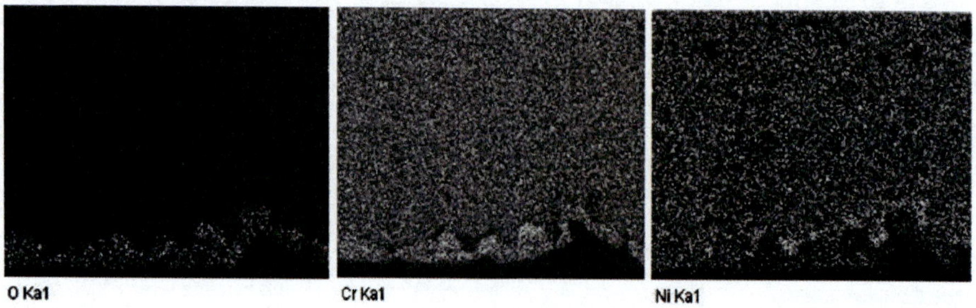

Fig. 11 Electron probe analysis; or X-ray mapping; inner tube surface

Table 3 Nominal composition of UNS S31035 (mass%)

Grades	C	Cr	Ni	Ti	Al	Ti + Al
Alloy 800	0.1 max	19.0–23.0	30.0–35.0	0.15–0.60	0.15–0.60	0.30–1.20
Alloy 800H	0.05–0.1	19.0–23.0	30.0–35.0	0.15–0.60	0.15–0.60	0.30–1.20
Alloy 800HT	0.06–0.1	19.0–23.0	30.0–35.0	0.25–0.60	0.25–0.60	0.85–1.20

reduction of the tube wall could be observed on the ring samples. Figure 10 shows the inner and the outer oxide layers after 16,000 h operation.

Figures 11, 12 show the X-ray mapping inside and outer side tube surfaces. There is a very thin Cr rich oxide on the inside and a bit thicker oxide with some deposits containing O, Al, Si, S, and K, and chromium and iron on the outside. There are small amount of small precipitates in the grains throughout the cross-section and very small precipitates along the grain boundaries near the outer surface of the tube.

5 Mechanical Behavior of Alloy 800HT at High Temperatures

The Alloy 800 series of alloys are a group of austenitic Fe–Ni–Cr stainless steels consisting of three alloys: Alloy 800, Alloy 800H and Alloy 800HT. They have similar chemical composition limits, but Alloy 800H and Alloy 800HT have more restricted chemistry (Table 3). They have significantly higher creep and rupture strength. The creep behavior of Alloy 800HT is also shown in Fig. 1. It has higher creep strength than others at temperatures higher than 750°C.

Alloy 800HT is a well developed material and provides a favorable combination of excellent creep properties, good resistance to high temperature oxidation, corrosion and carburization, and good structural stability at high temperatures [24]. One creep test with a stress of 25 MPa at 800°C for this material has been running for about 30 years, and is still running. This indicates that the material has both excellent creep strength and oxidation resistance at very high temperatures. This alloy has hence been used extensively in the power and process industries at temperatures between 500°C and 1,100°C. The material was recently recommended as a candidate material for the generation IV nuclear power plants in the temperature range above 750°C [25].

5.1 Structure Stability and Creep Strengthening Mechanism

At high temperatures, $M_{23}C_6$, Ni_3 (Ti, Al), designated as gamma prime (γ'), and Ti (C, N) are the main precipitates

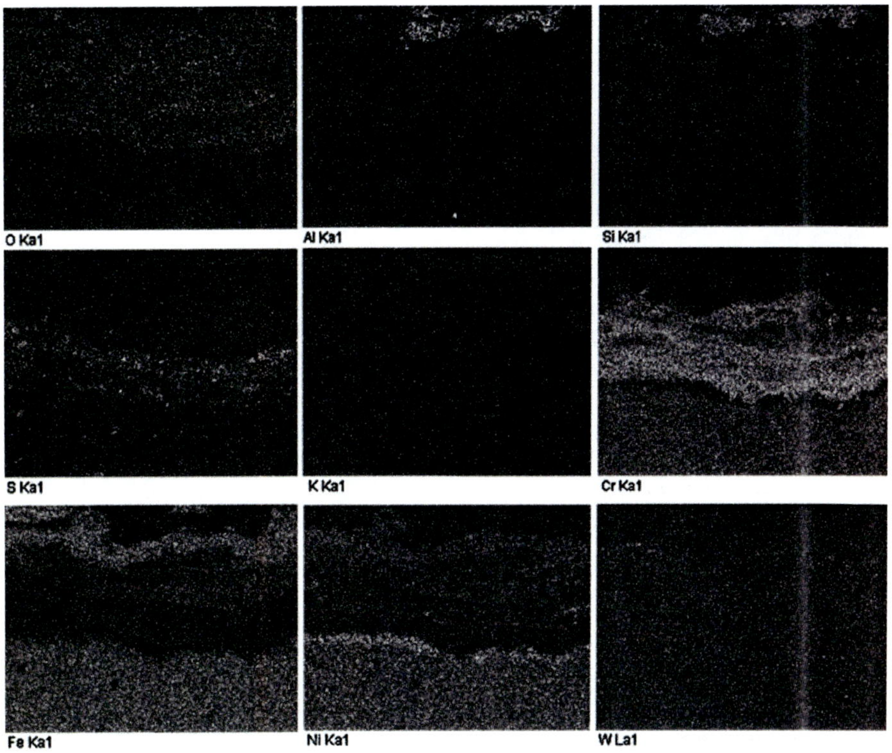

Fig. 12 Electron probe analysis or X-ray mapping; outer tube surface

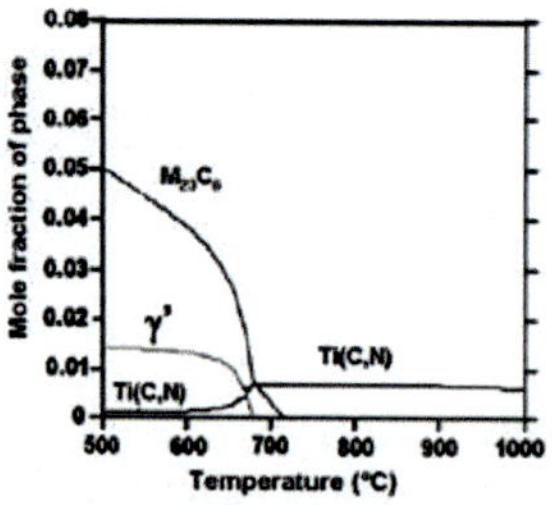

Fig. 13 Phase diagram of Alloy 800HT calculated by Thermo-Calc with the database TCFE3. Sigma phase is suppressed in this calculation since it is predicted by Thermo-Calc, but not observed in the material [26]

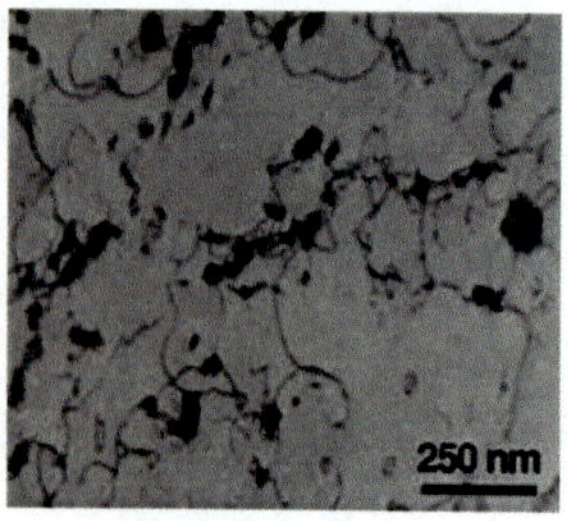

Fig. 15 Interaction between $M_{23}C_6$ precipitates and moving dislocations in the material crept with 140 MPa load at 600°C [28]

that contribute to the creep properties of Alloy 800HT (Fig. 13).

$M_{23}C_6$ precipitates can be observed both at the grain boundaries (Fig. 14a) and inside the grains (Fig. 14b) in Alloy 800HT materials aged between 600 and 800°C [26], but even in the material creep tested at 900°C [27]. This can be explained by the slow diffusion of C and N in Ti (C, N). These particles have a cube-to-cube orientation relationship with the matrix, but with a small mismatch between the matrix and the precipitates since the ratio between their lattice parameters ($M_{23}C_6$: $a = 1.06$ nm and austenite: $a = 0.36$ nm) is very close to three [26]. Elongated intragranular $M_{23}C_6$ precipitates can also be observed in some crept materials.

The contribution of the precipitation of $M_{23}C_6$ to the creep strength is that the precipitated carbides create a threshold stress on the moving dislocation. This leads to a reduction of their moving velocity and consequently creep strain rate [27]. Figure 15 shows the interactions between the $M_{23}C_6$ precipitates and the moving dislocations.

Ni_3 (Ti, Al) (γ') precipitates can be observed in the crept or aged material with dark-field imaging (Fig. 16a). These dense fine precipitates are homogeneously distributed in the matrix, but sometimes too small to be identified. Ni_3 (Ti, Al) has an ordered FCC structure with the space group $Pm\bar{3}m$. It can be found that this phase has a $(111)\gamma//(111)\gamma'$ orientation relationship with the matrix [26, 29]. These fine particles effectively interact with the moving dislocations and stop their movement as shown in Fig. 16b. Dislocation loops around γ' precipitates can be observed.

At 700°C, however, Ni_3 (Ti, Al) cannot precipitate homogeneously in the matrix. They precipitate heterogeneously on $M_{23}C_6$ or Ti (C, N) particles with different sizes and morphologies (Fig. 16c, d). No such phase can be observed in the materials crept at temperatures above 800°C.

Recently, the volume and the mean diameter of γ' precipitates in the materials aged at 600 and 650°C have been determined using EFTEM [26]. The jump-ratio images originating from the ionization edges of FeM, NiL, TiL, and AlL can be used to reveal the γ' phase as shown in Fig. 17.

Figure 18a shows the growth behavior of γ' precipitates with aging time. It follows the LSW equation [26]. It seems that those particles are still rather small at an aging time up to 100,000 h. The volume fraction of γ' precipitate increases in the early stage of the ageing up to about 2,000 h, and then reaches a plateau or little change at 600°C (Fig. 18b). This indicates an onset of coarsening. The volume fraction of this precipitate is lower at 650°C than at 600°C.

In a titanium stabilized austenitic stainless steel such as Alloy 800, fine carbon-rich Ti (C, N) precipitates with a size of 500 nm can sometime be observed in the solution annealed material. They may precipitate during the cooling after the solution annealing [26]. In the crept and aged material, the Ti (C, N) precipitates are much finer and only 10–100 nm in size (Fig. 19a).

Nitrogen uptake has been observed in the creep deformed material at 1,000°C both at the surface and in the center of the material. The microstructure reflects the change in equilibrium by the precipitation of large AlN particles at high temperature (Fig. 19b–d) [26]. This phenomenon has also been observed in other heat resistant austenitic stainless steel at high temperatures [30].

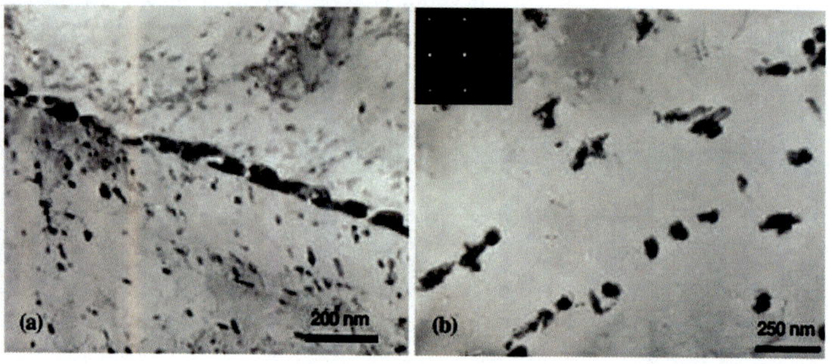

Fig. 14 **a** Grain boundary $M_{23}C_6$ precipitates in material crept with 160 MPa load at 600°C. **b** $M_{23}C_6$ precipitates inside grain in the material crept with 140 MPa load at 600°C [28]

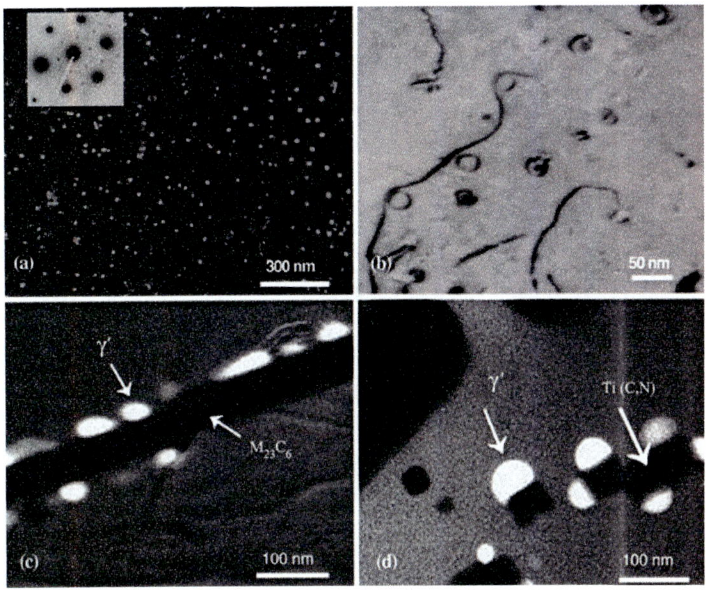

Fig. 16 **a** γ' precipitates in the material crept with 140 MPa at 600°C. **b** Interaction between $M_{23}C_6$ precipitates and moving dislocations in the crept material with 140 MPa at 600°C [28]. **c** γ' precipitates on $M_{23}C_6$ at 700°C. **d** γ' precipitates on Ti (C, N) at 700°C [26]

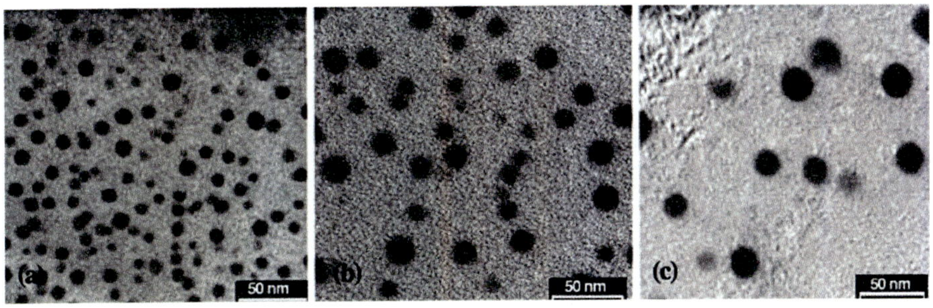

Fig. 17 EFTEM FeM jump-ratio images showing γ' precipitates after different aging time at 600°C to 650°C: **a** 600°C/4752 h. **b** 600°C/25165 h. **c** 600°C/85388 h[26]

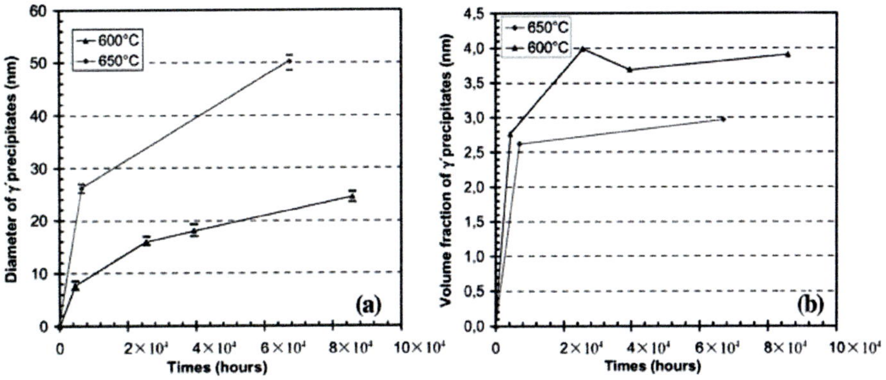

Fig. 18 Mean diameter and volume fraction of γ' precipitates plotted against aging time

5.2 Influence of Ti, Al and C on the Mechanical Behavior at Higher Temperature

As mentioned previously, the additions of C, Ti and Al in Alloy 800 series materials can greatly affect the creep behavior of the material. It is generally believed that the creep strength below 600°C is influenced mainly by these three alloying elements. However, the formation of a dense distribution of Ni_3 (Ti, Al) nanoprecipitates in the matrix by addition of Al and Ti is the main strengthening factors up to temperatures around 700°C [31–33]. At even higher temperatures, increase of C and Ti can improve the creep strength of the material [31]. Figure 20a shows the

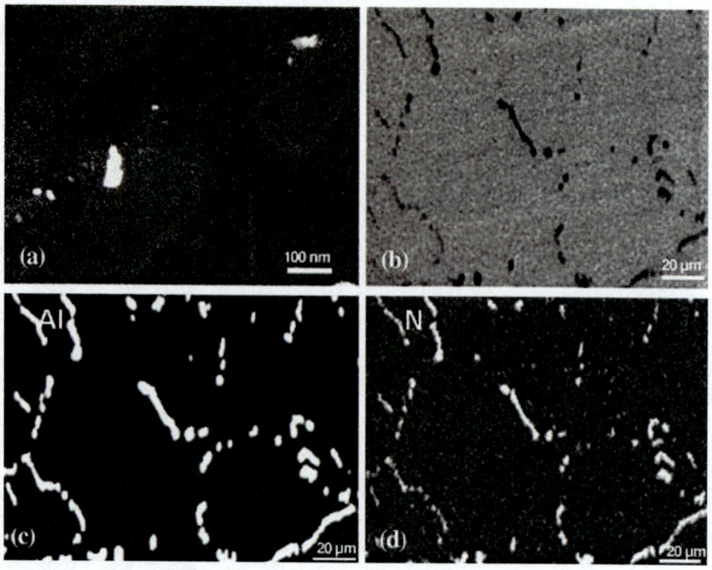

Fig. 19 a EFTEM FeM jump-ratio images showing Ti (C, N) precipitates in the material aged at 700°C for 62776 h. **b–d** EPMA point measurement showing the Al and N images in the material aged at 1,000°C for 347 h [26]

influence of carbon content on the creep strength at 700 and 900°C. It is obvious at 900°C but not at 700°C. Increase of the titanium content can significantly improve the creep strength at 900°C [31]. This might be attributed to the formation of TiC or Ti (C, N) precipitates. Titanium alone can cause a decrease of the creep strength when the titanium content is higher than 0.6%, but a combination of titanium and carbon can greatly increase the creep strength (Fig. 20b). Aluminum alone has a relatively small effect on the creep strength at 900°C.

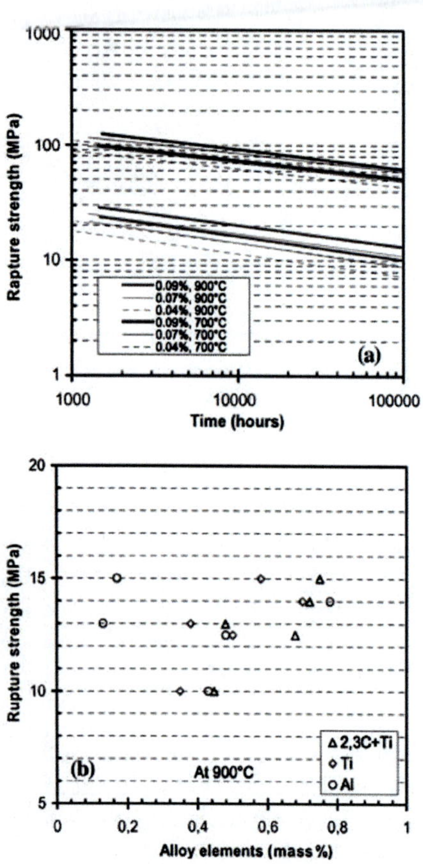

Fig. 20 Influence of alloying elements on the creep strength of Alloy 800 series material. **a** Influence of carbon. **b** Influence of titanium, aluminum and Ti + C

It is known that a fine grain size can lead to a low creep strength [34]. It is found that a smaller grain size can also lead to a higher creep rate at a given applied stress (Fig. 21). However, it has also been reported that the creep strength will not change much if the grain size is larger than ASTM 2–3; On the other hand, the ductility will decrease with a large grain size [35].

Cold deformation in the heat resistant material is always a concern. It is interesting to mention that a moderate cold deformation can reduce the creep rate (Fig. 21). Actually, a small cold work can increase the creep strength, but it strongly depends on the creep temperature [28]. At 700°C, the cold worked material up to a strain of 30% still shows higher creep strength than that of the annealed material. At 800°C, this strain is only 20%. At 900°C, the material with a cold work higher than 10% shows a less creep strength than an annealed material.

6 Composite Tube Products

Environmental protection is an international key issue. Great efforts are making globally to reduce the negative impacts on the nature as much as possible.

To reduce use of coal, oil and gas, more complex fuels such as different biological, industrial and municipal waste are increasingly used for steam and energy production.

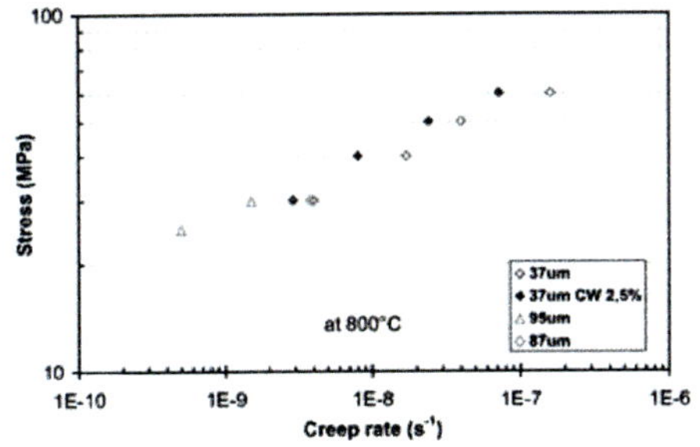

Fig. 21 Influence of grain size and cold work on the creep rate of Alloy 800HT material at 800°C

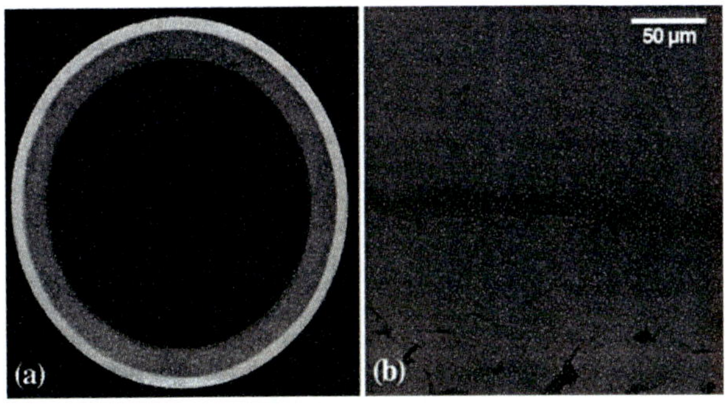

Fig. 22 Composite tube material. **a** Cross-section of a composite tube, white section is the corrosion resistant outer layer and the grey section is the carbon steel inner layer. **b** Microstructure of the bond zone in a composite tube. The corrosion resistant layer is above and the carbon steel is below the interface

Gasification of coal, pet coke, lignite, is another development striving to reduce the negative impact of using the world's scarce resources of oil and gas as feedstock for energy and chemicals production. These concepts or technologies are becoming more and more interesting worldwide [36].

The corrosive environment and lifetime of materials is a major concern when producing steam and energy in more complex flue gas compositions. For example in municipal waste incinerators the corrosion can be as high as 5 mm/year. This results in unscheduled standstills and considerable economical losses.

There are also design criteria that impose use of composite tubes instead of solid carbon or low alloyed material or solid higher alloyed stainless steel tubing. One example is in coal gasification syngas coolers with syngas on the inside and steam on the outside of the tubes. Composite tubes are the best option in this case as it combines corrosion resistance (Alloy 800 on the inside) with low thermal elongation of the outer low-alloyed material (SA213 T12). Full stainless tubes cannot be used in this case as the thermal elongation is too large for the vessel, a fire tube boiler. Full carbon steel tube could not be used as the material is not resistant enough against the inside synthetic gas.

To overcome these problems, composite tubes are used [37–43]. Composite tubes have been used in black liquor recovery boilers in the pulp and paper industry since the early 1970s. Main applications are in waterwalls and superheaters [37, 43]. With increased heat value as well as steam carbon or lower alloy tubing are less used in this type of boilers. Waste heat boilers for metal smelting industries are another application in which composite tubes have increased the service life of the steam boilers substantially.

A composite tube as shown in Fig. 22 consists of a material that fulfils the mechanical requirements and resists SCC from the steam side, and an other material that has very high corrosion resistance. The bond zone between the two components is a metallurgical bond of diffusion bonding. The effect of carbon diffusion can be seen in the microstructure of a composite tube bond (Fig. 22b).

Table 4 shows typical chemical compositions of the two different layers in composite tubes. One layer is mainly

Table 4 Typical chemical compositions of composite tube materials (mass%)

	Alloy	C	Si	Mn	Cr	Ni	Mo	Fe	Others
Inner layer	UNS N08800	≤008	≤015	≤02.0	20	32		Bal	Ti. Al
	SA210-A1	019	03	07				Bal	
	16Mo3	015	≤035	07				Bal	
	SA213-T22	010	0.25	0.5	225		1	Bal	
	SA213-T91	0.10	0.50	0.6	8.5	04	095	Bal	
Outer layer	TP 304L	<003	045	13	185	10		Bal	
	USN N 08028	0.030	0.5	0.6	27	31	35	Bal	Cu = 10
	Alloy 825 Mod	<0025	<05	08	20	38.5	25	Bal	Cu = 17 Ti ≤ 10
	Alloy 690	0.02	≤05	≤05	30	60		10	
	TP310	<0020	≤07	≤020	25	20	85	Bal	
	SA213-T12	0.15	0.25	0.5	0.9		0.5		

pressure vessel approved low-alloyed CrMo steels. The other layer material is either highly alloyed stainless steels or nickel based alloys. It has been found [36, 37, 40, 41] that the use of highly alloyed composite tube products greatly increases the corrosion resistance and makes the boiler more tolerant to corrosive environment, and thereby reduce the down time periods. Compared to ordinary carbon steel or low-alloyed pressure vessel steel, the composite tube materials increase lifetime by a factor of 2–20.

7 Conclusions

A new austenitic stainless steel grade, UNS S31035, has been developed intended for superheater and reheaters in advanced coal fired power boilers. Extrapolation from creep data with two methods gives a creep strength of (99 $\pm$ 3) MPa at 700°C for 100,000 h. The new grade has higher creep rupture strength than that of other austenitic stainless steels available today.

The properties of UNS S31035 make it a cost efficient option in high-efficient coal fired boilers for super-heaters and reheaters operating at metal temperatures from 650 up to 700°C. For such temperatures the other option is to use more costly nickel-base materials.

The creep strength is related to intragranular precipitates and nanoparticles acting as obstacles for dislocation movement.

Alloy 800 series of alloys have both excellent creep strength and oxidation resistance at temperature higher than 700°C. $M_{23}C_6$, Ni_3 (Ti, Al) and Ti (C, N) nanoparticles contribute to the high creep strength at high temperature. However, the alloy chemistry and other parameters like heat treatment will also affect the creep properties.

The use of composite tube products in highly alloyed materials in corrosive steam boilers can greatly increase its lifetime and efficiency of steam and power production.

Acknowledgments This chapter is published by permission of Sandvik Materials Technology. The support of Prof. Wijk O. and Mr. Lundström M. is gratefully acknowledged. The authors are also indebted to the co-authors in the reference list, especially Mr. Kjellström P. and Dr. Andersson M. for their contribution and collaborations. Fruitful discussions with Dr. Lundberg M. and Mr. Wilson A. on high temperature materials are also acknowledged.

References

1. R. Blum, R.W. Vanstone, C. Messelier-Gouze, *Proceedings 4th International Conference on Advances in Materials Technology for Fossil Power Plant*, 116 (2004)
2. P. Lacombe, B. Baroux, G. Beranger (ed.), *Stainless Steels*, Les Ulis, Editions de Physique (1993)
3. High Temperature Grades, Data Sheet. Sandvik Materials Technology (2000)
4. B. Rudolph, R.W. Vanstone, *Proceedings 4th International Conference on Advances in Materials Technology for Fossil Power Plant*, 317 (2004)
5. J. Bugge, S. Kjær, R. Blum, Energy **31**(10–11), 1437 (2006)
6. R. Rautio, S. Brua, *Proc. 4th Int. Conf. on Advances in Materials Technology for Fossil Power Plant*, 274 (2004)
7. Y.U. Kim, S.I. Kwun, J.H. Shim, D.B. Park, Y.W. Cho, Y.H. Chung, W.S. Jung, www.nims.go.jp/hrdg/USC/Proceeding/, *3rd Symposium on Heat Resistant Steels and Alloys for High Efficiency USC Power Plants*
8. S. Latha, M.D. Mathew, P. Parameswaran, K.B. Sankara, S.L. Mannan, Int. J. Pre. Vess. Pip. **85**, 866 (2008)
9. T. Sourmail, Mater. Sci. Technol. **17**, 1 (2001)
10. R. Viswanathan, J. Nutting (ed.), *Advanced Heat Resistant Steels for Power Generation* (IoM Communications Ltd., London, 1999)
11. K. Laha, J. Kyonob, N. Shinya, Scripta Mater. **56**, 915 (2007)
12. R.M. Davison, T.R. Laurin, J.D. Redmond, H. Watanabe, M. Semchyshen, Mater. Des. **3**, 111 (1986)
13. M. Boström, Sandvik Materials Technology 100156TE (2010)
14. Annex to International Standard ISO 6303, Method of Extrapolation Used in the Analysis of Creep Rupture Data, Parts 1–6 (1981)
15. Holdworth S.R. (ed.), Guidance for the Assessment of Creep Rupture, Creep Strain and Stress Relaxation Data, Data Validation and Assessment Procedures, ECCC WG1, European Collaborative Creep Committee, vol. 5 (1996)
16. R. Sandström, J. Test. Eval. **31**, 58 (2003)
17. M. Evans, J. Mater. Sci. **44**, 5842 (2009)
18. S.R. Holdworth, Key Engineering Materials (Switzerland) **171–174**, 1 (2000)
19. R. Sandström, L. Lindé, J. Test. Eval. **3**, 203 (1999)
20. Generic recommendations and guidance for the assessment of full size creep rupture datasets, ECCC recommendations, vol. 5 part 1a [issue 5] (2008)
21. J. Högberg, G. Chai, P. Kjellström, M. Boström, U. Forsberg, R. Sandström, Creep behavior of the newly developed advanced heat resistant austenitic stainless steel grade UNS S31035, PVP2010-25727, Bellevue Washington (2010)
22. VdTÜV material data sheet 546 and 550 (03/2007), and 555 (09/2008)
23. P. Nyblom, J. Högberg, M. Herrdin, U. Forsberg, UNS S31035 A new austenitic tube grade for use in coal fired boilers at material temperatures up to about 700°C, 09267, NACE Corrosion 2009, Atlanta (2009)
24. W. Betteridge, R. Krefeld, H. Krockel, S.J. Lloyd, M. Van de Voorde, C. Vivante (eds.), *Alloy 800* (North Holland Publishing, New York, 1978)
25. W. Ren, A Review of Alloy 800H for Applications in the Gen IV Nuclear Energy Systems, PVP2010-25727, Bellevue Washington (2010)
26. J. Erneman, J.O. Nilsson, H.O. Andrén, D. Torbjörk, Met. Trans. A **40**, 544 (2009)
27. E. El-Magd, G. Nicolini, M. Farag, Met. Trans. A **27**, 747 (1996)
28. Smith, Electron Metallography of Alloy 800, Sandvik, IT002841 (1978)
29. P. Liu, J.O. Nilsson, J. Mater. Sci. Tech. **6**, 764 (1990)
30. J. Erneman, M. Schwind, P. Liu, J.O. Nilsson, H.O. Andrén, J. Ågren, Acta Mater. **52**, 4337 (2004)
31. N.G. Persson, Mechanical properties of Alloy 800 above 600°C, *Proc. Of Inter. Conf. on Alloy 800*, Petten, The Netherlands (1978)
32. L. Egnell, Ductilite de fluage a rupture de lálloy 800, *Proc. 18e Colloque de Metallurie*, Saclay, Juin (1975)
33. T.H. Bassford, EC working group meeting in Brussel (1976)

34. C.E. Sessions, G. Grothe, A. Vaia, J.L.W. Wilson, Review of the behaviour of Alloy 800 for use in LMFBR steam generators, WNET-115 (1975)
35. R. Lagneberg, J. Iron Steel Inst. **297**, 1503 (1969)
36. B. Price, *Energy from Wastes* (Financial Times Energy Publishing, London, 1996) ISBN 1-85334-619-5
37. A. Wilson, S. Åsberg, Composite tubes for black liquor recovery boilers, Soodakattilapäivä, Helsinki (2003)
38. P. Pademakers, G. Grossmann, A. Karlsson, M. Montgomery, T. Eriksson, L. Nylof, *6th Liege Conference on Materials for Advanced Power Engineering* (1998)
39. J.R. Keiser, B. Taljat, X.L. Wang, *Proceedings of the 1998, International Chemistry Recovery Conference.* Tampa, p. 71 (1998)
40. A. Wilson, U. Forberg, J. Noble, CORROSION/97, paper no. 153 (1997)
41. U. Forsberg, M. Lundberg, L. Nylöf, A. Wilson, NACE-Italy seminar, materials and coating for environmental protection (1998)
42. A. Wilson, U. Forsberg, M. Lundberg, L. Nylöf (1999) *Stainless Steel World 1999 Conference*, vol. 2, p. 669 (1999)
43. J.R. Keiser, J.R. Kish, D.L. Singbeil, Corrosion Conference paper no. 10081 (2010)

Fig. 9 100 m walking beam cooling bed

Fig. 10 9 + 7 horizontal and vertical combined straightening

3.3 Optimization Study on High Straightness Rail

We have studied and optimized rail compound straightening technology and automatic complementary straightening technology and control model of flatness with the introduction system of compound straightening, automatic check and two-direction hydraulic straightening. Straightness can also be controlled by these research achievements. In order to cooperate with the operation of 9 + 7 new compound straightener, ANSYS software was adopted to study on the rail straightening technique. Main contents of the study were rail camber before straightening and the effect of straightening reduction to straightness after straightening. According to the research result, optimized straightening technique was formulated, and meet the requirements of 350 km/h high speed railway.

Table 2 Series of research contents on smelting process

No.	Research contents
1	Optimizing study on desulfurization process of hot metal
2	Research on efficient oxygen supply of BOF, argon supply of bottom blown, and dephosphorization technology
3	Research on deoxidation without aluminum of high purity molten steel, secondary refining, and control technique of inclusions
4	Research on vacuum degassing technology
5	Research on tundish metallurgy technology for high purity molten steel
6	Research on shielded casting technology for high purity molten steel
7	Research on mold flux of rail steel
8	Research on EMS of high purity molten steel
9	Research on control technology of secondary cooling for rail steel
10	Research on dynamic soft reduction at the final stage of solidification

Table 3 Technology study on high purify railway steel

No.	Research contents
1	Caliber design technique of universal rolling for rail high precision rolling
2	The control technology of microstrain in tandem rolling process
3	Guidance design technology of scuffing-free universal mill
4	Control technology of straight drawn steel
5	Control technique of low roller consume

3.4 Optimization Study on High Straightness Rail

Manufacture technique of bloom surface defect-free, rail surface decarbonization, and optimization of descaling process have been studied. In addition, the technology of preventing rail surface damnification in the process of rolling, straightening and transportation have been studied as well. So that there is almost no defect on surface.

3.5 Study on 100 m Rail Technology

Compared to 25 m rail, there are many advantages for 100 m rail. It has low welding cost, and because of fewer welded splices, the straightness of rail can be ensured easily. Therefore, the holistic ride of rail can be ensured easily accordingly. It is beneficial to safety of high speed train and make travel comfortable. Based on the above reasons, techniques of 100 m rail manufacture, vertical loading, and welding have been studied in Pangang. The problem of difficult to transport multitudinously and safely disappears. And the technique of 100 m rail manufacture can provide guarantee.

3.6 100 m Rail Heat Treatment Technology Research

Through independent development on-line heat treated equipment and technology are successfully developed. Product performance is stable and realized industrial production. Pangang is the only manufacturer who can produce heat treated rails in China. Main products are U75V, SS, PG4, etc.

4 Achievements of Rail Product Development

Based on advanced equipments and technologies research, Pangang developed series of high speed, 1300 MPa high strength heavy haul and exported pearlite rails, the production technology is improved compared with the year of 2000 and the products quality meet the world advanced level.

4.1 Series of High Speed Rails

Facing the opportunity of high speed railway development in our country, Pangang resolved several technologies, especially in domain of internal quality, corss-section dimension, straightness and surface quality. Pangang successfully developed rails with the speed of 250 and 350 km. Based on this, side track rails which made Pangang the only factory who can produce complete high speed rail products, overall quality of Pangang rail meet the world advanced level, as shown in Tables 4, 5, 6 and 7.

4.2 Development of 1300 MPa High Strength Heavy Haul Pearlite Rail

Pangang 1300 MPa super high strength rails which were used in heavy haul railways are the highest intensity level eutectoid pearlite steel which can be steadily produced at present in the world. This grade of rail still applied the design route of eutectoid pearlite structure and component design fully considered the features of Pangang such as resources and low cost. On the other hand, we adopted technology of on-line heat-treatment which is developed independently by Pangang and combining the technology with favorable characteristics of the steel. On this basis, through compound enhancement of the alloyed elements and heat-treatment, industrial production of 1300 MPa super high strength rail has come true.

Table 4 350 km rail steel continuous casting block sulphur print macro examination result

Item	Inner defect judgment (grade)					
	Centerline segregation	Central looseness	Central pipe	Centre crack	Middle crack	Angle inner crack
CC	0–1.0	0.5–1.0	0–1.0	0–0.5	0–0.5	0–0.5
Technology	≤1.0 (100%)	≤1.0 (100%)	≤1.0 (100%)	≤0.5 (100%)	≤0.5 (100%)	≤0.5 (100%)

Table 5 Contrast result before and after control technique

Item	A (grade)	B (grade)	C (grade)	D (grade)	Percent of pass (%)
Index requirement	≤2.0	≤1.0	≤1.0	≤1.0	98.1
Actual level	1.0–2.0	0.5–1.5	0.5–1.0	0.5–1.0,	98.1

Table 6 Cross-section dimension of Pangang rail

Item	Cross-section dimension of Pangang rail			350 km/h
	Range (mm)	Margin(mm)	Standard discrepancy	
Height of rail	$\frac{175.5-176.3}{175.96}$	0.8	0.140	176 ± 0.6
Width of rail head	$\frac{72.7-73.3}{73.04}$	0.6	0.063	73 ± 0.5
Width of rail foot	$\frac{149.2-150.2}{149.7}$	1.0	0.165	150 ± 1.0
Web thickness	$\frac{16.4-16.8}{16.59}$	0.4	0.074	$16.5^{+1.0}_{-0.5}$

Table 7 Main rail straightness index

Item	Pangang rail	350 km/h
Central 3 m straightness (mm/3 m)	$\frac{0.08\text{–}0.35}{0.22}$	≤0.3
Central 1 m straightness (mm/1 m)	$\frac{0.06\text{–}0.22}{0.17}$	≤0.2
Central 1.5 m straightness (mm/1.5 m)	$\frac{0.05\text{–}0.37}{0.21}$	≤0.45

Table 8 Performance index of Pangang 1300 MPa rail

Item	Performance index of Pangang 1300 MPa rail
Tensile strength R_m (MPa)	$\frac{1310\text{–}1370}{1345}$
Elongation (%)	$\frac{10\text{–}14}{11.6}$
Head surface hardness (HB)	370–415
Hardness 16 mm to surface) (HRC)	≥37

Table 9 Performance index of low alloy rail (AREMA) produced by Pangang

Item	Performance index of low alloy rail (AREMA)
Tensile strength R_m (MPa)	$\frac{1030\text{–}1080}{1058}$
Elongation (%)	$\frac{10\text{–}14}{10.2}$
Head surface hardness (HB)	$\frac{300\text{–}315}{306.0}$

Table 10 Performance index of standard strength (AREMA) produced by Pangang

Item	Performance index of standard strength rail (AREMA)
Tensile strength R_m (MPa)	$\frac{1010\text{–}1130}{1083.0}$
Elongation (%)	$\frac{10\text{–}12}{10.2}$
Head surface hardness (HB)	$\frac{301\text{–}326}{314.4}$

At present, 1300 MPa ultra high strength rail is widely used in Da-Qin line, Shi-Tai line and other heavy haul lines, and also exported to Brazil, Australia, etc. Service conditions showed that using effect of 1300 MPa high strength rail is perfect, there is almost no fracture caused by special damage. Compared with former U75V heat treated rail, the service life of 1300 MPa high strength rail is prolonged by 50% (Table 8).

4.3 Series of Heat Treated Rails

By using efficient and express strengthening method, rail head performances have been improved economically. At present, series of heat treated rails such as U75V, U71Mn, PG4, SS and LA have been developed. The specification is from 43 to 75 kg/m, any symmetrical rail can be produced which is used in freight and passenger and freight mixed lines, the service life of rail in small radius section is improved effectively.

4.4 Series of Exported Rails

Facing the vigorous market demand, Pangang successfully developed LA, SS rails according to USA national standard and 900A rails by modern equipment and technology at the beginning of 2005, and explored international markets for PG4 rail. There are more than ten specifications such as UIC54, UIC60, 115RE (AREMA), 136RE (AREMA), GB60, etc. At present, the amount of annual exportation is about 200,000 tons, the destination included USA, Brazil, Australia, New Zealand, Indonesia, Thailand and other countries and regions and entering first-class railway of USA, Pangang rails get a full affirmation. The performances of primary rail products are shown in Tables 9 and 10.

5 Conclusions

Aiming at insufficiency of equipments, Pangang developed high effective continuous casting and universal rolling technologies to acclimatize itself to rail development opportunities. Rail manufacturing practice is changed from die casting with 950 rolling line to universal rolling. Based on it, Pangang successfully developed high speed rails with the speed of 250 and 350 km, besides, the quality of series heavy haul rails with strength of 1300 MPa and export rails have been improved steadily. Pangang rail exported to USA, Brazil, Australia, Indonesia and other countries and regions, realizing production system innovation and new product development.

The Influence of Strong Magnetic Field on Alloy Carbide Precipitation in Fe-C-Mo Alloy

Tingping Hou, and Kaiming Wu

Abstract

The effect of 12 T strong magnetic field on molybdenum carbide precipitation in a Fe-C-Mo alloy was investigated by means of transmission electron microscopy and selected area electron diffraction technique. The sequence of molybdenum carbide precipitation in Fe-C-Mo alloy during tempering changes due to the effect of 12 T strong magnetic field. The precipitation of $(Fe, Mo)_6C$ was greatly promoted when a strong magnetic field was applied.

Keywords

Precipitation • Carbide • Microstructure • Magnetic field • Tempering

1 Introduction

The phenomenon that magnetic field exerts effects on phase transformations in metallic materials was first reported in the middle of the last century by Smoluchowski and Turner [1]. In recent years, results of studies of the effects of strong magnetic field on martensite [2], bainite [3], and ferrite [4] transformations, and precipitation [5–7] have been reported. It is generally accepted that strong magnetic field may have significant effects on ferromagnetic materials. Not only the difference of magnetic moment, but also the magnetocrystalline shape, magnetic anisotropy, induced magnetic anisotropy, and magnetostriction can affect the nucleation and growth rate, transformation kinetics, variants and microstructures of product phases [8, 9]. Molybdenum carbides are alloy carbides with a proportion of substitutional solutes of Fe replaced by Mo [10]. It is expected that the precipitation of molybdenum carbides can be influenced by strong magnetic fields. However, this has been paid less attention and less reported in the research literature than it is expected. The aim of the present study is to investigate the effects of 12 T magnetic field on molybdenum carbide precipitation of a Fe-C-Mo alloy during tempering at 530°C.

2 Experimental

In order to avoid the complication of interactive effects among multiple solutes, the alloy studied was prepared by vacuum induction melting, utilizing high purity electrolytic iron, graphite and molybdenum. The chemical composition is shown in Table 1. The alloy was homogenized at 1250°C for 48 h after hot forging, and then made into a square column of 7 × 7 × 25 mm (the dimensions were determined by the magnetic field heat treatment furnace). Specimens were reheated to 915°C and held for 30 min, and then followed by ice brine quenching to obtain a martensite microstructure for subsequent tempering. The specimens were heat treated under a magnetic field of 12 T at 530°C for 10 and 60 min. After heat treatment, specimens were polished and etched with Nital solution (3 vol.%) for TEM analysis.

Thin foils were prepared for carbide analysis by using transmission electron microscopy (TEM). Thin foil observation is crucial to define the orientation relationship

T. Hou · K. Wu (✉)
Institute of Advanced Steels and Welding Technology, Hubei Provincial Key Laboratory for Systems Science on Metallurgical Processing, Key Laboratory for Ferrous Metallurgy and Resources Utilization of Ministry of Education, Wuhan University of Science and Technology, Wuhan, 430081, China
e-mail: wukaiming2000@yahoo.com; wukaiming@wust.edu.cn

Y. Weng et al. (eds.), *Advanced Steels*, DOI: 10.1007/978-3-642-17665-4_51,

Table 1 Chemical composition of investigated steel (mass%)

C	Mo	Si	Mn	P	S
0.28	3.0	<0.004	<0.001	<0.003	<0.004

between the precipitates and the matrix. Thin foils were sliced from bulk specimens as 3 mm diameter discs to a thickness of ~500 μm. After slicing, the specimens were ground with silicon carbide paper to around 50 μm in thickness. Electropolishing was conducted at 40 V using a twin jet unit. The electrolyte consisted of 10% perchloric acid and 90% glacial acetic acid. The foils were examined in a JEM-2100F microscope operating at 200 kV.

3 Results

TEM micrographs of the carbides in the specimens tempered at 530°C for 10 and 60 min without the presence of magnetic field are shown in Fig. 1. The spherical particle was identified to be (Fe, Mo)$_3$C [10] and the lath-like particle (Fe, Mo)$_2$C [11].

The morphology and identification of M_6C is presented in Fig. 2. A strong magnetic field facilitates the precipitation of M_6C in the equiaxed morphology in Fe-C-Mo alloy [7]. Due to the similar crystal characteristics between M_6C and $M_{23}C_6$, it is difficult to distinguish the geometric configuration from the diffraction pattern, so a thorough understanding of the intensity of corresponding diffraction spots will aid in identification. M_6C in Fig. 2 belongs to the Fd3m space group including a d glide plane. So, only the spots satisfying the condition: $h + k = 4n$ $(n = 0, \pm1, \pm2,)$ in the diffraction patterns will appear. Accordingly, in Fig. 2d, {220} and {400} can be fully identified. In simulation, the average intensity of each spot in Fig. 2e is represented by its area, so {400} is about four times greater than {220} spots. At the same time, the {200} spots in the M_6C disappear. But in the <100> zone of $M_{23}C_6$, the difference between the {400} and {200} is relatively small.

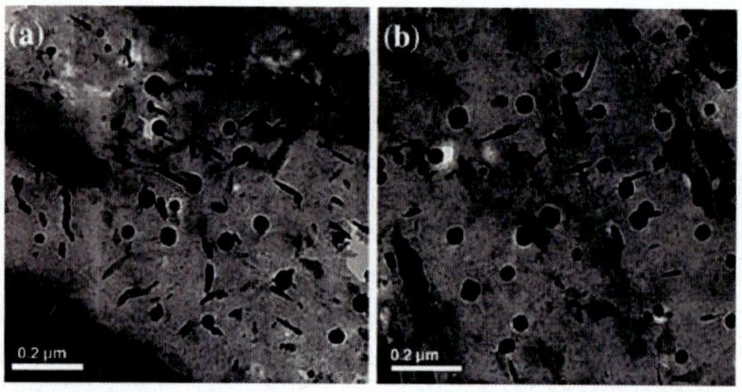

Fig. 1 TEM micrographs of carbides in the specimens kept at 530°C for **a** 10 min, **b** 60 min without the presence of magnetic field. The spherical particles are M_3C carbides and the lath-like particles M_2C carbides

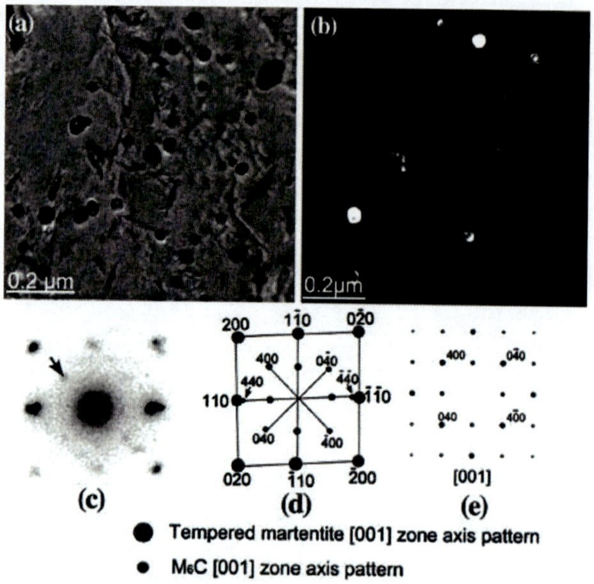

Fig. 2 Morphology, selected area electron diffraction (SAED) patterns, and the interpretation of a spherical type M_6C during magnetic tempering at 530°C for 60 min. **a** Bright field, **b** dark field, **c** SAED pattern of M_6C and the martensite matrix, **d** corresponding schematic key diagram, **e** simulation of M_6C. The *arrow* in (**c**) marks the diffraction spot of a dark pattern. The average intensity of each spot is represented by its area in simulation (**e**)

The precipitates of carbides formed at 530°C with and without the presence of 12 T magnetic field are summarized in Table 2. When there was no magnetic field, (Fe, Mo)$_2$C, (Fe, Mo)$_3$C were precipitated over 10 and 60 min. In contrast, in 12 T magnetic field, (Fe, Mo)$_2$C, (Fe, Mo)$_3$C and (Fe, Mo)$_6$C were precipitated at the above stages. It is obvious that (Fe, Mo)$_6$C was promoted under the strong magnetic field.

4 Discussion

There are still disagreements concerning the stability of some of the carbides such as M_3C, $M_{23}C_6$ and M_2C. Shtansky and Inden [11] recently reported that $M_{23}C_6$ is stable and Fe_2MoC is also stable. Zhou and Wu [7] reported that the stability of different kinds of carbides maybe changed due to high external magnetic field. The precipitation of (Fe, Mo)$_6$C was promoted when 12 T magnetic field was applied during isothermal holding at 570 and 610°C for various times.

Considering thermodynamics, (Fe, Mo)$_2$C, (Fe, Mo)$_3$C and (Fe, Mo)$_6$C are all ferromagnetic-like iron carbides. Iron carbides can be magnetized to some extent because the Gibbs free energy is lowered in relation to their

Table 2 Precipitation of molybdenum carbides in the specimens during tempering at 530°C with and without the presence of a 12 T magnetic field

	Without a 12 T magnetic field		With a 12 T magnetic field	
Time	10 min	60 min	10 min	60 min
Type	M_2C, M_3C	M_2C, M_3C	M_2C, M_3C, M_6C	M_2C, M_3C, M_6C
Carbide	$(Fe,Mo)_2C$, $(Fe,Mo)_3C$	$(Fe,Mo)_2C$, $(Fe,Mo)_3C$	$(Fe, Mo)_2C$, $(Fe, Mo)_3$ C, $(Fe, Mo)_6C$	$(Fe, Mo)_2C$, $(Fe, Mo)_3$ C, $(Fe, Mo)_6C$

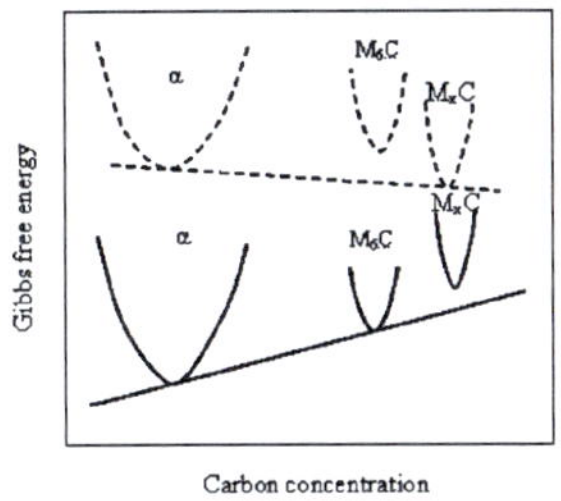

Fig. 3 Schematic illustration of Gibbs free energy vs. carbon concentration for α-ferrite and different carbides M_6C and M_xC M_2C and M_3C) with (*solid line*) and without (*dash line*) the presence of 12 T magnetic field [7]

magnetization. However, the extent of the decrease in Gibbs free energy of $(Fe, Mo)_2C$, $(Fe, Mo)_3C$ and $(Fe, Mo)_6C$ in 12 T magnetic field cannot be determined quantitatively owing to the lack of magnetic properties of molybdenum iron carbides. The results of Zhang et al. [12] regarding the precipitation of iron carbides show that carbides with lower carbon content have higher magnetizations, and that the amount of Gibbs free energy lowered in 14 T magnetic field for iron carbides with lower carbon content is larger than that for iron carbides with higher carbon content. Therefore, given the application of 12 T magnetic field, Gibbs free energy of molybdenum iron carbides can be lowered to different extents, as illustrated in Fig. 3. The molybdenum carbide M_6C has the lowest carbon content among M_2C, M_3C and M_6C. Therefore, the precipitation of $(Fe, Mo)_6C$ carbide is promoted by applying 12 T magnetic field.

5 Conclusions

The effect of 12 T high magnetic field on molybdenum carbide precipitation in a Fe-C-Mo alloy during tempering at 530°C for 10 and 60 min was studied. The following conclusions are drawn. $(Fe, Mo)_6C$ was precipitated in the shorter and longer tempering durations when 12 T magnetic field was applied, whereas it was not precipitated even at longer tempering duration without the presence of strong magnetic field. This indicates that the precipitation of $(Fe, Mo)_6C$ was greatly promoted by applying a strong magnetic field.

Acknowledgments The authors are grateful to the financial support for this work from the Ministry of Education (Grant no. NCET-05-0680), Natural Science Foundation of Hubei Province (Grant no. 2006ABB037) and Hubei Provincial Department of Education (Grant no. 200711001). The authors express their thanks to Professor M. Enomoto, Ibaraki University, Japan, for providing alloy specimen and to Electromagnetic Process Lab of the Key Laboratory of the Ministry of Education, Northeastern University for the help in magnetic field heat treatment.

References

1. R. Smoluchowski, R.W. Turner, Appl. Phys. **20**, 745 (1949)
2. K. Tomoyuki, F. Takashi, S. Toshio et al., Condens. Matter. **237–238**, 603 (1997)
3. H. Ohtsuka, Mater. Sci. Eng. A **438–440**, 136 (2006)
4. M. Enomoto, H. Guo, Y. Tazuke et al., Metall. Mater. Trans. A **32A**, 445 (2001)
5. C.T. Peters, A.P. Miodownik, Scripta Metall. **7**, 955 (1973)
6. Y. Watanabe, H. Nozaki, M. Kato et al., Acta Metall. **39**, 3161 (1991)
7. Z.N. Zhou, K.M. Wu, Scripta Mater. **61**, 670 (2009)
8. Y.D. Zhang, N. Gey, C.S. He et al., Acta Mater. **52**, 3467 (2004)
9. D.A. Molodov, P.J. Konijnenberg, Scripta Mater. **54**, 977 (2006)
10. D.V. Shtansky, G. Inden, Acta Mater. **45**, 2861 (1997)
11. S. Yamasaki, H.K.D.H. Bhadeshia, Mater. Sci. Technol. **19**, 723 (2003)
12. Y.D. Zhang, X. Zhao, N. Bozzolo et al., ISIJ Int. **45**, 913 (2005)